Henning Hopf

Classics in Hydrocarbon Chemistry

WILEY-VCH

Henning Hopf

Classics in Hydrocarbon Chemistry

Syntheses
Concepts
Perspectives

With a foreword by
W. von Eggers Doering

Weinheim · New York · Chichester
Brisbane · Singapore · Toronto

Prof. Dr. Henning Hopf
Institut für Organische Chemie
der TU Braunschweig
Hagenring 30
38106 Braunschweig
Germany

Library of Congress Card No.: applied for

A catalogue record for this book is available from the British Library.

A catalogue record for this publication is available from Die Deutsche Bibliothek.

© WILEY-VCH Verlag GmbH, D-69469 Weinheim (Federal Republic of Germany). 2000
Printed on acid-free and chlorine-free paper.
Composition: Data Source Systems, Romania. Printing: betz-druck gmbh, 64291 Darmstadt.
Bookbinding: J. Schäffer GmbH & Co. KG.
Printed in the Federal Republic of Germany.

Foreword

This uniquely original book brings together in delightful, coherent form many of the most important advances in organic chemistry of the last century. It can be read with great profit and enjoyment at several levels. At the most profound, its study enhances the reader's understanding of the uses and significances of synthesis in the growth of organic chemistry.

If the science of chemistry be the acquisition of intellectual control over the transformation of one structure of matter into another, its application to the synthesis of chemicals of use to humanity has been the popular measure of its success - pharmaceuticals, pigments, dyes, coatings, fabrics, plastics, tires, drugs, anesthethics, explosives, gases - an endless stream of not always positive 'goodies'. Quintessentially the province of organic chemistry, rational synthesis has been its power and glory for more than a century - the joy and agony, the thrill, of its practitioners.

In the beginning, the isolation of chemicals from natural sources provided an unceasing stimulus to the creation and development of the science. Structural hypotheses were deduced from a myriad of chemical transformations and degradations, in the course of which large numbers of new reactions were discovered and incorporated into the body of organic chemistry. Today, parenthetically, spectroscopic methods have almost entirely supplanted this classical approach, and therewith deprived the science of a nigh inexaustible source of unpremeditated discoveries. For decades, rational synthesis of a natural product was perceived as the final, definitive confirmation of its structural hypothesis.

Periodically, the successful synthesis of a natural product of ever more forbidding structure has served as a marker of the heights reached by the organic chemical enterprise. In their finest examples at the hands of ingenious, imaginative, and skilled practitioners of the science and the art, "recent advances in the synthesis of natural products" have served as encouragement to attack goals otherwise considered unachievable. To be sure, this eminently worthy endeavor has often been coupled to the irresistible attraction of a popular acclaim that follows the total synthesis of such as quinine, penicillin, cholesterol, cancer cures incredible in number - or simply to the childlike elation of bursting into a game and emerging the winner.

Concurrently, from the earliest days of the triumph of structural theory, synthesis of 'unnatural' products played a unique role for the *cognoscente* concerned with the advancement of intellectual control over organic chemistry. Questions, speculations and queries in profusion were raised that did not find answers among the vast collection of compounds already uncovered in Nature. Only from new, purposefully designed structures was clarification and enlightenment to be found. To answer the question, for example, how do cyclobutadiene and cyclooctatetraene, unknown at the time, compare to benzene, by setting off to find examples among the awesomely large number of compounds still lying undiscovered in the world of microorganisms, plants and animals was too ludicrous to contemplate! Only through synthesis were such new types of structures to be made available for exploration.

Unlike synthetic targets of natural origin, which were patently capable of existence and synthesis by some sequence of chemical processes available to Nature, targets of imaginary, unnatural structure presented quite different challenges. Some structural conceptions were so far removed from known types that their very existence as stable, isolable products could not be assured. Of others, it could not be assumed with confidence that any combination of chemical reactions known at the time might lead even in principle to a successful synthesis, let alone in quantities large enough to permit effective study.

Many a profound question in modern organic chemistry have found their essential answers in compounds of carbon and hydrogen alone, archetypes unadorned by perturbing elements. In *Classics in Hydrocarbon Chemistry*, Henning Hopf recreates the fascinating story of the central role played by hydrocarbons and their synthesis in new conceptual advances. We hear, among many fascinating examples, of adamantane, a 'classic' hydrocarbon in its multifarious impacts, if its extraordinary preparation be in essence more a tribute to J. Willard Gibbs than to the wiles of synthesis; of dodecahedrane, 'classic' not only in homage to Plato, but classic as an awesome triumph of the science and art of synthetic organic chemistry; of the incredible, strain-defying [1.1.1]propellane, classic in its revolutionizing impact on our conception of proper valence angles.

The staggering task of organizing the multitude of hydrocarbon targets that have been selected for synthesis over more than a century presents the author with a formidable challenge happily solved by generalizing a game that we all played during our introduction into the wonders of organic chemistry. There we deduced the constitutions and configurations of the isomeric heptanes and octanes in validation of the magnificent structural theory of the tetravalent, tetrahedral carbon atom. This game was in fact our introduction to the *Aufbau Prinzip*. We played it again with the first members of that delightful family - benzene, naphthalene, anthracene and phenanthrene. Here, the principle leads us systematically from basic structures to the most elaborate extensions and combinations, just as it has many of those who have undertaken the synthesis of its fascinating creations. The *Aufbau Prinzip* is not only a truly felicitous means of organizing the disparate collection of jewels on display in this treatise, it can be hoped to stimulate readers to create fancies of their own.

For me, and I believe for many readers, the underlying, unifying thread is 'significance'. From the start of my life in chemistry, the basic dilemma has involved choice. Why work on this project, rather than that one? Significant for the entire human race? For all scientists? All organic chemists? Answers are subjective. One person's 'highly significant' is another person's 'pedestrian'. By not being the product of several authors under a reigning editor, but by having only a single author of high accomplishment - a rare and sorely missed occasion these days - a consistent perspective, a consistent sense of chemical significance, emerges from '*Classics*' to enrich every page. From the small detail in a synthesis to the 'beauty' in a goal, from the inclusion of this example and the exclusion of another, the author's eclectic command of mechanistic and synthetic organic chemistry shines throughout. By thoughtfully - and critically - following a thread of 'significance', the reader is offered an opportunity to emerge enlightened and broadened, and with a self-developed sense of how the more important may be distinguished from the less important.

"Not for ambition or bread/or the strut and trade of charms/on the ivory stages" (nor for the New York Times!), but for the enlightenment, edification, and even amusement of the community of organic chemists have the synthetic works in this history been brought to fruition. Motivations may have run from a search for deeper understanding of basic organic chemistry to the construction of esthetically pleasing structures by chemical bonds! But for my grateful part, Henning Hopf has reaffirmed an enduring conviction that synthesis in the service of organic chemistry *per se* is a noble calling.

November 1999

William von Eggers Doering
Harvard University
Cambridge, Mass., USA

About the Author

Henning Hopf was born in 1940; he studied chemistry at the Universities of Göttingen and Wisconsin in Madison where he got his Ph. D. in 1967. After post-doctoral work at Marburg University he received his Habilitation at Karlruhe in 1972. Between 1974 and 1979 he was a faculty member at the University of Würzburg before becoming director of the Institute of Organic Chemistry at the Technical University of Braunschweig.

In his contributions to synthetic organic chemistry he is interested in the preparation and the study of the chemical properties of acetylenes, cumulenes, polyenes, cyclophanes, and polycyclic hydrocarbons, whereas his work in physical organic chemistry concentrates on the thermal behavior of hydrocarbons. Many of the hydrocarbons synthesized in his laboratories have become important reference compounds in structural chemistry.

Contents

1   Preface..1

2   From Simple Building Blocks to
     Complex Target Molecules and Multifaceted Reactions 5

3   Adamantane and other Cage Hydrocarbons............................23
3.1 Adamantane 23
3.2 Higher analogs of adamantane.. 29
3.3 Twistane... 30
3.4 Tetraasterane.. 34
     References.. 37

4   The Prismanes 41
4.1 Triprismane.. 42
4.2 [5]-Prismane.. 44
4.3 *En route* to hexaprismane 47
     References.. 50

5   The Platonic Hydrocarbons 53
5.1 Tetrahedrane 54
5.2 Cubane ([4]-prismane).. 60
5.3 Dodecahedrane 63
     References.. 75

6   Bridgehead-distorted Hydrocarbons 81
6.1 Propellanes.. 82
6.2 Fenestranes 88
6.3 *out,out, out,in* and *in,in* Hydrocarbons 94
     References.. 99

7   Alkenes 103
7.1 Linear conjugated polyenes 103
7.2 Cyclopropene, the smallest cycloolefin and how it can be used
     in hydrocarbon synthesis.. 112
7.3 Large-ring cycloalkenes by transition metal-catalyzed
     oligomerization of dienes... 118

7.4   Distorted olefins: *trans*-cycloalkenes, *anti*-Bredt-hydrocarbons,
       betweenanenes, and pyramidalized olefins... 122
7.5   Tetrakis-*tert*-butylethene - an exercise in preparative futility............... 138
       References... 142

8   Alkynes...151
8.1   Polyacetylenes - rods made of carbon 151
8.2   Angle-strained cycloalkynes... 156
8.3   Medium and large-ring alkynes 160
       References... 165

9   Allenes and Cumulenes...171
9.1   Acyclic allenes.. 171
9.2   Acyclic cumulenes... 178
9.3   Cyclic allenes.. 182
9.4   Cyclic cumulenes.. 188
       References... 190

10   The Annulenes..197
10.1   1,3-Cyclobutadiene 198
10.2   Benzene and its isomers.. 203
10.3   Cyclooctatetraene and the $(CH)_8$ isomers 209
10.4   The higher annulenes 218
10.5   Bridged annulenes 227
       References... 238

11   Cross-Conjugated and Related Hydrocarbons 251
11.1   The dendralenes 253
11.2   The fulvenes.. 260
11.3   The fulvalenes... 269
11.4   Pentalene, azulene, heptalene and other zero-bridged annulenes 277
11.5   The [*n*]radialenes 290
11.6   The [*n*]rotanes... 300
       References... 309

12   Leaving the π-Plane—Non-Planar Aromatic Compounds 321
12.1   The helicenes 323
12.2   The circulenes... 330
12.3   The cyclophanes 337
       References... 368

13   Three-dimensional Oligoolefins 379
13.1   Barrelene.. 380
13.2   Triptycene and the iptycenes 385
13.3   Bullvalene and semibullvalene 389
13.4   Triquinacene 397
13.5   Spiropolyenes, stellapolyenes and related hydrocarbons..................... 407

References.. 412

14   **Extended Systems–I. From Benzene to**
       Graphite Substructures 421
14.1   Linearly annelated polycyclic aromatic hydrocarbons 423
14.2   Angularly annelated polycyclic aromatic hydrocarbons...................... 426
14.3   Condensed aromatic hydrocarbons with ribbon structures 431
14.4   Graphite substructures 438
       References.. 443

15   **Extended Systems–II. Beyond the PAH Pattern 447**
15.1   From biphenylene to linear and angular [*n*]phenylenes....................... 447
15.2   Extended structures containing triple bonds 457
15.3   Building with cyclopropane rings.. 472
       References.. 480

16   **Classics in Hydrocarbon Synthesis—**
       Three Examples from Physical Organic Chemistry 485
16.1   The dehydrobenzenes 485
16.2   Non-Kekulé hydrocarbons.. 492
16.3   Bond fixation in benzene rings 501
       References.. 514

       Author Index 523

       Subject Index 527

1 Preface

The word *classic* has experienced an interesting metamorphosis over the years. Originally, the Latin word *classis*, meaning the assembled crowd, was used as a military term describing the Roman land and sea forces. It was also used in the Roman tax system which distinguished between five different tax classes, the fifth being of lowest rank. Many years later, during the 18th century, the word slowly changed its meaning and now referred to any group with particular characteristics, whether one was talking about its age, education or social standing.

Today the words *classic* and *classical* have a double meaning. On the one hand they refer to the "ancient Greek and Latin authors, or their works, or the culture and the architecture *etc*. of Greek and Roman antiquity generally".[1] On the other hand classic means "of the first class, of acknowledged excellence, remarkably typical, outstandingly important".[1]

It is with this double meaning that I am using the word classics in the title of this book. On the one hand, I want to discuss with the reader several older ('ancient', *i.e.* older than, say, 40 years in these rapidly changing times) hydrocarbon syntheses from the classical period of organic chemistry, which ended when the spectroscopic methods became routine in organic chemistry in the late fifties and early sixties of the 20th century. These hydrocarbon syntheses have had a lasting impression on the development of this part of the chemical sciences and belong securely to the canon of organic chemistry today.

On the other hand, I want to present more recent developments which certainly deserve to be called outstanding and which one day may become classical in the historic sense as well. Clearly, my selections in this latter category are much more subjective, because the newer syntheses and discoveries have not stood the test of time. Looking to other fields of human creativity should make one cautious—how often have the works of writers, composers, and painters been overlooked, dismissed and rejected only to become classics at some later time. It is rare that a masterpiece is recognized as a classic immediately as when Mozart's *Don Giovanni* was first performed in Prague in 1787. I thus excuse myself to all those whose work on hydrocarbons is not mentioned in this book. This is not only explained by my preferences and occasionally my ignorance—I also had to meet a clear page limit set by the editor of this series ("Not more than 500 pages"). Besides, the book has been written primarily for readers who are *not* specialists in hydrocarbon chemistry but for a more general readership which is interested and willing to learn about developments during the last half century in an important and fascinating area of organic chemistry. And, finally, when discussing certain groups of hydrocarbons I am, of course, attempting to retrace and cover the most important developments, but I am also trying to fire the reader's imagination and hoping that certain structures will trigger explorations from this starting point. After all, the advantage of modern electronically searchable data bases is that the information required for these explorations is at his or her fingertips.

It is challenging to write a text on important work in hydrocarbon chemistry at this point in time because the field is experiencing a deep-seated change. For the last three or four decades, a period in which most of the work discussed in this book originated, the main and often sole purpose was to *prepare* a certain hydrocarbon— for any reason which the authors decided to choose, reasons which will become clear in the course of this book. The more recent developments, however, are beginning to stress *function* over *synthesis*. In other words, rather than just making these molecules—which, as many of the following examples show, may be difficult enough or even impossible so far—they are increasingly becoming the starting points for the preparation of 'extended systems' which can perform certain tasks (*e.g.* conduct electricity when doped or serve as molecular wires or switches), which have special (*e.g.* non-linear) optical properties, or which form particular geometric shapes, *e.g.* spherical or tubular. Changes of paradigm are often just a *nouvelle vague*—in this instance, though, the use of the term seems to be justified. And similar developments can be noted in other parts of organic chemistry also, supramolecular chemistry[2] being a particularly illustrative example.

One of the biggest problems I encountered while writing this text was to keep the balance between a review on a certain class of hydrocarbons on the one hand, and a description of the history of ideas and concepts which were developed—or evolved— while trying to solve a certain synthetic or structural problem on the other. This is the eternal problem of the balance between breadth and depth of a text—all I can hope is that the readers judge my efforts with a kindly critical eye. Forced to decide where I personally would put the emphasis, I prefer breadth, plus the description or recognition of what I have called 'generic relationships' between classes of hydrocarbons or their reactions.

Besides the unintentional, there is a number of intentional omissions. To begin with, in the majority of the syntheses presented the book limits historical discussions to a minimum. It would certainly make very interesting reading to follow the discovery, structure elucidation and ultimately first synthesis of important hydrocarbons like methane, ethylene, acetylene or benzene.[3] No such historical excursions will be made here. This is a pity, because to follow the historic development of certain concepts and hypotheses gives insight into the human aspects of chemical research.

Furthermore, the book will not describe the synthesis of industrially important hydrocarbons. Because hydrocarbon chemistry forms the very basis of our technical civilization—thus answering the question of why one should study these compounds most convincingly—this is also a serious omission. But besides that the author is no expert in this area, there are—fortunately—excellent texts available on industrial hydrocarbon chemistry.[4-7]

And finally, the vast and quickly growing field of hydrocarbons as reactive intermediates cannot be considered here to the extent which it clearly deserves. No charged hydrocarbons will thus be discussed, which might be excusable on purist grounds because these species—carbocations and carbanions—almost always have a counter-ion which contains other atoms than carbon and hydrogen. But 'pure' reactive hydrocarbons such as radicals, diradicals, and carbenes will also largely be omitted (for exceptions see below)[8]—it is hoped that at some later time a volume on 'Classics in Reactive Intermediates' in this evolving series[9] will deal with these species. So that the compounds described are not restricted to those stable 'under normal laboratory conditions' reactive species such as cyclobutadiene, *anti*-Bredt hydrocarbons, small-ring alkynes, to name but a few, have, however, been included. To relieve the guilty conscience of the author in this particular area somewhat, a chapter on reactive hydrocarbon intermediates is included in which the preparation of the dehydrobenzenes and a selection of non-Kekulé hydrocarbons will be described as typical examples.

One of the molecules whose synthesis is described in this book, dodecahedrane, has been called "the Mount Everest of hydrocarbon chemistry".[10] Standing on such an insurmountable point—what can you do, where can you go from there? Certainly, you can look back to the pathways that led to success in this ultimate goal and compare their relative importance in organic chemistry, their elegance or lack of it, or look at the attempts which failed. But in contrast to the Himalayas, a view from *hydrocarbons'* Mt Everest offers a grandiose panorama of many more unclimbed peaks and of numerous valleys never entered by an explorer to be reached in the future.

A substantial section of this book was written during a sabbatical at Stanford University during the Fall of 1997. And I want to take the opportunity to thank John Brauman, Carl Djerassi and Barry Trost for their hospitality, the library staff for their help in making effective use of the facilities of the Chemistry Library and my graduate class for criticism of an early version of this text.

I am, furthermore, very grateful to many colleagues who provided me with information about their contributions to hydrocarbon chemistry and in particular to Virgil Boekelheide, Manfred Christl, Lutz Fitjer, Klaus Hafner, Dietmar Kuck, Goverdhan Mehta, Armin de Meijere, Klaus Müllen, Horst Prinzbach, Wolfram Sander, Sethuraman Sankararaman, Paul v. R. Schleyer, Günther Szeimies, Emmanuel Vogel, and Günther Wilke who read (and substantially improved!) selected chapters of this text. I especially thank Wolfgang Lüttke for a decade-long exchange of ideas on hydrocarbon chemistry (and many other matters). Cornelia Mlynek, Dr. Jörg Grunenberg, and Gabriele Salomon helped with the drawing of the formulae and typing of the text. I want to thank them for this important help, which, after all, was performed besides their "normal" daily duties. Finally, Ian Davies, considerably improved my English during the copy-editing process.

This book has been written by a chemist who has always derived joy and satisfaction from synthesizing hydrocarbons. Over the years, many of the molecules prepared in my laboratories have also attracted the attention of other chemists—spectroscopists, theoreticians, other preparative chemists—and much international cooperation and many friendships have developed. Today these personal relationships are as important to me as the actual synthetic work, and it is for this reason that I dedicate this monograph to international cooperation and friendship.

Henning Hopf   Braunschweig, Fall 1999

References

1. L. Brown (*Ed.*), *The New Shorter Oxford English Dictionary*, Vol. 1, Clarendon Press, Oxford, **1993**.
2. J. M. Lehn, *Supramolecular Chemistry—Concepts and Perspectives*, VCH, Weinheim, **1995**.
3. A rare example of a book describing the discovery and structure elucidation of many of the basic hydrocarbons has been published in the *Classic Researches in Organic Chemistry* Series H. Hart (*Ed.*), A. A. Baker, Jr, *Unsaturation in Organic Chemistry*, Houghton Mifflin, Boston, **1968**.
4. K. Weissermel, H.-J. Arpe, *Industrial Organic Chemistry*, 3rd ed., WILEY–VCH, Weinheim, **1997**.
5. *Ullmann's Encyclopedia of Industrial Chemistry*, H.-J. Arpe, E. Biekert, H. T. Davies, W. Gerhartz, H. Gerrens, W. Keim, J. L. McGuire, A. Mitsutani, H. Pilat, H. Reece, H. E. Simmons, E. Weise, R. Wirtz, H.-R. Wüthrich (*Eds.*), 6th ed.,WILEY–VCH, Weinheim, **1998**.
6. H.-G. Franck, J. W. Stadelhofer, *Industrial Aromatic Chemistry,* Springer, Berlin, **1988**.
7. R. E. Kirk, D. F. Othmer, *Encyclopedia of Chemical Technology*, 4th ed., J. Wiley and Sons, New York, **1991-1998**.
8. A good summary of recent work on radicals, carbenes, carbenoids, carbocations, and carbanions can be found in Houben-Weyl, *Methoden der Organischen Chemie*, Vol. E19, M. Regitz, B. Giese, M. Hanack (*Eds.*), Thieme, Stuttgart, **1989**.
9. K. C. Nicolaou, E. J. Sorensen, *Classics in Total Synthesis—Targets, Strategies, Methods*, VCH, Weinheim **1996,** E. M. Carreira, *Classics in Stereoselective Synthesis*, WILEY–VCH, in preparation.
10. See the review *Organic Chemistry* **1976** in *Nachr. Chemie, Labor, Technik*, **1977**, *25*, 59–70. Originally these yearly reviews of important developments in organic chemistry were written anonymously; the hydrocarbon section of the above summary was written by this author.

2 From Simple Building Blocks to Complex Target Molecules and Multifaceted Reactions

Hydrocarbons form the basis of organic chemistry and to a large extent the basis of organic reactivity as well. Although not the only binary compounds in organic chemistry—consider, for example, the chloro-[1] and the oxocarbons,[2] the cyanocarbons,[3] and the fully metalated (e.g. lithiated) hydrocarbons[4]—they are by far the most numerous and important. Hydrocarbons can be 'constructed' by a very simple *Aufbauprinzip* (built-up principle) that in its most elementary variant uses just three building units—the sp^3- (**1**), the sp^2- (**2**), and the sp-hybridized (**3**) carbon atom (Scheme 1).

Allowing only one additional carbon atom, and considering only molecules made up of covalently bonded atoms these three basic units translate into the C–C single (**4**), the C–C double (**5**), and the C–C triple bond (**6**) (Scheme 2).

How these very elementary bonding situations can be used to design—and later synthesize!—molecular edifices of often stunning beauty and bewildering complexity is the topic of a large part of this book. These molecular constructions occasionally have been created for their own sake; more often they are of great importance in many areas of basic and applied research.

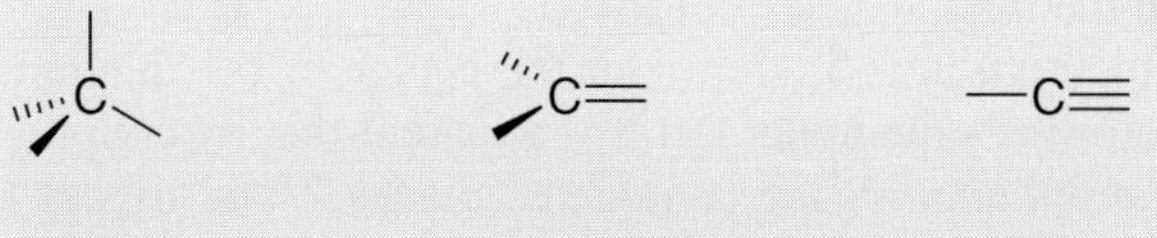

Scheme 1. The building blocks of hydrocarbon chemistry.

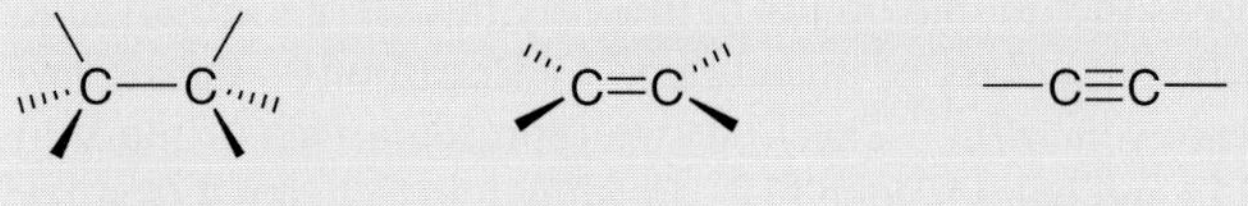

Scheme 2. Single, double and triple bonds between carbon atoms—the bricks of the molecular architect.

If a chemist were given just a carbon-carbon single bond as a repetitive unit (plus the appropriate number of hydrogen atoms to 'saturate' his construction at the end), what kind and types of molecular structures could he build? The simple answer, given in every elementary chemistry textbook, is shown in Scheme 3:

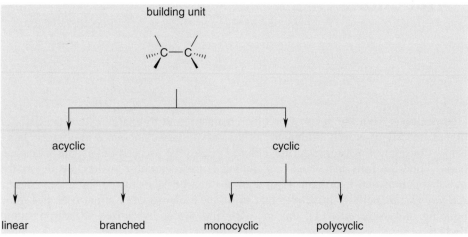

Scheme 3. Hydrocarbon structures that can be formed from C–C single bonds.

Whereas acyclic saturated hydrocarbons[5] will not be explicitly mentioned in this book, monocyclic[6] and especially polycyclic hydrocarbons will be discussed extensively. The reader not familiar with the wonders of polycyclic hydrocarbon chemistry—especially in the form of the so-called 'cage hydrocarbons'—will be surprised about the richness of structures covered by the simple term 'polycyclic' (see Chapters 3–5).

Whereas Scheme 3 does not show any structural detail, Scheme 4, which presents important results of building with the 'C=C-brick', does:
Six branches extend from the repetitive unit **5**, shown in the center of the diagram. Beginning with the arm pointing towards the upper left it is apparent that by coupling two and then three double bond units *linearly* the conjugated dienes **7** and trienes **8** are created, the first two members of the class of the *conjugated oligo-* and *polyenes* (see Chapter 7).

Preceding clockwise, two and then three double bonds are joined *circularly* to provide cyclobutadiene (**9**) and benzene (**10**), the first two members of the *[n]annulene* series (see Chapter 10). Whereas the double bonds in these hydrocarbons are arranged in an *endocyclic* fashion, in the next class of unsaturated hydrocarbons—the *radialenes*—they are all oriented *semicyclically*.[7] The first two representatives of the vinylogous series, [3]radialene (**11**) and [4]radialene (**12**) are shown in Scheme 4 (see Section 11.5). A hybrid between the two types of double bond arrangement just discussed is shown next with the *fulvenes*, which are made up of both *endo-* and *semicyclic* double bonds. The scheme shows triafulvene (**13**) and pentafulvene (**14**, usually just called 'fulvene') as the simplest members of this class of hydrocarbon (see Section 11.1). In contrast to the polyenes and the annulenes, the radialenes and the fulvenes are characterized by an arrangement of π-electrons which is called *cross-conjugated*.

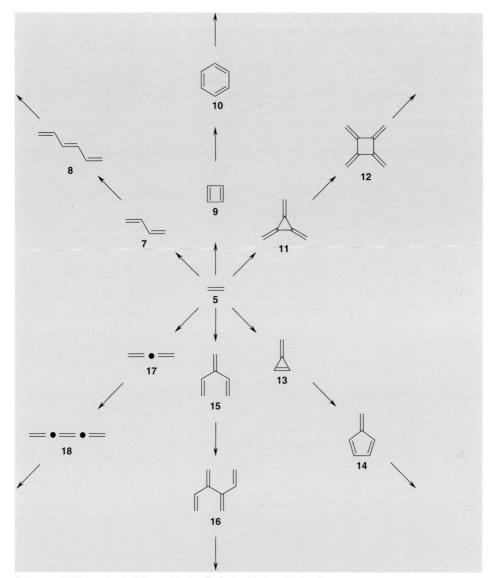

Scheme 4. Molecular building with the C–C double bond–simple structures.

Clearly, **13** is the simplest conceivable hydrocarbon of this type. Cross-conjugation is encountered also in the so-called *dendralenes* represented at the bottom of the scheme by the first two members of this particular series, [3]dendralene (**15**) and [4]dendralene (**16**) (see Section 11.4). Staudinger, who was interested in cross-conjugation during the early years of his career, has called molecules like **15** 'open fulvenes'[8]—because they can formally be generated by 'cutting through' certain single bonds of the corresponding 'closed' fulvenes.

In all examples discussed so far, the joining of the C=C blocks has been achieved by use of single bonds. If this task is taken over by a (then shared) carbon atom we can also include the cumulenic double bond systems in our scheme. This is shown with the first two members, allene (**17**, propadiene) and [3]cumulene (**18**, butatriene) in the last branch of the scheme (see Chapter 9).

The importance of these π-electron systems in organic chemistry differs widely. The polyolefins **7** and **8** and their higher vinylogs are subsystems of such diverse hydrocarbons as β-carotene (see Section 7.1) and polyacetylene[9] (see below) which can be regarded as archetypical examples from the *natural* and the *designed* world of organic chemistry. The overwhelming importance of aromatic hydrocarbons in both industry and fundamental research needs no further comment (or very detailed com-

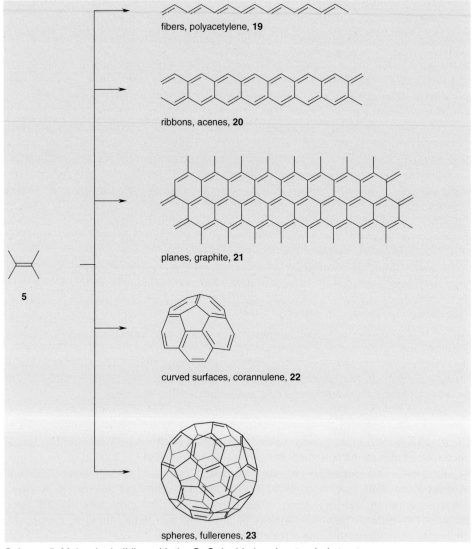

fibers, polyacetylene, **19**

ribbons, acenes, **20**

planes, graphite, **21**

curved surfaces, corannulene, **22**

spheres, fullerenes, **23**

5

Scheme 5. Molecular building with the C=C double bond–extended structures.

ment[10]). Although cross-conjugation is a phenomenon often encountered in *e.g.* dye-stuff chemistry, the cross-conjugated hydrocarbons—the fulvenes, radialenes, and dendralenes—are presently of far lesser practical importance than the first two classes of π-electron systems. This is also true for the cumulenic hydrocarbons and their derivatives[11].

Have we exhausted all conceivable combinations of double bonds with Scheme 4? Not at all!

On the next higher level of construction we can integrate the π-systems just generated into more complex 'extended π-systems' as illustrated in Scheme 5.

As one example—many more are possible and the reader is asked just to let her or his fantasy flow—consider polyacetylene **19**, which we can regard as the extension of the linear combination of double bonds to infinity.[9] We can also call **19** a fiber molecule and it has actually been likened to a 'molecular wire'. If two such fibers are connected by single bonds—as shown in **20**—ribbon structures arise. These are named *acenes* and the lower members of the series have been prepared as will be described later (Section 14.1). Continuing our *aufbau* work we can add an increasing number of polyacetylene fibers and generate molecular π-planes, which—when polymeric—are commonly called 'graphite' (**21**).

We must not stop here, either. If other than six-membered rings are included in our growing π-planes *non-planar structures* become possible, as exemplified by the curved surface of corannulene (**22**)[12] (see Section 12.2). Note that this hydrocarbon can also be regarded as a derivative of [5]radialene (see Section 11.5). As the planar arrangement of double bonds reaches its ultimate realization in **21**, the curved structures lead to molecular spheres, the fullerenes, of which the presently most prominent one, C_{60} (**23**), is shown in Scheme 5. At this point, however, we have left hydrocarbon chemistry and arrived in a new field—the new allotropes of carbon.[13] The close generic relationship between these two areas is also apparent from the different views of C_{60} shown in Scheme 6.

Depending on the 'resolution' of our view we can recognize numerous hydrocarbon substructures in **23**—relatively simple ones such as paracyclene (**24**), but also more complex ones such as corannulene (**21**) or tricyclopenta[*def;jkl;pqr*]triphenylene (**25**, sumanene). Many other hydrocarbon 'cut-outs' are possible—as an exercise the reader is asked to deconstruct C_{60} him/herself—and a substantial number has been prepared during the last few years. Clearly, a *designed* route to C_{60} will have to rely on hydrocarbon intermediates obtained by a rationally planned and executed synthetic sequence. In a text on important hydrocarbons, syntheses describing some of these 'steps towards C_{60}' must be included (see Section 12.2).

If triple bonds are the only allowed building units just one combination is possible—the molecular rod, $H-(C{\equiv}C)_n-H$ (Section 8.1). Cyclic variants are possible—and have actually been generated[14]—but they are again new forms of carbon ('cyclocarbon'), no longer hydrocarbons.

An endless number of hybrids is possible between the three basic systems **4** to **6**. A minute selection will be mentioned in this introductory chapter to show the reader what to expect later.

In the alkene field, hydrocarbons formally constructed from sp^2- and sp^3-hybridized carbon-atoms only are of great interest. They include highly strained olefins such as cyclopropene (**26**), the *trans*-cycloolefins, **27**, and the bi- or polycyclic *anti*-Bredt hydrocarbons, **28**, which we will discuss in detail in Chapter 7 (Scheme 7).

In molecules with more than one double bond these again must not lie in the same plane. All kinds of orientations are conceivable, the three-dimensional parallel alignment of barrelene (**29**) is just one (important) example (for more see Chapter 13).

Among molecules containing formal single and triple carbon-carbon bonds only the homologous series of the cycloalkynes is particularly noteworthy (Scheme 8).

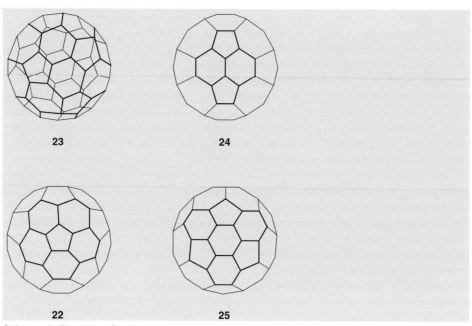

Scheme 6. Resolving C$_{60}$ into hydrocarbon substructures.

Scheme 7. A selection of alkenes built from **4** and **5** only.

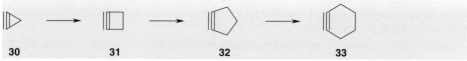

Scheme 8. The cycloalkynes: From which ring size onwards do they become isolable?

Beginning with the extremely strained cyclopropyne (**30**), when will we reach the stability limit which enables us to isolate the appropriate cycloalkynes? The answer will

be given in Section 8.2. It is not so long ago that a deliberate attempt to generate structures such as **30** and **31** would not have been undertaken by a serious organic chemist!

So far we have applied our simple *aufbau* principle to generate *hydrocarbon*
structures only. We have seen that it is quite easy to establish generic relationships—
family trees, so to speak—between many classes of hydrocarbons, thus creating order
in an area which on first sight might overpower us with its huge structural diversity.

A very similar back-of-the envelope approach[15] can also be employed to generate—and to discover!—lines of heritage between *hydrocarbon reactions*. Above we
have used the double bond, **5**, as one example to generate *structure patterns*; now we
will employ just one reaction to produce *reactivity patterns* in hydrocarbon chemistry.
The reaction which we select for this purpose is the *Cope* rearrangement (Scheme 9),
one of the most important and most thoroughly studied reactions in organic chemistry. In this pericyclic process, one bis allyl system **34** is converted into its isomer **35**
merely by heating. With R = H the process is *degenerate* and the reaction is an
automerization, with R ≠ H an equilibrium between the two valence isomers **34** and
35 is established in which one side might be strongly favored (thus making the process useful for synthesis).

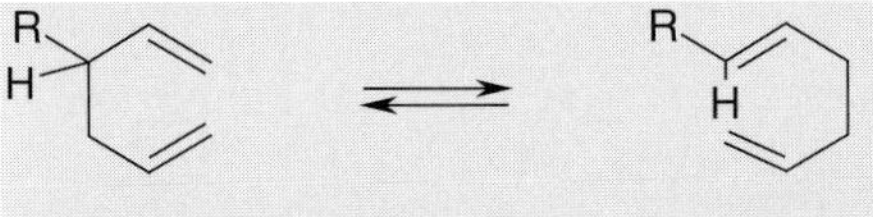

34   **35**

Scheme 9. The Cope rearrangement–one of the fundamental processes of organic chemistry.

Let us incorporate this rather simple arrangement of two double and three single
carbon-carbon bonds into more complex structures! Or, to put it another way, let us
take this motif and ask what variations of it are possible. Realizing that in **34/35** all
five building blocks can be replaced by structural elements with increasing unsaturation and/or complexity, a *rearrangement matrix*, shown in Scheme 10, results.

Without discussing every single entry of the matrix its underlying rationale is obvious: The role of the X–Y fragment may be played by a single, double, triple *etc.*
bond—all the way to the benzene ring, where we will deliberately stop (horizontal
variation). And the double bonds of the original Cope system can successively be replaced by triple bonds or allene groups, again ending our variations here to limit the
size of the matrix (vertical variations). Of the 42 combinations shown here many have
been studied and have become important in preparative organic chemistry—in addition
to the original Cope process (row 1, column 1, r1/c1),[16] these are the Bergman cyclization (r3/c2),[17] the divinylcyclopropane rearrangement (r1/c5),[18] the divinylcyclobutane isomerization (r1/c6),[19] the Saito-Meyers rearrangement (r5/c2),[17] the cyclization of 1,3-hexadien-5-yne to benzene (r2/c2),[20] the interconversion of 1,5-
hexadiyne to 3,4-bismethylenecyclobutene (r3/c1),[21] and several others.[22] Not all of
the combinations shown have been verified experimentally, though, leaving room for
discovery.

But as we could continue the building process with the C–C double bond—from
Scheme 4 to Scheme 5—we can incorporate the Cope process into increasingly complex—and again 'three-dimensional'—hydrocarbon frameworks (Scheme 11).

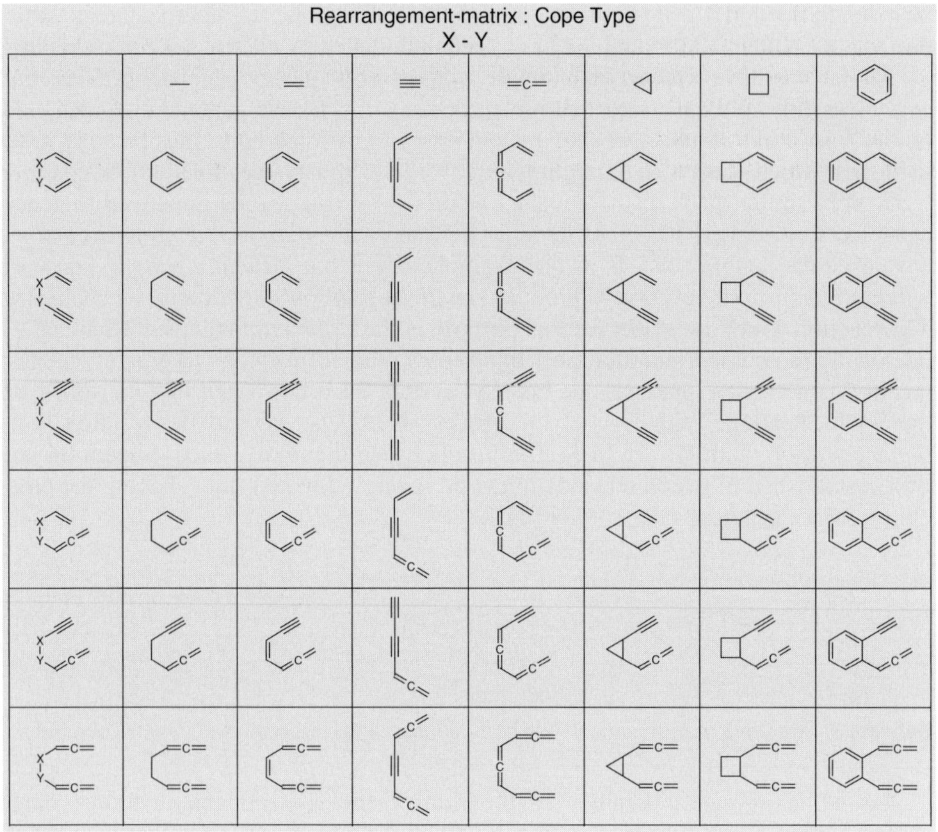

Scheme 10. A selection of hydrocarbons that in principle can undergo Cope-type rearrangements.

Beginning with divinylcyclopropane (**36**), which we have already encountered as one of the combinations in the initial Cope rearrangement matrix (Scheme 10), we can either connect its vinyl ends by a (then bridging) methylene group or we can short-circuit these ends—the resulting hydrocarbons are homotropylidene (**37**) and norcaradiene (**39**), respectively. Homotropylidene, a hydrocarbon noted for its fluxional behavior,[23] paved the way to the most celebrated fluxional molecule of them all, bullvalene (**38**),[24] a $C_{10}H_{10}$ molecule in which by way of repetitive Cope isomerizations any carbon and any hydrogen position can be interconverted into any other (at about 100 °C; see Section 13.3). The most characteristic property of norcaradiene (**39**), on the other hand, is its valence isomerization to the monocyclic hydrocarbon tropylidene (**42**, 1,3,5-cycloheptatriene).[25] When this Cope isomerization takes place in a derivative of **38** in which the bridgeheads are spanned by an additional 1,3-butadiene unit, hydrocarbon **40** results.[26] Norcaradiene-cycloheptatriene ring-opening of this produces the bridged aromatic hydrocarbon methano[10]-annulene (**41**),[27] a 10π-electron system which fulfils Hückel's rule and being aromatic on all counts (see Section 10.5). Tropylidene (**42**) itself can be used for (formal) construction higher vinylogs such as the doubly-bridged [14]annulene **43**[28] (see

Section 10.5).

Generic relationships such as this are typical of hydrocarbon chemistry and are a consequence of the *Aufbauprinzip*—although more often than not these connections become obvious only after extensive preparative work. For the future development of the field, heuristic thinking by which both complex products *and* processes are constructed from basic structure and reactivity 'modules' should, however, be employed more often.

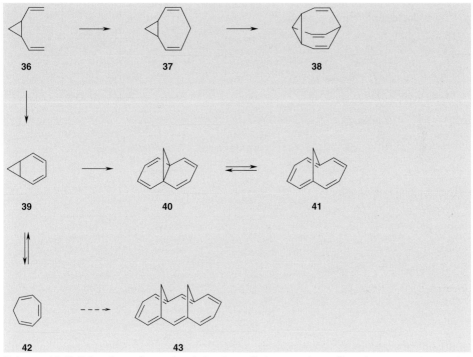

Scheme 11. Building three-dimensional structures that can undergo Cope-type rearrangements.

Before terminating our building-process with the Cope rearrangement we note that it cannot only be extemporized as shown in Schemes 10 and 11 but that we can also replace its hydrogen and carbon atoms by isovalent substitutents or atoms thus 'creating' such well-known processes as the Claisen rearrangement, the oxy-Cope rearrangement, the Claisen ester rearrangement, selecting these few examples more or less at random.[29] The Cope rearrangement is certainly an important organic reaction, but the principle it represents is of far greater significance for synthetic and mechanistic organic chemistry.

What we have just demonstrated for **34** and **35** we can also do with many other hydrocarbon reactions, both intra- and intermolecular processes. To fire the reader's imagination Scheme 12 summarizes (a small selection) of important intramolecular isomerizations which can—and to some extent have already been—subjected to similar permutations, as demonstrated in Schemes 10 and 11 above. All these processes have contributed substantially to the development of physical organic and theoretical organic chemistry.[30]

For mechanistic studies hydrocarbons are interesting substrates because no functional (meaning hetero-atom containing) group 'disturbs' the course of the process under consideration, *i.e.* does not make it more complex than it necessarily has to be. Needless to say, it is usually the polarizing effect of functional groups that accounts to a large extent for the richness and variability of organic reactions. Although it is a gross

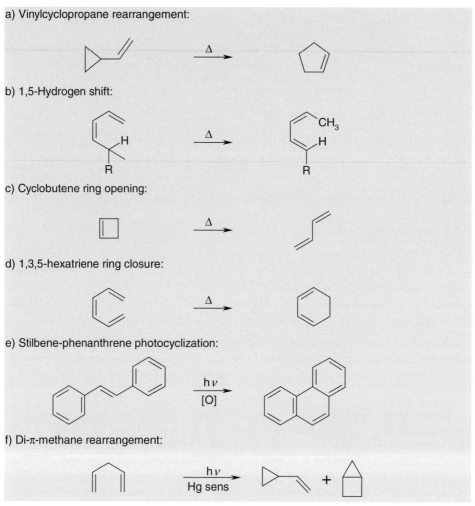

Scheme 12. A selection of intramolecular hydrocarbon reactions as prototypes for important chemical transformations.

oversimplification to say that problems of stereoselectivity (regio-, diastereo- and enantioselectivity) play no role in hydrocarbon chemistry—all the processes summarized in Scheme 12 have been studied carefully from the stereochemical viewpoint also—stereoselection and its generation is more a problem of (polar) functional group organic chemistry.

In our discussion of the significance of hydrocarbons in organic chemistry we have so far stressed structural and reactivity viewpoints and we have made the tacit assumption that in our various building exercises the bonding parameters of the building units **1** to **6** are unperturbed, *e.g.* that the angles in **1**, **2**, and **3** are 109.5, 120,

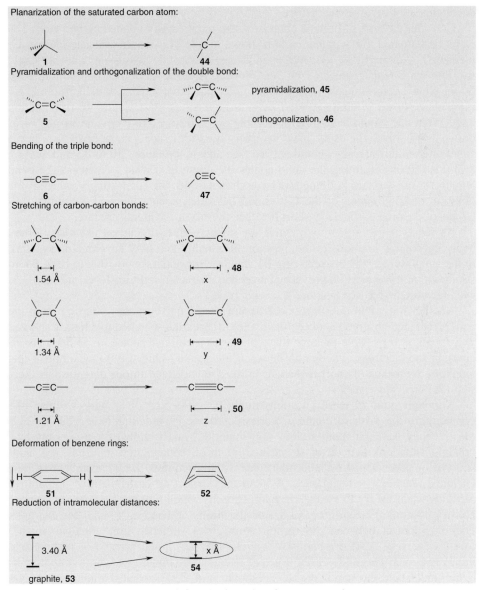

Planarization of the saturated carbon atom:

Pyramidalization and orthogonalization of the double bond:

pyramidalization, **45**

orthogonalization, **46**

Bending of the triple bond:

Stretching of carbon-carbon bonds:

Deformation of benzene rings:

Reduction of intramolecular distances:

graphite, **53**

Scheme 13. Ways to distort and deform hydrocarbon fragments and structures.

and 180°, and the carbon-carbon distances in **4**, **5**, and **6** are 1.54, 1.34, and 1.21 Å, respectively. Actually it is—and always has been since the days of Adolf von Baeyer and William Perkin, Jr—one of the main motives of chemists engaged in hydrocarbon work to violate and even break these standard bonding situations deliberately! How this can be accomplished for **1–6** is shown graphically in Scheme 13, which summarizes the most important of these bond-angle and bond-length deformations. Again, other molecular distortions can be conceived, and it is left to the reader to find more ways to arrange carbon and hydrogen atoms in three-dimensional space (and later prepare the 'designed' hydrocarbons and demonstrate their significance for the development of organic chemistry).

Consider the icon of organic chemistry, the tetrahedrally bonded carbon atom **1**— can it be flattened to the planar carbon shown in **44**? This would translate into planar methane if saturated with four hydrogen atoms, a molecule possessing what we could call *anti-van't Hoff* geometry (see Section 6.2). And the carbon-carbon double bond **5**—in what circumstances, *i.e.* in what molecular environment, will it become pyramidal, as shown in **45**, or twisted, even orthogonal as illustrated in **46** (see Section 7.4)? What one would have to do to a triple bond, **6**, to make it deviate strongly—as drawn in **47**—from the usual linear geometry is easier to see—its incorporation into a small or even normal ring should suffice (see above, structures **30** to **33** and Section 8.2). Instead of deforming the bond angles we might also want to increase the bond lengths of our standard building blocks as shown in **48–50**. Interestingly, whereas the question of lengthening of the C–C bond has often been addressed, systematic attempts of creating particularly short bonds between carbon atoms are rare.

As far as larger organic structures are concerned, hydrocarbon chemistry offers many interesting solutions to the question of whether and how far aromatic systems, in the simplest case the benzene ring **51**, can be distorted. Is it possible to bend **51** so strongly that it forsakes its proverbial aromatic character and transforms into a 1,3,5-cyclohexatriene (**52**, see Sections 12.3 and 16.3)?

Another important question concerns intramolecular distances. Starting from the 3.40 Å distance observed between the planes of graphite, sketched in a highly stylized way in **53**, is it possible to force these planes to a closer distance, as in **54**, thus increasing the electronic interaction between the planes? One way to accomplish this would be by means of short molecular bridges as indicated in our diagram (see Section 12.3 on cyclophanes).

Although some of these questions might have the ring of a sports(wo)man-like competition, the wish to create a world-record for a particular molecular arrangement,[6] they have far more serious and important implications. Firstly, to produce bonding situations that differ strongly from those encountered in usual, *e.g.* nonstrained or undeformed organic molecules usually requires the development of new chemical reactions or techniques. We can thus expect an enrichment of preparative methods when we try to synthesize hydrocarbons which will enable us to answer the above questions. Furthermore, to define the limits of bonding (when, for example, does the contact between the carbon atoms in a stretched bond finally vanish?) touches the very heart of chemistry, which, after all, is the science of making and breaking bonds. The study of deformed or even bizarre bonding situations can help us better understand the standard cases as shown in **4–6**. Theoretical studies on the nature of the chemical bond have always profited from, and been inspired strongly, by molecules with unusual bond lengths and bond angles. And, finally, deformed bond lengths and angles translate into unusual distributions of electrons which in turn causes surprising, often drastically enhanced, chemical reactivity.

There is one final reason why the study of hydrocarbons attracts many chemists—their wish to play is often fulfilled extremely well on this exciting playing ground of organic chemistry. Whether (the mostly male) practitioners speak of tinker toy chemistry, molecular Lego or Meccano sets, the connection to an earlier part of their lives is obvious enough. It has occasionally resulted in bizarre hydrocarbon structures, some of which are shown in Scheme 14 together with their amusing, but often highly descriptive trivial names.[31]

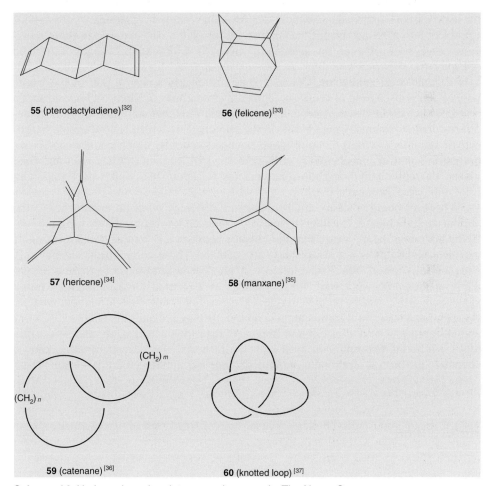

55 (pterodactyladiene)[32]   **56** (felicene)[33]

57 (hericene)[34]   **58** (manxane)[35]

(CH$_2$)$_m$

(CH$_2$)$_n$

59 (catenane)[36]   **60** (knotted loop)[37]

Scheme 14. Hydrocarbon chemistry as a playground—The Name Game.

The reader should not pass judgment on these molecular games too quickly, though! What often started as *fröhliche Wissenschaft,*[38] as a joyful undertaking in science—and often as *l'art pour l'art*—at some later stage was actually shown to be of considerable importance. The synthesis of molecules like the two interlocking rings **59**, a so-called catenane, and the knotted loop shown in **60** was originally performed mainly for topological reasons, certainly not with any kind of application in mind. Only a few years later it was discovered that nature had already prepared these amazing structures which are not held together by covalent bonds. In the early 1960ies single-stranded and double-stranded circular DNA was discovered[39] and it was shown that duplex circular DNA molecules are made up of two closed intact DNA rings which can appear in nature in different configurations, among them catena and knot structures—just as in **59** and **60**!

Turning now to detailed discussions of classical—in both meanings of the term (see Chapter 1)—hydrocarbon syntheses, I shall start with saturated, mostly polycyclic hydrocarbons—the adamantanes, the prismanes and the platonic hydrocarbons (Chapters 3 to 5). Since bridgehead carbon atoms play a decisive role in these molecules a separate chapter on bridgehead-distorted hydrocarbons seems warranted (Chapter 6). We then will look at various unsaturated hydrocarbons—alkenes, alkynes, cumulenes, annulenes (Chapters 7 to 10), clearly a central part of the whole book. After a discussion of cross-conjugated hydrocarbons (Chapter 11) our next step will take us out of the π-plane towards aromatic (Chapter 12) and olefinic three-dimensional π-systems (Chapter 13). In the two chapters which follow simple hydrocarbon subunits are used to build more extended systems, which could become important in nano- or picochemistry (Chapters 14, 15); we then throw a brief and final glance on reactive hydrocarbon intermediates (Chapter 16), which play important roles in physical organic chemistry.

Whenever possible I have tried to compare different synthetic approaches to the various target molecules without trying to presents all known syntheses for a particular hydrocarbon, being aware that the balance between a review and a discussion of chemical highlights is not always easy to maintain! These comparisons are not only pedagogically useful, but also are often very good demonstrations of progress in synthetic methodology very well. Retrosynthesis as a method for the design of future synthetic work[40] will be applied to selected cases, but is not used throughout the text. As useful and successful as this approach obviously is, it makes the chance discovery and the unforeseen result seem less important than they actually are, and sometimes looks a little bit like rationalization after the fact. As every experimentalist knows, chemistry is often cleverer and more intricate than our thinking about it—fortunately.

References

1.   Reviews: R. Stroh, W. Hahn in Houben-Weyl-Müller, *Methoden der Organischen Chemie*, Thieme, Stuttgart, **1962**, Vol. V/3, 503–1078; R. West, *Acc. Chem. Res.*, **1970**, *3*, 130–138; T. Chivers in *The Chemistry of the Carbon–Halogen Bond, Part 2*, S. Patai, (*Ed.*), J. Wiley & Sons, London, **1973**, 917–977.
2.   Reviews: R. West (*Ed.*) *Oxocarbons*, Academic Press, New York, N. Y., **1980**; F. Serratosa, *Acc. Chem. Res.*, **1983**, *16*, 170–176; G. Seitz, P. Imming, *Chem. Rev.*, **1992**, *92*,

1227-1260.

3. Reviews: E. Ciganek, W. J. Linn, O. W. Webster in *The Chemistry of the Cyano Group,* Z. Rappoport, *(Ed.)*, Interscience Publishers, New York, N. Y., **1970**, 423–638; R. Dworczak, H. Junek in *The Chemistry of triple-bonded functional groups, Supplement C2,* S. Patai, *(Ed.)*, J. Wiley, New York, N. Y., **1994**, 789–871.

4. Review: A. Maercker, M. Theis, *Top. Curr. Chem.,* **1987**, *138*, 1–61.

5. This may actually turn out not to be a mean task in a specific case. For example the preparation of the longest (non-polymeric) acyclic hydrocarbon known up to 1997, CH_3–$(CH_2)_{388}$–CH_3 required a multi-step sequence: I. Bidd, M. C. Whiting, *J. Chem. Soc. Chem. Commun.,* **1985**, 543–544.

6. The biggest saturated *monocyclic* ring system ever to be reported seems to be cyclooctaoctacontodictane, $C_{288}H_{576}$: K. S. Lee, G. Wegner, *Macromolec. Chem. Rapid Commun.,* **1985**, *6*, 203–208. For more of these world records in chemistry see H. J. Quadbeck-Seeger, R. Faust, G. Knaus, U. Siemeling, *Chemierekorde*, WILEY-VCH, Weinheim, **1997**; R. Faust, G. Knaus, U. Siemeling, *World Records in Chemistry*, WILEY-VCH, Weinheim, **1999**.

7. A. von Baeyer, *Ber. Dtsch. Chem. Ges.,* **1894**, *27*, 436–454, already distinguished three types of bonding situations for double bonds: they may be endocyclic, *i.e.* be incorporated into a ring—like in cyclohexene—or exocyclic, *i.e.* not part of a ring such as in vinylcyclohexane. The situation where one of the carbon atoms is part of the ring, and the other outside of the ring—as in methylenecyclohexane—was called semicyclic by him. Unfortunately this very clear differentiation never caught on in chemistry, since most chemists call the last bonding situation exocyclic. I will use von Baeyer's designation, which was rediscovered by G. Zinner (*Chemiker-Ztg.,* **1985**, *109*, 436) throughout this book and hope to make it more popular.

8. H. Staudinger, *Arbeitserinnerungen*, Hüthig Verlag, Heidelberg, **1961**, 15.

9. Review: H. Shirakawa, T. Masuda, K. Takeda in *The Chemistry of triple-bonded functional groups, Supplement C2,* Vol. 2, S. Patai *(Ed.)*, J. Wiley, New York, N. Y., **1991**, 945–1016.

10. Reviews on various aspects of industrial aromatic chemistry: H.-G. Franck, J. W. Stadelhofer, *Industrial Aromatic Chemistry*, Springer, Berlin, **1988**; M. Zander, *Polycyclische Aromaten*, Teubner Studienbücher, Teubner, Stuttgart, **1995**; G. A. Olah, A. Molnar, *Hydrocarbon Chemistry*, J. Wiley & Sons, New York, N. Y., **1995**; K. Weissermel, H.-J. Arpe, *Industrial Organic Chemistry*, 3rd ed., VCH, Weinheim, **1997**.

11. Reviews on allene chemistry: L. Brandsma, H. D. Verkruijsse, *Synthesis of Acetylenes, Allenes and Cumulenes*, Elsevier, Amsterdam, **1981**; S. R. Landor *(Ed.)*, *The Chemistry of the Allenes*, Academic Press, Vol. I-III, London, **1982**; H. F. Schuster, G. M. Coppola, *Allenes in Organic Synthesis*, Wiley-Interscience, New York, N. Y., **1984**.

12. Even if we were to stick to six-membered rings only, we could create three-dimensional π-systems; for example, if we would connect the ends of a sufficiently long acene ribbon, a torus-shaped π-structure would arise (see Section 14.3).

13. H. W. Kroto, J. R. Heath, S. C. O′Brien, R. F. Curl, R. E. Smalley, *Nature*, **1985**, *318*, 162–163; W. Krätschmer, L. D. Lamb, K. Fostiropolous, D. R. Huffman, *Nature*, **1990**, *347*, 354–358; H. W. Kroto, *Angew. Chem.*, **1997**, *109*, 1648–1664; *Angew. Chem. Int. Ed. Engl.*, **1997**, *36*, 1578–1593; R. E. Smalley, *Angew. Chem.*, **1997**, *109*, 1666–1673; *Angew. Chem. Int. Ed. Engl.*, **1997**, *36*, 1594–1601; Reviews: A. Hirsch, *Chemistry of the Fullerenes*, Thieme, New York, **1995**; A. Hirsch *(Ed.)*, *Fullerenes and Related Structures, Top. Curr. Chem.*, **1999**, Vol. 199.

14. F. Diederich, Y. Rubin, C. B. Knobler, R. L. Whetten, K. E. Schriver, K. N. Houk, Y. Li, *Science*, **1989**, *245*, 1088–1090; *cf.* F. Diederich in *Modern Acetylene Chemistry*, P. J.

Stang, F. Diederich (*Eds.*), VCH, Weinheim, **1995**, 443–471.

15. What is done here in a very elementary fashion and without requiring any special knowledge has been developed systematically employing the calculation powers of modern computers. Today a large number of expert systems for reaction prediction and synthesis design is available. Reviews: J. H. Winter, *Chemische Syntheseplanung*, Springer, Heidelberg, **1982**; J. Gasteiger, M. G. Hutchings, B. Christoph, L. Gann, Chr. Hiller, P. Löw, M. Marsili, H. Saller, K. Yuki, *Top. Curr. Chem.*, **1987**, *137*, 19–73; R. Herges, *J. Chem. Inf. Comput. Sci.*, **1990**, *30*, 377–383; R. Herges, Chr. Hoock, *Science*, **1992**, *255*, 711–713; A. V. Zeigarnik, D. Bonchev, *Chemical Reaction Networks*, CRC Press, Boca Raton, **1996**.

16. S. R. Wilson, *Org. React.*, **1993**, *43*, 93–250 and refs. to the earlier literature.

17. For a recent review on the Bergman and related rearrangements see K. K. Wang, *Chem. Rev.*, **1996**, *96*, 207–222.

18. The divincyclopropane rearrangement was discovered by E. Vogel, *Angew. Chem.* **1960**, *72*, 4–26; *cf.* E. Vogel, K.-H. Ott, K. Gajek, *Liebigs Ann. Chem.*, **1961**, *644*, 172–188; Review: T. Hudlicky, R. Fan, J. W. Reed, K. G. Gadamasetti, *Org. React.*, **1992**, *41*, 1–133.

19. E. Vogel, *Liebigs Ann. Chem.* **1958**, *615*, 1–14.

20. H. Hopf, H. Musso, *Angew. Chem.*, **1969**, *81*, 704; *Angew. Chem. Int. Ed. Engl.*, **1969**, *8*, 680; *cf.* U. Nüchter, H. Hopf, G. Zimmermann, *Liebigs Ann. Chem.*, **1997**, 1505–1515 and refs. cited therein.

21. W. D. Huntsman, H. J. Wristers, *J. Am. Chem. Soc.*, **1963**, *85*, 3308–3309; W. D. Huntsman, H. J. Wristers, *J. Am. Chem. Soc.*, **1967**, *89*, 342–347; Reviews: W. D. Huntsman in *The Chemistry of Ketenes, Allenes and Related Compounds, Part 2*, S. Patai, (*Ed.*), J. Wiley & Sons, Chichester, **1980**, 521–667; A. Viola, J. J. Collins, N. Filipp, *Tetrahedron*, **1981**, *37*, 3765–3811.

22. B. Engels, C. Lennartz, M. Hanrath, M. Schmittel, M. Strittmatter, *Angew. Chem.*, **1998**, *110*, 2067–2070; *Angew. Chem. Int. Ed. Engl.*, **1998**, *37*, 1960–1963; M. Schmittel, M. Keller, S. Kiau, M. Strittmatter, *Chem. Eur. J.*, **1997**, *3*, 807–816; N. Krause, M. Hohmann, *Synlett*, **1996**, 89–91; T. Gillmann, T. Hülsen, W. Massa, S. Wocadlo, *Synlett*, **1995**, 1257–1259.

23. W. v. E. Doering, B. M. Ferrier, E. T. Fossel, J. H. Hartenstein, M. Jones, G. Klumpp, R. M. Rubin, M. Saunders, *Tetrahedron*, **1967**, *23*, 3943–3963; W. v. E. Doering, W. R. Roth, *Angew. Chem.*, **1963**, *75*, 27–35; W. v. E. Doering, W. R. Roth, *Tetrahedron*, **1962**, *18*, 67–74; *cf.* G. Maier, *Valenzisomerisierungen*, Verlag Chemie, Weinheim, **1972**.

24. G. Schröder, *Angew. Chem.*, **1963**, *75*, 722–723; G. Schröder, *Chem. Ber.*, **1964**, *97*, 3140–3149; R. Merényi, J. F. M. Oth, G. Schröder, *Chem. Ber.*, **1964**, *97*, 3150–3161.

25. W. v. E. Doering, L. H. Knox, *J. Am. Chem. Soc.*, **1953**, *75*, 297–303; M. B. Rubin, *J. Am. Chem. Soc.*, **1981**, *103*, 7791–7792; Reviews: G. Maier, *Angew. Chem.*, **1967**, *79*, 446–458; *Angew. Chem. Int. Ed. Engl.*, **1967**, *6*, 402–414; W. Tochtermann, *Fortschr. Chem. Forsch.*, **1970**, *15*, 378–444; E. Vogel, W. Wiedemann, H. D. Roth, J. Eimer, H. Günther, *Liebigs Ann. Chem.*, **1972**, *759*, 1–36.

26. E. Vogel, H. D. Roth, *Angew. Chem.*, **1964**, *76*, 145; *Angew. Chem. Int. Ed. Engl.* **1964**, *3*, 228; *cf.* E. Vogel, W. Klug, *Org. Synth.*, **1974**, *54*, 11–18.

27. H. C. Dorn, C. S. Yannoni, H.-H. Limbach, E. Vogel, *J. Phys. Chem.*, **1994**, *98*, 11628–11629.

28. E. Vogel in *Current Trends in Organic Synthesis* H. Nozaki (*Ed.*), Pergamon Press, Oxford, **1983**, 379–400.

29. Reviews on the Claisen and Cope rearrangements: S. J. Rhoads, N. R. Raulins, *Org. React.*, **1975**, *22*, 1–252; F. E. Ziegler, *Acc. Chem. Res.*, **1977**, *10*, 227–232; G. B. Bennett,

Synthesis, **1977**, 589–606; S. Blechert, *Synthesis,* **1989**, 71–82.

30. Leading general references: J. M. Brown, *Thermolysis of Alicyclic Compounds* in MTP International Review of Science, Organic Chemistry, Series One, Vol. 5, W. Parker (*Ed.*), Butterworths, London, **1973**, 159–204: *cf.* D. Becker, N. C. Brodsky, MTP International Review of Science, Organic Chemistry, Series Two, Vol. 5, D. Ginsburg (*Ed.*), Butterworths, London, **1976**, 197–276; J. J. Gajewski, *Hydrocarbon Thermal Isomerizations*, Academic Press, New York, **1981**; G. Desimoni, G. Tacconi, A., Barco, G. P. Polini, *Natural Products Synthesis through Pericyclic Reactions*, American Chemical Society, ACS Monograph 180, Washington, D. C., **1983**. For reviews on: the vinylcyclopropane rearrangement see T. Hudlicky, T. M. Kutchan, S. M. Naqvi, *Org. React.,* **1985**, *33*, 247– 335; *cf.* Z. Goldschmidt, B. Crammer, *Chem. Soc. Rev.*, **1988**, *17*, 228–267; 1,5-hydrogen shifts see C. W. Spangler, *Chem. Rev.*, **1976**, *76*, 187–217; cyclobutene ring-opening processes see M. J. S. Dewar, C. Jie, *Acc. Chem. Res.*, **1992**, *25*, 537–543; W. R. Dolbier, Jr, H. Koroniak, K. N. Houk, C. Sheu, *Acc. Chem. Res.*, **1996**, *29*, 471–477; G. Mann, H. M. Muchall in Houben-Weyl, *Methods of Organic Chemistry*, Thieme, Stuttgart **1997**, Vol. E17f, A. de Meijere (*Ed.*), 667–684; electrocyclic reactions see E. N. Marvell, *Thermal Electrocyclic Reactions*, Academic Press, New York, **1980**; the stilbene-phenanthrene photocyclization see F. B. Mallory, C. W. Mallory, *Org. React.,* **1984**, *30*, 1–456; W. H. Laarhoven, *Org. Photochem.*, **1987**, *9*, 129–224; H. Meier, *Angew. Chem.*, **1992**, *104*, 1425–1446; *Angew. Chem. Int. Ed. Engl.*, **1992**, *31*, 1399–1420; the di-π-methane rearrangement see H. E. Zimmerman, D. Armesto, *Chem. Rev.,* **1996**, *96*, 3065– 3112 and lit. cit.

31. An amusing account of this playful attitude towards chemistry is described in A. Nickon, E. F. Silversmith, *Organic Chemistry: The Name Game*, Pergamon Press, New York, N. Y. **1987**.

32. *Pterodactyladienes* named after extinct flying reptiles known as *pterodactyli*; *cf.* R. Pettit, *Chem. Eng. News*, **1965**, August 23, 38–39; H.-D. Martin, M. Hekmann, *Chimia*, **1974**, *28*, 12–15; H.-D. Martin, M. Hekmann, *Angew. Chem.*, **1976**, *88*, 447–448; *Angew. Chem. Int. Ed. Engl.*, **1976**, *15*, 431–432; H.-D. Martin, M. Hekmann, *Tetrahedron Lett.,* **1978**, 1183–1186; H.-D. Martin, B. Mayer, M. Pütter, H. Höchstetter, *Angew. Chem.*, **1981**, *93*, 695–696; *Angew. Chem. Int. Ed. Engl.*, **1981**, *20*, 677–678.

33. *Felicene* named after the face of a smiling cat; *cf.* A. Gilbert, R. Walsh, *J. Am. Chem. Soc.*, **1976**, *98*, 1606–1607.

34. *Hericene* from the Latin word *hericeus* for hedgehog; *cf.* O. Pilet, P. Vogel, *Angew. Chem.*, **1980**, *92*, 1036–1037; *Angew. Chem. Int. Ed. Engl.,* **1980**, *19*, 1003–1004; O. Pilet, J.-L. Birbaum, P. Vogel, *Helv. Chim. Acta*, **1983**, *66*, 19–34. Review: P. Vogel in *Advances in Theoretically Interesting Molecules*, R. P. Thummel, (*Ed.*), JAI Press, Greenwich, CT, Vol. 1, **1989**, 201–355.

35. *Manxane* resembles the Isle of Man's coat of arms; *cf.* M. Doyle, W. Parker, P. A. Gunn, J. Martin, D. D. MacNicol, *Tetrahedron Lett.*, **1970**, 3619–3622.

36. *Catenanes* from the Latin word *catena* for chain; *cf.* E. Wasserman, *J. Am. Chem. Soc.*, **1960**, *82*, 4433–4434. For reviews and accounts on catenanes see: D. M. Walba, *Tetrahedron*, **1985**, *41*, 3161–3212; D. M. Walba, J. D. Armstrong III, A. E. Perry, R. H. Richards, T. C. Homan, R. C. Haltimanger, *Tetrahedron*, **1986**, *42*, 1883–1894; C. O. Dietrich-Buchecker, J.-P. Sauvage, *Chem. Rev.*, **1987**, *87*, 795–810; J.-P. Sauvage, *Acc. Chem. Res.*, **1990**, *23*, 319–327; C. O. Dietrich-Buchecker, J.-P. Sauvage, *Bioorg. Chem. Front.*, **1991**, *2*, 195–248; J.-C. Chambron, C. O. Dietrich-Buchecker, J.-P. Sauvage, *Top. Curr. Chem.*, **1993**, *165*, 131–162; F. Bickelhaupt, *J. Organomet. Chem.*, **1994**, *475*, 1– 14; D. B. Amabilino, J. F. Stoddart, *Chem. Rev.*, **1995**, *95*, 2725–2828; F. Vögtle, T. Dünnwald, T. Schmidt, *Acc. Chem. Res.*, **1996**, *29*, 451–460; A. C. Benniston, *Chem.*

Soc. Rev., **1996**, *25*, 427–435; M. Fujita, K. Ogura, *Coord. Chem. Rev.*, **1996**, *148*, 249-264; R. Jäger, F. Vögtle, *Angew. Chem.*, **1997**, *109*, 966–980; *Angew. Chem. Int. Ed. Engl.*, **1997**, *36*, 930-944; S. A. Nepogodiev, J. F. Stoddart, *Chem. Rev.*, **1998**, *98*, 1959–1976; D. G. Hamilton, J. E. Davies, L. Prodi, J. K. M. Sanders, *Chem. Eur. J.*, **1998**, *4*, 608–620; J.-C. Chambron, J.-P. Sauvage, *Chem. Eur. J.*, **1998**, *4*, 1362–1366.

37. G. Schill, *Catenanes, Rotaxanes, and Knots*, Academic Press, New York, N. Y., **1971**.

38. F. Nietzsche, *Die fröhliche Wissenschaft* (*La Gaya Sienza*), W. Goldmann, München, **1959**.

39. For a review on interlocked DNA structures see J. C. Wang, *Acc. Chem. Res.*, **1973**, *6*, 252–256;

40. E. J. Corey, X.-M. Cheng, *The Logic of Chemical Synthesis*, J. Wiley and Sons, New York, **1989**; St. Warren, *Organic Synthesis: The Disconnection Approach*, J. Wiley and Sons, Chichester, **1982**; J. Fuhrhop, G. Penzlin, *Organic Synthesis*, 2nd ed., VCH Publishers, Weinheim, **1994**.

3 Adamantane and other Cage Hydrocarbons

3.1 Adamantane

To begin a presentation of classic hydrocarbon syntheses with adamantane (**1**, Scheme 1) is justified on several grounds.

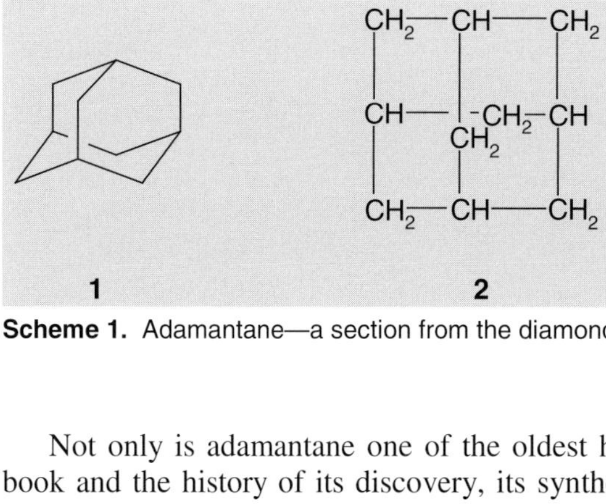

Scheme 1. Adamantane—a section from the diamond lattice.

Not only is adamantane one of the oldest hydrocarbons to be mentioned in this book and the history of its discovery, its syntheses and finally industrial production has all the hallmarks of a good story to catch and hold a readers attention. More important is the role which **1** played in the development and unfolding of organic chemistry.[1] Looking back along the winding road of organic chemistry, **1** can be recognized as a decisive signpost molecule marking the way not only to hydrocarbon chemistry as a separate field of study but also to an area of organic chemistry for which we still do not have an adequate term—*non-natural product chemistry*. Because this 'designer chemistry' is intimately connected to physical organic chemistry, **1** and its off-spring molecules are central to this branch of chemistry also.

That adamantane would attract the attention of chemists is easy to grasp: Its skeleton is a section of the defining element of organic chemistry—carbon—in its most beautiful and precious form—diamond. With its tell-tale name, adamante[2] reflects diamond's beauty in the high esthetic appeal of its symmetrical (point group T_d) molecular structure—although it tends to be forgotten that in days where computer-drawn, 'correct' three-dimensional structures are routinely used to represent molecules, the beauty of **1** is not easily recognized when looking at the early representations of this molecule, as shown in **2**.[3] In fact, adamantane—together with its 'sister molecule' urotropine—were the first cage compounds which forced chemists to accept three-dimensional structures of molecules.

Although **1** stands at the cross-roads of natural ('given') and non-natural ('designed, man-made, artifical') molecules it is a natural product itself. It was isolated in minute amounts (0.0004%) from a sample of petroleum collected near the village of Hodonín in Moravia by Landa and Machacek in 1933 who also proposed its name.[2] The constitution as shown in Scheme 1 was proposed by Lukes[4] on the basis of the unusual physical properties of the compound and of X-ray structural studies. The first synthesis, performed by Prelog and Seiwerth,[5] which put the early structural assignments on solid ground appeared eight years later and is reproduced in Scheme 2.

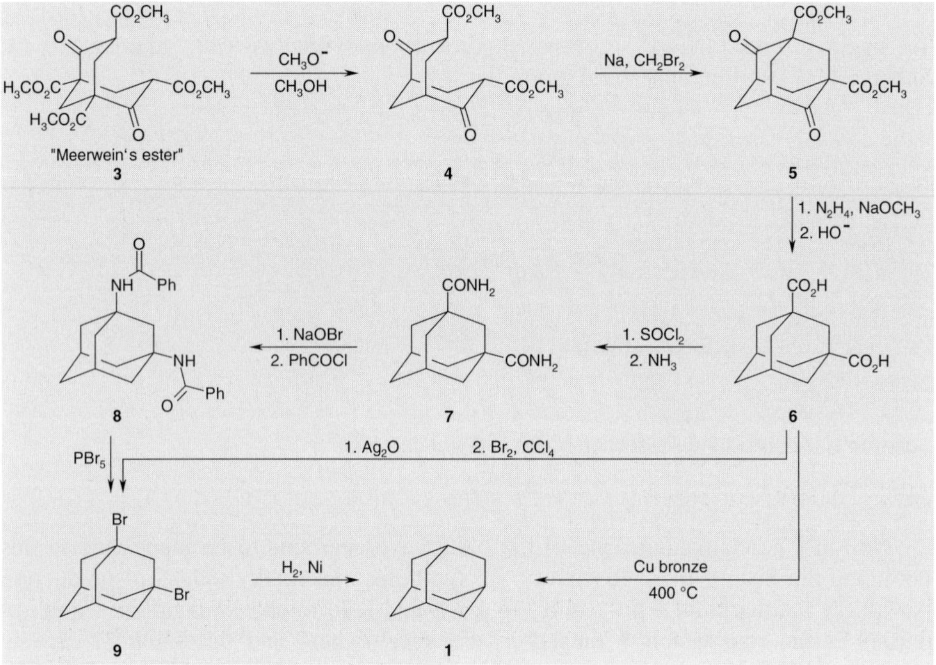

Scheme 2. The first synthesis of adamantane by Prelog (**1**).

The so-called Meerwein ester **3**, readily prepared by condensing formaldehyde with dimethyl malonate in the presence of piperidine, was employed as the starting material.[6] Although Böttger had already prepared the adamantane carbon framework from **3** by treating its disodium salt with dibromomethane, he was unable to produce the parent hydrocarbon from the various adamantane derivatives he had in hand.[7] Prelog circumvented these difficulties in the late part of the synthesis by using a less highly functionalized bicyclo[3.3.1]nonandione derivative—the diester **4**—which still allowed the bridging step to take place. When this was subjected to Böttger's conditions (a reaction which proceeds with varying and only moderate yields) the diester **5** resulted; this could be reduced by a Wolff–Kishner reduction and saponified to the 1,5-dicarboxylic acid **6**. Heating this with copper bronze at 400 °C lead to a compound identical with Landa's sample. Because the yield of the decarboxylation was miserable (2.4%) various alternatives were investigated to reduce **6** to **1**. Although these transformations of the derivatives **7–9**—which are shown in Scheme 2 without further comment because they all involve routine preparative chemistry—were successful, the total yield of **1** stayed in the mg range (0.3 % over all steps).[5]

A further simplification of the synthesis—again accompanied by increased yield (to *ca* 6% total) was devised by Stetter and coworkers.[8] It is reproduced in Scheme 3.

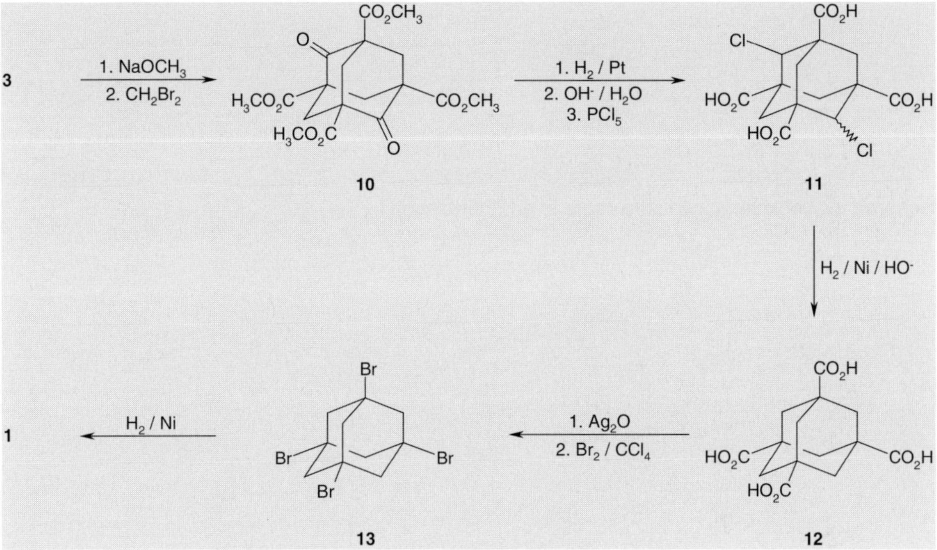

Scheme 3. Stetter's improved adamantane synthesis from **3**.

In this approach Böttger's original tetraester **10** was first converted into the dichloro derivative **11** as shown; this was then hydrogenated to the 1,3,5,7-tetraacid **12**. The latter, on Hunsdiecker degradation, provided the tetrabromide **13** which was finally reduced to the parent hydrocarbon.

Although **1** could now be prepared in 200-mg amounts, it remained a rare and precious, somewhat curious compound the chemistry of which could not be explored because of lack of material—a common fate in hydrocarbon chemistry, as we shall see in later chapters of this book for other compounds. The breakthrough came in 1957 when Schleyer published a one-page note on a discovery he had made while trying to isomerize *endo*-tetrahydrodicyclopentadiene **14** (which is easily obtained by catalytic hydrogenation of dicyclopentadiene) to its *exo* isomer **15**.[9] When **14** was heated under reflux overnight in the presence of 10% of its weight of aluminum trichloride the expected rearrangement indeed occurred and provided **15** in about 50% yield. It was, however, accompanied by approximately 12% of an isomeric hydrocarbon. Spectral comparison quickly showed that this was adamantane (**1**), which thus had been prepared literally overnight in only two steps form a commercially available precursor (Scheme 4).

Over the years numerous attempts were made to improve the yield of the **14→1** isomerization, and, indeed, conditions were found which gave **1** as the main product of the process. When gaseous **14** in dry hydrochloric acid was passed through a heated tube containing a chlorinated palladium-alumina catalyst[10] a 60% yield of **1** was realized. A practical laboratory synthesis, described in *Organic Syntheses*, yields **1** in 13–15%.[11]

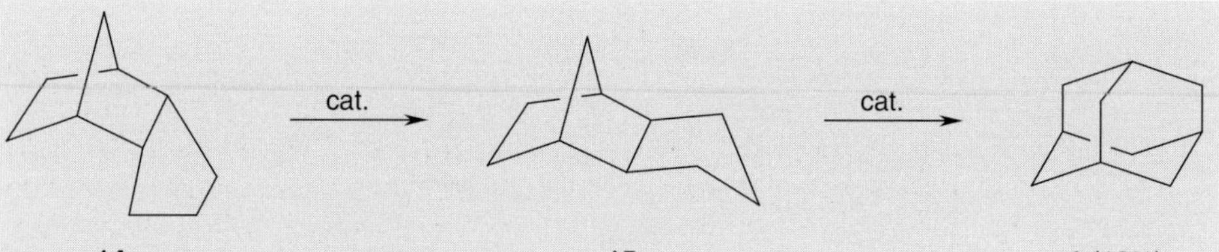

| 14 | 15 | 1 (12%) |

Scheme 4. Schleyer's two step route to adamantane (**1**).

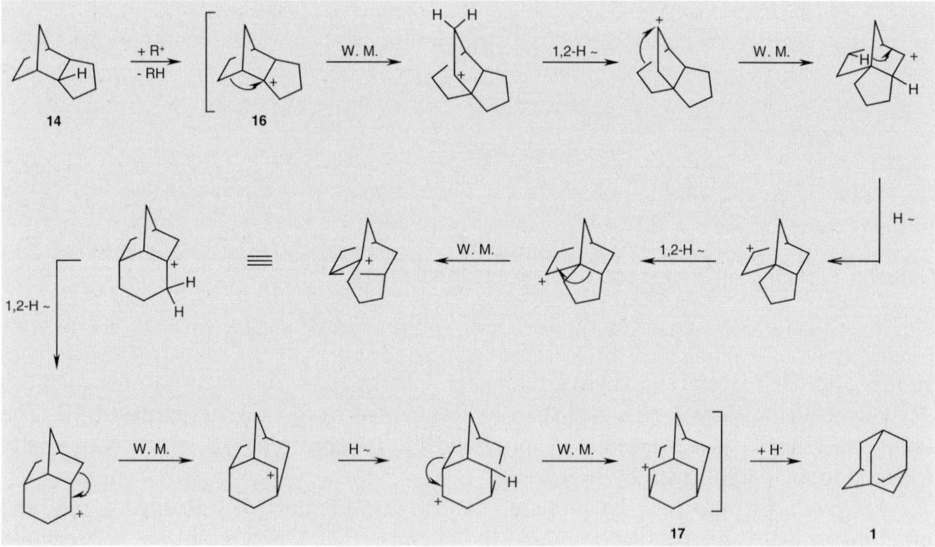

Scheme 5. A possible mechanism for the formation of adamantane (**1**) from *endo*-tetrahydrodicyclopentadiene (**14**, W.M. = Wagner–Meerwein rearrangement).

It is probably no exaggeration to state that no other experiment has had such far-reaching consequences for hydrocarbon chemistry as this—from the practical view-point—so simple process. With the possible exception of cyclooctatetraene (see Section 10.3), adamantane became the most important compound in modern hydrocarbon chemistry. Before these developments are discussed, two questions must be addressed. What is the mechanism of this deep-seated isomerization? Why does it proceed so readily and lead to **1** only or mostly? It is as easy to identify the force driving the formation of adamantane from **14** as it is difficult to answer the first question.

Since intermediates in the conversion of **14** to **1** could not be detected in the above rearrangement, the mechanism must remain speculative.

A general graph summarizing the rearrangement possibilities has been proposed; it shows that *ca* 3000 (2897 to be exact) independent pathways involving 1,2-alkyl shifts are conceivable to proceed from **14** to **1**, although not all of these are of the same importance and will take place with the same likelihood.[12] One of these connections—involving only 1,2-carbon (Wagner–Meerwein rearrangement) steps and various hydrogen shifts[13, 14]—is depicted in Scheme 5.

The isomerization starts by cation formation, *e.g.* to **16**, through intermolecular hydride transfer. Subsequently, these cations experience simple or multiple rearrangements, **16** → → → **17**, and finally the process is terminated by saturation of the cation (**17**) by hydride transfer. If the reaction is performed under heterogeneous conditions the product(s) is (are) distributed between the acid and the hydrocarbon phase.

The force driving the process is the high thermodynamic stability of adamantane. In fact, according to molecular mechanics calculations adamantane is the most stable $C_{10}H_{16}$ tricyclic hydrocarbon possible, it is the 'stabilomer'.[15, 16] The high thermodynamic stability of **1** results from the high degree of branching in the ring system and the nearly ideal positioning of all atoms in the whole molecular structure. According to X-ray and electron diffraction data the bond angles in **1** are 109.5° and the C–C bond lengths are 1.54 Å.[17] A selection of $C_{10}H_{16}$ hydrocarbons, all with a (calculated) higher heat of formation (in kcal mol^{-1}) than **1** is shown in a map of 'adamantaneland' in Scheme 6.[12,14]

This suggests that 'Everything should rearrange to adamantane'[18] and, indeed, for those isomers shown in Scheme 6, for which this dictum of Schleyer has been tested experimentally, it has turned out to be true. Thus *exo*-1,2-trimethylenenorbornane (**18**), twistane (**19**), 2,6-trimethylenenorbornane (**20**), protoadamantane (**21**) and tricyclo[5.2.1.0^{4,10}]decane (**22**) have all been isomerized to the $C_{10}H_{16}$ stablimomer: adamantane (**1**).[12–14, 19] As a matter of fact, even non-isomeric precursors such as cholesterol, nujol, cedrene, and caryophyllene have been isomerized to adamantane derivatives.[18]

The impact of Schleyer's discovery on organic chemistry was manifold and deep. First of all, it was made just at the right time—when physical organic chemistry was beginning to flourish and to become the cutting edge of organic chemistry in the early 1960ies. For mechanistic studies **1** was ideal, not only because it underwent many mechanistically important reactions itself, as we have just seen for one particular example, but also because with its rigid structure it offered a perfect platform or scaffold for all kinds of mechanistic studies, whether these involved bridgehead radicals, the possible introduction of double bonds at bridgeheads (*i.e.* the violation of Bredt's rule, see Section 7.4), or the precise orientation of functional groups in three-dimensional space *etc.* Furthermore, **1** and related cage compounds, *i.e.* molecules that can be defined as having three or more rings arranged topologically to enclose space in the center of a molecular structure became important reference and test systems for the predictive power of molecular mechanics calculations, the rapid development of which started at that time. In the same vein, **1** with its conformationally

completely fixed six-membered rings became an important molecule in conformational analysis, which was already experiencing widespread acceptance at that
time.[20] In preparative chemistry the question of functionalizing **1** gained growing
importance when it was discovered that 1-adamantylamine hydrochloride ('Symmetrel') had antiviral activity.[21]

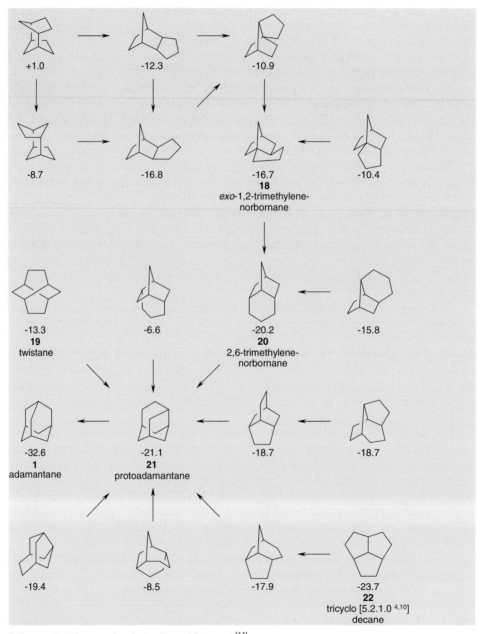

Scheme 6. Adamanteland as adopted from ref.[14]

3.2 Higher analogs of adamantane

Returning to hydrocarbon chemistry, the application of the Schleyer process and its variations to other, 'larger' hydrocarbons soon became the focus of research: If **1** is a 'monomer' unit of diamond, can higher oligomers, for example the dimer **23** and the trimer **24** be prepared by acid-catalyzed isomerisations? And if so, how far can this process be extended? (Scheme 7).

For such a project it is desirable to use hydrocarbon precursors which are isomeric with the final product and are at least moderately strained. The critical role of the starting material is nicely illustrated by the preparation of diamantane (**23**)[22] as shown in Scheme 8. The original preparation of **23** involved treatment of the [2+2]norbornene photodimer **25** with different Lewis acid catalysts. Depending on which isomer of **25** was used, yields between 1 and 10% of **23** were obtained.[23] When, however, precursor **26** was employed, the yield increased to 30%; unfortu-

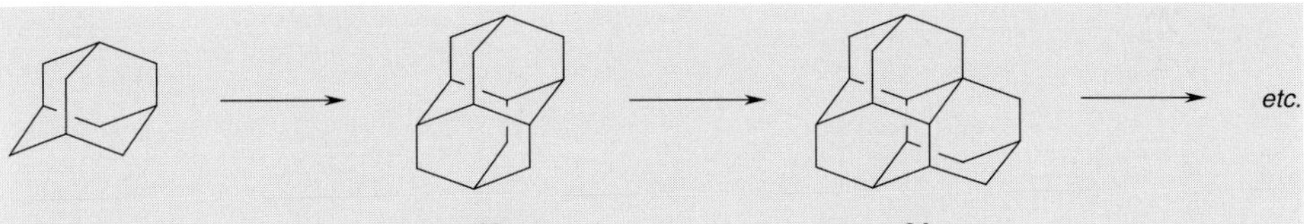

Scheme 7. From adamantane to the higher diamantanoid hydrocarbons.

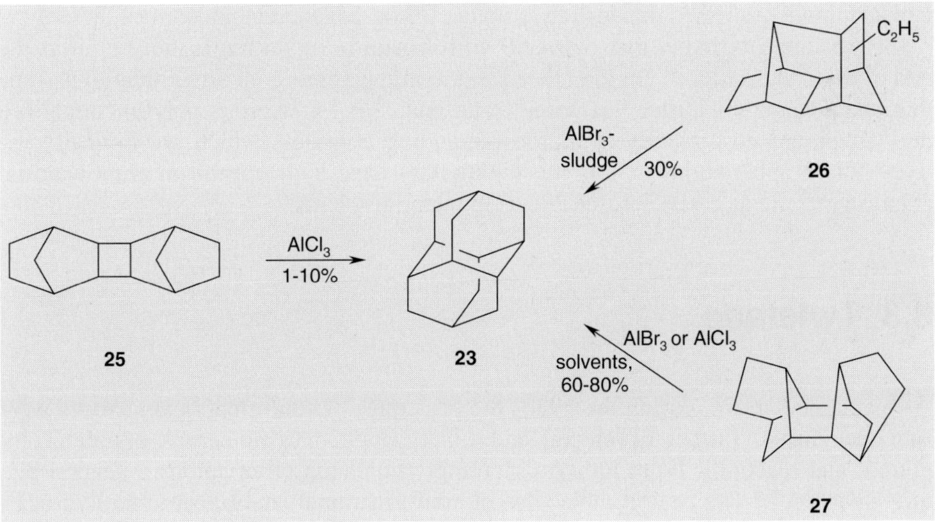

Scheme 8. Different routes to diamantane (**23**).

nately, this starting hydrocarbon is not readily available.[24] A further drastic increase in yield was accomplished when the so-called Binor-S, a dimer of norbornadiene, was first hydrogenated (cleavage of the cyclopropane rings), and the resulting tetrahydro Binor-S, **27**, was isomerized with aluminum tribromide in carbon disulfide. Diamantane was obtained in 65% yield,[24] and with aluminum trichloride in dichloromethane the yield was as high as 82%.[25]

Various routes to the third homolog of the diamantanoid series, triamantane (**24**), have been reported;[26] for example, the heptacyclooctadecanes **28** and **29** have been isomerized with most spectacular success by means of a complex catalyst system (Scheme 9):

Finally, the rearrangement process has also been employed in an attempt to syn-

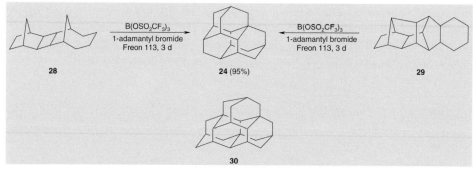

Scheme 9. A particularly successful route to triamantane (**24**).

thesize tetramantane, a hydrocarbon for which three structural isomers are conceivable, of which one, **30**, is shown in Scheme 9. Although interesting polycycles were isolated, none possessed the desired structure **30** or adamantanoid isomers thereof.[27]

It thus appears likely that diamond will remain a—sparkling!—holy grail at the end of this long route.[28] On the other hand, nothing is more incorrect than to assume that adamantane chemistry has come to an end. Since numerous polyfunctionalized derivatives are now readily available—including countless which are optically active—future applications of this most important cage hydrocarbon in combinatorial chemistry[29] are as easily foreseeable as in material science.

3.3 Twistane

As mentioned above, during the 1950s the principles of conformational analysis were first recognized, further developed and finally applied to numerous organic compounds and reactions. From hitherto seemingly unrelated observations a general picture emerged.[30] The central molecules of conformational analysis are two hydrocarbons: *n*-butane as the prototype for acyclic systems and cyclohexane (**31**) for cycloalkanes. For the latter, its ring inversion, which interchanges axial and equatorial substituents is the fundamental process. It was recognized that for this process to occur several intermediate conformations of different energy content must be passed, conformations that either represent local minima or transition states on the energy sur-

face. The minima are occupied by the chair form (**31**) and the twist-boat conformation (**33**), the transition states by the half-chair (**32**) and the boat conformation (**34**).
Whereas **32** and **34** lie, respectively, *ca* 11 and 7 kcal mol^{-1} above **31**, the energy difference between **33** and **31** is only 5.5 kcal mol^{-1} (Scheme 10):

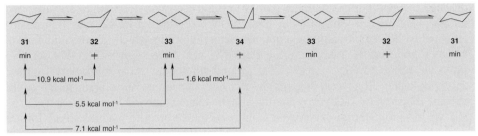

Scheme 10. Conformations involved in the inversion of cyclohexane (**31**).

Of course, the chair-chair inversion process shown in Scheme 10 is degenerate,
but when the degeneracy is lifted, *e.g.* by the introduction of a substituent, the process
becomes amenable to experimental study.

In the same way that adamantane (**1**) 'fixes' the chair form of cyclohexane, tricyclo[4.4.0.0^{3,8}]decane (**42**, twistane) fixes the twist-boat conformation. Twistane is a
twist-boat isomer of adamantane. The hydrocarbon is chiral (point group D_2) and is
formally composed of four twist-boat cyclohexane moieties with the same chirality.
Several syntheses for **42** have been reported, and we will discuss three of these approaches here which start from widely differing starting materials.

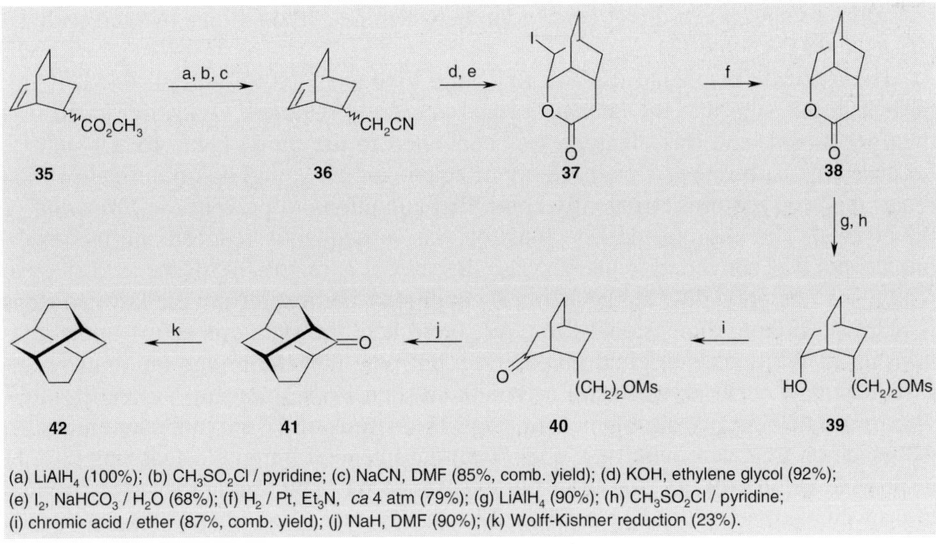

(a) LiAlH$_4$ (100%); (b) CH$_3$SO$_2$Cl / pyridine; (c) NaCN, DMF (85%, comb. yield); (d) KOH, ethylene glycol (92%);
(e) I$_2$, NaHCO$_3$ / H$_2$O (68%); (f) H$_2$ / Pt, Et$_3$N, *ca* 4 atm (79%); (g) LiAlH$_4$ (90%); (h) CH$_3$SO$_2$Cl / pyridine;
(i) chromic acid / ether (87%, comb. yield); (j) NaH, DMF (90%); (k) Wolff-Kishner reduction (23%).

Scheme 11. The first synthesis of twistane (**42**) by Whitlock.

Twistane was first prepared by Whitlock in 1962[31] by the route summarized in Scheme 11.

In the starting material **35**, easily prepared by a Diels–Alder addition of methyl acrylate to 1,3-cyclohexadiene, one of the features of the final product is already clearly discernible, a 'ethano bridge' connecting two 1,4-related atoms in a relatively highly functionalized six-membered ring. In fact, the 'remaining' part of the synthesis consisted in using these functional groups to build the other 1,4-bridge. Towards this end **35**, which is one carbon atom short of the 10 required in **42**, was first chain-elongated to **36**, employing conventional methodology in which the missing carbon atom is introduced by a Kolbe nitrile step. The next two steps—saponification, and iodolactonization to **37**—prepared the ground for the twistane-framework-forming step to be achieved several steps later. Because the iodine substituent in **37** could easily be removed by catalytic reduction and the lactone opened by a reduction step as well, we have transferred the oxygen atom from the carboxylic function to the former 2 position of **35**. Before the oxidation of the alcohol to a ketone was performed, how-ever, the primary OH-function of the reduction product of **38** was converted to the mesylate, providing **39**. This was then oxidized to the ketone **40** which is then ready for the second bridging process. Note that because of Bredt's rule (see Section 7.4) **40** can only be anionized at the methylene group, not at the bridgehead α-position. The enolate displaced the mesylate anion in an internal S_N2-process—which worked well, since the leaving group is bonded to a primary carbon atom—and the twistanone **41** was obtained. For the terminating step to twistane (**42**) the author used a Wolff-Kishner reduction; clearly there are several other possibilities. Not surprisingly, when twistane was treated with acid, it rearranged to adamantane (**1**), confirming that the latter "may be conceived as a bottomless pit into which many rearranging molecules may irreversibly fall".[12]

In the second twistane synthesis, reported a few years later,[32] the carbon atoms of the target molecule are all present in the starting material already, 2,7-dihydroxynaphthalene (**43**). In fact, what this synthesis accomplishes is the formation of a 'zero bridge', *i.e.* a direct connection between the carbon atoms marked with as-terisks in **43** (Scheme 12).

The aromatic bisphenol **43** was first fully hydrogenated to the diol **44**. It is im-portant in this step that the decalin formed is *cis*-configurated. Oxidation to **45** was then performed, and this diketone was converted to the mono ketal **46**. Obviously, because both carbonyl groups in **45** are of equal reactivity, and do not influence each other, this step requires stringently controlled conditions to prevent the formation of the bisketal. The unprotected keto function was subsequently reduced and the result-ing alcohol was converted to its mesylate. Because it is important for the next steps to really 'see' the prevailing stereochemical conditions, the result from the last two steps is redrawn in three-dimensional form, **47**. From here the synthesis shows some rela-tionship to Whitlock's original preparation, but note that the bridge building neces-sary must now occur between two carbon atoms in a 1,6-relationship—after all, in **47** we are still lacking two six-membered rings. Hydrolysis of **47** led to the ketomesylate **48** which on treatment with base underwent the intended intramolecular ring closure to provide ketone **49**, an isomer of the twistanone **41**. The termination step—reduc-tion of this ketone to the hydrocarbon **42**—was routine. Note that in **48** another α-methylene group is in principle available for ring closure. It seems that this pathway

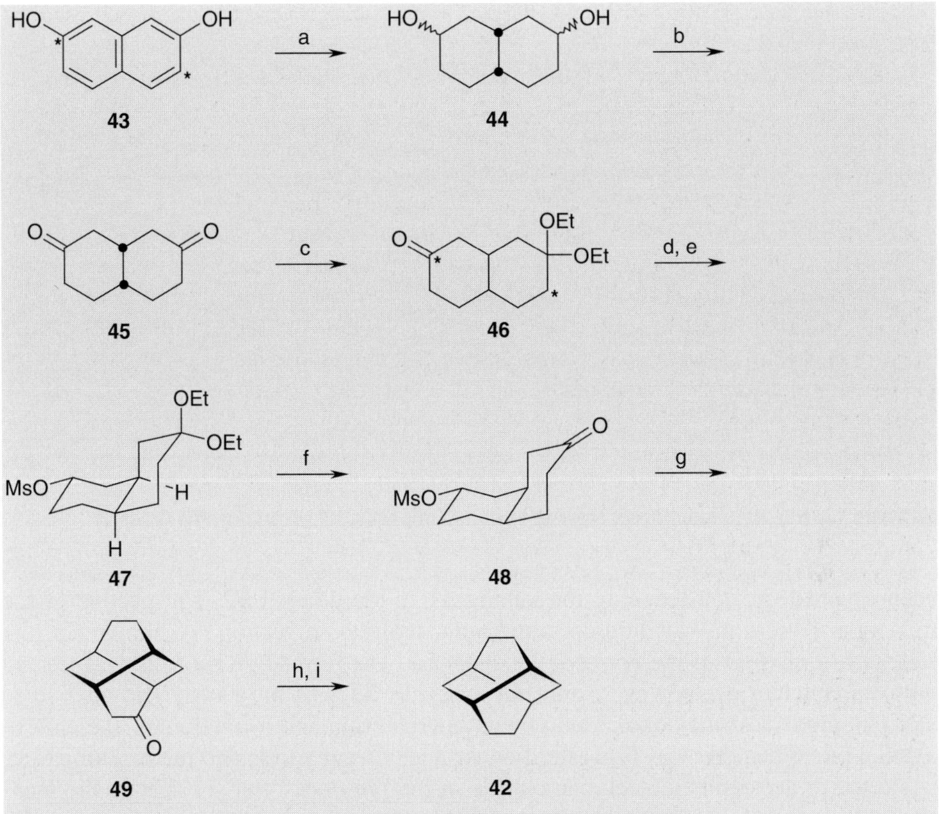

(a) Ra-Ni, H$_2$, EtOH, 150°C, 100 atm (63%); (b) CrO$_3$, H$_2$SO$_4$, H$_2$O (43%);
(c) (C$_2$H$_5$O)$_3$CH, p-TsOH, EtOH (65%); (d) Li, EtOH / THF, liq. NH$_3$ (100%);
(e) CH$_3$SO$_2$Cl, CH$_2$Cl$_2$, pyridine; (f) oxalic acid, H$_2$O (comb. yield 60%); (g)
NaH, dioxan (100%); (h) HSCH$_2$CH$_2$SH, HOAc, BF$_3$·Et$_2$O (100%); (i) Ra-Ni,
EtOH (34%)

Scheme 12. The synthesis of twistane (**42**) from an aromatic precursor by Deslongchamps and
Gauthier.

is not followed for stereoelectronic and strain reasons (*inter alia* formation of a four-
membered ring).

The last twistane synthesis to be discussed here[33] was carried out by Capraro and
Ganter when considerable knowledge on the behavior of cage hydrocarbons had already accumulated. It begins with a most unusual starting material which even on
second sight does not reveal the secret of its relation to **42**—the unusal ether **50**, a
substrate completely lacking a six-membered ring! (Scheme 13).

It was known to the authors, that on pyrolysis this readily available starting material undergoes a remarkable series of isomerizations which are terminated by an intramolecular cycloaddition to provide **51**.[34] Using modern terminology we would
call this a tandem reaction.[35] This adduct—which was formed in 50% yield—already shows a close resemblance to our final goal. In fact, formally we only have to
perform a 1,2-carbon shift to arrive at the twistane skeleton. In practice this was

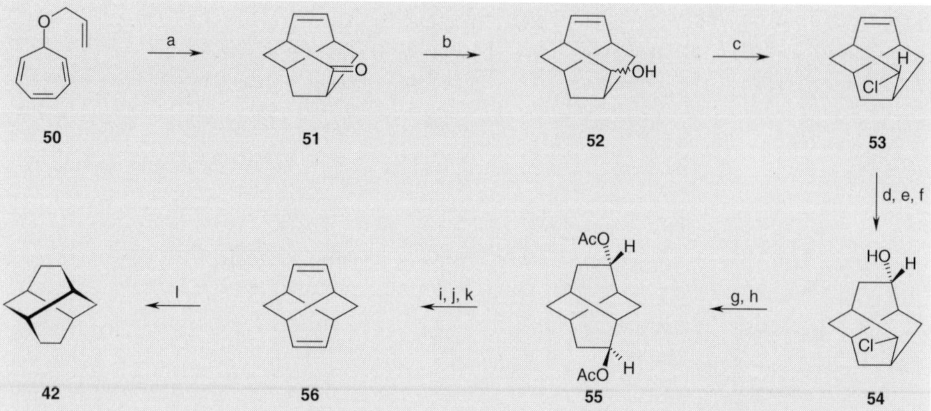

(a) 200 °C / 24 h (50%); (b) LiAlH$_4$ (100%); (c) SOCl$_2$ (> 90%); (d) Hg(OAc)$_2$, H$_2$O, THF; (e) NaBH$_4$; (f) aq. NaOH (comb. yield 60%); (g) Ac$_2$O; (h) AgOAc, Ac$_2$O; (i) LiAlH$_4$ (100%); (j) CH$_3$SO$_2$Cl (90%); (k) t-BuOK, DMSO (40%); (l) Pd / C, EtOH (71%)

Scheme 13. A twistane synthesis starting from a highly unusual precursor, the ether **50**.

accomplished by first reducing the ketone **51** to the alcohol **52**. Irrespective of the reduction conditions—a lithium aluminum hydride reduction is shown in the scheme—a mixture of the epimeric alcohols **52** was formed. When this was treated with thionylchloride, however, only the chloride **53** was produced. The next steps saw the removal of the double bond by oxymercuration and formation of the chloro-alcohol **54**. When this was first esterified with acetic anhydride and the resulting ester subjected to acetolysis in acetic anhydride in the presence of silver acetate the mentioned Wagner–Meerwein rearrangement did indeed take place and the bis acetate **55** was formed. Its reduction with lithium aluminum hydride then mesylation and, finally, β-elimination with potassium tert-butoxide afforded the remarkable (chiral) 4,9-twistadiene (**56**), which was effortlessly hydrogenated to **42**.

Other syntheses of **42** are known, and some were directed towards the preparation of optically active twistane.[33, 36–38] The absolute configuration of the (+)-enantiomer is 1(R),3(R),6(R),8(R). As with other rigid chiral hydrocarbons, twistane has been used as a test compound for theories of optical activity.[39]

Today twistane chemistry is a well-developed field. Many derivatives are known and their chemical behavior has been studied—although there can be no comparison with its all-chair isomer, adamantane (see above).[40]

3.4 Tetraasterane

As we have seen, the chair (**31**) and the twist-boat (**33**) are the 'building-blocks' of adamantane (**1**) and twistane (**42**), respectively. The cyclohexane boat conformation (**34**) is the constituent element of the asteranes, a homologous series of cage hydrocarbons which begins with triasterane (**57**) and then proceeds on to tetra- (**58**) and pentaasterane (**59**) (Scheme 14); in **58** the cyclohexane boat appears four times.[41]

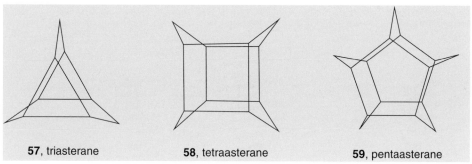

57, triasterane   58, tetraasterane   59, pentaasterane

Scheme 14. The asterane family.

Originally this series was designed to test the dependence of the structural and chemical properties of the asteranes on the bowsprit-flagpole interaction (a 1,4-interaction) in their cyclohexane rings. This transannular effect should be felt least in **57** and strongest in **59**.

The synthesis of **58**,[42, 43] summarized in Scheme 15, illustrates several important synthetic problems, often encountered during the preparation of highly symmetric cage molecules.

The synthesis begins with the photodimerization of 1,4-dihydrophthalic anhydride (**60**). This starting material is well chosen, because it can be prepared easily and it contains two double bonds of vastly different reactivity both of which will be used in due course. As expected, on irradiation solely the activated double bond in the anhydride ring was excited and the intended [2+2]cycloaddition set in. In principle, this could lead to the asterane skeleton directly, as shown by the photodimer **61**, which would be formed if the photoexcitation plus addition to the non-activated double bond occurred twice. And, in fact, **61** was formed. However, the yield was poor—it never exceeded 10%. As side products, isomerization, oxidation and reduction products of the starting material **60** were also produced, as were the *trans*-head-to-head and the *trans*-head-to-tail dimer of **60**, both of which cannot be used for the preparation of **58**, because their double bonds are not facing each other, as is required for the subsequent intramolecular photoaddition. The *cis*-head-to-tail dimer of **60** is evidently passed *en route* to **61**, and the *cis*-head-to-head product which would have involved initial cyclobutane formation at the unactivated double bond is either again a non-isolated intermediate or one which is not formed at all—the latter being more likely. The bottleneck of the whole synthesis occured at a very early step—a fault that could not be remedied at later stages. Because they provide at least two new bonds in an often stereocontrolled way in a one pot reaction, cycloadditions are obviously of great importance for the synthesis of cage hydrocarbons in particular and spherical structures in general. And we will thus see many more applications of this type of reaction in the course of this book. If they are employed early in a synthesis they must lead to

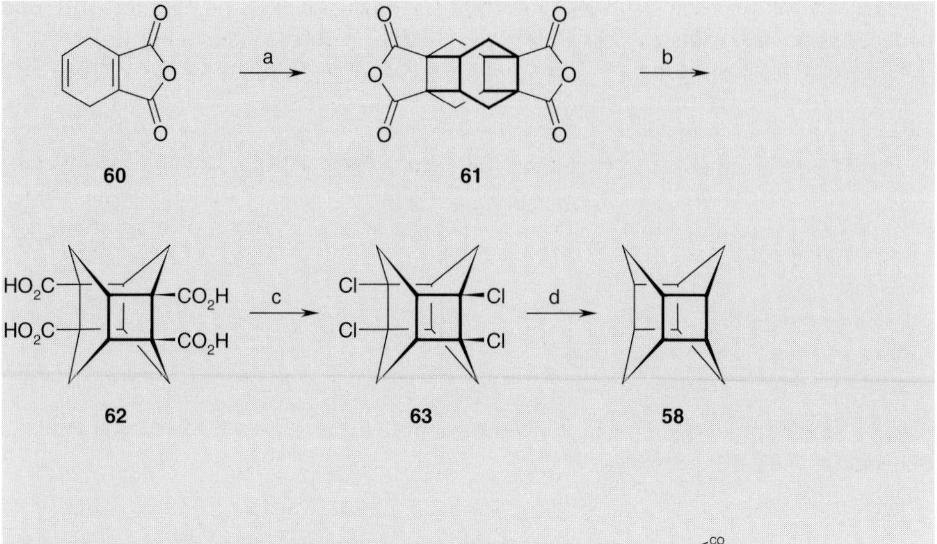

(a) hν, dioxan (9-10%); (b) dil. NaOH, then H$_2$SO$_4$ (86%); (c) Pb(OCOCH$_3$)$_4$ / [structure]

DMF / CH$_3$COOH (5:1) (20%); (d) Na / EtOH, reflux (74%)

Scheme 15. The synthesis of tetraasterane (**58**) by Musso.

the desired product in high yield. If they are essential at a later, or even at the last stage of a synthesis, yield is normally no longer the major concern.

The next steps to **58** look simple on paper, but they caused several practical problems. Because these reactions pose a general problem in hydrocarbon synthesis we will dwell on them here. The saponification of **61** to the tetraacid **62** proceeded well and in good yield. To convert this derivative, however, to **58** required extended experimentation and optimization. In general the direct decarboxylation of a tetraacid to a hydrocarbon employing any typical 19th-century reaction (*e.g.* high-temperature heating in the presence of copper bronze, see above, Section 3.1) cannot be recommended. Even if these processes give acceptable results for a monoacid, total yields are often poor when they have to be conducted several times in succession. Removal of these functional groups, introduced (as is often necessary) because they are needed to activate multiple carbon carbon bonds initially, must, therefore, occur in a stepwise manner. After the Hunsdiecker degradation and several of its variants had been unsuccessful, Grob decomposition finally enabled the preparation of the tetrachloro derivative **63**. This search for—and finally compliance with—often far less than optimal conditions is also typical in this area of hydrocarbon chemistry. General methods are not available and one must usually search for an individual solution. It does not add to 'planning safety' that multiple processes of this type can sometimes be performed successfully, as is shown by the last step, the reduction of **63** to the parent molecule, which proceeded in satisfactory yields.

Because of the above restrictions the chemistry of tetraasterane—and all other asteranes, for that matter—has not yet been developed to any significant extent. Not even the certainly very interesting Lewis acid-catalyzed isomerization of the highly strained **58** has yet been reported, nor the preparation of chiral derivatives of **58**. This hydrocarbon shares this fate with numerous other so-called 'interesting compounds' of this type. Interesting must always mean showing interesting, *e.g.* novel chemical, behavior. To discover that, one obviously needs enough material. Otherwise, the danger of unfulfilled promises and/or just the description of the bizarre is very real.

In closing we note that the synthesis of iceane (**64**) (Scheme 16), a cage molecule with a skeletal shape resembling that of ice I, the form that prevails under normal conditions, has also been described.[44–46] This cage hydrocarbon constitutes an interesting hybrid system because it is formally constructed out of two chair and three boat moieties.

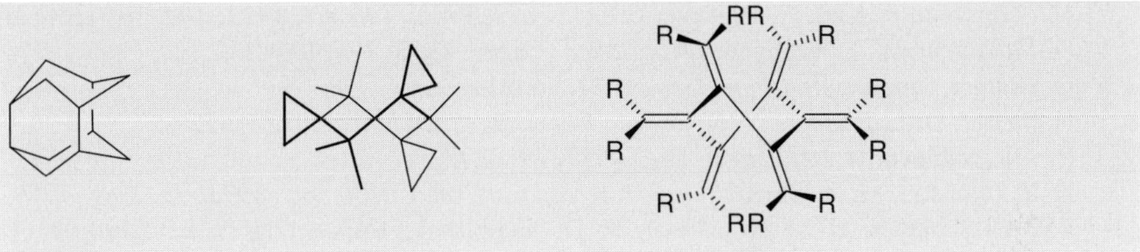

Scheme 16. Some other hydrocarbons containing non-chair cyclohexane rings.

The stabilization of otherwise unfavorable cyclohexane conformations can also be achieved by loading the ring with substituents. For example the trispirane **65** is a rare example of a cyclohexane ring with a pure twistboat conformation.[47, 48] Similar conformational distortions have also been observed for fully substituted [6]radialenes **66** (see Section 11.5).[49] To achieve non-chair cyclohexane conformations thus does not require the fixation of the ring in a rigid polycyclic skeleton.

References

1.   The chemistry of adamatane and other cage compounds has been reviewed many times: H. Stetter, *Angew. Chem.*, **1954**, *66*, 217–229; H. Stetter, *Angew. Chem.*, **1962**, *74*, 361–374; *Angew. Chem. Int. Ed. Engl.*, **1962**, *1*, 286–299; R. C. Fort, Jr, P. v. R. Schleyer, *Chem. Rev.*, **1964**, *64*, 277–300; G. Gelbard, *Ann. Chim.*, **1969**, *4*, 331–344; V. V. Sevost'yanova, M. M. Krayushkin, A. G. Yurchenko, *Russ. Chem. Rev.*, **1970**, *39*, 817–833; R. C. Bingham, P. v. R. Schleyer, *Fortschr. Chem. Forschung* (*Top. Curr. Chem.*), **1971**, *18*, 1–102; E. M. Engler, P. v. R. Schleyer in *MTP Internat. Rev. of Science, Organic Chemistry*, Series One, Butterworths, London, **1973**, Vol. 5, W. Parker (*Ed.*), 239–317; M. A. McKervey, *Chem. Soc. Rev.*, **1974**, *3*, 479–512; C. Ganter, *Top. Curr. Chem.*, **1976**, *67*, 15–106; R. C. Fort, Jr, *Adamantane. The Chemistry of Diamond Molecules*, M. Dekker, New York, N. Y., **1976**; A. P. Marchand, *Tetrahedron*, **1988**, *44*, 2377–2395; G. A. Olah (*Ed.*), *Cage Hydrocarbons*, J. Wiley and Sons, New York, N. Y., **1990**; E. Osawa, O. Yonemitsu (*Eds.*), *Carbocyclic Cage Compounds*, VCH Publishers, Inc., New

York, N.Y., **1992**.

2. The name adamantane is derived from the Greek for diamond (*adamas*), *adamant* referring to an impenetrably hard substance.

3. S. Landa, *Chem. Listy*, **1933**, *27*, 415; S. Landa, S. Machacek, J. Mzourek, *Chem. Listy*,
1933, *27*, 433; S. Landa, *Chim. Ind. (Paris)*, **1933**, 506; S. Landa, S. Machacek, *Coll.
Czech. Chem. Commun.*, **1933**, *5*, 1; S. Landa, *Petrol. Ztg.*, **1934**, *30*, 1; S. Landa, S.
Hala, *Chem. Listy*, **1957**, *51*, 2325; S. Landa, S. Hala, *Coll. Czech. Chem. Commun.*,
1959, *24*, 93–98.

4. This structure determination of adamanane is mentioned in refs. [3,5]; *cf.* E. Heilbronner,
F. A. Miller, *A Philatelic Ramble through Chemistry*, Verlag Helvetica Chimica Acta,
Basel, **1998**.

5. V. Prelog, R. Seiwerth, *Ber. Dtsch. Chem. Ges.*, **1941**, *74*, 1644–1648; V. Prelog, R.
Seiwerth, *Ber. Dtsch. Chem. Ges.*, **1941**, *74*, 1769–1772.

6. H. Meerwein, F. Kiel, G. Klösgen, E. Schoch, *J. Prakt. Chemie*, **1922**, *104*, 161–206; *cf.*
H. Meerwein, W. Schürmann, *Liebigs Ann. Chem.*, **1913**, *398*, 196–242.

7. O. Böttger, *Ber. Dtsch. Chem. Ges.*, **1937**, *70*, 314–325.

8. H. Stetter, O.-E. Bänder, W. Neumann, *Chem. Ber.*, **1956**, *89*, 1922–1926.

9. P. v. R. Schleyer, *J. Am. Chem. Soc.*, **1957**, *79*, 3292.

10. D. E. Johnston, M. A. McKervey, J. J. Rooney, *J. Am. Chem. Soc.*, **1971**, *93*, 2798–2799.

11. P. v. R. Schleyer, M. M. Donaldson, R. D. Nicholas, C. Cupas, *Org. Synth. Coll. Vol. V*,
J. Wiley and Sons, New York, N.Y., **1973**, 16.

12. H. W. Whitlock, Jr, M. W. Siefken, *J. Am. Chem. Soc.*, **1968**, *90*, 4929–4939.

13. Alternative mechanisms have also been discussed in the chemical literature, for a review
see T. S. Sorensen, S. M. Whitworth in *Cage Hydrocarbons*. G. A. Olah *(Ed.)*, J. Willey
& Sons, New York, N. Y., **1990**, 65–101.

14. The most recent review of the isomerization mechanisms of these $C_{10}H_{16}$ hydrocarbons is
discussed by C. Ganter in *Carbocyclic Cage Compounds* E. Osawa, Y. Yonemitsu,
(Eds.), VCH Publishers, New York, N. Y., **1992**, 293–317.

15. S. A. Godleski, P. v. R. Schleyer, E. Osawa, Y. Inamoto, Y. Fujikara, *J. Org. Chem.*,
1976, *41*, 2596–2605.

16. S. A. Godleski, P. v. R. Schleyer, E. Osawa, W. Todd Wipke, *Progr. Phys. Org. Chem.*,
1981, *13*, 63–117.

17. Despite its ideal structure parameters **1** is not a strain-free molecule, see R. H. Boyd, S.
N. Sanwal, S. Shary-Tehrany, D. McNally, *J. Phys. Chem.*, **1971**, *75*, 1264–1271; R. S.
Butler, A. S. Carson, P. G. Laye, W. V. Steele, *J. Chem. Thermodyn.*, **1971**, *3*, 277–280 .

18. M. Nomura, P. v. R. Schleyer, A. A. Arz, *J. Am. Chem. Soc.*, **1967**, *89*, 3657–3659.

19. E. M. Engler, M. Farcasiu, A. Sevin, J. M. Cense, P. v. R. Schleyer, *J. Am. Chem. Soc.*,
1973, *95*, 5769–5771.

20. Barton´s seminal paper which established conformational analysis had appeared in 1950:
D. H. R. Barton, *Experientia*, **1950**, *6*, 316.

21. W. L. Davies, R. R. Grunert, R. F. Hoff, J. W. McGahen, E. M. Neumayr, M. Paulshock,
J. C. Watts, T. R. Wood, E. C. Herrmann, C. E. Hoffman, *Science*, **1964**, *144*, 862.

22. Diamantane is also known as *congressane* since it served as a logo for a IUPAC con-
gress, held in London in 1963, where the chemical community was challenged to prepare
this hydrocarbon; *cf.* A. Nickon, E. F. Silversmith, *The Name Game*, Pergamon Press,
New York, N. Y., **1987**, p. 282. Diamantane has also been isolated from petroleum: S.
Hala, S. Landa, V. Hanus, *Angew. Chem.*, **1966**, *78*, 1060–1061; *Angew. Chem. Int. Ed.
Engl.*, **1966**, *5*, 1045–1046.

23. C. A. Cupas, P. v. R. Schleyer, D. J. Trecker, *J. Am. Chem. Soc.*, **1965**, *87*, 917–918; *cf.* I.
L. Karle, J. Karle, *J. Am. Chem. Soc.*, **1965**, *87*, 918–920.

24. T. M. Gund, V. Z. Williams, Jr, E. Osawa, P. v. R. Schleyer, *Tetrahedron Lett.*, **1970**, 3877–3880.

25. T. Courtney, D. E. Johnston, M. A. McKervey, J. J. Rooney, *J. Chem. Soc. Perkin* I, **1972**, 2691–2696.

26. Reviews: G. A. Olah in *Cage Hydrocarbons* G. A. Olah, (*Ed.*), J. Wiley & Sons, New York, N. Y., **1990**, 103–153; M. A. McKervey, *Chem. Soc. Review*, **1974**, *3*, 479–512.

27. P. v. R. Schleyer, E. Osawa, M. G. B. Drew, *J. Am. Chem. Soc.*, **1968**, *90*, 5034–5036.

28. Hydrocarbons are used as precursors for industrial diamond production, though, *e.g.* methane and acetylene are often used to make artificial diamonds; *cf.* R. H. Wentorf in *Kirk-Othmer, Encyclopedia of Chemical Technology*, J. Wiley & Sons, New York, N. Y., **1992**, 4th ed., Vol. 4, 1082–1096.

29. Reviews: F. Balkenhohl, Chr. v. d. Bussche-Hünnefeld, A. Lansky, Chr. Zechel, *Angew. Chem.*, **1996**, *108*, 2436–2488; *Angew. Chem. Int. Ed.*, **1996**, *35*, 2288–2337; K. S. Lam, M. Lebl, V. Krchnák, *Chem. Rev.*, **1997**, *97*, 411–448 for a selection of small, non-natural compounds which could be used in combinatorial chemistry.

30. E. L. Eliel, N. L. Allinger, S. J. Angyal, G. A. Morrsion, *Conformational Analysis*, Interscience Publishers, New York, N. Y., **1965;** W. J. Orville-Thomas, *Internal Rotations in Molecules*, J. Wiley and Sons, London, **1974**; E. L. Eliel, S. H. Wilen, *Stereochemistry of Organic Compounds*, J. Wiley and Sons, New York, N. Y., **1994**.

31. H. W. Whitlock, Jr, *J. Am. Chem. Soc.,* **1962**, *84*, 3412–3413.

32. J. Gauthier, P. Deslongchamps, *Can. J. Chem.*, **1967**, *45*, 297–300.

33. H.-G. Capraro, C. Ganter, *Helv. Chim. Acta*, **1976**, *59*, 97–100.

34. C. A. Cupas, W. Schumann, W. E. Heyd, *J. Am. Chem. Soc.*, **1970**, *92*, 3237–3239.

35. L. F. Tietze, U. Beifuss, *Angew. Chem.*, **1993**, *105*, 137–170; *Angew. Chem. Int. Ed. Engl.*, **1993**, *32*, 131–164; *cf.* Tse-Lok Ho, *Tandem Organic Reactions*, J. Wiley and Sons, New York, N. Y., **1992**.

36. For the first synthesis of optically active twistane by a route akin to Whitlock's original procedure see K. Adachi, K. Naemura, M. Nakazaki, *Tetrahedron Lett.*, **1968**, 5467–5470.

37. M. Tichý, J. Sicher, *Tetrahedron Lett.*, **1969**, 4609–4613; *cf.* M. Tichý, J. Sicher, *Collect. Czech. Chem. Comm.*, **1972**, *37*, 3106–3116.

38. H.-G. Capraro, C. Ganter, *Helv. Chim. Acta*, **1980**, *63*, 1347–1351.

39. Review: K. Naemura in *Carbocyclic Cage Compounds* E. Osawa, O. Yonemitsu (*Eds.*), VCH Publishers, Inc., New York, N. Y., **1992**, 61–90; *cf.* J. H. Brewster, *Tetrahedron Lett.*, **1972**, 4355–4358 and J. H. Brewster, *Topics Stereochem.*, **1967**, *2*, 1–72.

40. For the preparation of lower homologs of twistane, twistanone and twistandione, compounds derived from tricyclo[3.3.0.0^{3,7}]octane, see R. R. Sauers, K. W. Kelly, B. R. Sickles, *J. Org. Chem.*, **1972**, *37*, 537–543 and M. Nakazaki, K. Naemura, H. Harada, H. Narutaki, *J. Org. Chem.*, **1982**, *47*, 3470–3474.

41. Tetraasterane is not an isomer of adamantane (**1**) and twistane (**42**), though. It has two additional carbon atoms.

42. H.-M. Hutmacher, H.-G. Fritz, H. Musso, *Angew. Chem.*, **1975**, *87*, 174–175; *Angew. Chem. Int. Ed. Engl.*, **1975**, *14*, 180–181.

43. H.-G. Fritz, H.-M. Hutmacher, H. Musso, G. Ahlgren, B. Åkermark, R. Karlsson, *Chem. Ber.*, **1976**, *109*, 3781–3792.

44. C. A. Cupas, L. Hodakowski, *J. Am. Chem. Soc.*, **1974**, *96*, 4668–4669.

45. D. P. G. Hamon, G. F. Taylor, *Tetrahedron Lett.*, **1975**, 155–158; *cf.* D. P. G. Hamon, G. F. Taylor, *Austral. J. Chem.*, **1976**, *29*, 1721–1734.

46. It has been pointed out that the structure of **64** resembles more that of wurtzite, a crystal form of zinc sulfide, and should therefore be called wurtzitane: R. O. Klaus, H. Tobler,

C. Ganter, *Helv. Chim. Acta,* **1974**, *57*, 2517–2519.

47. L. Fitjer, H.-J. Scheuermann, U. Klages, D. Wehle, D. S. Stephenson, G. Binsch, *Chem. Ber.*, **1986**, *119*, 1144–1161; *cf.* K. Wulf, U. Klages, B. Rissom, L. Fitjer, *Tetrahedron*, **1997**, *53*, 6011–6018.
48. J. Weiser, O. Golan, L. Fitjer, S. E. Biali, *J. Org. Chem.*, **1996**, *61*, 8277–8284.
49. T. Sugimoto, Y. Misaki, T. Kajita, Z. Yoshida, Y. Kai, N. Kasai, *J. Am. Chem. Soc.*, **1987**, *109*, 4106–4107. When dodecamethyl-[6]radialene (R = CH$_3$ in **66**, a hydrocarbon which prefers a chair conformation, is irradiated in hexane with 254-nm light it isomerizes to a twistboat configurated isomer: Th. Höpfner, Ph. d. dissertation, Braunschweig, **1996**. For the generation of nearly planar cyclohexane rings see D. L. Mohler, K. P. C. Vollhardt, S. Wolff, *Angew. Chem.*, **1990**, *102*, 1200–1202; *Angew. Chem. Int. Ed. Engl.*, **1990**, *29*, 1151–1153; *cf.* H. Dodziuk, *Modern Conformational Analysis*, VCH Publishers, New York, N. Y., **1995**.

4 The Prismanes

The [n]-prismanes are a class of polyhedranes which have interested chemists ever since Ladenburg proposed the smallest member of this hydrocarbon family, [3]-prismane (**1**) as the structure for benzene more than a century ago.[1] Formally these cage compounds (Scheme 1) consist of an even number of methine units—their sum formula is thus $(CH)_{2n}$—positioned at the corners of a regular prism. Their high symmetry (D_{nh} point group, see below) has immediate esthetic appeal, and their preparation constitutes a considerable challenge which not only originates from their convex structure but also because they are expected to be highly strained. After all, the [n]-prismanes can be formally regarded as closed loops of all *cis*-fused cyclobutane units. In principle n, the order of the prismane under discussion, can vary from 3 to infinity. It has, however been predicted[2] that only the members with n up to approximately 12 will have planar n-membered rings.[3]

Extensive force field[4] and *ab initio* calculations[5] have been performed for the prismanes; they reflect the highly strained nature of these compounds and their unusual structural features (extended C–C bond lengths and deformed bond angles). Of the four prismanes shown in Scheme 1 Ladenburg benzene (**1**, [3]-prismane) is pre-

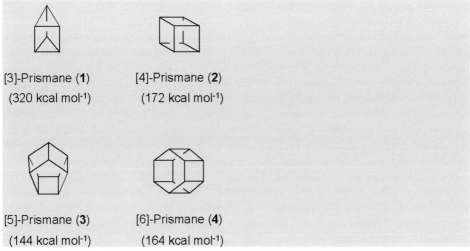

[3]-Prismane (**1**)
(320 kcal mol⁻¹)

[4]-Prismane (**2**)
(172 kcal mol⁻¹)

[5]-Prismane (**3**)
(144 kcal mol⁻¹)

[6]-Prismane (**4**)
(164 kcal mol⁻¹)

Scheme 1. The first four members of the prismane family and their calculated steric energies in kcal mol⁻¹.

dicted by MM2 calculations to have the highest steric energy. [4]-Prismane (**2**, cubane) is considerably less strained, and calculations predict that [5]-prismane (**3**, pentaprismane) has the minimum value for the whole series. Beginning with the next higher homolog, [6]-prismane (**4**, hexaprismane) a strain increase sets in which persists all the way to the higher members of the family. High strain, of course, must not necessarily cause high instability and difficulty or even impossibility in isolating an organic compound. Because a thermal $[_\pi 2_s + _\pi 2_s]$cleavage of a cyclobutane ring is forbidden on orbital-symmetry grounds,[6] the prismanes have high kinetic thermal stability. This is borne out, for example, by cubane (**2**) which withstands heating up to 200 °C[7] (see Section 10.3).

In this chapter the syntheses of [3]- (**1**) and [5]-prismane (**3**) and several attempts to prepare the still unknown [6]-prismane (**4**) will be presented. A discussion of [4]-prismane (**3**) is deferred until Chapter 5 because this unique hydrocarbon also belongs to the so-called *Platonic hydrocarbons*, the preparation of which I want to present collectively in a chapter of their own.

4.1 Triprismane

Retrosynthetically [3]-prismane (**1**, tetracyclo[2.2.0.0^{2,6}.0^{3,5}]hexane) might be decomposed in several ways. Focusing on the cyclobutane rings, these may be cleaved either 'horizontally' or 'vertically' (Scheme 2).

Whereas the former process leaves the three-membered rings intact, the latter destroys them. The product of the first cleavage is 3,3'-bicyclopropenyl (**5**); that of the second is Dewar benzene (**6**). Both hydrocarbons have been prepared (see Section 10.2), and it has been demonstrated that **5**, which is not easily synthesized, cannot be cyclized to **1**. Dewar benzene and several of its derivatives can, however, be converted to [3]-prismanes as we shall see later. Cyclobutane rings are commonly prepared by [2+2]photocycloadditions and we can hence expect this process to play a prominent role in the preparation of **1** and its higher homologs. Cyclopropane rings, on the other hand, are usually accessible by addition of carbenes to carbon-carbon double bonds.[8] Because of the particular carbon framework of **1** a direct carbene addition route is clearly impossible for the preparation of this hydrocarbon. A retro-carbene decomposition which breaks one cyclopropane and one cyclobutane bond is nevertheless conceivable and depicted in the lower half of Scheme 2. This would initially lead to the carbene **7**, which, by ring-opening of the second (strained) three-membered ring, would most likely isomerize to the cyclopentadienyl carbene **8**. Provided that this can be generated by decomposition of an appropriate precursor, *e.g.* a diazo compound, its cyclization to [3]-prismane, although formally possible as the dotted lines in **8** indicate, seems highly unlikely, though, because it has other options to stabilize itself. By a 1,4-addition to its butadiene moiety, for example, benzvalene (**9**) could be formed, and a 1,2-insertion process into the neighboring carbon–hydrogen bond could lead to fulvene (**10**). Although a practical route for the synthesis of **1** from either of these precursors is not recognizable on first sight, these C_6H_6 isomers should be taken into consideration as possible substrates for **1**, as should, of

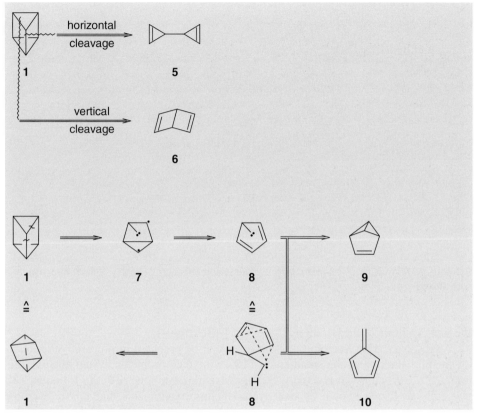

Scheme 2. Retrosynthesis of [3]-prismane (**1**).

course, the most stable of all possible hydrocarbons of this composition, benzene (see Section 10.2). As it turned out, **9** was the starting material for the first synthesis of [3]-prismane as we shall see below.

When liquid benzene was irradiated at 254 nm a complex mixture containing **9** and **10** resulted, but **1** could not be detected in the photolysates.[9] The situation changes, however, with benzene derivatives carrying bulky substituents. This was shown first by Viehe and co-workers[10] in a classical paper in which the term benzvalene was coined and which involved the photoisomerization of 1,2,3-trifluoro-4,5,6-tris-*tert*-butyl-benzene (**11**) to a mixture of the prismane derivative **12**, the Dewar benzene **13**, and **14**, the first benzvalene to be prepared (Scheme 3).

Later, this rearrangement process was not only observed for a hydrocarbon—**15**, which isomerized to **16**–**18**[11]—but also for numerous other highly substituted benzene derivatives.[12] The steric hindrance between the substituents of the substrates is a prerequisite for the success of the process, because it enforces non-planarity of the aromatic precursor and by this facilitates the formation of the Dewar benzene isomer (see Sections 10.2 and 12.3), which in a subsequent [2+2]photocycloaddition, closes to the prismane derivative (such as **12** and **16**, respectively).

As already hinted, benzvalene (**9**) turned out to be the C$_6$H$_6$ isomer of choice for

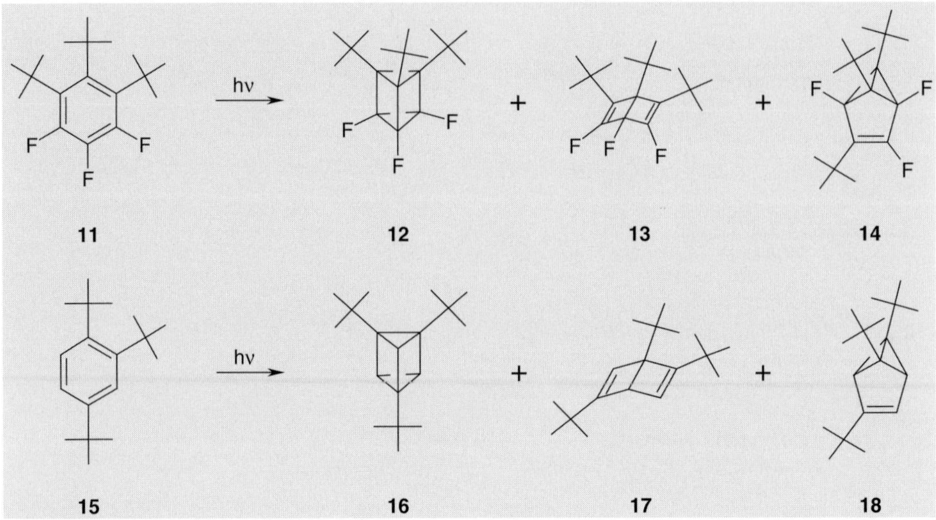

Scheme 3. Formation of [3]-prismanes by photoisomerization of highly substituted benzene derivatives.

the first synthesis of the parent hydrocarbon **1** (Scheme 4).

As discovered by Katz and co-workers **9** can be prepared in preparative amounts by treating lithium cyclopentadienide (**19**) with dichloromethane and methyl lithium.[9, 13, 14] When it was reacted with N-phenyl-triazolindione (**20**) the 1:1 adduct **23** was formed. It was shown by deuterium-labeling experiments that this remarkable process probably begins with the formation of the zwitterion **21**. Stabilization of this intermediate by σ-bridging then leads to the delocalized cation **22**, which can reclose to the observed adduct **23**, thus not only providing three of the four strained rings required in [3]-prismane, but also setting the stage for the ultimate ring closure. Hydrolysis followed by oxidation of **23** provided the azo compound **24** which on ultraviolet irradiation extruded nitrogen and yielded a complex photolysis mixture from which **1** could be isolated in 1.8% yield.[13, 15]

That Dewar benzene derivatives are plausible intermediates in the formation of [3]-prismanes was most convincingly demonstrated by a photoaddition experiment in which the parent molecule **6** was isomerized to **1**.[16] Although [3]-prismane was isolated in this experiment in 15% yield, its chemistry—and many of its physical properties—remain largely unexplored because of lack of material, a fate which **1** shares with many other hydrocarbons when they are made for the first time.

4.2 [5]-Prismane

Because both intramolecular [2+2]photoaddition (*cf.* **6**→**1**) and nitrogen extrusion (**24**→**1**) have been successfully employed in the synthesis of a prismane, **1**, we might

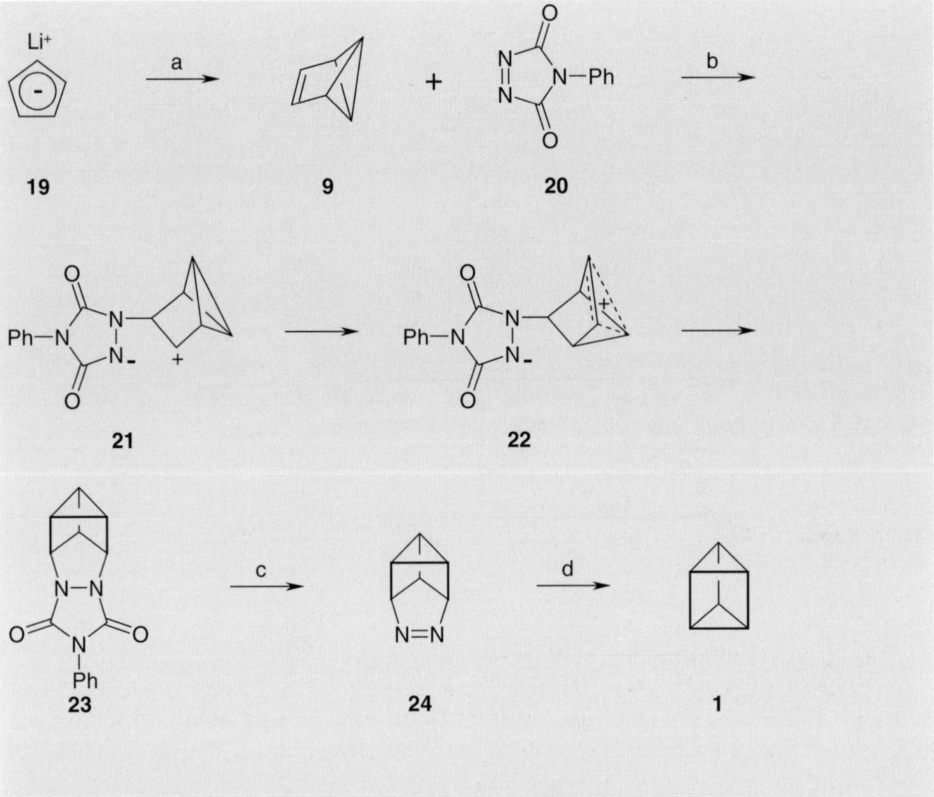

(a) CH₃Li, dichloromethane (29%); (b) ether/dioxan (50-60%);
(c) KOH, CH₃OH-H₂O (85:15), acidic CuCl₂, aq. NaOH (65%);
(d) hν, Pyrex (4-6%)

Scheme 4. The first synthesis of [3]-prismane (**1**) by Katz.

expect these reactions to work also for the preparation of [5]-prismane (**3**), even more so since this is the least strained member of the whole series.

Surprisingly, these expectations are not fulfilled. As summarized in Scheme 5 neither hypostrophene (**25**)[17] nor the mono- or the bis azo compounds, **26** and **27**, respectively, which are both obtainable from the C₁₀H₁₀ hydrocarbon basketene (see Section 10.4) can be converted into [5]-prismane (**3**).[18]

Situations like this are not uncommon in the synthesis of cage hydrocarbons and have been ascribed to strain effects, unfavorable electronic interactions and excessively large distances between functional groups (*i.e.* double bonds, see Section 7.2). They may be circumvented by not attempting to combine the correct number of carbon atoms which will eventually make up the desired target hydrocarbon at an early stage of the synthesis, but settle for a lesser goal, *i.e.* the assembly of *more* than the ultimately required carbon atoms and subjecting the thus generated precursor molecule to a *carbon-removing* process such as a ring-contraction. The attractiveness of this approach in this instance is enhanced by the observation, that homohypostrophene,

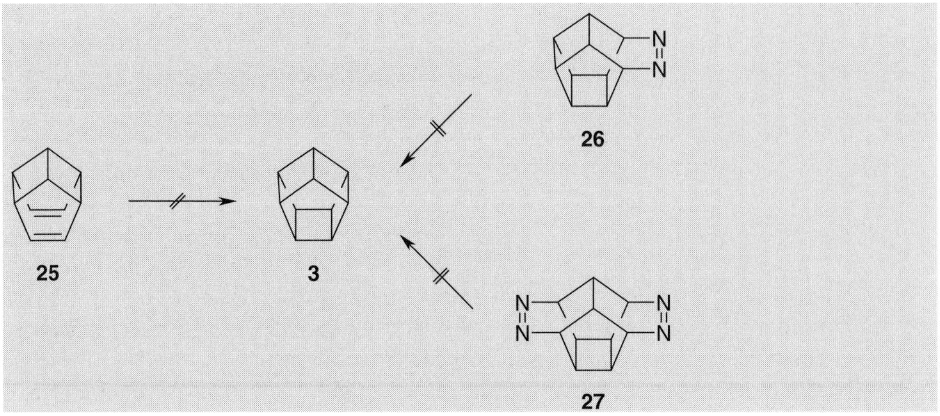

Scheme 5. Unsuccessful attempts to prepare [5]-prismane (**3**).

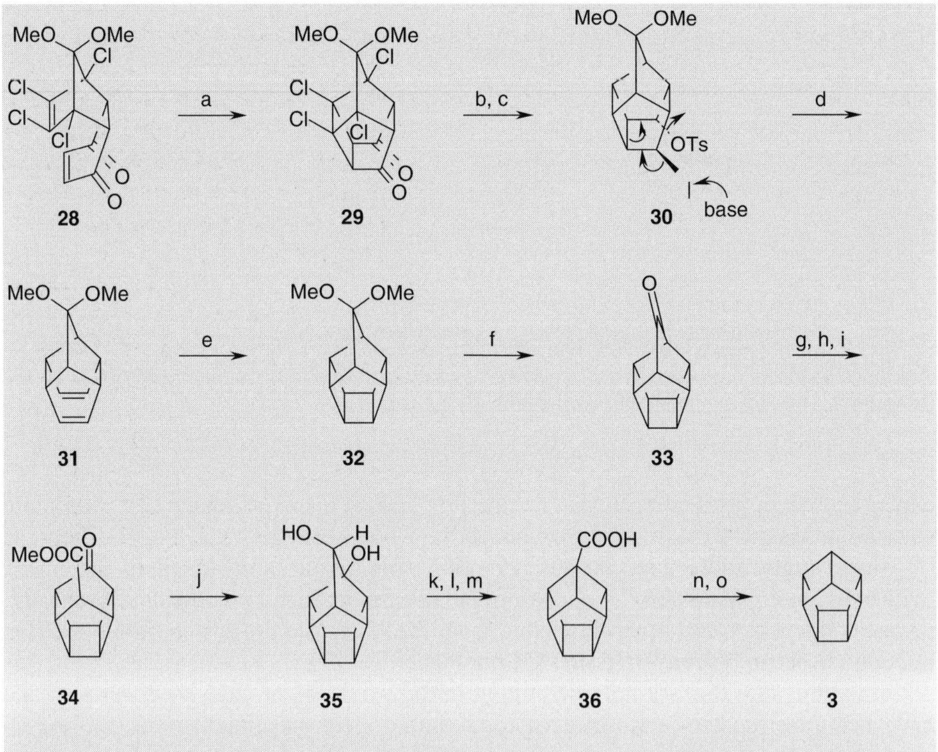

(a) hv, acetone; (b) Li-liq. NH$_3$, *t*-BuOH (H$_2$O); (c) 1. TsCl, py., 2. NaI, HMPA, 100°C (72%, comb. yield); (d) *t*-BuLi, ether (50%); (e) hv, acetone (92%); (f) 30% H$_2$SO$_4$, ether (95%); (g) *m*-CPBA, CH$_2$Cl$_2$ (90%); (h) KOH-H$_2$O, RuO$_4$, NaIO$_4$; (i) CH$_2$N$_2$, ether (90%, comb. yield); (j) Na-liq. NH$_3$ (83%); (k) Cl$_2$, Me$_2$S, CH$_2$Cl$_2$, Et$_3$N, py.; (l) TsCl, py.; (m) 20% aq. KOH (50%, comb. yield); (n) 1. ClCOCOCl, 2. *t*-BuOOH, py.; (o) 2,4,6-triisopropylnitrobenzene, 150°C (42%, comb. yield)

Scheme 6. The synthesis of [5]-prismane (**3**) by Eaton.

the hydrocarbon in which the single bond shared by the two cyclobutane rings in **25** is replaced by a methylene group, could be photochemically closed to homopentaprismane.[19] This extended-bridge concept (for other applications see Section 12.3 on cyclophanes) was exploited by Eaton and co-workers in the first synthesis of [5]-prismane,[20] which was published in 1981 (Scheme 6).

When the Diels–Alder adduct between *p*-benzoquinone and 1,2,3,4-tetrachloro-5,5-dimethoxycyclopentadiene (**28**) was irradiated, the anticipated [2+2]cycloaddition took place and provided **29** in which the two five-membered rings of the target **3** have been generated, but in which one zero bridge is missing and a second one is too large by one extra carbon atom. Reductive dechlorination, tosylation of the intermediate diol and partial replacement of the tosyl groups then led to **30**, which on base treatment lost its leaving groups but also of its four-membered ring. The resulting hyprostrophene derivative **31** could, however, be quickly recyclized photochemically to the homopentaprismane ketal **32**. Removal of the protecting groups led to the ketone **33**, which, although structurally so close to **3** still required nine steps, *i.e.* more than half of the total synthesis, to remove its bridging carbonyl group. To initiate this process, bridgehead functionalization, *e.g.* bromination, seems as a viable step; it is, however, prevented by the inability of **33** to enolize, because of the restrictions imposed on the system by Bredt's rule (see Section 7.4). The carbonyl bridge was, therefore, opened by oxidation—Baeyer-Villiger oxidation followed by rutheniumdioxide/sodium periodate oxidation (which operates on the open salt of the lactone)—to a keto acid, which on esterification with diazomethane led to the keto ester **34**. Acyloin condensation subsequently closed the bridge again and provided another homopentaprismane derivative, **35**. This derivative, however, had a functional group at one of its bridgeheads. After oxidation with the complex of dimethylsulfide and chlorine and the replacement of its hydroxyl substituent with tosylate as a better leaving group, everything was ready for a Favorskii ring contraction which, as hoped, led to the pentaprismane carboxylic acid **36**. Pentaprismane (**3**) was finally obtained by use of decarboxylation methodology already used successfully for the preparation of cubane (**2**, see Section 5.2).

Considering what has happened to the original diene component in the course of the synthesis it is easily seen that the perchlorocyclopentadiene ketal served as an equivalent of cyclobutadiene (see Section 10.1). As a result of realizing this relationship, a formal synthesis of [5]-prismane (**3**) has been developed which begins with the addition of cyclobutadiene itself to 4,4-dimethoxy-cyclohexa-2,5-diene-1-one, the mono acetal of *p*-benzoquinone.[21] Because the remaining steps to the crucial intermediate **34** are related to those presented in Scheme 6, the completion of this reaction sequence is left to the reader as an exercise.

4.3 *En route* to hexaprismane

The preparation of [6]-prismane (**4**) has not yet been accomplished. Obviously, with the number of four-membered rings increasing as we proceed through the series, the number of ways of deconstructing the target hydrocarbon also increases rapidly. The most obvious possibility consists in splitting hexaprismane symmetrically and totally into two halves—two benzene molecules! (Scheme 7). Determining the conditions under which these can be aligned in parallel fashion and are willing to give up their aromatic character is, however, a different matter.

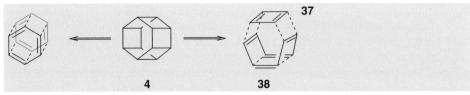

Scheme 7. Two ways to deconstruct [6]-prismane (**4**).

Another conceptually attractive route is provided by a cycloaddition between cyclobutadiene (**37**, see Section 10.1) and cyclooctatetraene (**38**, see Section 10.3), two readily available starting hydrocarbons, which, however, again 'only' have to be oriented in the right position (the 'capped' one shown in the scheme) for closure to hexaprismane. And there is another serious drawback which must be overcome if this route is to be successful—both **37** and **38** are known to react as 4π-components in cycloadditions, whereas [2+2]cycloadditions are required here. The use of equivalents for **37** and **38** is thus mandatory; and although this C_4+C_8-route to **4** has not yet been completed, it is of interest to discuss it in detail here, because it provides deeper insights into the building process by which polyhedral hydrocarbons may be prepared.

For **37** we have already encountered a synthetic equivalent, the tetrachlorocyclopentadiene acetal **39**.[20] For cyclooctatetraene (**38**), 1,5-cyclooctadiene (**40**, see Section 7.3) was introduced as an equivalent by Eaton and co-workers in their attempt to prepare **4**.[22, 23] On the basis of their studies, Mehta has embarked on a long journey to hexaprismane (Scheme 8) which—with the highly desired goal in immediate grasp—disappointingly failed at the very last step so far.[24]

As shown in the scheme, Diels–Alder addition between **39** and **40** led to the expected *endo* adduct **41**, which was converted into the *endo* peroxide **42** by first preparing a conjugated diene intermediate from **41** and then oxidizing it with singlet oxygen. When **42** was reduced with lithium aluminum hydride a bis diol was formed which provided a bis acetate with acetic anhydride. Evidently the latter can adopt a conformation which brings the two double bonds of the intermediate so close to each other that they could be closed photochemically. Transesterification then yielded the bis mesylate **43**, which by a substitution-elimination sequence was converted *via* **44** to the diene **45**, again set up for intramolecular photocycloaddition. After deketalization the resulting ketone **46** was ready to participate in a Favorskii ring contraction, which—after esterification—provided the secohexaprismane derivative **47**. Although this could be reduced, by functional group manipulations, to the hydrocarbon **48**, this one-bond-away precursor for **4** could not be coerced to give up as few as two hydrogens atoms.

A sizable number of other attempts towards the preparation of hexaprismane (**4**) has been reported during recent decades. Although all of them were unsuccessful,

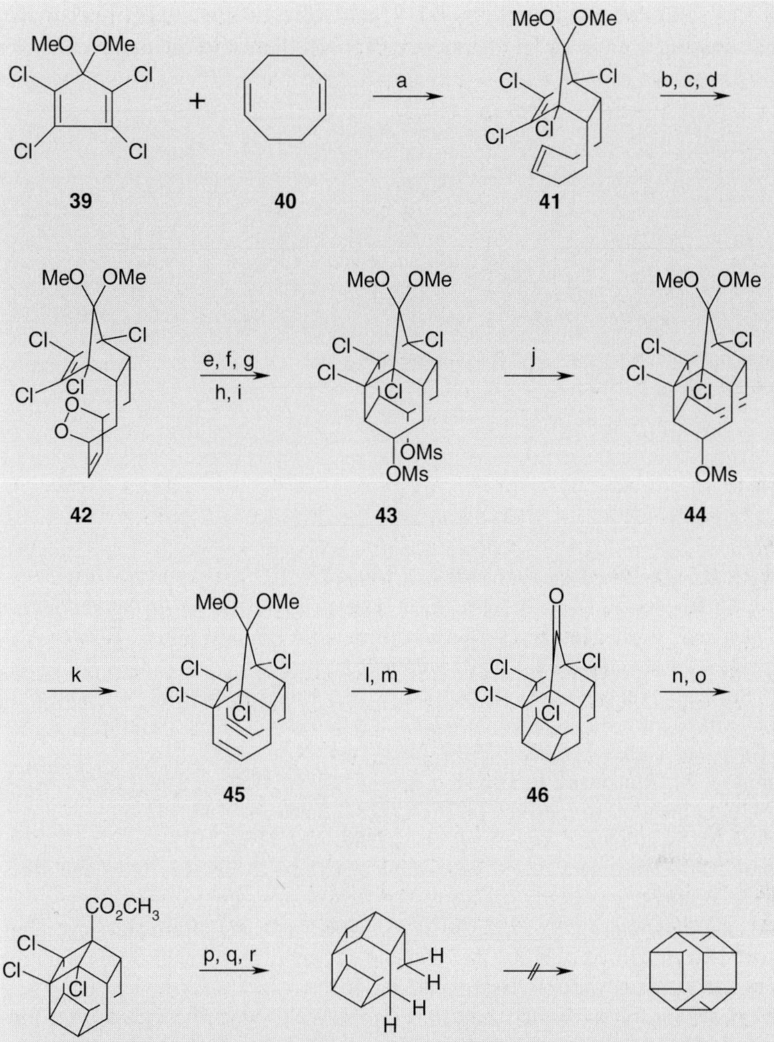

Scheme 8. *En route* to [6]-prismane: Mehta's approach (**4**).

(a) reflux, neat (84%); (b) NBS, AIBN, CCl$_4$ (100%, mixture of isomers); (c) DBU, DMSO, 20°C (71%); (d) hν, methylene blue, O$_2$, 7 d (70%); (e) LiAlH$_4$, ether (100%); (f) Ac$_2$O, pyridine (75%); (g) hν, acetophenone, benzene (73%); (h) aq. KOH, CH$_3$OH (quant.); (i) MsCl, pyridine (85%); (j) NaI, HMPA, 100°C (55%); (k) B$_2$H$_6$, THF, aq. NaOH (74%); (l) hν, acetone (60%); (m) H$_2$SO$_4$, CH$_2$Cl$_2$ (85%); (n) NaOH, toluene; (o) CH$_2$N$_2$, CH$_3$OH (comb. yield 85%); (p) aq. KOH, CH$_3$OH; (q) HgO, Br$_2$, CH$_2$Br$_2$ (comb. yield 82%); (r) Li, *tert*-BuOH, THF (32%)

they have not only extended our preparative knowledge but also led to a considerable number of interesting cage hydrocarbons (Scheme 9), among them the dienes **49**[25] and **50**[26] and the saturated hydrocarbons **51**[27] and **52** (1,4-bishomo[6]-prismane, garudane),[28] all bearing a more or less pronounced resemblance to the elusive **4**.

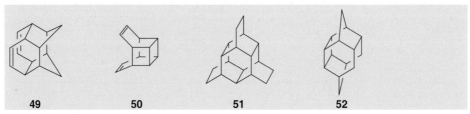

49   **50**   **51**   **52**

Scheme 9. Further potential precursors for [6]-prismane (**4**).

References

1.   A. Ladenburg, *Ber. Dtsch. Chem. Ges.*, **1869**, *2*, 140–142.
2.   E. Osawa, J. M. Rudzinski, D. A. Barbiric, E. D. Jemmis in *Strain and Its Implications in Organic Chemistry*, A. de Meijere, S. Blechert (*Eds.*), NATO, ASI Series, Kluwer Academic Pubishers, Dordrecht, **1989**, Vol. *273*, 259–261.
3.   Because of the resemblance to the coat of arms of Israel and Switzerland two isomers of [12]-prismane have been nick-named *israelane* and *helvetane*; *cf.* A. Nickon, E. F. Silversmith, *The Name Game*, Pergamon Press, New York, N. Y., **1987**, 87.
4.   E. D. Jemmis, J. M. Rudzinski, E. Osawa, *Chem. Express*, **1988**, *3*, 109–112; *cf.* M. A. Miller, J. M. Schulman, *J. Mol. Struct. (THEOCHEM)*, **1988**, *40*, 133–141.
5.   R. L. Disch, J. M. Schulman, *J. Am. Chem. Soc.*, **1988**, *110*, 2102–2105.
6.   R. B. Woodward, R. Hoffmann, *The Conservation of Orbital Symmetry*, Academic Press, New York, N. Y., **1971**.
7.   P. E. Eaton, *Tetrahedron,* **1979**, *35*, 2189–2223. Woodward and Hoffmann have compared the symmetry-imposed stability of **1** and **2** to an "angry tiger unable to break out of a paper cage": R. B. Woodward, R. Hoffmann, ref. [6].
8.   Houben-Weyl, *Methods of Organic Chemistry, Carbocyclic Three-Membered Ring Compounds* A. de Meijere (*Ed.*), Thieme Verlag, Stuttgart, **1997**, 4th ed., Vol 17.
9.   K. E. Wilzbach, J. S. Ritscher, L. Kaplan, *J. Am. Chem. Soc.*, **1967**, *89*, 1031–1032; *cf.* L. Kaplan, K. E. Wilzbach, *J. Am. Chem. Soc.*, **1967**, *89*, 1030–1031.
10.   H. G. Viehe, R. Merényi, J. F. M. Oth, J. R. Senders, P. Valange, *Angew. Chem.*, **1964**, *76*, 922; *Angew. Chem. Int. Ed. Engl.*, **1964**, *3*, 755.
11.   K. E. Wilzbach, L. Kaplan, *J. Am. Chem. Soc.*, **1965**, *87*, 4004–4006.
12.   For a recent review on prismanes see: G. Mehta, S. Padma in *Carbocyclic Cage Compounds* E. Osawa, O. Yonemitsu, (*Eds.*), VCH Publishers, Inc. New York, N. Y., **1992**, 183–215.
13.   T. J. Katz, N. Acton, *J. Am. Chem. Soc.,* **1973**, *95*, 2738–2739.
14.   H. R. Ward, J. S. Wishnok, *J. Am. Chem. Soc.*, **1968**, *90*, 1085–1086.
15.   N. J. Turro, C. A. Renner, W. H. Waddell, T. J. Katz, *J. Am. Chem. Soc.,* **1976**, *98*, 4320–4322.
16.   N. J. Turro, V. Ramamurthy, T. J. Katz, *Nouv. J. Chim.*, **1977**, *1*, 363–365.
17.   J. S. McKennis, L. Brener, J. S. Ward. R. Pettit, *J. Am. Chem. Soc.*, **1971**, *93*, 4957–4958;

cf. L. A. Paquette, R. F. Davis, D. R. James, *Tetrahedron Lett.*, **1974**, 1615–1618.

18. K. Shen, *J. Am. Chem. Soc.*, **1971**, *93*, 3064–3066; *cf.* E. L. Allred, B. R. Beck, *Tetrahedron Lett.*, **1974**, 437–440 and R. Askani, W. Schneider, *Chem. Ber.*, **1983**, *116*, 2366–2370.

19. P. E. Eaton, L. Cassar, R. A. Hudson, D. R. Hwang, *J. Org. Chem.*, **1976**, *41*, 1445–1448; *cf.* E. C. Smith, J. C. Barborak, *J. Org. Chem.*, **1976**, *41*, 1433–1437 and A. P. Marchand, T.-C. Chou, J. D. Ekstrand, D. van der Helm, *J. Org. Chem.*, **1976**, *41*, 1438–1444.

20. P. E. Eaton, Y. S. Or, S.J. Branca, *J. Am. Chem. Soc.*, **1981**, *103*, 2134–2136; *cf.* P. E. Eaton, Y. S. Or, S. J. Branca, B. K. R. Shankar, *Tetrahedron*, **1986**, *42*, 1621–1631.

21. W. G. Dauben A. F. Cunningham, Jr, *J. Org. Chem.*, **1983**, *48*, 2842–2847.

22. P. E. Eaton, U. R. Chakraborty, *J. Am. Chem. Soc.*, **1978**, *100*, 3634–3635.

23. G. Mehta, M. S. Nair, A. Srikrishna, *Ind. J. Chem.*, **1983**, *22B*, 959–963; *cf.* I. A. Akhtar, G. I. Fray, J. M. Yarrow, *J. Chem. Soc. (C)*, **1968**, 812–815.

24. G. Mehta, S. Padma, *Tetrahedron*, **1991**, *47*, 7783–7806.

25. A. Srikrishna, G. Sunderbabu, *J. Org. Chem.*, **1987**, *52*, 5037–5039.

26. N. C. Yang, M. G. Horner, *Tetrahedron Lett.*, **1986**, *27*, 543–546.

27. V. Boekelheide, R. A. Hollins, *J. Am. Chem. Soc.*, **1973**, *95*, 3201–3208; *cf.* G. Mehta, S. R. Shah, *Ind. J. Chem.*, **1990**, *29B*, 101–102.

28. G. Mehta, S. Padma, *J. Am. Chem. Soc.*, **1987**, *109*, 7230–7232; *cf.* G. Mehta, S. Padma, *Tetrahedron*, **1991**, *47*, 7807–7820. For further homologs and secologs of [6]-prismane see T.-C. Chou, Y.-L. Yeh, G.-H. Lin, *Tetrahedron Lett.*, **1996**, *37*, 8779–8782; T. D. Golobish, W. P. Daily, *Tetrahedron Lett.*, **1996**, *37*, 3239–3242.

5 The Platonic Hydrocarbons

According to the Greek philosopher Plato the matter surrounding us and out of which we are made is composed of four elements: fire, earth, air, and water. There is, furthermore, an immaterial element, not part of the physical world, but necessary to construct the 'heavenly matter' or 'ether' and responsible for the 'beautiful order' of the universe. The five elements consist of smallest units, to which characteristic regular polyhedra can be assigned—the tetrahedron (**1**, fire), the cube (**2**, earth), the octahedron (**3**, air), the icosahedron (**4**, water), and the pentagonal dodecahedron (**5**, ether), structures which have since become known as the Platonic bodies (Scheme 1).[1]

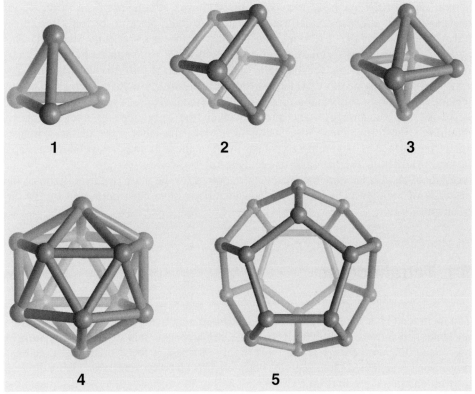

Scheme 1. The Platonic bodies.

That people are strongly attracted by the beauty, harmony and internal balance of these five structures is obvious to anybody who views our environment with open eyes, where we come across Platonic bodies in countless variations in pieces of art, in architecture, furniture, design, engineering *etc*. In many of nature's products we can also discover structures of Platonic symmetry.[2–7]

On the molecular level, *i.e.* when 'translating' these structures into molecular objects by replacing their vertices by atoms and their edges by bonds, it was realized relatively early in the history of chemistry that many inorganic compounds have structures corresponding to the Platonic bodies.[8] Platonic organic molecules, and in particular platonic hydrocarbons,[1] entered the stage of synthetic organic chemistry only in the early 1960s. Obviously, with carbon being tetravalent, the organic equivalents of **1** ('tetrahedrane'), **2** ('cubane'), and **5** ('dodecahedrane') only are possible synthetic targets. The octahedral structure **3** can 'at best' only be found in a new carbon allotrope, whereas an icosahedral molecule (**4**) cannot be constructed from tetravalent carbon atoms.

Besides the aesthetic appeal of **1**, **2**, and **5** their syntheses constitute a major challenge. Not only must a synthesis 'overcome' the extremely high strain in **1** and **2** (see below)—the strain energy per C–C bond amounts to *ca* 23 kcal mol^{-1} in **1** and to 14 kcal mol^{-1} in **2**[1]—but the closed surface of all three hydrocarbons constitutes a very difficult synthetic problem also. Imagine a stepwise approach in which the surfaces are gradually closed by forming one bond of the skeleton after the other. From a certain stage of the synthesis the increasingly convex structures of the synthetic intermediates will allow chemical transformations to take place at, or from one 'face' only—usually from the 'exterior', because this is the less sterically shielded. This geometric restraint will exclude a considerable number of synthetic transformations from further construction; for example, all those which require the departure of certain leaving groups or those which are stereoelectronically demanding as far as trajectories of entering and leaving groups are concerned, *viz*. S$_N$2-type processes. It will hence be a good strategy—and possibly even the only one—to avoid 'reagent-requiring' transformations altogether or at least in the latter steps of the synthesis when the surface is beginning to 'close in'. For these framework-producing steps, isomerizations and thermal and photochemical cycloadditions are very likely the methods of choice, because the 'reagents' required for them are just heat, light, or the presence of a catalyst. In fact, these considerations have been borne out by the experiment.

5.1 Tetrahedrane

Tetrahedrane (**1**) is the only platonic hydrocarbon which has not yet been prepared in unsubstituted form. Considering its extreme strain energy, this should not be surprising—according to theory **1** is the most highly strained of the formally saturated hydrocarbons (calculated strain energy between 126 and 140 kcal mol^{-1} depending on the computational method used).[9] When trying to estimate (or even calculate) the chances of preparing or isolating a highly strained compound, however, it is often kinetic, not thermodynamic stability which counts. High inherent strain of a hydro-carbon is *per se* no reason not to attempt to synthesize it.

Tetrahedrane, however, is unique, because its thermodynamic and its kinetic sta-bility are both very low, as we shall see. It is therefore mandatory to consider the in-troduction of stabilizing stubstituents from the very start of any synthetic project aimed at preparing a tetrahedrane sufficiently stable to enable study of its structural, spectroscopic, and chemical properties.

The classical substituent for stabilizing reactive species and normally unstable conformations of organic molecules is the *tert*-butyl group. This bulky substituent not only hinders or prevents the attack of 'outside' reagents on the molecular system un-der study, but can exert a pronounced 'internal' effect also. For tetrahedrane G. Maier has pointed out that the intramolecular repulsion between the four *tert*-butyl groups of the fully protected hydrocarbon, tetra-*tert*-butyl-tetrahedrane (**6**), is at a minimum when their mutual distance reaches a maximum (Scheme 2), a condition provided by the T_d symmetry of the tetrahedron.[10, 11]

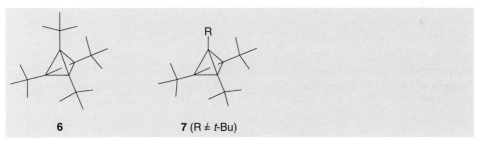

6   **7** (R ≠ *t*-Bu)

Scheme 2. Tetra-*tert*-butyl-tetrahedrane (**6**) and the 'corset effect'.

Any other conceivable arrangement of the substituents—which might also occur during the course of a chemical reaction—forces them into closer contact thus in-creasing the strain of **6**. This effect has been termed 'corset effect' and it has been noted that it must change, or even break down completely, when a *tert*-butyl group is removed or replaced by a smaller substituent as in **7**.[12]

These considerations have been fully supported by the decades-long race towards tetrahedrane, which was finally won by Maier in 1979.[13–15]

After extensive studies on the photochemical and thermal behavior of numerous *tert*-butylated cyclobutene-1,2-dicarboxylic anhydrides, cyclopentadienones and cy-clobutadienes (see Section 10.1), it was finally discovered that the photochemical be-havior of the fully *tert*-butylated cyclopentadienone **15** is completely different from that of lesser substitutend analogs, and that this unusual behavior was the decisive prerequisite for the preparation of **6**.

The synthesis of this first tetrahedrane began with the photoaddition of di-*tert*-butyl-acetylene (**8**) to *tert*-butyl maleic anhydride (**9**) (Scheme 3).

On irradiation, the resulting [2+2]adduct **10** lost carbon dioxide and was con-verted into a mixture of the two trisubstituted cyclopentadienones **11** and **12**. Bromine addition to the latter afforded the 1,4-dibromide **13** (see Section 10.3 for temporarily

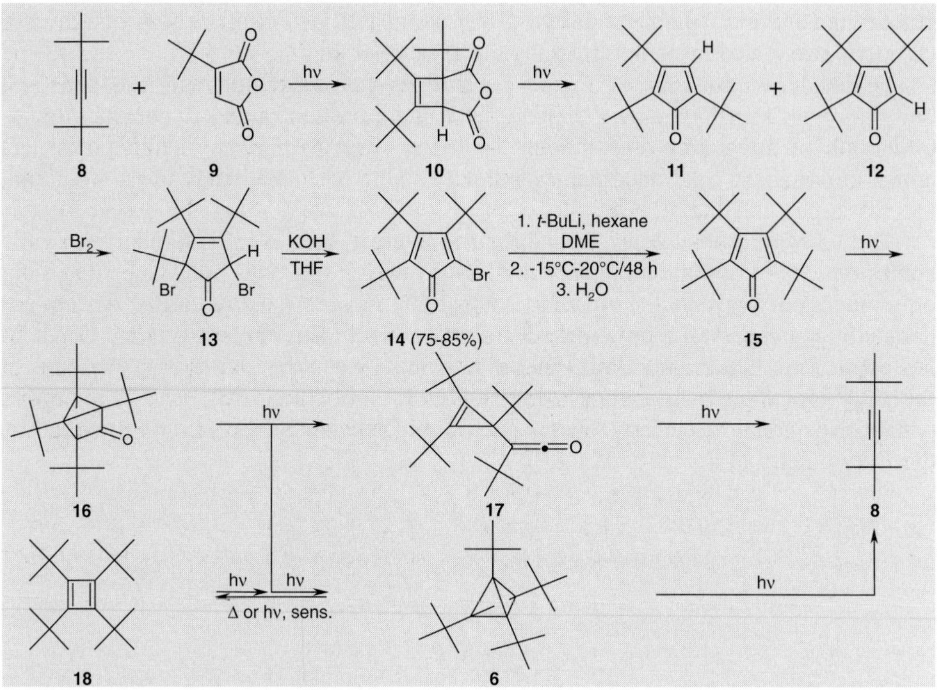

Scheme 3. The first synthesis of a tetrahedrane, tetra-*tert*-butyl- tetrahedrane (**6**) by Maier and co-workers.

'giving-up' one double bond during a hydrocarbon synthesis), which on treatment with potassium hydroxide was dehydrobrominated to **14**. The conversion of this vinylic bromide to the fully *tert*-butylated **15** required painstaking and extensive optimization, and, despite this, this antiaromatic ketone and several of its precursor molecules remain precious organic compounds which are not easily prepared. Excitation of the dienone isolated in an argon matrix with 254-nm light resulted in exclusive criss-cross addition to the tricyclopentanone **16** as observed by IR spectroscopy. Continued irradiation induced elimination of carbon monoxide as well as formation of the ketene **17** as shown by its typical IR band. The ketene is photo labile and on irradiation lost CO slowly. Irradiation in an argon matrix at 77 K enabled the isolation of the hydrocarbon products formed—the starting material **8** (which was the only end product if the photolysis was performed at room temperature) and the tetrahedrane **6**, a stable, colorless and crystalline material. Preparetively, **6** is obtained best when a matrix of **15** in 2,2-dimethyl-butane/*n*-pentane (8:3, the so-called Rigisolve) is irradiated at – 196 °C. Under these conditions the yield is 40% relative to transformed **15**. On heating, **6** isomerized to the fully substituted cyclobutadiene **18**, a process which may be reversed photochemically. Derivatives of **6** with deuterium- and ^{13}C-labeled *tert*-butyl substituents[16, 17] could be prepared by the same route, and both the ^{13}C NMR chemical shift and the ^{13}C–^{13}C coupling constant indicate that the bond between the *tert*-butyl groups and the neighboring carbon atoms must have a high degree of s-character, comparable, in fact, with that in *tert*-butylacetylene and in **8**.

Determination of the structure of **6** by X-ray structural analysis was not an easy task, because the room temperature form of the hydrocarbon consists of soft, ductile crystals. It has, however, been possible to obtain a low-temperature modification, which is stabilized by entrapped gases such as nitrogen or argon, which gave satisfactory diffraction patterns.[18, 19] The C–C bond of the tetrahedron of **6** is considerably shortened, 1.497 Å[19] compared with the usual 1.54 Å for a carbon–carbon single bond. As the deformation densities for the Ar clathrate of **6** show, the bending of the ring bonds amounts to *ca* 26°. From thermodynamic measurements[10, 11] and calculated thermochemical data[20] a strain enthalpy of 129.2 ± 2.1 kcal mol⁻¹ could be derived for **6**, corresponding to an astounding 21.5 kcal mol⁻¹ for each ring bond in the tetrahedron skeleton, making the likely assumption that the *tert*-butyl substituents do not contribute significantly to the strain present in **6**. This is the highest value ever found.[21] The chemical behavior of **6**, which will not be reviewed here,[10, 11] is quite rich, although only two reaction channels, protonation and oxidation, are open to this tetrahedrane.

Knowledge about the chemical and physical properties of **6** notwithstanding, it would still be desirable to have a less tedious and higher yield approach to this compound. An attractive precursor for **6** is the diazo compound **19**, which, by way of either

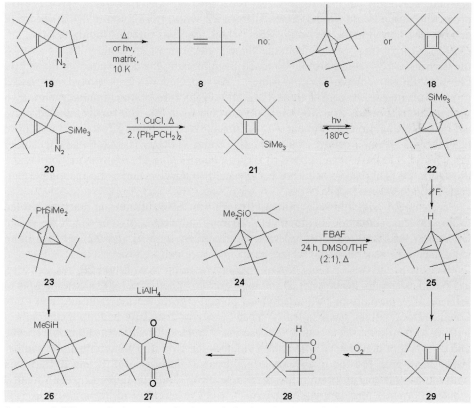

Scheme 4. Silicon-substituted tetrahedrane derivatives by the carbene route.

thermally or photochemically induced nitrogen loss, should give the corresponding carbene which by intramolecular addition could then yield the target **6**. Surprisingly, this route to tetrahedrane[22] could at first not be realized. Although **19** could be readily prepared, its photolysis in argon at 10 K or its flash vacuum thermolysis provided only the cleavage product **8**, neither **6** nor its valence isomer **18**[23] (Scheme 4).

When, however, the irradiation (λ = 254 nm) was performed in a Rigisolve matrix at 77 K **6** was produced in comfortable 60% yield, and even heating **19** in *n*-octane (100 °C) led to the Platonic hydrocarbon (20%).[24]

The usefulness of this approach was further illustrated by the conversion of the diazo compound **20** in which one of the *tert*-butyl substituents of **19** had been replaced by a trimethylsilyl group. Stirring this compound in dichloromethane at 0→20 °C in the presence of cuprous chloride led initially to a metal complex of the cyclobutadiene **21**, from which the pure organic ligand could subsequently be liberated by treatment with ethylene 1,2-bis(diphenylphosphino)ethane. When 'free' **21** was then irradiated in ether or cyclohexane with light of a wavelength ≥ 300 nm it criss-cross cyclized in quantitative yield to the trimethylsilyl tetrahedrane derivative **22**, the first new tetrahedrane to be prepared 11 years after the synthesis of **6**.[25, 26] Compound **22** is thermally more stable than the more symmetrical **6**, but on heating to 180 °C it also reverted to its cyclobutadiene precursor, **21**. The increased stability of **22** cannot be explained solely by steric considerations; although the corset effect is obviously working very well, an electronic effect must also be operative. Although the TMS protecting group could not be removed from **22** with fluoride ion under a variety of conditions to provide the derivative **25**,[26] there are hints that this highly reactive tetrahedrane might be produced from the silicon substituted tetrahedrane **24**, which—as the derivative **23**—could also be obtained by subjecting the corresponding diazo precursor to the above decomposition route.[27] As expected, the isopropoxydimethylsilyl group is a better leaving group than TMS and on fluoride treatment under rather drastic conditions it seems as if **25** was really formed from **24**. The finally isolated product in this desilylation experiment was the unsaturated diketone **27**; it could well have been formed by ring opening of the initially produced **25** to its cyclobutadiene isomer **29**, which could have been trapped by oxygen dissolved in the reaction medium to the dioxetane **28**. This also would not be expected to survive under the reaction conditions but rather open to **27**.

Treating **24** with lithium aluminum hydride caused a substitution reaction and led to **26**, the least stable tetrahedrane yet prepared, and a derivative in which we would not expect the corset effect to operate so effectively as in **6** and **22**, because more space has been generated into which steric hindrance can 'relieve' itself.

In view of the drastic reduction of stability in going from **6** and **22**, respectively, to **26** and especially **25** it is not surprising—in retrospect!—that all earlier attempts to prepare, *e.g.*, partially or fully methyl- or phenyl-substituted tetrahedranes and functionalized derivatives have failed or have been repudiated by authors trying subsequently to reproduce these results. This series of failures began with claims by Thorpe and co-workers in 1913,[28] who were evidently the first to try a tetrahedrane synthesis, to have prepared the methyl tricarboxylic acid derivative of **1**, and did not end with the postulation of a tetralithio-derivative of **1**[29] and its quenching with methyl iodide to provide tetramethyltetrahedrane.[30, 31] The moral of all these *illusions perdues* is, therefore, not to waste time and effort trying to prepare tetrahedranes in which the corset effect is necessarily weaker than in **6** or **22** but to be courageous enough to attempt the preparation of a molecule in which it is certainly completely absent—the parent hydrocarbon **1** itself! Of course, such an attempt would take the above experiences into account and be conducted under appropriate conditions such as in a matrix at very low temperatures. According to theoretical calculations[32] of the enthalpies of formation of various hydrocarbon isomers of **1** such as vinylacetylene, cyclobutadiene (Section 10.1), butatriene (Section 9.2) and triafulvene (Section 11.1), and also of reactive species isomeric with **1**, such as cyclopropenyl carbene, and of the transition states separating these minima, there is only one reasonable route to **1**, namely ring closure of radical **30** (Scheme 5). These calculations offer no advice as to the preparation of **30**, of course, but they illustrate that even this pathway might be difficult to follow, because the activation barriers of its ring closure to **1** and its rearrangement to the isomeric diradical **31** are approximately equal, but the latter is *ca* 28 kcal mol^{-1} more stable than its precursor **30**.

Because these calculations indicate only trends and since they refer to thermally induced processes in the gas phase, the experimentalist must not be discouraged by

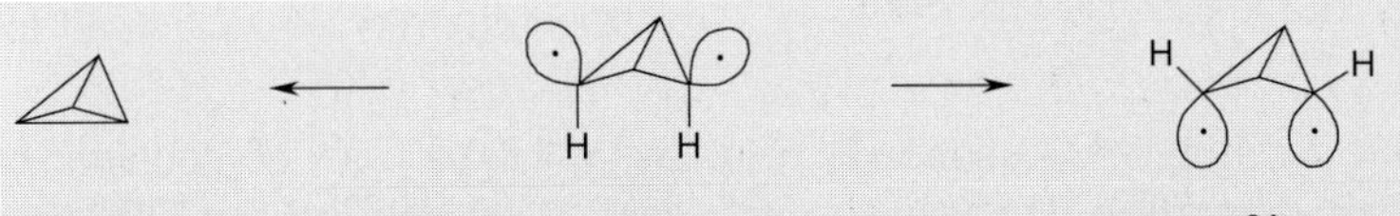

Scheme 5. A pathway to tetrahedrane (**1**) suggested by theoretical calculations.

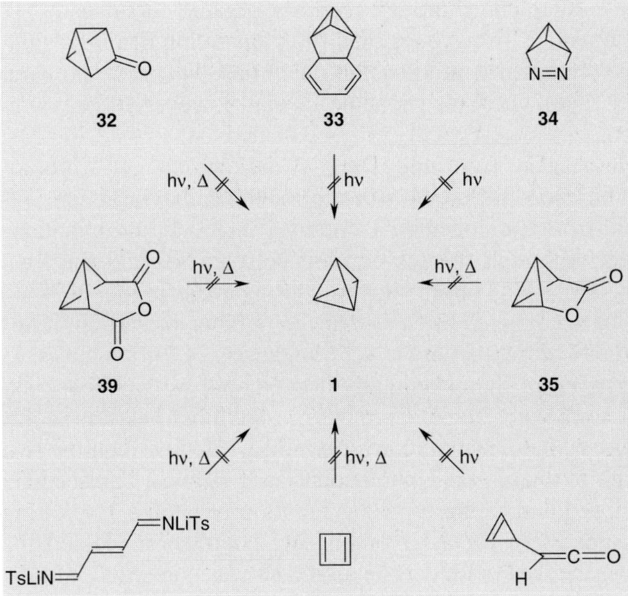

Scheme 6. A selection of unsuccessful attempts to prepare tetrahedrane (**1**) under low temperature matrix conditions.

them, especially because photochemically generated, *i.e.* electronically excited species could behave quite differently. Unfortunately, this hope has not yet been fulfilled, even though certain precursors seemed to be sure winners (Scheme 6).

Irrespective of the method tried, whether the decarbonylation of **32**[33] or the fragmentation of **33**, which would result in benzene as the other cleavage product,[34] the photochemical decomposition of **34** with nitrogen as an excellent leaving group,[35] the (formal) decarboxylation of **35**,[36] the decarbonylation and isomerization of the ketene **36**,[36] the 'back-isomerization' of cyclobutadiene (**37**),[10, 11] the generation of 2-butene-1,4-dicarbene, another reactive C_4H_4 species, from **38**,[37] or the decomposition of the anhydride **39**[2] is concerned—every attempt failed, leaving as the only consolation the knowledge of how *not* to make tetrahedrane.[10, 11] Knowing the ingenuity of preparative chemists, however, and the continuous development of new synthetic methods, it seems safe to predict that these results will not be the last in this challenging area of hydrocarbon chemistry.

5.2 Cubane ([4]-prismane)

Besides adamantane (Section 3.1) cubane (**2**, pentacyclo[4.2.0.0^{2,5}.0^{3,8}.0^{4,7}]octane) is the only other cage hydrocarbon which has found commercial application. This was not all obvious in 1964 when **2**, like so many other hydrocarbons (see, *e.g.*, adamantane: Section 3.1; cyclooctatetraene: Section 10.3; [1.1.1]propellane: Section 6.1; or [2.2]paracyclophane (Section 12.4), to name but a few of the hydrocarbons discussed in this book), made its rather humble entry into the world of organic chemistry.[38, 39] Since then numerous improvements[40–43] have enabled the preparation of this regular hydrocarbon polyhedron (O_h symmmetry) in kilogram quantities. The basic preparative sequence as developed by Eaton and Cole (Scheme 7) was, however, already present in the very first synthesis.

The synthesis of cubane started from the Diels–Alder adduct of 2-bromo-cyclopentadienone to itself, the *endo* dibromide **40**, the monomer having been prepared in *ca* 40% over-all yield from 2-cylopenten-1-one by a three-step bromination-dehydrobromination sequence. Although the relationship between **40** and the final product **2** is not obvious on first sight—after all, the starting material contains no four-membered ring—it can be recognized already that the number of carbon atoms ultimately required in **2** is already present from the very beginning of the synthesis. In fact, there are two more carbon atoms than eventually needed, and we can hence expect carbon-removal steps at some later stage. Certainly no additional carbon atoms are required from outside sources. Ketalization converted **40** to **41**, in which the two double bonds are close enough to enable [2+2]photoaddition to **42**—an intermediate with two cyclobutane rings closed and set up for the generation of a third. Exploiting its α-bromocyclopentanone subsystem for a Favorskii ring contraction yielded the carboxylic acid **43**. To remove the first extra carbon atom, **43** was converted into the perester **46**, which on heating in cumene at 150 °C decarboxylated to **45**, from which the remaining steps to **2** are clearly apparent—repetition of the ring-contraction process just presented (**45**→**48**→**49**→**2**). Short-cuts of this route are possible; they involve the photochemical cage-formation of **44** from **40**, and double Favorskii rearrangement of the former to the diacid **47** which was then again reductively decarboxylated to cubane (**2**).

The success of Eaton's synthesis was so overwhelming—especially when looking back at the developments which have occurred in this field during the last three decades—that a second route to **2**, that of Pettit,[44] which appeared nearly simultaneously with the first synthesis is all but forgotten today, even though it is as elegant as the Eaton approach (Scheme 8).

The decisive step here is the addition of cyclobutadiene—set free from its iron tricarbonyl complex **50** by oxidation (see Section 10.1)—to 2,5-dibromobenzoquinone (**51**). In the resulting [2+2]cycloadduct **52** the two double bonds are again close enough to form a four-membered ring on irradiation, providing the dibromodiketone **53**. One-pot double Favorskii ring contraction converted this to the diacid **54**—an isomer of **47**—which was converted to cubane by decarboxylating its *tert*-butyl perester.

Other cubane syntheses are known,[45, 46] but let us have a second look at Pettit's approach. Since obviously the dibromoquinone has been used as an equivalent of cy-

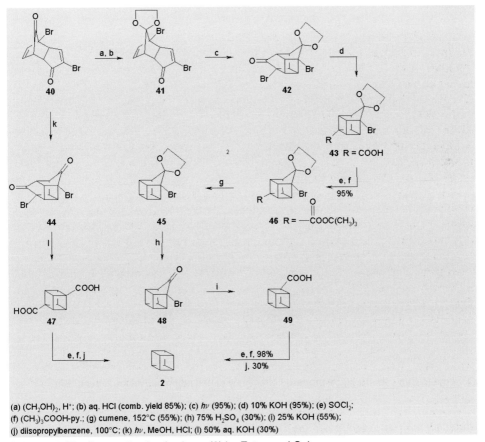

(a) (CH₂OH)₂, H⁺; (b) aq. HCl (comb. yield 85%); (c) *hν* (95%); (d) 10% KOH (95%); (e) SOCl₂;
(f) (CH₃)₃COOH-py.; (g) cumene, 152°C (55%); (h) 75% H₂SO₄ (30%); (i) 25% KOH (55%);
(j) diisopropylbenzene, 100°C; (k) *hν*, MeOH, HCl; (l) 50% aq. KOH (30%)

Scheme 7. The first synthesis of cubane (**2**) by Eaton and Cole.

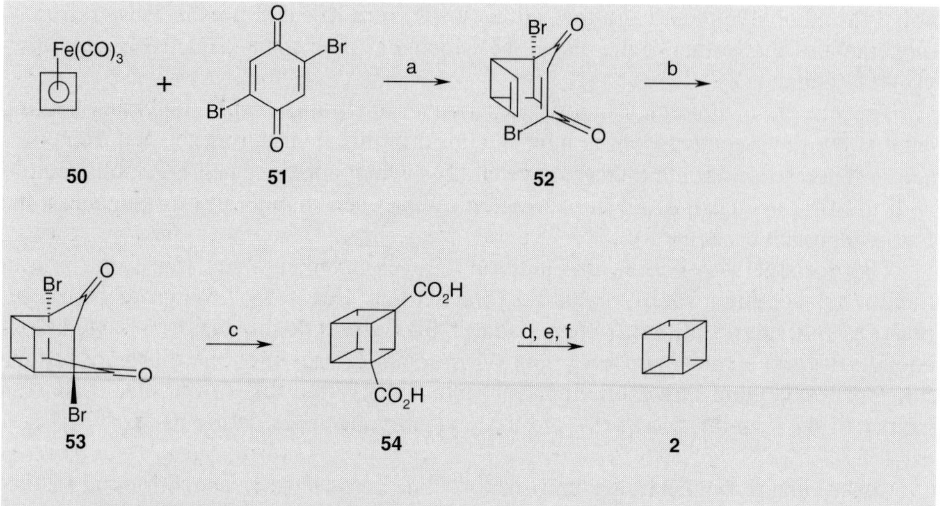

(a) Ce^{4+} (80%); (b) hν (90%); (c) aq. KOH (90%); (d) SOCl$_2$; (e) t-BuOOH;
(f) diisopropylbenzene, Δ

Scheme 8. Pettit's synthesis of cubane (**2**) from its 'half', cyclobutadiene (**37**).

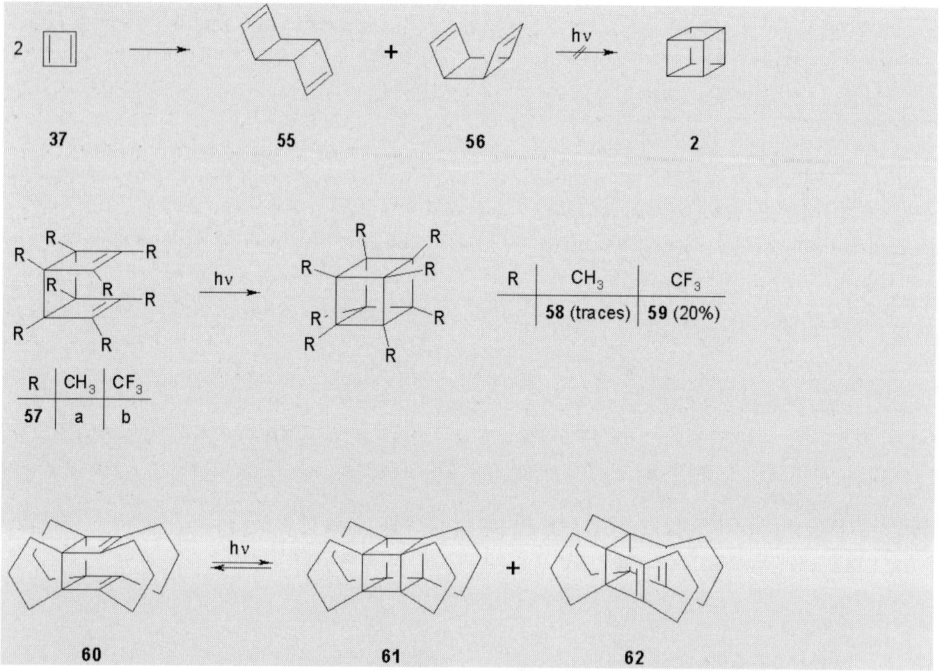

Scheme 9. Can cubane (**2**) be prepared directly by dimerization of cyclobutadiene (**37**)?

clobutadiene (see Section 4.3 on the (attempted) preparation of [6]-prismane, where
similar concepts have been discussed), and cyclobutadiene has been employed to pro-

vide one half of **2**, we might ask whether cubane could not be synthesized directly by dimerization of two molecules of cyclobutadiene (**37**, see Section 10.1) (Scheme 9).

Although cyclobutadiene readily dimerizes with formation of a new four-membered ring[47] (see Section 10.3), one of the products formed, **55**, has the wrong stereochemistry for a further cylization to **2**. What is worse is that the second component of this dimerization, the (correctly oriented) *syn* isomer, *syn*-tricyclo [4.2.0.0^{2,5}]octa-3,7-diene (**56**), does not photocylize under a variety of conditions. This inertness has been attributed to three causes. Firstly, there is a large difference in strain between substrate and product, the latter being far more strained than its open isomer. Secondly, the distance between the two parallel double bonds is relatively large (*ca* 3.05 Å, see below). Finally, because of the dominance of through-bond interaction in **56**, π_+ lies above π_- for both sets of MOs, making the reaction symmetry-forbidden. We would thus expect the process **56**→**2** to depend heavily on stereoelectronic and strain effects. This is indeed so. Although the permethyl derivative of **56**, **57a**, was, for a long time, thought not to photocyclize to octamethyl cubane (**58**)[48], this is actually not the case as was shown by a re-investigation of this process by Gleiter and co–worker.[49] Furthermore, photolysis of the perfluoro derivative **57b** caused photocylization to perfluorooctamethyl cubane **59** in moderate yield.[50]

The decisive role of strain effects in this route to cubane derivatives was demonstrated impressively by Gleiter and co–workers who showed that on irradiating **60**, a fully propano-bridged version of **56**, with a high-pressure mercury lamp in pentane at room temperature, a photoequilibrium is established which contains the starting material **60**, its cubane isomer **61** and the alternative *syn*-tricyclooctadiene derivative **62** in a 10:1:4 ratio.[51] The starting material is readily available and its X-structural analysis shows that the distance between the two double bonds has been reduced to 2.65 Å only. The short bridges in **60** not only bring the double bonds into closer proximity but also reduce the difference between the strain energies of the substrate and product, the large energy gap between **56** and **2** being responsible to a considerable extent for the failure of the photocyclization of the parent hydrocarbons (see above). From the viewpoint of the hydrocarbon chemist the cubane derivative **61** can either be regarded as a propellaprismane (see Sections 4 and 6.1) or a tetramer of cyclopentyne (Section 8.2); this formal consideration shows once again the close relationship between many of the hydrocarbons discussed in this book.

Today cubane chemistry is a well-developed and blossoming field. Only recently, methods such as the *ortho*-metalation of cubane amides, *trans*-metalation and *ortho*-magnesiation have ushered in a renaissance in this area of cage compound chemistry,[52–55] and the hydrocarbon, which started as a rare laboratory chemical, has become a cherished building block in material science, combinatorial chemistry and the preparation of high-energy materials (*vulgo* explosives) such as the polynitrocubanes.[55]

5.3 Dodecahedrane

With the number of C–C bonds exceeding those of tetrahedrane (**1**) and cubane (**2**) by factors of 5 and 2.5, respectively, the number of retrosynthetic pathways for dissection of dodecahedrane (**5**) into smaller precursor units is, of course, much higher than for the two other Platonic hydrocarbons. In principle, however, the number of assembly steps can be reduced significantly if these disconnections are performed in a way that provides building blocks in which the extraordinary high symmetry (point group I_h) of the target molecule is already beginning to emerge. The most important of these deconstruction strategies, all of which have been put to the experimental test, are assembled in Scheme 10.

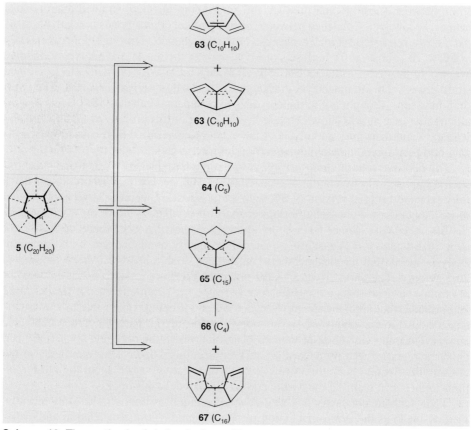

Scheme 10. The synthesis of dodecahedrane (**5**)–retrosyntheses leading to highly symmetrical precursors for a convergent synthesis.

Cleavage of the $C_{20}H_{20}$ hydrocarbon **5** along its 'equator' leads to two molecules of triquinacene (**63**, $C_{10}H_{10}$), whereas removal of a cyclopentanoid lid, **64**, provides a C_{15} fragment, **65**, called peristylane in the shown (saturated) form. In the third retrosynthetic cleavage a dodecahedron cap is cut-off to form a (branched) C_4 fragment, **66**, and the role of the remaining C_{16} unit could be played by the trisolefin **67**, hexaquinacene. Whereas **63** and **67** have the necessary functionality for the reverse, synthetic step, for all other proposed hydrocarbon precursor molecules appropriate functional groups would have to be introduced into the σ-framework.

Year-long intense efforts by the best hydrocarbon chemists showed that none of these approaches could be realized.[1, 56] How seriously Woodward and co-workers took the triquinacene route to dodecahedrane when they first synthesized **63** in 1964[57] (see Section 13.4) and discussed its possible use as a precursor for **5** is unknown. After all, in this dimerization process six double bonds, *i.e.* 12 carbon atoms, must be aligned properly in the transition state to enable formation of six new C–C single bonds. And this entropic disadvantage is augmented by a steric factor since all new bonds must be formed from the sterically highly shielded *endo* orientation of **63**. Consequently triquinacene has never been observed to form dodecahedrane under the action of heat, high pressure, irradiation, or transition metal catalysis.[56] Even if the two halves are preoriented relative to each other by either connecting them directly *via* a single bond as in Paquette's bivalvane attempt[58] or Roberts' triquinacenophane route[59] the entropic and steric obstacles could not be overcome.

In the second approach symbolized in Scheme 10 the number of 'cuts' has been reduced from six to five. Nevertheless, although Eaton and co-workers were able to connect a cyclo-C_5-lid to a peristylane bowl, the final closure of the corresponding 'capped' intermediate to the dodecahedrane skeleton failed.[60] And, finally, after hexaquinacene (**67**), a trisolefin also of interest in connection with long-range electronic interactions between non-conjugated π-systems (see Chapter 13), was prepared by Paquette and co-workers,[61] but could not be closed to dodecahedrane (**5**), convergent routes were abandoned altogether and replaced by serial strategies. After an enormous *tour de force* these were eventually successful, and the 'Mount Everest of hydrocarbon chemistry',[62] *viz.* **5**, was climbed by two completely independent and strategically very different routes.

The first dodecahedrane ever to be prepared was 1,16-dimethyldodecahedrane, prepared in a 19-step synthesis by Paquette and co-workers in 1982.[63, 64] Because the even greater success of synthesizing the parent hydrocarbon was reported only a short time later,[65, 66] I shall only discuss the synthesis of the unsubstituted dodecahedrane (**5**) here; this is also justified because the two synthetic pathways are almost identical.

The synthesis can be divided into three main section; after laying the foundations by preparing a 'cornerstone intermediate', the diester **75/76**, the number of carbon atoms ultimately required is assembled in the center section of the synthesis, which is then concluded by closing the dodecahedrane cage. Whereas the second part involves highly functionalized synthetic intermediates, the last part is characterized by stepwise defunctionalization.

The diester **76**, the fundament of the synthesis, was prepared as illustrated in Scheme 11 by the Diels–Alder addition of dimethyl acetylenedicarboxylate (**72**, E = $COOCH_3$) to the double diene 9,10-dihydrofulvalene (**69**, see Section 11.3).

The hydrocarbon **69** can be prepared either by flash vacuum pyrolysis of nickelocene (**68**) at 950 °C[67] or—more expediently—by oxidative coupling of sodium cyclopentadienide (**70**) with iodine.[68] In the subsequent, so-called domino Diels–Alder addition[69, 70] the dienophile **72** can approach the diene system **69** either from the periphery—as shown by the transition state **71**—or internally, between the two cyclopentadiene rings—see **73**. The former approach led to adduct **74**, whereas **76** was obtained by the latter. The two isomers were produced as a 1:1-mixture,[71] and when the whole sequence was optimized adduct **76** could be isolated in 15–20% yield.[72]

The relationship of **76** to dodecahedrane is not obvious when viewing this product

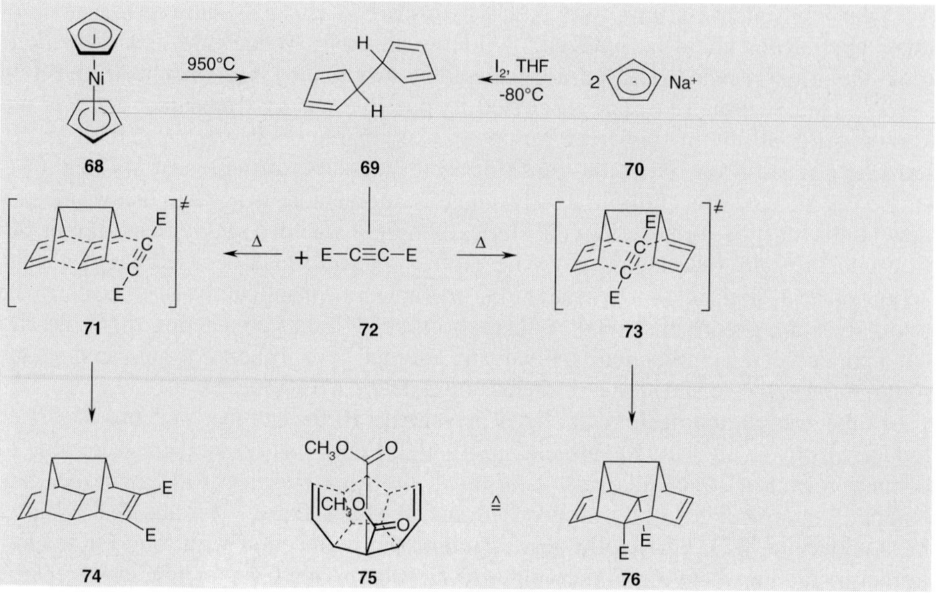

Scheme 11. Paquette's synthesis of dodecahedrane (**5**)—laying the foundations (E = COOCH₃ in all structures).

in the projection shown in the scheme. Redrawing the diester as in **75**, however, reveals that this symmetric intermediate (C_{2v} symmetry) already contains four five-membered rings connected in a way that fixes the six methine hydrogen atoms in an all-*cis* relationship. Furthermore, the diester already contains 14 of the ultimately required 20 carbon atoms of dodecahedrane (**5**). The above projection also clearly shows that the central C–C bond, originally constituting the central σ-bond of the dienophile **72**, must be removed at some time later in the synthesis. For the ensuing steps, however, it was retained, to take advantage of the norbornenyl character that it confers on the two halves of the molecule, thus guaranteeing high stereochemical control in several later transformations.

In the second section of the synthesis the number of carbon atoms was brought to completion. The transformations required towards this end are easier to visualize in structural formulas of type **76**, and I shall hence change to 'spherical structures' à la **75** only after the appropriate C_{20} intermediate has been reached (Scheme 12).

A suitable precursor for the addition of further carbon sources is the cross-corner diketone **80**. This was obtained highly efficiently (68% overall yield) by iodolactonization of **76**, saponification of the resulting iodohydrin **77**, with sodium methoxide in methanol, to **78**, oxidation of this diol with Jones reagent to **79**, and reductive removal of the iodine substituents of the latter compound with zinc–copper couple and ammonium chloride in methanol. For introduction of the still missing carbon atoms, **80** was treated with diphenylcyclopropylsulfonium ylid (**81**). This reagent, introduced by Trost [73] converted **80** initially into oxaspiropentanes which isomerized to the bis(spirocaclobutanone) **82** under the acidic work-up conditions. Baeyer–Villiger oxidation of **82** finally yielded the dilactone diester **83** in quantitative yield.

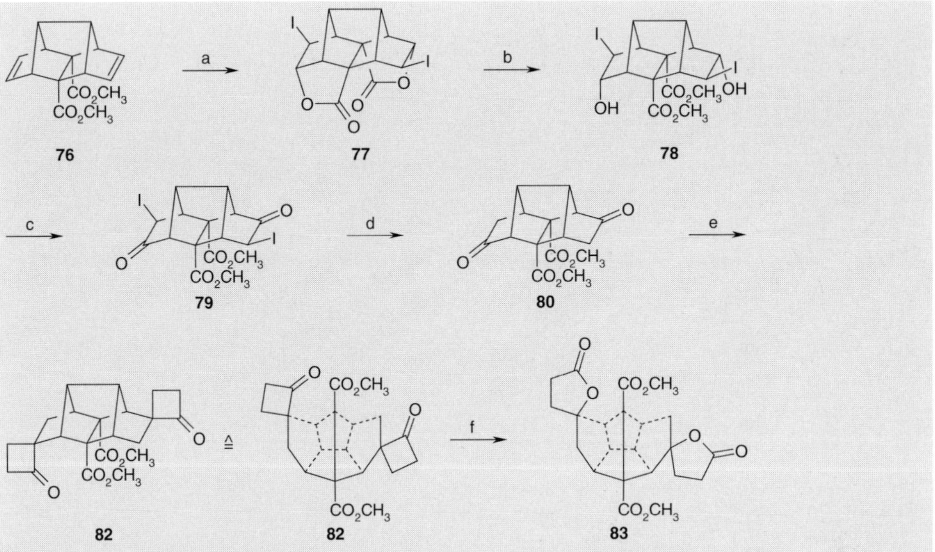

(a) KOH, aq. CH₃OH, then I₂, NaHCO₃ (94%); (b) NaOH, CH₃OH (quant.); (c) Na₂Cr₂O₇, H₂SO₄ (92%); (d) Zn / Cu, CH₃OH (78%); (e) + ▷—ŠPh₂ (**81**) (77%); (f) H₂O₂, CH₃OH (100%)

Scheme 12. Paquette's synthesis of dodecahedrane (**5**)—the completion of the number of carbon atoms.

Although these steps have not increased the number of methine groups (6) and five-membered rings (4) compared with **76**, the number of carbon atoms is now complete and the stage is set for the construction of new five-membered rings by suitable manipulation of the existing carbon framework. This was accomplished in the third part of the synthesis as summarized in Scheme 13.

Before discussing the final transformations it must again be pointed out that because of the spherical geometry of the target molecule and many of its precursor molecules the generation of nearly every methine carbon must be accomplished in a contrathermodynamic manner, *i.e.* with the much smaller hydrogen atom oriented towards the exterior and the larger functional groups towards the interior of the (growing) sphere.

When **83** was treated with phosphorous pentoxide in methanesulfonic acid its spirocyclic functionality was transposed into laterally fused cyclopentenone rings as shown in **84**. Not surprisingly catalytic hydrogenation (Pd/C) occurred from the 'outside' (convex side) of **84** furnishing the diketo diester **85**, which marks the half-way point of the synthesis because 10 of the 20 CH-groups and 6 of the 12 cyclopentane rings have been prepared. With the spherical structure emerging step by step stereochemical alternatives other than reagent attack from the convex surface become less and less likely, yes, even impossible. Thus sodium borohydride reduction of **85** converted it directly into the 'closed' bis-lactone **86**. When this was opened by treatment with hydrochloric acid in methanol the dichloro diester **87** was produced. Although this transformation sounds simple it actually required extensive optimization work and at times even threatened the whole approach.

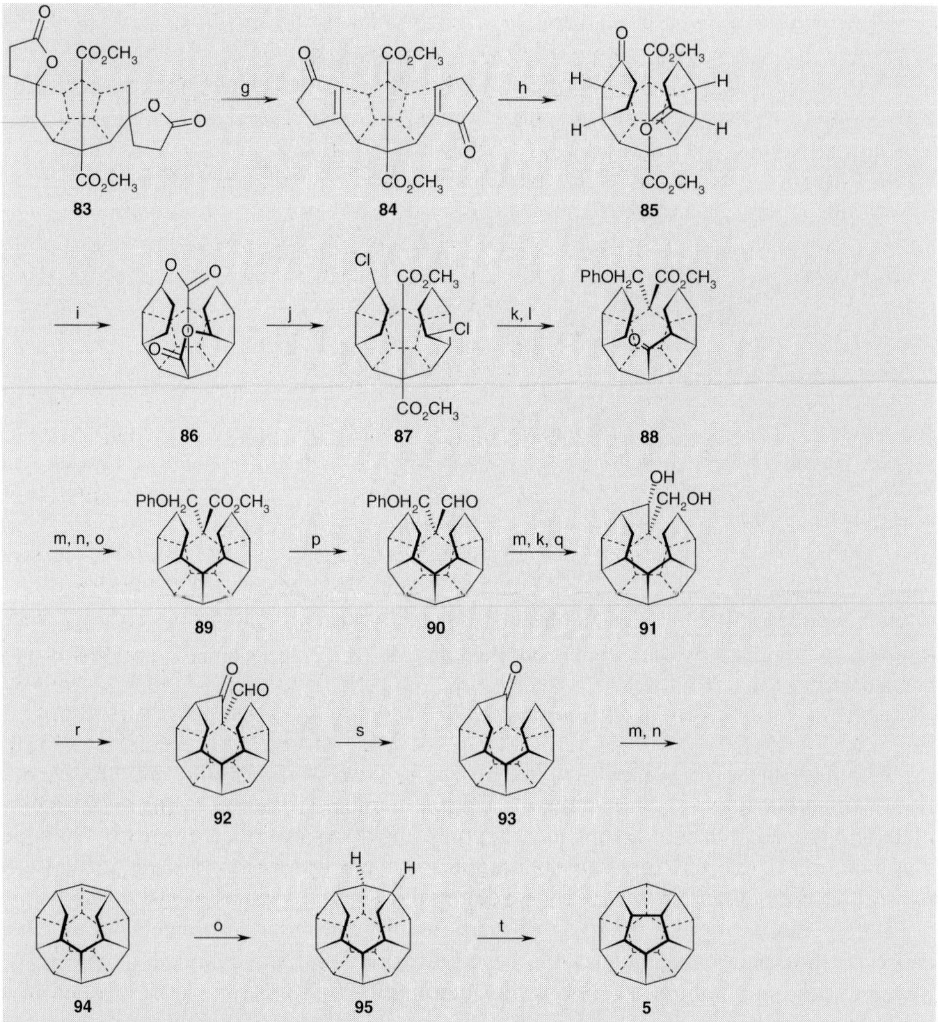

(g) P_4O_{10}, H_3CSO_3H (83%); (h) H_2, Pd / C, EtOAc (100%); (i) $NaBH_4$, CH_3OH (81%); (j) HCl / CH_3OH (62%); (k) Li / NH_3; (l) $PhOCH_2Cl$ (comb. yield 48%); (m) hv; (n) TosOH; (o) HN=NH; (p) Dibal-H; (q) H_3O^+; (r) PCC; (s) KOH, EtOH (37%); (t) Pd / C, 250°C (40-50%)

Scheme 13. Paquette's synthesis of dodecahedrane (**5**)–the closing of the dodecahedrane cage.

The next five-membered ring was produced when **87** was converted into the tetra-seco ketoester **88** by treatment with lithium in liquid ammonia in the presence of chloromethylphenylether as a trapping reagent. As the comparison of the two structures shows, several deep-seated changes in the functionality of **87** occur during this crucial step, the mechanism of which has not been fully established. Not only has one five-membered ring been closed under concomitant loss of chloride, but the second chlorine substituent has also been removed, a primarily generated dianion has been trapped by the added electrophile, and the central auxiliary C–C bond has been severed. It should be noted that **87** also served as the pivotal intermediate in the synthesis of 1,6-dimethyldodecahedrane by changing the trapping reagent to methyl iodide.

The eighth cyclopentane ring was introduced by subjecting **88** to what could be called a homo-Norrish photocyclization. This process initially led to the corresponding Norrish alcohol, which was dehydrated with *p*-toluenesulfonic acid. Diimine reduction then yielded the triseco derivative **89**. After this had been reduced to the aldehyde **90** with diisobutylaluminum hydride a second Norrish cyclization was conducted and the protecting groups were removed by Birch-type reduction followed by acid hydrolysis. From the resulting diol **91**, a C_{21}-intermediate, the extra carbon atom was removed by PCC-oxidation to **92** and base treatment of the latter (retrograde aldol reaction, loss of formate). The thus obtained ketone **93** is ideally set up for still another Norrish photocyclization, and when the alcohol formed was again dehydrated with *p*-toluenesulfonic acid, the monoseco olefin **94** resulted. Diimine hydrogenation as above converted **94** into *seco*-dodecahedrane **95**, and the crowning dehydrogenation to **5** was accomplished with H_2-presaturated 10% palladium on carbon at 250 °C—a 23 step, most difficult journey, more then once threatened by frustration and failure had come to a happy and successful ending!

Although Paquette's dodecahedrane synthesis provided enough material for study of the spectroscopic, structural and selected chemical properties of **5** (see below), a shorter sequence was clearly desirable. Recapitulating the adamantane success story (see Section 3.1) an isomerization approach starting from another $C_{20}H_{20}$ hydrocarbon immediately comes to mind, especially because it had been predicted by force-field calculations that **5** is the stabilomer, *i.e.* the most stable isomer among its class.[74] Beginning with attempts to isomerize the *anti* dimer of basketene (see Section 13.3), **96**,[75] all these experiments aimed at either **5** or its dimethyl derivatives **99**, failed, irrespective of whether other saturated polycyclic precursors such as **98**, **100-102** or various multiply-bridged cyclophanes, **97** and **103** were used as precursors.[76] In the end it was Prinzbach and co-workers who demonstrated that dodecahedrane can be prepared by a thermodynamically controlled isomerization of another $C_{20}H_{20}$ hydrocarbon, pagodane (**104**) (Scheme 14).

The name for this highly symmetric polyquinane (D_{2h} symmetry) was coined because its structure reminded the authors of an oriental temple and its reflection in water.[77–80] Like dodecahedrane **104** contains 12 multifused cyclopentane rings. To convert it to **5** formally two cyclobutane bonds have to be cleaved hydrogenolytically; inversion of configuration at these centers has to take place accompanied by an opening of the molecular sphere; and finally ring closure has to be accomplished by oxidative C–C bond formation between opposing methylene bridges as summarized in structure **105**. Energetically this is a favorable isomerization, force-field calculations having shown the heat of formation of **104** to be more than 40 kcal mol^{-1} higher than that of the stabilomer **5**.[74] Structurally **104** is also of interest because it has substructures which are dealt with in other chapters of this book; *i.e.* the central cyclobutane ring forms part of several propellanes (see Section 6.1) and each of the quaternary carbon atoms is the center of a [5.5.5.5]fenestrane unit (see Section 6.2).

Prinzbach's synthesis of pagodane (**104**) begins with the hexachloride isodrin (**106**). This substrate, formerly used as an insecticide, already contains 12 carbon atoms, which are arranged in four properly oriented cyclopentane rings—in practice this molecule already represents one-half of the pagodane structure. The route from

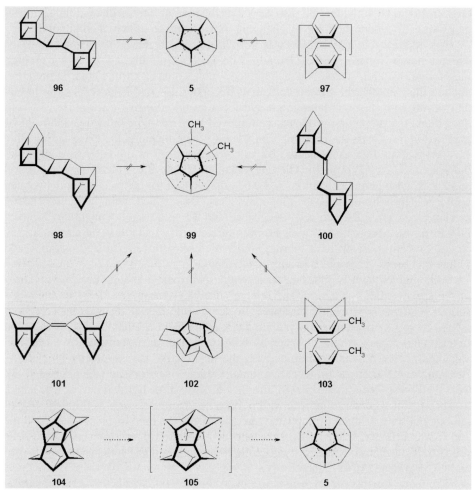

Scheme 14. Dodecahedranes by thermodynamically controlled isomerizations?

106 to **104** can be roughly divided into two halves: in the first half of the sequence nearly all the carbon atoms which eventually will form the skeleton of the target molecule are assembled (Scheme 15), whereas the second half of the synthesis creates the 'mirror image' of the pagoda.

To add the needed additional carbon atoms to **106** a double benzannelation furnishing **113** (a bridged [3.3]orthocylophane intermediate, see Section 12.3) was first performed by reacting isodrin successively with two equivalents of the highly reactive diene tetrachlorothiophene-1,1-dioxide (**107**), which—as the comparison of starting material and product **113** shows—functions as an equivalent of cyclobutadiene (see Section 10.1). In the actual synthesis **106** and **107** first underwent Diels–Alder addition to the mono adduct **108** which spontaneously split off sulfur dioxide and provided the intermediate **109**. Under the reaction conditions (reflux in carbon tetrachloride) this readily aromatized to **110** (note the internal hydrogen atom transfer from

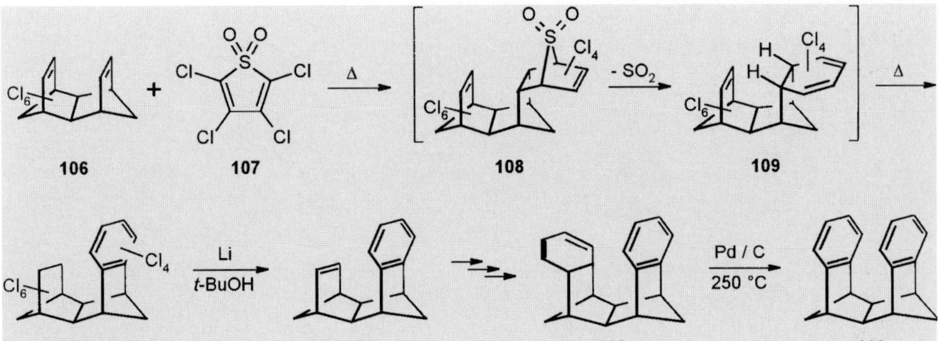

Scheme 15. Prinzbach's synthesis of dodecahedrane (**5**)–assembling the carbon framework of pagodane (**104**).

the six-membered ring to the etheno bridge), and when this was dehalogenated with lithium in *tert*-butanol the mono-annelated hydrocarbon **111** was obtained in an 95 % over-all yield.[78, 81] Although subjecting **111** to the same protocol yielded hydrocarbon **112** effortlessly, the aromatization of the newly generated six-membered ring turned out to be difficult because the two inner hydrogen atoms which have to be removed for that purpose are effectively shielded by the already present first benzene ring. After extensive variation of methods and conditions **113** could be obtained from **112** by palladium-catalyzed dehydrogenation at 250 °C, the yield of the **111**→**113**-interconversion amounting to *ca* 35%.

In the first step of the second half of the pagodane synthesis the floor of the pagoda was completed by an intramolecular [2+2]photocycloaddition. Thus irradiation of **113** with monochromatic 254-nm light established a photoequilibrium to which the photoproduct **114** contributed in *ca* 30%. The tetraene is thermally very stable (the activation energy for the back reaction to **113** was determined to be 38 kcal mol^{-1}), ring-opening processes destroying the just closed four-membered ring need not, therefore, be feared during the subsequent transformations. The next step, addition of maleic anhydride (**115**) to **114** in benzene under reflux, is similar to the domino Diels–Alder addition of dimethyl acetylenedicarboxylate (**72**) to dihydrofulvalene (**69**) in Paquette's dodecahedrane synthesis (see above). As illustrated in Scheme 16 the dienophile approaches **114** from the least hindered direction (*i.e.* away from the *endo*-oriented methylene hydrogen substituent of the roof of the diene). [2+4]Cycloaddition then created a new dienophilic double bond which is ideally positioned to enable a second, now internal, Diels–Alder addition to occur. The primary 1:1 adduct was, in fact, so reactive that it could not be detected. In the resulting anhydride **116**, formed stereospecifically and in quantitative yield, the structural relationship to pagodane cannot be overlooked. Still, 'image' and 'mirror image' do not yet match—there are two extra carbon atoms in the lower half of the carbon framework of the future pagodane. In other words, the twenty carbon atoms of the precursor molecules **113** and **114** are not the twenty carbon atoms of **104**; to generate the pagodane skeleton one of the original carbon atoms of **114** must be removed as must one of the carbon atoms introduced by the Diels–Alder addition with **115**.

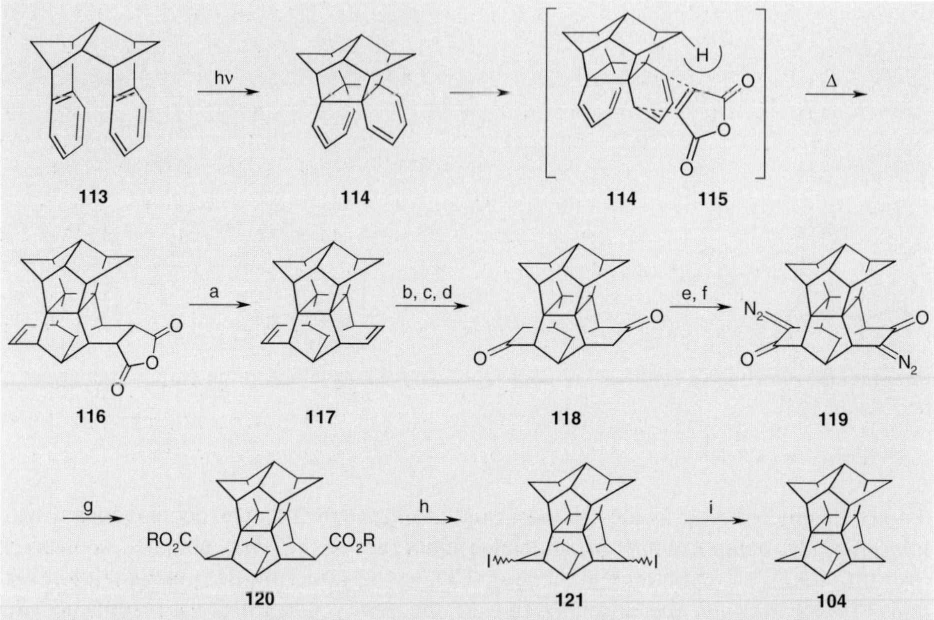

(a) Cu$_2$O, bipyridyl, H$_2$O; (b) B$_2$H$_6$ / THF, 0°C, quinoline, 100-150°C (78%); (c) NaOH, H$_2$O$_2$ (comb. yield 96%); (d) CrO$_3$, acetone, 0°C (97%); (e) HCOOCH$_3$ / NaH, THF, room temp.; (f) p-TsN$_3$ / NEt$_3$, room temp. (comb. yield 82%); (g) MeOH, hν, room temp. (95%); (h) Pb(OAc)$_4$, I$_2$, CCl$_4$, hν (80%); (i) Na-K, THF, then t-BuOH (100%)

Scheme 16. Prinzbach's synthesis of dodecahedrane (**5**)–the preparation of pagodane (**104**).

The required double ring-contraction/carbon number reduction was initiated by degradation of **116** to the diene **117** by cuprous oxide-induced decarboxylation in boiling quinoline. This was next converted to the diketone **118** by standard procedures—hydroboration, oxidation—and subsequent formylation (activation of the diketone) and diazo-group transfer then led directly in a one-pot operation to the bis(diazoketone) **119**. When methanolic solutions of this intermediate were subjected to the photochemical Wolff ring contraction, uniform conversion to the diester **120** occurred. The thermodynamically unfavorable *syn,syn*-stereochemistry—proven by X-ray structural analysis—of the ester groups in **120** probably results from the kinetically controlled *anti*-capture of methanol by the sterically less shielded bottom side of the ketenes formed as intermediates. To remove the two surplus carbon atoms, **120** was first saponified and the resulting diacid iododecarboxylated by irradiation in the presence of iodine and lead tetraacetate. Complete reduction to pagodane (**104**) was finally achieved by reducing the obtained diiodide **121** (mixture of isomers) with sodium-potassium alloy in *tert*-butanol.

The synthesis of pagodane from isodrin (**106**)[82] required a total of 45 functional changes which could be concentrated into 14 one-pot manipulations. The overall yield was 24% implying an average yield of *ca* 90% per step or 97% per functional group change. The structure of **104** was established by the usual spectroscopic methods, the ^{1}H-^{13}C coupling constants of 139 and 141 Hz for the bridgehead carbon hydrogen bonds demonstrating their pronounced s-type character, as is typical for many strained hydrocarbons. According to X-ray structural analysis[83] the so-called lateral cyclo-butane bonds (C1–C2) are stretched to 1.57 Å—a promising property in view of their intended cleavage in the **104**→**5**-transformation.

Although **104** rearranged upon exposure to catalytic amounts of Lewis acids and strong protic acids, no dodecahedrane could be detected among the isomerization products. Fortunately, noble metal catalyzed isomerizations on—*inter alia*—Pt/Re/Al$_2$O$_3$, Rh/C, Pd/Al$_2$O$_3$, whether performed in the condensed or the gas phase, did indeed close **104** to dodecahedrane (**5**), thus demonstrating after all that the summit of the Mt Everest of hydrocarbon chemistry can be reached by an alternative route.[84] Yields were only in the 2–8% range and **5** had to be separated and purified from the product mixture by painstaking purification steps. A way out of the dilemma of having prepared an 'immediate' precursor of **5** but not being able to use it as effectively as expected in dodecahedrane chemistry, was, however, found, by realizing that the strained central cyclobutane ring of **104** is part of several small-ring propellane units known to add electrophiles or radicals across their central C–C bond (see Section 6.1).

Hence, when pagodane was treated with elemental bromine with simultaneous irradiation 1,4-addition took place and the dibromide **122** was obtained as a single stereoisomer in quantitative yield. The process is reversible under photochemical and thermal conditions, but heating **122** in the presence of zinc in DMF furnished the ring-opened nonacyclic diene **123** in 89% yield (Scheme 17). Note that this hydro-carbon is a retro[2+2]product of **104**, and as such could in principle be produced from pagodane by thermal ring opening. When **104** was heated at 750 °C, however, only one monomeric product could be identified—naphthalene, produced in 70% yield.[85] Bissecododecahedradiene **123** is a remarkable cage compound in its own right. It not only has the reduced reactivity of a hyperstable olefin[86] but has also has been con-verted into unusually persistent pagodane radical cations[87] and di-cations.[83, 88] Cyclopropanation of **123** with dichlorocarbene under phase transfer conditions yielded the corresponding bis-adduct[89] and when this was dehalogenated with so-dium/*tert*-butanol in THF the cyclopropanated pagodane **124** was obtained;[84] this turned out to be a valuable intermediate for various dodecahedranes, including the parent hydrocarbon **5**.

Under carefully adjusted reaction conditions 30% of **125** could be obtained from **124** *via* the dimethyl pagodane intermediate **126** and by demethylation the parent hydrocarbon **5** was even formed in 50% yield. As a further dimethyldodecahedrane the isomer **127** was also produced in small amounts, **128** evidently serving as an in-termediate.[79]

The high symmetry of dodecahedrane (point group I_h), is reflected by all its spec-troscopic and physical properties. With $\delta = 3.38$ the dodecahedryl protons absorb at relatively low-field, presumably because of strong deshielding caused by the three adjoining, fully eclipsed C–H bonds. The ^{13}C–H coupling constant is in the expected range for a strained hydrocarbon (134.5 Hz, see above).[66] The infrared and the Ra-man spectra of **5** are both characterized by a small number of absorption bands. With 430 ± 10 °C the melting point of dodecahedrane is unusually high. The interesting question concerning the size of the cavity of **5** was answered after successful X-ray crystallographic structure determination—the intracavity distance is 4.310-4.317 Å.[90] The C–C bonds of the hydrocarbon (1.535–1.541 Å) are somewhat shorter than

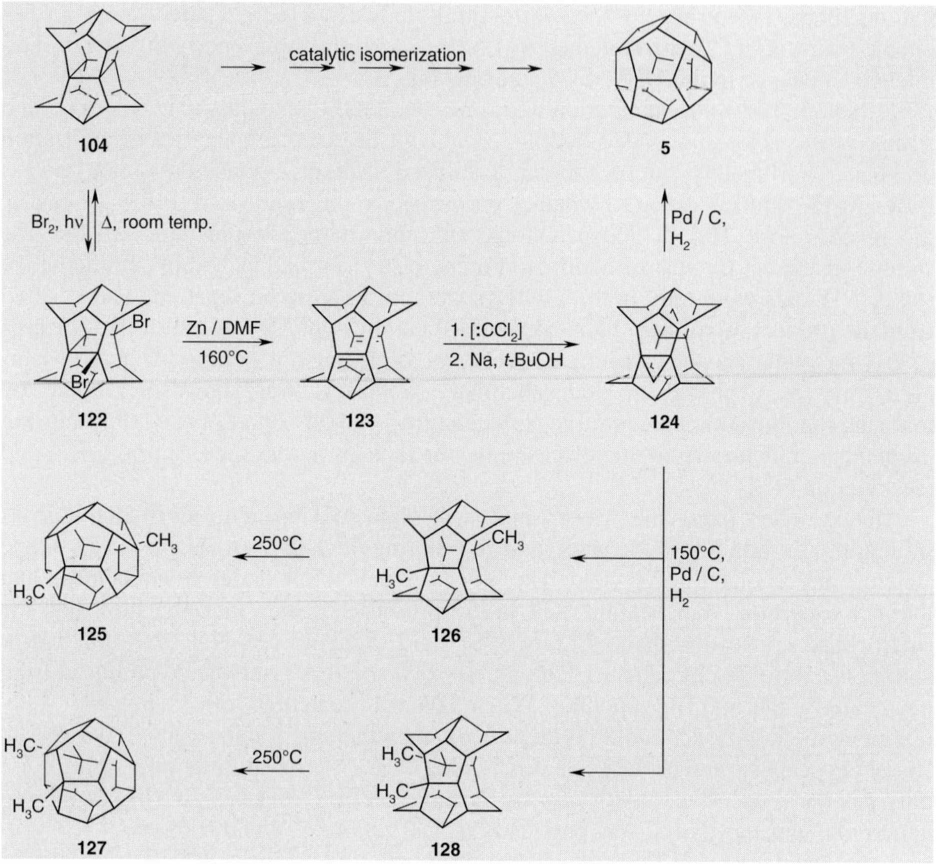

Scheme 17. Prinzbach's synthesis of dodecahedrane (**5**)–closing the pagodane cage.

the 1.546 Å value determined for cyclopentane. The C–C–C bond angles, which range from 107.7 to 108.1° conform very well to the value expected for perfect do-decahedral symmetry (108°). On the basis of the experimentally determined enthalpy of formation of 1,6-bis(methoxycarbonyl)dodecahedrane a heat of formation ($\Delta H^\circ_f(g)$) of 18.2 kcal mol^{-1} and a strain energy of 61.4 ± 1.1 kcal mol^{-1} have been derived for the parent hydrocarbon.[91]

Recent improvements of both Paquette's[92] and Prinzbach's dodecahedrane syn-thesis[93] have enabled systematic investigation of its chemical reactivity and led to the synthesis of numerous dodecahedrane derivatives. As shown by the representative collection in Scheme 18 these include a large variety of monosubstituted derivatives such as the monohalides **129** (Hal = F, Cl, Br), the hydrocarbons **130** (C_xH_y = vinyl, phenyl, ethyl *etc.*), dodecahedranes with polar functional groups **131** (functional group F = CHO, COOH, COOCH$_3$, CONH$_2$, NH$_2$, *etc.*) and bisfunctionalized do-decahedranes **132** ($F^1 = F^2$ = Br, OCH$_3$, COOCH$_3$; F^1 = Br, F^2= OCH$_3$ *etc.*). Poly-functionalization with four and even six functional groups has also been achieved.[93]

Of particular interest is the introduction of C–C double bonds on the surface of **5**.

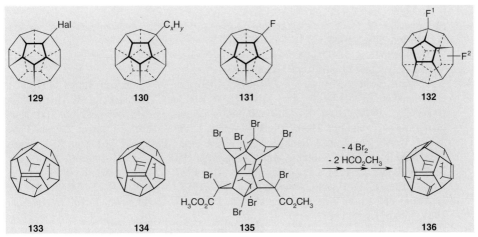

Scheme 18. A selection of functionalized dodecahedranes.

On the one hand its rigid spherical geometry enforces pyramidalization at the sp^2-centers (see Section 7.4) and on the other hand successive dehydrogenation could in principle lead to a continous series of dodecahedrenes, which ultimately would end in eicosadehydrododecahedrane—fullerene C_{20}! Indeed the first steps towards this goal, the preparation of the monoolefin **133**[56, 94] and some of its derivatives,[95] and of the diene **134**[94, 96] have been reported, and the hexaene **136** has left its fingerprint in the MS fragmentation pattern of the highly brominated pagodane diester **135**.[93] More recent studies with highly functionalized dodecahedranes even indicate that species containing only two residual hydrogen atoms, *i.e.* $C_{20}H_2$, can be generated under MS conditions.[97]

Although the chemistry of dodecahedrane is by no means as 'easy' as that of adamantane (see Section 3.1), and this hydrocarbon and its derivatives are definitely 'not for everybody'[93] it clearly has come of age.

References

1.   W. Grahn, *Chem. in uns. Zeit*, **1981**, *15*, 52–61; *cf.* P. Friedländer, *Platon als Atomphysiker* in P. Friedländer, *Platon*, Berlin, **1964**, 3rd ed., Vol. I, 260–275; L. R. MacGillivray, J. L. Atwood, *Angew. Chem.*, **1999**, *111*, 1080–1096; *Angew. Chem. Int. Ed.*, **1999**, *38*, 1018–1033.
2.   I. Bernal, W. C. Hamilton, *Symmetry*, W. H. Freedman and Comp., San Francisco, **1972**.
3.   P. Pearce, S. Pearce, *Polyhedra Primer*, Van Nostrand Reinhold Comp. New York, N. Y., **1978**.
4.   I. Hargittai, M. Hargittai, *Symmetry through the Eyes of a Chemist*, VCH Verlagsgesellschaft, Weinheim, **1986**.
5.   R. Wille (*Ed.*), *Symmetrie in Geistes- und Naturwissenschaft*, Springer-Verlag, Heidel-

berg, **1988**.

6.   M. Senechal, G. Fleck (*Eds.*), *Shaping Space - a Polyhedral Approach*, Birkhäuser Verlag, Boston-Basel, **1988**.

7.   E. Heilbronner, J. D. Dunitz, *Reflections on Symmetry in Chemistry...and Elsewhere*, Verlag Helvetica Chimica Acta, Basel, **1993**.

8.   A. F. Wells, *Structural Inorganic Chemistry*, Clarendon Press, Oxford, 4th ed., **1975**.

9.   Calculated strain energies for **1** by MNDO: 136 kcal mol^{-1} (A. Schweig, W. Thiel, *J. Am. Chem. Soc., ***1979**, *101*, 4742–4743), ab initio calculations: 132–140 kcal mol^{-1} (H. Kollmar, *J. Am. Chem. Soc.*, **1980**, *102*, 2617–2621; K. B. Wiberg, *J. Comput. Chem.*, **1984**, *5*, 197; B. A. Hess, Jr, L. J. Schaad, *J. Am. Chem. Soc.*, **1985**, *107*, 865–866; R. L. Disch, J. M. Schulman, M. L. Sabon, *J. Am. Chem. Soc.*, **1985**, *107*, 1904–1906; K. B. Wiberg, R. F. W. Bader, C. D. H. Lau, *J. Am. Chem. Soc.*, **1987**, *109*, 985–1001; K. B. Wiberg, R. F. W. Bader, C. D. H. Lau, *J. Am. Chem. Soc.*, **1987**, *109*, 1001–1012); MINDO/3: 126 kcal mol^{-1} (H. Kollmar, F. Carrion, M. J. S. Dewar, R. C. Bingham, *J. Am. Chem. Soc.*, **1981**, *103*, 5292–5305).

10. Review: G. Maier, *Angew. Chem.*, **1988**, *100*, 317–341; *Angew. Chem. Int. Ed. Engl.*, **1988**, *27*, 309–332.

11. G. Maier in *Cage Hydrocarbons* G. A. Olah (*Ed.*), J. Wiley and Sons, New York, N. Y., **1990**, 219–259.

12. Although this is a powerful image, a corset actually functions differently. It is tubular structure with two open ends at either side: any hindrance between the matter in its interior is thus reduced by 'shifting' it, at least partially, towards the 'outside', where it, in fact, might fulfill various tactic and strategic functions. Hydrocarbon **6** does not have these options.

13. G. Maier, S. Pfriem, *Angew. Chem.*, **1978**, *90*, 551–552; *Angew. Chem. Int. Ed. Engl.*, **1978**, *17*, 519–520.

14. G. Maier, S. Pfriem, U. Schäfer, R. Matusch, *Angew. Chem.*, **1978**, *90*, 552–553; *Angew. Chem. Int. Ed. Engl.*, **1978**, *17*, 520–521.

15. G. Maier, S. Pfriem, U. Schäfer, K.-D. Malsch, R. Matusch, *Chem. Ber.*, **1981**, *114*, 3965–3987.

16. L. H. Franz, Ph. d. Dissertation, University of Giessen, **1982**, *cf.* ref.[10, 11]

17. T. Loerzer, R. Machinek, W. Lüttke, L. H. Franz, K.-D. Malsch, G. Maier, *Angew. Chem.*, **1983**, *95*, 914; *Angew. Chem. Int. Ed. Engl.*, **1983**, *22*, 878; *cf.* G. Maier, S. Pfriem, K.-D. Malsch, H.-O. Kalinowski, K. Dehnicke, *Chem. Ber.*, **1981**, *114*, 3988–3996.

18. H. Irngartinger, A. Goldmann, R. Jahn, M. Nixdorf, H. Rodewald, G. Maier, K.-D. Malsch, R. Emrich, *Angew. Chem.*, **1984**, *96*, 967–968; *Angew. Chem. Int. Ed. Engl.*, **1984**, *23*, 993–994.

19. H. Irngartinger, R. Jahn, G. Maier, R. Emrich, *Angew. Chem.*, **1987**, *99*, 356–357; *Angew. Chem. Int. Ed. Engl.*, **1987**, *26*, 356–357.

20. P. v. R. Schleyer, J. E. Williams, K. R. Blanchard, *J. Am. Chem. Soc.*, **1970**, *92*, 2377–2386.

21. The strain energy per bond in [1.1.1]propellane (see Section 6.1) is 15 kcal mol^{-1}. M. D. Newton, J. M. Schulman, *J. Am. Chem. Soc.*, **1972**, *94*, 773–778; *cf.* K. B. Wiberg, *Angew. Chem.*, **1986**, *98*, 312–322; *Angew. Chem. Int. Ed. Engl.*, **1986**, *25*, 312–322.

22. S. Masamune, N. Nakamura, M. Suda, H. Ona, *J. Am. Chem. Soc.*, **1973**, *95*, 8481–8483; *cf.* P. Eisenbarth, M. Regitz, *Chem. Ber.*, **1982**, *115*, 3796–3810.

23. G. Maier, K. A. Reuter, L. H. Franz, H. P. Reisenauer, *Tetrahedron Lett.*, **1985**, *26*, 1845–1848; *cf.* G. Maier, I. Bauer, D. Born, H.-O. Kalinowski, *Angew. Chem.*, **1986**, *98*, 1132–1134; *Angew. Chem. Int. Ed. Engl.*, **1986**, *25*, 1093–1095.

24. G. Maier, F. Fleischer, *Tetrahedron Lett.*, **1991**, *32*, 57–60; *cf.* G. Maier, F. Fleischer, *Liebigs Ann. Chem.*, **1995**, 169–172; G. Maier, H. Rang, R. Emrich, *Liebigs Ann. Chem.*, **1995**, 153–160.

25. G. Maier, D. Born, *Angew. Chem.*, **1989**, *101*, 1085–1087; *Angew. Chem. Int. Ed. Engl.*, **1989**, *28*, 1050–1052.

26. G. Maier, D. Born, I. Bauer, R. Wolf, R. Boese, D. Cremer, *Chem. Ber.*, **1994**, *127*, 173–189; *cf.* G. Maier, R. Wolf, F. Fleischer, *GIT Zeitschrift*, **1992**, 36. For the photoisomerization of 1,2,3-tri-*tert*-butyl-4-(trimethylgermyl)cyclobutadiene to the corresponding tri-*tert*-butyl(trimethylgermyl)tetrahedrane see G. Maier, R. Wolf, H.-O. Kalinowski, *Chem. Ber.*, **1994**, *127*, 201–204.

27. G. Maier, R. Wolf, H.-O. Kalinowski, R. Boese, *Chem. Ber.*, **1994**, *127*, 191–200; *cf.* G. Maier, R. Wolf, H.-O. Kalinowski, *Angew. Chem.*, **1992**, *104*, 764–766; *Angew. Chem. Int. Ed. Engl.*, **1992**, *31*, 738–740.

28. R. M. Beesley, J. F. Thorpe, *Proc. Chem. Soc.*, **1913**, *29*, 346–347; *cf.* R. M. Beesley, J. F. Thorpe, *J. Chem. Soc.*, **1920**, *117*, 591–620.

29. G. Rauscher, T. Clark, D. Poppinger, P. v. R. Schleyer, *Angew. Chem.*, **1978**, *90*, 306–307; *Angew. Chem. Int. Ed. Engl.*, **1978**, *17*, 276–277.

30. N. S. Zefirov, V. N. Kirin, N. M. Yur'eva, A. S. Koz'min, N. S. Kulikov, Y. N. Luzikov, *Tetrahedron Lett.*, **1979**, 1925–1926.

31. A summary of these attempts is given by N. S. Zefirov, A. S. Koz'min, A. V. Abramenkov, *Russ. Chem. Rev.*, **1978**, *47*, 163–171 who cite a publication on the ester of Thorpe's tricarboxylic acid with cellulose as having useful properties for industrial applications. For modern attempts to prepare peralkylated tetrahedranes **7** with R ≠ *tert*-butyl see: G. Maier, H. P. Reisenauer, *Chem. Ber.*, **1981**, *114*, 3959–3964; G. Maier, L. H. Franz, *Liebigs Ann. Chem.*, **1995**, 139–145; G. Maier, L. H. Franz, R. Boese, *Liebigs Ann. Chem.*, **1995**, 147–151; G. Maier, F. Fleischer, H.-O. Kalinowski, *Liebigs Ann. Chem.*, **1995**, 173–186.

32. H. Kollmar, F. Carrion, M. J. S. Dewar, R. C. Bingham, *J. Am. Chem. Soc.*, **1981**, *103*, 5292–5305.

33. G. Maier, M. Hoppe, H. P. Reisenauer, *Angew. Chem.*, **1983**, *95*, 1009–1011; *Angew. Chem. Int. Ed. Engl.*, **1983**, *22*, 990–992. For a review on **32** and its derivatives see P. Dowd, H. Irngartinger, *Chem. Rev.*, **1989**, *89*, 985–996.

34. M. Christl, S. Freund, *Chem. Ber.*, **1985**, *118*, 979–999.

35. D. A. Kaisaki, D. A. Dougherty, *Tetrahedron Lett.*, **1987**, *28*, 5263–5266.

36. G. Maier, M. Hoppe, K. Lanz, H. P. Reisenauer, *Tetrahedron Lett.*, **1984**, *25*, 5645–5648.

37. L. B. Rodewald, H.-K. Lee, *J. Am. Chem. Soc.*, **1973**, *95*, 623–624; L. B. Rodewald, H.-K. Lee, *J. Am. Chem. Soc.*, **1973**, *95*, 3084; *cf.* B. Wolf, Ph. d. dissertation, University of Giessen, **1985**.

38. P. E. Eaton, T. W. Cole, Jr, *J. Am. Chem. Soc.*, **1964**, *86*, 962–964.

39. P. E. Eaton, T. W. Cole, Jr, *J. Am. Chem. Soc.*, **1964**, *86*, 3157–3158.

40. N. B. Chapman, J. M. Key, K. J. Toyne, *J. Org. Chem.*, **1970**, *35*, 3860–3867.

41. T.-Y. Luh, L. M. Stock, *J. Am. Chem. Soc.*, **1974**, *96*, 3712–3713; *cf.* A. J. H. Klunder, B. Zwanenburg, *Tetrahedron*, **1972**, *28*, 4131–4138.

42. R. S. Abeywickrema, E. W. Della, *J. Org. Chem.*, **1980**, *45*, 4226–4229; *cf.* E. W. Della, J. Tsanaktsidis, *Austr. J. Chem.*, **1986**, *39*, 2061–2066.

43. M. Bliese, J. Tsanaktsidis, *Austr. J. Chem.*, **1997**, *50*, 189–192 and references therein.

44. J. C. Barborak, L. Watts, R. Pettit, *J. Am. Chem. Soc.*, **1966**, *88*, 1328–1329.

45. P. E. Eaton, T. W. Cole, Jr, *J. Chem. Soc. Chem. Commun.*, **1970**, 1493–1494.

46. C. G. Chin, H. W. Cuts, S. Masamune, *J. Chem. Soc. Chem. Commun.*, **1966**, 880–881.

47. M. Avram, J. G. Dinulescu, E. Marcia, G. Mateescu, E. Sliam, C. D. Nenitzescu, *Chem.*

Ber., **1964**, *97*, 382–389.

48. D. Seebach, *Angew. Chem.*, **1965**, *7*, 119–129; *Angew. Chem. Int. Ed. Engl.*, **1965**, *4*, 121–131; *cf.* C. E. Berkoff, R. C. Cookson, J. Hudec, D. W. Jones, R. O. Williams, *J. Chem. Soc.*, **1965**, 194–200.
49. R. Gleiter, St. Brand, *Tetrahedron Lett.*, **1994**, *35*, 4969–4972.
50. L. F. Pelosi, W. T. Miller, *J. Am. Chem. Soc.*, **1976**, *98*, 4311–4312.
51. R. Gleiter, M. Karcher, *Angew. Chem.*, **1988**, *100*, 851–852; *Angew. Chem. Int. Ed. Engl.*, **1988**, *27*, 840–841; *cf.* R. Gleiter, *Angew. Chem.*, **1992**, *104*, 29–46; *Angew. Chem. Int. Ed. Engl.*, **1992**, *31*, 27–44.
52. G. W. Griffin, A. P. Marchand, *Chem. Rev.*, **1989**, *89*, 997–1010.
53. Review: H. Higuchi, I. Ueda in *Carbocyclic Cage Compounds* E. Osawa, O. Yonemitsu (*Eds.*), VCH Publishers, New York, N. Y., **1992**, 217–247.
54. R. E. Hormann, *Aldr. Chim. Acta*, **1996**, *29*, 31–38.
55. A. Bashir-Hashemi, G. Doyle, *Aldr. Chim. Acta*, **1996**, *29*, 43–49.
56. Reviews: L. A. Paquette, *Chem. Rev.*, **1989**, *89*, 1051–1065; L. A. Paquette in *Strategies and Tactics in Organic Synthesis* Th. Lindberg (*Ed.*), Academic Press, Orlando, **1984**, 175–200; L. A. Paquette, *Top. Curr. Chem.*, **1979**, *79*, 43–165; P. E. Eaton, *Tetrahedron*, **1979**, *35*, 2189–2223.
57. R. B. Woodward, T. Fukunaga, R. C. Kelly, *J. Am. Chem. Soc.*, **1964**, *86*, 3162–3164.
58. L. A. Paquette, I. Itoh, W. B. Farnham, *J. Am. Chem. Soc.*, **1975**, *97*, 7280–7285; *cf.* L. A. Paquette, I. Itoh, K. B. Lipkowitz, *J. Org. Chem.*, **1976**, *41*, 3524–3529.
59. W. P. Roberts, G. Shoham, *Tetrahedron Lett.*, **1981**, *22*, 4895–4898. The phane concept actually does not apply to this case, since phanes are bridged aromatic compounds and triquinacene is not an aromatic subsystem.
60. P. E. Eaton, R. H. Mueller, *J. Am. Chem. Soc.*, **1972**, *94*, 1014–1016; *cf.* P. E. Eaton, R. H. Mueller, G. R. Carlson, D. A. Cullinson, G. F. Cooper, T.-C. Chou, E.-P. Krebs, *J. Am. Chem. Soc.*, **1977**, *99*, 2751–2767; P. E. Eaton, G. D. Andrews, E.-P. Krebs, A. Kumai, *J. Org. Chem.*, **1979**, *44*, 2824–2834.
61. L. A. Paquette, R. A. Snow, J. L. Muthard, T. Cynkowski, *J. Am. Chem. Soc.*, **1978**, *100*, 1600–1602; L. A. Paquette, R. A. Snow, J. L. Muthard, T. Cynkowski, *J. Am. Chem. Soc.*, **1979**, *101*, 6991–6996; G. G. Christoph, J. L. Muthard, M. C. Böhm, R. Gleiter, *J. Am. Chem. Soc.*, **1978**, *100*, 7782–7784.
62. See the review *Organic Chemistry, **1976*** in *Nachr. Chem. Labor, Technik*, **1977**, 59–70.
63. L. A. Paquette, D. W. Balogh, *J. Am. Chem. Soc.*, **1982**, *104*, 774–783; *cf.* L. A. Paquette, D. W. Balogh, J. F. Blount, *J. Am. Chem. Soc.*, **1981**, *103*, 228–230; G. G. Christoph, P. Engel, R. Usha, D. W. Balogh, L. A. Paquette, *J. Am. Chem. Soc.*, **1982**, *104*, 784–791.
64. L. A. Paquette, D. W. Balogh, R. Usha, D. Kountz, G. G. Christoph, *Science*, **1981**, *211*, 575.
65. L. A. Paquette, R. J. Ternansky, D. W. Balogh, *J. Am. Chem. Soc.*, **1982**, *104*, 4502–4503.
66. R. J. Ternansky, D. W. Balogh, L. A. Paquette, *J. Am. Chem. Soc.*, **1982**, *104*, 4503–4504; *cf.* L. A. Paquette, R. J. Ternansky, D. W. Balogh, W. J. Taylor, *J. Am. Chem. Soc.*, **1983**, *105*, 5441–5446; L. A Paquette, R. J. Ternansky, D. W. Balogh, G. Kentgen, *J. Am. Chem. Soc.*, **1983**, *105*, 5446–5450; L. A. Paquette, M. J. Wyvratt, O. Schallner, J. L. Muthard, W. J. Begley, R. M. Blankenship, D. W. Balogh, *J. Org. Chem.*, **1979**, *44*, 3616–3630.
67. E. Hedaya, D. W. McNeil, P. Schissel, D. J. McAdoo, *J. Am. Chem. Soc.*, **1968**, *90*, 5284–5286.
68. W. v. E. Doering in *Theoretical Organic Chemistry - The Kekulé- Symposium*, Butterworths, London, **1959**, 45.

69. For an overview of intramolecular Diels–Alder additions see: G. Brieger, J. M. Bennett, *Chem. Rev.*, **1980**, *80*, 63–97.

70. L. A. Paquette, M. J. Wyvratt, H. C. Berk, R. E. Moerck, *J. Am. Chem. Soc.*, **1978**, *100*, 5845–5855.

71. L. A. Paquette, M. J. Wyvratt, *J. Am. Chem. Soc.*, **1974**, *96*, 4671–4673. The 2:1-adduct **76** has also been called the "pincer" adduct;[70] for further uses of the stereoselective construction of polycyclic ring systems using the Domino pincer Diels–Alder reaction see M. Lautens, E. Fillion, *J. Org. Chem.*, **1997**, *62*, 4416–4427 and refs. quoted.

72. D. W. McNeil, B. R. Vogt, J. J. Sudol, S. Theodoropulos, E. Hedaya, *J. Am. Chem. Soc.*, **1974**, *96*, 4673–4674.

73. B. M. Trost, M. J. Bogdanowicz, *J. Am. Chem. Soc.*, **1973**, *95*, 5298–5307; B. M. Trost, M. J. Bogdanowicz, *J. Am. Chem. Soc.*, **1973**, *95*, 5321–5334 and previous papers in this series.

74. S. A. Godleski, P. v. R. Schleyer, E. Osawa, Y. Inamoto, Y. Fujikura, *J. Org. Chem.*, **1976**, *41*, 2596–2605; *cf.* T. Iizuka, M. Imai, N. Tanaka, T. Kann, E. Osawa, *Science Reports of the Faculty of Education*, Gunma University, **1981**, *30*, 5; *Chem. Abstr.*, **1982**, *97*, 126567m.

75. N. J. Jones, W. D. Deadman, E. LeGoff, *Tetrahedron Lett.*, **1973**, 2087–2088.

76. P. Grubmüller, Ph. d. thesis, University of Erlangen-Nürnberg, **1979**.

77. W.-D. Fessner, H. Prinzbach, G. Rihs, *Tetrahedron Lett.*, **1983**, *24*, 5857–5860.

78. W.-D. Fessner, G. Sedelmeier, P. R. Spurr, G. Rihs, H. Prinzbach, *J. Am. Chem. Soc.*, **1987**, *109*, 4626–4642.

79. W.-D. Fessner, H. Prinzbach in *Cage Hydrocarbons* G.A. Olah (*Ed.*), J. Wiley and Sons, Inc., New York, N. Y., **1990**, 353–405; *cf.* H. Prinzbach, W.-D. Fesssner in *Organic Synthesis: Modern Trends* O. Chizhov (*Ed.*), Blackwell Scientific, Oxford, **1987**, 23.

80. Review: H. Prinzbach, K. Weber, *Angew. Chem.*, **1994**, *106*, 2329–2348; *Angew. Chem. Int. Ed. Engl.*, **1994**, *33*, 2239–2258.

81. The hydrocarbon **111** has previously been prepared in lower yield (38%) from **106** using the dimethyl acetal of tetrachlorocyclopentadienone as the diene component: K. Mackenzie, *J. Chem. Soc.*, **1965**, 4646–4653.

82. Interestingly isodrin (**106**) also served as the starting material in Woodward`s synthesis of triquinacene (**63**),[57] which could not—as we have seen above—be dimerized to dodecahedrane (**5**).

83. G. K. S. Prakash, V. V. Krishnamurthy, R. Herges, R. Bau, H. Yuan, G. A. Olah, W.-D. Fessner, H. Prinzbach, *J. Am. Chem. Soc.*, **1988**, *110*, 7764–7772.

84. W.-D. Fessner, B. A. R. C. Murty, J. Wörth, D. Hunkler, H. Fritz, H. Prinzbach, W. D. Roth, P. v. R. Schleyer, A. B. McEwen, W. F. Maier, *Angew. Chem.*, **1987**, *99*, 484–486; *Angew. Chem. Int. Ed. Engl.*, **1987**, *26*, 452–454.

85. W.-D. Fessner, B. A. R. C. Murty, H. Prinzbach, *Angew. Chem.*, **1987**, *99*, 482–484; *Angew. Chem. Int. Ed. Engl.*, **1987**, *26*, 451–453.

86. According to *Schleyer* hyperstable olefins are olefins with strain energies smaller than that of the corresponding saturated hydrocarbons: W. F. Maier, P. v. R. Schleyer, *J. Am. Chem. Soc.*, **1981**, *103*, 1891–1900; A. B. McEwen, P. v. R. Schleyer, *J. Am. Chem. Soc.*, **1986**, *108*, 3951–3960. A considerable number of these olefins has in the meantime been reported (for a review see W. Luef, R. Keese, *Top. Stereochemistry*, **1991**, *20*, 231–318), including hyperstable bridgehead olefins: H. Kukuk, E. Proksch, A. de Meijere, *Angew. Chem.*, **1982**, *94*, 304: *Angew. Chem. Int. Ed. Engl.*, **1982**, *21*, 306; H. Hopf, R. Savinsky, P. G. Jones, I. Dix, B. Ahrens, *Liebigs Ann./Recueil*, **1997**, 1499–1504.

87. H. Prinzbach, B. A. R. C. Murty, W.-D. Fessner, J. Mortensen, J. Heinze, G. Gescheidt, F. Gerson, *Angew. Chem.*, **1987**, *99*, 488–490; *Angew. Chem. Int. Ed. Engl.*, **1987**, *26*,

457–459.

88. G. K. S. Prakash, V. V. Krishnamurthy, R. Herges, R. Bau, H. Yuan, G. A. Olah, W.-D. Fessner, H. Prinzbach, *J. Am. Chem. Soc.*, **1986**, *108*, 836–838.

89. P. R. Spurr, B. A. R. C. Murty, W.-D. Fessner, H. Fritz, H. Prinzbach, *Angew. Chem.*, **1987**, *99*, 486–488; *Angew. Chem. Int. Ed. Engl.*, **1987**, *26*, 455–457.

90. J. C. Galluci, C. W. Doecke, L. A. Paquette, *J. Am. Chem. Soc.*, **1986**, *108*, 1343–1344. For the X-ray structural analysis of 1,16-dimethyldodecahedrane see G. G. Christoph, P. Engel, R. Usha, D. W. Balogh, L. A. Paquette, *J. Am. Chem. Soc.*, **1982**, *104*, 784–791.

91. H. D. Beckhaus, Chr. Rüchardt, D. R. Lagerwall, L. A. Paquette, F. Wahl, H. Prinzbach, *J. Am. Chem. Soc.*, **1994**, *116*, 11775–11778; *cf.* the correction in *J. Am. Chem. Soc.*, **1995**, *117*, 8885. For calculated enthalpies of formation see J. M. Schulman, R. L. Disch, *J. Am. Chem. Soc.*, **1984**, *106*, 1202–1204 and refs. cited.

92. L. A. Paquette, J. C. Weber, T. Kobayashi, Y. Miyahara, *J. Am. Chem. Soc.*, **1988**, *110*, 8591–8599; *cf.* G. A. Olah, G. K. S. Prakash, W.-D. Fessner, T. Kobayashi, L. A. Paquette, *J. Am. Chem. Soc.*, **1988**, *110*, 8599–8605.

93. M. Bertau, J. Leonhardt, A. Weiler, K. Weber, H. Prinzbach, *Chem. Eur. J.*, **1996**, *2*, 570–579.

94. J.-P. Melder, K. Webert, A. Weiler, E. Sackers, H. Fritz, D. Hunkler, H. Prinzbach, *Res. Chem. Intermed.*, **1996**, *22*, 667–702. With 43.5° dodecahedrene (**133**) possesses a highly out of plane bent C-C double bond: R. C. Haddon, *Science*, **1993**, *261*, 1545.

95. J.-P. Melder, R. Pinkos, H. Fritz, H. Prinzbach, *Angew. Chem.*, **1989**, *101*, 314–319; *Angew. Chem. Int. Ed. Engl.*, **1989**, *28*, 305–310.

96. J.-P. Melder, R. Pinkos, H. Fritz, H. Prinzbach, *Angew. Chem.*, **1990**, *102*, 105–109; *Angew. Chem. Int. Ed. Engl.*, **1990**, *29*, 95–99.

97. K. Scheumann, E. Sackers, M. Bertau, J. Leonhardt, D. Hunkler, H. Fritz, J. Wörth, H. Prinzbach, *J. Chem. Soc.*, *Perkin Trans. 2*, **1998**, 1195–1210.

6 Bridgehead-distorted Hydrocarbons

Although the bond angles of many of the bridgehead carbons atoms incorporated in the cage systems discussed in Chapters 3 to 5 span a considerable range and sometimes differ drastically from the ideal 109.5 ° angle found in, *e.g.*, adamantane, they have one property in common—the four bonds extending from these carbon atoms point into *both* hemispheres. In other words, in all these compounds the tetrahedral stereochemistry of our original building unit, the sp^3-hybridized carbon atom (**1**) can always be recognized, even in the most highly distorted hydrocarbons.

The question might, however, be asked, whether or not it is possible to deform **1** to such an extent that all of its four bonds point into *one* hemisphere only, as shown by structure **2** in Scheme 1.

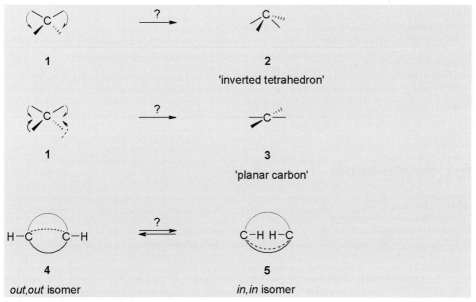

Scheme 1. Inverted and planar carbon atoms.

One way to reach this state of 'inverted tetrahedral' geometry would involve the displacement of two bonds out of their original hemisphere all the way towards the side of the molecule which contains the other two bonds—as the arrows in the **1**→**2** transformation indicate. If, on the other hand, the 'down movement' of the in-plane bonds in **1** were accompanied by an 'up movement' of its out-of-plane bonds, could a situation be reached in which all bonds lie in one plane—the carbon atom having been 'planarized' as in **3**?[1]

Because the concept of the tetrahedral carbon atom is one of the fundaments on which organic chemistry is built—irrespective of whether one considers its stereochemical, dynamic or synthetic aspects—the problem of violating this geometry is clearly very important. That this deformation does not come 'naturally' is amply illustrated by the millions of organic compounds constructed from carbon atoms of type **1**. Although the configurational stability of the tetrahedral carbon atom is a prerequisite of life as we know it, the question might be raised whether one cannot construct, *e.g.*, bicyclic systems which could undergo inversion from an *out, out* (**4**) to an *in, in* isomer (**5**). If this reaction could be shown to occur what would be its mechanism?

The stability of the tetrahedral over the planar arrangement is supported by theoretical calculations. Thus, the energy difference between square-planar and tetrahedral methane far exceeds the typical carbon-hydrogen and carbon-carbon bond strengths. In other words, hydrocarbons would rather 'decompose' than adopt this particular geometry.[2–6]

It should, however, be pointed out that these considerations apply to hydrocarbons only. Theoretical and experimental evidence shows that planar carbon is, in fact, preferred by several metalorganic systems, in which metal atoms are bonded directly to that particular carbon atom.[7–9] A discussion of these most interesting 'anti-van't-Hoff molecules' is beyond the scope of this book.

We will begin our discussion of the above questions by looking at the so-called propellanes first. Not only has this group of compounds been studied far more thoroughly than the molecules discussed below in Sections 6.2 and 6.3, but there are also many review articles available which should enable the reader to familiarize him/herself quickly with the vast literature.[10–16]

6.1 Propellanes

Ginsburg, who coined the term propellanes,[10, 11] defined these compounds as tricyclic systems conjoined in or by a carbon-carbon single bond. They thus have the general structure **6**; the smallest conceivable representative is [1.1.1]propellane (**7**, Scheme 2).

The next higher homolog is [2.1.1]propellane (**8**) with which—on further homologation—branching may set in, leading to either [2.2.1]propellane (**9**) or [3.1.1]propellane (**10**). The inverted geometry discussed above can evidently only be expected for propellanes in which *m*, *n*, and *o* are sufficiently small;[17] this seems to be so for molecules not exceeding the [3.2.1] and the [4.1.1] skeletons.

Although [1.1.1]propellane (**7**) seems at first sight to be prohibitively strained, it is actually a stable compound which has been prepared by different routes (Scheme 3).

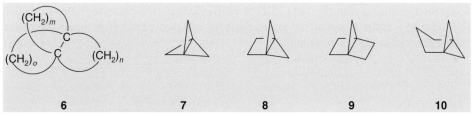

Scheme 2. The small-ring propellanes.

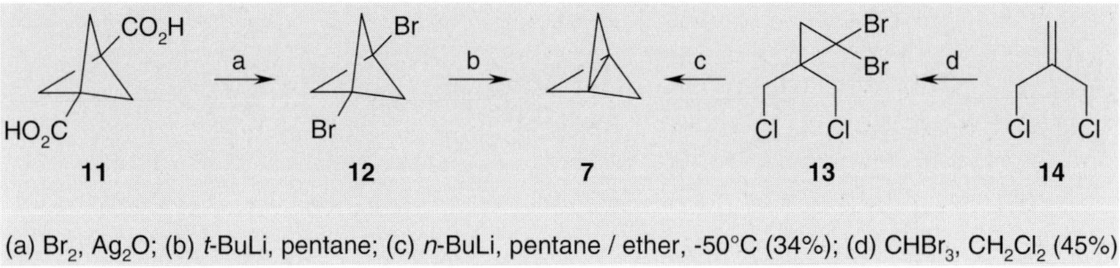

(a) Br$_2$, Ag$_2$O; (b) *t*-BuLi, pentane; (c) *n*-BuLi, pentane / ether, -50°C (34%); (d) CHBr$_3$, CH$_2$Cl$_2$ (45%)

Scheme 3. Preparation of [1.1.1]propellane (**7**), the smallest propellane.

In the original synthesis by Wiberg and co-workers[18] bicyclic dicarboxylic acid **11** was first converted into the dibromide **12** by Hunsdiecker degradation. When the latter was dehalogenated by treatment with *tert*-butyllithium in pentane the propellane was produced as the only isolable hydrocarbon. The success of the transformation rests largely on the high strain energy of the precursor molecule **12**—only a relatively small additional strain increase is required to close the central 'zero bridge'. Because **11** was not readily available at that time,[19] **7** remained a hydrocarbon the chemical behavior of which could not be investigated to any extent. As with adamantane (see Section 3.1) and cyclooctatetraene (Section 10.3)—to name but two of many hydrocarbons which initially could only be prepared in small or even minute amounts—the situation changed for the better dramatically when a ready access route became available. This breakthrough was accomplished by Szeimies and co-workers[20,21] who showed that [1.1.1]propellane could be obtained in only two steps from the commercial product **14**. Addition of dibromocarbene to **14** first furnished the tetrahalide **13**; this reacted with butyllithium to produce **7** in good yield, 1-bromo-3-chloromethyl-bicyclo[1.1.0]butane being passed *en route*. This synthesis made **7** available in multi-gram quantities. Today **7** is a well-investigated compound; its structure has been determined by neutron diffraction,[22] and, of course, its other spectroscopic data, including its photo electron spectrum[23] have been recorded.[17] The hydrocarbon has a rich and often surprising chemical behavior with the transformations often occurring at the central bond.[12, 14–16] The use of **7** as a monomer for the construction of the so-called staffanes will be described in Section 15.3. Finally, several hydrocarbons have been prepared which incorporate **7** has a subunit, *i.e.* these molecules contain an even greater number of bridges than [1.1.1]propellane.[24–27]

The next higher homolog of **7**, [2.1.1]propellane (**8**) has been obtained from 1,4-diiodobicyclo[2.1.1]hexane (**20**),[28] and it is of interest to follow the whole synthetic sequence beginning with propargyl alcohol (**15**) (Scheme 4), because it illustrates that the route to a suitable precursor molecule from which ultimate C–C bond formation can be attempted can be quite lengthy.

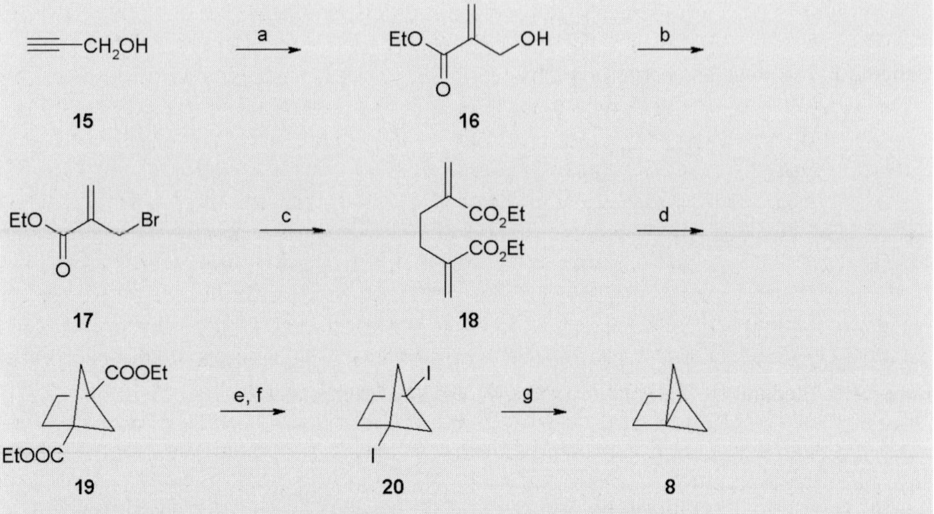

(a) Ni(CO)$_4$, EtOH, HOAc, H$_2$SO$_4$; (b) PBr$_3$; (c) Ni(CO)$_4$ (70%); (d) CH$_3$CN, Ph$_2$CO, hν (ca 30%); (e) KOH (79%); (f) t-BuOI, hν (60%); (g) K$_{vapour}$, 110°C

Scheme 4. The synthesis of [2.1.1]propellane (**8**).

The acrylic ester **16** was first prepared by reaction of **15** with nickel carbonyl in ethanol-acetic acid followed by acidic work-up. Because compound **16** contains just half the carbon atoms required in the target molecule, a dimerization reaction is an obvious possibility. This was achieved by converting **16** to the allyl bromide **17** and dimerizing it with nickel carbonyl to the 1,5-hexadiene derivative **18**. On triplet-sensitized photolysis this gave the bicyclic diester **19**, which was hydrolyzed to the corresponding diacid. Although this derivative could be transformed to the dibromide by a Hunsdiecker reaction, conversion of the latter to the diiodide **20**, which is re-quired to accomplish the last (dehalogenation) step, failed. In the end, the diacid was converted directly to **20** by use of potassium *tert*-butoxide and iodine under con-comitant irradiation.

When **20** was exposed to potassium vapor at 110 °C in a sonicated stream of ni-trogen the [2.1.1]propellane was formed.[29] Proof of the structure of this highly reac-tive hydrocarbon rests largely on the measurement of its infrared spectrum, which was obtained in a matrix. Trapping experiments with bromine—which successfully produce bridgehead dibromides from some of the higher propellanes (see below)—failed with this compound. Hydrocarbon **8** is thus considerably more reactive than the smallest propellane **7**. As with this latter hydrocarbon, more complex structures have been synthesized which incorporate **8** as a subsystem.[30]

[2.2.1]Propellane (**9**) can be compared in many respects with its lower homolog **8**.

It can be obtained similarly from 1,4-diiodonorbornane by treatment with potassium in the gas phase[28] and it has been trapped at 20 K in an argon matrix; other conceptually similar approaches have also been described.[31–33] It is also highly reactive—it polymerizes when the matrix is warmed above 50 K—but in contrast to **8** it yields the corresponding bridgehead dibromide when bromine is admitted.

Beginning with **10** the propellanes become increasingly stable. It is, for example, possible to heat 1,5-dibromo-bicyclo[3.1.1]heptane with sodium in triglyme under reflux (at 216 °C!) and remove **10** by distillation as it is formed—and in an excellent 75% yield.[34] A sizable number of [3.1.1]- and [4.1.1]propellane derivatives is known, and Diels–Alder addition to bicyclo[1.1.0]butene derivatives is the method of choice for their preparation[35] (see Section 7.4).

The high reactivity of **8** and **9** has been traced to their high strain energies, calculated to be 104 and 102 kcal mol^{-1} as compared to the corresponding bicyclic compounds.[36] Free-radical addition across the central bond of the hydrocarbon is a highly exothermic process. Although [1.1.1]propellane (**7**) is equally strained (calculated strain energy 102 kcal mol^{-1}) it does not polymerize in solution (even up to temperatures of *ca* 100 °C)[37] because the radical formed by addition to the conjoining bond is destabilized both by the geometry at the bridgehead and the strain inherent in the bicyclo[1.1.1]pentane framework. Polymerization of **10** sets in only at much higher temperatures.[34] These surprising reactivity differences are also apparent from the (calculated) heats of hydrogenation of the small-ring propellanes.[38] Thus for **8** and **9** energies of –73 and –99 kcal mol^{-1}, respectively, were calculated, whereas only –39 kcal mol^{-1} is obtained for **7**, comparable with the –37 kcal mol^{-1} for cyclobutane and –38 kcal mol^{-1} for cyclopropane.

According to these calculations [2.2.2]propellane (**29b**) should possess an energy of hydrogenolysis of *ca* –93 kcal mol^{-1}, *i.e.* it should be a very reactive propellane again. This is indeed so. Because the synthesis of the first derivative of this hydrocarbon,[39] the amide **29a**, has several interesting preparative features it is presented here in detail (Scheme 5).

The construction of the polycyclic framework began with a [2+2]photocycloaddition of ethylene to the α,β-unsaturated keto acetate **21**. On β-elimination of acetic acid this provided a new unsaturated ketone, **23**, which could be subjected to the same photoaddition, leading to **24**. This ketone is already a propellane, *i.e.* the crucial central bond has been generated in the early part of the synthesis. What follows are essentially two ring-contraction processes, performed very similarly. Diazo-group transfer converted **24** into the diazo ketone **25**, which, on photolysis, Wolff rearranged to the ketene **26**. The first carbon atom was cleaved off by ozonolysis and the resulting ketone **27** was converted to **28** by a second diazo-group transfer. When this diazoketone was irradiated in the presence of dimethylamine it ring-contracted again and the ketene, which is very likely produced as an intermediate, was trapped to provide the [2.2.2]propellane **29a**. This amide is not very stable at room temperature: It underwent thermal cycloreversion with a half-life of just 1 h and provided the two possible isomers **30** and **31**, the ease of the cleavage process having previously been predicted by theoretical considerations.[40] No generalizations about the reactive behavior of the propellanes described here can presently be made, since the reactivity trends in the series differ significantly for different organic reactions (polymerization, halogen addition, addition of electrophiles *etc.*).[12, 14]

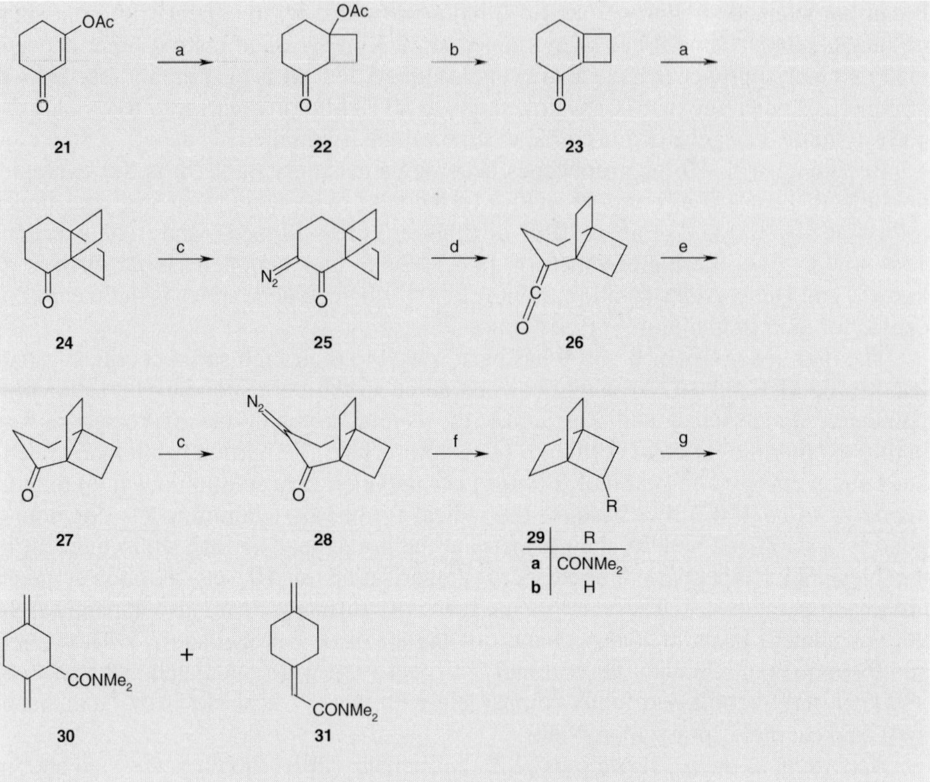

(a) H$_2$C=CH$_2$, hv, -70°C; (b) *tert*-BuOK, *tert*-BuOH (85%); (c) HCO$_2$CH$_3$, ‾OCH$_3$, TsN$_3$ (80%); (d) hv, CH$_2$Cl$_2$, -70°C; (e) O$_3$, CH$_2$Cl$_2$ (comb. yield 45%); (f) hv, Me$_2$NH (40%); (g) 20°C, CDCl$_3$

Scheme 5. The preparation of the first [2.2.2]propellane (**29a**).

To select 'classical syntheses' from the numerous routes leading to the higher propellanes[10–14] is difficult, all the more because for many of these compounds different preparative routes have been devised. Still, certain reaction types are particularly useful for the preparation of the higher propellanes, which often have highly symmetrical structures and hence have considerable esthetic appeal.

Inter- and intramolecular carbene additions have often been used for the preparation of [*m.n.*1]propellanes. For example, [3.2.1]propellane (**33**) can be obtained directly by methylenation of the olefin **32**[41] or *via* the dichlorocarbene adduct of **32**.[42, 43] With a half-life of 20 h at 195 °C, **33** demonstrates impressively how a reduction in strain increases stability. An interesting intramolecular carbene addition is provided by the pyrolysis of the tosyl hydrazone salt **34** which yields the adamantane-derived propellane **35** on heating[44] (Scheme 6).

The formation of three-membered rings by solvolysis-type processes is demonstrated by two routes to [4.4.1]propellane (**38**). The first—starting from **36** (Scheme 7) and also constituting the first synthesis reported for this hydrocarbon[45] exploits an intramolecular displacement process, in which the cyclopropane ring is formed by participation of a double bond in a homoallylic position. The unsaturated propellane produced, **37**, was readily hydrogenated to the saturated compound **38**. In the second route, this propellane was formed on diazotation of the primary amine **39**.[46]

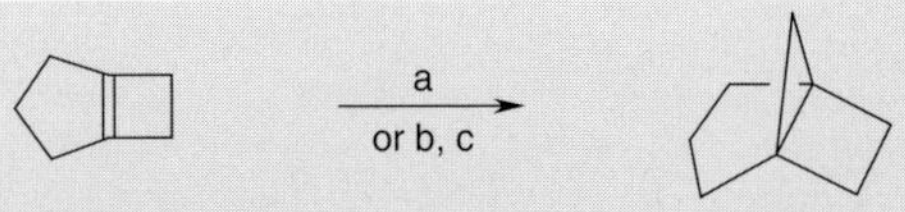

32 **33**

(a) CH_2N_2, ether (60%); (b) CCl_3CO_2Et / $NaOCH_3$ (40%); (c) Na, $NH_{3\,liq.}$ (30%)

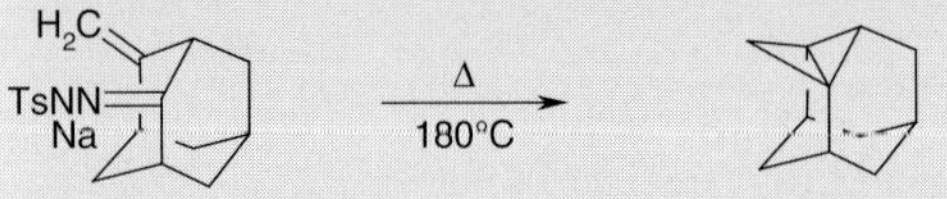

34 **35** (70%)

Scheme 6. Synthesis of some higher [*m.n.*1]propellanes.

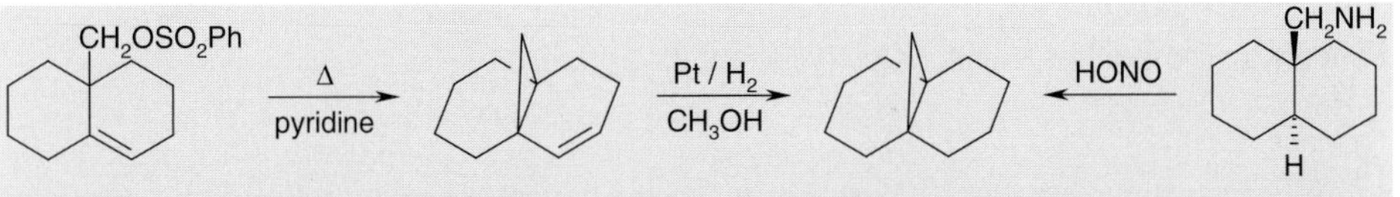

36 **37** **38** **39**

Scheme 7. Synthesis of propellanes by solvolysis reactions.

Finally, the Diels–Alder reaction is the major synthetic reaction for the preparation of numerous propellanes containing six-membered rings, as illustrated in Scheme 8 for the preparation of [4.4.4]propellane (**48**). The ring-construction process was initiated with the [2+4]cycloaddition of 1,3-butadiene (**40**) to acetylene dicarboxylic acid (**41**), a classic reaction, first described by Alder in 1938.[47] The 2:1 adduct obtained was converted to the anhydride **42**, which is already a propellane and one of the central molecules of all of propellane chemistry.[48] By conventional chain elongation reactions this was converted to the half ester **43** which in turn was transformed into **44** by Arndt–Eistert homologation. When **44** was subjected to a Dieckmann condensation the enol ester **45** was formed and this—after saponification—was readily decarboxylated to the unsaturated ketone **46** from which **48** was obtained *via* the saturated ketone **47** by catalytic hydrogenation followed by Wolff-Kishner reduction[49] (Scheme 8).

In a second synthesis by the same authors[49] **47** was prepared by first thermally decomposing the diazoketone **49** in the presence of copper sulfate and subsequently reducing the cyclopropane produced as an intermediate. According to X-ray structural analysis the three six-membered rings in **48** are somewhat flattened cyclohexane chairs, the molecule has D_3 symmetry and is chiral.[10]

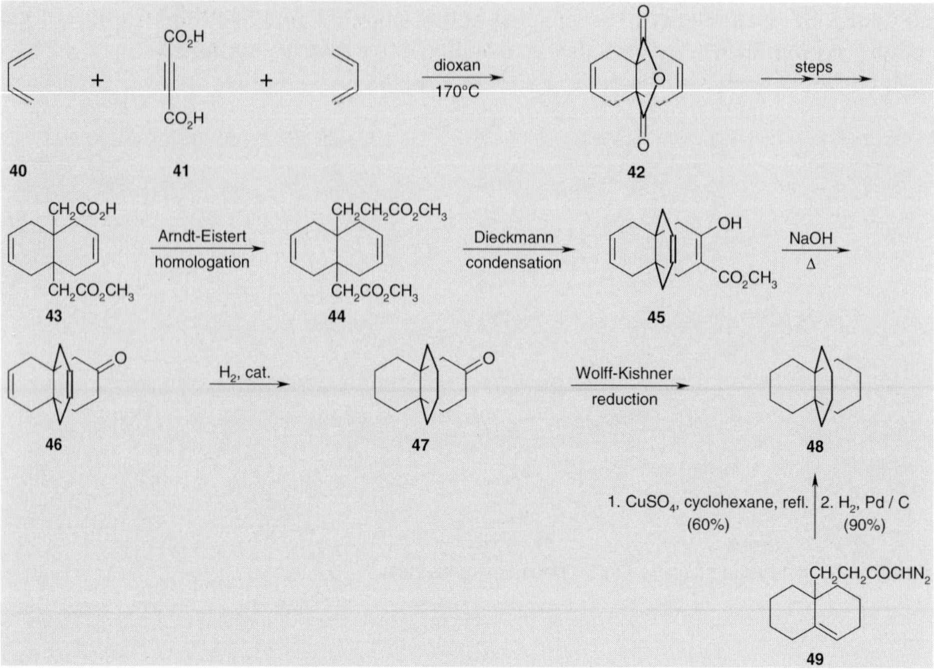

Scheme 8. The synthesis of [4.4.4]propellane (**48**).

Whereas the propellanes with short bridges were used to fathom the limits of distortion of the tetrahedral carbon atom, their homologs with extended bridges—especially when these are functionalized—have turned out to be interesting model compounds for stereochemical investigations of another type. Because many of these propellanes have curved overall geometries, they enable the study of problems related to the selectivity of *exo-* and *endo*-attack on concave molecules.[50]

6.2 Fenestranes

Because molecules represented by the general structure **50** are, on first sight, reminiscent of a lattice window (especially when $m = n = p = q = 4$), the compounds have been called fenestranes.[51, 52] According to theoretical calculations, the central carbon atom in these polycyclic systems can actually be flat,[3, 53–58] if the bridges fulfil certain requirements. In the saturated series the [3.5.3.5]- (**51**) and the [4.4.4.4]fenestrane (**52**) are attractive target molecules. According to *ab initio* calculations the strain energy of **51** should be about 150 kcal mol^{-1},[59] slightly lower than the 160 kcal mol^{-1} calculated for **52**.[60] The calculated bond angle across the central carbon atom is *ca* 130°, indicating that flattening has set in. Among the unsaturated fenestranes the [5.5.5.5]system **53** and its lower and higher vinylogs are of special interest. Formally these molecules correspond to annulenes into whose very center a carbon atom has been placed; in **53** a [12]annulene ring (see Section 10.4) forms the perimeter.[3] If the rim of the molecule is planar, the central carbon atom should be planar also (Scheme 9).

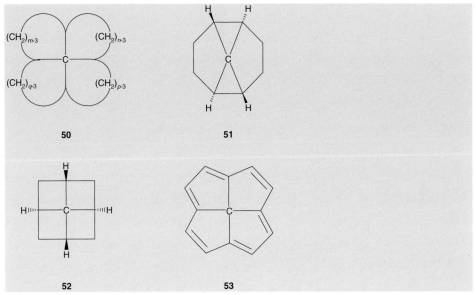

Scheme 9. A selection of fenestranes.

Although a sizable number of molecules with the general structure **50** is known—even several natural products with this unusual polycyclic framework have been isolated and synthesized[52]—none of the molecules **51–53** has yet been synthesized. Derivatives of **53** have, however, been prepared, and a selection of syntheses leading to this unique fusion of four five-membered rings will be described below. The [5.5.5.5]hydrocarbons are presently the most thoroughly studied fenestranes although higher and lower homologs (see below) have been described including benzoannelated congeners.[52] In all three cases discussed here, alternate syntheses have been published by the same authors.

The synthesis of the parent molecule, all-*cis*-[5.5.5.5]fenestrane (**64**)[61] begins with the tricyclic ketone **54** (Scheme 10), readily available from dicyclopentadiene. On Cu(I)-catalyzed 1,4-addition of 3-benzyloxy-1-propyllithium followed by borohydride reduction this was converted to the *endo* alcohol **55**, which, in turn, was protected by treatment with ethyl vinyl ether to provide **56**. The next step—potassium permanganate oxidation—was critical, because it generated the functionality at the bicyclo[3.3.0]skeleton where it is needed for the subsequent steps. As intended, **57** with a formyl substituent and a hemiacetal grouping (produced by reaction of the other 'half' of the original C–C double bond and the hydroxyl group provided by hydrolysis of the acetal function) was formed; on treatment with base under phase transfer conditions this epimerized to **58**.

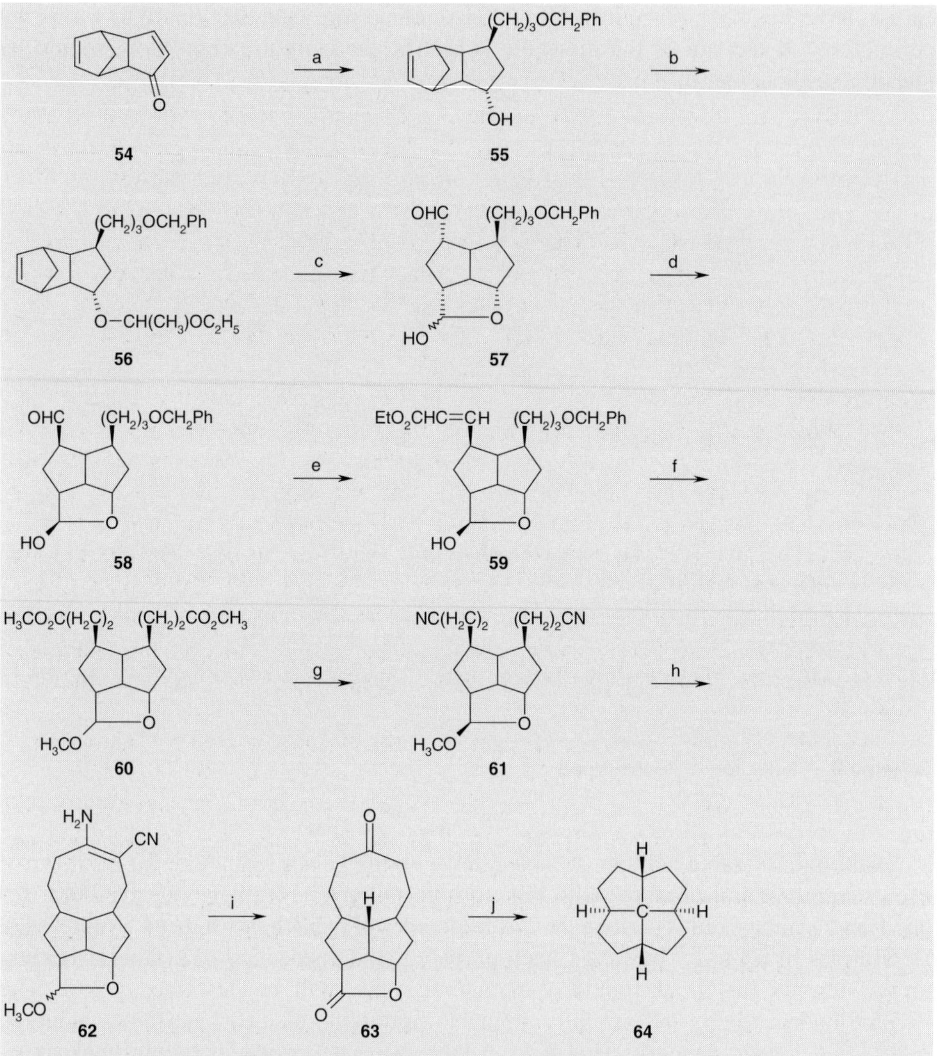

(a) 1. PhCH$_2$O(CH$_2$)$_3$Li, CuI, 2. NaBH$_4$; (b) H$^+$, ⌣o⌣; (c) KMnO$_4$; (d) base, PTC;
(e) (EtO)$_2$P(O)CH$_2$CO$_2$Et, n-BuLi (56%); (f) 1. H$_2$, Pd / C, 2. MeOH / p-TsOH, 3. RuCl$_3$,
4. MeOH / H$^+$ (42%); (g) 1. NH$_3$, 2. - H$_2$O (72%); (h) HMDS, n-BuLi, THF (59%); (i) 1. H$_2$O,
H$^+$, 2. Jones ox. (20%); (j) H$_2$, Pd / C, 320°C / 5 h (40%)

Scheme 10. The synthesis of all-cis-[5.5.5.5]fenestrane (**64**) by Keese and co–workers.

The next steps served to effect chain elongation at the aldehyde position and to convert the benzyl ether group into an ester. This was accomplished by Horner reaction of **58** with ethyl (diethoxyphosphoryl)acetate (**58**→**59**) followed by the redox chemistry shown in the scheme (**59**→**60**). Because Dieckmann cyclization of the resulting diester **60** gave only poor yields (5%), even when performed at high dilution, resort was taken to the Thorpe–Ziegler cyclization for which **60** had to be converted into the dinitrile **61**. When this was treated with hexamethyldisilazane (HMDS) in the presence of butyllithium in tetrahydrofuran cyclization to the β-enamino nitrile **62** occurred in satisfactory yield. Oxidation of the latter under reflux with Jones reagent provided the cyclic ketone **63** which was the required intermediate for a final ring closure. By transannular carbene insertion into the central carbon–hydrogen bond the two five-membered rings still missing and the fenestrane constituting quaternary carbon atom should be generated; a reductive decarboxylation step should then remove the lactone group. Both processes had, in fact, been employed by the authors in their original synthesis of **64**.[62, 63] Because, surprisingly, the carbene generated from **63** failed to insert, this ketone was subjected directly to a remarkable reduction reaction used previously.[63] Heating of **63** over palladium on charcoal in the presence of hydrogen at 320 °C provided, in 40% yield, the fenestrane **64**, identical in all its properties with the hydrocarbon previously prepared.[63]

Although the preparation of **64** constitutes a remarkable achievement, the synthesis is a *tour de force* because it depends almost exclusively on reactions which only provide one new carbon-carbon bond per step. It is one of the lessons to be learned from hydrocarbon synthesis that whenever possible processes should be employed which provide more than one bond per step, and that these reactions—cycloadditions, tandem reactions *inter alia*—should not only be used *once* during the synthesis but as *often as possible*. In other words, the high symmetry which is typical of many hydrocarbon targets should emerge early in and persist throughout the course of the synthesis already, not appear in the last step. Early symmetrization is also important with regard to the development of general synthetic strategies, not just one route leading to a specific product.

These concepts are well illustrated by studying the preparation of two other [5.5.5.5]fenestranes, the tetraene **75** and its tetrabenzo derivative **87**.[64]

The key intermediate in the synthesis of **75** by Cook and co-workers[65, 66] is the diketone **70** (Scheme 11). This was prepared by use of the highly efficient (and large scale) route to polyquinanes[67–70] by reaction of 1,2-dicarbonyl compounds with dimethyl 3-oxoglutarate (**68**). This route, the so-called Weiss reaction, currently the simplest to 1,5-disubstituted *cis*-bicyclo[3.3.0]octan-3,7-diones, has been used successfully in a growing number of syntheses of natural and non-natural products.

The dicarbonyl compound employed for the preparation of **75** was the cyclopentene-3-glyoxal **67**, present as a hydrate. This intermediate can be readily prepared from the commercially available cyclopentene acid **65** in four steps *via* the phosphazine **66**. The first multi-bond forming construction step was performed by treating a freshly prepared solution of **67** with the keto diester **68**. The 2:1-condensation product **69** was formed in excellent yield—four new bonds having been prepared in a one-pot operation! Saponification then yielded the desired diketone **70**. The role of the carbon–carbon double bond, the function of which might not have been apparent from the beginning of the synthesis, now becomes clear—it is a not only a latent 1,2-diol (oxidation of **70** to **71**) but also embodies the carbon atoms for a future diacid, **72**, obtained by Jones oxidation from **71**. Acid-catalyzed cyclization—again a process delivering more than one new carbon bond and a reaction known from previous work—subsequently provided the tetraketone **73**, already a derivative of the parent hydrocarbon **64**. The X-ray structure of this compound has been determined; it shows slight flattening at the center carbon atom with bond angles between *ca* 115 and 118°. The tetraol **74** could be obtained as a mixture of isomers by diborane reduction of **73**;

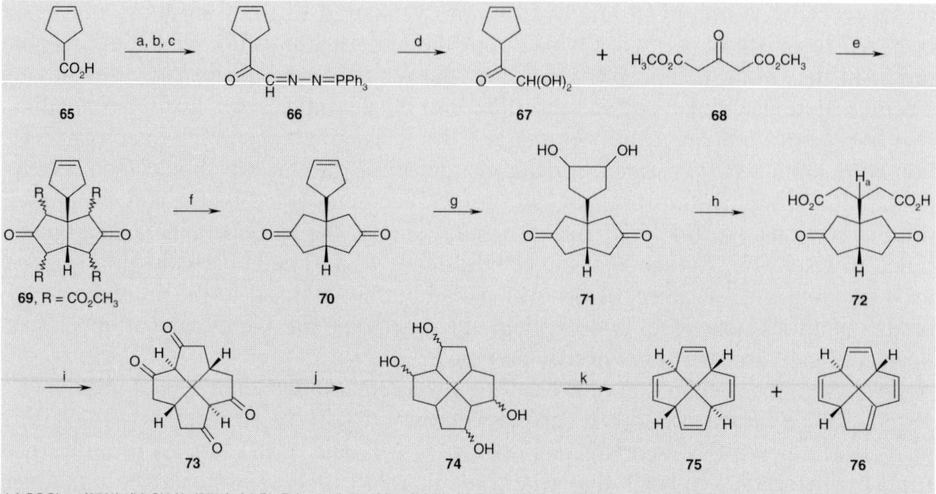

(a) SOCl$_2$, Δ (85%); (b) CH$_2$N$_2$ (87%); (c) Ph$_3$P / anhydrous ether (93%); (d) THF, NaNO$_2$, HCl, 0°C; (e) pH = 8.3 (>90%); (f) CH$_3$CO$_2$H, HCl, 87°C (>90%); (g) OsO$_4$, [structure] (>92%); (h) Jones`reagent, 1.4 M (70%); (i) NSA, diglyme, cumene, Δ (>70%); (j) B$_2$H$_6$, THF, 0°C / 2 d (92%); (k) HMPA, reflux, 2 d

Scheme 11. The synthesis of [5.5.5.5]fenestratetraene **75** by Cook and co–workers.

when these isomers were heated under reflux in HMPA for two days, the [5.5.5.5]fenestratetraene (**75**, 'staurane-2,5,8,11-tetraene') was produced. The bridge-head isomer **76** was isolated as a side product.

The synthesis of the tetrabenzoannelated fenestrane, **87**, also termed fenestrin-dane, due to Kuck and co-workers[71–73] and summarized in Scheme 12, is comparable in elegance with the route to **75**. It is presented here not only because it again stresses the importance of symmetrical precursor molecules, but also because **87** has turned out to be an important starting molecule for the preparation of the so-called centro-polyquinanes, hydrocarbons with unique topological features.[74]

The synthetic sequence begins with a two-bond forming process, the condensation of 1,3-indandione (**77**) with dibenzylideneacetone (**78**) to the triketone **79**. Interest-ingly, whereas the flattened central carbon in **64** is produced in the last step of the synthesis, and that in **75** at approximately half way, the central quaternary carbon atom is generated here at the very beginning of the whole sequence, and at a time when only one five-membered ring is present, this having been contributed by the simple precursor **77**. The spiro compound **79** was next reduced to the diol **80** (mix-ture of isomers) which, when subjected to a double Friedel–Crafts cyclization, led to **81**—a fenestrane skeleton has thus been prepared in three short, high-yield processes! Two simple steps—oxidation to the ketone **82** and bromination to the α,α'-dibromide **83** followed, and when these latter were treated with base not only a Favorskii ring contraction but also a β-elimination took place, yielding the tribenzoannelated car-boxylic acid **84**. After this had been decarboxylated to **85** the fourth benzene ring was introduced by Diels–Alder addition of tetrachlorothiophenedioxide, followed by de-hydrochlorination-dechlorination to the target hydrocarbon **87**. According to X-ray structure determination the flattening of the center carbon atom of this compound is comparable with that of **73**—C1–C13–C7 angle is 116.5°. Interestingly, bridgehead-

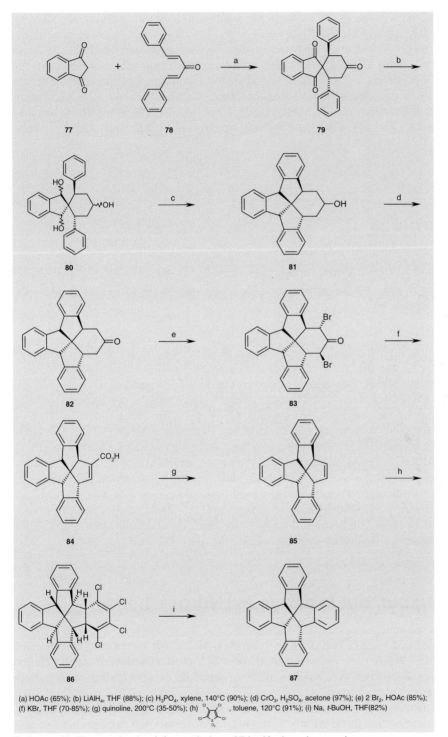

(a) HOAc (65%); (b) LiAlH₄, THF (88%); (c) H₃PO₄, xylene, 140°C (90%); (d) CrO₃, H₂SO₄, acetone (97%); (e) 2 Br₂, HOAc (85%); (f) KBr, THF (70-85%); (g) quinoline, 200°C (35-50%); (h) [structure], toluene, 120°C (91%); (i) Na, t-BuOH, THF(82%)

Scheme 12. The synthesis of fenestrindane **87** by Kuck and co-workers.

substituted derivatives of **87** such as the tetrabromide show significantly widened C1–C13–C7 angles (121.5°).[52]

Both larger and smaller fenestrane skeletons have been described. In the larger systems, *e.g.* with a [5.5.5.7]skeleton, hardly any flattening effect is expected, but they are of interest because some diterpenes contain this unusual carbon framework.[75–77] In the smaller fenestranes, however, flattening should increase. This is indeed so, as shown by X-ray structural analysis of the [4.4.5.5] amide **88** (Scheme 13 the C1–C11–C6- and C3–C11–C9-bond angles are 128.2 and 123.0°, respectively.[78,79]

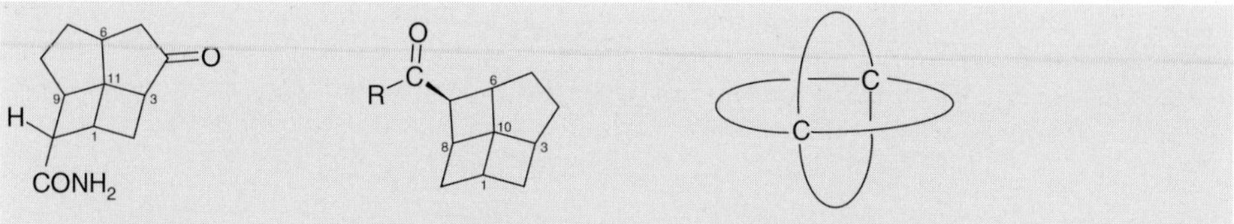

Scheme 13. Pronounced carbon flattening in the fenestranes **88** and **89** and the general structure of a paddlane, **90**.

The most highly flattened central carbon atom known to date in a fenestrane has been observed for **89**, a derivative of the currently smallest representative of these tetracyclic molecules. The central angles in this [4.4.4.5]fenestrane C1–C10–C6 and C3–C10–C8 are 128.3 and 129.2°, respectively.[80] Interestingly, certain bonds leading to the core of the molecule are shortened, *e.g.* C3–C10 and C6–C10 = 1.49 Å, whereas perimeter bonds for the cyclobutane rings are extended (1.57 Å). These changes in bond distance have been predicted correctly by MNDO calculations.[52, 81]

To close this section on ways of deforming the saturated carbon atom within hydrocarbon frameworks the so-called paddlanes, of general structure **90**, should be mentioned. Although far less is known about these molecules than about propellanes and fenestranes these systems have been prepared.[82–91] If flat, these molecules could be labeled as 'double fenestranes' or as propellanes in which the central C–C bond has been extended by further atomic centers.

6.3 *out,out, out,in* and *in,in* Hydrocarbons

The general structures of two of these molecules, the *out,out* and *in,in* hydrocarbons **4** and **5,** have already been presented in Scheme 1. If they interconvert, a hydrocarbon with an *out,in* structure (see below) could, in principle, serve as a transition state or intermediate.

Although, as this section will show, hydrocarbons of this type have been prepared, *in,out* isomerism has so far mostly been studied for heterocyclic compounds, in particular for diazamacrobicyclics, *i.e.* molecules in which nitrogen atoms serve as bridgeheads in **4** and **5**. In fact, the term *in,out* isomerism was coined by Simmons and Park in 1968[92–94] when they first synthesized these bicyclic diamines. These authors also proposed the term 'homeomorphic isomerism' for the process of inter-conversion (**4** ⇄ **5**) of these isomers. The development of cryptand chemistry by Lehn [95–98] and later by many others strongly focused research effort in this area on molecules containing heteroatoms (mostly nitrogen and oxygen). In the form of their quaternary salts the above diamines can exist in all three possible stereoisomeric forms, these having been produced *via* reversible protonation. This approach is clearly not possible for hydrocarbons; but before presenting the leading preparative work in this area we will address the problem of *in,out* isomerism from standpoint of molecular mechanics. The preferred conformations of a large number of bicyclic hy-drocarbons, ranging from bicyclo[3.2.2]nonane to bicyclo[6.6.6]eicosane,[99] have been calculated by application of stochastic search procedures.[100,101] These calcula-tions enabled the location of all stable conformations of these hydrocarbons and pre-diction of their thermodynamic stabilities. Some results are shown in Scheme 14.

Bicyclic Hydrocarbon	Isomer		
	out,out	*in,out*	*in,in*
bicyclo [3.2.2] nonane	24.25	81.43	
bicyclo [3.3.2] decane	29.95	66.78	130.17
bicyclo [5.3.1] undecane	26.19	37.52	63.93
bicyclo [4.3.3] dodecane	48.60	55.80	93.45
bicyclo [4.4.3] tridecane	58.35	54.81	82.43
bicyclo [5.3.3] tridecane	56.46	52.43	78.20
bicyclo [5.4.2] tridecane	48.00	46.51	67.33
bicyclo [6.5.1] tetradecane	42.16	40.18	42.44
bicyclo [7.5.1] pentadecane	41.83	41.77	41.48
bicyclo [5.5.5] heptadecane	60.83	54.16	49.78
bicyclo [6.6.6] eicosane	47.42	43.62	36.40

Scheme 14. Steric energies (kcal mol^{-1}) of the lowest-energy conformations of selected bicyclic hydrocarbons calculated by the MM2 method.

As expected, and known from long experience, the short-bridged molecules all are *out,out* isomers—the strain of the alternative structures would be too high. When, however, the length of the bridges is increased a change sets in. For example, for bi-cyclo[4.4.3]tridecane the *in,out* isomer should be the most stable; for the *out,out* iso-mer non-bonding interactions between the methylene groups of the bridges cause in-creasing strain, whereas the *in,in* structure still cannot compete. Continuing through the homologous series the situation begins to change, though, and with bicy-clo[7.5.1]pentadecane the interesting situation is reached where all three isomers have almost the same strain energy. Finally, for bicyclo[5.5.5]heptadecane and bicy-clo[6.6.6]eicosane the *in,in* hydrocarbon is the most favored. *In,out* and *in,in* hydro-carbons should thus be obtainable, and this is indeed so.

The first compounds of this type to be synthesized were **93** (Scheme 15) and **99** and **101** (Scheme 16, below), which were reported simultaneously. To prepare the [8.2.2] derivative **93** the cyclic diene **91** was treated with perfluoro-2-butyne (**92**).[102, 103] Cyclic dienes containing at least one *trans* double bond often undergo [2+2]cycloadditions. This does not occur with **91**, a known compound readily prepared from *trans*-cyclododecene. The competing process is also held at bay by employing the powerful dienophile **92**, which participates in free-radical [2+2]cycloadditions only sluggishly. Later the Diels–Alder approach to *in,out* bicyclic compounds was extended to even smaller cyclodienes and other dienophiles yielding products which were amenable to X-ray structural analysis. These studies unequivocally established the *in,out* structure of the cycloadducts.[104–108]

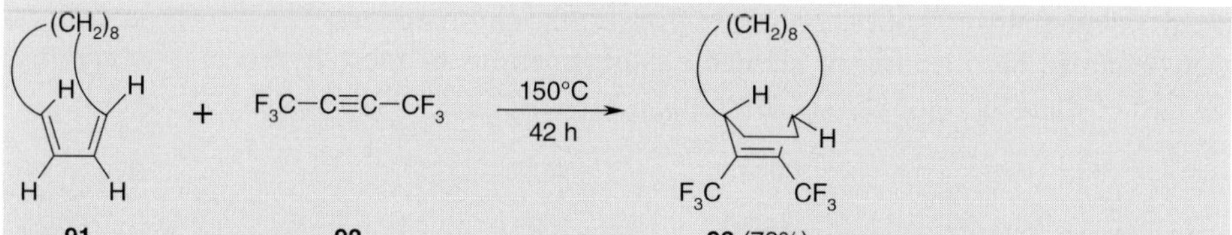

91 **92** **93** (78%)

Scheme 15. Preparation of *in,out* bicyclic compounds by Diels–Alder addition.

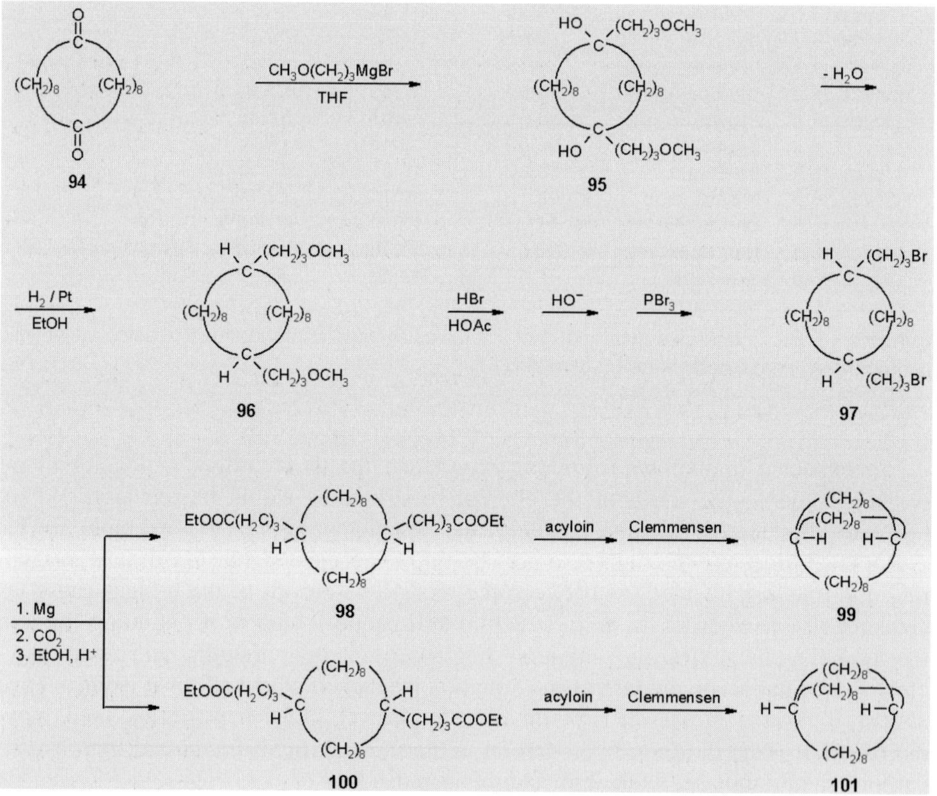

94 **95** **96** **97** **98** **99** **100** **101**

Scheme 16. Preparation of *in,out* bicyclic hydrocarbons by Simmons and Park.

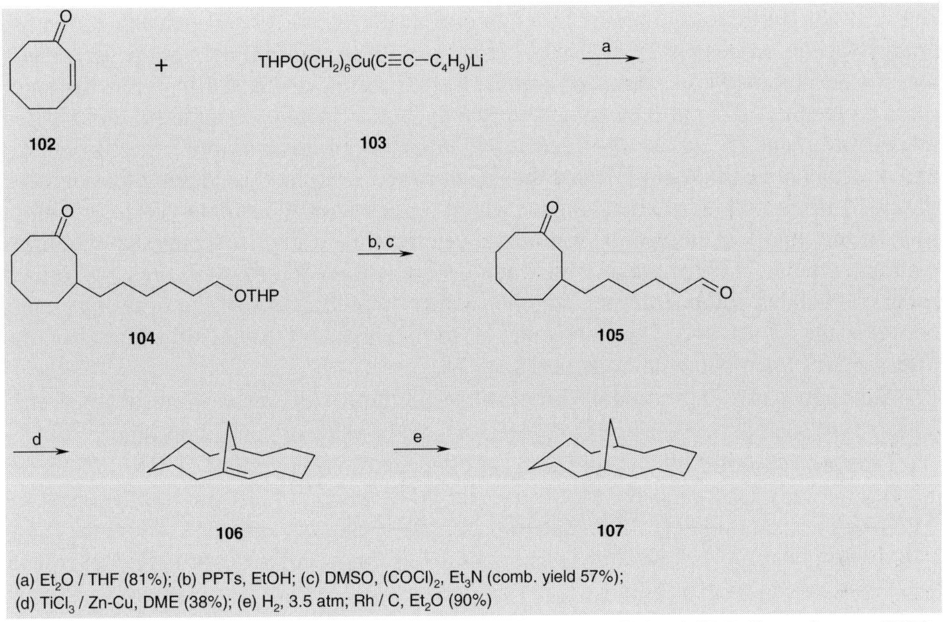
+
THPO(CH2)6Cu(C≡C–C4H9)Li
a
102
103
OTHP
b, c
104
105
d
e
106
107
(a) Et2O / THF (81%); (b) PPTs, EtOH; (c) DMSO, (COCl)2, Et3N (comb. yield 57%);
(d) TiCl3 / Zn-Cu, DME (38%); (e) H2, 3.5 atm; Rh / C, Et2O (90%)

Scheme 17. Synthesis of the *out,out*, *out,in* and *in,in* isomers of bicyclo[6.5.1]tetradecane (**107**).

In the preparation of the bicyclo[8.8.8]hexacosanes **99** and **101** the critical steps were acyloin condensations. Starting with 1,10-cyclooctadecadione (**94**) the first few steps involved chain extension to the dibromide **97** *via* **95** and **96**, by conventional methods. The dibromide **97** could then by separated into its *trans* and *cis* isomers by fractional crystallization. The *cis* isomer was converted to the diester **98** which was then subjected to an acyloin condensation then a Clemmensen reduction. Although this cyclization could in principle yield both the *in,in* and the *out,out* hydrocarbons, spectroscopic data indicated only **99** was formed.[109]

Likewise, the *trans* isomer of **97** was converted into the *in,out* hydrocarbon, **101**, as the only isomer to be expected in this case.

The most thorough study of *in,out* isomerism in hydrocarbons yet published[110] involves bicyclo[6.5.1]tetradecane (**107**), for which—as can be seen in Scheme 14—the calculated strain energies of the three isomers are also nearly equal, making **107** a particularly interesting goal. The hydrocarbon was prepared by Saunders and Krause by the sequence summarized in Scheme 17.

Starting with cycloocten-3-one (**102**) the required carbon atoms were first assembled by conjugate addition of the mixed cuprate **103** to **102**. The resulting saturated ketone **104** was transformed by removal of the THP-protecting group and Swern oxidation of the resulting alcohol to the dicarbonyl compound **105**. When the latter was ring-closed by McMurry coupling the bridgehead olefin (see Section 7.4) **106** resulted; this, a hyperstable olefin (see Section 7.4), could be reduced only under forcing conditions to the target molecule **107**. It was shown by ^{13}C NMR spectroscopy not only that all three isomers had been formed but that the precursor olefin **106** also comprised a mixture of two *in,out* isomers. Because these could be separated chromatographically, temperature dependent NMR studies could be performed. They showed that equilibration occurs at temperatures above 80 °C and that the activation energy for the process is 28 kcal mol^{-1} (log A = 14.0), the first activation parameters ever to be obtained for a homeomorphic isomerization process. Likewise, the saturated molecules **107** could be subjected to a dynamic NMR investigation; this yielded an activation energy for *out,out-in,in* interconversion of 24 kcal mol^{-1} (log A = 15.5) and a value above 30 kcal mol^{-1} for the *in,out* process, no linebroadening having been observed up to 100 °C. A mechanism which agrees with these data certainly cannot involve any planarization of carbon atoms (see above). Rather it seems that a conformational change is involved in which one bridge is threaded through the ring formed by the remainder of the structure. In other words by pulling one of the bridging chains between the other two, the molecule is turned inside out, as illustrated by the **108** ⇄ **109** interchange in Scheme 18.

Although a rare but certainly interesting structural and stereochemical phenomenon, *in,out* isomerism has also been observed in a number of natural products.[115]

The route to *in,out* bicyclic olefins as applied for the preparation of **106** in Scheme 17 was actually first used by McMurry and co–workers[111, 112] for the synthesis of **111** from the ketoaldeyde **110** (Scheme 19); these authors also performed the catalytic hydrogenation to **112** and later generalized this method to produce several of these bridgehead olefins which served as precursor for novel carbocations with bent three-center, two-electron C–H–C bonds.[113]

The highly strained *in*-[3^{4,10}][7]metacyclophane (**113**) was produced in 11% yield by pyrolyzing the corresponding trissulfone (see Section 12.3).[114] Its endocyclic hydrogen atom absorbs at δ = – 4.08 and according to calculations is only 1.78 Å above the plane of the facing benzene ring.

Although a rare but certainly interesting structural and stereochemical phenomenon, *in,out* isomerism has also been observed in a number of natural products.[115]

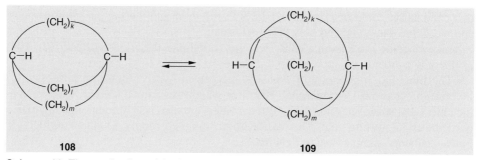

108   **109**

Scheme 18. The mechanism of the homeomorphic isomerization.

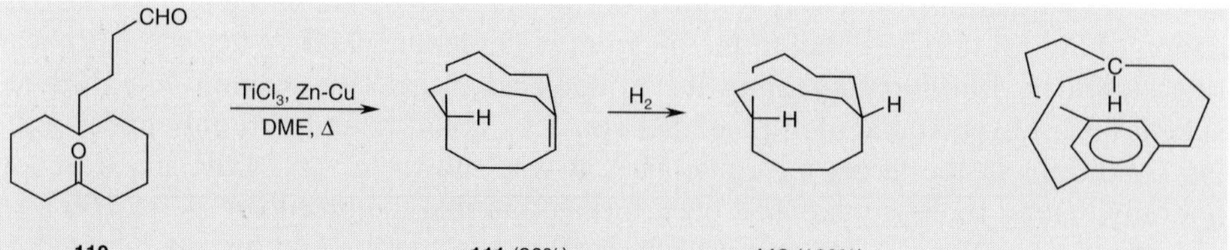

110   **111** (30%)   **112** (100%)   **113**

Scheme 19. Olefinic and aromatic hydrocarbons with inside C–H bonds.

References

1. Formally the same planar arrangement can be reached by a twist motion in which two opposing bond angles remain constant, whereas a second pair opens up and a third one closes. For reviews on the planarization of the tetrahedral geometry of organic compounds see R. Keese in *Organic Synthesis: Modern Trends* O. Chizhov (*Ed.*), Blackwells, Oxford, **1987**, 43–52; W. Luef, R. Keese in *Advances in Strain in Organic Chemistry*, Vol. 3, B. Halton, (*Ed.*), JAI Press, Greenwich, CT, **1993**, 229–267; H. Dodziuk, *J. Mol. Struct.*, **1990**, *239*, 167–172; H. Dodziuk, *Top. Stereochem.*, **1994**, *21*, 351–380.

2. R. Hoffmann, R. W. Alder, C. F. Wilcox, Jr, *J. Am. Chem. Soc.*, **1970**, *92*, 4992–4993; *cf.* R. Hoffmann, *Pure Appl. Chem.*, **1971**, *28*, 181–194.

3. J. N. Murrell, J. B. Pedlay, S. Durmaz, *J. Chem. Soc. Faraday Trans. 2*, **1973**, *69*, 1370–1380.

4. J. B. Collins, J. D. Dill, E. D. Jemmis, Y. Apeloig, P. v. R. Schleyer, R. Seeger, J. A. Pople, *J. Am. Chem. Soc.*, **1976**, *98*, 5419–5427; K. Sorger, P. v. R. Schleyer, *J. Mol. Struct.*, **1995**, *338*, 317–346.

5. K. B. Wiberg, G. B. Ellison, J. J. Wendoloski, *J. Am. Chem. Soc.*, **1976**, *98*, 1212–1218.

6. D. C. Crans, J. P. Snyder, *J. Am. Chem. Soc.*, **1980**, *102*, 7152–7154.

7. W. Luef, R. Keese, H.-B. Bürgi, *Helv. Chim. Acta*, **1987**, *70*, 534–542.

8. S. Harder, J. Boersma, L. Brandsma, A. van Heteren, J. A. Kanters, W. Bauer, P. v. R. Schleyer, *J. Am. Chem. Soc.*, **1988**, *110*, 7802–7806; *cf.* H. Dietrich, W. Mahdi, W. Storch, *J. Organomet. Chem.*, **1989**, *349*, 1–10.

9. Reviews: D. Röttger, G. Erker, *Angew. Chem.*, **1997**, *109*, 840–856; *Angew. Chem. Int. Ed. Engl.*, **1997**, *36*, 812–827; R. Choukroun, P. Cassoux, *Acc. Chem. Res.*, **1999**, *32*, 494–502.

10. D. Ginsburg, *Acc. Chem. Res.*, **1969**, *2*, 121–128; *cf.* D. Ginsburg, *Acc. Chem. Res.*, **1972**, *5*, 249–256.

11. D. Ginsburg, *Propellanes, Structure and Reactions*, Verlag Chemie, Weinheim, **1975**; D. Ginsburg, *Propellanes, Sequel I*, Technion, Haifa, **1981**; D. Ginsburg, *Propellanes, Sequel II*, Technion, Haifa, **1985**; *cf.* D. Ginsburg in *International Review of Science, Organic Chemistry*, Butterworths, London, **1976**, Series 2, Vol. 5, 369–415. For a more recent review of propellane chemistry see Y. Tobe in *Carbocyclic Cage Compounds* E. Osawa, O. Yonemitsu (*Eds.*), VCH Publishers, New York, N. Y., **1992**, 125–153.

12. K. B. Wiberg, *Acc. Chem. Res.*, **1984**, *17*, 379–386.

13. D. Ginsburg, *Top. Curr. Chem.*, **1987**, *137*, 1–17.

14. K. B. Wiberg, *Chem. Rev.*, **1989**, *98*, 975–983.

15. Review of the chemistry of [1.1.1]propellanes: P. Kaszynski, J. Michl in *The Chemistry of the Cyclopropyl Group*, Vol. 2, Z. Rappoport, (*Ed.*), J. Wiley and Sons, Chichester, **1995**, 773–812.

16. W. C. Agosta in *The Chemistry of Alkanes and Cycloalkanes*, S. Patai, Z. Rappoport, (*Eds.*), J. Wiley and Sons, Chichester, **1992**, 927–962.

17. It has been pointed out[12] that inverted geometries may in some cases be found in the absence of the structural constraints indicated by the general structure **1**. In bicyclo[1.1.0]butane and its derivatives for example the bridgehead C–H bond is 11.5° below the plane formed by the carbon atoms 1, 2, and 4: K. W. Cox, M. D. Harmony, G. Nelson, K. B. Wiberg, *J. Chem. Phys.*, **1969**, *50*, 1976–1980.

18. K. B. Wiberg, F. H. Walker, *J. Am. Chem. Soc.*, **1982**, *104*, 5239–5240.

19. D. E. Applequist, T. L. Renken, J. W. Wheeler, *J. Org. Chem.*, **1982**, *47*, 4985–4995.

20. K. Semmler, G. Szeimies, J. Belzner, *J. Am. Chem. Soc.*, **1985**, *107*, 6410–6411.

21. J. Belzner, U. Bunz, K. Semmler, G. Szeimies, K. Opitz, A.-D. Schlüter, *Chem. Ber.,* **1989,** *122,* 397–398; F. Alber, G. Szeimies, *Chem. Ber.,* **1992,** *125,* 757–758; M. Werner, D. S. Stephenson, G. Szeimies, *Liebigs Ann. Chem.,* **1996,** 1705–1715; M. Kenndoff, A. Singer, G. Szeimies, *J. prakt. Chem.,* **1997,** *339,* 217–232; *cf.* K. M. Lynch, W. P. Dailey, *Org. Syntheses,* **1997,** *75,* 98–105.

22. L. Hedberg, K. Hedberg, *J. Am. Chem. Soc.,* **1985,** *107,* 7257–7260.

23. E. Honegger, H. Huber, E. Heilbronner, W. P. Dailey, K. B. Wiberg, *J. Am. Chem. Soc.,* **1985,** *107,* 7172–7174.

24. J. Belzner, G. Szeimies, *Tetrahedron Lett.,* **1986,** *27,* 5839–5842.

25. J. Belzner, G. Gareiß, K. Polborn, W. Schmid, K. Semmler, G. Szeimies, *Chem. Ber.,* **1989,** *122,* 1509–1529.

26. J. Belzner, G. Szeimies, *Tetrahedron Lett.,* **1987,** *28,* 3099–3102.

27. G. Kottirsch, K. Polborn, G. Szeimies, *J. Am. Chem. Soc.,* **1988,** *110,* 5588–5590.

28. K. B. Wiberg, F. H. Walker, W. E. Pratt, J. Michl, *J. Am. Chem. Soc.,* **1983,** *105,* 3638–3641.

29. F. H. Walker, K. B. Wiberg, J. Michl, *J. Am. Chem. Soc.,* **1982,** *104,* 2056–2057.

30. J. Morf, G. Szeimies, *Tetrahedron Lett.,* **1986,** *27,* 5363–5366.

31. C. F. Wilcox, Jr, C. Leung, *J. Org. Chem.,* **1968,** *33,* 877–880.

32. K. B. Wiberg, W. F. Bailey, M. E. Jason, *J. Org. Chem.,* **1976,** *41,* 2711–2714.

33. W. F. Carrol, Jr, D. G. Peters, *J. Am. Chem. Soc.,* **1980,** *102,* 4127–4134.

34. P. G. Gassman, G. S. Proehl, *J. Am. Chem. Soc.,* **1980,** *102,* 6862–6863.

35. A.-D. Schlüter, H. Harnisch, J. Harnisch, U. Szeimies-Seebach, G. Szeimies, *Chem. Ber.,* **1985,** *118,* 3513–3528.

36. K. B. Wiberg, H. A. Cannon, W. E. Pratt, *J. Am. Chem. Soc.,* **1979,** *101,* 6970–6972.

37. Neat **7** is, however, moderately stable at room temperature, provided air is excluded. Otherwise a considerable amount of polymeric material begins to build up in a few hours.

38. K. B. Wiberg, *J. Am. Chem. Soc.,* **1983,** *105,* 1227–1233.

39. P. E. Eaton, G. H. Temme III, *J. Am. Chem. Soc.,* **1973,** *95,* 7508–7510; *cf.* J. V. Silverton, G. W. A. Milne, P. E. Eaton, K. Nyi, G. Temme III, *J. Am. Chem. Soc.,* **1974,** *96,* 7429–7432.

40. W. D. Stohrer, R. Hoffmann, *J. Am. Chem. Soc.,* **1972,** *94,* 779–786.

41. P. G. Gassman, A. Topp, J. W. Keller, *Tetrahedron Lett.,* **1969,** 1093–1095.

42. K. B. Wiberg, G. J. Burgmaier, P. Warner, *J. Am. Chem. Soc.,* **1971,** *93,* 246–247.

43. K. B. Wiberg, G. J. Burgmaier, *J. Am. Chem. Soc.,* **1972,** *94,* 7396–7401.

44. K. Mlinaric-Majerski, Z. Majerski, *J. Am. Chem. Soc.,* **1980,** *102,* 1418–1419; *cf.* K. Mlinaric-Majerski, Z. Majerski, *J. Am. Chem. Soc.,* **1983,** *105,* 7389–7395.

45. J. W. Rowe, A. Melera, D. Arigoni, O. Jeger, L. Ruzicka, *Helv. Chim. Acta,* **1957,** *40,* 1–12. This process has also been obeserved to occur in steroid derivatives: A. Zürcher, O. Jeger, L. Ruzicka, *Helv. Chim. Acta,* **1954,** *37,* 2145–2152.

46. W. G. Dauben, P. Laug, *Tetrahedron Lett.,* **1962,** 453–456.

47. K. Alder, K. H. Backendorf, *Ber. Dtsch. Chem. Ges.,* **1938,** *71,* 2199–2209.

48. J. Altman, E. Babad, J. Itzchaki, D. Ginsburg, *Tetrahedron,* Supplement 8, **1966,** 279–304.

49. J. Altman, D. Becker, D. Ginsburg, H. J. E. Loewenthal, *Tetrahedron Lett.,* **1967,** 757–758.

50. D. Ginsburg, *Acc. Chem. Res.,* **1974,** *7,* 286–293.

51. The term fenestrane (from the Latin *fenestra* for window) was introduced by S. Georgian, M. Saltzman, *Tetrahedron Lett.,* **1972,** 4315–4317. The hydrocarbon **51** has also been called 'windowpane': *cf.* ref.[55]

52. Reviews on fenestranes: A. Greenberg, J. F. Liebman, *Strained Organic Molecules,* Academic Press, New York, N.Y., **1978,** Chapter 6, *p.* 342; R. Keese, *Nachr. Chem. Techn. Lab.,* **1982,** *30,* 844–849; B. Rao Venepalli, W. C. Agosta, *Chem. Rev.,* **1987,** *87,* 399--

410; K. Krohn, *Nachr. Chem. Techn. Lab.*, **1987**, *35*, 264–266; R. Keese, W. Luef, J. Mani, S. Schüttel, M. Schmid, C. Zhang in *Strain and Its Implications in Organic Chemistry* A. de Meijere, S. Blechert, (*Eds.*), NATO ASI Series, Kluwer Academic Publishers, Dordrecht, **1989**, 77–107; W. C. Agosta in *The Chemistry of Alkanes and Cycloalkanes*. S. Patai, Z. Rappoport, (*Eds.*) J. Wiley and Sons, Chichester, **1992**, 927–962; X. Fu, G. Kubiak, W. Zhang, W. C. Han, A. K. Gupta, J. M. Cook, *Tetrahedron*, **1993**, *49*, 1511–1524; D. Kuck in *Theoretically Interesting Molecules, Vol. 4*, R. P. Thummel, (*Ed.*), JAI Press, Greenwich, CT, **1998**, 81–155.

53. R. C. Bingham, M. J. S. Dewar, D. H. Ho, *J. Am. Chem. Soc.*, **1975**, *97*, 1294–1301.
54. M. C. Böhm, R. Gleiter, P. Schang, *Tetrahedron Lett.*, **1979**, 2575–2578.
55. K. B. Wiberg, L. K. Olli, N. Golembeski, R. D. Adams, *J. Am. Chem. Soc.*, **1980**, *102*, 7467–7475.
56. J. Chandrasekhar, E.-U. Würthwein, P. v. R. Schleyer, *Tetrahedron*, **1981**, *37*, 921–927.
57. P. Gund, T. M. Gund, *J. Am. Chem. Soc.*, **1981**, *103*, 4458–4465.
58. K. B. Wiberg, J. J. Wendoloski, *J. Am. Chem. Soc.*, **1982**, *104*, 5679–5686.
59. K. B. Wiberg, *J. Org. Chem.*, **1985**, *50*, 5285–5291; *cf.* L. Skattebøl, *J. Org. Chem.*, **1966**, *31*, 2789–2794.
60. J. M. Schulman, M. L. Sabio, R. L. Disch, *J. Am. Chem. Soc.*, **1983**, *105*, 743–744.
61. M. Luyten, R. Keese, *Helv. Chim. Acta*, **1984**, *67*, 2242–2245; *cf.* H. Schori, B. B. Patil, R. Keese, *Tetrahedron*, **1981**, *37*, 4457–4463.
62. R. Keese, A. Pfenninger, A. Roesle, *Helv. Chim. Acta*, **1978**, *62*, 326–334.
63. M. Luyten, R. Keese, *Angew. Chem.*, **1984**, *96*, 358–359; *Angew. Chem. Int. Ed. Engl.*, **1984**, *23*, 390–391. For still another [5.5.5.5]fenestrane synthesis see J. Brunvoll, R. Guidetti-Grept, I. Hargittai, R. Keese, *Helv. Chim. Acta*, **1993**, *76*, 2838–2846.
64. These fenestranes have also been called stauranes from the Greek word for cross, *stauros*; *cf.* ref.[67]
65. M. N. Deshpande, M. Jawdosiuk, G. Kubiak, M. Venkatachalam, U. Weiss, J. M. Cook, *J. Am. Chem. Soc.*, **1985**, *107*, 4786–4788.
66. M. Venkatachalam, M. N. Deshpande, M. Jawdosiuk, G. Kubiak, S. Wehrli, J. M. Cook, *Tetrahedron*, **1986**, *42*, 1597–1605.
67. R. Mitschka, J. M. Cook, U. Weiss, *J. Am. Chem. Soc.*, **1978**, *100*, 3973–3974.
68. R. Mitschka, J. Oehldrich, K. Takahashi, J. M. Cook, U. Weiss, J. Silverton, *Tetrahedron*, **1981**, *37*, 4521–4542.
69. W. C. Han, K. Takahashi, J. M. Cook, U. Weiss, J. V. Silverton, *J. Am. Chem. Soc.*, **1982**, *104*, 318–321.
70. M. N. Deshpande, S. Wehrli, M. Jawdosiuk, J. T. Guy, Jr, D. W. Bennett, J. M. Cook, M. R. Depp, U. Weiss, *J. Org. Chem.*, **1986**, *51*, 2436–2444.
71. D. Kuck, H. Bögge, *J. Am. Chem. Soc.*, **1986**, *108*, 8107–8109; D. Kuck, A. Schuster, B. Paisdor, D. Gestmann, *J. Chem. Soc. Perkin I*, **1995**, 721–732; M. Seifert, D. Kuck, *Tetrahedron*, **1996**, *52*, 13167–13180; D. Kuck, R. A. Krause, D. Gestmann, F. Posteher, A. Schuster, *Tetrahedron*, **1998**, *54*, 5247–5258.
72. D. Kuck, A. Schuster, *Angew. Chem.*, **1988**, *100*, 1222–1224; *Angew. Chem. Int. Ed. Engl.*, **1988**, *27*, 1192–1194; *cf.* D. Kuck, A. Schuster, R. A. Krause, *J. Org. Chem.*, **1991**, *56*, 3472–3475.
73. D. Kuck, *Chem. Ber.*, **1994**, *127*, 409–425; *cf.* J. Tellenbröker, D. Kuck, *Angew. Chem.*, **1999**, *111*, 1000–1004; *Angew. Chem. Int. Ed. Engl.*, **1999**, *38*, 919–922.
74. D. Kuck, *Liebigs Ann./Recueil*, **1997**, 1043–1057.
75. T. Tsunoda, M. Amaike, U. S. F. Tambunan, Y. Fujise, S. Ito, *Tetrahedron Lett.*, **1987**, *28*, 2537–2540.
76. M. T. Crimmins, L. D. Gould, *J. Am. Chem. Soc.*, **1987**, *109*, 6199–6200.

77. P. A. Wender, T. W. von Geldern, B. H. Levine, *J. Am. Chem. Soc.*, **1988**, *110*, 4858–4860.
78. V. B. Rao, S. Wolff, W. C. Agosta, *Tetrahedron*, **1986**, *42*, 1549–1553.
79. W. G. Dauben, J. Pesti, C. H. Cummins, unpublished results.
80. V. B. Rao, C. F. George, S. Wolff, W. C. Agosta, *J. Am. Chem. Soc.*, **1985**, *107*, 5732–5739.
81. W. Luef, R. Keese, *Helv. Chim. Acta*, **1987**, *70*, 543–553 and references quoted therein.
82. C. F. H. Allen, J. A. VanAllan, *J. Org. Chem.*, **1953**, *18*, 882–894.
83. T. Mori, K. Kimoto, M. Kawanisi, H. Nozaki, *Tetrahedron Lett.,* **1969**, 3653–3656.
84. E. H. Hahn, H. Bohm, D. Ginsburg, *Tetrahedron Lett.,* **1973**, 507–510.
85. C. W. Thornber, *J. Chem. Soc. Chem. Commun.*, **1973**, 238.
86. R. Helder, H. Wynberg, *Tetrahedron Lett.*, **1973**, 4321–4324.
87. F. Vögtle, P. K. T. Mew, *Angew. Chem.*, **1978**, *90*, 58–60; *Angew. Chem. Int. Ed. Engl.*, **1978**, *17*, 60–62.
88. K. B. Wiberg, M. J. O'Donnell, *J. Am. Chem. Soc.*, **1979**, *101*, 6660–6666.
89. P. Warner, B.-L. Chen, C. A. Bronski, B. A. Karcher, R. A. Jacobson, *Tetrahedron Lett.,* **1981**, *22*, 375–376.
90. P. E. Eaton, B. D. Leipzig, *J. Am. Chem. Soc.*, **1983**, *105*, 1656–1658.
91. K.-L. Noble, H. Hopf, L. Ernst, *Chem. Ber.*, **1984**, *117*, 455–473.
92. H. E. Simmons, C. H. Park, *J. Am. Chem. Soc.*, **1968**, *90*, 2428–2429.
93. C. H. Park, H. E. Simmons, *J. Am. Chem. Soc*, **1968**, *90*, 2429–2431.
94. H. E. Simmons, C. H. Park, R. T. Uyeda, M. F. Habibi, *Trans. N.Y. Acad. Sci. Ser. II*, **1970**, *32*, 521.
95. B. Dietrich, J.-M. Lehn, J. P. Sauvage, *Tetrahedron Lett.,* **1969**, 2885–2888.
96. B. Dietrich, J.-M. Lehn, J. P. Sauvage, *Tetrahedron Lett.,* **1969**, 2889–2892.
97. J.-M. Lehn, *Acc. Chem. Res.*, **1978**, *11*, 49–57.
98. J.-M. Lehn, *Angew. Chem.*, **1988**, *100*, 91–116; *Angew. Chem. Int. Ed. Engl.*, **1988**, *27*, 89–114.
99. M. Saunders, *J. Am. Chem. Soc.*, **1987**, *109*, 3150–3152.
100. M. Saunders, *J. Comput. Chem.*, **1989**, *10*, 203.
101. N. Weinberg, S. Wolfe, *J. Am. Chem. Soc.*, **1994**, *116*, 9860–9868 and references cited therein.
102. P. G. Gassman, R. P. Thummel, *J. Am. Chem. Soc.*, **1972**, *94*, 7183–7184.
103. P. G. Gassman, S. R. Korn, R. P. Thummel, *J. Am. Chem. Soc.*, **1974**, *96*, 6948–6955.
104. P. G. Gassman, S. R. Korn, T. F. Bailey, T. H. Johnson, J. Finer, J. Clardy, *Tetrahedron Lett.*, **1979**, 3401–3404.
105. P. G. Gassman, T. F. Bailey, R. C. Hoye, *J. Org. Chem.*, **1980**, *45*, 2923–2924.
106. P. G. Gassman, R. C. Hoye, *J. Am. Chem. Soc.*, **1981**, *103*, 215–217.
107. P. G. Gassman, R. C. Hoye, *J. Am. Chem. Soc.*, **1981**, *103*, 2496–2498.
108. P. G. Gassman, R. C. Hoye, *J. Am. Chem. Soc.*, **1981**, *103*, 2498–2500.
109. C. H. Park, H. E. Simmons, *J. Am. Chem. Soc.,* **1972**, *94*, 7184–7186.
110. M. Saunders, N. Krause, *J. Am. Chem. Soc.,* **1990**, *112*, 1791–1795.
111. J. E. McMurry, C. N. Hodge, *J. Am. Chem. Soc.*, **1984**, *106*, 6450–6451.
112. J. E. McMurry, T. Lectka, C. N. Hodge, *J. Am. Chem. Soc.*, **1989**, *111*, 8867–8872.
113. J. E. McMurry, T. Lectka, *J. Am. Chem. Soc.*, **1993**, *115*, 10167–10173.
114. R. A. Pascal, Jr, R. B. Grossman, D. Van Engen, *J. Am. Chem. Soc.*, **1987**, *109*, 6878–6880; *cf.* R. A. Pascal, Jr, C. G. Winans, D. Van Engen, *J. Am. Chem. Soc.*, **1989**, *111*, 3007–3010.
115. For a comprehensive review on *in,out* isomerism see R. W. Alder, St. P. East, *Chem. Rev.*, **1996**, *96*, 2097–2111.

7 Alkenes

The fundamental role of the carbon-carbon double-bond in organic chemistry has already been referred to in Chapter 2, in particular when we used this important functional group as a building block ('molecular brick') for the construction of numerous basic π-systems (*cf.* Chapter 2, Schemes 4 and 5). Review literature thus abounds,[1–7] and the main problem in dealing with alkenes in a monograph such as this one is one of selection. The following five sections are thought to represent the major lines of development of alkene chemistry during recent decades; in these sections I not only want to discuss important questions and problems of current olefin chemistry, but I also want to show how these problems are rooted in older, more classical times. To follow the development of an idea or a concept through time is—unfortunately—only rarely used in the teaching of chemistry nowadays. All too often modern chemistry is presented as an unhistorical success story, which overemphasizes the contribution of the present generation. It is for this reason that I have also included a section describing *failures only*: the numerous, often highly imaginative, but so far *in toto* unsuccessful attempts to prepare a formally quite simple hydrocarbon—tetrakis-*tert*-butylethene. May this section have a sobering effect and yet stimulate the reader not to give up too early!

7.1 Linear conjugated polyenes

Hydrocarbons with long, conjugated polyene chains have played a significant role in the development of preparative and theoretical organic chemistry, ever since the structure of β-carotene (**1**; Scheme 1) was established by Karrer in 1930.[8] There can, in fact, hardly be a better example of a classic in hydrocarbon synthesis than that of **1** (see below). Conjugation is the hall mark of many terpenoids, and to add just one more example lycopene (**2**) must be mentioned, the red coloring matter of the tomato and various fruits.[9–11]

Although these preparative achievements were accomplished several decades ago, conjugated polyene systems are still of great interest in natural products chemistry, as illustrated by, *e.g.*, amphoterecin B,[12] a macrolide which contains a conjugated heptaene substructure or linearmycin A1, a polyene antibiotic isolated from *Streptomyces* which contains a conjugated penta- and a tetraene subsystem.[13] As mentioned in Chapter 2, however, extended polyene systems are also important in unnatural (de-

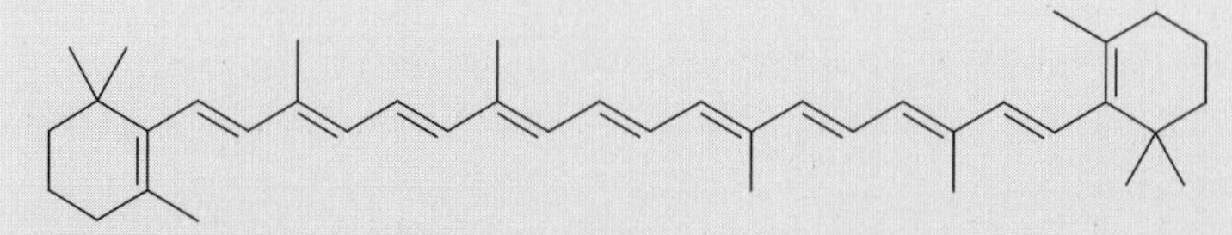

Scheme 1. Two important naturally occurring linearly conjugated hydrocarbons: β-carotene (**1**) and lycopene (**2**).

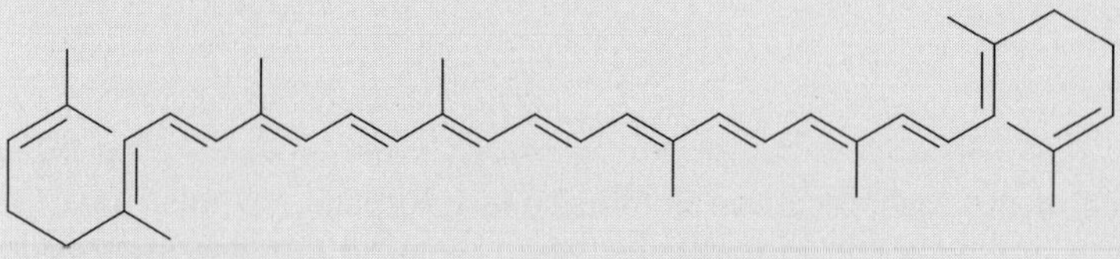

signed) products chemistry, where long π-systems are discussed as subsystems of polyacetylene and in connection with so-called molecular wires, which could be important *structural* and *functional* elements in the evolving fields of molecular electronics and nanochemistry.[14, 15] With these (potential) applications in mind, this section will concentrate largely on the preparation of long polyene chains which are 'disturbed' as little as possible by extensive branching, functional groups—either at the end of the π-chain or at some interior position—annelation of aromatic rings, and interruption of the chain by aromatic systems.[16]

Because polyene chemistry is one of the most successful examples of a building block approach in chemistry, one should be able to synthesize more highly substituted derivatives when the problem of preparing the parent systems has been mastered. We will thus begin our discussion of the linear polyenes with the preparation of the unsubstituted hydrocarbons. Although a lot of important—and preparatively very difficult—work has been conducted on these systems, it should be stressed from the very beginning that many of these classical studies no longer fulfil our modern quality demands, especially as far as purity is concerned. This is primarily because of the very high instability of the unsubstituted polyenes and the separation techniques which were available to the early workers in this field, and certainly has nothing to do with a lack of quality in synthesis design. From 1,3,5-hexatriene onwards the conjugated parent systems become increasingly difficult to handle—they are easily oxidized, they polymerize readily, particularly in the presence of air, they undergo cycloadditions and isomerization reactions with numerous reagents and lose their stereochemical integrity quickly ((*E*)/(*Z*) isomerization). As a result, the number of determinations of the structures of polyene hydrocarbons by X-ray analysis is very small.[17] Even today the X-ray structure of such simple polyenes as (*Z*)-1,3,5-hexatriene is unknown;[18] for the next higher vinylog the structure in the crystalline state has been reported for all-*trans*-1,3,5,7-octatetraene only.[19] No information is available on still more extended unsubstituted polyenes—for the experimentalist an unsatisfactory and astonishing situation considering the important role these basic systems have played in, *e.g.*, the development of molecular orbital theory.[20]

The preparation of the parent polyenes and their terminally substituted alkyl and phenyl derivatives nicely reflects progress in C–C double-bond-forming reactions during recent decades.

On the basis of the observation that 1-hexen-5-yne can be isomerized to 1,3,5-hexatriene by treatment with base—a delocalization of electrons in the literal meaning of the term—Sondheimer and co-workers were able to develop a general route to the parent polyenes.[21] For example the C_{10} hydrocarbon **3** could be isomerized to all-*trans*-decapentaene (**4**) by treatment with potassium *tert*-butylate in *tert*-butanol as shown in Scheme 2. Quantitative analysis by UV spectroscopy indicated that the pentaene had been produced in *ca.* 16% yield. When the hydrocarbon was actually isolated, only 9% survived, reflecting the high reactivity of **4**.

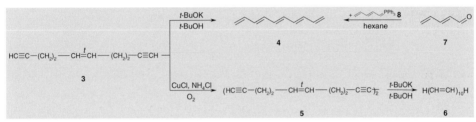

Scheme 2. Routes to H-(CH=CH)$_n$-H hydrocarbons.

Glaser coupling of the substrate **3** led to its dimer **5**, which—in turn—could be isomerized to the decaene **6**, which was identified by qualitative UV data. When the appropriate precursor for a pentadecaene was rearranged, no conclusive evidence for the formation of this product could be obtained, the method thus having passed its limit of applicability. Other routes to **4**, which polymerizes rapidly even in the crystalline state at low temperatures, have been described, among them the Wittig reaction between 2,4-pentadienal (**7**) and the ylid **8**.[22]

Introduction of methyl substituents into the α and ω positions results in a pronounced increase in the stability of the polyenes. For example, 2,4,6,8,10,12-tetradecahexaene, the first colored aliphatic hydrocarbon to be prepared, can be sublimed *in vacuo* and melts at 205 °C![23] For the preparation of these dimethyl derivatives two general approaches have turned out to be particularly valuable—the reduction of 2-butyn-1,4-diols with lithium aluminum hydride to conjugated dienes, and the Wittig reaction.

The first approach was introduced by Whiting and co-workers[24] and is shown in general form in Scheme 3.

The required acetylenic diol **12** was assembled from two unsaturated aldehyde components, **9** and **11**—substrates which have been known for a long time[25]—and acetylene (**10**) as the center part of the future hydrocarbon. Intermediate **12** was then reduced at low temperatures to the dimethyl polyene **13**. Less symmetric routes to the acetylenic diol are possible, as shown by the coupling of the pentaene aldehyde **14** with the Grignard reagent **15** to provide **16**, which on reduction with lithium aluminum

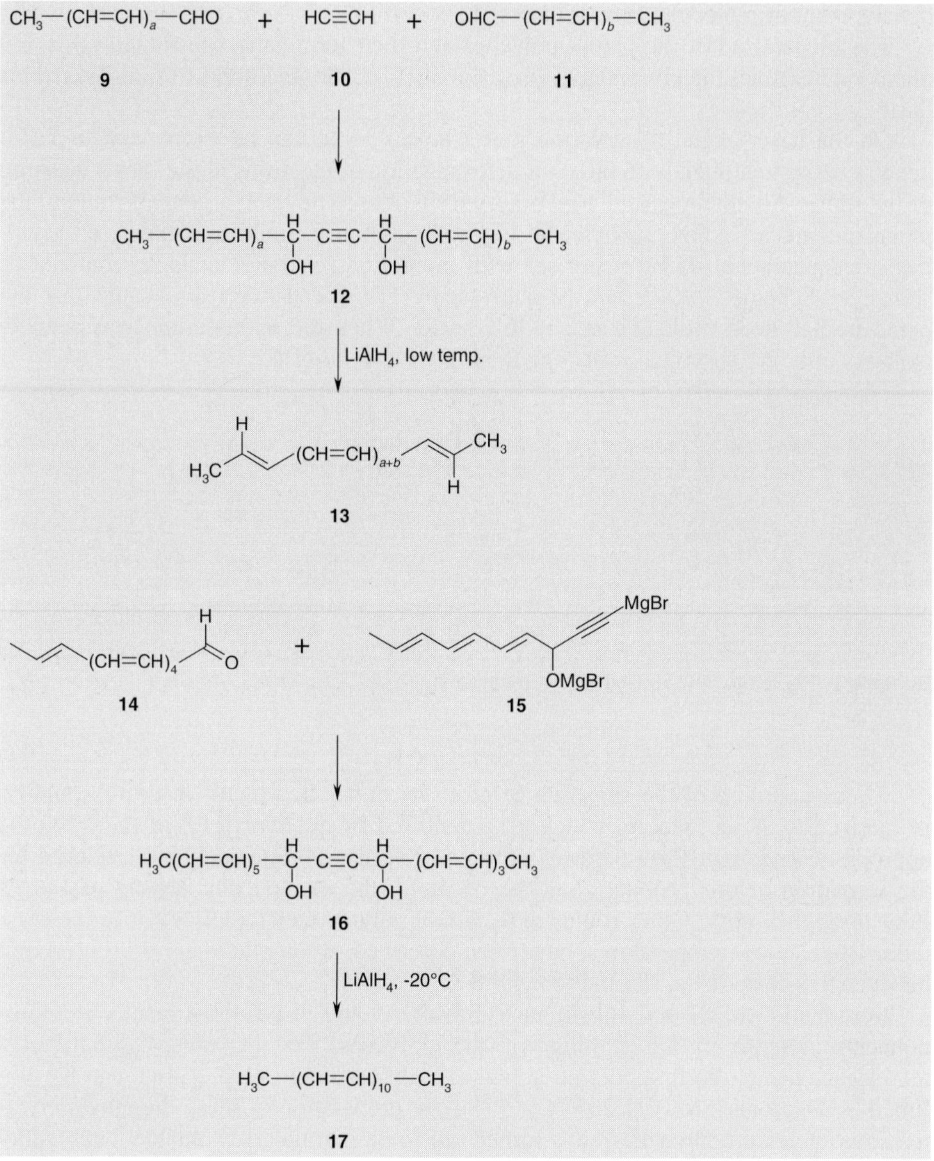
CH3—(CH=CH)a—CHO + HC≡CH + OHC—(CH=CH)b—CH3
9
10
11
CH3—(CH=CH)a—C(H)(OH)—C≡C—C(H)(OH)—(CH=CH)b—CH3
12
LiAlH4, low temp.
(CH=CH)a+b
13
(CH=CH)4
14
MgBr
OMgBr
15
H3C(CH=CH)5—C(H)(OH)—C≡C—C(H)(OH)—(CH=CH)3CH3
16
LiAlH4, -20°C
H3C—(CH=CH)10—CH3
17

Scheme 3. Routes to α,ω-dimethyl polyenes–the reduction of acetylenic diols.

hydride was transformed to the decaene **17** in low yield. Extending this approach beyond **17** gave inconclusive results or failed.

The application of the Wittig reaction for the preparation of these long-chain polyenes was introduced by Bohlmann and co-workers.[26] To prepare the decapentaene derivative **17**, for example, they treated the already highly unsaturated dialdehyde **18** with two equivalents of the 'extended' ylid **19**. The resulting diacetylene **20** was then Lindlar-hydrogenated and the resulting isomer mixture of polyenes isomerized to the most stable all-*trans* isomer **17** by treatment with a trace of iodine in benzene (Scheme 4).

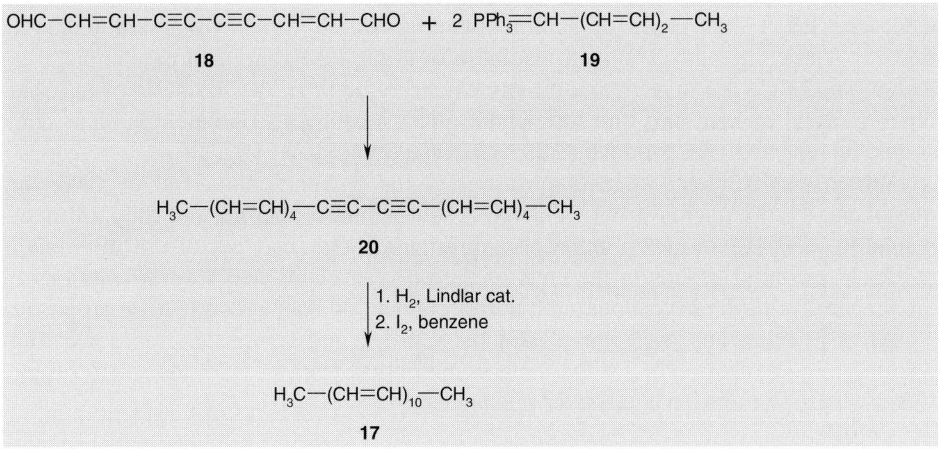
OHC–CH=CH–C≡C–C≡C–CH=CH–CHO + 2 PPh3=CH–(CH=CH)2–CH3
18
19
H3C–(CH=CH)4–C≡C–C≡C–(CH=CH)4–CH3
20
1. H2, Lindlar cat.
2. I2, benzene
H3C–(CH=CH)10–CH3
17

Scheme 4. Routes to α,ω-dimethyl polyenes—the Wittig approach.

Alternatively, these unstable and very poorly soluble π-chains have been prepared by coupling double Wittig reagents with different monoaldehydes; the yield is occasionally excellent.[27]

The first molecular wires—although this term was not used at the time—were prepared by R. Kuhn and his co-workers, who developed a general route to α,ω-diphenyl polyenes with up to 15 consecutive double bonds.[23] To achieve this goal—more than 60 years ago!—two basic strategies were developed. For the 'smaller' polyolefins (up to eight double bonds) the so-called lead oxide route proved to be invaluable (Scheme 5).

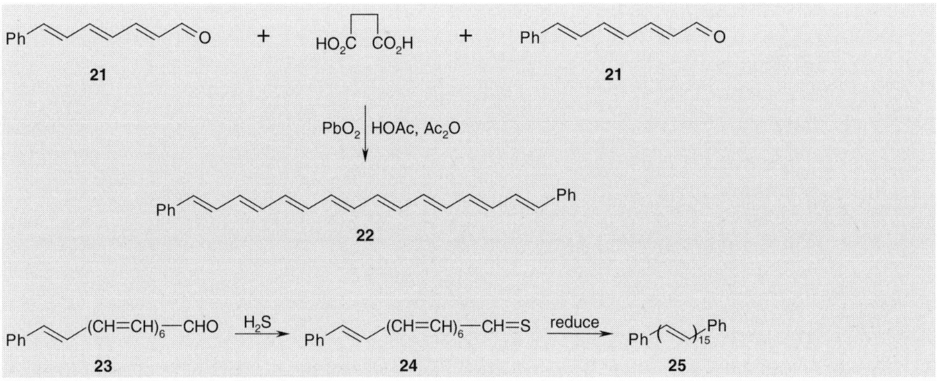
Ph
O
+
HO2C CO2H
+
Ph
O
21
21
PbO2
HOAc, Ac2O
Ph
Ph
22
Ph
(CH=CH)6–CHO
H2S
Ph
(CH=CH)6–CH=S
reduce
Ph
Ph
15
23
24
25

Scheme 5. Kuhn's routes to α,ω-diphenylpolyenes.

In this method, which reached its limit with the octatetraene **22**,[28] two molecules of 7-phenyl-heptatrienal (**21**) were condensed with succinic acid in acetic acid-acetic anhydride in the presence of lead oxide. To prepare still longer polyenes appropriate aldehyde precursors such as **23** were first converted into their thio derivatives **24** by treatment with hydrogen sulfide. When the latter were desulfurized by treatment with piperidine or other amines or with metals, dimerization took place yielding the pentadecaene **25** a dark-green crystalline material which forms violet colored solutions.[29] The compound is almost insoluble in organic solvents and its chemical behavior was hence not investigated to any extent at the time. On the basis of the solubilizing effect of medium- and long-chain alkyl substituents, Kuhn's classical work should be repeated and extended to these derivatives.

Whereas substituents at inner positions of the polyene chain tend to make this non-planar,[17] the phenyl groups change its electronic properties and they are detrimental to solubility in the common organic solvents. One way out of this dilemma is to attach *tert*-butyl groups to the ends of the polyene chain, and several solutions of the preparative problems encountered in this connection have recently been proposed.

An interesting approach, developed by Schrock and co-workers, uses the ring-opening-oligomerization of 7,8-bis(trifluoromethyl)tricyclo[4.2.2.0^{2,5}]deca-3,7,9-triene (**26**) by use of a tungsten catalyst **27** (Scheme 6).[30]

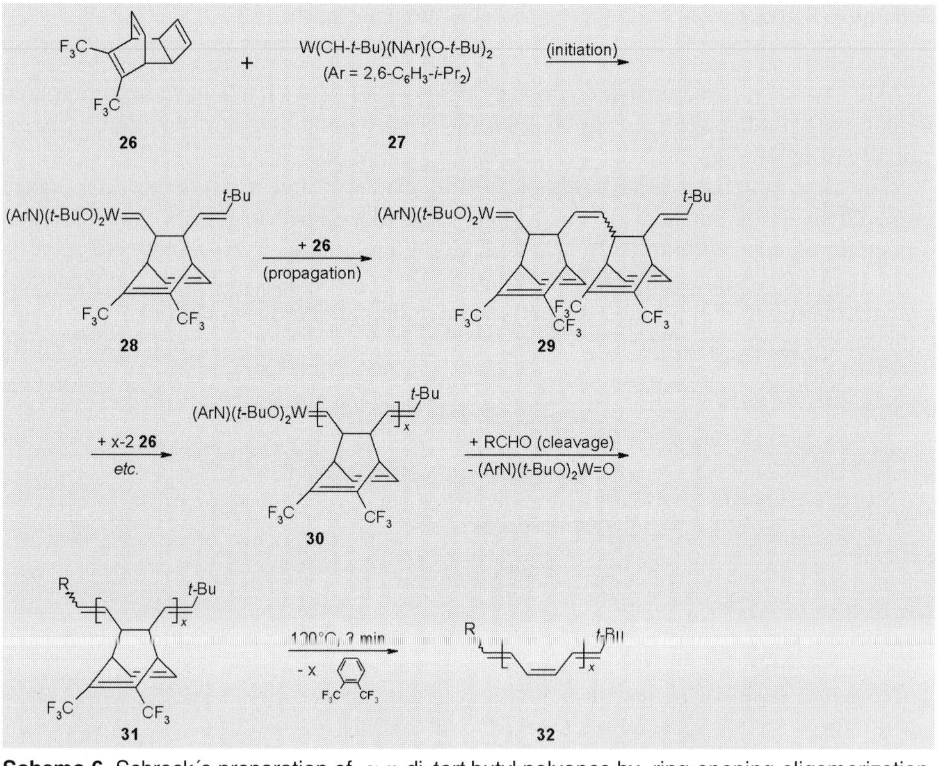

Scheme 6. Schrock's preparation of α,ω-di-*tert*-butyl polyenes by ring-opening-oligomerization.

The process is initiated by formation of tungsten complex **28**. When this was reacted with further **26** propagation set in, finally yielding the oligomers **30**. These could be end-capped with an aldehyde such as pivaldehyde, furnishing the metal free oligomers **31**. Short-time pyrolysis at relatively low temperatures resulted in removal of 1,2-bis-trifluormethyl benzene as a leaving group and the *tert*-butyl-capped polyene **32** was formed. Although this method could be extended as far as the decapentaenes, it yielded polydisperse mixtures which must be subjected to chromatographic separation for the isolation of the various pure polyenes.

Stepwise approaches which combine efficiency with complete control of every step are thus required. One recent solution, that of Müllen and co-workers, which illustrates the building block approach most convincingly, sets out from *tert*-butylacetylene (**33**) (Scheme 7).[31]

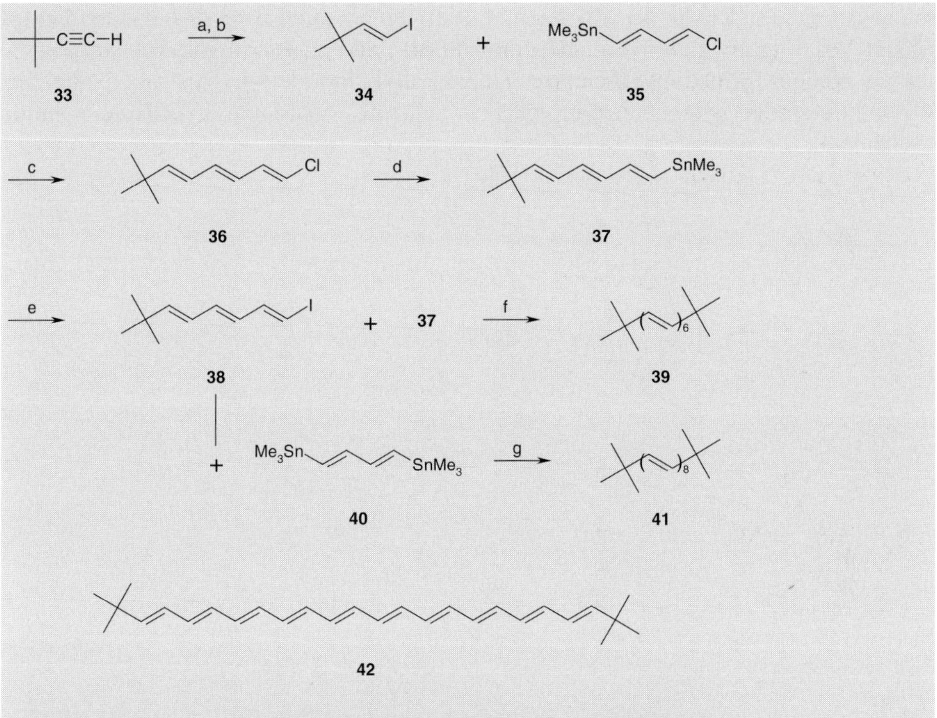

(a) DIBAH / hexane; (b) I$_2$, THF, -50°C (comb. yield 67%); (c) PdCl$_2$(CH$_3$CN)$_2$, DMF, 20°C (64%); (d) Me$_3$SnLi, THF, 0°C (56%); (e) I$_2$ (64%); (f) PdCl$_2$(CH$_3$CN)$_2$, DMF, 20°C (82%); (g) (39%)

Scheme 7. Preparation of α,ω-di-*tert*-butyl polyenes by Stille coupling.

In the first step **33** was transformed into the vinyl iodide **34** by treatment with DIBAH followed by trapping of the formed vinyl alane with iodine. When **34** was subjected to Stille coupling with the chlorodiene **35**—which can be prepared from 1,4-dichloro-1,3-butadiene—the extended chloride **36** was formed in good yields. This was converted to the trimethylstannone **37**, which not only served as a precursor for the iodide **38** but was also a coupling partner in the next Stille reaction step: The coupling of the latter with **37** provided the hexaene **39** in excellent yield, and reaction with the bis-stannyl derivative **40**, also available from 1,4-dichloro-1,3-butadiene, yielded the octaene **41**, with the (most stable) all-*trans* isomer again formed in a final isomerization step, which this time was performed by irradiation with light of wavelength > 345 nm. In fact, this last coupling reaction also provided small amounts of di-*tert*-butylicosadecaene (**42**) *via* partial exchange of iodine and trimethyltin groups. Because these derivatives are again only poorly soluble in organic solvents the method was extended to the introduction of longer (solubilizing) alkyl groups which substantially increased the solubility of the corresponding derivatives.

An iterative strategy was also employed by Hopf and co-workers for the preparation of polyenes which are fully *tert*-butylated at their ends, a protection which makes the corresponding hydrocarbons not only more stable but also increases their solubility; all derivatives of this type prepared so far are, furthermore, solids, so their structures can be determined by X-ray structural analysis.[32] The required capping unit is the unsaturated aldehyde **45** which could easily be prepared from di-*tert*-butyl ketone (**43**) by reacting it with vinyl magnesium chloride, and rearranging and converting the tertiary alcohol formed into the corresponding allyl chloride **44**. This was oxidized to **45** by treatment with 2-nitro-propane in aqueous potassium hydroxide solution (Scheme 8).

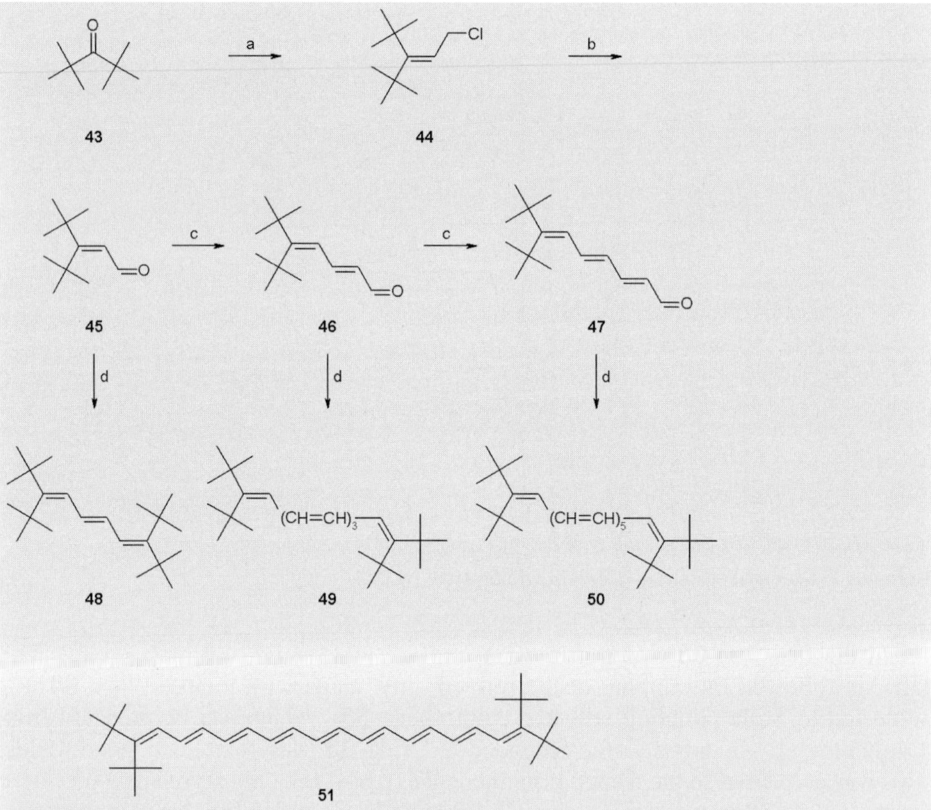

(a) H$_2$C=CHMgBr, THF; SOCl$_2$, pyridine (comb. yield 55%); (b) KOH, 2-nitro-propane, isopropanol (58%); (c) BrPh$_3$PCH$_2$CH(OR)$_2$, *tert*-BuOK, THF (85%); (d) TiCl$_4$, Zn, pyridine, THF (50-60% all-*trans* isomer)

Scheme 8. The preparation of polyenes fully *tert*-butylated at their ends.

McMurry coupling of **45** led to a (readily separable) mixture of the trienes **48**, chain extension by conventional Wittig methodology provided **46**. This again could be reductively dimerized to the pentaene **49** or chain-extended to **47**, which effortlessly led to heptaene **50**. Going through the whole cycle yet again provided **51**. Because the route described in Scheme 7 leads to molecules with an even number of double bonds, and the approach presented in Scheme 8 to odd-numbered systems, the two methods complement each other.

As these few examples show, it is possible to prepare conjugated polyenes with

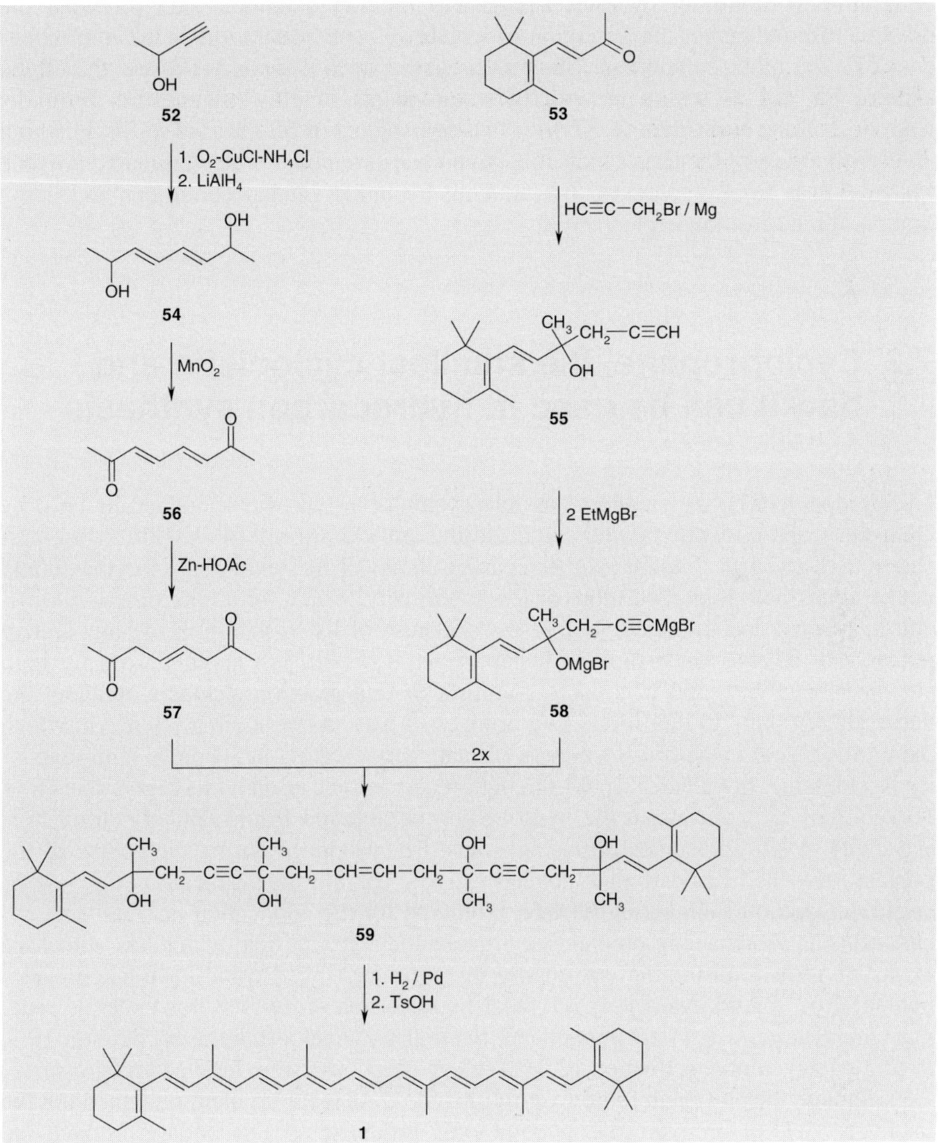

Scheme 9. The first synthesis of β-carotene (**1**) by Karrer and Eugster.

considerably more double bonds than are found in β-carotene (**1**). Despite this, β-carotene and its 'half', Vitamin A aldehyde (retinal), remain the most important molecules of this general structure and so I believe that nothing could pay tribute better to the numerous researchers who have prepared β-carotene in dozens of syntheses[2] than to repeat the original route to hydrocarbon **1** here which was published by Karrer and Eugster half a century ago (Scheme 9).[33]

In the converging approach two building blocks were prepared first—the diketone **57**, and the acetylide **58**. The former was obtained from the propargyl alcohol **52** by dimerizing it oxidatively and reducing the diacetylenic diol obtained to the diol **54** with lithium aluminum hydride. Manganese dioxide oxidation then provided the diketone **56**, the center diene section of which was subsequently reduced to a monoene, **57**. The other component, **58**, was prepared from β-ionone (**53**) *via* the allylic alcohol **55** and its treatment with two equivalents of ethyl magnesium bromide. Coupling of one equivalent of **57** with two equivalents of **58** then led to **59**, in which all carbon atoms of the target molecule have been assembled. The conjugated polyene system **1** was finally generated by reducing **59** under Lindlar conditions and dehydration of the resulting heptaene diol.

7.2 Cyclopropene, the smallest cycloolefin and how it can be used in hydrocarbon synthesis

Cyclopropene (**61**), the smallest possible cycloalkene, was first obtained in 1926 by Demyanov and co-workers who subjected the ammonium salt **60** to Hofmann elimination (Scheme 10);[34] there had been earlier claims of its synthesis,[35] but they could not be confirmed. Regarded more or less a curiosity, which for a long time was difficult to prepare and to handle (see below) because of its 'instability', the preparative potential of **61** was not exploited. Instability is, however, just another word for high chemical reactivity, and it is up to chemists to find reaction channels enabling the harnessing of this 'driving force' in a productive way. After all, in a science in which the ultimate goal is the transformation of matter by creating new bonds, high reactivity is a blessing, not a curse. In **61** the high reactivity is caused by its high strain ($E_s = $ 56 kcal mol^{-1}),[36] and when the hydrocarbon is prepared from relatively strain free, especially acyclic precursors this strain must be introduced during the course of the preparation—the hydrocarbon is 'loaded', so to speak, during synthesis.

In this section I am taking **61** as a prototype for the application of small-ring cycloolefins as substrates in preparative organic chemistry, a field which has witnessed explosive growth during the last quarter of a century.[37–44] I could have taken cyclobutene also.[45] And again it is a typical hydrocarbon story (see, for example, adamantane, Section 3.1, [1.1.1]propellane, Section 6.1, cyclooctatetraene, Section 10.3, or [2.2]paracyclophane, Section 12.3) in which great oaks from little acorns grow.

Although the Hofmann route was optimized,[46] its main problem remained not the mediocre yield (1-propyne (**62**) is commonly produced as a secondary product, see below), but the starting materials, which also often have to be prepared by multi-step

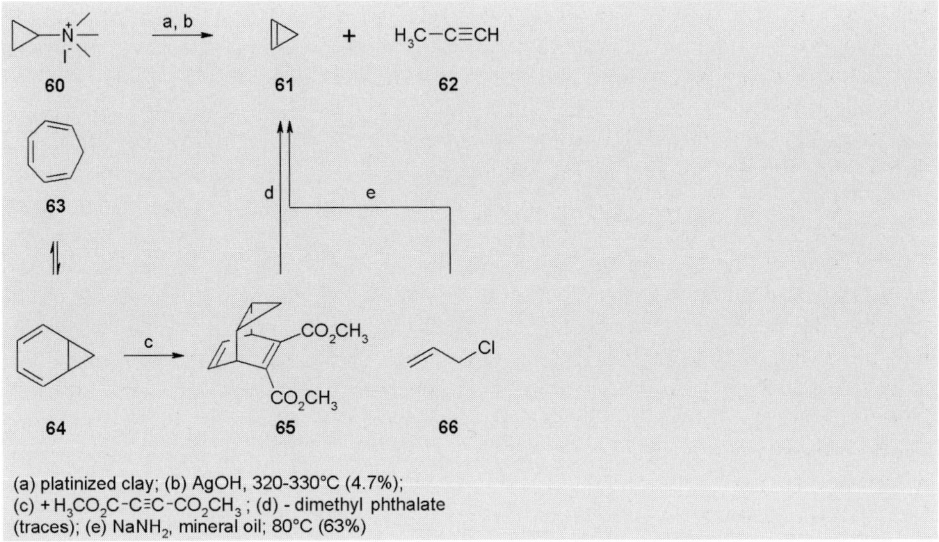

(a) platinized clay; (b) AgOH, 320-330°C (4.7%);
(c) + H$_3$CO$_2$C-C≡C-CO$_2$CH$_3$; (d) - dimethyl phthalate
(traces); (e) NaNH$_2$, mineral oil; 80°C (63%)

Scheme 10. Different routes to the smallest cycloalkene, cyclopropene (**61**).

routes, and problems related to the handling of the product. For example when cyclo-propyl bromide is dehydrobrominated with potassium *tert*-butoxide supported on sil-ica gel, cyclopropene is produced in 75% yield when working on a 50 mg scale yet the yield drops to a meager 14% when the β-elimination is conducted with more ma-terial.[47] Likewise, subjecting the Diels–Alder adduct **65**, obtained by adding di-methyl acetylenedicarboxylate to the valence isomer pair **63/64**, to a Alder–Rickert cleavage yielded dimethy phthalate in quantitative yields, but **61** only in traces;[48, 49] evidently **61** does not survive the conditions of its generation. In fact, when an at-tempt was made to purify **61** by fractional distillation, it rapidly polymerized at –36 °C, its boiling point at atmospheric pressure.[35] The best route to **61** currently consists in the dehydrochlorination of allyl chloride (**66**) with base. With sodium amide as base yields as high as 63% have been achieved[50] if the product is removed quickly from the reaction zone. The process also works well for the preparation of substituted cyclopropenes although the choice of the base is critical. For example, when 2-butenyl chloride was treated with lithium amide in dioxan 3-methyl-cyclopropene was obtained in excellent yield and purity; merely switching to sodium amide in di-*n*-butyl ether led, however, to methylenecyclopropane.[51]

The obvious route of adding methylene to a triple bond affords no viable method for generating **61**; it has, however, been employed successfully for the preparation of substituted cyclopropenes, although stopping the process at the mono addition stage constitutes a problem.[52]

Before discussing some typical reactions of **61** we want to look at this molecule more closely. Its structural formula is simple, but we can discover much from it if we focus on certain parts (Scheme 11). Molecular formulas often resemble paintings which also yield more of their 'secrets' on closer examination.

The hydrocarbon has four locations where it can react: the C=C double bond; the olefinic C–H bonds; the methylene group; and the carbon-carbon single bond which

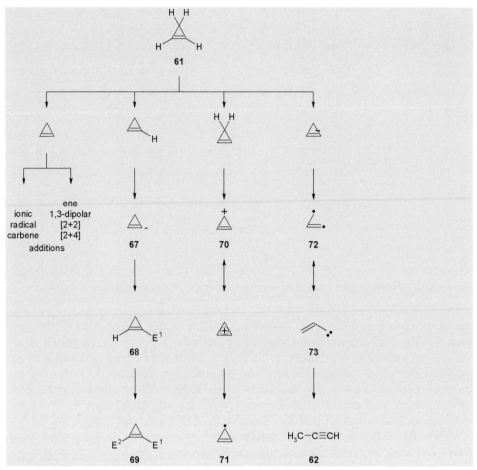

Scheme 11. How to 'read' a simple molecule—cyclopropene (**61**).

connects the saturated and the unsaturated part of the molecule. As far as the double bond is concerned we would, of course, expect addition reactions to occur, whether these are ionic, radical, or cycloadditions *etc*. Because the molecule is strained, we would expect these additions to occur more readily than for strain-free olefinic model compounds. As far as the olefinic C–H bonds are concerned, **61** can be regarded as a 'homoacetylene' and we would thus expect to see increased C,H acidity at this position. This is confirmed experimentally, as we shall see, and is preparatively very useful for the functionalization of **61** *via* its carbanion **67** (stepwise introduction of one and two electrophilic groups E^1 and E^2, to **68** and **69**, respectively). The methylene group is of interest because by heterolytic bond cleavage (hydride abstraction) we could form an aromatic cation, the cyclopropenium cation (**70**), as the smallest conceivable compound obeying Hückel's rule (see Chapter 10). Whereas homolytic bond cleavage at this position seems feasible (formation of the radical **71**), the removal of a proton is highly unlikely because the anti-aromatic cyclopropenyl anion (a cyclic 4π-electron system) would have to be left behind. C-C bond rupture to **72** destroys the three-membered ring, but furnishes a most useful intermediate, vinyl carbene (**73**), the excess energy of which can be used to drive follow-up reactions, for example the already mentioned isomerizition of **61** to propyne (**62**). In principle, therefore, hydrocarbon **61** could be incorporated into products in closed (cyclic) or open form.

Numerous polar addition reactions to **61** and its derivatives have been performed with a plethora of reagents, and provide ready access to functionalized cyclopropanes.[39] The cycloaddition reactions of cyclopropenes are particularly useful in hydrocarbon chemistry.

The tendency of **61** to polymerize has been noted early, and although polymerization to **75** could be a free-radical-initiated process,[48] an alternative, beginning with its known[53] ene reaction to 3-cyclopropyl-cyclopropene (**74**), is also worth considering (Scheme 12).

Cyclopropenes can undergo [2+2]cycloadditions under a variety of conditions (photochemical, thermal, Lewis-acid or metal activation) either to each other or to

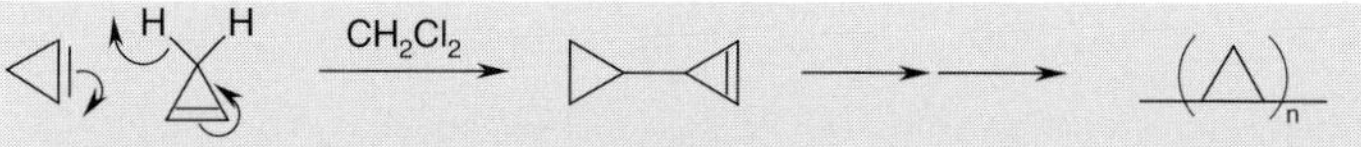

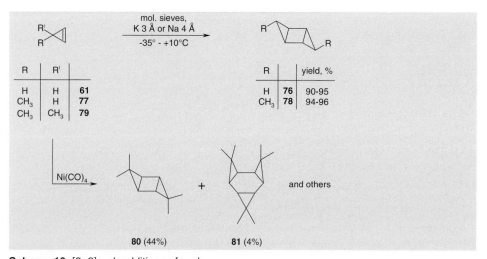

Scheme 12. Di- and polymerization of cyclopropene (**61**) by the ene reaction.

Scheme 13. [2+2]cycloadditions of cyclopropenes.

other olefins, furnishing structurally interesting highly strained polycyclics. For example, when **61** was absorbed on molecular sieves at low temperatures and the reaction mixture then slowly warmed, the dimer **76** was produced in quantitative yield; with the monomethyl derivative **77** the dimer **78** is formed[54] (Scheme 13).

In the presence of various metal catalysts cyclopropenes di- and trimerize. With nickel carbonyl 3,3-dimethyl-cyclopropene (**79**) yields **80** and **81**[55] the latter be-coming the exclusive product in the presence of tetrakis-triphenylphosphine palladium.[56] Considering the close chemical analogy between the cyclopropane ring and the carbon-carbon double bond, these di- and trimers can be called bis-homo-cyclobutadiene and tris-homobenzene, respectively. They are discussed later in connection with the problem of *homoaromaticity* (see Section 13.4).

Cyclopropene (**61**) has become a popular dienophile in Diels–Alder additions, its reactivity being roughly comparable with that of maleic anhydride. With 1,3-butadiene it provided norcarene (**82**),[48, 57] substituted dienes such as isoprene led to the expected adducts, *e.g.* **83**,[58] and with 1,3-cyclopentadiene the tricyclic adduct **84** was formed.[48, 50] The drastically enhanced reactivity of **61** compared with that of ethylene itself is nicely demonstrated by the cycloaddition of the latter to cyclopenta-diene, which requires not only high temperatures 190–220 °C but high pressure also (200–400 atm)[59] (Scheme 14).

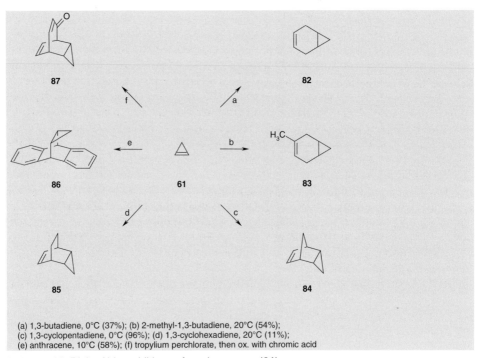

(a) 1,3-butadiene, 0°C (37%); (b) 2-methyl-1,3-butadiene, 20°C (54%);
(c) 1,3-cyclopentadiene, 0°C (96%); (d) 1,3-cyclohexadiene, 20°C (11%);
(e) anthracene, 10°C (58%); (f) tropylium perchlorate, then ox. with chromic acid

Scheme 14. Diels–Alder additions of cyclopropene (**61**).

With 1,4-cyclohexadiene and anthracene the 1:1 adducts **85**[60] and **86**[61] were produced, and tropylium perchlorate initially provided an alcohol which may be oxidized to the α,β-unsaturated ketone **87**.[62] Concluding this brief excursion into peri-cyclic addition chemistry it should be mentioned that **61** and its derivatives have been extensively employed both in hetero Diels–Alder[39, 63] and 1,3-dipolar cycloaddition reactions.[39, 48, 64, 65]

The cyclopropene-homoacetylene analogy is also demonstrated by the comparable C,H acidity of the two hydrocarbons (pK_a of **61** 29,[66] pK_a of acetylene 25) and the resulting ready deprotonation of **61** to the carbanion **67** (Scheme 11). In practice cyclopropene can be metalated easily by treatment with lithium, sodium or potassium amide in liquid ammonia. The resulting metal organic compound **88** (Scheme 15) can, of course, be quenched by the addition of electrophiles,[39] but it can also undergo self-trapping by still non-metalated starting material providing the interesting dimer bicyclopropylidene (**90**, cyclopropylidenecyclopropane) (see Section 15.3) and the trimer **91**.

The products are formed by the addition of **88** to **61** to provide a 2-cyclopropenyl-1-metallocyclopropane which subsequently isomerizes to the more stable 2-cyclopropyl-1-metallocyclopropene isomer **89**. This reacts with the starting **61** to give 1,2-(dicyclopropyl)cyclopropene (**91**) or can be protonated to 1-cyclopropylcyclopropene which stabilizes itself to **90**. Under controlled conditions this highly strained olefin (see Section 11.6) can be obtained in 70% yield.[67, 68]

Hydride abstraction—or, more generally, the removal of a negatively charged substituent[39]—from vastly different cyclopropene precursors is the method of choice for preparation of cyclopropenium cations. Although the parent system **70** was obtained (as the hexachloroantimonate salt) by a different route,[69] many cyclopropene-derived hydrocarbons have been charged by hydride removal, in most cases with the perchlorate or tetrafluoroborate salts of the trityl cation as the hydride acceptor.[39] Cyclopropenyl radicals have, on the other hand, been generated *inter alia* by one-electron reduction processes, which can be performed either by electrolysis or by the use of conventional chemical reductants such as zinc or magnesium powder.[39]

In the very first experiments aimed at the synthesis of **61** it was noted that the product mixture always contained propyne (**62**) as a byproduct. When **61** was deliberately subjected to gas phase pyrolysis it was shown that **62** is the sole reaction product.[48] For a long time it was assumed that this isomerization proceeds *via* the propene-1,3-diyl intermediate **72**, formed by ring-opening of **61** and subsequent 1,2-hydrogen shift.[70] Recently, however, it was shown that an alternative exists; this involves the propenylidene intermediate **92** formed by ring opening of **61** with a synchronous hydrogen transfer;[71] again the process would be terminated by a hydrogen transfer step (Scheme 16).

If such an unsaturated carbene is, indeed, generated initially, *ab initio* calculations suggest that its formation might be reversible.[72] That this is so was shown by pyrolyzing the substituted cyclopropene **93** at 200 °C. Not only was the expected

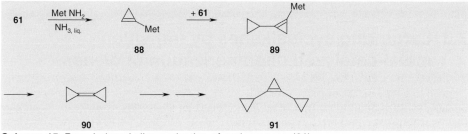

Scheme 15. Base-induced oligomerization of cyclopropene (**61**).

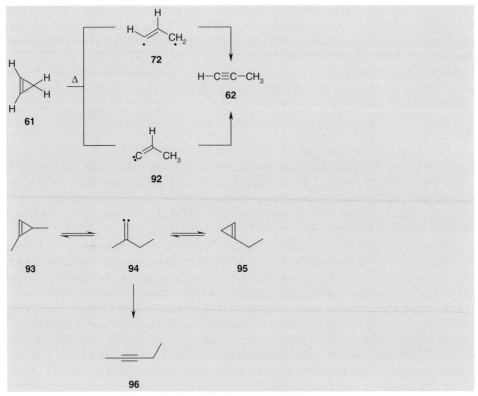

Scheme 16. The thermal isomerization of cyclopropene (**61**).

acetylene **96** produced, but the rearranged cyclopropene **95** also. Clearly there are two means of insertion of **94** into a neighboring C–H bond.[73]

In summary, the chemistry of a molecule as structurally simple as **61** is far richer than one might, *a priori*, have assumed. If, furthermore, we incorporate **61** as a building block into more complex systems,[74] numerous other structure patterns arise which lead to novel and often surprising chemical and physical behavior. Among the many hydrocarbons with cyclopropene subunits,[75] triafulvene (Section 11.2), several (substituted) fulvalenes (Section 11.3), cyclopropabenzene (Section 16.3) and bicyclo[1.1.0]-1(3)-butene (see below) will be discussed in this book.

7.3 Large-ring cycloalkenes by transition metal-catalyzed oligomerization of dienes

One of the most significant reactions of hydrocarbon chemistry, leading to the largest hydrocarbons ever prepared, is the polymerization of ethylene (**97**) to polyethylene (**98**) (Scheme 17).

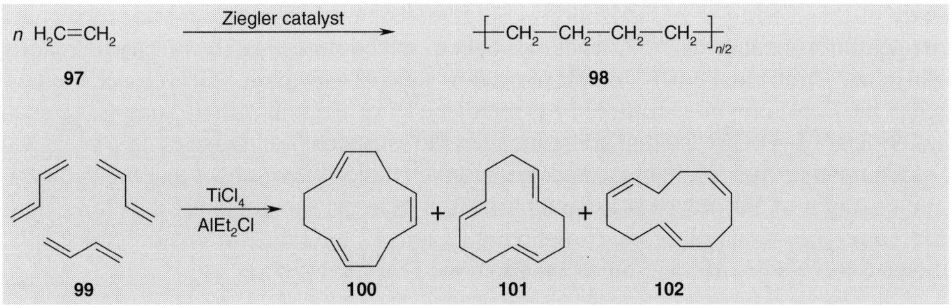

Scheme 17. Ziegler's synthesis of polyethylene (**98**) and Wilke's synthesis of the cyclododecatrienes **100–102**.

The variant discovered by Ziegler, in which this process is conducted at normal pressure and low temperature (70 °C) in the presence of metal-organic catalyst systems such as the famous triethylaluminum-titanium(IV) chloride combination, the original Ziegler catalyst, in an inert solvent such as *iso*-dodecane, became of particularly huge commercial importance.[76–78] The largely unbranched paraffin chains in **98** led to a very high crystallinity (up to 80%) in this form of polyethylene, for which relative molecular masses of up to 6 million have been observed. Low-pressure polyethylene polymerisates with molecular masses between 50 and 100 000 are commonly used commercially.

Although this reaction can also be performed with other monoolefins such as propene (preparation of isotactic polypropylene according to Natta[79]) and 1-butene, Wilke discovered in the mid 1950s that the simplest conjugated diene, 1,3-butadiene (**99**), behaves differently. With a catalyst prepared from titanium(IV) chloride and diethylaluminumchloride (Ti/Al-ratio 1:4.5), which rapidly converted **97** into **98**, the diene **99** was cyclotrimerized to *trans,trans,cis*-1,5,9-cyclododecatriene (**100**) in yields better than 80%.[80] With other transition metal catalysts the yield of the cyclododecatrienes could be increased to over 96%, and the all-*trans* isomer **101** and the *cis,cis,trans* hydrocarbon **102** (always a minor component) could be obtained.[81] Because traditionally medium-sized and large carbocyclic ring systems are difficult to prepare,[82] this very easy and efficient access to twelve-membered ring compounds constituted a real breakthrough and quickly became a classic. This was even more so because it could be shown that the method is not restricted to cyclooligomerization but can also be performed as a cyclo-co-oligomerization with numerous other olefins, dienes, acetylenes *etc.* (see below). Metal-mediated carbon-carbon bond formation reactions are of great importance in hydrocarbon synthesis, as we shall see, *e.g.*, in Chapters 14 and 15 where we will exploit metal-catalyzed coupling processes to construct very large extended hydrocarbon systems. Mechanistically the cyclotrimerization is of a type different from these reactions, and can better be compared with Reppe's 1940 cyclooctatetraene synthesis[83] (see Section 10.3) in which, *e.g.*, nickel catalysts act as templates for the synthesis of cyclic hydrocarbons from (small) acyclic building blocks. Hydrocarbons such as **101** are not only of industrial importance (preparation of nylon and vestamide) but can also serve as starting materials for a variety of bi- and polycyclic ring systems.[84]

In subsequent studies Wilke and co-workers discovered that the scope and reac-

tivity of these catalytic trimerizations could be enhanced by use of homogeneous ze-
rovalent nickel catalysts.[81, 85] These so-called 'naked nickel catalysts' can be readily
generated from nickel(0) complexes with ligands such as Ni(cyclooctadiene)$_2$,
Ni(COD)$_2$, that can be substituted by butadiene[81] or from nickel(II) complexes such
as [Ni(acac)$_2$]-Al(OEt)Et$_3$ that are reduced in the presence of butadiene.

Comprehensive mechanistic and preparative studies have shown that the cyclooligomerization of **99** proceeds as summarized in Scheme 18 and that by 'catalyst tuning' (*inter alia* variation of the complexing ligands L) the ring-forming process can be
directed in widely different directions (Scheme 18).[86, 87]

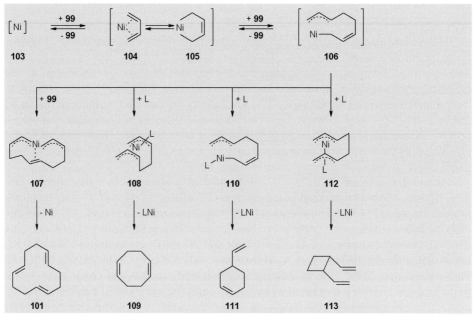

Scheme 18. The mechanism of the cyclooligomerization of 1,3-butadiene (**99**).

The process begins with the loss of the ligands of a nickel catalyst system **103**
such as the nickel complex of **101**[88] or Ni(COD)$_2$ under the influence of butadiene
and formation of a non-isolable intermediate, which has been formulated either as a
η^4 (**104**) or η^2 (**105**) system. At low temperatures (–20 to –30 °C) this reacts with
further **99** to provide the nickel complex **106**, a crucial manifold in the cyclooligomerization process. In the presence of ligands L such as P(C$_6$H$_5$)$_3$, P(OC$_6$H$_4$-*o*-
C$_6$H$_5$)$_3$ or P(cyclohexyl)$_3$, intermediates such as **108**, **110** or **112** could be isolated
and with further 1,3-butadiene **106** provided **107**; all these structures have been elucidated by NMR spectroscopy.[89, 90] By reductive elimination these nickel complexes
finally ring-close to **101**, 1,5-cyclooctadiene (**109**, COD), 4-vinyl-1-cyclohexene
(**111**), and *cis*-1,2-divinylcyclobutane (**113**),[91, 92] making all these hydrocarbons
available in preparative and even industrial amounts from one common precursor, **99**.
The superiority of this approach becomes particularly impressive when it is compared

with conventional hydrocarbon synthesis. For example, **113**, an important model compound in connection with the study of the Cope rearrangement,[93] was first prepared by Vogel starting from bicyclo[4.2.0]oct-3-ene (**116**), a hydrocarbon originally synthesized by Alder and co-workers[94] from the cycloadduct **115** between 1,3-butadiene (**99**) and maleic anhydride (**114**) (Scheme 19).[95]

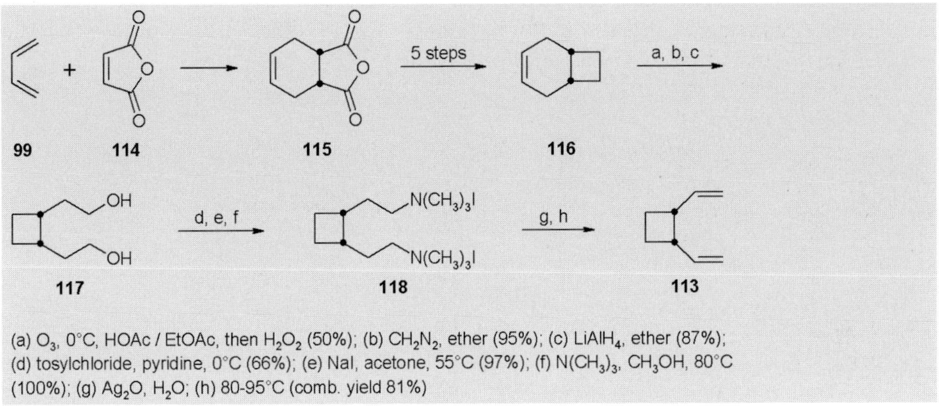

(a) O_3, 0°C, HOAc / EtOAc, then H_2O_2 (50%); (b) CH_2N_2, ether (95%); (c) LiAlH$_4$, ether (87%); (d) tosylchloride, pyridine, 0°C (66%); (e) NaI, acetone, 55°C (97%); (f) N(CH$_3$)$_3$, CH$_3$OH, 80°C (100%); (g) Ag$_2$O, H$_2$O; (h) 80-95°C (comb. yield 81%)

Scheme 19. The first synthesis of *cis*-1,2-divinylcyclobutane (**113**) by Vogel.

Ring-opening of **116** by ozonolysis followed by oxidative work-up first led to *cis*-1,2-cyclobutanediacetic acid; this was converted to the *cis*-diol **117** by esterification with diazomethane and lithium aluminum hydride reduction of the diester produced. The *cis*-diol was then transformed into the corresponding diiodide by treatment of the bis tosylate with sodium iodide in acetone, and when the dihalide was heated with trimethylamine at 80 °C the bis ammonium salt **118** resulted. Conventional Hofmann elimination finally provided the desired hydrocarbon **113**. Although most steps of this route proceed well—the poorest yield is obtained in the ozonisation step—the synthesis is lengthy and can thus not compete with the Ni-mediated dimerization of **99** as described above. The *trans* isomer of **113** is most easily prepared by triplet-sensitized photodimerization of **99**, a [2+2]cycloaddition in which **113** is generated only in small amounts.[96]

As already mentioned mixed oligomerizations of butadiene with other unsaturated hydrocarbons are also possible under these conditions, and the examples assembled in Scheme 20 are, again, of exemplary nature only.

For example, 1,3-butadiene (**99**) and ethylene react in the presence of naked nickel or nickel ligand catalytic systems to provide—via the intermediate complex **119**—*cis,trans*-1,5-cyclododecadiene (**120**) in yields of up to 80%.[97] Among the co-oligomerizations with acetylenes the reactions with cyclic diynes (see Section 8.3) such as cyclotetradeca-1,6-diyne or cyclododecayne are of particular interest.[98] With the former acetylene either a mono-, **121**, or a bis-, **122**, adduct could be prepared, whereas the latter furnished the bicyclic triene **123**. When these hydrocarbons were partially hydrogenated their central, tetrasubstituted double bonds survived; they could, however, be cleaved by ozonolysis thus enabling the preparation of the very large ring diketones **124–126**.

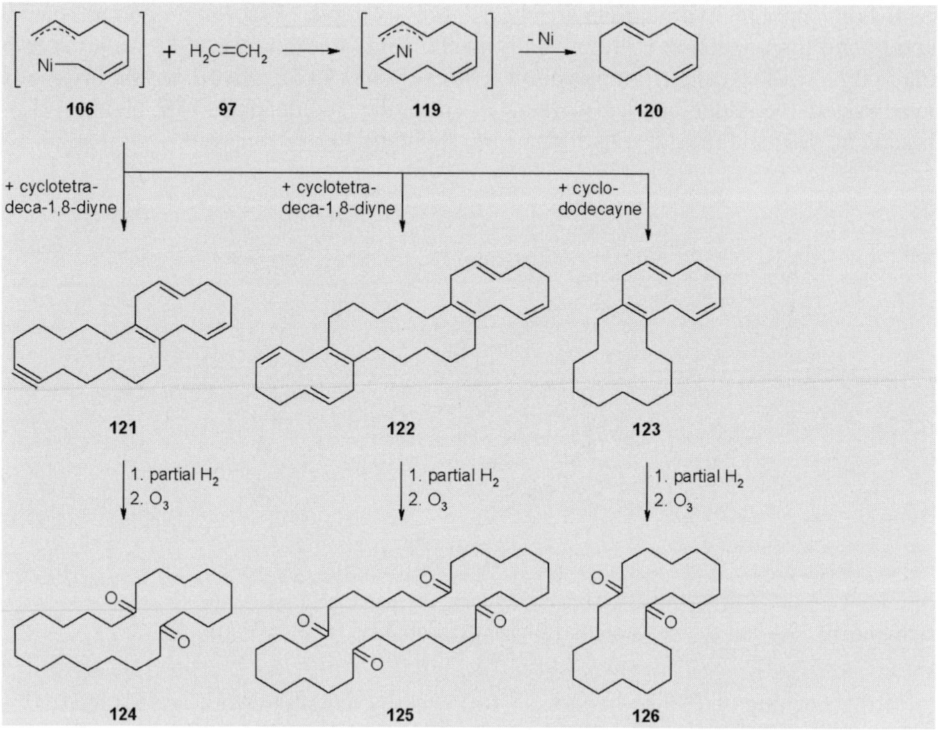

Scheme 20. Cyclo-co-oligomerization of 1,3-butadiene (**99**) with alkenes and acetylenes.

7.4 Distorted olefins: *trans*-cycloalkenes, *anti*-Bredt-hydrocarbons, betweenanenes, and pyramidalized olefins

As mentioned in Chapter 2—and in every introductory organic chemistry textbook—in ethylene in its ground state all six atoms lie in one plane, the bond angles are near 120° and the carbon–carbon distance is 1.34 Å. When the olefinic double bond carries a hydrocarbon substituent at either end, the (*E*)-configurated diasteromer (**128**) is usually thermodynamically more stable than the (*Z*) isomer (**127**, Scheme 21).

If, however, the olefinic carbon atoms are connected by molecular bridges—in the simplest case by polymethylene chains—the situation rapidly becomes more complicated. Whereas planar *cis* double bonds can be incorporated into carbocyclic rings of any size (see **129**) this is no longer so for the *trans* isomer **130**. Although we would not expect the *trans* isomer of, e.g., cyclopropene (**61**) or cyclobutene to exist—not to speak of their isolability—we can envisage that from a certain chain length onwards the *trans* isomer **130** could not only be prepared but actually be the more stable of the two forms.

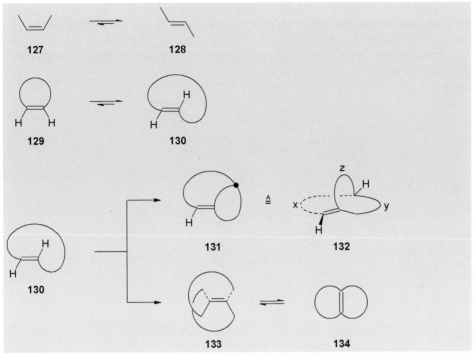

Scheme 21. *trans*-Cycloolefins, *anti*-Bredt-hydrocarbons and betweenanenes.

If we introduce an additional bridge (of any chain length, *i.e.* beginning with zero) into **130** this could have its second bridgehead in any position of the first-formed bridging unit as symbolized by **131** or, alternatively, it could 'bite back' to the double bond as shown by the general structure **133**. Compounds of type **131**, redrawn more realistically in **132**, are called *anti*-Bredt hydrocarbons, after J. Bredt, one of the pioneers in this area, whereas **133** is a so-called betweenanene, its '*trans*' double bond being sandwiched 'between' two (in the simplest case saturated) ring systems. If **133** undergoes *cis–trans* isomerization it would be converted into **134**, which formally consists of two '*cis*' configurated cycloalkene rings, and represents a class of compounds long known in organic chemistry (see, however, below for examples in which the molecular bridges become very small).

For small bridging elements in particular we would expect severe deviations of the usual planar double-bond geometry in **130, 132,** and **133**. For example, in **132** it is apparent that the substituents at the non-brigdehead olefinic carbon atom lie in a different plane from the bonds extending from the bridgehead carbon atom of the double bond. We must therefore ask how much an olefinic double bond can be distorted. In principle the deformations might occur in the π-plane and involve the bond angles and bond distances or there might be out-of-plane distortions, with the substituents moving perpendicularly to the molecular plane. The total deformation space of ethylene (and its derivatives) can be described systematically by group theory;[99] however, we shall be concerned here with two types of non-planar deformations only

(Scheme 22): the twisting of the double bond as represented by **135**, which in its most extreme form would lead to complete orthogonalization and uncoupling of the olefin's p-orbitals and the symmetric out-of-plane bending, and **136**, in which all four substituents point into one hemisphere only and which leads to pyramidalization of the double bond.

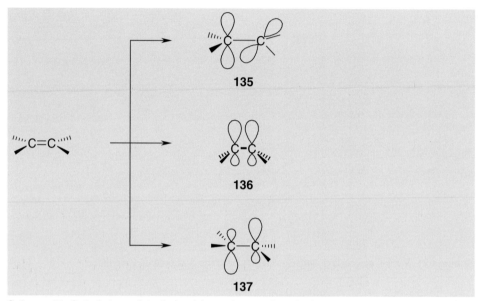

135

136

137

Scheme 22. Out-of-plane distortions of the carbon–carbon double bond.

Other deformations are possible, as illustrated by **137**, which shows the antisymmetric out-of-plane bending of a double bond; combinations of the twisting and the symmetric and antisymmetric bending modes are also conceivable.[99, 100]

In describing distorted olefinic systems I shall follow the systematic order given in Scheme 21, not the historic development[101–103] and hence begin with the *trans* cycloolefins **130**.

The *trans* cycloalkenes are not only of interest because of the large amount of deformation and hence high strain expected for the lower members of the series but also because of their inherent chirality. These hydrocarbons have a plane of chirality like substituted cyclophanes (see Section 12.3) and metallocenes, and that they usually occur as racemates was recognized as early as 1952.[104, 105]

The borderline molecule in the *trans* cycloalkene series is *trans*-cyclooctene (**140**), a hydrocarbon which can be generated under normal laboratory conditions by applying classical methods of olefin chemistry. This was shown first by Ziegler,[106] who subjected the ammonium salt **138** to a Hofmann elimination and obtained a mixture of *cis*- (**139**) and *trans*-cyclooctene (**140**). This work was repeated and confirmed by Cope and co-workers, using the newly discovered amine-oxide elimination reaction that later should bear his name; it was shown that the elimination takes place in very good yield (89 %) and that the product mixture contains 40% of the *cis* isomer and 60% of the *trans* (Scheme 23); the latter could be isolated by extraction with 20% aqueous silver nitrate solution, which dissolved the *trans* form but not the *cis*.[107]

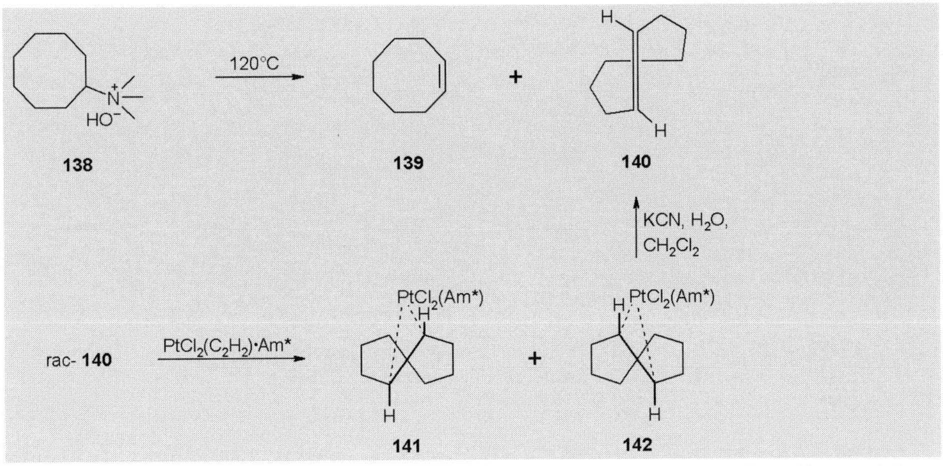

Scheme 23. *trans*-Cyclooctene (**140**), the first *trans* cyclolefin to be prepared, and its resolution.

To resolve *trans*-cyclooctene Cope and co-workers treated *rac*-**140** with a platinum complex prepared from Zeise's salt and (+)-α-methylbenzylamine (Am* in Scheme 23) to obtain the diastereomeric metal complexes **141** and **142**. After separation by recrystallisation, these adducts were decomposed by treatment with potassium cyanide in aqeuos solution. The enantiomer shown in Scheme 23 is (*R*)-configurated.[108] With an activation energy of $\Delta H^{\neq} = 34.7$ kcal mol^{-1}[109] the racemization barrier of **140** is considerably lower than the (*E*)/(*Z*) barrier for 2-butene ($\Delta H^{\neq} = 62.2$ kcal mol^{-1})[110] indicating the strained nature of *trans*-cyclooctene. This strain manifests itself as a twisted double bond; Traetteberg showed by gas-phase electron diffraction measurements that the torsion angle is 136°.[111]

The higher homologs of **140**, *trans*-cyclononene[109, 112] and *trans*-cyclo-decene[109, 113] are also known and have also been resolved. They are, however, easily racemized at low temperatures by rotating the polymethylene-chain loop around the two faces of the π-bond, a motion that also interconverts the enantiomers. For example, the half-life for optically active *trans*-cyclononene is only 6 s at 30 °C.[109, 112] This compares with a full hour for the racemization of **140** at 183.9 °C,[109] indicating that this conformational change is not possible for the smaller *trans* cycloolefins; these, of course, become less and less stable with increasing strain and hence cannot be subjected to the usual resolving conditions. There is, however, abundant chemical and spectroscopic proof of their existence.

trans-Cycloheptene (**145**) was first prepared by a Corey–Winter fragmentation of the *trans*-1,2-cycloheptane thionocarbonate **143**[114] which on treatment with trimethyl phosphite first yielded the carbene intermediate **144** which subsequently lost carbon dioxide (Scheme 24).

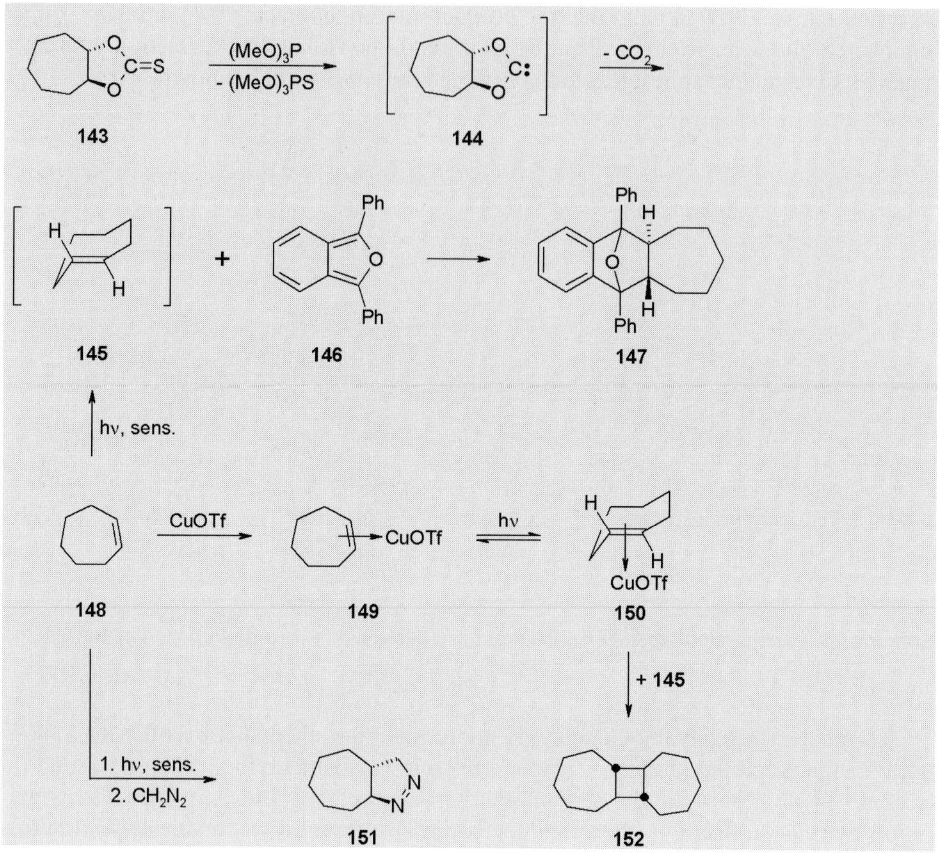

Scheme 24. The preparation of *trans*-cycloheptene (**145**).

Because of its high reactivity, **145** could not be isolated in these first experiments but was trapped with 2,5-diphenyl-3,4-isobenzofuran (**146**) as the *trans*-configurated Diels–Alder adduct **147**. Later it was shown that **145** can be generated directly from its *cis* isomer **148** under various conditions. For example, in the presence of copper(I) triflate a *cis*-metal complex **149** was produced initially; on irradiation this was converted into its *trans* isomer **150**.[115] Not only could this reversible process be studied in great detail by NMR techniques,[116] **150** be isolated,[117] and **145** could be liberated from it by treatment with trimethyl phosphite at low temperature;[118] there is also chemical evidence. Addition of methanol under acidic conditions to photochemically generated **145**[119] led to the expected trapping product as did the addition of a second molecule of **145** to provide the doubly *trans*-fused dimer **152** (although *cis*-fused dimers are thermodynamically more stable).[115] The best way to generate **145** from **148** consists in its irradiation in methanol at −78 °C in the presence of methyl benzoate, a triplet sensitizer.[120–122] By measurement of the kinetic stability of **145** at different (low) temperatures, a lifetime of 47 s was extrapolated for this highly strained compound at room temperature and an Arrhenius activation energy of *ca* 17 kcal mol^{-1} was calculated for the *trans-cis* reisomerization process.[122] *trans*-Cycloheptene generated in this way can also be trapped with diazomethane to provide the 1,3-dipolar addition product **151**,[122] and it can be observed by the modern spectroscopic techniques. Its UV spectrum suggest a substantial twist of the double bond, but the vinyl H,H coupling constant implies an angle of *ca* 180° for the vinyl protons.[123] According to molecular mechanics (PMI) calculations **145** has 25.6° twist and 23.4° pyramidalization angles.[122–124] Nowadays *trans*-cycloheptene is a well understood hydrocarbon and a good example of the combined use of chemical, physical, and computational techniques to study a highly reactive molecule. *cis-trans* Isomerizations of this type are not limited to the parent hydrocarbons but have also been observed for several α,β-unsaturated cycloalkenones.[125–128]

Not surprisingly, several experiments aimed at generating the next lower homolog in the series, *trans*-cyclohexene, were performed analogously to the preparation of **140** and **145**. Thus, a dimer corresponding to **152** was produced when *cis*-cyclohexene was irradiated in the presence of copper(I) triflate,[115] and a trapping experiment in the presence of 1,3-butadiene afforded *trans*-Δ²-octalin in surprisingly high yield (> 50%).[129] The most thorough—and revealing—investigation so far, however, was performed with a cyclohexene derivative, 1-phenyl-cyclohexene (**153**) (Scheme 25).

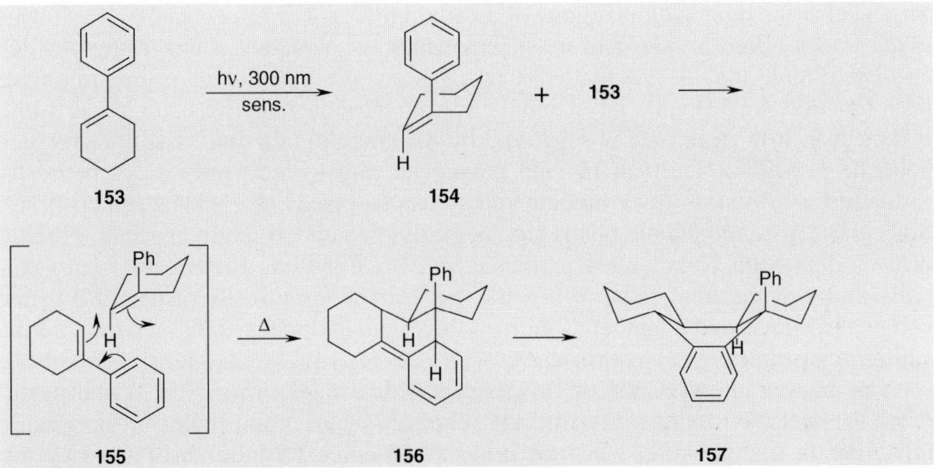

Scheme 25. The generation of a *trans*-cyclohexene, **154**.

When **153** was subjected to laser photolysis at room temperature in methanol at λ = 265 nm a new isomer was formed which, according to its spectroscopic properties, must be **154**. The compound quickly reverted to the starting material, and it has been estimated that the Arrhenius activation energy for this step amounts to a mere 7.5 kcal mol⁻¹. The estimated lifetime of **154** at room temperature is 9 μs.[130]

When, furthermore, **153** was photolyzed in methanol at −75 °C a dimer was formed the structure of which, **157**, was established by X-ray structural analysis.[131, 132] The most likely reaction pathway involves initial isomerization to *trans*-1-phenylcyclohexene (**154**) which is reactive enough to 'use' the π-electrons of the phenyl substituent of **153** for a Diels–Alder addition as shown in **155**. The primary product **156** then recovers its aromatic character by a hydrogen shift yielding the isolated dimer **157**.

trans-Cyclopentenes have not received any serious consideration and no evidence for the formation of *trans*-cyclopentenone was obtained in photochemical studies on cyclic enones.[128]

trans-Configured double bonds have also been introduced into a variety of monocylic ring systems which already contain double bonds, among the isolable hydrocarbons are the remarkable *cis,trans*-1,5-cyclooctadiene, which has also been prepared in an optically active form[133] and 1,3-*trans*-5-cycloheptatriene (*trans*-tropylidene) which has been postulated as a reactive intermediate in the thermal conversion of homobenzvalene to bicyclo[3.2.0]-hepta-2,6-diene.[134]

How does the instability of *trans*-cyclooctenes with a ring size below eight manifest itself in more complex, multi-bridged hydrocarbon systems? As shown in Scheme 21 introduction of an additional bridge into **130** which connects one double bond carbon atom with another carbon atom of the original ring—making it a second bridgehead—generates the bridgehead olefins **131**. As a results of several decades of intensive work in the terpene field, Bredt postulated in the early 20th century that 'based on our conception of the position of atoms in space, in the systems of the camphene and pinene series, as well as in similarly constituted compounds, a C=C bond cannot be placed at the branching positions of a carbon bridge (the bridgeheads).[135,136] This is the famous Bredt's rule, and molecules which are violating it are fittingly called *anti*-Bredt molecules. It is a matter of fact that countless violations of this rule have been reported since Bredt's days, and the rule as formulated above is no longer valid. If, however, it is rephrased as suggested by Szeimies,[137] the rule is still useful as a heuristic principle: 'In small bi- and polycyclic ring systems only a geometrically distorted C=C bond of lower binding energy can be placed at a bridgehead position'. Since these distorted double bonds are not easily prepared, it might hence be difficult or even impossible to prepare a particular *anti*-Bredt system. Bredt was at least partially aware of the limitations of his rule, because he specifically restricted it to the certain terpenoids and compounds derived therefrom and realized that it would not be valid for molecules with larger bridges.[138–140]

The modern interpretation of Bredt's rule is due to Wiseman,[141, 142] who recognized the decisive role of a *trans*-cycloalkene subsystem in molecules of the general type **132**. In such a hydrocarbon the bridges x, y and z form the three rings xy, xz, and yz. The C=C bond is *exo*- (better: *semi*-, see Chapter 2) cyclic with respect to ring yz and *endo*-cyclic with respect to rings xy and xz. In the latter ring this bond is *cis* configured, whereas in ring xy it must assume a *trans* configuration. According to Wiseman the strain of the bridgehead olefin under discussion is closely related to the strain of the corresponding *trans*-cycloolefin ring which makes up part of its structure. From what we have learned above about the stability and reactivity of the (monocyclic) *trans* olefins **130**, we can hence conclude that *anti*-Bredt hydrocarbons can be isolated if the *trans* olefin ring size exceeds or is equal to eight, and that they will be increasingly difficult to make and to handle once this critical ring size is passed. The Wiseman model has been confirmed numerous times, and I can present only a small selection of successful syntheses of *anti*-Bredt olefins here.[143] The different synthetic strategies which have been employed successfully to reach this goal are particularly well demonstrated by bicyclo[3.3.1]-1(2)-nonene (**159**) for which a sizable number of syntheses has been reported; some of these are summarized in Scheme 26.

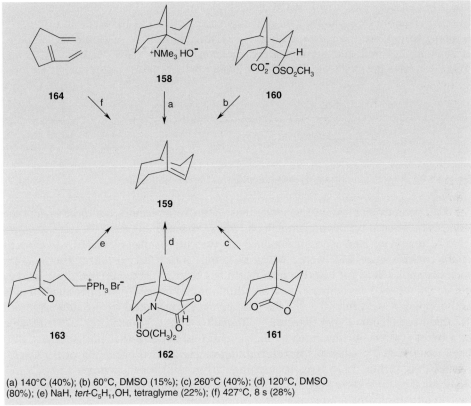

(a) 140°C (40%); (b) 60°C, DMSO (15%); (c) 260°C (40%); (d) 120°C, DMSO (80%); (e) NaH, *tert*-C$_5$H$_{11}$OH, tetraglyme (22%); (f) 427°C, 8 s (28%)

Scheme 26. Various routes to the *anti*-Bredt hydrocarbon bicyclo[3.3.1]-1(2)-nonene (**159**).

Many of these syntheses start from precursors which already possess the complete σ-framework. Thus in the first syntheses of **159**, published simultaneously by Wiseman and Marshall and their co-workers in 1967, either the Hofmann base **158** was subjected to β-elimination[141, 142] or the salt **160** to fragmentation.[144] The latter process also yielded the lactone **161** which could be decarboxylated separately by gas phase pyrolysis. 1,2-Elimination was also successful for 1-chloro-bicyclo[3.3.1]-nonane[145] and fragmentation of the sulfoximine **162**.[146] The bicyclic framework of **159** could, however, also be constructed by an intramolecular Wittig reaction of the cyclohexanone derivate **163**,[147, 148] and even, particularly elegantly, from an acyclic precursor, the triene **164**, by an intramolecular Diels–Alder addition.[149] Today **159** is not only a thoroughly investigated hydrocarbon,[137, 143] it has even been prepared in optically active form and its absolute configuration has been assigned;[150] several of its isomers have been prepared,[137, 147] and X-ray structural investigations have been reported for some of its derivatives.[143] Although these compounds are more reactive than unstrained cycloolefins, working with them poses no special problems.

As predicted from the Wiseman model there is a pronounced reactivity increase when the 'core cycloalkene' is made smaller; *anti*-Bredt hydrocarbons containing a *trans*-cycloheptene ring have been prepared—again in significant numbers[137, 140]— but they can only rarely be isolated as such and more often have been identified by trapping experiments, usually with diphenylisobenzofuran (**146**) or furan or by self-trapping to provide the corresponding [2+2]cycloadducts. To illustrate the range of compounds covered, some typical representatives, **165**,[151] **166**,[152] **167**,[153–156] and **168**,[7–157] have been collected in Scheme 27.

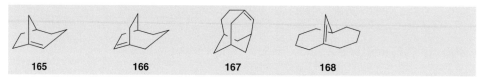

| 165 | 166 | 167 | 168 |

Scheme 27. A selection of *trans*-cycloheptene-derived *anti*-Bredt olefins.

The methods employed to generate these reactive molecules are more or less the same as those used for the preparation of **159** (*cf.* Scheme 26).

The reactivity increase is maintained for the still smaller bicyclic hydrocarbons (*trans*-cyclohexenes, as it were), which have only a fleeting existence. The most famous examples here are 1-norbornene which has been generated by treating 1-iodo-2-chloro (or bromo) norbornane with alkyllithium compounds or with sodium amide and trapping it with furan[158, 159] and 1-adamantene, produced by dehalogenation of 1,2-diiodo-adamantane and detected by trapping with 1,3-butadiene.[160–162] Double *anti*-Bredt systems, as illustrated by the general formula **169** (Scheme 28), have also been described;[140] either [2+4]cycloaddition or partial hydrogenation of paracyclophanes (see Section 12.3) have been particularly useful for their preparation.[163, 164] Some of the 'three-dimensional olefins' discussed later (Chapter 13) formally also belong into this hydrocarbon category.

Returning to our bridging scheme above (Scheme 21), two other structure patterns deserve further comment. In **131** the newly introduced bridge may not contain any carbon atom, *i.e.* it may be a zero bridge giving rise to hydrocarbons of the general structure **170** (Scheme 28); many representatives of this group of bridgehead olefins have been generated or proposed as reaction intermediates.[137]

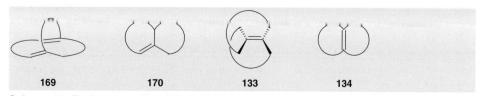

| 169 | 170 | 133 | 134 |

Scheme 28. Further types of *anti*-Bredt olefins.

Finally, there is the group of the fully-bridged olefins **133,** the betweenanenes, and **134,** a pair of rather unusual diasteromers.

A molecular system with the general structure **133** had already been mentioned by Cahn, Ingold, and Prelog in the extension of their stereochemical nomenclature system to planar chiral molecules.[165] It took more than a decade, however, before the first betweenanenes were actually synthesized. Two quite general approaches, which were developed almost simultaneously, have turned out to be particularly useful for the preparation of these hydrocarbons.

The rather short route reported by Nakazaki[166–169] begins with either cyclododecyne (**171a**) or cyclodecyne (**171b**), to which two equivalents of 1,3-butadiene were added exploiting the methodology discussed above (Section 7.3). The resulting bicyclic olefins **172**, representatives of the fully-bridged bicyclics **134**, were hydrogenated with the highly substituted central double bond surviving the reduction process (Scheme 29).

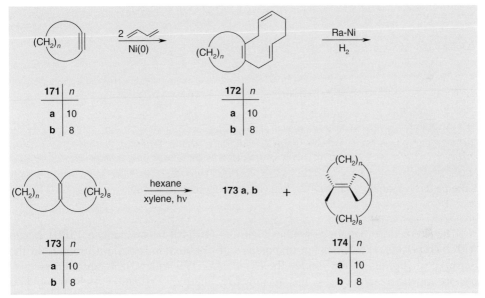

Scheme 29. A three-step route to [10][8]- (**174a**) and [8][8]-betweenanene (**174b**).

Xylene-sensitized photoisomerization with a low-pressure mercury lamp in hexane subsequently led to the desired hydrocarbons **174** which were produced as a 2:1 mixture with the starting materials **173**. In other words, the photoequilibration which was so successful for the synthesis of several *trans*-cycloolefins (see above) works well in this case also. To prepare analytically pure betweenanenes **174** a 'chemical' separation was applied: Whereas the *cis* isomers **173** reacted with dichlorocarbene, no addition took place to the 'bridge-protected'—see structure **133**—internal double bond of **174**.

The second route to betweenanenes, that of Marshall and co–workers, who also coined the name,[170] was more elaborate in its original form and relied on several classical steps including chain-elongation of an appropriately configured cyclododecane derivative and ultimate ring closure by an acyloin condensation;[171–173] it has, however, been replaced by a new approach[174] in which modern organic reactions are used (Scheme 30) and which offers the additional advantage of enabling the synthesis of optically active betweenanenes of far higher enantiopurity than that previously reported for, *e.g.*, the [8][8]-homolog.[168]

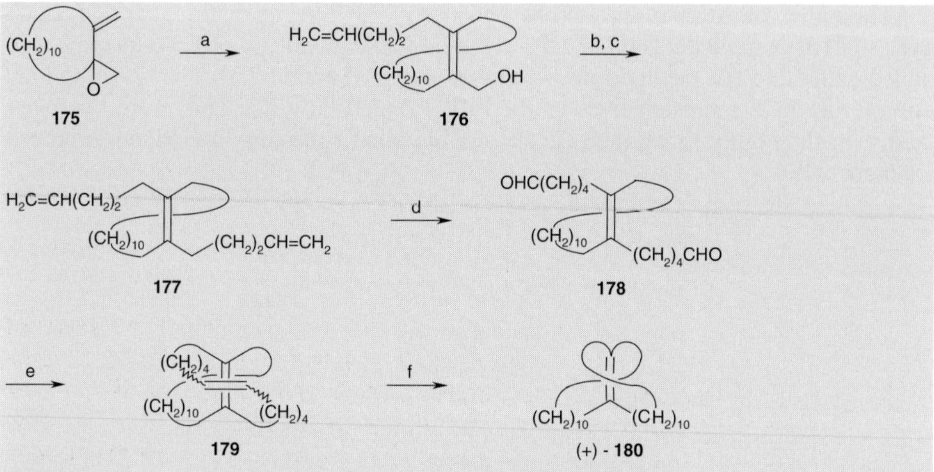

(a) H₂C=CH(CH₂)₂MgBr, CuI, Me₂S, THF (98%); (b) Sharpless oxidation (78%);
(c) ClPO(OEt)₂, pyridine, then H₂C=CH(CH₂)₂MgBr, CuI (90%); (d) (siam)₂BH, THF; H₂O₂,
NaOH; (e) CrO₃, pyridine, CH₂Cl₂ (comb. yield 82%); (f) Mc Murry coupling (*trans : cis* =
4 : 1), then cat. H₂

Scheme 30. Preparation of optically active [10][10]-betweenanene (**180**).

The stereodifferentiating synthesis of (+)-[10][10]-betweenanene (**180**) began with the epoxide **175** which on treatment with 3-butenylmagnesium bromide in the presence of cuprous iodide yielded the (*E*) alcohol **176** with > 98% diastereoselectivity. Its kinetic resolution by the asymmetric Sharpless-oxidation yielded (+)-**176** (with 95% ee) which was subsequently transformed *via* the triene **177** and the derived bis primary diol into the corresponding dextrorotatory bis aldehyde **178**. When this was ring-closed by McMurry coupling a 4:1 mixture of the (*E*) and (*Z*) dienes **179** resulted. Catalytic hydrogenation—again only the less shielded 'outer' double bond reacted—finally led to (+)-[10][10]-betweenanene (**180**) which was shown to be (*R*)-configurated by chemical correlation and CD measurements.

Whereas the limits of the betweenanene concept, *i.e.* the lengths the molecular bridges in these hydrocarbons may be shortened, have not yet been explored, this is not so for the, at first sight, much simpler and more familiar hydrocarbons **134** (Scheme 31). Hydrocarbons such as bicyclo[4.4.0]-1(6)-decene ('octalin')[175] have been known for a very long time and the olefin **181** ('7,8-dehydropentalane'), with its two 'merged' five-membered rings could be prepared by more or less conventional methods.[176] Continuing our shortening process, however, we might expect some very unusually structured and highly strained double bond systems indeed.

For example, in **182** will the bond angles be reduced from the usual 120° for the

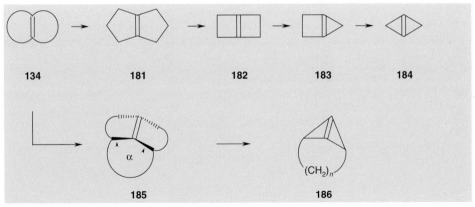

Scheme 31. *En route* to bicyclo[1.1.0]-1(3)-butene (**184**).

sp^2-hybridized carbon atom to only 90°? Or to a mere 60° in the bicyclic olefin **184**, the smallest member of this series, which can also be regarded as a homo-cyclopropyne (see Section 8.2), and which we formally could reach *via* **183**? On the other hand, if *en route* to these smaller homologs of **134** not only in-plane-deformation occurs (see above), but also out-of-plane bending, *e.g.* pyramidalization, should then not an additional bridge, as symbolized by the general structure **185**, support the pyramidalization and at the same time protect the presumably extremely reactive deformed double bond from outside attack from at least one of its faces? The exploration of this latter idea has led to some remarkable, highly distorted olefins among them the bridged hydrocarbons **186**, which are derivatives of the ultimate **184**.

After several unsuccessful attempts[177, 178] bicyclo[2.2.0]-1(4)-hexene (**182**) was finally generated by gas phase decomposition of the tosylhydrazone salt **187**[179] by Wiberg and co-workers (Scheme 32).

In the original work the successful preparation of **182** was inferred from the [2+4]adduct **188** which it yields in the presence of cyclopentadiene as a trapping reagent and by the formation of the dimer **190**, the hydrocarbon evidently being produced by initial ring opening of **182** to 1,2-bismethylenecyclobutane (**189**)—which is also isolated as a side product—and Diels–Alder addition of the latter to its closed isomer **182**. Later experiments showed that the short-lived **182** (half-life < 10 s at –23 °C) can be isolated as the bis(triphenylphosphine) platinum complex[180] and, furthermore, that a most convenient route to this bicyclic olefin consists in the controlled electroreduction of 1-bromo-4-chloro-bicyclo[2.2.0]hexane in dimethyl formamide at –20 °C.[181,182] According to its Raman and infrared spectra, recorded at –190 °C, **182** is a planar hydrocarbon with D_{2h} symmetry, *i.e.* there is no pyramidalization of the double bond under these conditions.[183] Because the distortion has thus occurred exclusively in the π-plane and not significantly influenced the overlap between the p-orbitals, the high reactivity of **182** is primarily a consequence of its highly strained σ-framework. When **190** was heated, ring-opening of its cyclobutene ring took place and diene **191** was formed.[179] Note that the hydrocarbons **188**, **190**, and **191** are also propellanes; a route to this interesting class of hydrocarbons (see Chapter 6) is

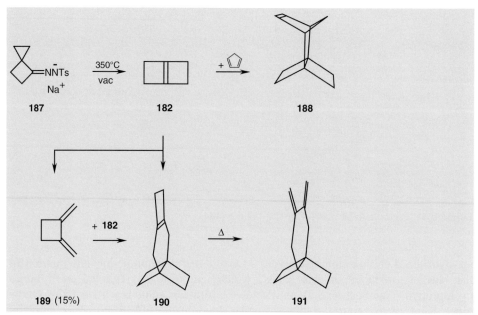

Scheme 32. The first preparation of bicyclo[2.2.0]-1(4)-hexene (**182**) by Wiberg and co-workers.

thus also possible *via* strained polycyclic olefins. We shall return to the high reactivity of **182** in connection with its most surprising dimerization behavior[184] at the end of this section.

The lower homologs of **182**, the hydrocarbons **183** and **184**, apparently have not been prepared. They have, however, been the subject of extensive theoretical calculations. According to *ab initio* calculation on the STO-3G and the 4-31G levels there is pyramidalization in either of these bridgehead olefins—129.3° in **183**[185] and 128.3° and 132.5° (STO-3G and 4-31G level, respectively) for **184**.[185, 186] On the other hand, derivatives of **183** and **184** could be generated by applying the bridging concept illustrated by **185** (see below).

That *syn* pyramidalization can be introduced into monoolefins of type **134** with comparatively large bridges was demonstrated by Borden[187] with the hydrocarbon **195**, a 1,2-benzobridged version of **181** (Scheme 33). To generate this distorted olefin the readily available diketone **192** was first reduced to the vicinal diol **193** with zinc amalgam. Because the thiocarbonate of **193** could not be induced to undergo a Corey–Winter fragmentation it was converted to the dimethylamino dioxolane **194** by treatment with the dimethyl acetal of dimethyl formamide. Pyrolysis of **194** under reflux in tetraglyme in the presence of one equivalent of acetic acid caused fragmentation and generation of **195**, as judged from its trapping product with diphenylisobenzofuran (**146**) and its dimerization (self-trapping) to the cyclobutane derivative **196**, the structure of which was determined by X-ray structural analysis.

In later experiments the parent hydrocarbon of **195**, *i.e.* that with an ethano rather than a 1,2-benzobridge could be obtained by flash vacuum pyrolysis at 450 °C of a precursor molecule in which the future double bond was masked by a lactone ring, the

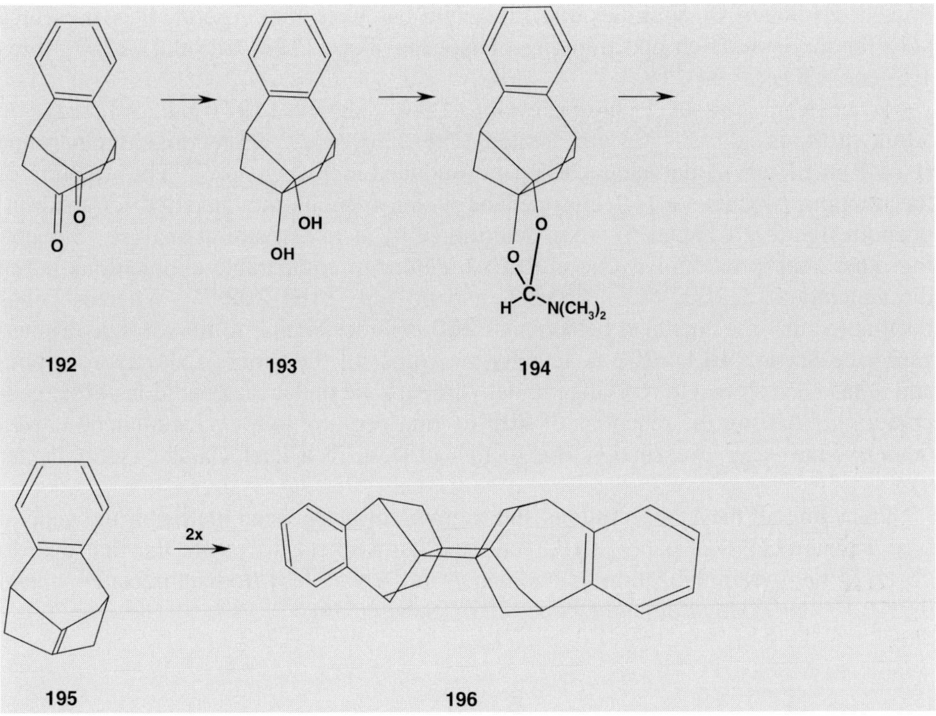

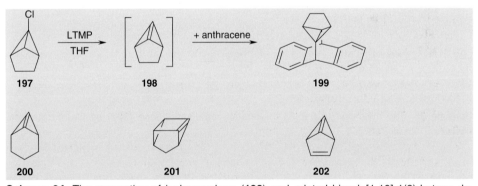

Scheme 33. The preparation of the *syn*-pyramidalized hydrocarbon **195** by Borden and co–workers.

Scheme 34. The generation of isobenzvalene (**198**) and related bicyclo[1.10]-1(3)-butene derivatives by Szeimies and co-workers.

experiment also enabling matrix isolation of this hydrocarbon.[188, 189] The out-of-plane angle of this bridgehead olefin (see α in **185**) was calculated to be 119° by Schleyer and co-workers,[190] *i.e.* the olefinic carbon atoms in this molecule are nearly sp^3-hybridized, making the rapid [2+2]dimerization understandable.

As far as derivatives of **182** and **183** are concerned a remarkable general approach developed by Szeimies and his group has been most useful. It is presented here for the generation and trapping of isobenzvalene (**198**, tricyclo[3.1.0.0^{2,6}]hex-1(6)-ene) (Scheme 34).[191]

When a mixture of 1-chlorotricyclo[3.1.0.0^{2,6}]hexane (**197**) and anthracene in tetrahydrofuran at –20 °C was treated with lithium 2,2,6,6-tetramethylpiperidide (LTMP) the Diels–Alder adduct **199** was produced in 36% yield.[191] The structure of the trapping product—a [4.1.1]propellane which exhibits the 'inverted tetrahedron' phenomenon (see Chapter 6)—was determined by X-ray structural analysis. Subjecting other appropriate polycyclic precursor halides to comparable eliminations led to the generation of **200**,[192–196] **201**,[196] and possibly even **202**.[197] Whereas it has been noted that the chemical behavior of **200** is rather similar to that of dehydrobenzene (see Section 16.1), **202** is actually an isomer of the former C$_6$H$_4$ hydrocarbon, and it has been shown to rearrange to dehydrobenzene under the conditions of its generation.[197] Among the reactions of **200** its ring-opening to a cyclocumulene is particularly intriguing; we shall return to this process in a later chapter (see Chapter 9).[198]

Preceding all these observations and trapping of bridgehead olefins of the general type **134** are classical papers by Greene and co–workers who were the first ever to observe the pyramidalization phenomenon on 9,9'-didehydrodianthracene[199] and 9,9',10,10'-tetraedehydrodianthracene (**207**) (Scheme 35).[200]

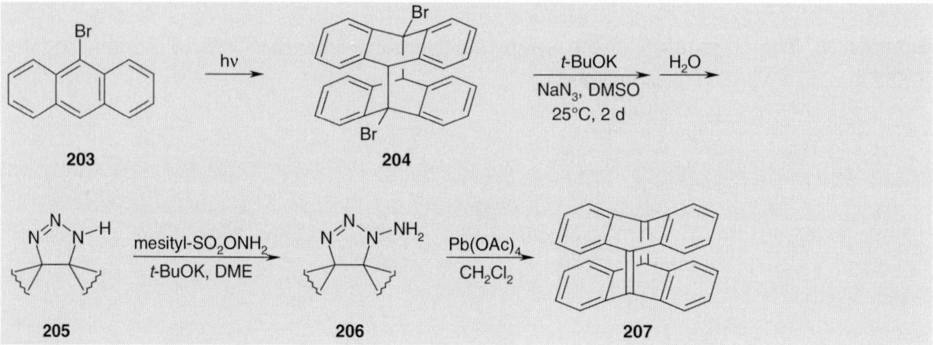

Scheme 35. The synthesis of 9,9',10,10'-tetradehydrodianthracene (**207**) by Greene and co-workers.

The synthesis of this bisolefin begins with the photodimerization of 9-bromoanthracene (**203**) to 9,10'-dibromodianthracene (**204**), which on treatment with potassium *tert*-butoxide in the presence of sodium azide provided the bistriazoline **205** after work-up. In this reaction the bridgehead double bonds have already been produced—most likely in a two-step manner—but do not survive the elimination conditions. When **205** was treated with hydroxylamine *O*-mesitylene sulfonate the bis-*N*-aminotriazoline derivative **206** was formed. Lead tetraacetate oxidation of this intermediate finally furnished the desired hydrocarbon **207**. According to an X-ray structural analysis its olefinic bridgeheads lie 0.5 Å outside of the plane defined by the four attached carbon atoms; this corresponds to a pyramidalization angle of

19.7°.[201] The remarkable use of **207** for the construction of picotubes will be addressed later (Section 14.3).

When the benzene rings of **207** are replaced by ethano bridges the hydrocarbon **211** (tricyclo[4.2.2.2^{2,5}]dodeca-1,5-diene) results; this is one of the simplest conceivable molecules with two parallel double bonds in exactly defined orientation. This hydrocarbon has been called the simplest member of the series of the so-called superphanes (see Section 12.3). This is, however, a misnomer, because phanes by definition must either have aromatic or antiaromatic subunits.

Diene **211** is a dimer of the highly reactive bicyclo[2.2.0]hex-1(4)-ene (**182**, see above, Scheme 32) from which it has been obtained by leaving **182** to stand in dimethyl formamide or pentane at room temperature[202] (Scheme 36).

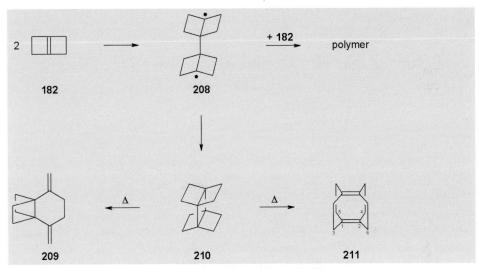

Scheme 36. The dimerization of bicyclo[2.2.0]hex-1(4)-ene (**182**).

At relatively high concentrations (> 0.01 M) a polymer forms from **182**, while in dilute solutions (0.002 M) the main product is **211**, which is accompanied by small amounts of the ring-opened hydrocarbon **209**. The branching points of these transformations are the diradical **208** and the pentacyclic propellane **210**, which consists exclusively of merged cyclobutane rings, and which so far has escaped isolation. The structure of **211** was determined by the usual spectroscopic methods and by X-ray crystallography.[203] The distance between the parallel double bonds amounts to 2.40 Å only—much lower than the distance between the layers of graphite (3.4 Å) or the benzene rings of [2.2]paracyclophane (ca 3 Å, see Section 12.3). The olefinic carbon atoms C1 and C2 (see formula **211**) lie 40 pm out of the plane defined by C3, C4, C5, and C6 and the dihedral angles between the substituent and each of the planes defined by C1, C3, C5 and C2, C4, and C6 are 27.3°. The dihedral angle between the C1, C2, C3, C6 and the C1, C2, C4, C5 planes is 36.5°—for unstrained double bonds these deformation angles would be zero, of course.

7.5 Tetrakis-*tert*-butylethene - an exercise in preparative futility

Many hydrocarbon syntheses have failed—some only in the last step (see, *e.g.*, [6]-prismane, Section 4.3). Although it is always useful to analyze ones mistakes, in preparative chemistry all too often such rationalization has something futile about it, because the reasons a particular approach did not work as anticipated are too manifold. Poor analysis of the problem at hand can certainly be the decisive reason for failure, but so also can be poor craftsmanship of the experimenter, unavailable methodology at a certain point in time *etc.*—all these can make insurmountable obstacles to reach a certain target molecule. A molecule illustrating this situation particularly well is tetrakis-*tert*-butylethene (**219**), which has so far thwarted all attempts to conquer it (Scheme 37).

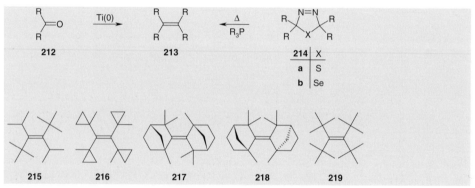

Scheme 37. Successful methods to prepare fully substituted alkenes.

Two routes stand out for the preparation of fully substituted alkenes **213**—McMurry coupling,[204] in which a ketone **212** is reductively dimerized by treating it with zero valent titanium, and the Barton–Kellog fragmentation in which either Δ^3-1,3,4-thiodiazolines **214a** or their selenium analogs **214b** are first thermally decomposed to the corresponding episulfides or -selenides which on treatment with, *e.g.*, triphenylphosphine are dechalcogenated and provide the olefin **213**. The substrates **214** are usually easily prepared (see below), yields are good, and the whole fragmentation process can also be carried out in one step.[205–206] Although these methods have been employed successfully for the preparation of many olefins carrying voluminous groups—among them tetraisopropylethene,[207, 208] 1,2-di-*tert*-butyl-1,2-diethylethene,[209] 1,2-di-*tert*-butyl-1,2-diisopropylethene (**215**),[210] tetrakis-(1-methylcyclopropyl)ethene (**216**),[211] and *syn*- or (*E*)- (**217**) and *anti*- or (*Z*)-2,2'-bisfenchylidene (**218**),[212] these routes failed for the tetra-*tert*-butylated hydrocarbon **219**.

Neither McMurry coupling of ketone **220**[213] nor the debromination of **221** with magnesium[214] or the low-temperature photodecomposition of the diazo compound

222[215] nor, finally the attempt to combine **222** with **223** (or the corresponding selone)[216]—a reaction which has been used successfully numerous times[205, 206, 217]—led to the olefin **219** (Scheme 38).

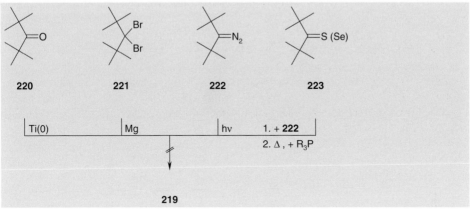

Scheme 38. Failed routes to **219**—*tert*-butylated starting materials and late generation of the double bond.

In every case the starting materials find a way to avoid coupling to **219**—by giving reduced products (from **220** and **221**) or side-products formed from the intermediate carbene by C,H-insertion and rearrangement (**222**), or by not reacting with each other at all (**222**, **223**).

In all these attempts the voluminous substituents of the target molecule **219** are already present in the substrates and the formation of the central double bond is deferred to the last step of the synthesis. As the results show, the intramolecular steric repulsion between the four *tert*-butyl groups cannot be overcome. One might, however, also proceed exactly the other way round by trying to form the double bond first and then generate the *tert*-butyl substituents late in the synthesis from some appropriate precursor group. When this approach is followed the large groups are 'smuggled' into the evolving hydrocarbon structure and it is reasonable to attempt this process by making the substituent as small as possible initially without, of course, making too many transformation steps necessary at a later stage. This 'tied-back' approach is particularly well illustrated by the triene **227**, which is readily prepared[218, 219] from the building blocks **224** and **225** *via* **226** by the Barton–Kellog process (Scheme 39).

Although the endocyclic double bonds could be oxidized[217, 219]—for example by ozonization under carefully controlled conditions as shown in the Scheme—the thus generated formyl substituents could not be transformed further in the desired direction. Their lithium aluminum hydride reduction provided the bis hemiacetal **231**, not the expected tetra alcohol **229**. Clearly, the reaction begins as expected, but as soon as one aldehyde group has been reduced to the primary alcohol function this is forced on to the neighboring, still unreduced substituent to form a hemiacetal. Very likely strain release contributes to the driving force of this process. In a remarkable decarbonylation reaction **228** is transformed into the oxaadamantanoid structure **230** on

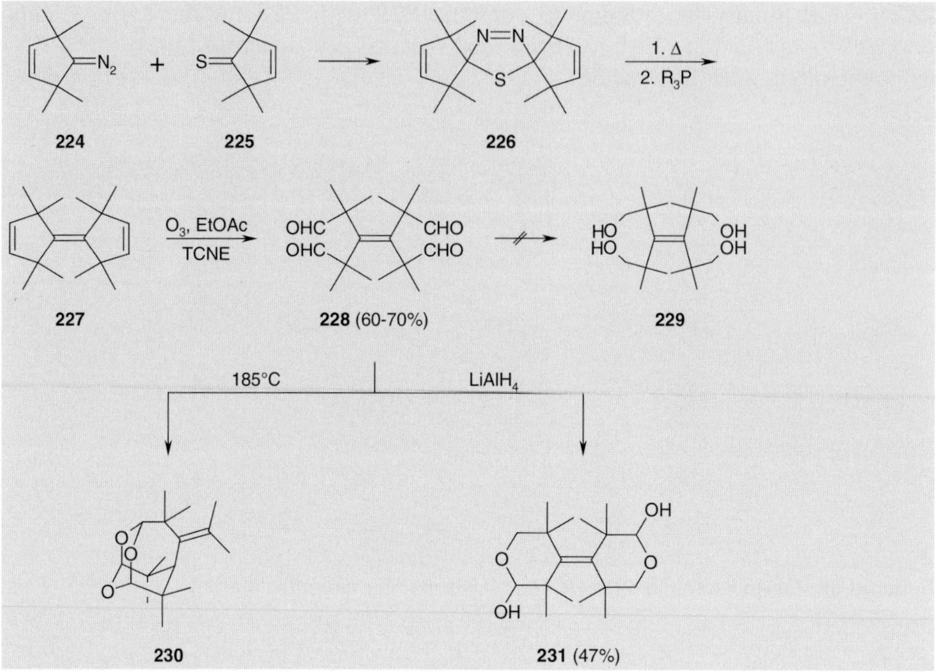

Scheme 39. Failed routes to **219**: The 'tied-back' approach—late generation of the *tert*-butyl groups.

heating.[220] The aldehyde **228** is already strongly distorted—with a dihedral angle of 28.6° it shows the strongest twisting of all purely aliphatic crowded alkenes;[217, 221] its double bond is extended to a length of 1.36 Å.

Many other 'tied-back' approaches have been attempted, all have failed,[222–224] including an attempt to hydrogenate the imaginatively tied-back olefin **216** which could be prepared in good yield from the thiadiazoline **232**. The cyclopropylmethyl group, known to be sterically less demanding than a *tert*-butyl-substituent (of which it is a 'short-circuited' version), offers its strain as an additional bonus. Even so, the saturation of this octadehydro derivative to its parent **219** could not be achieved, as had to be accepted by Lüttke and co-workers[211] (Scheme 40).

The acetyl function is another proto-*tert*-butyl group, and it has been shown that it can be alkylated to the latter by reaction with dimethyl titanium dichloride.[225] Again, when applied to the tetraketone **233**, this often so successful reductive methylation fails, providing the bicyclic ether **234** instead. And obviously by a process similar to the formation of **230** from **228**—the reaction begins as expected (introduction of the first methyl groups) but then breaks out in another direction when further strain increase would become unbearable.[226] Finally—and as a tribute to organic chemists' imagination and tenacity—a completely different route to conquer this *Eigennordwand* of hydrocarbon chemistry was started by Krebs and co–workers.[219] *tert*-Butyllithium was first added to the angle-strained thiacycloheptyne **235** (see Section 8.2) to provide the vinyl lithium derivative **236**. On hydrolysis, this was converted to **237**, from which hydrogenation over Raney nickel liberated the hidden *tert*-butyl

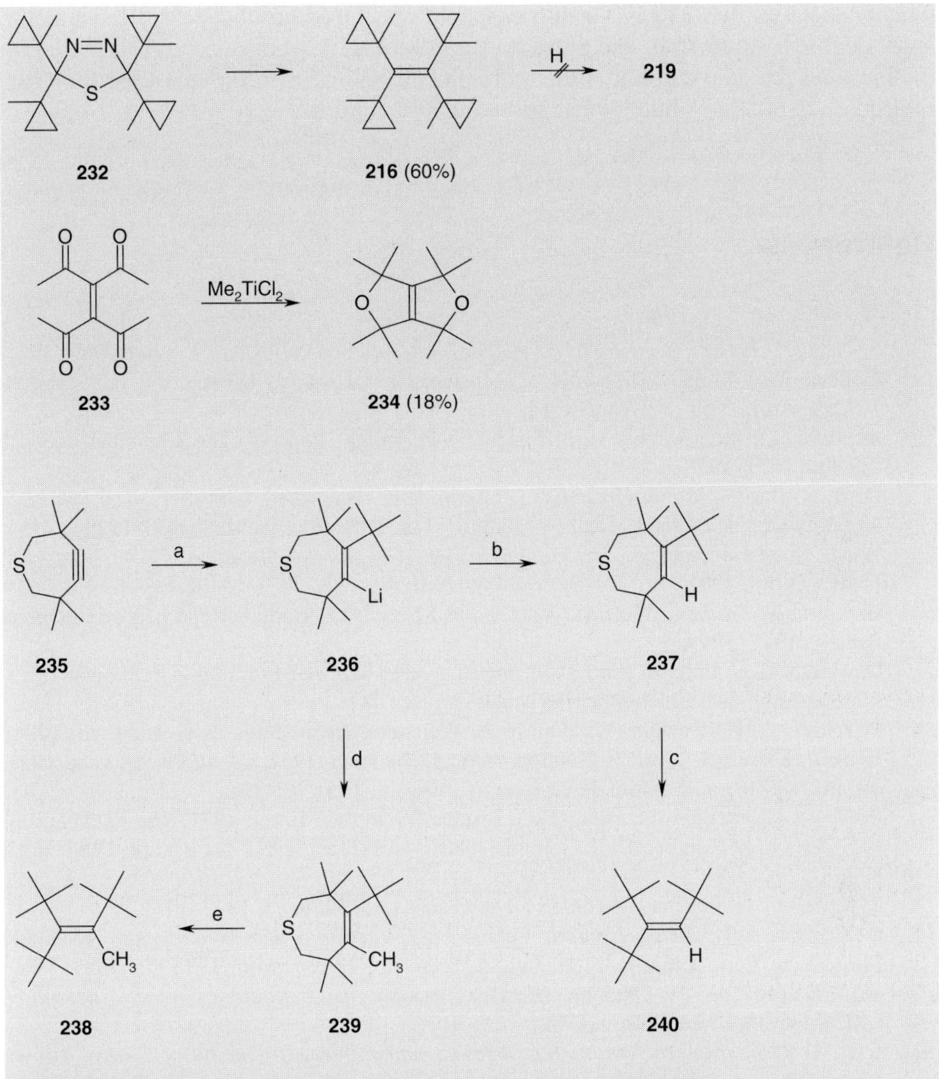

(a) *t*-BuLi (70%); (b) H$_2$O; (c) H$_2$, Ra-Ni (48%); (d) CH$_3$I (30%); (e) H$_2$, Ra-Ni (74%)

Scheme 40. ...yet more failures to prepare tetrakis-*tert*-butylethene (**219**).

substituents and yielded tris-*tert*-butylethene (**240**) in acceptable yield. Although **236** could be trapped also by quenching with methyl iodide, and the resulting **239** could be reduced to the per-substituted olefin **238**, reaction of **236** with a *tert*-butyl-group-delivering electrophile has not yet been accomplished.

In the face of all these failures and disappointments, the question must be asked why the different explorers of this clearly very hostile molecular environment just do not give up and apply their expertise to more hopeful target molecules. One answer is given by yet another theoretical evaluation of the tetrakis-*tert*-butylethene (**219**) problem by *ab initio* and density functional calculations at the BLYP/DZd level which concludes that **219** is a stable molecule with an S_0 ground state, D_2 symmetry, a 45° double bond torsion, and a strain energy near 93 kcal mol^{-1}.[227] With these predictions and the conviction that the requirements can be met, the quest for the 'grail molecule' tetrakis-*tert*-butylethene must and will continue!

References

1. S. Patai (*Ed.*), *The Chemistry of the Alkenes*, Interscience Publishers, London, **1964**.
2. Houben-Weyl-Müller, *Methoden der Organischen Chemie*, Thieme Verlag, Stuttgart, **1972**, 4. Aufl., Vols. V/1b and V/1d.
3. *Aliphatic Chemistry*, Vols. 1–5, Specialist Periodical Reports, The Chemical Society, London, **1971–1977**.
4. D. H. R. Barton, W. D. Ollis (*Eds.*), *Comprehensive Organic Chemistry, The Synthesis and Reactions of Organic Compounds*, Vols. 1–6, Pergamon Press, Oxford, **1979**.
5. B. M. Trost, I. Fleming (*Eds.*), *Comprehensive Organic Synthesis*, Vols. 1–9, Pergamon Press, Oxford, **1991**.
6. *General and Synthetic Methods*, Vols. 1–16, Specialist Periodical Reports, The Chemical Society, **1977–1994**.
7. G. Mehta, H. S. Prakash Rao, *The Chemistry of Dienes and Polyenes* Z.Rappoport (*Ed.*), J. Wiley and Sons, Chichester, **1997**, Vol. 1, 359–480.
8. P. Karrer, A. Helfenstein, W. Wehrli, A. Wettstein, *Helv. Chim. Acta*, **1930**, *13*, 1084–1099; P. Karrer, R. Morf, K. Schöpp, *Helv. Chim. Acta*, **1931**, *14*, 1036–1040 and **1931**, *14*, 1431–1436; cf. R. Kuhn, E. Lederer, *Naturwiss.*, **1931**, *19*, 306.
9. Reviews: *Carotenoids* O. Isler, (*Ed.*), Birkhäuser Verlag, Basel, **1971**; *The Retinoids* M. B. Sporn, A. B. Roberts, D. S. Goodman (*Eds.*), Acadamic Press, New York, **1984**, Vols. I and II.
10. P. Karrer, C. H. Eugster, E. Tobler, *Helv. Chim. Acta*, **1950**, *33*, 1349–1352.
11. L. Zechmeister, *cis-trans-Isomeric Carotenoids, Vitamin A and Arylpolyenes*, Academic Press, New York, **1962**; R. S. H. Liu, A. E. Asato, *Tetrahedron*, **1984**, *40*, 1931–1969; M. I. Dawson, W. H. Okamura (*Eds.*), *Chemistry and Biology of Synthetic Retinoids*, CRC Press, Inc., Boca Raton, **1990**.
12. I. M. Tereshin, *Polyene Antibiotics, Present and Future*, University of Tokyo, Tokyo, Japan, **1976**.
13. S. Sakuda, U. Guce-Bigol, M. Itoh, T. Nishimura, Y. Yamada, *Tetrahedron Lett.*, **1995**, *36*, 2777–2780.
14. J. M. Lehn, *Angew. Chem.*, **1990**, *101*, 1347–1362; *Angew. Chem. Int. Ed. Engl.*, **1990**, *29*, 1304–1319.
15. A. J. Heeger, in *Handbook of Conducting Polymers* T. A. Skotheim, (*Ed.*), Marcel Dekker, New York, N. Y., **1986**, Vol. 2, 729–756.
16. In other words, making some of the double bonds of the polyene chain part of an aromatic subunit–a well-known stabilizing effect in polyene chemistry. Formally this exclusion is also justified since the resulting hydrocarbons belong to the group of cross-conjugated systems to be discussed in Chapter 11.
17. J. Benet-Buchholz, R. Boese, Th. Haumann, M. Traetteberg, in *The Chemistry of Dienes and Polyenes* Z. Rappoport (*Ed.*), John Wiley and Sons, Chichester, **1997**, Vol. 1, 25–65.
18. R. Boese, H. Hopf, unpublished results.

19. Of all octatetraene isomers this is by far the most thoroughly investigated one, for a review see: B. E. Kohler, *Chem. Rev.*, **1993**, *93*, 41–54.

20. V. Branchadell, M. Sodupe, A. Oliva, J. Bertran in *The Chemistry of Dienes and Polyenes* Z. Rappoport (*Ed.*), J. Wiley and Sons, Chichester, **1997**, Vol. 1, 1–23.

21. F. Sondheimer, D. A. Ben-Efraim, R. Wolovsky, *J. Am. Chem. Soc.*, **1961**, *83*, 1675–1681; *cf.* F. Sondheimer, D. A. Ben–Efraim, Y. Gaoni, *J. Am. Chem. Soc.*, **1961**, *83*, 1682–1685.

22. C. W. Spangler, D. A. Little, *J. Chem. Soc. Perkin Trans.* I, **1982**, 2379–2385.

23. R. Kuhn, *Angew. Chem.*, **1937**, *50*, 703–718 and references therein.

24. P. Nayler, M. C. Whiting, *J. Chem. Soc.*, **1955**, 3037–3047.

25. I. Heilbron, E. H. R. Jones, R. A. Raphael, *J. Chem. Soc.,* **1944**, 136–139.

26. F. Bohlman, H.-J. Mannhardt, *Chem. Ber.*, **1956**, *89*, 1307–1315.

27. C. W. Spangler, R. A. Rathunde, *J. Chem. Soc. Chem. Commun.*, **1989**, 26–27.

28. R. Kuhn, A. Winterstein, *Helv. Chim. Acta*, **1928**, *11*, 87–116.

29. R. Kuhn, K. Wallenfels, quoted in ref.[23]; *cf. D. R. P.* 683 030 (**1937**); *D. R. P. Org. Chem.* **3** 311.

30. K. Knoll, R. Schrock, *J. Am. Chem. Soc.*, **1989**, *111*, 7989–8004.

31. A. Kiehl, A. Eberhardt, K. Müllen, *Lieb. Ann Chem.*, **1995**, 223–230.

32. H. Hopf, C. Mlynek, D. Fischer, *Eur. J. Org. Chem.*, **2000**, in press.

33. P. Karrer, C. H. Eugster, *Helv. Chim. Acta,* **1950**, *33*, 1172–1174; *cf.* P. Karrer, C. H. Eugster, *Compt. rend.*, **1950**, *230*, 1920–1921. β-Carotene (**1**) was synthesized practically simultaneously by Karrer and by Inhoffen, whose original short communication arrived at the respective editorial office one week later than the contribution of the Swiss chemists: H. H. Inhoffen, F. Bohlmann, K. Bartram, H. Pommer, *Chem. Ztg.*, **1950**, *74*, 285; *cf.* H. H. Inhoffen, F. Bohlmann, K. Bartram, G. Rummert, H. Pommer, *Liebigs Ann. Chem.*, **1950**, *570*, 54–69; H. H. Inhoffen, H. Siemer, *Fortschr. Chem. Org. Naturstoffe*, **1952**, *9*, 1–40.

34. N. Y. Demyanov, M. N. Doyarenko, *Bull. Acad. Sci. Russ.*, **1922**, *16*, 297; *cf.* N. Y. Demyanov, M. N. Doyarenko, *Ber. Dtsch. Chem. Ges.*, **1923**, *56*, 2200–2207.

35. A review on the early history of cyclopropene (covering the chemical literature until *ca.*, **1963**) has been published: F. L. Carter, V. L. Frampton, *Chem. Rev.*, **1964**, *64*, 497–525.

36. K. B. Wiberg, W. J. Bartley, F. P. Lossing, *J. Am. Chem. Soc.*, **1962**, *84*, 3980–3981.

37. G. L. Closs in *Advances in Alicyclic Chemistry* H. Hart, G. J. Karabatsos, (*Eds.*), Academic Press, New York, N. Y., **1966**, Vol. 1, 53–127.

38. M. L. Deem, *Synthesis*, **1972**, 675–691.

39. M. L. Deem, *Synthesis*, **1982**, 701–716.

40. B. Halton, M. G. Banwell, *The Chemistry of Functional Groups, The Chemistry of the Cyclopropyl Group* S. Patai (*Ed.*), J. Wiley and Sons, Chichester, **1987**, Vol. 2, *pp.* 1224.

41. M. S. Baird, *Top. Curr. Chem.*, **1988**, *144*, 137–209.

42. M. S. Baird, *Advances in Strain in Organic Chemistry* B. Halton (*Ed.*), The JAI Press, Greenwich, **1991**, Vol. 1, 65–116.

43. M. S. Baird, I. G. Bolesov, *The Chemistry of Functional Groups, Supplement D2, The Chemistry of Halides, Pseudohalides and Azides* S. Patai, Z. Rappoport (*Eds.*), John Wiley and Sons, Chichester, **1995**, 1351–1394.

44. Houben-Weyl, *Methods of Organic Chemistry, Carbocyclic Three- and Four-membered Ring Compounds* A. de Meijere (*Ed.*), Thieme Verlag, Stuttgart, **1997**, Vol. E17d.

45. Houben-Weyl, *Methods of Organic Chemistry, Carbocyclic Three- and Four-membered Ring Compounds* A. de Meijere (*Ed.*), Thieme Verlag, Stuttgart, **1997**, Vol. E17f.

46. M. J. Schlatter, *J. Am. Chem. Soc.,* **1941**, *63*, 1733–1737.

47. J. M. Denis, R. Niamayoua, M. Vata, A. Lablache-Combier, *Tetrahedron Lett.*, **1980**, *21*,

515–518. For the elimination of 1-bromo-2-ethoxy-cyclopropane with magenium in tetrahydrofuran see A. I. D'yachenko, N. M. Abramova, T. Y. Rudashevskaya, O. A. Nesmeyanova, O. M. Nefedov, *Izv. Akad. Nauk SSR, Ser. Khim*, **1982**, 1193; *Bull. Acad. Sci. USSR, Div. Chem. Sci.*, **1982**, *31*, 1066.

48. K. B. Wiberg, W. J. Bartley, *J. Am. Chem. Soc.*, **1960**, *82*, 6375–6380.
49. K. Alder, G. Jacobs, *Chem. Ber.*, **1953**, *86*, 1528–1539; *cf.* K. Alder, K. Kaiser, M. Schumacher, *Liebigs Ann. Chem.*, **1957**, *602*, 80–93.
50. G. L. Closs, K. D. Krantz, *J. Org. Chem.*, **1966**, *31*, 638.
51. R. Koster, S. Arora, P. Binger, *Liebigs Ann. Chem.*, **1973**, 1219–1235; *cf.* R. Koester, S. Arora, P. Binger, *Angew. Chem.*, **1970**, *82*, 839–840; *Angew. Chem. Int. Ed. Engl.*, **1970**, *9*, 810–811.
52. W. v. E. Doering, T. Mole, *Tetrahedron,* **1960**, *10*, 65–70.
53. P. Dowd, A. Gold, *Tetrahedron Lett.*, **1969**, 85–86.
54. A. J. Schipperijn, J. Lukas, *Tetrahedron Lett.*, **1972**, 231–232.
55. P. Binger, A. Brinkmann, *Chem. Ber.*, **1978**, *111*, 2689–2695.
56. P. Binger, G. Schroth, J. McMeeking, *Angew. Chem.*, **1974**, *86*, 518; *Angew. Chem. Int. Ed. Engl.,* **1974**, *13*, 465.
57. By employing *E*- and *Z*-1-deuterio-1,3-butadiene this cycloaddition has been shown to take place as an *endo*-process: J. E. Baldwin, V. P. Reddy, *J. Org. Chem.*, **1989**, *54*, 5264–5267.
58. V. V. Plemenkov, V. A. Breus, *Zh. Org. Khim.*, **1974**, *10*, 1656–1658; *J. Org. Chem. USSR*, **1974**, *10*, 1672–1674.
59. L. M. Joshel, L. W. Butz, *J. Am. Chem. Soc.*, **1941**, *63*, 3350–3351.
60. G. R. Wenzinger, J. A. Ors, *J. Org. Chem.,* **1974**, *39*, 2060–2063.
61. V. V. Plemenkov, L. A. Yanykino, *Zh. Org. Khim.*, **1970**, *6*, 2041–2044; *J. Org. Chem. USSR*, **1970**, *6*, 2049–2052.
62. S. Itô, I. Itoh, I. Saito, A. Mori, *Tetrahedron Lett.*, **1974**, 3887–3888.
63. R. W. LaRochelle, B. M. Trost, *Chem. Commun.*, **1970**, 1353–1354; *cf.* R. Srinivasan, *J. Am. Chem. Soc.*, **1967**, *89*, 4812–4813.
64. D. F. Eaton, R. G. Bergman, G. S. Hammond, *J. Am. Chem. Soc.*, **1972**, *94*, 1351–1353.
65. P. G. Gassman, W. J. Greenlee, *J. Am. Chem. Soc.*, **1973**, *95*, 980–982; *cf.* J. P. Visser, P. Smael, *Tetrahedron Lett.*, **1973**, 1139–1140.
66. pk$_a$ of **61**: Houben-Weyl, *Methods of Organic Chemistry, Carbocyclic Three- and Four-membered Ring Compounds* A. de Meijere (*Ed.*), Thieme Verlag, Stuttgart, **1997**, Vol. E17/d, p. 2760. The high degree of s-character of the carbon-carbon double bond in **61** can also be inferred from its spectral properties, see J. D. Dunitz, H. G. Feldman, V. Schomaker, *J. Chem. Phys.*, **1952**, *20*, 1708–1709 for its electron diffraction spectrum and P. H. Kasai, R. J. Meyers, D. F. Eggers, Jr, K. B. Wiberg, *J. Chem. Phys.*, **1959**, *30*, 512–516 for its microwave spectrum.
67. A. J. Schipperijn, P. Smael, *Recl. Trav. Chim. Pays-Bas*, **1973**, *92*, 1121–1133.
68. A. J. Schipperijn, *Recl. Trav. Chim. Pays-Bas*, **1971**, *90*, 1110–1112.; *cf.* L. Fitjer, J.-M. Conia, *Angew. Chem.*, **1973**, *85*, 347–349; *Angew. Chem. Int. Ed. Engl.*, **1973**, *12*, 332–334.
69. R. Breslow, J. T. Groves, *J. Am. Chem. Soc.*, **1970**, *92*, 984–987.
70. H. Hopf, G. Wachholz, R. Walsh, *Chem. Ber.*, **1985**, *118*, 3579–3587 and references therein.
71. R. Walsh, C. Wolf, S. Untiedt, A. de Meijere, *J. Chem. Soc. Chem. Commun.*, **1992**, 422–424; *cf.* H. Hopf, A. Plagens, R. Walsh, *Liebigs Ann. Chem.,* **1996**, 825–835.
72. M. Yoshimine, J. Pacansky, N. Honjou, *J. Am. Chem. Soc.*, **1989**, *111*, 4198–4209 and references therein; *cf.* W. Graf v. d. Schulenburg, N. Goldberg, *J. Chem. Soc. Chem.*

Commun., **1998**, 2761–2762.

73. H. Hopf, W. Graf v. d. Schulenburg, R. Walsh, *Angew. Chem.,* **1997**, *109*, 415–417; *Angew. Chem. Int. Ed Engl.,* **1997**, *36*, 381–383; *cf.* I. R. Likhotvorik, D. M. Brown, M. Jones, Jr, *J. Am. Chem. Soc.*, **1994**, *116*, 6175–6178; H. Hopf, W. Graf v. d. Schulenburg, R. Walsh, *Angew. Chem.*, **1999**, *111*, 1200–1203; *Angew. Chem. Int. Ed Engl.*, **1999**, *38*, 1128–1130.

74. Natural products containing cyclopropene rings have been known for a long time. For a review see S. Beckmann, H. Geiger in Houben-Weyl, *Methoden der Organischen Chemie*, Thieme Verlag, Stuttgart, **1971**, Vol. IV/4, 445–478.

75. For a review of many of these hydrocarbons see W. E. Billups, M. M. Haley, G.-A. Lee, *Chem. Rev.*, **1989**, *89*, 1147–1159.

76. K. Ziegler, E. Holzkamp, H. Breil, H. Martin, *Angew. Chem.*, **1955**, *67*, 541–547.

77. K. Ziegler, *Angew. Chem.*, **1959**, *71*, 623–625; *cf.* K. Ziegler, H. Martin, *Makromolekulare Chem.*, **1955**, *18/19*, 186–194.

78. K. Ziegler, *Angew. Chem.*, **1964**, *76*, 545–553; *Angew. Chem. Int. Ed. Engl.*, **1964**, *3*

79. G. Natta, *Angew. Chem.*, **1964**, *76*, 553–566; *Angew. Chem. Int. Ed. Engl.*, **1964**, *3*

80. G. Wilke, *Angew. Chem.*, **1957**, *69*, 397–398; *cf.* H. W. B. Reed, *J. Chem. Soc.* (*London*), **1954**, 1931–1941. The Ti/Al-ratio is critical in these reactions: For lower ratios (1:0.5 to 1.0) the (linear) polymerization is preferred over cyclooligomerization: S. E. Horne, Jr, F. P. Kiehl, I. I. Shipman, V. L. Folt, C. F. Gibbs, *Ind. Eng. Chem.*, **1956**, *48*, 784

81. Review: G. Wilke, *Angew. Chem.*, **1963**, *75*, 10–20

82. G. Mann, G. Haufe, *Chemistry of Alicyclic Compounds*, Elsevier, Amsterdam, **1989**.

83. W. Reppe, O. Schlichting, K. Klager, T. Toepel, *Liebigs Ann. Chem.*, **1948**, *560*, 1–92.

84. F. Turecek, V. Hanus, P. Sedomera, H. Antropiusová, K. Mach, *Tetrahedron*, **1979**, *35*, 1463–1467.

85. P. W. Jolly, G. Wilke, *The Organic Chemistry of Nickel*, Academic Press, New York, N. Y., **1975**; *cf.* V. M. Akhmedov, M. T. Anthony, M. L. H. Green, D. Young, *J. Chem. Soc. Dalton Trans.*, **1975**, 1412–1419.

86. G. Wilke, A. Eckerle in *Applied Homogeneous Catalysis by Organonmetallic Compounds* B. Cornils, W. A. Hermann (*Eds.*), VCH, Weinheim, Vol. 2, **1996**, 358–373; *cf.* P. Heimbach, *Angew. Chem.*, **1973**, *85*, 1035–1049; *Angew. Chem. Int. Ed. Engl.*, **1973**, *12*, 975–989.

87. Review: G. Wilke, *Angew. Chem.*, **1988**, *100*, 190–211; *Angew. Chem. Int. Ed. Engl.*, **1988**, *27*, 185–206.

88. K. Jonas, P. Heimbach, G. Wilke, *Angew. Chem.*, **1968**, *80*, 1033; *Angew. Chem. Int. Ed. Engl.*, **1968**, *7*, 949; *cf.* P. W. Jolly, I. Tkatchenko, G. Wilke, *Angew. Chem.*, **1971**, *83*, 328–329; *Angew. Chem. Int. Ed. Engl.*, **1971**, *10*, 328–329.

89. B. Bogdanovic, P. Heimbach, M. Kröner, G. Wilke, E. G. Hoffmann, J. Brandt, *Liebigs Ann. Chem.*, **1969**, *727*, 143–160.

90. B. Henc, P. W. Jolly, R. Salz, G. Wilke, R. Benn, E. G. Hoffmann, R. Mynott, G. Schroth, K. Seevogel, J. C. Sekutowski, C. Krüger, *J. Organomet. Chem.*, **1980**, *191*, 425–448; *cf.* R. Benn, B. Büssemeier, S. Holle, P. W. Jolly, R. Mynott, I. Tkatchenko, G. Wilke, *J. Organomet. Chem.*, **1985**, *279*, 63–68.

91. P. Heimbach, W. Brenner, *Angew. Chem.*, **1967**, *79*, 813–814; *Angew. Chem. Int. Ed. Engl.*, **1967**, *6*, 800; *cf.* W. Brenner, P. Heimbach, H. Hey, E. W. Müller, G. Wilke, *Liebigs Ann. Chem.*, **1969**, *727*, 161–182. For the catalytic dimerization of dienes with other catalyst systems, *e.g.* nitrosocarbonyl transition metal compounds see J. P. Candlin, W. H. Janes, *J. Chem. Soc.* (C), **1968**, 1856–1860.

92. Butadiene (**99**) has also been dimerized to an odd-numbered ring system, 2-methylenevinylcylopentane, with a nickel catalyst: J. Kiji, K. Masui, J. Furukawa, *Tetra-*

hedron Lett., **1970**, 2561–2564; *cf.* J. Kiji, K. Masui, J. Furukawa, *J. Chem. Soc. Chem. Commun.*, **1970**, 1310–1311.

93. E. Vogel, *Angew. Chem.*, **1962**, *74*, 829–839.

94. K. Alder, H. A. Dortmann, *Chem. Ber.*, **1954**, *87*, 1492–1498.

95. E. Vogel, *Liebigs Ann. Chem.*, **1958**, *615*, 1–14.

96. W. G. Herkstroeter, A. A. Lamola, G. S. Hammond, *J. Am. Chem. Soc.*, **1964**, *86*, 4537-4540 and references cited therein; *cf.* C. D. Deboer, N. J. Turro, G. S. Hammond, *Org. Synth.*, **1967**, *47*, 64–68; W. L. Dilling, *Chem. Rev.*, **1969**, *69*, 845–877.

97. P. Heimbach, G. Wilke, *Liebigs Ann. Chem.*, **1969**, *727*, 183–193. For the mixed cyclo-oligomerization with bicyclo[2.2.1]hept-2-ene see P. Heimbach, R.-V. Meyer, G. Wilke, *Liebigs Ann. Chem.*, **1975**, 743–751.

98. P. Heimbach, W. Brenner, *Angew. Chem.*, **1966**, *78*, 983–984; *Angew. Chem. Int. Ed. Engl.*, **1966**, *5*, 961–962. For the co-oligomerization of butadiene and 2-butyne to 4,5-dimethyl-1-*cis*,4-*cis*,7-*trans*-cyclododecatriene (formed in over 95% yield) see W. Brenner, P. Heimbach, G. Wilke, *Liebigs Ann Chem.*, **1969**, *727*, 194–207; *cf.* P. Heimbach, R. Schimpf, *Angew. Chem.*, **1968**, *80*, 704–705; *Angew. Chem. Int. Ed. Engl.*, **1968**, *7*, 727–728; W. Brenner, P. Heimbach, K.-J. Ploner, F. Thömel, *Angew. Chem.*, **1969**, *81*, 744–745; *Angew. Chem. Int. Ed. Engl.*, **1969**, *8*, 753–754. The Ni-catalyzed mixed cyclooligomerization of 1,3-butadiene with allene has also been described: P. Heimbach, H. Selbeck, E. Troxler, *Angew. Chem.*, **1971**, *83*, 731–732; *Angew. Chem. Int. Ed. Engl.*, **1971**, *10*, 659–660.

99. O. Ermer, *Aspekte von Kraftfeldberechnungen*, W. Baur Verlag, München, **1981**.

100. W. Luef, R. Keese, *Topics in Stereochemistry*, E. L. Eliel, S. H. Wilen (*Eds.*), **1995**, *20*, 231–318.

101. Review of the early history of Bredt's rule: G. L. Buchanan, *Chem. Soc. Rev.*, **1974**, *3*, 41–63 and G. Köbrich, *Angew. Chem.*, **1973**, *85*, 494–503; *Angew. Chem. Int. Ed. Engl.*, **1973**, *12*, 464–473.

102. J. A. Marshall, *Acc. Chem. Res.*, **1980**, *13*, 213–218.

103. A. Greenberg, J. F. Liebman, *Strained Organic Molecules*, Academic Press, Inc., New York, N. Y. **1978**, 90–133.

104. A. T. Blomquist, L. H. Liu, J. C. Bohrer, *J. Am. Chem. Soc.*, **1952**, *74*, 3643–3647.

105. For a summary of work on chiral cycloolefins see: M. Nakazaki, K. Yamamoto, K. Naemura, *Topics in Current Chemistry*, Springer-Verlag, Berlin, **1984**, Vol. 125, 1–25.

106. K. Ziegler, H. Wilms, *Liebigs Ann. Chem.*, **1950**, *567*, 1–43, *cf.* K. Ziegler, H. Wilms, *Naturwiss.*, **1948**, *35*, 157–158.

107. A. C. Cope, R. A. Pike, C. F. Spencer, *J. Am. Chem. Soc.*, **1953**, *75*, 3212–3215.

108. A. C. Cope, C. R. Ganellin, H. W. Johnson, Jr, T.V. van Anken, H. J. S. Winkler, *J. Am. Chem. Soc.*, **1963**, *85*, 3276–3279; *cf.* A. C. Cope, C. R. Ganellin, H. W. Johnson, Jr, *J. Am. Chem. Soc.*, **1962**, *84*, 3191–3192.

109. A. C. Cope, B. A. Pawson, *J. Am. Chem. Soc.*, **1965**, *87*, 3649–3651, *cf.* A. C. Cope, K. Banholzer, H. Keller, B. A. Pawson, J. J. Whang, H. J. S. Winkler, *J. Am. Chem. Soc.*, **1965**, *87*, 3644–3649.

110. R. B. Cundall, T. F. Palmer, *Trans. Faraday Soc.*, **1961**, *57*, 1936–1941.

111. M. Traetteberg, *Acta Chem Scand. Ser.* B, **1975**, *29*, 29–36; *cf.* O. Ermer, *Angew. Chem.*, **1974**, *86*, 672–673; *Angew. Chem. Int. Ed. Engl.*, **1974**, *13*, 604–605 and O. Ermer, *Tetrahedron*, **1975**, *31*, 1849–1854.

112. For other racemizations of this type see: J. A. Marshall, V. H. Audia, T. M. Jenson, W. C. Guida, *Tetrahedron*, **1986**, *42*, 1703–1709.

113. G. Binsch, J. D. Roberts, *J. Am. Chem. Soc.*, **1965**, 87, 5157–5162.

114. E. J. Corey, F. A. Carey, R. A. E. Winter, *J. Am. Chem. Soc.*, **1965**, *87*, 934–935.

115. R. G. Salomon, K. Folting, W. E. Streib, J. K. Kochi, *J. Am. Chem. Soc.*, **1974**, *96*, 1145–1152.
116. G. M. Wallraff, R. H. Boyd, J. Michl, *J. Am. Chem. Soc.,* **1983**, *105*, 4550–4555.
117. J. T. M. Evers, A. Mackor, *Recl. Trav. Chim. Pays-Bas*, **1979**, *98*, 423–424.
118. G. M. Wallraff, J. Michl, *J. Org. Chem.*, **1986**, *51*, 1794–1800.
119. P. J. Kropp, *J. Am. Chem. Soc.*, **1969**, *91*, 5783–5791; for a review see J. A. Marshall, *Acc. Chem. Res.*, **1969**, *2*, 33–40.
120. Y. Inoue, T. Takamuku, H. Sakurai, *J. Chem. Soc. Perkin II*, **1977**, 1635–1642.
121. Y. Inoue, T. Ueoka, T. Kuroda, T. Hakushi, *J. Chem. Soc. Chem. Commun.*, **1981**, 1031–1033.
122. Y. Inoue, T. Ueoka, T. Kuroda, T. Hakushi, *J. Chem. Soc. Perkin II*, **1983**, 983–988.
123. M. Squillacote, A. Bergman, J. de Felippis, *Tetrahedron Lett.*, **1989**, *30*, 6508–6808.
124. N. L. Allinger, J. T. Sprague, *J. Am. Chem. Soc.*, **1972**, *94*, 5734–5747 for calculations of the structures and energies of some cycloalkenes by the force field method.
125. E. J. Corey, M. Tada, R. LaMahieu, L. Libit, *J. Am. Chem. Soc.,* **1965**, *87*, 2051–2052.
126. P. E. Eaton, K. Lin, *J. Am. Chem. Soc.*, **1965**, *87*, 2052–2054.
127. H. Hart, B. Chen., M. Jeffers, *J. Org. Chem.*, **1979**, *44*, 2722–2726.
128. P. E. Eaton, *Acc. Chem. Res.*, **1968**, *1*, 50–57.
129. J. Th. Evers, A. Mackor, *Tetrahedron Lett.*, **1978**, 2317–2320.
130. R. Bonneau, J. Joussot-Dubien, L. Salem, A. J. Yarwood, *J. Am. Chem. Soc.,* **1976**, *98*, 4329–4330.
131. W. G. Dauben, H. C. H. A. van Riel, C. Hauw, F. Leroy, J. Joussot-Dubien, R. Bonneau, *J. Am. Chem. Soc.*, **1979**, *101*, 1901–1903.
132. W. G. Dauben, H. C. H. A. van Riel, J. D. Robbins, G. Wagner, *J. Am. Chem. Soc.*, **1979**, *101*, 6383–6389.
133. A. C. Cope, C. F. Howell, A. Knowles, *J. Am. Chem. Soc.*, **1962**, *84*, 3190–3191; *cf.* ref.[106]
134. M. Christl, G. Brüntrup, *Angew. Chem.*, **1974**, *86*, 197; *Angew. Chem. Int. Ed. Engl.*, **1974**, *13*, 208; *cf.* R. M. Coates, L. A. Last, *J. Am. Chem. Soc.*, **1983**, *105*, 7322–7326.
135. J. Bredt, J. Houben, P. Levy, *Ber. Dtsch. Chem. Ges.*, **1902**, *35*, 1286–1291.
136. J. Bredt, *Liebigs Ann. Chem.*, **1924**, *437*, 1–13.
137. G. Szeimies in *Reactive Intermediates* R. A. Abramovitch (*Ed.*), Plenum Press, New York, N. Y., **1983**, Vol. 3, 299–366.
138. J. Bredt, *Ann. Acad. Scient. Fennicae,* **1927**, *19A*, 3.
139. For historical reviews of Bredt's rule see ref.[7–101] and F.S. Fawcett, *Chem. Rev.*, **1950**, *47*, 219–274.
140. For modern reviews on Bredt's rule see ref.[137] and P. M. Warner, *Chem. Rev.*, **1989**, *89*, 1067–1093; *cf.* J. McMurry, *Chem. Rev.*, **1989**, *89*, 1513–1524.; R. Keese, *Angew. Chem.*, **1975**, *87*, 568–578; *Angew. Chem. Int. Ed. Engl.*, **1975**, *14*, 528–538.
141. J. R. Wiseman, *J. Am. Chem. Soc.*, **1967**, *89*, 5966–5968.
142. J. R. Wiseman, W. A. Pletcher, *J. Am. Chem. Soc.*, **1970**, *92*, 956–962.
143. A recent compilation and analysis of structural data of distorted bridgehead olefins and amides has been published by T. G. Lease, K. J. Shea in *Advances in Theoretically Interesting Molecules* R. P. Thummel (*Ed.*), JAI Press, Inc., Greenwich, CT, **1992**, Vol. 2, 99–112.
144. J. A. Marshall, H. Faubl, *J. Am. Chem. Soc.*, **1967**, *98*, 5965–5966; *cf.* J. A. Marshall, H. Faubl, *J. Am. Chem. Soc.*, **1970**, *92*, 948–955.
145. K. B. Becker, R. W. Pfluger, *Tetrahedron Lett.*, **1979**, 3713–3716.
146. M. Kim, J. D. White, *J. Am. Chem. Soc.*, **1977**, *99*, 1172–1180.
147. K. B. Becker, *Helv. Chim. Acta*, **1970**, *60*, 81–93; *cf.* K. B. Becker, *Helv. Chim. Acta,*

1970, *60*, 94–102. The intramolecular Wittig reaction is conceptually superior to some of the other approaches to *anti*-Bredt olefins since it can be performed in a directed manner, whereas eliminations like, *e.g.*, the Hofmann elimination can lead to mixtures of regioisomers in certain cases.

148. W. G. Dauben, J. Ipaktschi, *J. Am. Chem. Soc.*, **1973**, *95*, 5088–5089.

149. K. J. Shea, S. Wise, *J. Am. Chem. Soc.*, **1978**, *100*, 6519–6521; *cf.* K. J. Shea, S. Wise, *Tetrahedron Lett.*, **1979**, 1011–1014 and ref.[143].

150. M. Nakazaki, K. Naemura, S. Nakahara, *J. Org. Chem.*, **1979**, *44*, 2438–2441, have shown that the levorotatory enantiomer has the S-configuration.

151. J. A. Chong, J. R. Wiseman, *J. Am. Chem. Soc.*, **1972**, *94*, 8627–8629.

152. J. R. Wiseman, J. A. Chong, *J. Am. Chem. Soc.*, **1969**, *91*, 7775–7777.

153. B. L. Adams, P. Kovacic, *J. Am. Chem. Soc.*, **1973**, *95*, 8206–8207.

154. B. L. Adams, P. Kovacic, *J. Am. Chem. Soc.*, **1974**, *96*, 7014–7018.

155. B. L. Adams, P. Kovacic, *J. Org. Chem.*, **1974**, *39*, 3090–3094.

156. M. Farcasiu, D. Farcasiu, R. T. Coulin, M. Jones, Jr, P. v. R. Schleyer, *J. Am. Chem. Soc.*, **1973**, *95*, 8207–8209; *cf.* A. D. Wolf, M. Jones, Jr, *J. Am. Chem. Soc.*, **1973**, *95*, 8209–8210.

157. K. B. Becker, J. L. Chappuis, *Helv. Chim. Acta*, **1979**, *62*, 34–43.

158. R. Keese, E.-P. Krebs, *Angew. Chem.*, **1971**, *83*, 254–256; *Angew. Chem. Int. Ed. Engl.*, **1971**, *10*, 262–264.

159. R. Keese, E.-P. Krebs, *Angew. Chem.*, **1972**, *84*, 540–542; *Angew. Chem. Int. Ed. Engl.*, **1972**, *11*, 518–520.

160. W. Burns, D. Grant, M. A. McKervey, G. Step, *J. Chem. Soc. Perkin I*, **1976**, 234–238.

161. W. Burns, M. A. McKervey, *J. Chem. Soc. Chem. Commun.*, **1974**, 858–859.

162. D. Lenoir, *Tetrahedron Lett.*, **1972**, 4049–4052; *cf.* D. Lenoir, J. Firl, *Liebigs Ann. Chem.*, **1974**, 1467–1473.

163. W. G. L. Aalbersberg, K. P. C. Vollhardt, *Tetrahedron Lett.*, **1979**, 1939–1942; A. E. Murad, H. Hopf, *Chem. Ber.* **1980**, *113*, 2358–2371.

164. H. Hopf, R. Savinsky, P. G. Jones, I. Dix, B. Ahrens, *Liebigs Ann./Recueil*, **1997**, 1499–1504.

165. R. S. Cahn, C. K. Ingold, V. Prelog, *Angew. Chem.*, **1966**, *78*, 413–447; *Angew. Chem. Int. Ed. Engl.*, **1966**, *5*, 385–419.

166. M. Nakazaki, K. Yamamoto, J. Yanagi, *J. Chem. Soc. Chem. Commun.*, **1977**, 346–347.

167. M. Nakazaki, K. Yamamoto, J. Yanagi, *J. Am. Chem. Soc.*, **1979**, *101*, 147–151.

168. M. Nakazaki, K. Yamamoto, M. Maeda, *J. Chem. Soc. Chem. Commun.*, **1980**, 294–295.

169. M. Nakazaki, K. Yamamoto, M. Maeda, *J. Org. Chem.*, **1980**, *45*, 3229–3232.

170. J. A. Marshall, M. Lewellyn, *J. Am. Chem. Soc.*, **1977**, *99*, 3508–3510.

171. J. A. Marshall, R. E. Bierenbaum, K.-H. Chung, *Tetrahedron Lett.*, **1979**, 2081–2084.

172. J. A. Marshall, T. H. Black, R. L. Shone, *Tetrahedron Lett.*, **1979**, 4737–4740.

173. J. A. Marshall, T. H. Black, *J. Am. Chem. Soc.*, **1980**, *102*, 7581–7583; *cf.* J. A. Marshall, *Acc. Chem. Res.*, **1980**, *13*, 213–218 for a review on *trans*-cycloalkenes and [*a.b*]-betweenanenes.

174. J. A. Marshall, K. E. Flynn, *J. Am. Chem. Soc.*, **1983**, *105*, 3360–3362.

175. W. Hückel, R. Danneel, A. Schwartz, A. Gercke, *Liebigs Ann. Chem.*, **1929**, *474*, 121–144 and references cited therein.

176. E. Vogel, *Chem. Ber.*, **1952**, *85*, 25–29.

177. E. F. Kiefer, C. H. Tanna, *J. Am. Chem. Soc.*, **1969**, *91*, 4478–4480.

178. J. J. Gajewski, C. N. Shih, *J. Am. Chem. Soc.*, **1970**, *92*, 4457–4458; *cf.* K. B. Wiberg, G. J. Burgmaier, *J. Am. Chem. Soc.*, **1972**, *94*, 7396–7401.

179. K. B. Wiberg, G. J. Burgmaier, P. Warner, *J. Am. Chem. Soc.*, **1971**, *93*, 246–247.

180. M. E. Jason, J. A. McGinnety, K. B. Wiberg, *J. Am. Chem. Soc.*, **1974**, *96*, 6531–6532; *cf.* M. E. Jason, J. A. McGinnety, *Inorg. Chem.*, **1975**, *14*, 3025–3029.

181. J. Casanova, H. R. Rogers, *J. Org. Chem.*, **1974**, *39*, 3803.

182. K. B. Wiberg, W. F. Bailey, M. E. Jason, *J. Org. Chem.*, **1974**, *39*, 3803–3804.

183. J. Casanova, J. Bragin, F. D. Cotrell, *J. Am. Chem. Soc.*, **1978**, *100*, 2264–2265.

184. K. B. Wiberg, M. E. Jason, *J. Am. Chem. Soc.*, **1976**, *98*, 3393–3395. For a review of the chemistry of **182** see K. B. Wiberg, M. G. Matturro, P. J. Okarma, M. E. Jason, W. P. Dailey, G. J. Burgmaier, W. F. Bailey, P. Warner, *Tetrahedron*, **1986**, *42*, 1895–1902.

185. H.-U. Wagner, G. Szeimies, J. Chandrasekhar, P. v. R. Schleyer, J. A. Pople, J. St. Binkley, *J. Am. Chem. Soc.*, **1978**, *100*, 1210–1213.

186. W. J. Hehre, J. A. Pople, *J. Am. Chem. Soc.*, **1975**, *97*, 6941–6955.

187. R. Greenhouse, W. Th. Borden, K. Hirotsu, J. Clardy, *J. Am. Chem. Soc.*, **1977**, *99*, 1664–1666; *cf.* W. Th. Borden, *Chem. Rev.*, **1989**, *89*, 1095–1109.

188. G. E. Renzoni, T.-K. Yin, F. Miyake, W. Th. Borden, *Tetrahedron*, **1986**, *42*, 1581–1584.

189. A chiral variant of this type of hydrocarbon in which two allylic positions are spanned by a propano bridged has been synthesized by a route similar to the one used for the **192→195**-interconversion: R. Greenhouse, W. Th. Borden, T. Ravindranathan, K. Hirotsu, J. Clardy, *J. Am. Chem. Soc.*, **1977**, *99*, 6955–6961; *cf.* W. Th. Borden, T. Ravindranathan, *J. Org. Chem.*, **1971**, *36*, 4125–4127.

190. W. F. Maier, P. v. R. Schleyer, *J. Am. Chem. Soc.*, **1981**, *103*, 1891–1900.

191. U. Szeimies-Seebach, J. Harnisch, G. Szeimies, M. Van Meerssche, G. Germain, J.-P. Declerq, *Angew. Chem.*, **1978**, *90*, 904–905; *Angew. Chem. Int. Ed. Engl.*, **1978**, *17*, 848–850.

192. G. Szeimies, J. Harnisch, O. Baumgärtel, *J. Am. Chem. Soc.*, **1977**, *99*, 5183–5184.

193. G. Szeimies, J. Harnisch, K.-H. Stadler, *Tetrahedron Lett.*, **1978**, 243–246.

194. U. Szeimies-Seebach, G. Szeimies, *J. Am. Chem. Soc.*, **1978**, *100*, 3966–3967.

195. J. Harnisch, H. Legner, U. Szeimies-Seebach, G. Szeimies, *Tetrahedron Lett.*, **1978**, 3683–3686.

196. J. Harnisch, O. Baumgärtel, G. Szeimies, M. Van Meerssche, G. Germain, J.-P. Declerq, *J. Am. Chem. Soc.*, **1979**, *101*, 3370–3371.

197. A.-D. Schlüter, J. Belzner, H. Heywang, G. Szeimies, *Tetrahedron Lett.*, **1983**, *24*, 891–894.

198. H.-G. Zoch, G. Szeimies, R. Römer, G. Germain, J.-P. Declerq, *Chem. Ber.*, **1983**, *116*, 2285–2310; *cf.* H.-G. Zoch, G. Szeimies, R. Römer, R. Schmitt, *Angew. Chem.*, **1981**, *93*, 894–895; *Angew. Chem. Int. Ed. Engl.*, **1981**, *20*, 877–878.

199. N. M. Weinshenker, F. D. Greene, *J. Am. Chem. Soc.*, **1968**, *90*, 506.

200. R. L. Viavattene, F. D. Greene, L. D. Cheung, R. Majeste, L. M. Trefonas, *J. Am. Chem. Soc.*, **1974**, *96*, 4342–4343. The original synthesis has been improved by R. Herges and H. Neumann, unpubl. results; *cf.* H. Neumann, Ph. d. dissertation, Erlangen, **1995**.

201. Apparently all known pyramidalized olefins are syn-pyramidalized (*cf.* structure **136** *vs.* **137**, Scheme 22).

202. K. B. Wiberg, M. G. Matturro, P. J. Okarma, M. E. Jason, *J. Am. Chem. Soc.*, **1984**, *106*, 2194–2200; *cf.* K. B. Wiberg, M. G. Matturro, R. D. Adams, *J. Am. Chem. Soc.*, **1981**, *103*, 1600–1602).

203. K. B. Wiberg, R. D. Adams, P. J. Okarma, M. G. Matturro, B. Segmuller, *J. Am. Chem. Soc.*, **1984**, *106*, 2200–2206.

204. Reviews: J. E. McMurry, *Chem. Rev.*, **1989**, *89*, 1513–1524; D. Lenoir, *Synthesis*, **1989**, 883–897; A. Fürstner, B. Bogdanovic, *Angew. Chem.*, **1996**, *108*, 2583–2609; *Angew. Chem. Int. Ed. Engl.*, **1996**, *35*, 2442–2469.

205. D. H. R. Barton, B. J. Willis, *J. Chem. Soc. Perkin Trans. I*, **1972**, 305–310.

206. R. M. Kellog, S. Wassenaar, J. Buter, *Tetrahedron Lett.*, **1970**, 4689–4692; *cf.* R. M. Kellog, S. Wassenaar, *Tetrahedron Lett.*, **1970**, 1987–1990 and R. M. Kellog, *J. Org. Chem.*, **1972**, *37*, 4045–4060.
207. D. S. Bomse, T. H. Morton, *Tetrahedron Lett.*, **1975**, 781–784.
208. R. F. Langler, T. T. Tidwell, *Tetrahedron Lett.*, **1975**, 777–780.
209. D. Lenoir, D. Malwitz, B. Meyer, *Tetrahedron Lett.*, **1984**, *25*, 2965–2968.
210. W. v. E. Doering, W. R. Roth, F. Bauer, R. Breuckmann, T. Ebbrecht, M. Herbold, R. Schmidt, H.-W. Lennartz, D. Lenoir, R. Boese, *Chem. Ber.*, **1989**, *122*, 1263–1275.
211. T. Loerzer, R. Gerke, W. Lüttke, *Tetrahedron Lett.*, **1983**, *24*, 5861–5864; *cf.* J. Deuter, H. Rodewald, H. Irngartinger, T. Loerzer, W. Lüttke, *Tetrahedron Lett.*, **1984**, *26*, 1031–1034.
212. P. R. Brooks, R. Bishop, J. A. Counter, E. R. T. Tiekink, *J. Org. Chem.*, **1994**, *59*, 1365–1368 and references cited therein.
213. D. Lenoir, H. Burghard, *J. Chem. Res.*, **1980** (S), 396–397.
214. H. O. Kalinowski, E. Röcker, G. Maier, *Org. Magn. Res.*, **1983**, *21*, 64–66.
215. J. E. Gano, R. H. Wettach, M. S. Platz, V. P. Senthilnathan, *J. Am. Chem. Soc.*, **1982**, *104*, 2326–2327; *cf.* H. Bock, B. Berkner, B. Hierholzer, D. Jaculi, *Helv. Chim. Acta*, **1992**, *75*, 1798–1815.
216. D. H. R. Barton, F. S. Guziec, Jr, I. Shahak, *J. Chem. Soc. Perkin Trans I*, **1974**, 1794–1799; *cf.* T. G. Back, D. H. R. Barton, M. R. Britten-Kelly, F. S. Guziec, Jr, *J. Chem. Soc. Perkin Trans I*, **1976**, 2079–2098.
217. A. Krebs, B. Kaletta, W.-U. Nickel, W. Rüger, L. Tikwe, *Tetrahedron*, **1986**, *42*, 1693–1702.
218. A. Krebs, W. Rüger, W.-U. Nickel, *Tetrahedron Lett.*, **1981**, *22*, 4937–4940; *cf.* E. R. Cullen, F. S. Guzciec, Jr, C. J. Murphy, *J. Org. Chem.*, **1982**, *47*, 3563–3566.
219. A. Krebs, W. Born, B. Kaletta, W.-U. Nickel, W. Rüger, *Tetrahedron Lett.*, **1983**, *24*, 4821–4823.
220. E. R. Cullen, F. S. Guziec, Jr, *J. Org. Chem.*, **1986**, *51*, 1212–1216.
221. A. Krebs, W.-U. Nickel, L. Tikwe, J. Kopf, *Tetrahedron Lett.*, **1985**, *26*, 1639–1642.
222. F. S. Guziec, Jr, C. J. Murphy, *J. Org. Chem.*, **1980**, *45*, 2890–2893.
223. E. R. Cullen, F. S. Guziec, Jr, M. I. Holander, C. J. Murphy, *Tetrahedron Lett.*, **1981**, *22*, 4563–4567.
224. P. J. Garratt, D. Payne, D. Tocher, *J. Org. Chem.*, **1990**, *55*, 1909–1915.
225. Review: M. T. Reetz, *Organotitanium Reagents in Organic Synthesis*, Springer-Verlag, Berlin, **1986**.
226. J. Dannheim, W. Grahn, H. Hopf, C. Parrodi, *Chem. Ber.*, **1987**, *120*, 871–872.
227. H. M. Sulzbach, E. Bolton, D. Lenoir, P. v. R. Schleyer, H. F. Schaefer III, *J. Am. Chem. Soc.*, **1996**, *118*, 9908–9914.

8 Alkynes

Like olefin chemistry, acetylene chemistry is a huge field, and as in the previous chapter (Chapter 7) I can only present a brief overview of alkynes here, and will not even attempt to cover all important aspects of the chemistry of acetylenic hydrocarbons. Because of its importance to organic chemistry in general, alkyne chemistry has been, and continues to be, regularly reviewed,[1–15] the most recent monograph having appeared as late as 1996.[15]

In this chapter I shall thus address three questions only: the preparation of polyacetylenes, the incorporation of the carbon-carbon triple bond into (relatively small) cyclic hydrocarbons, and, finally, the preparation of larger ring systems which can also contain more than one triple bond. In these latter cases we will not look at cyclically conjugated hydrocarbons because these are discussed in the chapter on annulenes (Chapter 10). Hydrocarbons with more than two terminal acetylene groups will also be deferred to a later chapter (Chapter 15) in which I shall show how these alkynes can be used as starting materials for the construction of extended acyclic and cyclic π-systems. Because the triple bond precludes branching, the structural possibilities of building with this unit are obviously far smaller than for the C–C double bond (see, *e.g.*, Chapters 7, 13). This becomes particularly obvious if the triple bond is the only repeating unit available. In such circumstances the only class of compounds which can be constructed are the linear conjugated polyacetylenes.

8.1 Polyacetylenes - rods made of carbon

Interest in polyacetylenes originally stems from the discovery of a large number of these compounds in nature. By the end of the 1980s the group of Bohlmann alone, admittedly the most prolific in this field, had isolated approximately 700 acetylenes from *Compositae* and *Umbelliferae*.[7, 9, 14] Although many of these unusual natural products are functionalized or contain hetero atoms, a sizable number of hydrocarbons has also been isolated; among these is the pentaynene **1** (Scheme 1), which is widespread in *Compositae* and also serves as a precursor for many other acetylenic compounds.

Because polyacetylenes, especially the parent hydrocarbons described in this section, are very reactive, sometimes even explosive compounds, it is surprising that molecules such as **1** survive in plants at all. One reason could be that in *Compositae*,

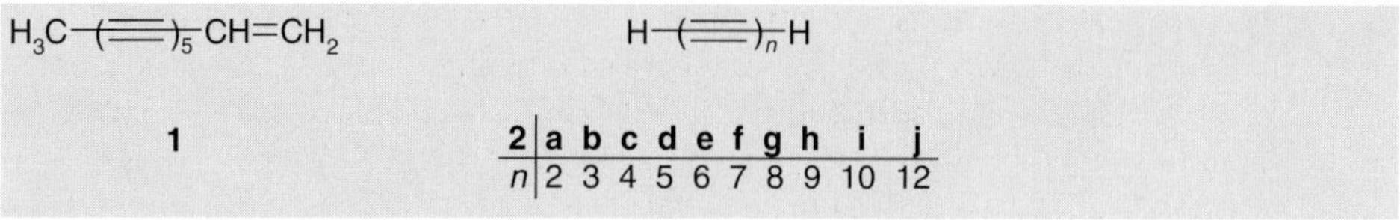

$$H_3C-(\!\!\equiv\!\!)_5-CH=CH_2 \qquad\qquad H-(\!\!\equiv\!\!)_n-H$$

1

2	a	b	c	d	e	f	g	h	i	j
n	2	3	4	5	6	7	8	9	10	12

Scheme 1. Polyacetylenes–natural and man-made carbon rods.

for example, they are usually concentrated in the so-called oil-channels, *i.e.* they are kept in solution all the time, not neat.[7]

The important observation that certain diacetylene derivatives can be polymerized to provide conjugated polyenepolyyne chains in the solid state, which are of interest in connection with organic conducting materials, opened up a second field of intense research on polyacetylenes.[16]

And finally, and most recently, polyacetylenes have been discussed as possible precursors for C_{60} and other fullerenes.[17, 18]

The homologous series **2** with *n* up to 12—with the only exception of $n = 11$—has been prepared. From triacetylene (**2b**) onwards these hydrocarbons become increasingly unstable, although it has been possible to isolate tetra- (**2c**) and even pentaacetylene (**2d**) as solids at low temperatures (see below). The higher ethynylogs all have to be studied in solution, the dodecamer (**2j**), can for example be handled at room temperature in dilute methanol solution.

Some of these hydrocarbons have been known for a very long time. Diacetylene (**2a**), for example, was first prepared by Adolf von Baeyer in his famous studies on 'explosive diamonds' which should play an important role in the formulation of his strain theory.[19] He prepared it by treatment of the bis ammonium salt of diacetylene dicarboxylic acid with cuprous chloride in aqueous ammonia then decomposition of the resulting copper acetylide with concentrated potassium cyanide solution.

The next higher homolog, triacetylene (**2b**) was only obtained 65 years later, when Hunsmann treated 1,6-dibromo-2,4-hexadiyne with a solution of silver nitrate in ammonia and then liberated the free hydrocarbon from the formed bis silver salt by treatment with hydrogen sulfide.[20] Several conceptionally related approaches to **2b** were reported nearly simultaneously[21, 22] as were the first syntheses of octatetrayne (**2c**) and decapentayne (**2d**).[23]

Rather than tracing the historic development of this field, I will now present a general concept enabling the (mostly first) preparation of many polyacetylenes; it was developed and executed by Walton and co-workers in 1972, in a study which certainly deserves to be called classic.[24] This investigation is also one of the best examples of a building block approach in acetylenic chemistry.

For the construction of di- and polyacetylenes, the copper mediated oxidative coupling of terminal acetylenes in the presence of air, originally described by Glaser[25] and later modified by other authors,[1, 8–12, 26] is the most versatile method; it has been used on numerous occasions for the preparation of acyclic and cyclic acetylenes containing at least one conjugated diyne unit (see also Chapters 10, 15). When, however, this technique is employed directly for the synthesis of terminal polyacetylenes uncontrolled chain growth occurs, because the products are more reactive towards further coupling than their precursor molecules. This problem can be prevented by the use of the triethylsilyl protective group which has the additional advantage that it is stable under the conditions of the so-called Hay coupling,[27] which turned out to be the method of choice for the preparation of the hydrocarbons **2**. In this variation of the Glaser coupling the complex of copper(I) chloride with *N,N,N',N'*-tetramethylethylenediamine (TMEDA) is used as a catalyst; having completed the coupling step the protective group is readily removed by treatment of the products with dilute alkali at room temperature. To prepare the carbon rods, Walton used both homo and mixed couplings, as shown in Schemes 2 and 3.

For the preparation of the hydrocarbons with an even number of triple bonds (**2**, *n* = 4, 6, 8, 12) the examples presented in Scheme 2 are typical.

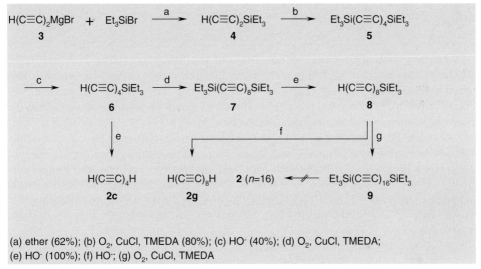

(a) ether (62%); (b) O_2, CuCl, TMEDA (80%); (c) HO$^-$ (40%); (d) O_2, CuCl, TMEDA; (e) HO$^-$ (100%); (f) HO$^-$; (g) O_2, CuCl, TMEDA

Scheme 2. Preparation of carbon rods with an even number of triple bonds.

The Grignard reagent prepared from diacetylene, **3**, was first converted to the monoprotected diyne **4** by quenching with triethylsilyl bromide. Hay coupling next provided the bis-protected tetrayne **5**, which was desilylated completely to **2c** on treatment with very dilute aqueous methanolic alkali. Clearly, this must be a stepwise process, and because the rate of cleavage of **5** is approximately twice that of the monosilylated intermediate **6**, the latter accumulates in the course of the deprotection. If, therefore, this reaction was terminated at a stage when the concentration of **6** had reached its maximum (as determined by ultraviolet spectroscopy, the interruption being accomplished by an acid quench) the 'half-protected' tetrayne **6** could be separated from **5** and **2c** in moderate yield. An additional coupling-deprotection cycle *via* **7** and **8** led to the free octayne **2g**, and because this process could be interrupted at the monoprotected stage again, at **8**, a precursor was at hand for the preparation of **9**. Although doubly blocked, this is a highly reactive compound, and it was impossible to obtain it in pure form. With its 32 consecutive carbon atoms, however, it is the longest fully characterized rod-system prepared to date. That the preparative approach seems to reach its limits at this point can also be inferred from the observation that the deprotection of **9** to the hydrocarbon **2**, with $n = 16$, failed.[28] Starting from the monosilylated derivative of **2b**, **2j** could be obtained by an analogous series of steps.

Mixed Hay oxidative couplings have been used to synthesize **2d**, **f**, and **h**; the route to **2d** is typical (Scheme 3).

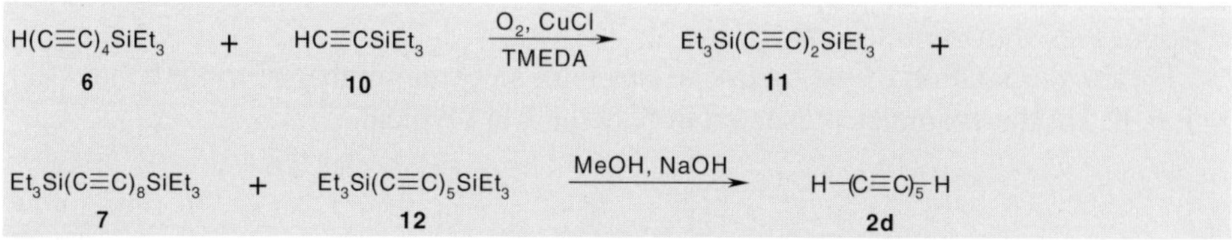

H(C≡C)₄SiEt₃   +   HC≡CSiEt₃ $\xrightarrow[\text{TMEDA}]{\text{O}_2, \text{CuCl}}$ Et₃Si(C≡C)₂SiEt₃   +

    6   **10**   **11**

Et₃Si(C≡C)₈SiEt₃   +   Et₃Si(C≡C)₅SiEt₃ $\xrightarrow{\text{MeOH, NaOH}}$ H—(C≡C)₅—H

    7   **12**   **2d**

Scheme 3. Preparation of carbon rods with an odd number of triple bonds.

Coupling of **6** with a twelvefold excess of triethylsilylacetylene (**10**) led to **12** which was readily deprotected to the desired **2d**. The use of excess **10**, which is less reactive than **6** in coupling reactions, minimizes the formation of the symmetrical coupling product **7**. This and the also produced **11** could be readily removed from the reaction mixture by chromatographic separation.

As expected, substitution of the terminal hydrogen atoms in **2** by the bulky *tert*-butyl or by aryl groups results in a dramatic increase of stability, enabling the use of established methods of conventional acetylene chemistry to prepare the appropriate hydrocarbons. Thus Bohlmann was able[29] to prepare the heptayne **17** from the bis-acetylenic aldehyde **13** and the bis Grignard reagent of **2a**, **14**, as shown in Scheme 4.

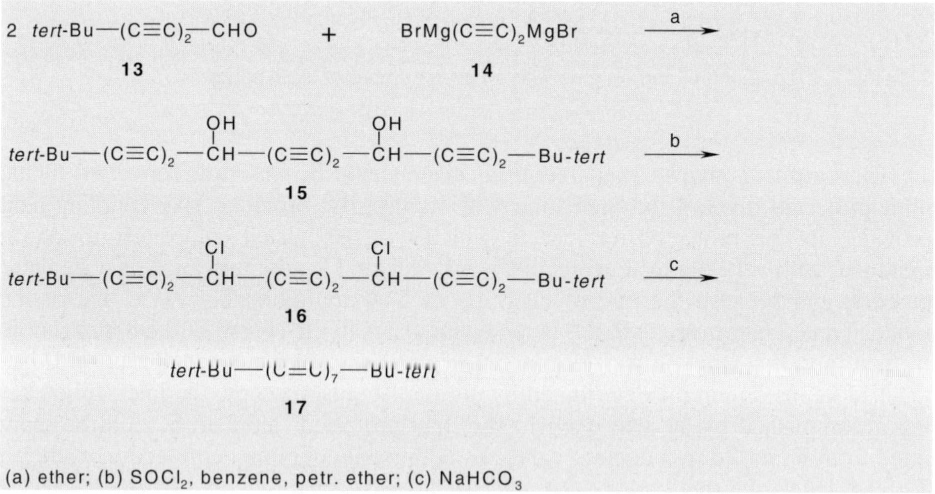

2 *tert*-Bu—(C≡C)₂—CHO   +   BrMg(C≡C)₂MgBr $\xrightarrow{a}$

       13   **14**

                  OH   OH

tert-Bu—(C≡C)₂—CH—(C≡C)₂—CH—(C≡C)₂—Bu-*tert* $\xrightarrow{b}$

                              15

                  Cl   Cl

tert-Bu—(C≡C)₂—CH—(C≡C)₂—CH—(C≡C)₂—Bu-*tert* $\xrightarrow{c}$

                              16

        tert-Bu—(C≡C)₇—Bu-*tert*

                       17

(a) ether; (b) SOCl₂, benzene, petr. ether; (c) NaHCO₃

Scheme 4. Bohlmann's route to more stable polyacetylenes...

The initially formed diol **15** was converted with thionyl chloride to the dichloride **16** which was easily dehydrochlorinated by sodium bicarbonate. The resulting **17**, a crystalline solid, only began to decompose when heated above 150 °C! In a similar vein, the decaacetylene **21** was prepared by Jones and co–workers[30] starting with **13** again but coupling it with the mono Grignard compound **18** derived from 1,4-pentadiyn-3-ol (Scheme 5).

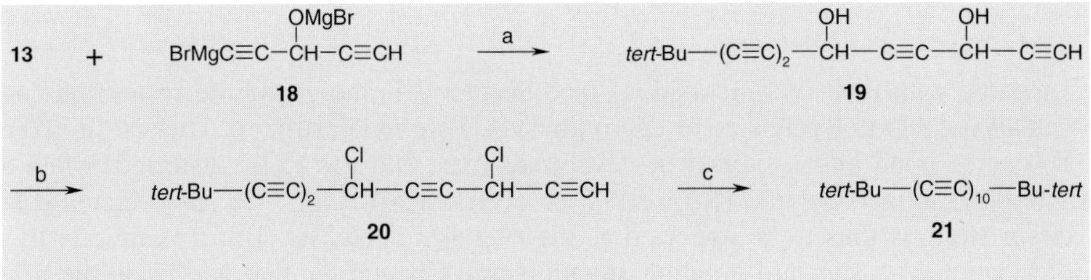

(a) (58%); (b) SOCl$_2$, pyridine; (c) Cu(OAc)$_2$, pyridine (comb. yield 39%)

Scheme 5. ...and an extension of this approach by Jones.

The resulting diol **19** was converted to its dichloride **20**, which on treatment with copper(II) acetate and pyridine, furnished **21**. The 'record holder' in this series is the bis *tert*-butyl derivative **22** (Scheme 6), obtained by a methodology similar to that used for the preparation of the higher polyacetylenes **2** (see above)[31]—a hydrocarbon that can be heated to nearly 50 °C. The stabilizing effect of the *tert*-butyl groups has been ascribed to their preventing too close approach of the polyyne chains to each other.

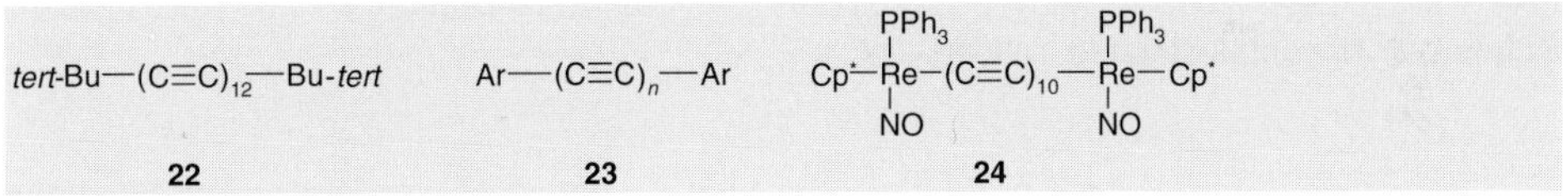

Scheme 6. More carbon rod systems.

Numerous aryl-substituted polyacetylenes **23** are known, and for their preparation the combination of dehydrohalogenation and oxidative coupling has often turned out to be the method of choice.[32–34]

As already mentioned, interest in linear carbon chains built only of acetylenic carbon atoms has recently been reawakened in connection with fullerene formation. In this context it is also of interest that in the presence of capping groups (*tert*-butyl, trifluoromethyl) laser-based synthetic techniques similar to those normally applied to generate fullerenes enable the production from graphite of thermally stable acetylenic carbon species with chain lengths in excess of 300 carbon atoms. These compounds have been called the "sp carbon allotrope";[35] individual compounds have apparently not been isolated and characterized from the complex reaction mixture, however. Another recent renaissance of this area stems from the possible use of polyacetylenes as molecular wires, connecting, *e.g.*, metal centers. Such a molecule is the bis rhenium complex **24** in which the exchange of 'electronic information' along the carbon wire can be studied.[36]

8.2 Angle-strained cycloalkynes

Compared with their olefinic analogs (see Section 7.1) the geometric restrictions on cycloalkynes up to a ring size of approximately 7 are much stricter. After all, it is the 180-degree bond angle, typical of acyclic alkynes, that has to be distorted, when a triple bond is incorporated into a small or even normal ring.[37] A steep increase in Baeyer strain is thus to be expected as the ring size decreases. But it is exactly this deformation of a standard bonding situation (see Chapter 2), that has made the title hydrocarbons attractive study objects, both for the preparative and the theoretical chemist.

As for the small-ring cycloalkenes we shall proceed (Scheme 7) from the (formally) simple cyclopropyne (**25**) through the homologous series all the way to the first 'barely isolable' cycloalkyne, cycloheptyne (**29**), which, as we shall see, still has only a very short half-life. Stable at room temperature are cyclooctyne and its higher homologs.

25   **26**   **27**   **28**   **29**

Scheme 7. The angle-strained cycloalkynes.

On our way, which does not follow the historic development and which has been reviewed several times,[3, 38–43] we shall also consider stabilization of the parent hydrocarbons by either annelation or by protecting alkyl groups, whereas the—well-known—stabilizing effect of exchanging methylene groups for hetero atoms will only be mentioned in selected cases.

For the lower members of the series, **25**[44] and **26**,[45, 46] so far only theoretical data exist—experimental attempts to generate **26** failed.[47, 48] Theoretical calculations suggest that the singlet of **25** is a transition state for the degenerate rearrangement of propadienylidene lying about 45 kcal mol^{-1} above the energy of this simplest cumulenic carbene system.[44, 49]

Solid experimental ground is reached with cyclopentyne (**27**), which has been generated by various routes; three of them, which represent conceptually very different approaches, are summarized in Scheme 8.

Thus the geminal dibromide **30** on treatment with phenyl lithium is debrominated, and it is reasonable to assume that this step results in the formation of the vinylidene carbene **33**. This, in turn, can undergo the well-known vinylidenecarbene→acetylene rearrangement[50] to provide **27**. That the cycloalkyne has actually been generated was

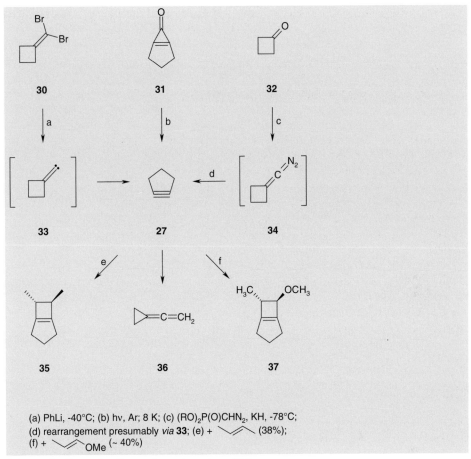

(a) PhLi, -40°C; (b) hν, Ar; 8 K; (c) (RO)$_2$P(O)CHN$_2$, KH, -78°C;
(d) rearrangement presumably *via* **33**; (e) + ∕∖∕∖ (38%);
(f) + ∕∖∕∖OMe (~ 40%)

Scheme 8. Several pathways to cyclopentyne (**27**).

shown by different trapping experiments.[51] Thus, in the presence of *trans*-2-butene the bicyclic olefin **35** was formed. When the cyclopropenone **31**, itself produced by photodecomposition of 2,6-diazocyclohexanone in an argon matrix, was further irradiated at 8 K it lost carbon monoxide, and again **27** was produced.[52] Originally, only a secondary product of **27**, the allene **36**, could be observed; however, in later experiments this 1,3-carbon migration process could be controlled and the IR spectrum of **27** could be recorded in matrix.[53] Finally, treatment of cyclobutanone (**32**) with dialkyl (diazomethyl) phophonates in the presence of potassium hydride at low temperature first led to the diazo compound **34**, which by way of nitrogen loss was converted to **27**, presumably *via* **33** also. A trapping experiment in this case with *trans*-1-methoxypropene yielded the *trans* cycloadduct **37**.[54]

Cyclopentyne can be stabilized in various ways,[55] one possibility is presented in Scheme 9 which shows **27** as a substructure of the aromatic hydrocarbon acenaphthyne (**40**).

Thus photoextrusion of either nitrogen or nitrogen and carbon monoxide from the

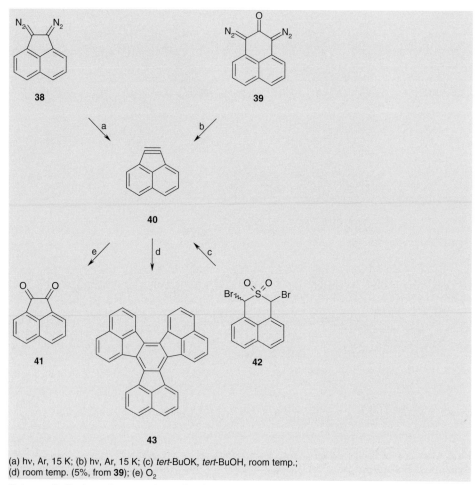

(a) hν, Ar, 15 K; (b) hν, Ar, 15 K; (c) *tert*-BuOK, *tert*-BuOH, room temp.;
(d) room temp. (5%, from **39**); (e) O$_2$

Scheme 9. Cyclopentyne condensed to an aromatic system.

precursor molecules **38** and **39**, respectively, in a matrix led to **40**[52] as did subjecting the dibromosulfone **42** to the conditions of the Ramberg-Bäcklund rearrangement.[56] Chemical proof of the intermediate formation of **40** was again provided by trapping experiments—addition of oxygen led to the quinone **41** and self-trapping to a trimer of **40**, decacyclene (**43**).

Among the angle strained cycloalkynes cyclohexyne (**20**) has received by far the greatest attention, ever since Wittig, Roberts, Blomquist and others opened up this area of acetylene chemistry between 1950 and 1960.[57] Since this classical work has been summarized several times[3, 6, 9–12, 43] only the most important routes to **28** will be presented here (Scheme 10).

In the debromination of 1,2-dibromocyclohexene (**44**)[58] and the dehydrochlorination of 1-chlorocyclohexene (**45**)[59] the relationship between the substrate and the desired product **28** is particularly obvious. The oxidation of **46** can be rationalized by invoking a biscarbene intermediate,[60] and the last two approaches could proceed *via*

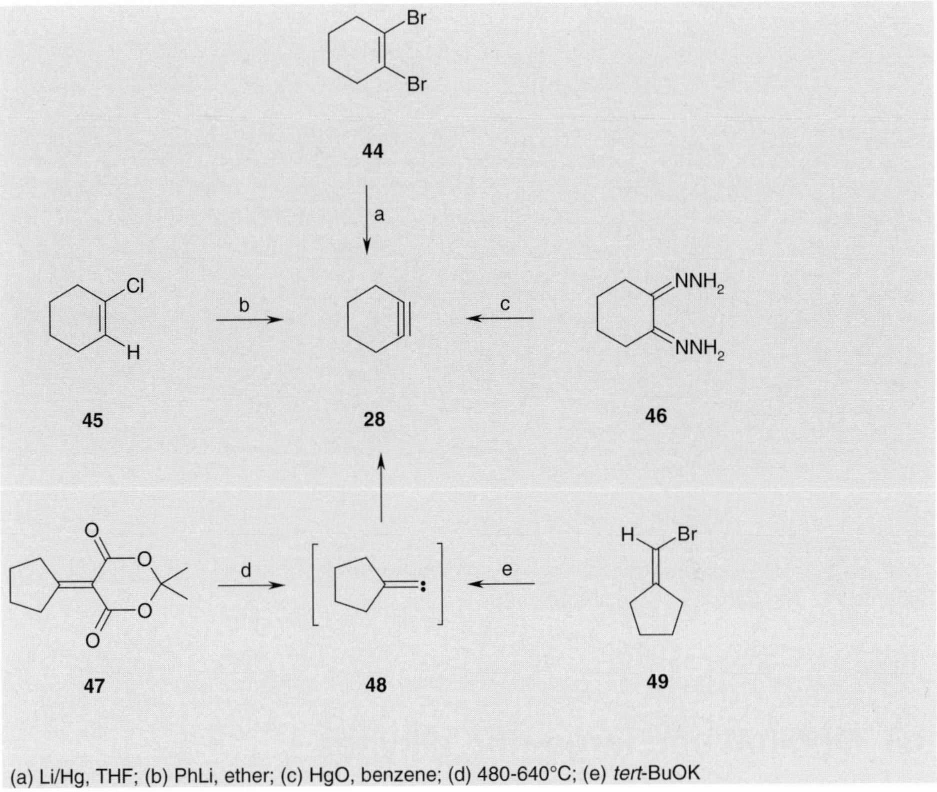
Br
Br
44
a
Cl
H
b
c
NNH2
NNH2
45
28
46
O
O
O
O
d
e
H
Br
47
48
49
(a) Li/Hg, THF; (b) PhLi, ether; (c) HgO, benzene; (d) 480-640°C; (e) tert-BuOK

Scheme 10. The generation of cyclohexyne (**28**) by different routes.

48, the higher homolog of **33**. As shown, this unsaturated carbene is either formed by flash-vacuum pyrolysis of the condensation product between cyclopentanone and Meldrum's acid, **47**,[61] or by dehydrobromination of the monobromide **49**.[62]

Again, because the cycloalkyne is too reactive to be isolated its mode of formation and its structure proof rest on matrix isolations studies or on trapping experiments; some typical examples are summarized in Scheme 11.

Thus Diels–Alder addition of tetracyclone (**50**) provided—after CO extrusion—the tetralin derivative **51**;[62] stabilizing the reactive triple bond in **28** with tris(triphenylphosphine)platinum(0) (**52**) allowed isolation of the metal complex **53**[63, 64] and trimerization furnished dodecahydrotriphenylene (**54**).[56, 58] Neither generation nor proof of existence of cycloheptyne (**29**) differ in principle from the methods employed for the preparation of its lower homologs; with this hydrocarbon, however, we are beginning to reach the shores of stability. In dilute dichloromethane solution at –25 °C **29** has a half-life of less than a minute, but at –78 °C this has already increased to 1 h.[65] Recalling the discussion on the stabilizing effect of *tert*-butyl groups on polyacetylenes (see above, Section 8.1), it does not come as a surprise that α-*gem*-dimethylation of a strained cycloalkyne also increases its stability—3,3,7,7-tetramethylcycloheptyne is a stable hydrocarbon at room temperature,[66, 67] as are the higher homologs of **25**-**29**.

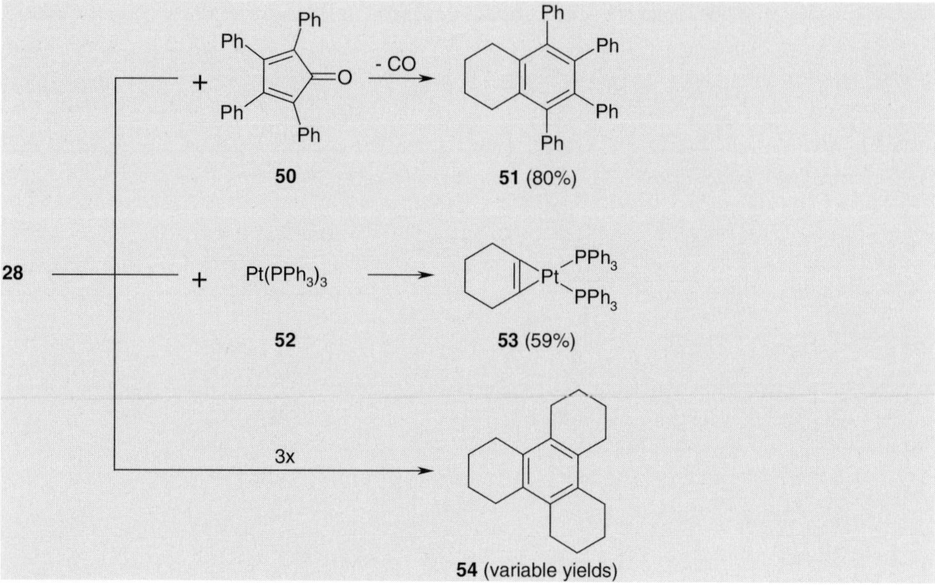

Scheme 11. The trapping of cyclohexyne (**28**).

8.3 Medium and large-ring alkynes

The medium-ring acetylenes (*i.e.* the 8- to 11-membered ring compounds) with one triple bond are all isolable compounds which can be prepared by the commonly used methods of acetylene chemistry, *e.g.* base-induced elimination of the appropriate 1,1- or 1,2-dihalocycloalkanes or 1-halocycloalkenes, oxidation of 1,2-bishydrazones *etc.*[6, 9-12] When basic reagents are employed in these processes secondary reactions, *viz.* isomerizations to cycloallenes or 1,3-cycloalkadienes can occur.

More reactive hydrocarbons are obtained when additional unsaturation is introduced into the ring systems. A typical example is provided by 1,5-cyclooctadiyne (**56**) which was first prepared by the dimerization of butatriene (**55**).[68] A much better yield of this interesting hydrocarbon was obtained, when the dibromide **57** was subjected to treatment with base in the presence of a crown ether (Scheme 12).[69]

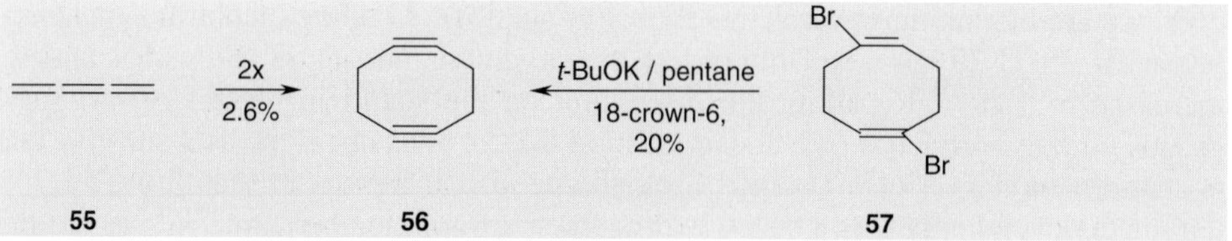

Scheme 12. Two routes to 1,5-cyclooctadiyne (**56**).

Benzo derivatives of **56** are known in which the 'ethano bridges' of the hydrocarbon have been replaced by benzene rings; because these are fully conjugated molecules they will be discussed later in the chapter on annulenes (Chapter 10).

Higher homologs of **56** have been known for decades; originally they were prepared by treating α,ω-dibromides (**58**) with a mixture of sodium acetylide and disodium acetylide in liquid ammonia.[70–72] In a shorter approach, however, the dibromides **58** can also be treated with bis salts of α,ω-diacetylenes, **59**, providing the cyclic hydrocarbons, *e.g.* **61** and **62** often in surprisingly high yields (up to *ca* 60%) (Scheme 13).

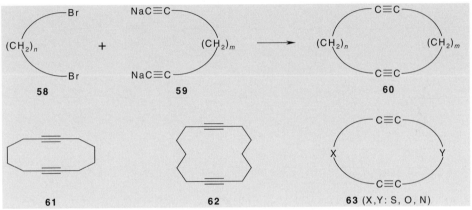

Scheme 13. The synthesis of macrocyclic non-conjugated diacetylenes.

The scope of this approach is illustrated not only by its application to the synthesis of hydrocarbons of type **60** but also by the preparation of various heterocyclic variants of general structure **63**.[73] Furthermore, nucleophilic substitution is also the method of choice when yet higher unsaturated macro rings are required. For example the diacetylenes **70–72**[74, 75] were prepared by Gleiter and co–workers from the dihalides **64** and **65** by first extended these to the acyclic diacetylenes **66** and **67**, respectively;[76] these could finally be ring-closed to the target molecules by coupling either with the dibromide **68** or the diiodide **69** (Scheme 14).

Rather than creating the macro ring in the last step it is sometimes advantageous to begin the synthesis with the complete ring already and subsequently introduce the triple bonds. As shown in Scheme 15 this approach has been successfully used to convert 1,6-cyclodecadione (**73**) into 1,6-cyclodecadiyne (**76**).[77]

Towards this end **73** was first transformed into the bis semicarbazone **74**. When this was treated with selenium dioxide it was converted to the bis selenadiazole **75**, which, on pyrolysis in the presence of copper powder, fragmented to the desired decadiyne **76**. The latter process is well-known in acetylene chemistry and has been applied many times.[78, 79]

The preparation of these medium or large-ring diacetylenes has been highlighted here, not only because these hydrocarbons are important model compounds in connection with a better understanding of the mode of action of the endiyne antibiot-

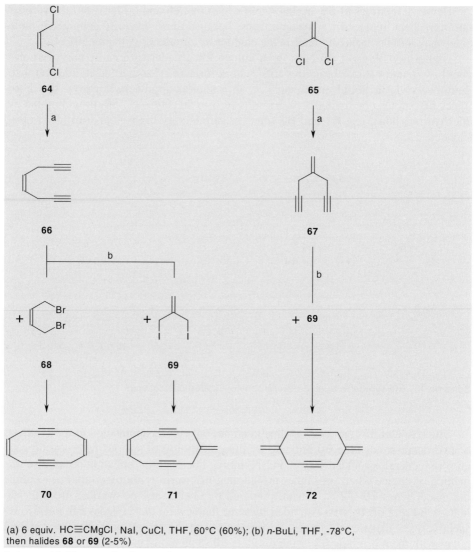

(a) 6 equiv. HC≡CMgCl, NaI, CuCl, THF, 60°C (60%); (b) *n*-BuLi, THF, -78°C,
then halides **68** or **69** (2-5%)

Scheme 14. Macrocyclic diynes with additional double bonds.

ics,[73, 80] but they also have interesting structural and spectroscopic behavior (interaction between the triple bonds through space) and have been used as precursors for a range of new compounds of the superphane type,[73] as will be discussed in Section 12.3.

When two unsaturated structural elements are separated by a methylene group we describe this as 'skipped' bonding. This not only occurs in important natural products like arachidonic acid, the parent compound of the prostaglandins, but also in man-designed molecules. An interesting situation arises when in cyclic compounds double and/or triple bonds are separated by sp^3-hybridized carbon atoms, because the mole-

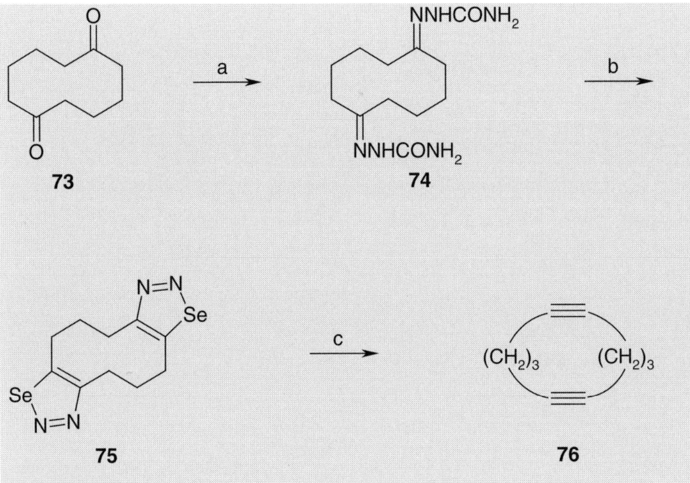

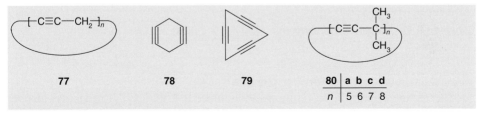

(a) $H_2NNHCONH_2$, EtOH (94%); (b) SeO_2, dioxan (36%); (c) Δ, Cu (52%)

Scheme 15. The conversion of a medium-sized precursor into a medium-sized diyne.

cules thus generated can show the phenomenon of *homoconjugation*[81] (see Section 13.4). As far as acetylenes are concerned this situation has recently been realized in the so-called [*n*]pericyclynes,[82, 83] the parent systems of which possess the general structure **77** (Scheme 16).

80	a	b	c	d
n	5	6	7	8

77   **78**   **79**   **80**

Scheme 16. The [*n*]pericyclynes.

That the monomer cyclopropyne (**25**) and the dimer 1,4-cyclohexadiyne (**78**) have not been generated is not surprising in view of the extreme reactivity of these highly angle-strained hydrocarbons (see Section 8.2, above). However, the trimer **79** and any of the higher homologs thereof are also presently unknown, most probably because of the inherent strain of these molecules and the increased reactivity of the bare methylene bridging unit which is, after all, activated by two flanking triple bonds. Substituting its hydrogen atoms by blocking methyl groups, however, results in the corresponding permethylated [*n*]pericyclynes, which are isolable and stable hydrocarbons; the whole series **80a–d** has been prepared.[83]

The preparative methodology employed—another example of the building-block approach so typical of polyacetylene chemistry—is illustrated in Scheme 17 which

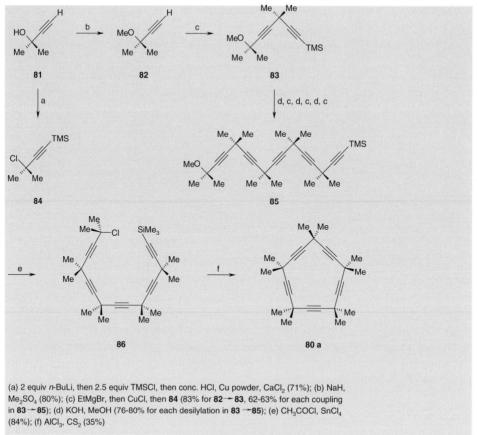

(a) 2 equiv *n*-BuLi, then 2.5 equiv TMSCl, then conc. HCl, Cu powder, CaCl₂ (71%); (b) NaH, Me₂SO₄ (80%); (c) EtMgBr, then CuCl, then **84** (83% for **82** → **83**, 62-63% for each coupling in **83** → **85**); (d) KOH, MeOH (76-80% for each desilylation in **83** → **85**); (e) CH₃COCl, SnCl₄ (84%); (f) AlCl₃, CS₂ (35%)

Scheme 17. The preparation of permethyl [5]pericyclyne (**80a**).

presents one route to the [5]pericyclyne **80a**[82, 84].

The monomer unit of the future cyclooligomer is 1,1-dimethyl-propargyl alcohol (**81**) which was converted either into its ether **82** or to the protected chloride **84**, the coupling unit for subsequent chain extension to the open-chain dimer **83** and—after several repetitive coupling cycles—to the acyclic pentamer **85**. On treatment with acetyl chloride in the presence of tin tetrachloride this was converted to the chloride **86** which was finally cyclized to the hydrocarbon **80a**.[84] The homologs **80b–c** were prepared by analogous convergent routes,[85] and with the preparative knowledge accumulated in the course of these studies variations of the building blocks could be explored. For example, when 1,3-diyne spacers were inserted between the permethylated corners, 'exploded pericyclynes' such as **87** (and its higher homologs) could be synthesized.[86] A combination of building blocks containing one or two triple bonds led to homoconjugated mixed polyalkynes/diyne macrocycles such as the box-like structure **88**,[87] and exploded [*n*]rotanes (see Section 11.6). The completely spirocyclopropanated macrocyclic oligodiacetylenes **89** were obtained[88] (Scheme 18) from 1,1-diethynyl-cyclopropane[89] as the starting monomer system.

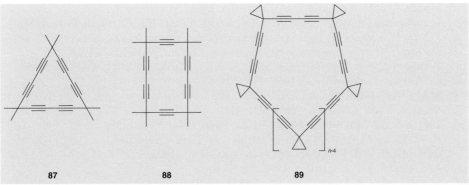

Scheme 18. A selection of extended [*n*]pericyclynes.

Large-ring polyacetylenes in which triple bond sections alternate with saturated units other than methylene are finally of great importance in annulene chemistry, a fascinating chapter of hydrocarbon chemistry which will be addressed in Section 10.4.

References

1. R. A. Raphael, *Acetylenic Compounds in Organic Synthesis*, Butterworths Scientific Publications, London, **1955**.
2. W. Ziegenbein, *Äthinylierung und Alkinylierung*, Verlag Chemie, Weinheim, **1963**.
3. R. W. Hoffmann, *Dehydrobenzene and Cycloalkynes*, Academic Press, New York, N. Y., **1967**.
4. T. F. Rutledge, *Acetylenic Compounds: Preparation and Substitution Reactions*, Reinhold Book Corporation, New York, N. Y., **1968**.
5. T. F. Rutledge, *Acetylenes and Allenes: Additions, Cyclization and Polymerization Reactions*, Reinhold Book Corporation, New York, N. Y., **1969**.
6. H. G. Viehe (*Ed.*), *Chemistry of Acetylenes*, Marcel Dekker, New York, N.Y., **1969**.
7. F. Bohlmann, T. Burkhardt, C. Zdero, *Naturally Occurring Acetylenes*, Academic Press, London, **1973**.
8. M. F. Shostakovskii, A. V. Bogdanova, *The Chemistry of Diacetylenes*, John Wiley and Sons, New York, N. Y., **1974**; *Polydiacetylenes* H-J. Cantow, (*Ed.*), Springer-Verlag, Berlin, **1984**; *Polydiacetylenes: Synthesis, Structure and Electronic Properties* D. Bloor, R. R. Chance (*Eds.*), Martinus Nijhoff Publishers, Dordrecht, **1985**.
9. S. Patai (*Ed.*), *The Chemistry of the Carbon-Carbon Triple Bond*, Part I, John Wiley and Sons, Chichester, **1978**.
10. S. Patai (*Ed.*), *The Chemistry of the Carbon-Carbon Triple Bond*, Part 2, John Wiley and Sons, Chichester, **1980**.
11. Houben-Weyl, *Methoden der Organischen Chemie, Alkine, Di- und Polyine, Allene und Cumulene*, Vol. V/2a, Georg Thieme Verlag, Stuttgart, **1981**.
12. S. Patai, Z. Rappoport (*Eds.*), *The Chemistry of Triple-bonded Functional Groups*, Sup-

plement C, John Wiley and Sons, Chichester, **1983**.

13. L. Brandsma, *Preparative Acetylenic Chemistry*, 2nd ed., Elsevier, Amsterdam, **1988**.

14. J. Lam, H. Breteler, T. Arnason, L. Hansen (*Eds.*), *Chemistry and Biology of Naturally Occurring Acetylenes and Related Compounds*, Elsevier, Amsterdam, **1988**.

15. F. Diederich, P. J. Stang (*Eds.*), *Modern Acetylene Chemistry*, VCH, Weinheim, **1996**.

16. G. Wegner, *Z. Naturforsch.*, **1969**, *24b*, 824–832; G. Wegner, *Angew. Chem.*, **1971**, *83*, 368; *Angew. Chem. Int. Ed. Engl.*, **1971**, *10*, 355; U. Jonas, K. Shah, S. Norvez, D. H. Charych, *J. Am. Chem. Soc.*, **1999**, *121*, 4580–4588; Review: G. Wegner, *Angew. Chem.*, **1981**, *93*, 352–371; *Angew. Chem. Int. Ed. Engl.*, **1981**, *20*, 361–380.

17. H. W. Kroto, D. R. M. Walton in *Carbocyclic Cage Compounds*, E. Osawa, O. Yonemitsu (*Eds.*), VCH Publishers, Inc. New York, N. Y., **1992**.

18. N. S. Goroff, *Acc. Chem. Res.*, **1996**, *29*, 77–83.

19. A. v. Baeyer, *Ber. Dtsch. Chem. Ges.*, **1885**, *18*, 2269–2281.

20. W. Hunsmann, *Chem. Ber.*, **1950**, *83*, 213–217.

21. F. Bohlmann, *Chem. Ber.*, **1951**, *84*, 785–794.

22. J. B. Armitage, C. L. Cook, E. R. H. Jones, M. C. Whiting, *J. Chem. Soc.*, **1952**, 2010–2014.

23. C. L. Cook, E. R. H. Jones, M. C. Whiting, *J. Chem. Soc.*, **1952**, 2883–2891. It does not depreciate these early accomplishments when one states that the purity of the original samples probably was not very high. On the other hand the preparative expertise required to first obtain these polyacetylenes in analytically pure form should also not be underrated: E. Kloster-Jensen, *Angew. Chem.*, **1972**, *84*, 483–485; *Angew. Chem. Int. Ed. Engl.*, **1972**, *11*, 438–439.

24. R. Eastmond, T. R. Johnson, D. R. M. Walton, *Tetrahedron*, **1972**, *28*, 4601–4616; *cf.* R. Eastwood, D. R. M. Walton, *J. Chem. Soc. Chem. Commun.*, **1968**, 204–205.

25. C. Glaser, *Ber. Dtsch. Chem. Ges.*, **1869**, *2*, 422–424; Review: P. Cadiot, W. Chodkiewicz in *Chemistry of Acetylenes* H. G. Viehe (*Ed.*), Marcel Dekker, New York, N. Y., **1969**, 597–647.

26. G. Eglinton, W. McRae, *Adv. Org. Chem.*, **1963**, *4*, 225: Review: W. G. Nigh in *Oxidation in Organic Chemistry* W.S. Trahanovsky, (*Ed.*), Academic Press, New York, N.Y., Part B, **1978**, 11–31.

27. A. S. Hay, *J. Org. Chem.*, **1962**, *27*, 3320–3321.

28. Looking at the calculated elemental analysis of this compound (C 99.48%, H 0.52%), the hexadecayne may be called 'linear carbon with a small impurity of hydrogen'.

29. F. Bohlmann, *Chem. Ber.*, **1953**, *86*, 657–667.

30. E. R. H. Jones, H. H. Lee, M. C. Whiting, *J. Chem. Soc.*, **1960**, 3483–3489.

31. T. R. Johnson, D. R. M. Walton, *Tetrahedron*, **1972**, *28*, 5221–5236.

32. H. H. Schlubach, V. Franzen, *Liebigs Ann. Chem.*, **1951**, *572*, 116–121.

33. S. Akiyama, M. Nakagawa, *Bull. Chem. Soc. Japan*, **1967**, *40*, 340.

34. M. Nakagawa, S. Akiyama, K. Nakasuji, K. Nishimoto, *Tetrahedron*, **1971**, *27*, 5401–5418 and refs. cited.

35. R. J. Lagow, J. J. Kampa, H.-C. Wei, S. L. Batle, J. W. Genge, D. A. Lude, C. J. Harper, R. Bau, R. C. Stevens, J. F. Haw, E. Munson, *Science*, **1995**, *267*, 362–366 and the extensively cited literature.

36. T. Bartik, B. Bartik, M. Brady, R. Dembinski, J. A. Gladysz, *Angew. Chem.*, **1996**, *108*, 467–469; *Angew. Chem. Int. Ed. Engl.*, **1996**, *35*, 414–417.

37. Krebs[41] has proposed to call these hydrocarbons 'angle-strained cycloalkynes'; I follow this suggestion since 'small-ring cycloalkynes' would include only the three- and the four-membered ring systems and hence not consider that the strain-induced reactivity increase 'lasts' all the way up to the seven-membered ring.

38. A. Krebs in H. G. Viehe (*Ed.*), *Chemistry of Acetylenes*, Marcel Dekker, New York, N. Y., **1969**, 987- 1062.
39. H. Meier, *Synthesis*, **1972**, 235–253.
40. M. Nakagawa in S. Patai (*Ed.*), *The Chemistry of the Carbon-Carbon Triple Bond*, J. Wiley and Sons, New York, N. Y., **1978**, 635–712.
41. A. Krebs, J. Wilke, *Top. Curr. Chem.*, **1983**, *109*, 189–233.
42. R. P. Johnson in *Molecular Structure and Energetics* J. Liebman, A. Greenberg (*Eds.*), VCH Publishers, Deerfield Beach, **1986**, Vol. 3, 85–140.
43. C. Wentrup, *Reactive Molecules*, Wiley-Interscience, New York, N. Y., **1984**.
44. R. S. Grev, H. F. Schaefer III, *J. Chem. Phys.*, **1984**, *80*, 3552–3555 and lit. quoted; *cf.* D. L. Michalopoulos, M. E. Geusic, P. R. R. Langridge-Smith, R. E. Smalley, *J. Chem. Phys.*, **1984**, *80*, 3556–3560.
45. G. Fitzgerald, P. Saxe, H. F. Schaefer, III, *J. Am. Chem. Soc.*, **1983**, *105*, 690–695.
46. H. Kollmar, F. Carrion, M. J. S. Dewar, R. C. Bingham, *J. Am. Chem. Soc.*, **1981**, *103*, 5292–5303.
47. L. K. Montgomery, J. D. Roberts, *J. Am. Chem. Soc.*, **1960**, *82*, 4750–4751.
48. G. Wittig, E. R. Wilson, *Chem. Ber.*, **1965**, *98*, 451–457. It should be mentioned though, that recently 3-silacyclopropyne has been prepared and characterized in a matrix: G. Maier, H. P. Reisenauer, H. Pacl, *Angew. Chem.*, **1994**, *106*, 1347–1349; *Angew. Chem. Int. Ed. Engl.*, **1994**, *33*, 1248–1250; *cf.* G. Maier, H. Pacl, H. P. Reisenauer, A. Meudt, R. Janoshek, *J. Am. Chem. Soc.*, **1995**, *117*, 12712–12720.
49. G. Fitzgerald, H. F. Schaefer III, *Isr. J. Chem.*, **1983**, *23*, 93–96.
50. R. F. C. Brown, F. W. Eastwood, K. J. Harrington, G. L. McMullen, *Austr. J. Chem.*, **1974**, *27*, 2393–2402; R. F. C. Brown, K. J. Harrington, G. L. McMullen, *J. Chem. Soc., Chem. Commun.*, **1974**, 123–124; Review: R. F. C. Brown, *Pyrolytic Methods in Organic Chemistry*, Academic Press, New York, N. Y., **1980,** 124.
51. L. Fitjer, S. Modaressi, *Tetrahedron Lett.*, **1983**, *24*, 5495–5498; *cf.* L. Fitjer, U. Klie- bisch, D. Wehle, S. Modaressi, *Tetrahedron Lett.*, **1982**, *23*, 1661–1664.
52. O. L. Chapman, J. Gano, P. R. West, M. Regitz, G. Maas, *J. Am. Chem. Soc.*, **1981**, *103*, 7033–7036.
53. As reported by R. P. Johnson in ref.[42], 102.
54. J. C. Gilbert, M. E. Baze, *J. Am. Chem. Soc.,* **1984**, *106*, 1885–1886: *cf.* J. C. Gilbert, M. E. Baze, *J. Am. Chem. Soc.*, **1983**, *105*, 664–665; J. C. Gilbert, D.-R. Hou, J. W. Grimme, *J. Org. Chem.*, **1999**, *64*, 1529–1534.
55. For the incorporation of cyclopentyne into a bicyclic framework (norbornyne) see P. G. Gassman, I. Gennick*, J. Am. Chem. Soc.*, **1980**, *102*, 6863–6864 and references therein. For the intermediate generation of bicyclo[2.2.1]hept-2-en-5-yne see T. Kitamura, M. Kotani, T. Yokoyama, Y. Fujiwara, K. Mori, *J. Org. Chem.*, **1999**, *64*, 680–691. For sta- bilization provided by a hetero atom in the ring plus neighboring *gem*-dimethyl substitu- ents (4-thia-3,3,5,5-tetramethylcyclopentyne) see J. M. Bolster, R. M. Kellog, *J. Am. Chem. Soc.*, **1981**, *103*, 2868–2869.
56. J. Nakayama, E. Ohshima, A. Ishii, M. Hoshino, *J. Org. Chem.*, **1983**, *48*, 60–65 and lit. quoted.
57. Attempts to prepare cycloalkynes at all are much older, see *e.g.* M. Blumenthal, *Ber. Dtsch. Chem. Ges.*, **1874**, *7*, 1092–1095.
58. G. Wittig, U. Mayer, *Chem. Ber.*, **1963**, *96*, 342–348.
59. G. Wittig, G. Harborth, *Chem. Ber.*, **1944**, 77, 306–314; *cf.* F. Scardiglia, J. D. Roberts, *Tetrahedron,* **1957**, *1*, 343–343.
60. G. Wittig, A. Krebs, *Chem. Ber.*, **1961**, 94, 3260–3275; *cf.* G. Wittig, R. Pohlke, *Chem. Ber.*, **1961**, *94*, 3276–3286.

61. G. J. Baxter, R. F. C. Brown, *Austr. J. Chem.*, **1978**, *31*, 327–339. The pyrolysis of the lower homolog of **47** is also described in this publication. Although products are formed which could originate from **27**, the authors also discuss reasonable alternatives not involving this strained cycloalkyne.

62. K. L. Erickson, J. Wolinsky, *J. Am. Chem. Soc.*, **1965**, *87*, 1142–1143.

63. M. A. Bennett, G. B. Robertson, P. O. Whimp, T. Yoshida, *J. Am. Chem. Soc.*, **1971**, *93*, 3797–3798.

64. G. B. Robertson, P. O. Whimp, *J. Am. Chem. Soc.*, **1975**, *97*, 1051–1059.

65. G. Wittig, J. Meske-Schüller, *Liebigs Ann. Chem.*, **1968**, *711*, 65–75; *cf.* F. G. Willey, *Angew. Chem.*, **1964**, *76*, 144; *Angew. Chem. Int. Ed. Engl.*, **1964**, *3*, 138.

66. A. Krebs, H. Kimling, *Angew. Chem.*, **1971**, *83*, 540–541; *Angew. Chem. Int. Ed. Engl.*, **1971**, *10*, 509–510.

67. S. F. Karaev, A. Krebs, *Tetrahedron Lett.*, **1973**, 2853–2854.

68. E. Kloster-Jensen, J. Wirz, *Angew. Chem.*, **1973**, *85*, 723; *Angew. Chem. Int. Ed. Engl.*, **1973**, *12*, 671; *cf.* E. Kloster-Jensen, J. Wirz, *Helv. Chim. Acta*, **1975**, *58*, 162–177.

69. H. Detert, B. Rose, W. Mayer, H. Meier, *Chem. Ber.* **1994**, *127*, 1529–1532. Reviews: H. Meier, N. Hanold, T. Molz, H. J. Bissinger, H. Kolshorn, J. Zountsas, *Tetrahedron*, **1986**, *42*, 1711–1719; H. Meier in *Advances in Strain in Organic Chemistry* B. J. Halton, (*Ed.*), The JAI Press, Greenwich, CT, Vol. 1, **1991**, 215–272; R. Gleiter, R. Merger, *Modern Acetylene Chemistry* F. Diederich, P. J. Stang (*Eds.*), VCH, Weinheim **1995**, 285–319.

70. J. H. Wotiz, R. F. Adam, C. G. Parsons, *J. Am. Chem. Soc.*, **1961,** *83*, 373–376.

71. J. Dale, A. J. Hubert, G. S. D. King, *J. Chem. Soc.*, **1963**, 73–86; *cf.* J. Dale, *J. Chem. Soc.*, **1963**, 93–111.

72. G. Schill, U. Keller, *Synthesis*, **1972**, 621–627; *cf.* G. Schill, E. Logemann, H. Fritz, *Chem. Ber.*, **1976**, *109*, 497–502.

73. Reviews: R. Gleiter, *Angew. Chem.*, **1992**, *104*, 29–46; *Angew. Chem. Int. Ed. Engl.*, **1992**, *31*, 27–44; R. Gleiter, R. Merger in *Modern Acetylene Chemistry* F. Diederich, P. J. Stang (*Eds.*), VCH Verlagsgesellschaft, Weinheim, **1995**, 285–319.

74. R. Gleiter, R. Merger, *Tetrahedron Lett.*, **1989**, *30*, 7183–7186.

75. R. Gleiter, R. Merger, *Tetrahedron Lett.*, **1990**, *31*, 1845–1848.

76. B. E. Looker, F. Sondheimer, *Tetrahedron,* **1971**, *27*, 2567–2571; *cf.* H. Hopf, T. Lehrich, *Tetrahedron Lett.*, **1987**, *28*, 2697–2700.

77. R. Gleiter, M. Karcher, R. Jahn, H. Irngartinger, *Chem. Ber.*, **1988**, *121*, 735–740; *cf.* R. Gleiter, D. Kratz, V. Schehlmann, *Tetrahedron Lett.,* **1988**, *29*, 2813–2816.

78. I. Lalezari, A. Shafiee, M. Yalpani, *Angew. Chem.*, **1970**, *82*, 484; *Angew. Chem. Int. Ed. Engl.*, **1970**, *9*, 464–465.

79. H. Meier, H. Petersen, *Synthesis,* **1978**, 596–598 Review: H. Meier in *Advances in Strain in Organic Chemistry* B. Halton (*Ed.*), The JAI Press, Greenwich, CT, Vol. 1, **1991**, 215–272

80. Review: K. C. Nicolaou, A. L. Smith in *Modern Acetylene Chemistry* F. Diederich, P. J. Stang (*Eds.*). VCH, Weinheim **1996**, 203–283.

81. K. N. Houk R. W. Gandour, R. W. Strozier, N. G. Rondan, L. A. Paquette, *J. Am. Chem. Soc.*, **1979**, *101*, 6797–6802 and references therein; *cf.* K. A. Klingensmith, W. Püttmann, E. Vogel, J. Michl, *J. Am. Chem. Soc.*, **1983**, *105*, 3375–3380.

82. L. T. Scott, G. J. DeCicco, J. L. Hyun, G. Reinhardt, *J. Am. Chem. Soc.*, **1983**, *105*, 7760–7761.

83. For a review see: L. T. Scott, M. J. Cooney in *Modern Acetylene Chemistry* F. Diederich, P. J. Stang (*Eds.*), VCH Verlagsgesellschaft, Weinheim, **1995**, 321–351.

84. For the preparation of large ring cycloalkynes by intramolecular Friedel-Crafts acylation of trimethylsilyl acetylenes see K. Utimoto, M. Tanaka, M. Kitai, H. Nozaki, *Tetrahe-*

dron Lett., **1978**, 2301–204.

85. L. T. Scott, G. J. DeCicco, J. L. Hyun, G. Reinhardt, *J. Am. Chem. Soc.*, **1985**, *107*, 6546–6555; *cf*. K. N. Houk, L. T. Scott, N. G. Rondan, D. C. Spellmeyer, G. Reinhardt, J. L. Hyun, G. J. DeCicco, R. Weiss, M. H. M. Chen, L. S. Bass, J. Clardy, F. S. Jørgensen, T. A. Eaton, V. Sarkozi, C. M. Petit, L. Ng, K. D. Jordan, *J. Am. Chem. Soc.*, **1985**, *107*, 6556–6562.

86. L. T. Scott, M. J. Cooney, D. Johnels, *J. Am. Chem. Soc.*, **1990**, *112*, 4054–4055.

87. M. J. Cooney, Ph. d. Dissertation, University of Nevada, Reno, **1993**.

88. A. de Meijere, S. Kozhushkov, Th. Haumann, R. Boese, C. Puls, M. J. Cooney, L. T. Scott, *Chem. Eur. J.*, **1995**, *1*, 124–131; S. Kozhushkov, Th. Haumann, R. Boese, B. Knierim, S. Scheib, P. Bäuerle, A. de Meijere, *Angew. Chem.*, **1995**, *107*, 859–861; *Angew. Chem. Int. Ed. Engl.*, **1995**, *34*, 781–783; *cf*. L. T. Scott, M. J. Cooney, C. Otte, C. Puls, Th. Haumann, R. Boese, P. J Carroll, A. B. Smith III, A. de Meijere, *J. Am. Chem. Soc.*, **1994**, *116*, 10275–10283 and references cited therein. For a recent review on pericyclynes see: S. I. Kozhushkov, A. de Meijere, *Top. Curr. Chem.*, **1999**, *201*, 1–42.

89. N. S. Zefirov, S. I. Kozhushkov, T. S. Kuznetsova, R. Gleiter, M. Eckert-Maksic, *Zh. Org. Khim.*, **1986**, *22*, 110–121; *J. Org. Chem. USSR*, **1986**, *22*, 95–105.

9 Allenes and Cumulenes

According to Richard Kuhn, a pioneer in this area of hydrocarbon chemistry as well (see also Section 7.1) cumulenes are compounds with *n* carbon atoms with (*n*-1) double bonds between them, *n* being equal or exceeding 3.[1] For the smallest member of the series and its derivatives the name *allenes* is commonly employed. Although the first allenes were described as early as 1864 already,[2] and this name was proposed for propadiene in 1875,[3] these highly unsaturated compounds, which even were shown to occur in nature (as constituents of pyrethrum flowers) at a relatively early date,[4–6] remained a curiosity for almost a century. Cumulenes were primarily studied for structural, *viz.* stereochemical, not preparative reasons. In 1875 van´t Hoff had predicted that in cumulenes with an even number of double bonds the four substituents must be arranged in two perpendicular planes, whereas for the odd-membered series all substituents must be in the same plane.[7] Provided the rotational barrier about the double bonds is sufficiently high—which is normally true for the lower cumulenic hydrocarbons—van´t Hoff´s proposals implied that the allenes, pentatetraenes *etc.* must be chiral if they carry one substituent different from hydrogen at both ends, whereas the butatrienes, hexapentaenes *etc.* should show (*E*)/(*Z*) isomerism when appropriately substituted. These predictions were confirmed in 1935[8, 9] and 1954, respectively,[10] when the first optically active allenes and *cis/trans*-butatrienes were described (see below).

The rapid progress in synthetic methodology which began after the second World War, and especially since chromatographic separation and spectroscopic identification methods became routine in chemical laboratories, had a strong influence on cumulene chemistry also. Today, cumulenes, and in particular allenes, are readily available, versatile starting materials and intermediates in organic synthesis because their double bonds can be used in all types of addition processes, their (acidic) hydrogen atoms can easily be replaced by functional groups, and their axial chirality is increasingly exploited in stereoselective synthesis.[11]

9.1 Acyclic allenes

Because allenes constitute a special type of dienes, it is not surprising that many of the olefin-forming processes have been employed to prepare these simplest cumulenic systems. Initially allenes were largely obtained from suitable olefinic precursors by elimination reactions such as dehalogenations with metals, dehydrohalogenations with bases, or dehydrations with the aid of acids or catalysts. These procedures, which have been reviewed in detail,[11] and will be illustrated by several examples below, are valuable for the preparation of simple and stable allenes and various functionalized systems. Quite often, however, the relatively harsh reaction conditions induce secondary processes such as isomerization, addition or polymerization. Preparative work which appeared before the introduction of the modern analytical techniques should thus be viewed critically. Later, rearrangement and addition processes (*e.g.* to 1,3-enynes) became of importance, and finally specific methods which yield allenes in high yield and purity were introduced, among them the Wittig reaction and the so-called carbene or Doering-Moore-Skattebøl (DMS) method.

In the Wittig route to allenes **3** either a phosphorane **1** is reacted with a ketene **2** or a vinylphosphorane **4** with a ketone **5** (Scheme 1).

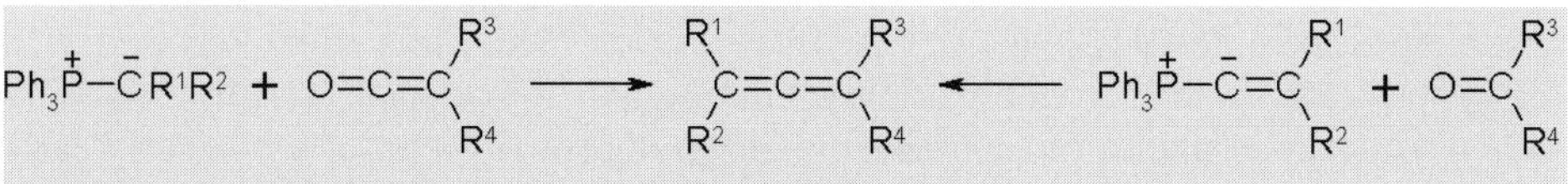

Scheme 1. Direct routes to allenes: The Wittig reaction.

Both methods suffer from the disadvantage that one of the starting materials— either the ketene **2** or the vinylphosphorane **4**—is often difficult to obtain and/or to handle. Although the method has been widely used to prepare allene derivatives such as acids, esters, or ketones, it is of limited use in hydrocarbon chemistry. In fact, one of the starting materials here is often the easily available diphenylketene and the route thus provides 1,1-diphenylallenes (usually 1,1-diphenyl-3,3-dialkylallenes).[12] Occasionally the ketene is generated *in situ* as will be described later.

In the DMS route, discovered by Doering in 1958,[13] dibromocarbene **7** (normally prepared by base-induced α-elimination of bromoform) is first added to an olefin **6** to provide a *gem*-dibromocyclopropane derivative **8**. When this is treated with alkali amalgam, halogen-metal exchange takes place and leads to the intermediate **9**, which—on loss of alkalibromide—provides the cyclopropylidene intermediate **10**

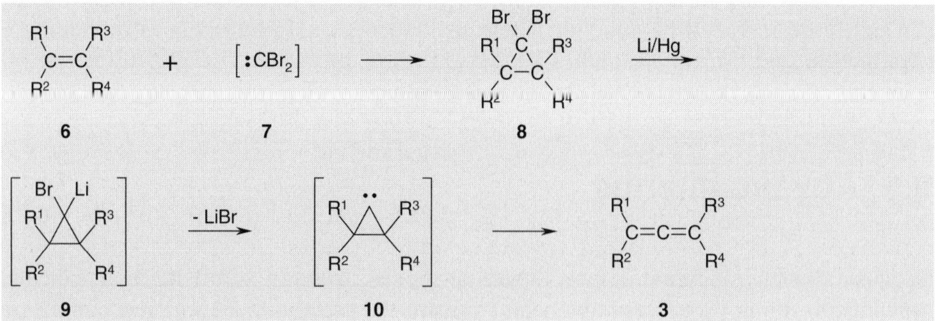

Scheme 2. Direct routes to allenes: The DMS-method.

(also known as carbenacyclopropane or cyclopropacarbene). Ring-opening of this highly strained species finally leads to the allene **3**[14] (Scheme 2).

Later Moore[15] and Skattebøl[16] replaced the amalgam by lithium alkyls; *n*-butyl lithium, in particular, is commonly used today for the debromination of **8**. When starting material and product in Scheme 2 are compared, it is seen that the DMS method achieves an over-all insertion of a carbon atom into a carbon-carbon double bond.[17]

Among all methods for the preparation of allenes the carbene route has the greatest general applicability; it has been used for the preparation of a very large variety of acyclic, mono-, bi- and polycyclic allenic hydrocarbons and it tolerates many functional groups. It has, furthermore, proven its value for the synthesis of higher cumulenes also (see Section 9.2, below).

For the preparation of the parent hydrocarbon propadiene (allene, **12**, R = H) and several of its simple alkyl derivatives (**12**, R ≠ H), a classical elimination reaction, such as treatment of the unsaturated vicinal dihalides **11** with zinc, is often simpler and preparatively superior (Scheme 3).

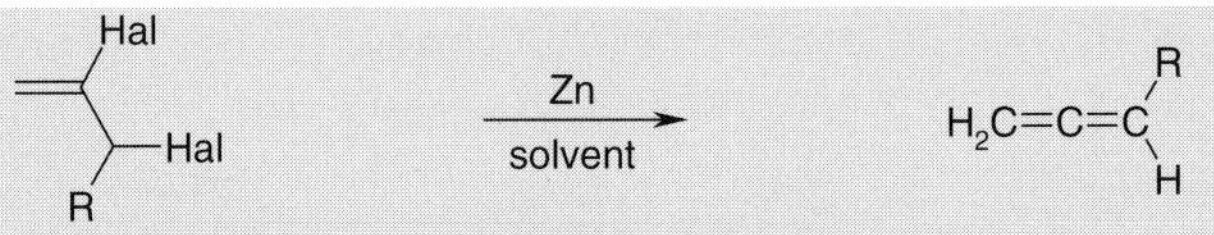

11
(Hal = Cl, Br; R = H, Et, Pr, *i*-Pr, *n*-Bu, *i*-Bu)

12

Scheme 3. Allene and alkyllallenes **12** by dehalogenation of *vic*-dihalides **11**.

Thus, in what amounts to a variation of the classical Gustavson and Demjanoff synthesis,[18] propadiene was obtained in 98% yield and free from olefinic or acetylenic (propyne) impurities by heating 2-chloro-allylchloride under reflux in butyl or isopentyl acetate in the presence of zinc.[19] Although the substrates **11** are not always easily available, an important advantage of this elimination is that the position of the olefinic double bond remains unchanged.

Among the tetraalkyl allenes tetrakis-*tert*-butyl-allene (**14**), obtained by either direct or indirect dehydration of the allyl alcohol **13**, is certainly one of the most unusual allenes ever prepared[20] (Scheme 4).

The hydrocarbon does not react with ozone at room temperature, *m*-chloroperbenzoic acid in dichloromethane under reflux, or bromine in boiling carbon tetrachloride, or with other reagents which normally attack allenes vigorously. The reason for this inertness has been ascribed both to the shielding effects of the four bulky substituents towards external reagents and to the increase in steric hindrance between the *tert*-butyl groups should a reagent bind to the central carbon atom of **14**.

For the preparation of arylallenes, dehydrohalogenation is often the method of choice especially when the resulting product cannot rearrange to its acetylenic isomer because it lacks an abstractable hydrogen atom. The preparation of tetraphenylallene (**16**)[21] and bis-biphenyleneallene (**18**)[22] from the precursor olefins **15** and **17** are representative examples (Scheme 5).

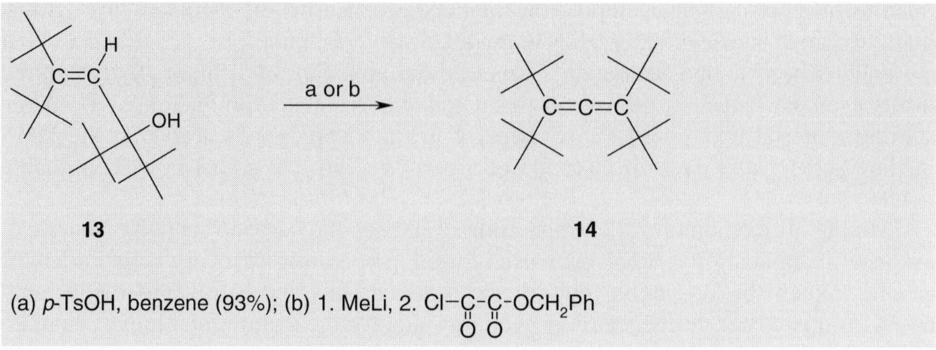

(a) p-TsOH, benzene (93%); (b) 1. MeLi, 2. Cl−C−C−OCH₂Ph
                                              ‖ ‖
                                              O O

Scheme 4. Tetrakis-*tert*-butyl-allene (**14**)—an inert allene.

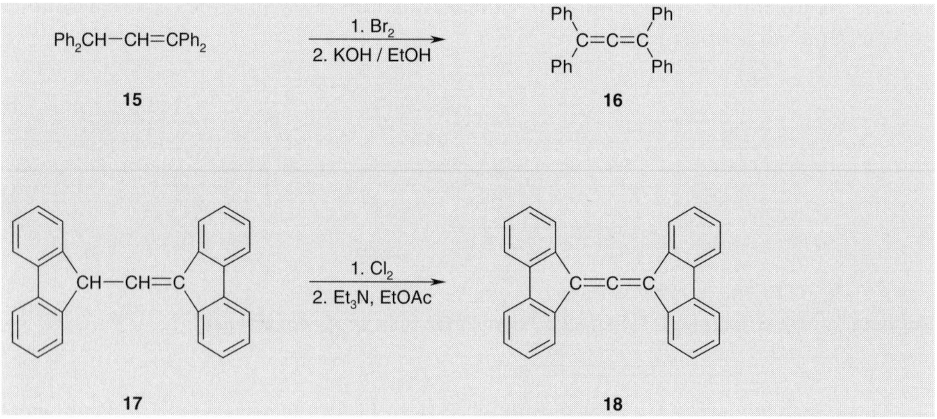

Scheme 5. Preparation of arylallenes by dehydrohalogenation.

Tetraphenylallene (**16**) can also be used as an example to demonstrate the application of the Wittig reaction in allene synthesis. The hydrocarbon has been prepared by heating either diphenylketene with triphenyl(diphenylmethylidene)phosphorane[12, 23] or benzophenone with (2,2-diphenylvinylidene)triphenylphosphorane.[24]

In an interesting Wittig-type route to tetrasubstituted allenes with cycloalkyl or alkyl substituents, the cumulenic carbon atom of the target molecule is provided by carbondioxide, the 'exchange' of its oxygen atoms takes place stepwise by first preparing a betaine intermediate from carbondioxide and the ylid. Heating this zwitterionic intermediate then generates a ketene *in situ* which is finally quenched by a second molecule of the Wittig reagent, which is present in excess.[25]

Among the acyclic allenic hydrocarbons carrying unsaturated substituents (vinyl, ethynyl, allenyl groups *etc.*) the vinylallenes and the bisallenes have been studied most thoroughly. The parent hydrocarbon of the vinylallenes, 1,2,4-pentatriene (**21**), is the smallest molecule conceivable that at the same time contains a cumulenic and a conjugated diene system. Since the former preferentially participates in [2+2]- and the latter in [2+4]cycloaddition reactions (Diels–Alder additions) the question arises, whether the allene or the conjugated diene character of **21** dominates the chemical behavior. The hydrocarbon was first prepared by E. R. H. Jones and co-workers, in 70% yield, by treating *trans*-5-chloro-3-penten-1-yne (**19**) with zinc–copper couple in butanol (Scheme 6).[26]

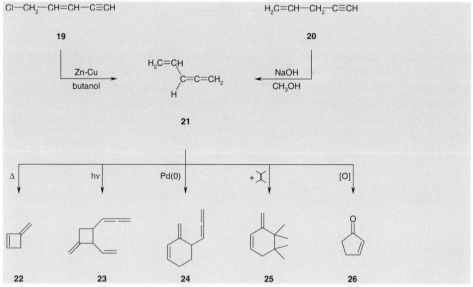

Scheme 6. Vinylallene **21**–the smallest hydrocarbon containing a cumulene and a conjugated diene unit.

Alternatively, **21** can be prepared in 80% yield by base-catalyzed isomerization of the non-conjugated precursor 1-penten-4-yne (**20**),[27] exploiting the propargyl to allene isomerization, a process which has been used very often in allene chemistry, but sometimes suffers from the disadvantage that the isomerization cannot be stopped at the cumulenic stage, because its conjugated diene isomer—if structurally possible—is thermodynamically more stable. Both processes have been generalized,[27] the zinc–copper reagent can be replaced by magnesium,[28] and substituted vinyl allenes result when the starting enyne **19** is treated with Grignard reagents.[29] Some representative applications of conjugated vinyllallenes in organic synthesis are also summarized in Scheme 6. Heating **21** in the gas phase causes ring closure to 3-methylenecyclobutene (**22**),[30] a (reversible) electrocyclization which is driven by the high endothermicity of the allene group. In 1,3-butadiene, which lacks this feature, the equilibrium lies on the side of the open structure, *i.e.* the ring strain of the cyclobutene cannot be overcome. In the photodimerization of **21**, which leads to the highly functionalized cyclobutane **23**, one molecule of the starting material reacts with its vinyl, the other with its allenyl end.[30] And in the palladium(0)-catalyzed dimerization the complex π-system of **21** is used in yet another manner, the dimer **24** now being the main reaction product.[31] Formally **24** is the result of a Diels–Alder addition of **21** with itself, and this addition mode is also observed with numerous typical dienophiles (formation of

25) such as dimethyl acetylenedicarboxylate,[26] maleimide,[26] α,ß-unsaturated ketones,[32, 33] and various quinones, the latter process providing easy access to substituted naphtho- and anthraquinones.[26, 34] Oxidation of vinylallenes with peracids led
to conjugated cyclopentenones **26**[35, 36] and alkyl substituted vinylallenes are capable
of undergoing thermal [1.5]sigmatropic hydrogen-shift reactions to 1,3,5-trienes, the
driving force of the process again being provided by the formation of a fully conjugated system. This rearrangement has been employed very successfully in the vitamin
A and vitamin D field.[37]

Replacement of the vinyl group in **21** by a second allenyl moiety leads to conjugated bisallenes, another group of cumulenic hydrocarbons with the potential to participate in both [2+2]- and [2+4]cycloadditions. The parent system, 1,2,4,5-
hexatetraene (**29**), has been prepared in 40–50% yield by copper(I)-catalyzed coupling of allenyl magnesium bromide (**27**) with propargyl bromide (**28**).[38] The reaction is accompanied by the formation of 1,2-hexadien-5-yne (**30**), presumably formed
by direct (S$_N$2-type) substitution of **28** with **27** (Scheme 7).

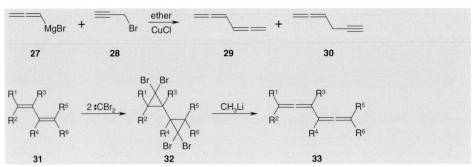

Scheme 7. Preparation of conjugated bisallenes.

Although metal-catalyzed coupling has frequently been used for the preparation of
bisallenes,[39–43] it cannot compete with the DMS method, which usually provides the
desired products free from their propargyl isomers and in better yields. Alkyl- and
aryl-substituted or cyclic conjugated dienes **31** are used as starting materials; these
are first converted into the bis adducts **32** by addition of dibromocarbene; treatment
with methyl lithium subsequently yields the bisallenes **33**, often in excellent yields
(>90 %).[44–47] With dienophiles the conjugated bisallenes react as 1,3-butadiene derivatives;[48] one of these [2+4]cycloadditions, leading to [2.2]paracyclophane derivatives,[49] will be described in detail in Section 12.4.

When two dibromocyclopropane rings are not joined directly as in **32**, but linked
by a polymethylene chain, as shown by **34** in Scheme 8, reduction with methyl lithium leads to α,ω-diallenes **35** in 83 (n = 3) and 86% yield (n = 4), as expected.

With an ethano linker, however, besides the desired **35** (n = 2, 68%), the bridged
spiro hydrocarbon **36** is also formed (28%),[50] presumably by the addition of a 'not
yet opened' cyclopropylidene unit to an 'already open' allene group as indicated by **37**.

1,2,6,7-Octatetraene (**35**, n = 2), on heating at 310 °C in the gas phase, underwent

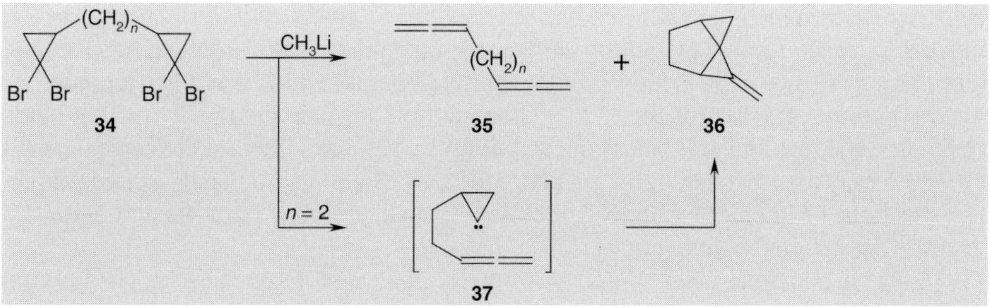

Scheme 8. Preparation of nonconjugated bisallenes.

an intramolecular allene cycloaddition to the tetramethylene ethane derivative **38**[51–54] (see Section 16.2), which ring-opened to 3,4-dimethylene-1,5-kexadiene (**39**) or [4]dendralene, Scheme 9, a member of the cross-conjugated hydrocarbons discussed in Section 11.4.

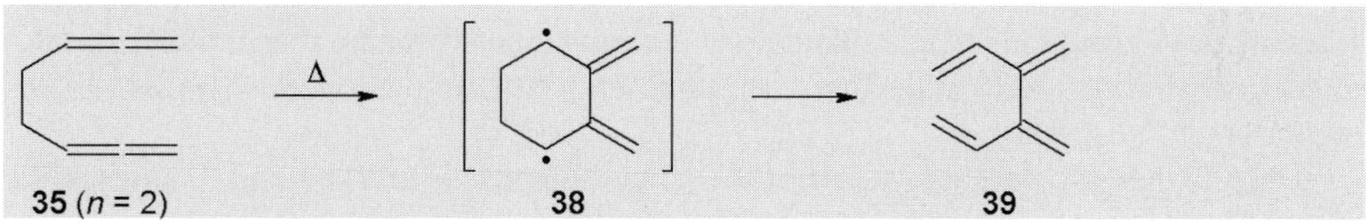

Scheme 9. Acyclic cross-conjugated hydrocarbons from α,ω-bisallenes.

Bisallenes in which two allene moieties are separated by a double bond or an aromatic ring as in **41** (not isolated)[55] and **44**[56] have been generated by base-catalyzed rearrangement of their propargyl isomers **40** and **43** (Scheme 10). These conjugated bisallenes readily cyclized to the *ortho*-quinoid hydrocarbons **42** and **45**, respectively, which may either dimerize or be trapped by added dienophiles (see Sections 12.3 and 16.2), oxygen *etc.*

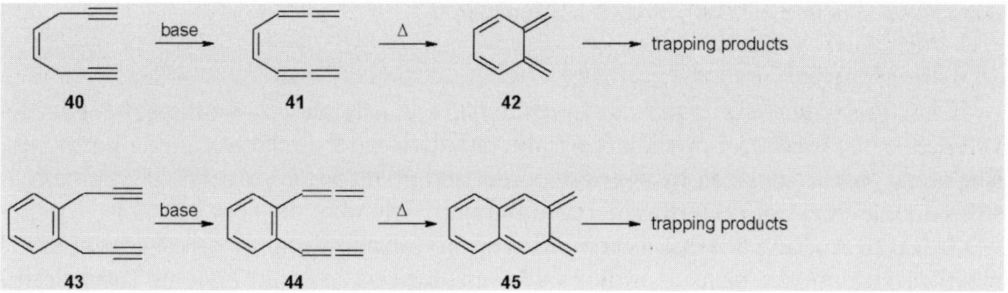

Scheme 10. *o*-Xylylenes from α,ω-bisallenes.

Optically active alkylallenes[57] have been prepared either by resolving a racemic mixture of the allene hydrocarbon or by converting optically active precursor molecules by stereoselective syntheses into the desired products. The former approach is illustrated by the kinetic resolution of racemic alkylallenes by partial asymmetric hydroboration with (+)-*sym*-tetraisopinocampheyldiborane, which is readily prepared by hydroboration of (-)-α-pinene.[58, 59] This method provided optically active alkyl (methyl, ethyl, propyl, *tert*-butyl) allenes in up to 50% ee. Even higher optical yields (70–80% ee) were realized in the second approach when, *e.g.*, optically active propargyl alcohols or their derivatives (acetates, mesylates *etc.*) were treated with hydrides[60] or various organocuprates.[61–63]

9.2 Acyclic cumulenes

Compared with the oligomeric polyenes (Section 7.1) and polyacetylenes (Section 8.1) far less is known about the higher cumulenes. As was true of their olefinic and acetylenic counterparts, the stability of the cumulogs of allene decreases quickly with growing chain length, and the introduction of the usual stabilizing groups (bulky substituents, aryl groups) is hence mandatory if one wants to prepare these highly unsaturated hydrocarbons and study their chemical and structural properties under normal laboratory conditions.

Despite this, the parent systems of the [*n*]cumulenes, with *n* = 3 and 4, and several still higher cumulenes—'protected' by polyalkylation—are known. Usually, however, stable [*n*]cumulenes with *n* > 4 are stabilized by aryl substituents (see below).

1,2,3-Butatriene (**47**) was first prepared unequivocally by Schubert et al. in 1952[64] by debromination of 1,4-dibromo-2-butyne (**46**) with zinc dust in diethylene glycol diethyl ether (Scheme 11).

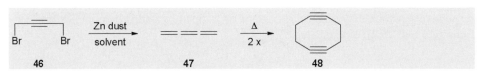

Scheme 11. The first synthesis of 1,2,3-butatriene (**47**).

Later this route was improved by Brandsma, Arens and co–workers,[65] who developed it into a general procedure for the preparation of aliphatic cumulenic trienes. The metal can be replaced by sodium iodide, and polar protic solvents, e.g. dimethylsulfoxide, are best. It is, furthermore, important to remove the very unstable cumulenes from the reaction mixture as soon as they are formed by applying a vacuum to the reaction flask. Under these conditions **47** was obtained in *ca* 80% yield. The reactive behavior of butatriene and its derivatives has been studied,[64–66] and its highly remarkable thermal dimerization to 1,5-cyclooctadiyne (**48**) has been discovered by Kloster-Jensen and Wirz[67] (see Section 8.3).

Gas-phase pyrolysis—which usually avoids the presence of other reagents or solvents altogether and thus keeps secondary reactions at bay—has been employed very successfully for the preparation of the higher cumulenes also. Thus Alder–Rickert cleavage of 7-isopropylidene-5,6-dimethylene-bicyclo[2.2.1]hept-2-ene (**49**), 5,6-dimethylene-7-oxa-bicyclo[2.2.1]hept-2-ene (**50**), and 7,8-dimethylenebicyclo[2.2.2] octa-2,5-diene (**51**) (Scheme 12) all yield **47** when pyrolyzed at temperatures above 500 °C.[68]

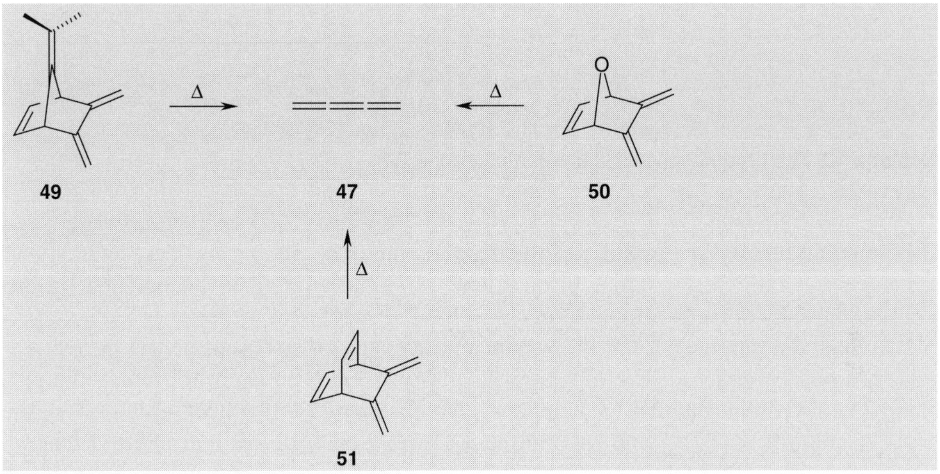

Scheme 12. Preparation of 1,2,3-butatriene (**47**) by retro-Diels–Alder reaction.

Most known alkyl-substituted [3]cumulenes have been obtained by methods involving carbenoid intermediates. The DMS procedure has been used to synthesize 4-methyl-1,2,3-pentatriene,[69, 70] 4-methyl-1,2,3-hexatriene,[69] *cis-* and *trans*-1,4-dimethylbutatriene (*cis-* and *trans*-2,3,4-hexatriene),[71] and tetramethylbutatriene;[72] the starting materials were always prepared by addition of dibromocarbene to an allene. Tetramethylbutatriene was also formed when 1,1-dibromo-2,2-dimethylethylene was metalated with *n*-butyl lithium in tetrahydrofuran at –60 °C, and the reaction temperature then slowly raised. Presumably the cumulene was formed by dimerization of the corresponding vinylidene carbene in this case.[73, 74] In an interesting catalytic process, *tert*-butylacetylene was dimerized to *cis-* and *trans*-1,4-di-*tert*-butyl-butatriene with dihydridocarbonyltris(triphenylphosphine)ruthenium serving as the catalyst.[75]

[*n*]Cumulenes with an even number of double bonds cannot be prepared by dimerization of unsaturated carbenes or by α,ω-elimination of unsaturated precursor molecules with an even number of carbon atoms. Special methods are hence required to obtain this particular class of hydrocarbon. This is well illustrated for [4]cumulene, (**55**, 1,2,3,4-pentatetraene) prepared first by Ripoll in 1976 (Scheme 13).[76, 77]

Treatment of the dibromoketone **52** with excess methyl lithium not only converted the dibromocyclopropane ring of the substrate to an allene group, but also the keto function to a tertiary alcohol, **53**. When the latter was exposed to aluminum oxide under carefully controlled conditions dehydration occurred, yielding the vinylallene **54**. Flash vacuum pyrolysis (700 °C, 10^{-3} torr) split off the 'leaving group' an-

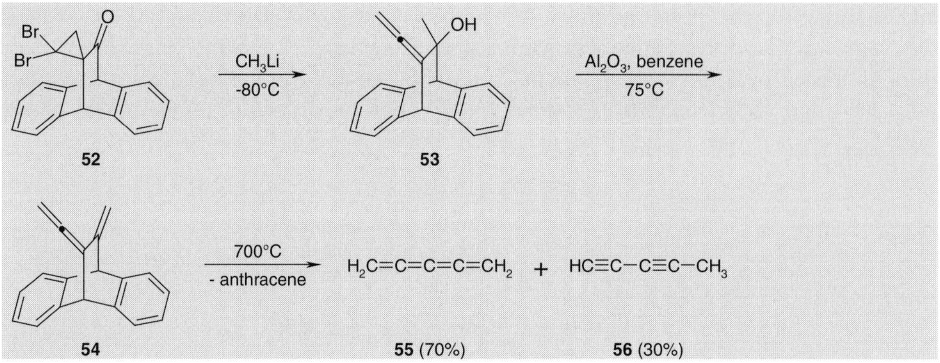

Scheme 13. The preparation of 1,2,3,4-pentatetraene (**55**).

thracene and furnished a mixture of the desired cumulene **55** and its acetylenic isomer **56** in 85% yield. Surprisingly, [4]cumulene is stable enough to be purified by gas chromatography at room temperature.

Neither the parent 1,2,3,4,5-hexapentaene nor any of its simple alkyl derivatives have yet been prepared. Several routes have, however, been used to prepare the permethylated hydrocarbon **58** (Scheme 14), which *inter alia* has been obtained by the carbene route from **57**[78] and by treatment of the dichloride **59** with methyl magnesium bromide in ether.[72]

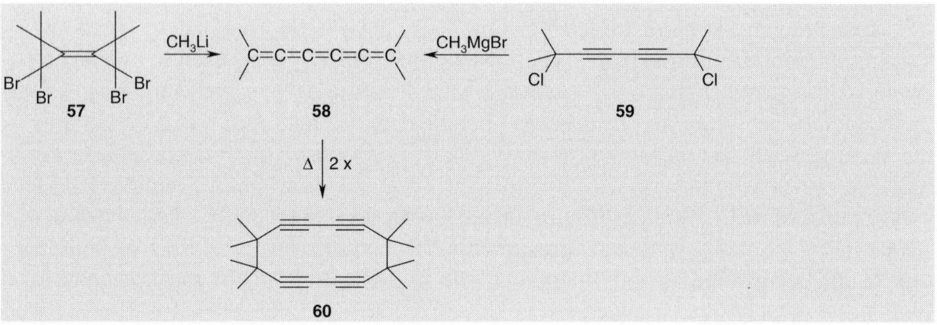

Scheme 14. Preparation of permethyl 1,2,3,4,5-hexapentaene (**58**).

Like its lower cumolog 1,2,3-butatriene (see above), **58** undergoes a thermal dimerization, in this instance furnishing the 1,3,7,9-cyclododecatetrayne **60**.[79]

The stabilizing effect of exhaustive substitution by *tert*-butyl groups is dramatically underlined by the cumulated pentaene **62** which is either accessible from the propargyl acetate **61** *via* a carbene intermediate[80] or by dehalogenation of the vicinal dibromide **63** with zinc (Scheme 15).[80–81] This [5]cumulene is a solid, which melts at 188 °C!

Still, there is a limit to the shielding effect of bulky groups in cumulene chemistry also. Although the aliphatic octaheptaene **64**, prepared by reduction of the corre-

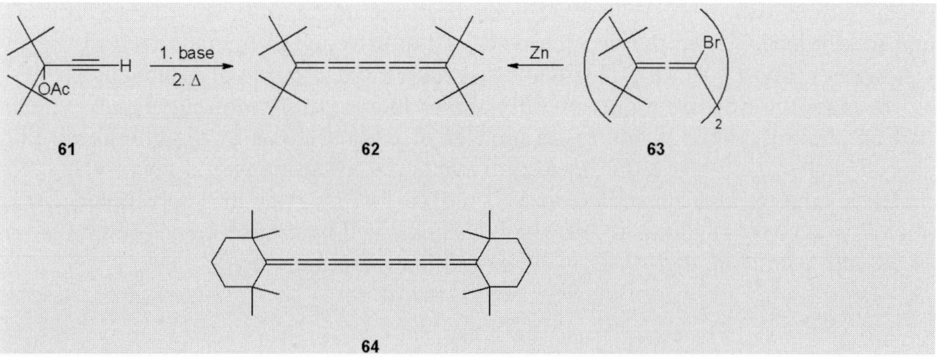

Scheme 15. A selection of higher [*n*]cumulenes protected by peralkylation.

sponding triyne-α,ω-diol as described in the following paragraph for arylsubstituted [*n*]cumulenes could be isolated in the form of orange crystals, these decompose within a few seconds at room temperature,[82] and the still larger decanonaene, identically substituted, could only be characterized by the UV spectrum of its colored solution.[83]

Almost all of the other fully characterized higher cumulenic hydrocarbons are completely substituted by aryl groups. The classical route leading to these hydrocarbons is shown in Scheme 16 in general form.

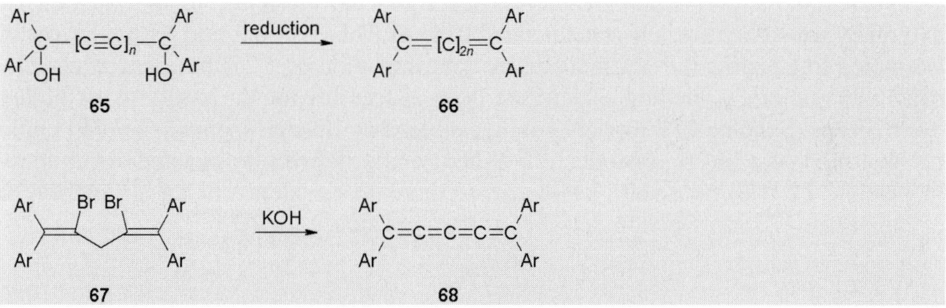

Scheme 16. A general route to the higher arylsubstituted cumulenes.

The starting diols **65** are readily obtained from the corresponding ketones and acetylenes by traditional Reppe chemistry. The reduction is usually performed by treating the α,ω-diols with stannous chloride-ether-HCl, potassium iodide-sulfuric acid-ethanol, phosphorous tribromide in pyridine or phosphorous triiodide in triethylamine. The aromatic substituents must not be identical and they can be connected to each other (as when, *e.g.*, fluorenone is used as a starting ketone). Not surprisingly, the most thoroughly studied group of compounds in this class are the butatrienes, for which van't Hoff's prediction that they should exist as *cis/trans* isomers (see above) was confirmed by Kuhn and Scholler in 1954.[10, 84]

The diols **65** required for the preparation of various hexapentaenes (**66**, *n* = 2) and octaheptaenes (**66**, $n = 3$) were prepared from the appropriate ketones and di-[85-87] and triacetylene,[88] and the decatetrayne-1,10-diols required for the synthesis of de-canonaenes (**66**, $n = 4$) were obtained by oxidative dimerization of pentadiynols.[82, 83]

Because the **65**→**66** reduction only allows the preparation of cumulenes with an odd number of double bonds (even number of carbon atoms in the cumulene unit) special routes again had to be devised to synthesize even-numbered polyarylated cu-mulenes. This is also illustrated in Scheme 16 for the fully aryl-substituted (Ar = phenyl, *p*-anisyl) [4]cumulene **68** which was obtained by double dehydrobromination of **67** with potassium hydroxide in ethanol-dimethyl formamide.[89]

9.3 Cyclic allenes

The cycloallenes[90] resemble their acetylenic isomers (see Section 8.2) in many respects. The dependence of their stability and isolability on ring size is comparable and originally rather similar methods were employed to prepare them, *i.e.* mostly 1,2-elimination reactions. According to molecular models, 1,2-cyclodecadiene and its higher homologs should contain undeformed allene units and hence be relatively strain-free. With decreasing ring size the allene group should be more and more bent, and eventually become planar. The molecular orbitals of such a planarized structure are best described as an allyl system, which is orthogonal to an sp^2-hybrid orbital.[91]

No reliable experimental evidence concerning the existence of the presumably extremely unstable 1,2-cyclopentadiene (**70**) is available. In the mid 1930s Favorskii attempted to generate this cycloallene by debromination of 1,2-dibromocyclopentene (**69**) with sodium, a method which had been successful for the synthesis of higher cycloallenes (Scheme 17); its only conjugated isomer 1,3-cyclopentadiene (**72**) could be identified as a reaction product.[92] When Wittig dehydrobrominated 1-bromocy-clopentene **71** with potassium *tert*-butoxide there was evidence of the generation of cyclopentyne (**73**, see Section 8.2), but again no evidence for **70**.[93]

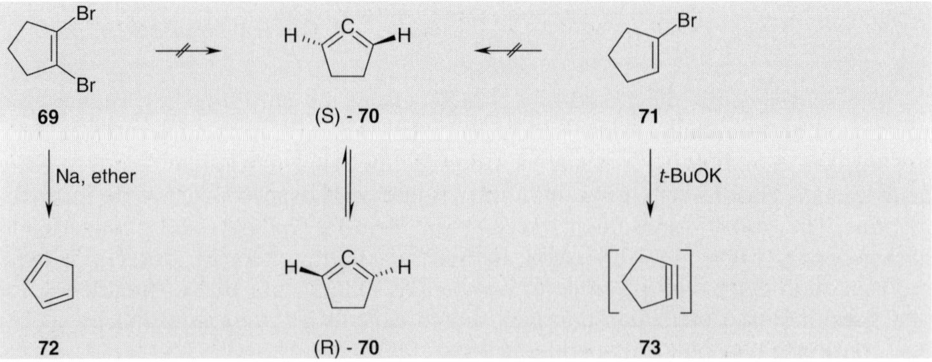

Scheme 17. Attempts to prepare 1,2-cyclopentadiene (**70**).

According to *ab initio* calculations **70** should still be chiral, the two allenic hydrogen atoms being bent out of the molecular plane by *ca* 21°. The calculated racemization barrier of 4.9 kcal mol^{-1}[94] is far lower than the experimentally determined inversion barrier of a typical acyclic allene such as 1,3-dimethylallene.[95]

Compared to this scarce information, the chemistry of the next higher homolog, 1,2-cyclohexadiene (**77**), is very rich. After several early attempts to prepare it, which largely yielded nonvolatile oligomeric hydrocarbons,[92, 96, 97] the first unequivocal demonstration of the existence of **77** was reported by Wittig and co-workers in 1966,[98] who treated 1-bromocyclohexene (**74**) with potassium *tert*-butoxide in dimethylsulfoxide, and isolated the allene dimer **75**, produced from **77** by [2+2]cycloaddition (allene dimerization, Scheme 18[99]).

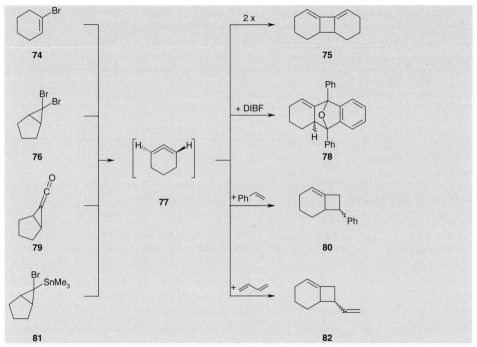

Scheme 18. Generation and trapping of 1,2-cyclohexadiene (**77**), the smallest cycloallene.

Trapping of **77** with diphenylisobenzofuran (DIBF) led to two isomeric cycloadducts **78**, interception with styrene yielded adduct **80**,[100] and 1,3-butadiene provided largely the [2+2]cycloadduct **82** (accompanied by a small amount of the Diels–Alder adduct of butadiene to **77**).[101] Some of these trapping experiments can in principle also be rationalized by postulating cyclohexyne as an intermediate—its cyloaddition being followed by an isomerization; this alternative was, however, ruled out by detailed labeling studies.[102]

The most efficient way to prepare **77** is again the carbene route (from **76**),[100] and carbene intermediates are probably also involved in two studies in which **77** was matrix-trapped at low temperatures, enabling its IR spectrum to be recorded. Pyrolysis of the ketene **79** caused decarbonylation and formation of **77**,[103] as did the heating of the precursor molecule **81**, which split off trimethyltin bromide at 500 °C.[104] Whatever the exact electronic structure of **77**, there is experimental evidence that this intermediate is chiral if generated from an appropriate substrate molecule.[105]

1,2-Cycloheptadiene (**83**; Scheme 19) has been generated by 1,2-elimination reactions[97, 106] from various precursors, and although it could be trapped[105] and stabilized by metal complex formation,[107, 108] no spectral data have yet been recorded.

Beginning with 1,2-cyclooctadiene (**84**), first prepared unambiguously by Landor and Ball in 1961 by dehydrochlorination of 1-chlorocyclooctene with base,[97, 109] and today best prepared by the DMS method from the dibromocarbene adduct of cycloheptene,[110] the stability limit is slowly reached: On the one hand this cycloallene readily dimerizes and undergoes the typical trapping reactions (see above); on the other, cold, dilute solutions of the compound could be obtained and subjected to rapid spectroscopic analysis.[107] As expected, introduction of alkyl groups at the allene function of **84** increases its kinetic stability, the 1-*tert*-butyl derivative surviving purification by gas chromatography.[111]

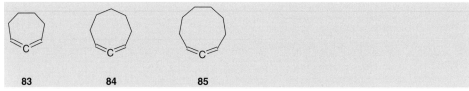

Scheme 19. Reaching the stability limit—the medium-sized cycloallenes.

Finally, 1,2-cyclononadiene (**85**), originally prepared by Blomquist and co-workers in 1951 by dehalogenation of 1-bromo-2-chlorocyclononene[112] and later by Skattebøl in multigram quantities from cyclooctene, by means of the DMS-method,[113] is a more or less 'normal' allene, which only dimerizes at 150 °C. According to electron diffraction[114] and X-ray structural analysis[115] the allene group of **85** is slightly (*ca* 10°) bent from linearity; the hydrocarbon has also been obtained in optically active form.[59]

Although other preparative methods are known[111] the reductive dehalogenation of the bicyclic 1,1-dihalocyclopropanes (DMS method) is the most general and efficient route to the cycloallenes beyond **85**.[111] The older dehalogenations or dehydrohalogenations of medium-sized cycloalkyldihalides or cycloalkylhalides often yield product mixtures which contain all possible isomers, *e.g.* the desired cycloallene, the corresponding cycloalkyne and the conjugated cyclodiene. Chromatographic techniques are required for purification.

Considering the high reactivity of the cycloallenes with ring-sizes smaller than nine it is surprising that it has been possible to increase the degree of unsaturation of these compounds even further by introducing additional double and triple bonds or a second allene group. A particularly impressive example is 1,2,4-cyclohexatriene (**87**) or *isobenzene*. This cyclic C_6H_6 hydrocarbon (see Section 10.2) was first produced by

Christl and co-workers from the fluorobromo- or dibromocarbene adduct of cyclo-pentadiene (**86**, Hal = F or Br) by treatment with methyl lithium[116] (Scheme 20).

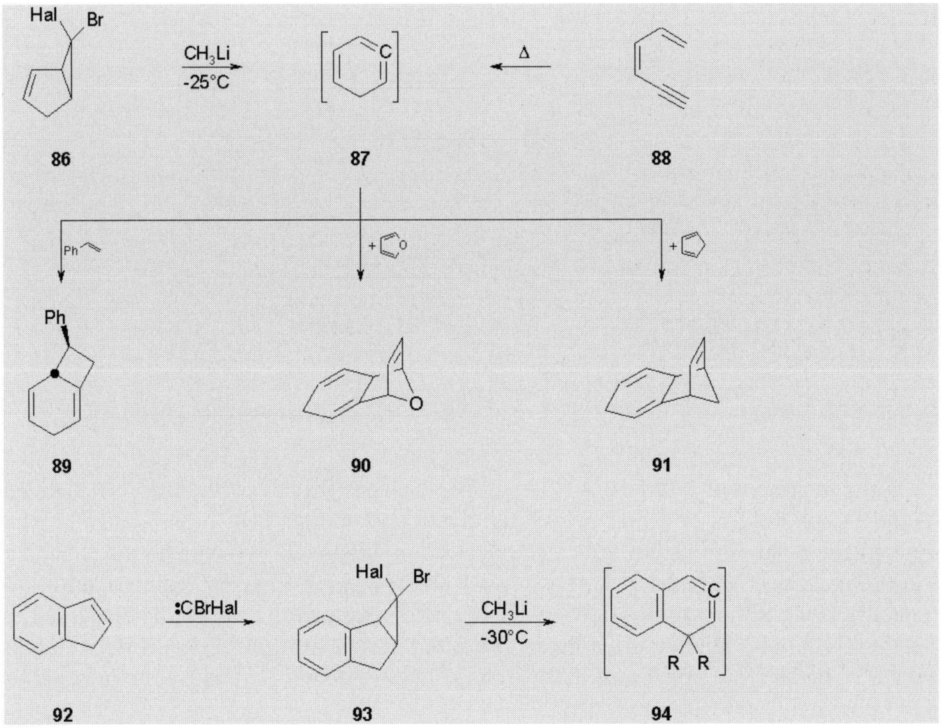

Scheme 20. Routes to 1,2,4-cyclohexatriene (**87**)–a cyclic isomer of benzene.

The intermediate formation of isobenzene during these dehalogenations has been inferred from trapping experiments with—*inter alia*—styrene, furan, and cyclopenta-diene, leading to the cycloadducts **89–91**.[117] Subjecting indene (**92**) to the same se-quence of steps provided the benzannelated derivative of **87**, isonaphthalene **94** (R = H).[116] In fact, the dimethyl derivative of this compound (**94**, R = CH$_3$), prepared by base-induced dehydrobromination of 3-bromo-1,1-dimethyl-1,2-dihydronaphthalene was the first example of a 1,2,4-cyclohexatriene derivative to be reported.[118] More recently Hopf, Zimmermann and co–workers have obtained **87** by thermal electrocy-clization of *cis*-1,3-hexadien-5-yne (**88**),[119] and finally isobenzene derivatives have been postulated as the initial products of the Diels–Alder addition of activated acety-lenes and vinylacetylenes.[120]

A theoretical study of the electronic nature of **87** has predicted that the allene struc-ture—as shown in the scheme—should be more stable than a diradical structure in which the allene π-bond between C2 and C3 is uncoupled.[121] This latter structure is, however, implied by results from an investigation of the thermochemistry of the **88→87** cyclization.[122]

Among the hydrocarbons derived from 1,2-cycloheptadiene no system has been studied more thoroughly than 1,2,4,6-cycloheptatetraene (**96**), probably the most studied cyclic cumulene. The hydrocarbon plays a central role on the C_7H_6 potential energy surface and has been generated by numerous routes, a selection of which is summarized in Scheme 21.[123]

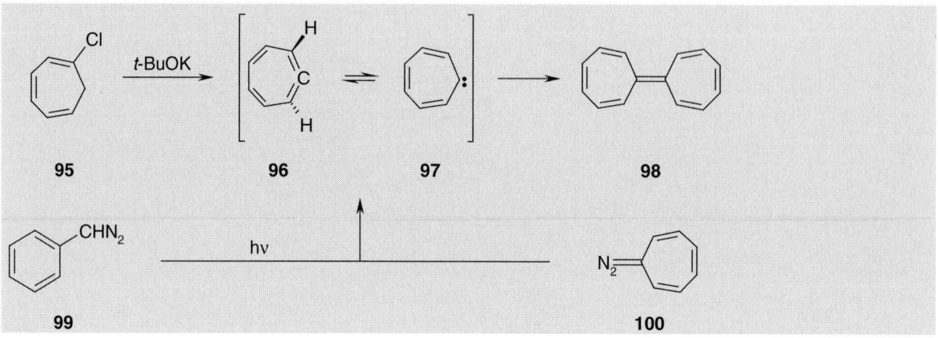

Scheme 21. Generation of 1,2,4,6-cycloheptatetaene (**96**), an important C_7H_6 intermediate.

Thus treatment of 1-chloro-1,3,5-cycloheptatriene (**95**)[124] with base provided **96** as did the photolysis of either one of the diazo compounds **99**[125] and **100**.[126] The cycloallene is in equilibrium with cycloheptatrienylidene **97**[127] which can either be trapped by added reagents such as styrene or undergo 'self-trapping' (dimerization) to heptafulvalene **98**, a representative of the cross-conjugated hydrocarbons discussed in Section 11.3. The allenic structure of **96** is supported by molecular orbital calculations,[128] IR and UV spectra of the matrix-isolated species,[127] and by the generation and trapping of optically active **96**.[129]

Depending on ring-size, monocyclic bisallenes are either reactive intermediates or can be isolated under normal laboratory conditions. For example, in the eight-membered ring series 1,2,4,6,7-cyclooctapentaene[130] and 1,2,4,5-cyclooctatetraene,[131] a 'bis-allenic isomer' of cyclooctatetraene (see Section 10.3), occur only as reactive intermediates in thermal isomerizations of various acetylenes. For ten-membered rings the situation is more subtle, and the stability and isolability of the cyclo-bis-allenes depends on the position of the two allene moieties relative to each other. Whereas 1,2,4,5-cyclodecatetraene (**101**), formed by methyl lithium treatment of the bis-dibromocarbene adduct of 1,3-cyclooctadiene rapidly cyclized to the bismethylenecyclobutene derivative **102**,[132] in a reaction which is typical for conjugated bisallenes,[131] the 'skipped' isomer **103**, which presumably is less strained, is isolable[133] (Scheme 22).

Cyclic bisallenes can, in principle, exist as *meso* or *d,l* diastereoisomers, as shown in **104** and **105**, respectively, for the 1,2,6,7-cyclodecatetraene diastereomeric pair. In practice, ring-opening of the bis-dibromocarbene adduct of 1,5-cyclooctadiene yielded only the *meso* compound **104**[16] as shown by an X-ray structure determination.[134] According to molecular models, the *d,l* isomer **105** does not seem to be prohibitively strained.

Another interesting cyclo-bis-allene, 1,2,4,6,7,9-cyclodecahexaene (**107**) was produced when the bis-dibromocarbene adduct of cyclooctatetraene **106** was dehalo-

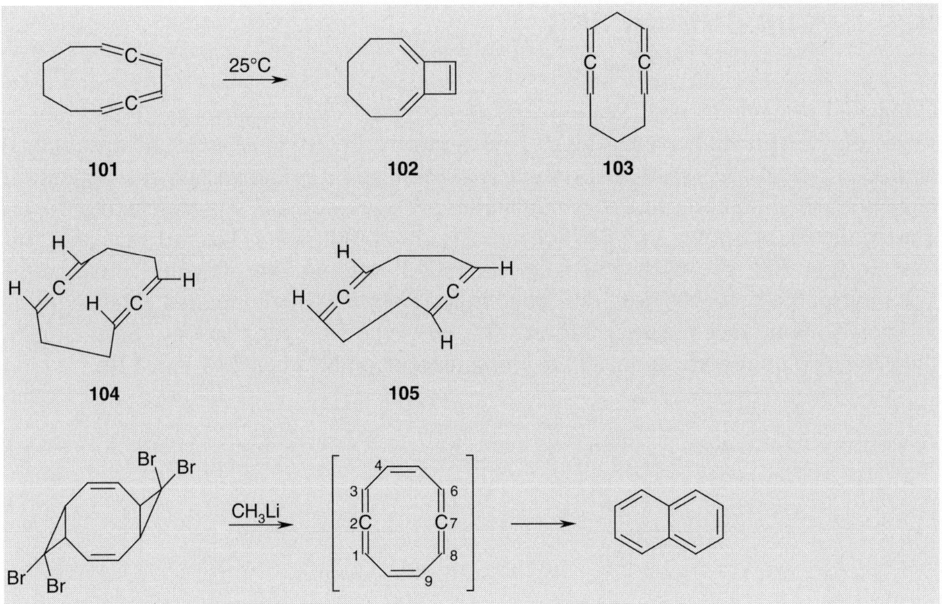

101 102 103

104 105

106 107 108

Scheme 22. A selection of ten-membered cyclo-bis-allenes.

genated with two equivalents of methyl or *n*-butyl lithium.[54, 135] At temperatures above –48 °C **106** furnished a complex product mixture from which naphthalene **108** could be separated as the main component. The most likely intermediate for its formation is the bis-allene **107**.

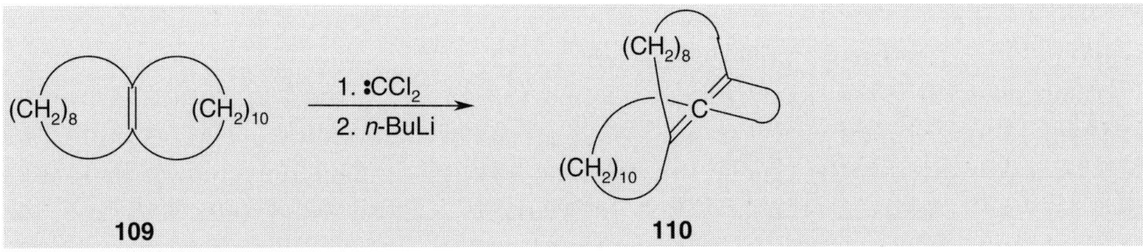

109 110

Scheme 23. Incorporation of the allene group into a doubly-bridged cycloallene.

Just as alkenes can be incorporated into more complex (poly)cyclic frameworks leading to *anti*-Bredt olefins or betweenanenes (Section 7.4), the allene group can also be part of a bi- or polycyclic skeleton. A case in point is provided by the first doubly-bridged allene **110** prepared from **109** by Nakazaki and co–workers in 1982.[136] When the dehalogenation-carbene rearrangement of the dichlorocyclopropane intermediate was performed with a sparteine-butyl lithium complex, optically active **110** was obtained.

9.4 Cyclic cumulenes

Compared to cycloallenes (Section 9.3, above) 1,2,3-cycloalkatrienes (**111**)—not to speak of the still higher cumulogs!—are a poorly investigated class of unsaturated hydrocarbons. As we shall see, only a few examples are known and their chemical behavior and structural properties remain largely unexplored. For example, their *trans* diastereomers, as shown in **112** (Scheme 24) are chiral, just as the *trans* cycloolefins (see Section 7.4), of which **112** is an 'extended' version. No optically active *trans*-1,2,3-cycloalkatriene has been described in the chemical literature, and hence nothing is known about racemization barriers for this type of hydrocarbons. Surprisingly, there are also no reports on the *cis-trans* isomerization between **111** and **112**.

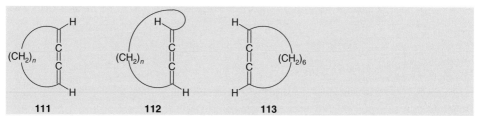

Scheme 24. Structural properties of cyclo[3]cumulenes.

According to molecular models the linear 1,2,3-butatriene unit can be incorporated strain-free into a ten-membered ring, *i.e.* the stability limit in this hydrocarbon series should be reached with 1,2,3-cyclodecatriene (**113**). Smaller rings will contain a progressively bent cumulene unit and show increased strain and reduced lifetimes.[137] These assumptions are generally supported by experimental evidence.

The smallest cyclo[3]cumulene hydrocarbon[138] described to date is 1,2,3-cyclohexatriene (**117**). This highly reactive species was prepared by Johnson and co-workers[139] in 1990 as summarized in Scheme 25.

Starting with 2-bromo-cyclohexen-3-one (**114**), the bromo substituent was first exchanged for a trimethylsilyl group by conventional methodology. The resulting **115** was then converted to the crucial intermediate **116**, a conjugated cyclohexadiene with two leaving groups at C2 and C3, respectively. These were removed, and the [3]cumulene **117** generated, by treatment with cesium fluoride, exploiting a technique which has also been applied to the synthesis of 1,2-cyclohexadiene and other strained cycloallenes. Proof of the generation of the intermediate **117** rests on trapping it with diphenylisobenzofuran **118** to form the cycloadduct **119**.

The next higher homolog, 1,2,3-cycloheptatriene (**122**), was first prepared by Szeimies and co-workers on 1981[140] (Scheme 26).

Fluoride-induced elimination from **120** led to the pyramidalized olefin **121** (see Section 7.5) which can be trapped directly by various dienophiles, but which can also ring-open to 1,2,3-cycloheptatriene (**122**), possibly *via* a vinylidene carbene intermediate. The short-lived **122** could be intercepted by numerous cyclic and acyclic dienes including 1,3-cyclohexadiene and **118**, as shown in Scheme 26. Interestingly, when a

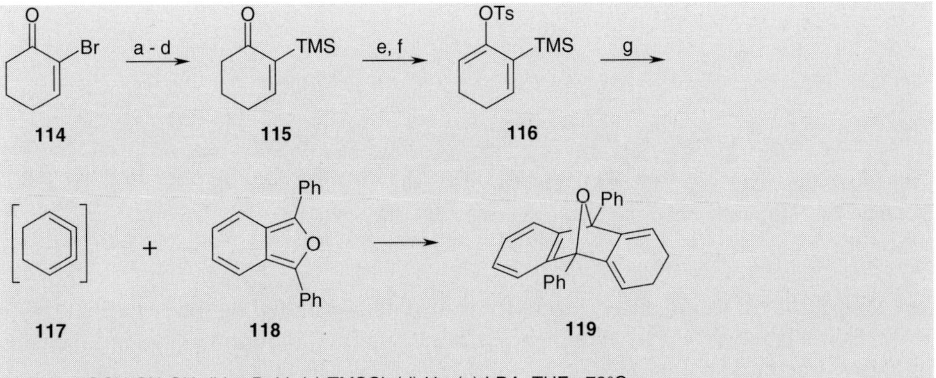

(a) H⁺, HOCH₂CH₂OH; (b) *n*-BuLi; (c) TMSCl; (d) H⁺; (e) LDA, THF, -78°C;
(f) PhN(OTf)₂; (g) CsF, DMSO, 25°C

Scheme 25. Preparation of 1,2,3-cyclohexatriene (**117**), the smallest cyclo[3]cumulene hydro-carbon.

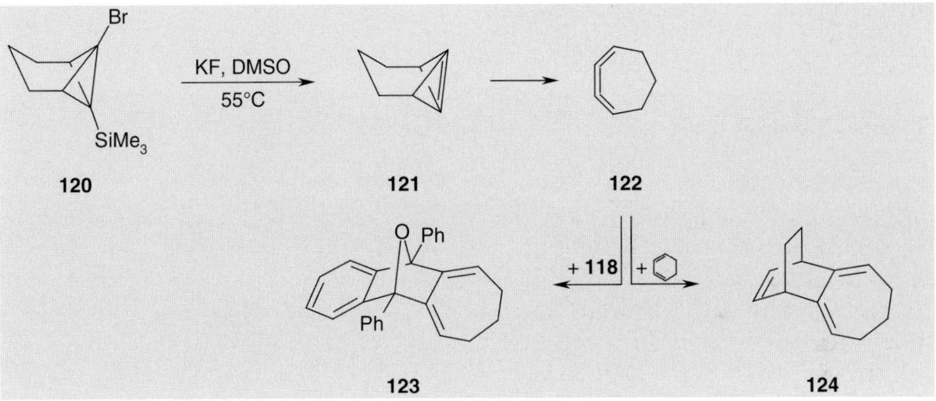

Scheme 26. Generation and trapping of 1,2,3-cycloheptatriene (**122**).

similar approach was applied to the next lower homolog of **120**, *i.e.* a benzvalene derivative (see Section 10.2), no **117** could be generated.[141] The dimerization of **122** to a bridged [4]radialene derivative will be discussed in Section 11.5.

Proceeding through the homologous series, 1,2,3-cyclooctatriene was prepared by an approach similar to that used to obtain **117**,[142] and for the still larger ring systems the DMS route is again superior.

Both cycloheptene and cyclooctene (**125**, *n* = 5 and 6) were taken through two sequential DMS ring-expansions, providing 1,2,3-cyclononatriene (**127**) (*n* = 5)[137] and 1,2,3-cyclodecatriene (**127**) (*n* = 6),[143] the latter being, in 1967, the first cy-clo[3]cumulene to be prepared; in both cases the corresponding cycloallenes **126** are passed *en route* (Scheme 27).

Cyclo[3]cumulenes with still larger rings have been obtained by the same method.[144] To avoid dimerization, cyclononatriene **127** (*n* = 5) must be handled at

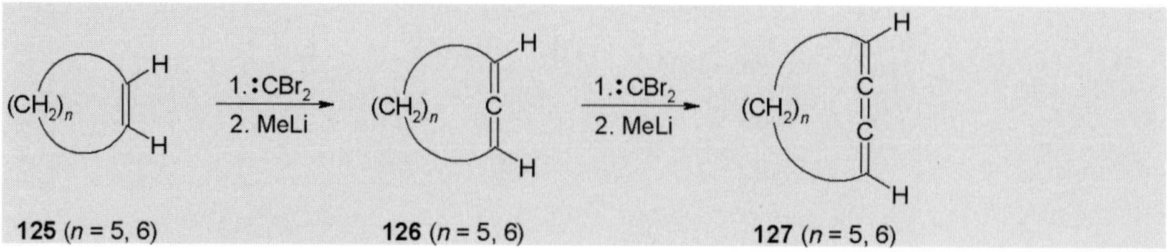

125 (*n* = 5, 6)   **126** (*n* = 5, 6)   **127** (*n* = 5, 6)

Scheme 27. Preparation of cyclo[3]cumulenes by the DMS method.

low temperatures; its enhanced reactivity is also demonstrated by ready trapping with Wilkinson's catalyst $(C_6H_5)_3PRhCl$ to yield a crystalline rhodium complex amenable to X-ray structural analysis.[145]

To conclude this chapter, it is noted that the [3]cumulenic analog of **110**, the doubly bridged [3]cumulene **129**, has been prepared by an intramolecular Wittig–Horner reaction from the phosphonate **128**[146] and that the longest known cyclocumulenes— incorporating six and five consecutive double bonds!—are the paracyclophane **130**[147] and the 1,2,3,4,5-tetradecapentaene **131**,[148] which are both stabilized by *tert*-butyl protective groups (Scheme 28).

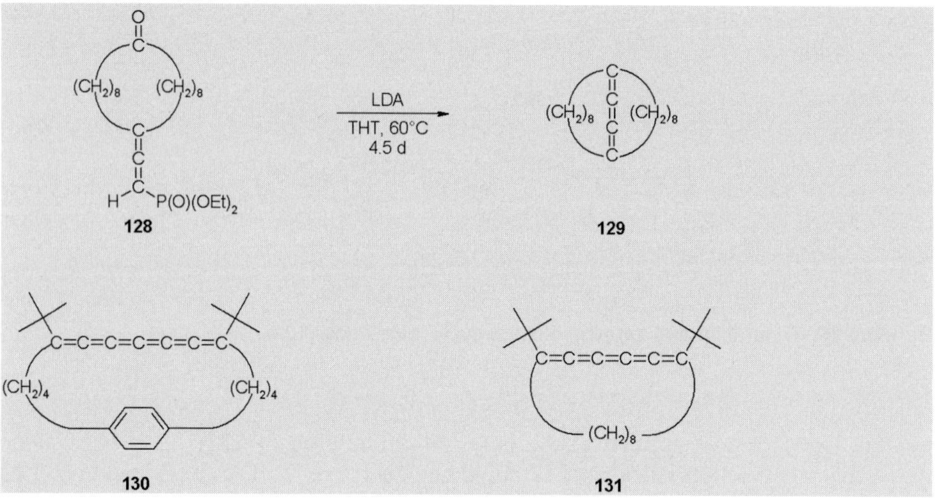

128   **129**

130   **131**

Scheme 28. A selection of mono- and doubly-bridged cyclocumulenes.

References

1.   R. Kuhn, K. Wallenfels, *Ber. Dtsch. Chem. Ges.*, **1938**, *71*, 783–790.
2.   F. Reboul, *Ann. Chem.*, **1864**, *131*, 238.
3.   L. Henry, *Ber. Dtsch. Chem. Ges.*, **1875**, *8*, 398–416.
4.   H. Staudinger, L. Ruzicka, *Helv. Chim. Acta*, **1924**, *7*, 177–201.

5. F. B. LaForge, F. Acree, Jr, *J. Org. Chem.*, **1942**, *7*, 416–418.

6. W. D. Celmer, I. A. Solomons, *J. Am. Chem. Soc.*, **1952**, *74*, 1870–1871. For reviews on naturally occurring allenes see S. R. Landor in S. R. Landor (*Ed.*), *The Chemistry of the Allenes*, Academic Press, London, **1982**, 679–707; C. H. Robinson, D. F. Covey in S. Patai (*Ed.*), *The Chemistry of Ketenes, Allenes and Related Compounds*, J. Wiley and Sons, Chichester, **1980**, Vol. I, 451–485.

7. J. H. van 't Hoff, *La Chimie dans L'Espace*, Bazendijk, Rotterdam, **1875**; *Die Lagerung der Atome im Raume*, F. Vieweg & Söhne, Braunschweig, **1877**.

8. P. Maitland, W. H. Mills, *Nature*, **1935**, *135*, 994.

9. E. P. Kohler, J. T. Walker, M. Tishler, *J. Am. Chem. Soc.*, **1935**, *57*, 1743–1745.

10. R. Kuhn, K. L. Scholler, *Chem. Ber.*, **1954**, *87*, 598–611.

11. Review literature: H. Fischer in *The Chemistry of Alkenes* (S. Patai, *ed.*), J. Wiley and Sons, Ltd. London, **1964**, 1025–1159; A. A. Petrov, A. V. Fedorova, *Russ. Chem. Rev.*, **1964**, *33*, 1–13; K. Griesbaum, *Angew. Chem.*, **1966**, *78*, 953–966; *Angew. Chem. Int. Ed. Engl.*, **1966**, *5*, 933–946; M. V. Mavrov, V. F. Kucherov, *Russ. Chem. Rev.*, **1967**, *36*, 233–249; D. R. Taylor, *Chem. Rev.*, **1967**, *67*, 317–359; T. F. Rutledge, *Acetylenes and Allenes*, Reinhold Book Corporation, New York, **1969**; S. R. Sandler, W. Karo, *Organic Functional Group Preparations*, Academic Press, New York, **1971**, Vol. 2, 1 ; R. Rossi, P. Diversi, *Synthesis*, **1973**, 25–36; M. Murray in Houben-Weyl, *Methoden der Organischen Chemie* E. Müller (*Ed.*), G. Thieme Verlag, Stuttgart, **1977**, Vol. 5/2a, 963–1076; H. Hopf in *The Chemistry of Functional Groups, The Chemistry of Ketenes, Allenes, and Related Compounds*, Part, 2 S. Patai (*Ed.*), John Wiley and Sons, Chichester, **1980**, 779–901; S. R. Landor (*Ed.*), *The Chemistry of the Allenes*, Academic Press, London, **1982**, Vol. I–III; H. F. Schuster, G. M. Coppola, *Allenes in Organic Synthesis*; J. Wiley and Sons, New York, **1984**; D. J. Pasto, *Tetrahedron*, **1984**, *40*, 2805–2827.

12. G. Wittig, A. Haag, *Chem. Ber.*, **1963**, *96*, 1535–1543.

13. W. v. E. Doering, P. M. LaFlamme, *Tetrahedron*, **1958**, *2*, 75–79; *cf.* W. v. E. Doering, A. K. Hoffman, *J. Am. Chem. Soc.*, **1954**, *76*, 6162–6165.

14. For the mechanism of the DMS reaction see H. Hopf in *The Chemistry of Functional Groups, The Chemistry of Ketenes, Allenes, and Related Compounds*, Part 2, S. Patai (*Ed.*), John Wiley and Sons, Chichester, **1980**, 785–786.

15. W. R. Moore, H. R. Ward, *J. Org. Chem.*, **1960**, *25*, 2073; *cf.* W. R. Moore, H. R. Ward, *J. Org. Chem.*, **1962**, *27*, 4179–4181.

16. L. Skattebøl, *Tetrahedron Lett.*, **1961**, 167–172.

17. This 'carbon insertion' becomes particularly obvious in a direct alkene to allene conversion which by-passes the intermediate **8** altogether and consists in the treatment of an olefin with one equivalent of carbon tetrabromide and two equivalents of methyllithium in ether at low temperatures: K. G. Untch, D. J. Martin, N. T. Castellucci, *J. Org. Chem.*, **1965**, *30*, 3572–3573.

18. G. Gustavson, N. Demjanoff, *J. Prakt. Chem.*, **1888**, *38*, 201–207.

19. Ya. M. Slobodin, A. P. Khitrov, *J. Gen. Chem. USSR*, **1961**, *31*, 3680–3681; *cf.* M. Bouis, *Ann. Chim. Paris,* **1928**, *9*, 402–465.

20. R. Bolze, H. Eierdanz, K. Schlüter, W. Massa, W. Grahn, A. Berndt; *Angew. Chem.*, **1982**, *94*, 926–927; *Angew. Suppl.*, **1982**, 2039–2049.

21. D. Vorländer, C. Siebert, *Ber. Dtsch. Chem. Ges.*, **1906**, *39*, 1024–1035.

22. H. Fischer, H. Fischer, *Chem. Ber.*, **1964**, *97*, 2975–2986; *cf.* F. Koelsch, *J. Am. Chem. Soc.*, **1933**, *55*, 3394–3399.

23. G. Kuscher, Dissertation, ETH Zürich, **1922**.

24. H. Gilman, R. A. Tomasi, *J. Org. Chem.*, **1962**, *27*, 3647–3650.

25. H. J. Bestmann, Th. Denzel, H. Salbaum, *Tetrahedron Lett.*, **1974**, 1275–1276.

26. E. H. R. Jones, H. H. Lee, M. C. Whiting, *J. Chem. Soc.*, **1960**, 341–346.
27. J. Grimaldi, M. Bertrand, *Bull. Soc. Chim. France*, **1971**, 947–957; *cf.* J. Grimaldi, M. Bertrand, *Bull. Soc. Chim. France*, **1971**, 4316–4320.
28. J. P. Dulcère, J. Goré, M. Roumestant, *Tetrahedron Lett.*, **1972**, 4465–4468.
29. J. P. Dulcère, J. Goré, M. L. Roumestant, *Bull. Soc. Chim. France*, **1974**, 1119–1123; *cf.* J. Goré, J. P. Dulcère, *J. Chem. Soc. Chem. Commun.*, **1972**, 866–867.
30. R. Schneider, H. Siegel, H. Hopf, *Liebigs Ann. Chem.*, **1981**, 1812–1825.
31. H. Siegel, H. Hopf, A. Germer, P. Binger, *Chem. Ber.*, **1978**, *111*, 3113–3118.
32. M. Bertrand, J. Grimaldi, B. Waegell, *J. Chem. Soc. Chem. Commun.*, **1968**, 1141–1142.
33. M. Bertrand, J. Grimaldi, B. Waegell, *Bull. Soc. Chim. France*, **1971**, 962–973.
34. F. Bohlmann, H.-J. Förster, C.-H. Förster, *Liebigs Ann. Chem.*, **1976**, 1487–1513.
35. J. Grimaldi, M. Bertrand, *Tetrahedron Lett.*, **1969**, 3269–3272.
36. J. Grimaldi, M. Bertrand, *Bull. Soc. Chim. France*, **1971**, 957–962; M. Bertrand, J. P. Dulcère, J. Grimaldi, M. Malacria, *C. R. Acad. Sci.*, Paris, **1974**, *279*, 805–806.
37. For a review see W. H. Okamura, *Acc. Chem. Res.*, **1983**, *16*, 81–88.
38. H. Hopf, *Angew. Chem.*, **1970**, *82*, 703; *Angew. Chem. Int. Ed. Engl.*, **1970**, *9*, 732.
39. F. Toda, Y. Takehira, *J. Chem. Soc. Chem. Commun.*, **1975**, 174.
40. F. Toda, Y. Takahara, *Bull. Chem. Soc. Japan*, **1976**, *49*, 2515–2517.
41. F. Toda, M. Ohi, *J. Chem. Soc. Chem. Commun.*, **1975**, 506.
42. E. Ghera, S. Shoua, *Tetrahedron Lett.*, **1974**, 3843–3846.
43. F. Coulomb-Delbecq, J. Goré, *Bull. Soc. Chim. France*, **1976**, 541–549.
44. K. Kleveland, L. Skattebøl, *J. Chem. Soc. Chem. Commun.*, **1973**, 432–433.
45. K. Kleveland, L. Skattebøl, *Acta Chem. Scand.*, **1975**, *B29*, 191–196; *cf.* K. Kleveland, L. Skattebøl, *Acta Chem. Scand.*, **1975**, *B29*, 827–830 and references cited.
46. H. Hopf, P. Blickle, *Tetrahedron Lett.*, **1978**, 449–452.
47. R. F. Heldeweg, H. Hogeveen, E. P. Schudde, *J. Org. Chem.*, **1978**, *43*, 1916–1920.
48. H. Hopf, G. Schön, *Liebigs Ann. Chem.*, **1981**, 165–180.
49. H. Hopf, *Angew. Chem.*, **1972**, *84*, 471–472; *Angew. Chem. Int. Ed. Engl.*, **1972**, *11*, 419–420.
50. L. Skattebøl, *J. Org. Chem.*, **1966**, *31*, 2789–2794. 1,2,5,6-Heptatetraene (diallenylmethane, **35**, *n* = 1) cannot be prepared by this method: K. J. Drachenberg, H. Hopf, *Tetrahedron Lett.*, **1974**, 3267–3270.
51. L. Skattebøl, S. Solomon, *J. Am. Chem. Soc.*, **1965**, *87*, 4506–4513.
52. W. R. Roth, M. Heiber, G. Erker, *Angew. Chem.*, **1973**, *85*, 511–512; *Angew. Chem. Int. Ed. Engl.*, **1973**, *12*, 504–505; *cf.* W. R. Roth, G. Erker, *Angew. Chem.*, **1973**, *85*, 510–511; *Angew. Chem. Int. Ed. Engl.*, **1973**, *12*, 503; W. R. Roth, G. Erker, *Angew. Chem.*, **1987**, *85*, 512; *Angew. Chem. Int. Ed. Engl.*, **1973**, *12*, 505.
53. W. Grimme, H. J. Rother, *Angew. Chem.*, **1973**, *85*, 512–514; *Angew. Chem. Int. Ed. Engl.*, **1973**, *12*, 505–506.
54. For a cyclic variant of this interesting process see. E. V. Demlow, G. C. Ezimora, *Tetrahedron Lett.*, **1970**, 4047–4050.
55. D. A. Ben Efraim, F. Sondheimer, *Tetrahedron Lett.*, **1963**, 313–315.
56. C. W. Bowes, D. F. Montecalvo, F. Sondheimer, *Tetrahedron Lett.*, **1973**, 3181–3184.
57. The first chiral allenes to be resolved into their enantiomers were not hydrocarbons but allene carboxylic acids, see refs.[8] and.[9] For reviews on the chirality of allenes see W. Runge in S. Patai (*Ed.*), *The Chemistry of Ketenes, Allenes and Related Compounds*, J. Wiley and Sons, Chichester, **1980**, Vol. 1, 98–154; W. Runge in S. R. Landor (*Ed.*), *The Chemistry of the Allenes*, Academic Press, London, **1982**, 579–678.
58. G. Zweifel, H. C. Brown, *J. Am. Chem. Soc.*, **1964**, *86*, 393–397.
59. W. L. Waters, M. C. Caserio, *Tetrahedron Lett.*, **1968**, 5233–5236.

60. A. Claesson, L.-I. Olsson, *J. Am. Chem. Soc.*, **1979**, *101*, 7302–7311.
61. J. L. Luche, E. Barreiro, J. M. Dollat, P. Crabbé, *Tetrahedron Lett.*, **1975**, 4615–4618.
62. W. M. Jones, J. W. Wilson, Jr, F. B. Tutwiler, *J. Am. Chem. Soc.*, **1963**, *85*, 3309–3010.
63. A. Claesson, L.-I. Olsson, *J. Chem. Soc. Chem. Commun.*, **1979**, 524–525.
64. N. M. Schubert, T. M. Liddicoet, W. N. Lanka, *J. Am. Chem. Soc.*, **1952**, *74*, 569; *cf.* N. M. Schubert, T. M. Liddicoet, W. N. Lanka, *J. Am. Chem. Soc.*, **1954**, *76*, 1929–1932. Earlier hints on the generation of 1,2,3-butatriene can be found in the patent literature: W. H. Carothers, G. J. Berchet, U. S. patent 2,136,178 (Nov. 8, **1937**).
65. P. P. Montijn, L. Brandsma, A. F. Arens, *Rec. Trav. Chim.*, **1967**, *86*, 129–146.
66. D. Mirejovsky, J. F. Arens, W. Drenth, *Rec. Trav. Chim.*, **1976**, *95*, 270–273.
67. E. Kloster-Jensen, J. Wirz, *Helv. Chim. Acta*, **1975**, *58*, 162–177; *cf.* E. Kloster-Jensen, J. Wirz, *Angew. Chem.*, **1973**, *85*, 723; *Angew. Chem. Int. Ed. Engl.*, **1973**, *12*, 671.
68. W. R. Roth, H. Humbert, G. Wegener, G. Erker, H.-D. Exner, *Chem. Ber.*, **1975**, *108*, 1655–1658.
69. W. J. Ball, S. R. Landor, N. Punja, *J. Chem. Soc. (C)*, **1967**, 194–197.
70. T. L. Jacobs, P. Prempree, *J. Am. Chem. Soc.*, **1967**, *89*, 6177–6182.
71. W. R. Roth, H.-D. Exner, *Chem. Ber.*, **1976**, *109*, 1158–1162.
72. L. Skattebøl, *Tetrahedron*, **1965**, *21*, 1357–1367.
73. G. Köbrich, H. Heinemann, W. Zündorf, *Tetrahedron*, **1967**, *23*, 565–584; *cf.* W. Krestinsky, *Ber. Dtsch. Chem. Ges.*, **1926**, *59*, 1930–1936.
74. G. Köbrich, W. Drischel, *Angew. Chem.*, **1965**, *77*, 95–96; *Angew. Chem. Int. Ed. Engl.*, **1965**, *4*, 74–75; G. Köbrich, W. Drischel, *Tetrahedron*, **1966**, *22*, 2621–2636.
75. H. Yamazaki, *J. Chem. Soc. Chem. Commun.*, **1976**, 841–842.
76. J. L. Ripoll, *J. Chem. Soc. Chem. Commun.*, **1976**, 235–236.
77. J. L. Ripoll, A. Thuillier, *Tetrahedron*, **1977**, *33*, 1333–1336.
78. L. Skattebøl, *Tetrahedron Lett.*, **1965**, 2175–2179.
79. L. T. Scott, G. J. DeCicco, *Tetrahedron Lett.*, **1976**, 2663–2666.
80. H. D. Hartzler, *J. Am. Chem. Soc.*, **1966**, *88*, 3155–3156; *cf.* H. D. Hartzler, *J. Am. Chem. Soc.*, **1971**, *93*, 4527–4531.
81. T. Nagi, T. Kaneda, H. Mizuno, T. Toyoda, Y. Sakata, S. Misumi, *Bull. Chem. Soc. Japan*, **1974**, *47*, 2398–2405.
82. F. Bohlmann, K. Kieslich, *Chem. Ber.*, **1954**, *87*, 1363–1372.
83. F. Bohlmann, K. Kieslich, *Abhandl. Braunschweig. Wiss. Ges.*, **1957**, *9*, 147.
84. Actually the first demonstration of geometrical isomerism for [3]cumulenes involved not a hydrocarbon but the dinitro derivative of bis-[diphenylene]-butatriene.
85. R. Kuhn, D. Blum, *Chem. Ber.*, **1959**, *92*, 1483–1714; *cf.* K. Brand, F. Kercher, *Ber. Dtsch. Chem. Ges.*, **1921**, *54*, 2007–2017.
86. R. K. Kuhn, K. Wallenfalls, *Chem. Ber.*, **1938**, *71*, 1510–1512; *cf.* R. Kuhn, H. Krauch, *Chem. Ber.*, **1955**, *88*, 309–315.
87. P. Cadiot, *Ann. Chim. Paris*, **1956**, 214–272; *cf.* W. Chodkiewicz, *Ann. Chim. Paris*, **1957**, 819–869.
88. R. Kuhn, H. Zahn, *Chem. Ber.*, **1951**, *84*, 566–570.
89. R. Kuhn, H. Fischer, H. Fischer, *Chem. Ber.*, **1964**, *97*, 1760–1766.
90. Reviews: R. P. Johnson in *Molecular Structure and Energetics*, J. F. Liebman, A. Greenberg, (*Eds.*), VCH Publishers, Inc., Deerfield Beach, FL, **1986**, Vol. 3, 85–140; R. P. Johnson, *Chem. Rev.*, **1989**, *89*, 1111–1124; R. P. Johnson in *Advances in Theoretically Interesting Molecules* R. P. Thummel (*Ed.*), JAI Press, Inc., Greenwich, CT, **1989**, 401–436.
91. B. Lam, R. P. Johnson, *J. Am. Chem. Soc.*, **1983**, *105*, 7479–7483.
92. A. E. Favorskii, *J. Gen. Chem. USSR*, **1936**, *6*, 720 ; *cf.* A. E. Favorskii, *Bull. Soc. Chim.*

France, **1936**, *5*, 1727–1732.
93. G. Wittig, J. Heyn, *Liebigs Ann. Chem.*, **1972**, *756*, 1–13.
94. R. J. Angus, Jr, M. W. Schmidt, R. P. Johnson, *J. Am. Chem. Soc.*, **1985**, *107*, 532–537 and lit. cited.
95. From the rate of racemization of 1,3-dimethyl- and 1,3-di-tert-butylallene in the gas phase the rotational barriers of the C–C double bonds were determined to be 46.2 and 46.9 kcal mol^{-1}, respectively: W. R. Roth, G. Ruf, Ph. W. Ford, *Chem. Ber.*, **1974**, *107*, 48–52.
96. N. A. Domnin, *J. Gen. Chem. USSR*, **1945**, *15*, 461; *cf.* N. A. Domnin, *J. Gen. Chem. USSR*, **1940**, *10*, 1939.
97. W. J. Ball, S. R. Landor, *Proc. Chem. Soc. London*, **1961**, 143–144; *cf.* W. J. Ball. S. R. Landor, *J. Chem. Soc.*, **1962**, 2298–2304.
98. G. Wittig, P. Fritze, *Angew. Chem.*, **1966**, *78*, 905; *Angew. Chem. Int. Ed. Engl.*, **1966**, *5*, 846; *cf.* G. Wittig, P. Fritze, *Liebigs Ann. Chem.*, **1968**, *711*, 82–87.
99. For a review see H. Hopf in S. R. Landor (*Ed.*), *The Chemistry of the Allenes*, Academic Press, London, **1982**, 525–562.
100. W. R. Moore, W. R. Moser, *J. Am. Chem. Soc.*, **1970**, *92*, 5469–5474.
101. M. Christl, M. Schreck, *Chem. Ber.*, **1987**, *120*, 915–920; *cf.* M. Christl, M. Schreck, *Angew. Chem.*, **1987**, *99*, 474–475; *Angew. Chem. Int. Ed. Engl.*, **1987**, *26*, 449–450.
102. A. T. Bottini, F. P. Corson, R. Fitzgerald, K. A. Frost II, *Tetrahedron*, **1972**, *28*, 4883–4904.
103. C. Wentrup, G. Gross, A. Maquestiau, R. Flammang, *Angew. Chem.*, **1983**, *95*, 551; *Angew. Chem. Int. Ed. Engl.*, **1983**, *22*, 542.
104. A. Runge, W. Sander, *Tetrahedron Lett.*, **1986**, *27*, 5835–5838.
105. M. Balci, W. M. Jones, *J. Am. Chem. Soc.*, **1980**, *102*, 7607–7608.
106. G. Wittig, J. Meske-Schüller, *Liebigs Ann. Chem.*, **1968**, *711*, 76–81.
107. J. P. Visser, J. E. Ramakers, *J. Chem. Soc. Chem. Commun.*, **1972**, 178–179.
108. S. M. Oon, A. E. Koziol, W. M. Jones, G. J. Palenik, *J. Chem. Soc. Chem. Commun.*, **1987**, 491–492.
109. G. Wittig, H.-L. Dorsch, J. Meske-Schüller, *Liebigs Ann. Chem.*, **1968**, *711*, 55–64.
110. E. T. Marquis, P. D. Gardner, *Tetrahedron Lett.*, **1966**, 2793–2798.
111. J. D. Price, R. P. Johnson, *Tetrahedron Lett.*, **1986**, *27*, 4679–4682.
112. A. T. Blomquist, R. E. Burge, Jr, L. H. Liu, J. C. Bohrer, A. C. Sucsy, J. Kleis, *J. Am. Chem. Soc.*, **1951**, *73*, 5510–5511.
113. L. Skattebøl, S. Solomon, *Organic Syntheses*, J. Wiley and Sons, New York, N. Y., **1973**, *Coll. Vol. V*, 306–310.
114. M. Traetteberg, P. Bakken, A. Almenningen, *J. Mol. Struct.*, **1981**, *70*, 287–295.
115. J. L. Luche, J. C. Damiano, P. Crabbé, C. Cohen-Addad, J. Lajzerowicz, *Tetrahedron*, **1977**, *33*, 961–964.
116. M. Christl, M. Braun, G. Müller, *Angew. Chem.*, **1992**, *104*, 471–473; *Angew. Chem. Int. Ed. Engl.*, **1992**, *31*, 473–475; *cf.* M. Christl, M. Braun in *Strain and its Implications in Organic Chemistry* A. de Meijere, S. Blechert, (*Eds.*), NATO ASI Series, Kluwer Academic Publishers, Dordrecht, **1989**, 121–131. For the generation of an oxa-variant of **94**, 3δ^2-chromene (2,3-didehydro-2*H*-1-benzopyran), see M. Christl, St. Drinkuth, *Eur. J. Org. Chem.*, **1998**, 237–241.
117. With cyclopentadiene, **87** also yields the isomer of **91** in which the methylene and the etheno bridge in the bicyclo[2.2.1]part of the molecule are interchanged.
118. B. Miller, X. Shi, *J. Am. Chem. Soc.*, **1987**, *109*, 578–579.
119. H. Hopf, H. Berger, G. Zimmermann, U. Nüchter, P. G. Jones, I. Dix, *Angew. Chem.*, **1997**, *109*, 1236–1238; *Angew. Chem. Int. Ed. Engl.*, **1997**, *36*, 1187–1190.

120. R. L. Danheiser, A. E. Gould, R. F. de la Pradilla, A. L. Helgason, *J. Org. Chem.*, **1994**, *59*, 5514–5515; *cf.* R. C. Burrell, K. J. Daoust, A. Z. Bradley, K. J. Di Ricio, R. P. Johnson, *J. Am. Chem. Soc.*, **1996**, *118*, 4218–4219.

121. R. Janoschek, *Angew. Chem.*, **1992**, *104*, 473–475; *Angew. Chem. Int. Ed. Engl.*, **1992**, *31*, 476–478.

122. W. R. Roth, H. Hopf, C. Horn, *Chem. Ber.*, **1994**, *127*, 1765–1779.

123. For a review see R. A. Moss, M. Jones, Jr in *Reactive Intermediates* R. A. Moss, M. Jones, Jr, (*Eds.*), John Wiley and Sons, New York, N. Y., **1985**, Vol. 3, 91–92.

124. B. L. Duell, W. M. Jones, *J. Org. Chem.*, **1978**, *43*, 4901–4903; *cf.* K. Untch, *Int. Symposium on the Chemistry of Non-Benzenoid Aromatic Compounds*, Sendai, Japan, **1970**.

125. P. R. West, O. L. Chapman, J.-P. LeRoux, *J. Am. Chem. Soc.*, **1982**, *104*, 1779–1782.

126. E. E. Waali, W. M. Jones, *J. Am. Chem. Soc.*, **1973**, *95*, 8114–8118.

127. R. J. McMahon, C. J. Abelt, O. L. Chapman, J. W. Johnson, C. L. Kreil, J.-P. LeRoux, A. M. Mooring, P. R. West, *J. Am. Chem. Soc.*, **1987**, *109*, 2456–2469 and refs. cited.

128. C. L. Janssen, H. F. Schaefer III, *J. Am. Chem. Soc.*, **1987**, *109*, 5030–5031 and refs. therein.

129. J. W. Harris, W. M. Jones, *J. Am. Chem. Soc.*, **1982**, *104*, 7329–7330.

130. G. H. Mitchell, F. Sondheimer, *J. Am. Chem. Soc.*, **1969**, *91*, 7520–7521.

131. H. Hopf, L. Eisenhuth, *Chem. Ber.*, **1975**, *108*, 2635–2648.

132. S. Masamune, C. G. Chin, K. Hojo, R. T. Seidner, *J. Am. Chem. Soc.*, **1967**, *89*, 4804–4805.

133. M. S. Baird, C. B. Reese, *Tetrahedron*, **1976**, *32*, 2153–2156.

134. H. Irngartinger, H. Jäger, *Tetrahedron Lett.*, **1976**, 3595–3596.

135. E. V. Dehmlow, M. Lissel, *Liebigs Ann. Chem.*, **1979**, 181–193. For a discussion of the **107**→**108** cyclization from the orbital symmetry viewpoint, see R. B. Woodward, R. Hoffmann, *The Conservation of Orbital Symmetry,* Verlag Chemie, Weinheim, **1971**, 63–64.

136. M. Nakazaki, K. Yamamoto, M. Maeda, O. Sato, T. Tsutsui, *J. Org. Chem.*, **1982**, *47*, 1435–1438.

137. For MNDO calculations on the homologous series **111** with *n* = 1–6, see R. O. Angus, Jr, R. P. Johnson, *J. Org. Chem.*, **1984**, *49*, 2880–2883.

138. Heterocyclic [3]cumulenes apparently tolerate strain easier. Thus 3,4-didehydrothiophene has been generated and trapped: X.-S. Ye, W.-K. Li, H. N. C. Wong, *J. Am. Chem. Soc.*, **1996**, *118*, 2511–2512; *cf.* J.-H. Liu, H.-W. Chan, F. Xue, Q.-G. Wang, T. C. W. Mak, H. N. C. Wong, *J. Org. Chem.*, **1999**, *64*, 1630–1634.

139. W. C. Shakespeare, R. P. Johnson, *J. Am. Chem. Soc.*, **1990**, *112*, 8578–8579. A bicyclic derivative of **117**, bicyclo[2.2.0]nona-1,2,3-triene, is possibly formed by intramolecular Diels–Alder addition when 1-nonen-6,8-diyne is subjected to flash vacuum pyrolysis: R. C. Burrell, K. J. Daoust, A. Z. Bradley, K. J. Di Rico, R. P. Johnson, *J. Am. Chem. Soc.*, **1996**, *118*, 4218–4219.

140. H.-G. Zoch, G. Szeimies, R. Romert, R. Schmitt, *Angew. Chem.* **1981**, *93*, 894–895; *Angew. Chem. Int. Ed. Engl.*, **1981**, *20*, 877–878.

141. A.-D. Schlüter, J. Belzner, U. Heywang, G. Szeimies, *Tetrahedron Lett.*, **1983**, *24*, 891–894.

142. S. Hernandez, M. M. Kirchhoff, S. G. Swartz, Jr, R. P. Johnson, *Tetrahedron Lett.*, **1996**, *37*, 4907–4910; *cf.* H. N. C. Wong, T. Chan, F. Sondheimer, *Tetrahedron Lett.*, **1978**, *7*, 667–670 for an early attempt to prepare a derivative of 1,2,3-cyclooctatriene.

143. W. R. Moore, T. M. Ozretich, *Tetrahedron Lett.*, **1967**, 3205–3207.

144. H. W. Schmitt, Ph. d. Dissertation, University of Braunschweig, **1986**.

145. R. O. Angus, Jr, M. N. Janakiraman, R. A. Jacobson, R. P. Johnson, *Organometallics*,

1987, *6*, 1909–1992.

146. R. S. Macomber, T. C. Hemling, *J. Am. Chem. Soc.*, **1986**, *108*, 343–344.

147. T. Negi, T. Kaneda, H. Mizuno, T. Toyoda, Y. Sakata, S. Misumi, *Bull. Chem. Soc. Japan*, **1974**, *47*, 2398–2405.

148. T. Negi, T. Kaneda, Y. Sakata, S. Misumi, *Chem. Lett.*, **1972**, 703–706.

10 The Annulenes

[n]Annulenes are fully conjugated monocyclic hydrocarbons of the general formula $(C_2H_2)_n$, **1**, with $n \geq 2$. Although the first three members of this vinylogous series could thus be called [4]- (**2**), [6]- (**3**) and [8]annulene (**4**), respectively, the traditional names for these hydrocarbons, 1,3-cyclobutadiene, benzene, and cyclooctatetraene, never had to fear to be replaced. Only from $n = 5$ onwards is the annulene nomenclature used, *i.e.* hydrocarbon **5** is [10]annulene *etc.* (Scheme 1).

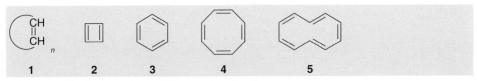

Scheme 1. The first members of the annulene family.

The annulenes are of utmost importance in synthetic, structural, mechanistic, theoretical and industrial chemistry, whether they are employed as starting materials, reaction intermediates, solvents or reference compounds. The phenomenon of aromaticity, which is best illustrated using annulenes as model compounds, is probably the single most general and important principle of all of organic chemistry, especially because it can be applied not only to molecules in their ground but also for characterization of transition states. No other class of hydrocarbons has inspired chemists more than the molecules symbolized by the simple structure **1**. And with the possible exception of the H_2O formula of water no other chemical structure has more been associated by the general public with the chemical sciences than the hexagon of benzene. The hydrocarbons **1** have exert an undiminished attraction on practical and theoretical chemists ever since Fararady discovered benzene in 1825. In fact, the history of organic chemistry in the 19th century is to a large extent the history of aromatic compounds and the establishment of the concept of aromaticity. There are countless monographs and reviews on benzene, aromatics, and annulenes in general, covering all aspects of these hydrocarbons from their synthesis to their role in the ecosphere.[1–25] Rather than making a—hopeless—effort to be comprehensive I have tried to trace the most important developments in this area, with the intention of showing that even after more than a century after entering the chemical stage some annulenes have not lost their attractiveness in preparative and theoretical chemistry. This is particular well demonstrated by the smallest of the [n]annulenes, 1,3-cyclobutadiene (**2**).

10.1 1,3-Cyclobutadiene

The first attempts to prepare 1,3-cyclobutadiene or derivatives thereof go back to the early days of organic chemistry. Both W. H. Perkin, Jr[26] and A. Kekulé[27] attempted to synthesize derivatives of **2**; the first systematic effort to obtain the hydrocarbon was that of R. Willstätter and his students.[28, 29] Willstätter was particularly interested in this problem because he wanted to establish the validity of Thiele's theory of partial valencies.[30] Whereas he was successful in preparing the higher vinylog of benzene, cyclooctatetraene (**4**, see Section 10.3) he failed with the synthesis of **2**. Sporadic attempts to prepare **2**—all of them unsuccessful[4]—appeared in the chemical literature until a seminal paper by Longuet-Higgins and Orgel in 1956 suggested that transition metal complexes of cyclobutadienes should be stable.[31] Three years later, Criegee and Schröder proved this proposal to be correct,[32] initiating a renaissance in cyclobutadiene chemistry which finally led to the solution of the century-old problem of synthesizing this hydrocarbon which, although structurally so simple, was so important for the development of the theory of aromaticity.[4, 33]

The first isolable derivative of **2**, the metal complex **7**, was obtained by treating the dichloride **6** with nickel tetracarbonyl in different organic solvents (Scheme 2).[32]

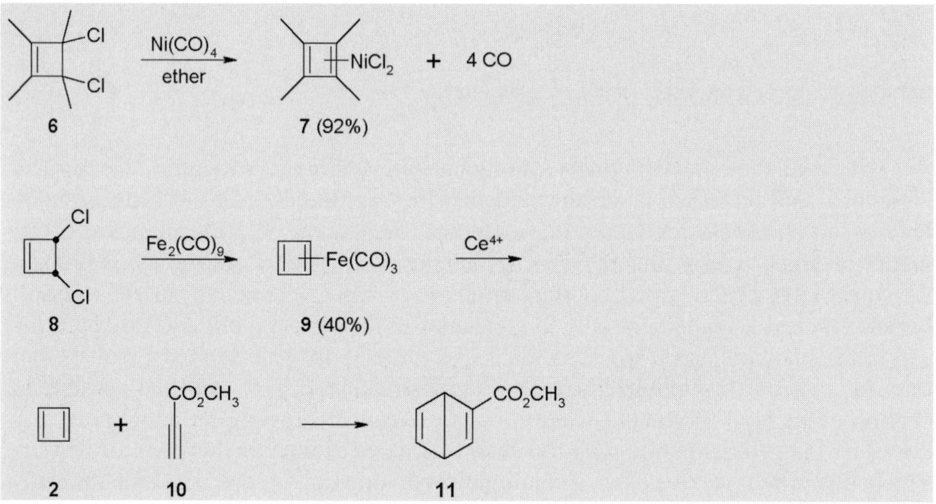

Scheme 2. The first stable cyclobutadienes-transition metal complexes.

In ether the yield of the diamagnetic complex, which crystallizes from chloroform as shiny, nearly black needles, was close to quantitative (92%). Several years later Pettit and co–workers prepared the iron tricarbonyl complex **9** of the parent hydrocarbon **2** by a similar approach using *cis*-3,4-dichlorocyclobutene (**8**) as the starting material.[34] Although the organic ligand can be released from these metal complexes by heating, a milder and more effective way of generating free cyclobutadiene consists in the oxidation of **7** and **9**, diammonium hexanitrocerate(IV) ('cerium ammonium nitrate', CAN) being the oxidant most often used. When this decomposition is per-formed in the presence of a trapping reagent such as methyl propiolate (**10**) cycload-ducts such as the Dewar benzene derivative **11** (see Section 10.2) result.[35]

That **2** is, in fact, generated as a transient intermediate in the oxidation, *i.e.* that the iron carbonyl is not involved in the cycloaddition process, was shown either by em-ploying optically active derivatives of **9**[36, 37] or by the so-called three-phase test in which the 'releasing' cyclobutadiene complex and the 'accepting' dieneophile were anchored to a polymeric support.[38]

Having shown that **2** and its simple derivatives are capable of existence, subse-quent developments to harness this highly reactive hydrocarbon followed two major lines. On the one hand techniques were developed to generate **2** from appropriate pre-cursors by thermolysis or photolysis and then either trap it in an inert matrix at low temperatures or to produce it in the gas phase—*i.e.* at high dilution—thus enabling recording its IR, UV, and photoelecton spectra. On the other hand, substituents were introduced into the kinetically unstable **2**, which either protected it sterically or changed its electronic properties (push-pull-cyclobutadienes[39]).

The matrix isolation of **2** was first achieved in 1973 by Chapman[40, 41] and by Krantz[42] who photolyzed α-pyrone (**12**) in argon at 8–10 K with 245-nm light. (Scheme 3).

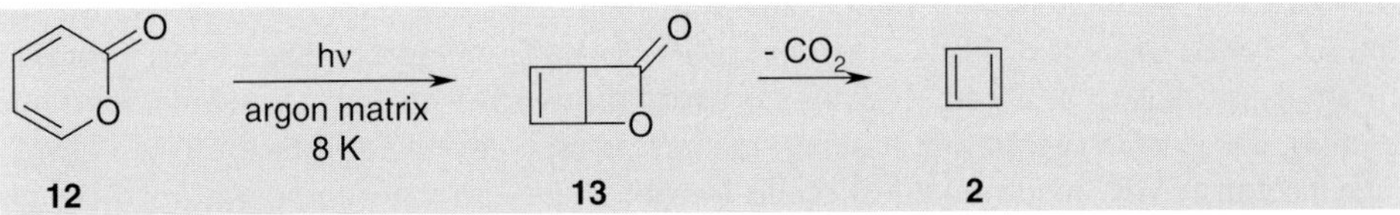

Scheme 3. The first generation of cyclobutadiene (**2**) in a matrix.

The primary step of this decarboxylation consists in the photoisomerization of **12** to 2-oxabicyclo[2.2.0]hex-5-en-3-one (**13**),[43] which subsequently fragments into **2** and carbon dioxide. Numerous other cycloreversions were subsequently used to gen-erate **2**, as shown by the representative examples summarized in Scheme 4.

Thus photolysis of cyclobutene dicarboxylic acid anhydride (**14**) at −196 °C led to **2**, with extrusion of carbon dioxide and carbon monoxide.[44] This method has been employed to prepare not only the parent hydrocarbon and several alkyl derivatives,[44] but also various chloro-[45] and fluorocyclobutadienes.[46] In what amounts to the cleanest route to **2**, tricyclo[2.1.0.0^{2,5}]pentan-3-one (**15**) has been photodecar-bonylated (λ_{max} = 254 nm) in an argon matrix at 10 K.[47] Clearly, **2** cannot be the primary product in this process, but in contrast to the tetrakis-*tert*-butyl derivative of **15** (see below and Section 5.1) no tetrahedrane could be detected in this case. Hetero-analogous Dewar benzenes such as **16**—generated as a reactive intermediate by oxi-dation of the corresponding bicyclic hydrazine derivative—have been used to prepare **2** (extrusion of N$_2$) [48, 49] as have the cyclic ether **17**,[50] the addition product of di-methyl acetylenedicarboxylate to cyclooctatetraene **18** (see Section 10.3),[51] and the heterocycle **19**,[44] all of which, on photolysis, decomposed into **2** and a stable aro-matic or heteroaromatic fragment.

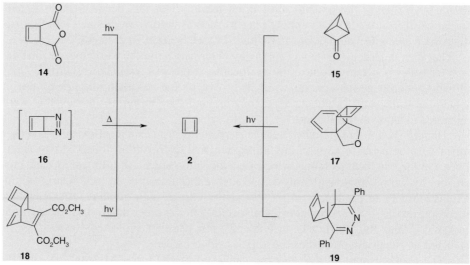

Scheme 4. The preparation of cyclobutadiene (**2**) by different cycloreversions.

Among the physical data of **2**[33] its NMR spectra are particularly difficult to measure. Nevertheless, by employing highly sophisticated experimental techniques the ^{13}C NMR spectra of [1,2-^{13}C$_2$]- and [1,4-^{13}C$_2$]cyclobutadiene have been recorded in an argon matrix at 25 K.[52] These investigations showed that cyclobutadiene has rectangular geometry and that the two valence isomers are in equilibrium (see below). The proton NMR spectrum of **2** could be obtained—at room temperature!—by replacing the low temperature matrix by a container molecule, the so-called hemicarcerand **20**, which encloses and protects the highly reactive species in its interior.[53] When the precursor molecules **12** and **20** are refluxed in chlorobenzene the lactone can slip into the hollow center of **20**, thus forming the encapsulated system **21** (Scheme 5).

At room temperature, however, the 'exits' of **21** are too narrow to allow **12** to leave the molecular prison again. Irradiation of **21** with a xenon lamp first caused electrocyclization to **22** and subsequently decarboxylation to **23**—just as in the matrix experiment (see above, Scheme 3). According to the ^{1}H NMR spectrum the cyclobutadiene in **23** exists as stable singlet.

A second route to cyclobutadiene and its derivatives involves ring expansion of cyclopropenylcarbenes which have been generated, for example, by decomposition of the diazo ketones **24** followed by a Wolff rearrangement and decarbonylation of the resulting ketene intermediate (Scheme 6).[54]

Highly substituted diazomethanes such as **26** are the substrates of choice for the preparation of cyclobutadienes carrying bulky substituents, including tetrakis-*tert*-butylcyclobutadiene (**25**)[55–57] which we have already encountered as the thermal isomerization product of tetrakis-*tert*-butyl tetrahedrane (see Section 5.1).

The obvious route to cyclobutadienes by dimerization of alkynes is attractive on first sight, but meets with practical difficulties at least as far as the actual isolation of stable derivatives is concerned. Both acetylenic hydrocarbons[57, 58] and functionalized

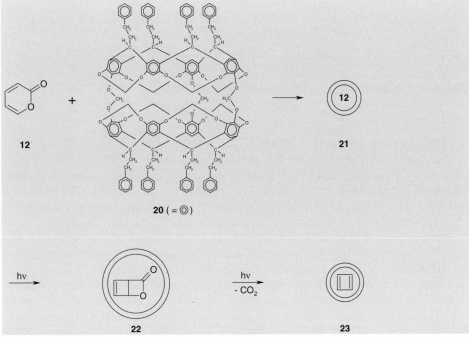

Scheme 5. Trapping of cyclobutadiene (**2**) in a molecular container.

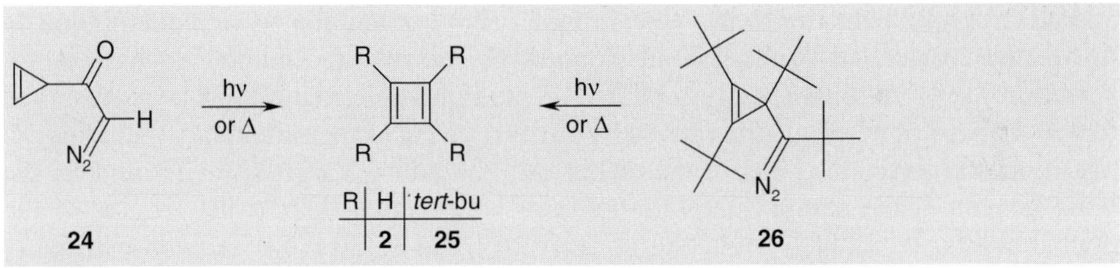

Scheme 6. Cyclobutadienes by ring expansion of cyclopropenyl carbenes.

alkynes such as cyanoacetylene[60] undergo [2+2]cycloadditions, but the resulting cyclobutadiene derivatives can be detected solely by spectroscopic measurements[58, 59] or trapping experiments[60] (see below). Only for the sterically shielded cycloheptynes **27** (X = CH$_2$ and S) (see Section 8.2), and exploiting the trick of transition metal stabilization and subsequent liberation of the cyclobutadiene generated (see above), can a [2+2]acetylene dimerization leading to isolable products be realized, in this case to **29**, which was formed *via* the palladium complex **28**. The dithia derivative **29** (X = S), prepared by Krebs and co-workers in 1972 was the first kinetically stabilized and electronically undisturbed [4]annulene to be reported (Scheme 7).[61]

Later the parent hydrocarbon **29** (X = CH$_2$) was prepared in the same laboratory.[62]

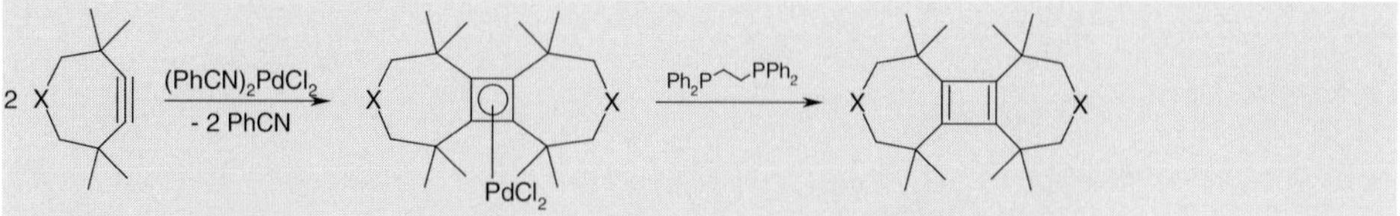

Scheme 7. The first kinetically stable cyclobutadiene derivatives.

The intense research effort devoted to the solution of the cyclobutadiene problem is—of course—explained by the role this particular annulene plays in the theory of aromaticity. According to Erich Hückel's famous rule[63] aromatic compounds must have $(4n+2)$ π-electrons (n = 0, 1, 2, 3 *etc.*). Clearly, cyclobutadiene does not meet this criterion. In fact, just as benzene (**3**) is the aromatic hydrocarbon *par excellence*, cyclobutadiene is the epitome of the *anti*-aromatic molecules, *i.e.* those possessing $4n$ π-electrons (n = 1, 2 *etc.*). Simple Hückel MO theory predicts that **2** should be a square planar, triplet diradical with zero resonance energy; its highest occupied molecular orbitals should be a pair of degenerate non-bonding MOs. For a square structure (D_{4h} symmetry), which obviously cannot show bond alternation between single and double bonds, IR selection rules allow four fundamental vibrations. If, on the other hand, **2** prefers a rectangular, planar structure (D_{2h} symmetry) the degeneracy of the non-bonding MOs is lifted and electron pairing in a bonding molecular orbital becomes possible, *i.e.* rectangular cyclobutadiene should be a singlet, with bond alternation. Its IR spectrum should consist of seven fundamentals, and the high-energy HOMO and the low-energy LUMO of such a species would imply both nucleophilic and electrophilic behavior to be typical of **2**.[64] Experimental evidence unambigously proves the correctness of the second alternative. The required number of lines is observed in the vibrational spectrum of **2**,[65, 66] and bond alternation has been observed for several cyclobutadiene derivatives by various physical measurements including X-ray structural analysis. Thus in the stable cyclobutadienes **29** (X = CH$_2$ and S) the bond lenghts of the single bonds lie between 1.597 and 1.602 Å; the lengths of the double bonds are 1.344 Å in the sulfur derivative and 1.339 Å in the hydrocarbon.[67, 10]

A cyclic arrangement of single and double bonds implies, furthermore, that valence tautomerism should be possible for appropriately substituted cyclobutadienes. In the energy profile (Scheme 8) for interconversion of the two rectangular forms of cyclobutadiene, **30** and **32**, the square structure **31** (actually the square singlet structure which is more stable than the square triplet form[69, 70]) would then correspond to the transition state. For 1,2-dideuteriocyclobutadiene, generated from bisdeuterated (at its bridgeheads) **16**, an energy of activation ($\Delta H^{\ddagger}$) between 1.6 and 10 kcal mol^{-1} and a strongly negative entropy of activation $\Delta S^{\ddagger}$ in the range of -17 to -32 e.u. were determined for the isomerization between the two valence isomers.[49, 71] For the valence isomerization betwen the highly substituted tautomers **33** and **34** an energy barrier ($\Delta G^{\ddagger}$) of 5.8 kcal mol^{-1} has been measured by use of variable temperature NMR spectroscopy.[57, 72]

Replacing the isopropoxydimethylsilyl substituent in **33/34** by a dimethylphenylsilyl group caused a decrease in the isomerization barrier ($\Delta G^{\ddagger}$ = 4.5 kcal mol^{-1}).[57, 72]

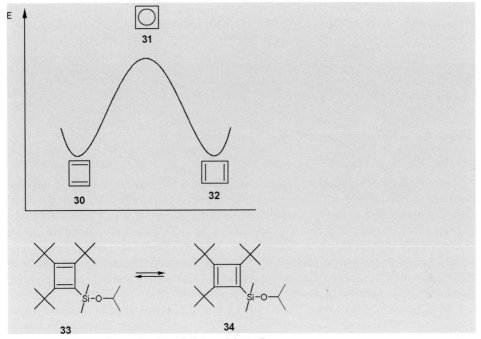

Scheme 8. Valence isomerization of 1,3-cyclobutadiene.

10.2 Benzene and its isomers

Formally six carbon and six hydrogen atoms can be combined to form a total number of 217 C_6H_6 isomers.[73] Among these, one stands out—the stabilomer benzene (**3**). Its isolation, synthesis and structure determination were among the main problems and achievements of organic chemistry of the 19th century. The benzene story is mentioned in every textbook of organic chemistry and has been discussed at length from the historical viewpoint also.[74, 75] Nothing better illustrates the importance of this hydrocarbon for the chemists of the 19th century than the famous *Benzolfest* which took place in 1890, 25 years after Kekulé's structure proposal for **3**. The guest list of this event reads like an almanac of chemical nobility.[76] Benzene and its derivatives can be prepared by ring-forming reactions from small open chains compounds, whether these are alkanes, alkenes or alkynes by use of thermal, photochemical and catalytic processes; transition metal-catalyzed processes are particularly useful. Alternatively, starting materials which already contain a ring can be used for benzene syntheses; the ring can be the same size or smaller or larger.[77]

Among the C_6H_6 hydrocarbons six belong to a special class—those in which every carbon atom bears one hydrogen atom, *i.e.* to the group of valence isomers of the brutto formula $(CH)_6$ (Scheme 9). Besides **3** these are Dewar benzene (bicyclo[2.2.0]hexa-2,5-diene, **35**), triprismane (tetracyclo[2.2.0.0^{2,6}.0^{3,5}]hexane, **36**, see

Section 4.1), benzvalene (tricyclo[3.1.0.^{2,6}]hex-3-en, **37**, see below), 3,3'-bicylo-propenyl (**38**, see below) and Claus benzene (**39**) (Scheme 9).

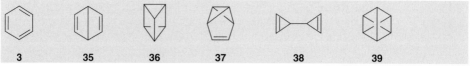

| 3 | 35 | 36 | 37 | 38 | 39 |

Scheme 9. The (CH)$_6$-hydrocarbon family.

Although other alternatives for the structure of **3** were discussed before the connectivity of the Kekulé structure was shown to be correct, the (CH)$_6$-hydrocarbons played a special role in this area of hydrocarbon chemistry since its 19th-century proponents believed that the chemical properties of the compound called benzene known at that time could best be reconciled with the structures given in Scheme 9. This misunderstanding originated in part from not taking the three-dimensional structures of these hydrocarbons into account; this nevertheless remained a challenge to the preparative skills of organic chemists for more than a century.

With the exception of **39**, all the hydrocarbons in Scheme 9 have since been prepared. That the missing Claus benzene with its 'crossed' single bonds ever will be synthesized seems highly unlikely in view of the extremely unfavorable bond angles and lengths, and hence prohibitively high internal strain, of this molecule.

Dewar benzene was first prepared by van Tamelen und Pappas[78] from phthalic acid (**40**) as shown in Scheme 10.

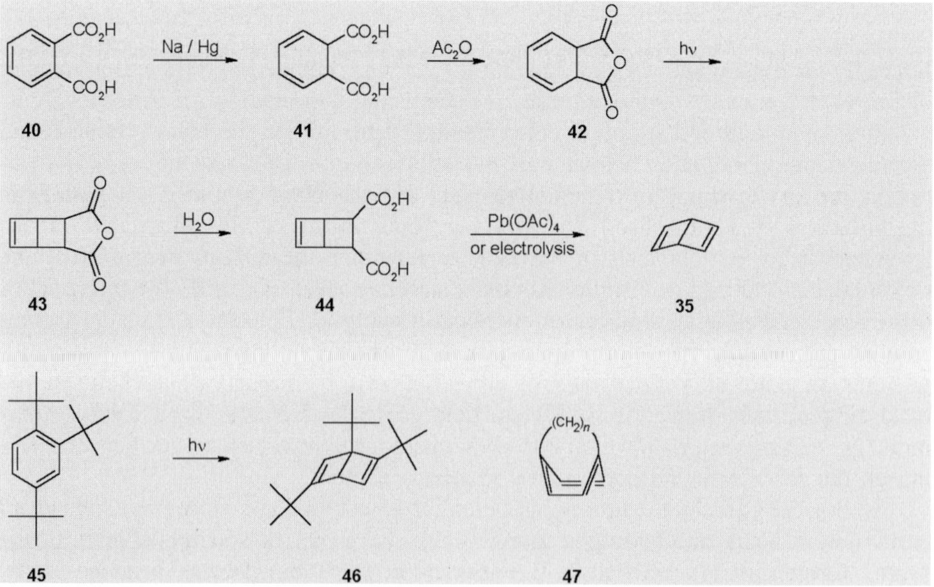

Scheme 10. The first synthesis of Dewar benzene (**35**) by van Tamelen and Pappas.

Partial reduction of **40** provided the 1,3-cyclohexadiene derivative **41**, which, after conversion into its anhydride **42**, was photocyclized to the anhydride **43**, which already contains the complete σ-skeleton and one of the double bonds of the target molecule. After **43** had been hydrolyzed to the diacid **44** this was decarboxylated either with lead tetraacetate or electrolytically. In both cases Dewar benzene **35** was obtained in *ca* 20% yield. Actually **35** was not the first Dewar benzene hydrocarbon to be reported—the same research group had already described the photoisomerization of the sterically hindered benzene derivative **45** to the tris-*tert*-butyl derivative **46**.[79] The driving force for this isomerization is provided by release of internal strain—the steric hindrance between the *ortho*-substituted *tert*-butyl substituents of **45** is higher than in the product **46**, in which the non-planar carbon framework results in a larger distance between the voluminous substituents. Strain-release-driven isomerizations of this type have since then often been used for the preparation of Dewar benzene derivatives, especially in the case of short-bridged cyclophanes (general structure **47**), which fulfil the prerequisite of a non-planar benzene ring, as will be discussed in Section 12.3.

That cyclobutadienes can be trapped by triple bond dienophiles to yield Dewar benzene derivatives has already been mentioned (see Section 10.1, above). In a particularly elegant experiment, carried out by Schäfer and Hellmann in 1967, these higly reactive species are evidently passed on the way from 2-butyne (**48**) to hexamethyl Dewar benzene (**51**) (Scheme 11).[80]

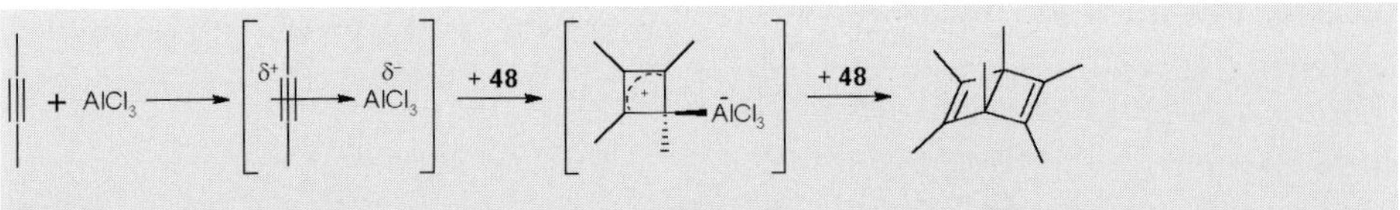

48   **49**   **50**   **51**

Scheme 11. The preparation of hexamethyl Dewar benzene (**51**) by trimerization of 2-butyne (**48**).

Treatment of **48** with a AlCl₃ catalyst led to the Dewar benzene **51** in yields between 60 and 70%! It is likely that the trimerization begins with polarization of the starting alkyne by the catalyst and formation of the π-complex **49**. A second molecule of **48** adds to this, providing the homoaromatic complex **50**, which finally is intercepted by a third 2-butyne molecule to yield the stable **51**.

The close relationship between Dewar benzene (**35**) and triprismane (**36**) has already been referred to in Section 4.1 when we discussed the synthesis of the latter hydrocarbon. This tetracyclic (CH)₆-hydrocarbon was, however, originally not prepared by intramolecular [2+2]photoaddition from **35**[81] but from benzvalene (**37**), the next in our list of benzene valence isomers.

Preceded by the discovery and characterization of several highly substituted benzvalene derivatives[82–84]—the name was coined by Viehe who obtained a trifluoro-tris-*tert*-butyl derivative of **37**, the first benzvalene ever to be prepared, by thermal trimerization of *tert*-butyl-fluoroacetylene[82]—the parent hydrocarbon was discovered in minute amounts by Wilzbach and co-workers in 1965 as a photoiso-merization product of benzene (**3**) (Scheme 12).[85]

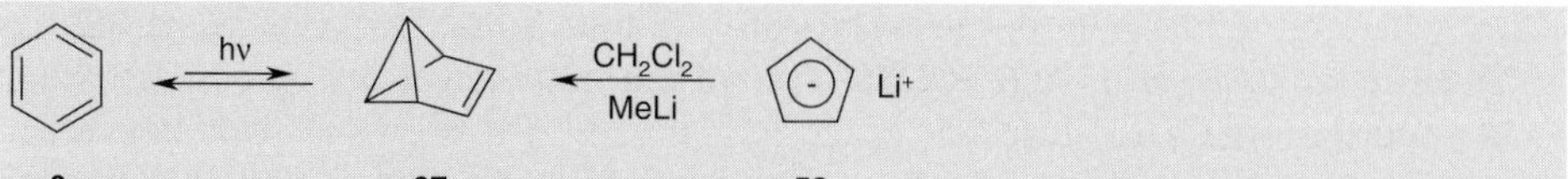

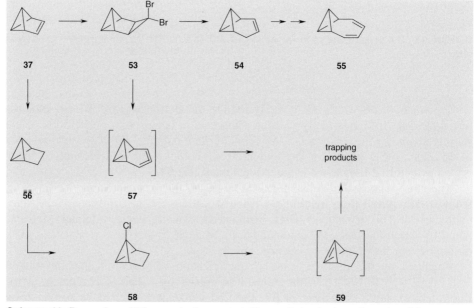

Scheme 12. The preparation of benzvalene (**37**).

As so often in hydrocarbon chemistry, however, (see, for example, adamantane ((Section 3.1), cyclooctatetraene (**4**, Section 10.3), [2.2]paracyclophane (Section 12.3), [1.1.1]propellane (Section 6.1)) the yield was so low (maximum 1%) that to study the chemical behavior of this interesting hydrocarbon was unthinkable. This changed—again a typical hydrocarbon success story—fundamentally when Katz and co–workers reported that multigram quantities of **37** (containing only a few percent of **3** as a side product) could be obtained by reacting lithium cyclopentadienide (**52**) with dichloromethane and methyl lithium in dimethyl ether.[86]

Not only a hydrocarbon with interesting structural and spectroscopic properties,[87] benzvalene (**37**) became an important substrate in small-ring chemistry enabling the preparation of numerous other polycyclic hydrocarbons which without it would have been very difficult to synthesize.[87] A small selection of these hydrocarbons, largely prepared by Christl and his co-workers, is summarized in Scheme 13.

Scheme 13. Benzvalene (**37**) as a starting material in small-ring hydrocarbon chemistry.

Thus carbene addition to **37** provided adducts such as the *gem*-dibromide **53** which could be converted[88] to homobenzvalene (**54**),[89] itself an important relay molecule to, for example, octavalene (**55**), one of the (CH)$_8$ isomers (see Section 10.3).[90] Among the highly reactive hydrocarbons generated from **37** the cycloallene **57** (see Section 9.3)[91] and the pyramidalized olefin **59** (see Section 7.5)—which is an isomer of benzvalene—are noteworthy. The latter has been synthesized from **37** *via* the saturated hydrocarbon **56**[92] and the bridgehead chloride **58**.[93]

In view of the very large number of cyclopropene derivatives known (see Section 7.2) it is surprising, that 3,3'-bicyclopropenyl (**38**) was the last (CH)$_6$ isomer to be prepared, although it has been estimated that this hydrocarbon should be the thermo-dynamically least stable of all benzene valence isomers. The parent molecule **38** was prepared in 1989 by Billups and Haley from the dichloride **60** by dechlorosilylation with fluoride at room temperature applying the so-called VGSR (vacuum-gas-solid reaction) technique. Because **38** decomposes rapidly at −10 °C it is essential to re-move it quickly from the reaction zone and trap it for further characterization[94] (Scheme 14).

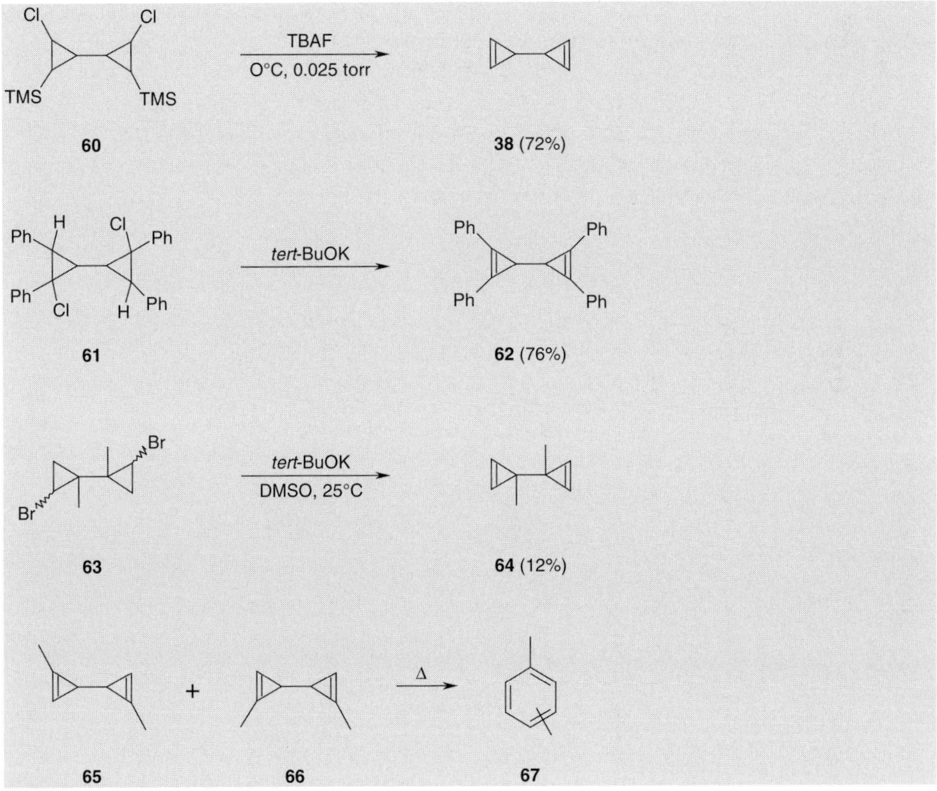

Scheme 14. The preparation of 3,3'-bicyclopropenyl derivatives by β-elimination reactions.

ß-Eliminations have also been successfully employed for the synthesis of substi-tuted bicyclopropenyl derivatives. Thus the tetraphenyl derivative **62** was prepared from **61** by treatment with potassium *tert*-butoxide[95] and the dimethyl derivative **64** was obtained by dehydrohalogenating **63**.[96] Both derivatives are thermally unstable—whereas **62** isomerized to a mixture of 1,2,4,5- and 1,2,3,4-tetraphenylbenzene at 135 °C, **64** underwent a [3,3]sigmatropic rearrangement to **65** and **66** before these three isomers all find their global minima in the isomeric xylenes **67**.[97, 98] Fully substituted 3,3'-bicyclopropenyl derivatives have finally been obtained by reductive coupling of triphenyl- and tris-*tert*-butylcyclopropenylium ions on the one hand,[99] and of 1,2,3-tris-trifluoromethyl-3-iodocyclopropenes on the other.[100]

Benzene (**3**) and its valence isomers are not solely interconnected by photochemical[101] and thermal processes—there are many pathways connecting the (CH)$_6$-family to many of the other 211 C$_6$H$_6$ hydrocarbons. Thus irradiation of liquid benzene with 254-nm light causes excitation to the S$_1$ state which can collapse to the diradical **68**. This species can either close directly to benzvalene (**37**, see above) or isomerize by 1,2-hydrogen shifts to fulvene (**71**, see Section 11.2). When, however, the irradiation is repeated in the gas phase, **3** rearranges to **71** or to a mixture of the ring-cleavage products *cis*- and *trans*-1,3-hexadiene-5-yne (**70**).[102] Heating the latter acyclic ben-

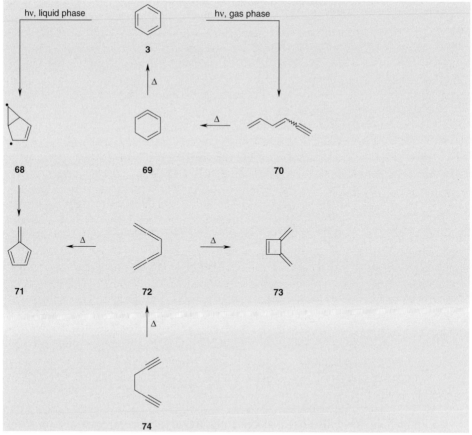

Scheme 15. Selected pathways connecting different C$_6$H$_6$ isomers.

zene isomers resulted in electrocyclization to 1,2,4-cyclohexatriene (**69**, isobenzene, see Section 9.3), which by a subsequent hydrogen shift rearomatizes to **3** (Scheme 15).[103]

The monocyclic hydrocarbon **71** can also be accessed by subjecting the open chain C$_6$H$_6$ hydrocarbon 1,5-hexadiyne (**74**) to gas phase pyrolysis.[104] The first step of the transformation consists in a Cope-type isomerization leading to the conjugated bisallene **72** (see Section 9.1) which subsequently either electrocyclizes to 3,4-bismethylene-cyclobutene (**73**) or undergoes 1,5-bridging followed by C,H-insertion of the thus generated carbene intermediate to yield **71**.[105] It should be pointed out that this map of 'C$_6$H$_6$-land' (see the map for 'adamantane-land' in Section 3.1) shows only a very small selection of all the presently known connecting pathways. Obviously, the landscape which this map represents becomes richer and more complex with every newly discovered C$_6$H$_6$ hydrocarbon.

10.3 Cyclooctatetraene and the (CH)$_8$ isomers

Allowing only C–C single, double and triple bonds as well as C–H bonds for 'saturation' there are 7426 ways to combine eight carbon and eight hydrogen atoms.[106] These isomers can again be classified according to structural types; for example, there are 122 possible acyclic C$_8$H$_8$ isomers. As for the C$_6$H$_6$ series (see Section 10.2, above) the largest scientific effort has been devoted to those isomers in which every carbon atom carries one hydrogen substituent only, *i.e.* to the (CH)$_8$ valence isomers. The 21 possible combinations in this family are derived from 17 different constitutional formulas, *i.e.* they include four pairs of stereoisomers.[107–109] This class of hydrocarbons includes cyclooctatetraene (**4**, see below), bicyclo[4.2.0]octa-2,4,7-triene (**75**, see below), barrelene (**76**, see Section 13.1), *syn*-and *anti*-tricyclo[4.2.0.0^{2,5}]octa-3,7-diene (**77**, see below), octavalene (**55**, see Section 10.2, above), semibullvalene (**78**, see Section 13.3), tricyclo[3.3.0.0^{2,6}]octa-3,8-diene (**79**, see below), cubane (**80**, see Section 5.2), cuneane (**81**, pentacyclo[3.3.0.0^{2,4}.0^{3,7}.0^{6,8}]octane, see below), and octabisvalene (**82**, pentacyclo[5.1.0.0^{2,4}.0^{3,5}.0^{6,8}]octane, see below) (Scheme 16), some of which are discussed in other chapters of this book, where they fit better for structural and synthetic reasons.

Among the (CH)$_8$ hydrocarbons cyclooctatetraene (**4**) is not only the most thoroughly studied but has also been used most often as a starting material, especially in hydrocarbon chemistry, as we shall see. Cubane (**75**), although very extensively studied and derivatized in many ways during the last two decades (see Section 5.2) cannot yet challenge the number one position of cyclooctatetraene.[2, 8, 110] And while we are comparing the general importance of various hydrocarbons—there is only one other hydrocarbon (at least of those mentioned in this book) which can compete with cyclooctatetraene—adamantane (see Sections 3.1 and 3.2). The history of these two central molecules is remarkably similar—originally only prepared in minute amounts by lengthy synthetic sequences, thus making comprehensive studies of their chemical behavior impossible, the situation changed 'overnight' as a result of serendipitous discoveries making cyclooctatetraene and adamantane commercial products available on the ton scale.[111]

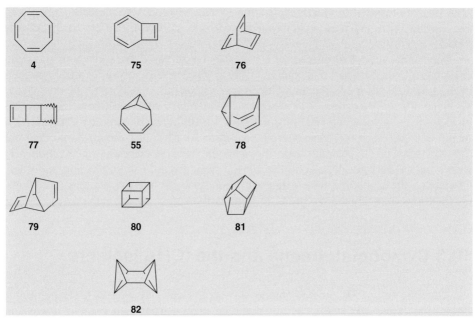

Scheme 16. A selection of important (CH)$_8$ hydrocarbons.

In their classic synthesis Willstätter and co–workers used *pseudo*-pelleterine (**83**), an alkaloid from the bark of the pomegranate tree, as the starting material for **4** (Scheme 17).[112] Except for the eight-membered ring of carbon atoms, substrate and product have nothing in common, so **83** had to be subjected to a long series of elimination steps. A crucial intermediate was 1,3,5-cyclooctatriene (**87**) which was prepared by reducing **83** first to the alcohol **84**, dehydrating this to the monoolefin **85** and subjecting the latter to exhaustive methylation and Hofmann elimination, the last two steps being performed twice in succession.

Bromination of the triene **87** subsequently yielded a mixture of dibromides which on treatment with dimethylamine provided the bis amine **89**. These transformations lead to 'loss' of one of the ultimately desired double bonds but also create the prerequisite for the final step—the double Hofmann elimination of **89** to **4**. This temporary sacrifice of double bonds was later used quite often for the synthesis of unsaturated hydrocarbons. An example from annulene chemistry will be discussed in Section 10.5. Although the cyclooctatetraene prepared by this route probably contained *ca* 30 % of styrene as a byproduct, the original reaction sequence was reproduced nearly four decades later by Cope and Overberger[113] thus confirming Willstätter's original claim. Whereas these early studies clearly showed that **4** was not an 'aromatic compound with extended ring-size', the hydrocarbon remained a rare and precious laboratory chemical. This unsatisfactory situation only changed when Reppe and co-workers discovered in 1940 that acetylene (**90**) can be catalytically cyclotetramerized in the presence of various nickel(II) salts such as nickel cyanide—the catalyst used preferentially—or the nickel salts of 2,4-pentanedione, ethyl acetoacetate, or salicylic aldehyde (Scheme 18).[114] From now on cyclooctatetraene (**4**) was a commercial[111]

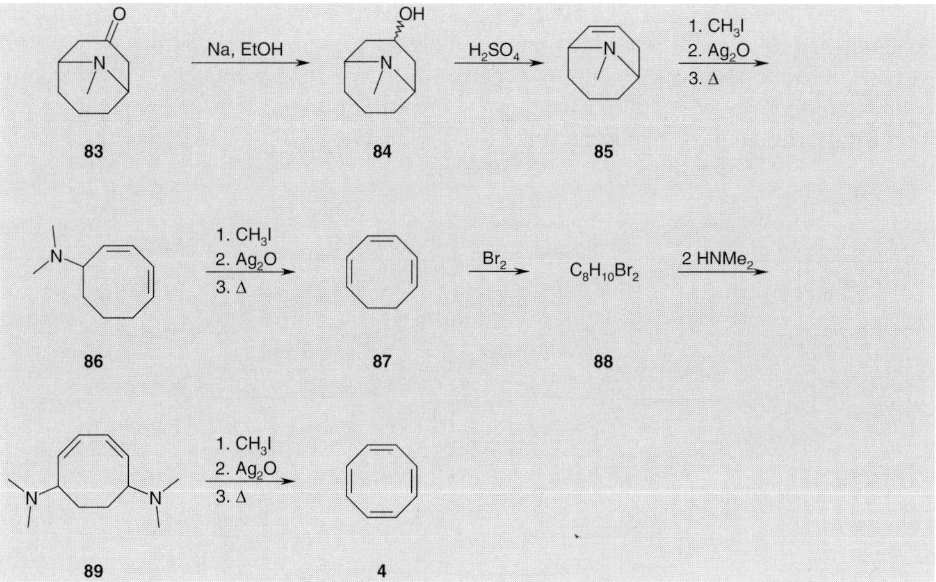

Scheme 17. The first synthesis of cyclooctatetraene (**4**) by Willstätter.

product, the chemical and physical properties of which could be studied on a very large scale. It was shown that the hydrocarbon behaves as a typical cyclopolyolefin and that it has the tub- or boat-like structure **91** shown in the scheme.

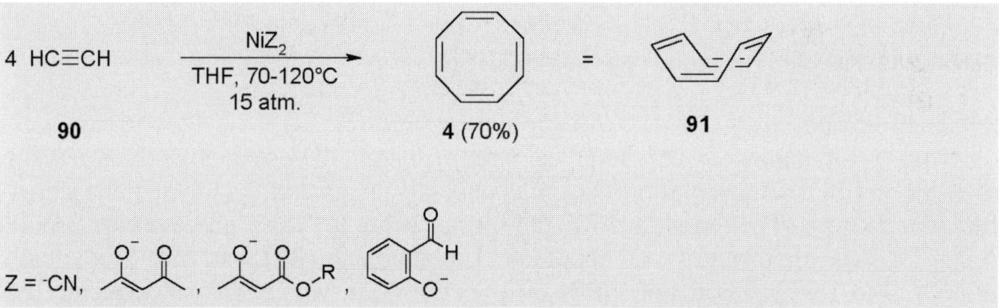

Scheme 18. Cyclooctatetraene (**4**) from acetylene (**90**)—the Reppe synthesis.

In the Reppe synthesis, besides **4** (produced in yields up to 70%), benzene, styrene, 1-phenyl-1,3-butadiene, and some other cyclic and acyclic oligomers of **90** are formed. When a second acetylenic component is added, whether terminal or non-terminal, further functionalized or not, the cyclo-co-oligomerization leads to derivatives of cyclooctatetraene.[2, 8, 110] Although the yields are usually low (10–20%), the synthetic transformation which has been accomplished more than offsets this disadvantage, especially because the starting materials are usually inexpensive.

The Reppe synthesis is by far the most important route to cyclooctatetraenes, but is not the only one, of course. Treating *cis*-3,4-dichloro-1-cyclobutene (**8**) with sodium amalgam led to *syn*-**77**, whereas lithium amalgam provides *anti*-**77**, the two hydro-carbons being one of the mentioned pairs of $(CH)_8$ stereoisomers.[115] Both com-pounds are thermolabile and on heating[115] or in the presence of metal salts (see be-low) isomerized to **4**[116] (Scheme 19).

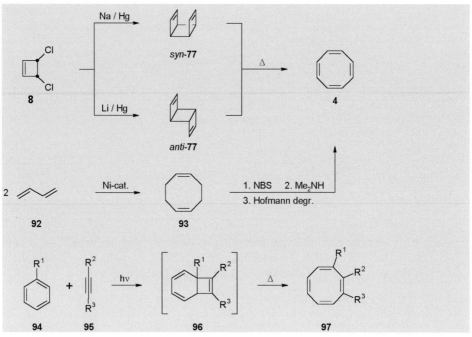

Scheme 19. Alternative routes to cyclooctatetraene (**4**).

That cyclobutadiene is very likely an intermediate in these reductive dimerizations is supported by the observation that independently generated cyclobutadiene dimer-izes to a mixture of *syn*- and *anti*-**77**, the former being the main product (see Section 5.2).[115, 117] In another $C_4 + C_4$ approach, 1,3-butadiene (**92**) was first catalytically dimerized to 1,5-cyclooctadiene (**93**) (see Section 7.3), which was subsequently de-hydrogenated to **4** *via* NBS-bromination, double dimethylamine substitution and Hofmann degradation.[118] Finally, photocycloaddition of substituted benzenes **94** to various alkynes **95** led to the [2+2]photoadducts **96**, which as derivatives of the bicy-clic valence isomer of cyclooctatetraene, **75**, ring-opened (see below) to the substi-tuted or functionalized cyclooctatetraenes **97**. Of course, a complete cyclooctatetraene nucleus can also be used as the starting material, and numerous alkyl and aryl cy-clooctatetraenes have been obtained *via* metal organic derivatives of cyclooctatet-raene, whether these are Grignard or alkali metal organic compounds.[2, 8, 110]

As with the lower vinylogs (see Section 10.2), isomerization reactions which con-nect the various $(CH)_8$ hydrocarbons have been studied very intensely both for mechanistic and preparative reasons. From the vast literature I preferentially consider those processes which either lead to (CH)$_8$ hydrocarbons the syntheses of which have not been described yet or are not presented in other sections of this book, or in which novel preparative procedures are employed. Various more complete '(CH)$_8$ maps' can be found in the review literature.[108, 109] The connections between the isomers are usually established by thermal, photochemically or metal-catalyzed rearrangements. Thermal interconversions are summarized in Scheme 20 which also demonstrates the central role played by cyclooctatetraene (**4**), the most thermodynamically stable isomer of this family.[109]

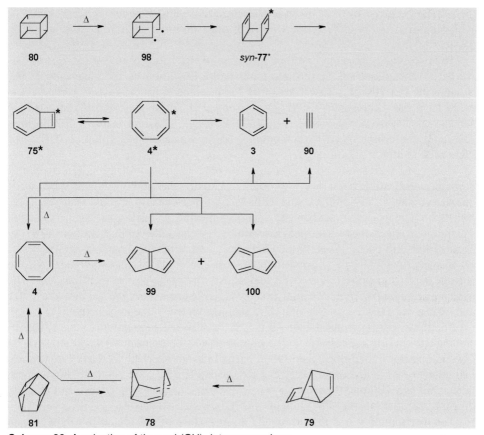

Scheme 20. A selection of thermal (CH)$_8$ interconversions.

Beginning with cubane (**80**, see Section 5.2), the (CH)$_8$ hydrocarbon with the highest heat of formation and strain energy, this pentacyclic isomer only begins to decompose above 200 °C, yielding benzene (**3**), cyclooctatetraene (**4**), acetylene (**90**) and the dihydropentalenes **99** and **100**[119] (see Section 11.4). The product composition is pressure-dependent which has been interpreted as an indication of the involvement of vibrationally excited species. The isomerization is initiated by rate-determining homolysis of **80** to the diradical **98**. Because this intermediate still has a cage structure, the bond cleavage would not be expected to contribute significantly to strain release, thus explaining the thermal stability of **80**. Opening of a second cyclobutane ring could then lead to vibrationally excited *syn*-**77** (*syn*-**77***), from which the immediate precursors of the isolated products, the vibrationally excited species **75*** and **4***, could be produced. Note that **99** and **100** are not $(CH)_8$ isomers any more, *i.e.* that here a door to the huge group (see above) of non-valence isomers of cyclooctatetraene has been opened. Cuneane (**81**), another non-olefinic and highly strained $(CH)_8$ hydrocarbon, produced **4** and semibullvalene (**78**) on gas phase thermolysis. The **4/78**-ratio is pressure-dependent, making vibrationally excited intermediates very likely again.[121] In contrast, the thermal rearrangement of tricyclo[3.3.0.0^{2,6}]octa-3,7-diene (**79**) occurred even at room temperature and yielded semibullvalene (**78**),[122, 123] which in turn thermally isomerized to **4** under high-temperature conditions (200–360 °C).[124]

The pyrolytic behavior of cyclooctatetraene (**4**) itself has been studied very extensively.[109] Depending on the reaction temperature it isomerizes and cleaves to **3**, **90**, styrene, **99** and **100** and their 1,5 isomer (high temperature products). A classical—preparative and mechanistic—study by Huisgen et al. has shown[125] that at *ca* 80 °C **4** and its valence isomer bicyclo[4.2.0]octa-2,4,7-triene (**75**) are in rapid equilibrium. The latter could be trapped by dienophiles, and it was established that *ca* 0.01 % of **75** is present in **4** at 100 °C.

Compared with the thermal behavior, the photochemistry of the $(CH)_8$ hydrocarbons has been studied far less thoroughly. Usually the photochemical work is of qualitative nature, *i.e.* product studies have been performed carefully, but relatively little is known about the mechanistic details of the various photoprocesses. A notable exception is the famous synthesis of semibullvalene (**78**) from barrelene (**76**) by Zimmerman and co–workers[126] (see Section 13.3). Whereas the direct photolysis of **76** in methylcyclohexane at room temperature gave only cyclooctatetraene (**4**), the acetone-sensitized process led to **78** and **4**, the former being the main product. The mechanism of this rearrangement, which presumably takes place *via* the diradicals **101** and **102**, is the classic example of a di-π-methane rearrangement (Scheme 21).[127]

Photolysis of the thermolabile **79** at –60 °C also led to semibullvalene (**78**), accompanied by **4** (in 1:2-ratio).[128] The photoproducts obtained from cyclooctatetraene (**4**) under various conditions correspond partially to the thermal products (see above), *i.e.* among other hydrocarbons benzene (**3**), acetylene (**90**), the isomeric dihydropentalenes **99** and **100** and their not shown 1,5 isomer have been identified. In this case, however, semibullvalene (**78**) and cuneane (**81**) were also detected.[129–131]

The disappointing inability of *syn*-**77** to ring-close to cubane (**80**)—shown as an inset in Scheme 21—has already been referred to (Section 5.2). As explained earlier, this inertness has been attributed to various factors including the high strain of the desired photoproduct **80**, the (too large) distance between the double bonds in *syn*-**77**, and unfavorable orbital interactions.[132]

Transition metal-catalyzed isomerizations of strained hydrocarbons are mechanistically interesting and often very useful preparatively. In the $(CH)_8$ family, the silver or palladium salt-catalyzed rearrangement of cubane (**80**) into cuneane (**81**), which was first synthesized by Eaton et al. by this route,[133] is most remarkable, because it transforms four cyclobutane rings into two cyclopropane and two cyclopentane rings in a one-pot operation (Scheme 22).

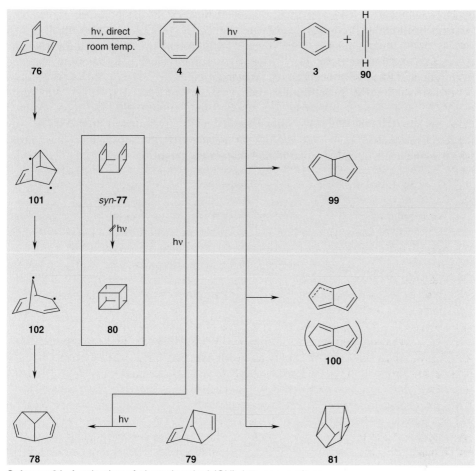

Scheme 21. A selection of photochemical (CH)$_8$ interconversions.

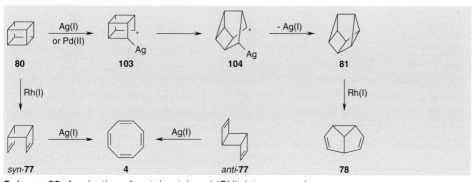

Scheme 22. A selection of metal-catalyzed (CH)$_8$ interconversions.

It has been postulated that the Ag(I)-catalyzed process is initiated by electrophilic attack of Ag(I) on one of the strained cyclobutane bonds (formation of **103**), followed by rearrangement to **104** and finally demetalation.[134] In the presence of catalytic amounts of rhodium(I) complexes **80** ring-opened to *syn*-**77**[135] and treatment of this and its *anti*-isomer, *anti*-**77**, with silver tetrafluoroborate converted these tricyclic dienes into cyclooctatetraene (**4**).[136] Finally, semibullvalene (**78**) has been obtained from cuneane (**81**) by Rh(I)-catalyzed isomerization. [133]

Octabisvalene (**82**) is the third saturated member of the (CH)₈ family. According to MM2 calculations it should be less strained than cubane (**80**) and more strained than cuneane (**81**), the other two fully saturated alternatives. It was prepared last—by Rücker, Prinzbach et al. in 1988[137, 138]—by a route differing substantially from those which we have encountered so far in this chapter (Scheme 23).

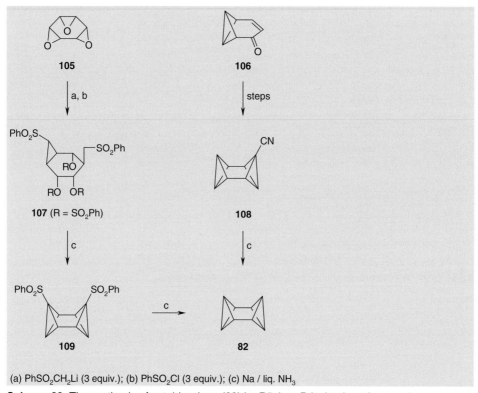

(a) PhSO₂CH₂Li (3 equiv.); (b) PhSO₂Cl (3 equiv.); (c) Na / liq. NH₃

Scheme 23. The synthesis of octabisvalene (**82**) by Rücker, Prinzbach and co–workers.

cis-Benzenetrioxide (**105**) was first converted to **107** by treatment with three equivalents of the lithium salt of phenylmethyl sulfone; this was then esterified with phenyl sulfonyl chloride (overall yield 84%). When **107** was reacted with the same base or potassium methylsulfinylmethide in dimethyl sulfoxide triple-cyclization took place and **109** was formed in 29% yield. Synthesis of the parent hydrocarbon **82** (20%) was achieved by reducing **109** in liquid ammonia with excess sodium at –78 °C, followed by quenching with solid ammonium chloride. Alternatively, tricyclo[4.1.0.0^{2,7}]hept-4-en-3-one (**106**) could be transformed in five steps, with an over-all yield of 10%, into the nitrile **108**. Again, reduction with sodium in liquid ammonia did not effect the octabisvalene skeleton, and **82** was obtained in 32% yield.

Closing this section on (CH)$_8$ hydrocarbons we return briefly to cyclooctatetraene (**4**). The fascination of this particular valence isomer stems not only from its chemical properties, especially its manifold use in hydrocarbon synthesis, but also from its unique structural properties. Its non-planar, tub-like geometry gives rise to an interesting dynamic behavior characterized by bond-switching and ring inversion processes.[139] For better comparision of **4** with its lower and higher vinylogs, however, it would be desirable to 'planarize' this archetypical non-planar molecule.[140] In particular it would be of interest to record the ^{1}H NMR spectrum of such a planar cyclooctatetraene derivative, because its olefinic hydrogen atoms should absorb at relatively high field: They should show a *paratropic* shift compared with the typical low-field resonances registered for the (*diatropic*) [4*n*+2]-annulenes.

One way of planarizing **4** consists in the replacement of at least one of its double bonds by an acetylene function. Such a dehydro[8]annulene would be a member of a whole series of 'dehydroaromatics', which begins with *ortho*-dehydrobenzene (see Section 16.1) and proceeds all the way to the higher dehydroannulenes which will be discussed in later sections of this chapter.

Cyclooctatrienyne (**111**), the smallest dehydro[8]annulene, was, in fact, generated by Krebs from bromocyclooctatetraene (**110**) by treatment with potassium *tert*-butoxide in ether at room temperature, as shown in Scheme 24.[141]

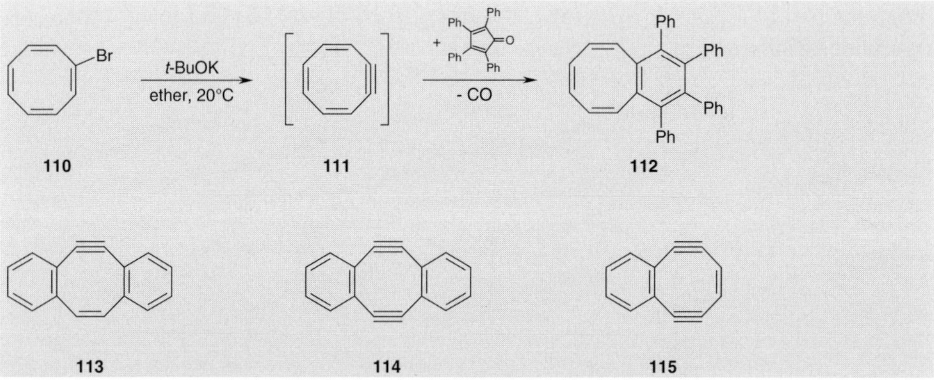

Scheme 24. Dehydro[8]annulenes—planar derivatives of cyclooctatetraene.

That the highly reactive **111** was formed during this elimination reaction was shown by various trapping experiments; for example, with tetraphenylcyclopentadienone the Diels–Alder adduct **112** was formed in good yield after decarbonylation of the primary adduct. Unfortunately, **111** was too reactive to enable measurement of its NMR spectra. When this dehydroannulene is stabilized by annelation with benzene rings, however, the desired data became available. For example, dehydrobromination of 5,6-dibromo-5,6-dihydrodibenzo[*a,e*]cyclooctene or 5-bromodibenzo[*a,e*]cyclooctene with potassium *tert*-butoxide in tetrahydrofuran[142] yielded the dibenzodehydro[8]annulene derivative **113** as a relatively unstable, planar[143] hydrocarbon, which nevertheless could be subjected to ^{1}H NMR analysis. Its olefinic protons absorb at $\delta = 5.45$, distinctly higher than the chemical shifts of appropriate non-planar model compounds—*i.e.* **113** is a paratropic hydrocarbon, as required by theory. A further increase in stability was realized when a second benzene ring was annelated as shown in **114**; this hydrocarbon, which again is planar according to X-ray analysis[144] only begins to decompose when heated above 110 °C. Of course, there are no eight-membered ring protons in **114** to test its paratropicity. For **115**, however, the simplest known, isolable—but quite unstable—planar neutral dehydro[8]annulene,[145] the olefinic protons absorb at $\delta = 4.93$. The signals of the corresponding olefinic protons of the precursor of **115**, 5,10-dibromobenzocyclooctatetraene, a non-planar compound, appear at $\delta = 5.75$, *i.e.* **115** shows the expected paratropic shift.

10.4 The higher annulenes

1,3,5,7,9-Cyclodecapentaene is the first hydrocarbon of general structure **1** with which the term annulene is normally associated. Although it formally fulfils the requirements of the Hückel rule—with $n = 2$ it contains 10 π-electrons—it is not at all a higher vinylog of benzene. The problem begins with the configuration of the double bonds in the hydrocarbon—whereas in benzene any double bond configuration other than all-*cis* would lead to prohibitively high strain, this is no longer so for the larger ring systems. For example the cyclodecapentaene could exist in an all-*cis*- (**116**), a mono-*trans*- (**117**), the di-*trans*- (**5**), the tri-*trans*-di-*cis*- (**118**) and many other configurations (Scheme 25).

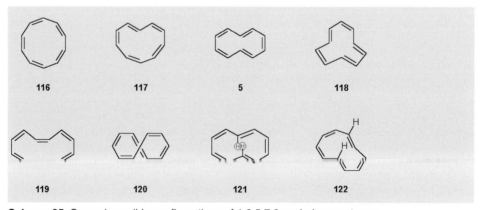

Scheme 25. Several possible configurations of 1,3,5,7,9-cyclodecapentaene.

It is *a priori* not obvious what the three dimensional structures of these alternatives would be. For a planar **116** its internal angles with 144° would differ significantly from the normal double bond angle. In fact, MINDO/2[146] and *ab initio* calculations[147] suggest that **116** would prefer the non-planar configuration shown in **119**, with alternating double bonds (D_5h symmetry) and no π-bond delocalization as required for an aromatic compound ($D_{10}h$ symmetry). Diastereomer **117** also prefers a non-planar structure,[146, 147] **120**, and when we draw **5** without its hydrogen substituents we overlook a most severe steric interaction—for a planar structure with 120° bond angles the internal hydrogen atoms must occupy the same space, as shown by structure **121**. Because this is physically impossible, the π-perimeter of the hydrocarbon must be twisted, in the extreme case as illustrated in **122**. This chairlike deformation, in turn, would result in poorer π-orbital overlap and hence reduced stability of the diastereomer. The deliberate removal of the steric interference by replacement of the inner hydrogen atoms by a methylene bridge, conceived and realized by Vogel, constitutes one of the most far-reaching ideas of annulene chemistry as we shall see in Section 10.5.

After numerous failures and some hopeful exploratory studies,[148] [10]annulene was finally prepared by Masamune in an outstanding investigation the craftsmanship of which deserves to be discussed in detail[149] (Scheme 26).

Irradiation of *trans*-9,10-dihydronaphthalene (**123**) at –70 °C provided all-*cis*-[10]annulene (**116**) which thermally isomerized at –10 °C to the *cis*-dihydronaphthalene **124**. Subjecting the latter to low-temperature photolysis led to the mono-*trans*-isomer **117**—which thermally rearranges to **123** at –25 °C—and to the tetracyclic diene **125**.

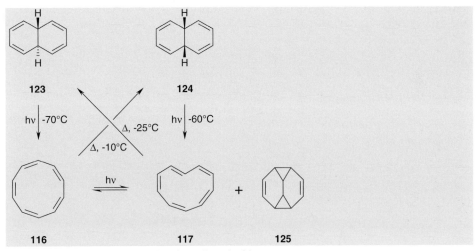

Scheme 26. The preparation of [10]annulene by Masamune.

In a preparative experiment, performed at –60 °C, the mixture of photoisomers obtained from **124** was separated by chromatography on alumina at –80 °C, and the instable [10]annulenes **116** and **117** were isolated as crystalline products! At –100 °C a photoequilibrium, favoring **117**, is established between the two diastereomers. As shown by NMR spectroscopy, the all-*cis*-isomer **116** is a non-planar hydrocarbon, which equilibrates between two non-equivalent conformations. In the mono-*trans*-isomer **117** the *trans*-double bond seems to migrate around the ring at a rate too fast to be studied by ^{1}H NMR spectroscopy at –100 °C. The spectra of both isomers identify them as typical polyenes, *i.e.* the combined effect of bond-angle deformation and non-bonded steric interactions counteracts any gain in stabilization that might have been achieved by delocalization.

The valence isomers and the interconversions shown in Scheme 26 are actually only a small selection of a much larger and much more intricate (CH)$_{10}$ map.[107, 108] Altogether 71 different constitutional formulas (so-called planar graphs) can be drawn for the (CH)$_{10}$ family,[150] and many of these hydrocarbons have been prepared during the last decades, including such famous 'named' structures as Nenitzescu's hydrocarbon (**126**, see Chapter 13),[151] bullvalene (**127**, see Section 13.3),[152] triquinacene (**128**, see Section 13.4),[153] basketene (**129**),[154] prebullvalene (**130**),[155] bicyclo[6.2.0]deca-2,4,6,9-tetraene (**131**),[156] 9,10-dihydrofulvalene (**132**, see Section 5.3),[157] isobullvalene (**133**),[158] snoutene (**134**),[159] hypostrophene (**135**, see Sec-

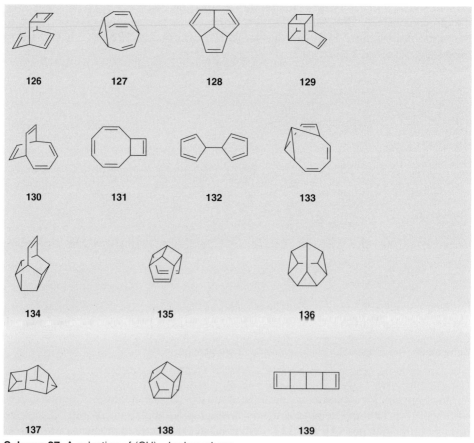

Scheme 27. A selection of (CH)$_{10}$ hydrocarbons.

tion 4.2),[160] diademane or mitrane (**136**, see Section 13.4),[161] barretane (**137**, see Section 13.4),[162] pentaprismane (**138**, see Section 4.2),[163] and pterodactyladiene (**139**, see Chapter 2)[164] (Scheme 27).

We have already encountered several of these hydrocarbons as starting materials or synthetic intermediates in earlier chapters of this book and will return to other representatives in a different context at a later stage. Clearly, these (CH)$_{10}$ isomers are not only related to each other,[107, 108] but also to other areas of hydrocarbon chemistry, thus demonstrating their important integrative role.

The synthesis of [12]annulene (**144**, cyclododecahexaene) by Schröder and co–workers underlines not only the central role of cyclooctatetraene (**4**) as a substrate for many important hydrocarbon syntheses —see above and *e.g.*, Section 13.4—it is also an interesting example of annulene-to-annulene ring expansions[165, 166] (Scheme 28).

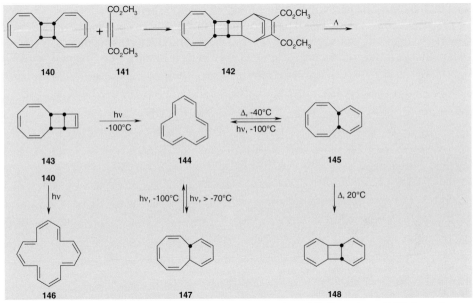

Scheme 28. Schröder's syntheses of [12]annulene (**144**) and [16]annulene (**146**) from the cyclooctatetraene dimer **140**.

Starting with the cyclooctatetraene dimer **140** cycloaddition of dimethyl acetylenedicarboxylate (**141**) first yielded the cycloadduct **142**, which on thermolysis fragmented into dimethyl phthalate and the (CH)$_{12}$ hydrocarbon *syn*-tricyclo[8.2.0.0^{2,9}]dodeca-3,5,7,11-tetraene (**143**). On photolysis at -100 °C this ring-opened to the thermo- and photolabile [12]annulene (**144**) in yields up to 80%. Not surprisingly, this latter hydrocarbon participates in various equilibria (**144** ⇄ **145**, **144** ⇄ **147**) involving other (CH)$_{12}$ valence isomers.[167] Leaving the protective low-temperature environment caused **145** to ring-close to the benzene dimer **148**, and, indeed, when this was warmed above room temperature it dissociated into two molecules of benzene (**3**). NMR investigations show that there is bond alternation in

144—as expected for a 4n-cyclopolyolefin—and that rapid interconversion of two conformations occurs.[168] Irradiation of the dimer **140** also enabled a convenient synthesis of [16]annulene (**146**, yield 60%),[165] a higher [n]annulene which will be readdressed below.

Annelation of benzene rings to the [12]annulene nucleus results in drastic stability increases; hydrocarbons of this type could be obtained by Wittig reaction between various dialdehydes and bis Wittig reagents.[169, 170]

Starting in the mid-1950s Sondheimer and co–workers began to develop an approach to the fully conjugated cyclopolyolefins **1** with $n > 12$, called *annulenes* from the Latin word for ring,[171] which quickly became the most general and important route to this class of nonbenzenoid aromatics. Together with several of the bridged annulenes discussed below, the combination of Sondheimer's annulene chemistry, Hückel molecular orbital theory, and the newly introduced NMR spectroscopy into organic chemistry ushered in a new era of aromatic chemistry, comparable in its impact with the period approximately one hundred years earlier.

The numerous annulenes prepared by the Sondheimer group have been reviewed countless times[172] and are part of every—even introductory—chemistry textbook today. Rather than reproducing these syntheses in their totality, only the most typical

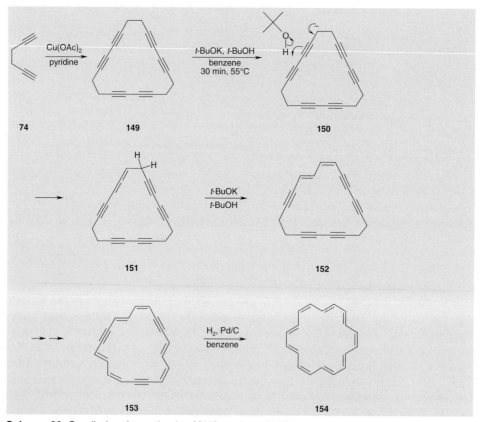

Scheme 29. Sondheimer's synthesis of [18]annulene (**154**).

and consequential will be presented here.

The central molecule of this part of annulene chemistry is [18]annulene (**154**) which was prepared in three steps from 1,5-hexadiyne (**74**) as shown in Scheme 29.

The first step consisted in the formation of the 18-membered ring system **149** by oxidative trimerization of bipropargyl **74** using the Eglinton coupling process (copper(II) acetate in pyridine).[173] The cyclotrimerization was the bottleneck of the synthesis, because it took place in only 6% yield, although the poor yield was partially offset by the isolation of several other cyclic coupling products of **74**, which could be used as substrates for the synthesis of still higher annulenes (see below). In the second step the π-electrons of **149** were 'delocalized' in the literal meaning of the term, by base-catalyzed isomerization. It has long been known that under the influence of base **74** stabilizes itself to 1,3-hexadien-5-yne (**70**),[174] and that this process involves 1,2-hexadien-5-yne (propargylallene) as an intermediate.[175] Applying this mechanism to **149** led—*via* the carbanion **150**—to the mono-rearranged hydrocarbon **151**, which with its doubly-activated methylene group readily isomerized to the partially conjugated hydrocarbon **152**. Twofold repetition of this rearrangement provided the fully conjugated **153** (yield *ca* 50%), which is—not surprisingly—produced in the form of different isomers. The structure shown in **153** is, however, the major isomer obtained. Note that this hydrocarbon can be regarded as an aryne, *i.e.* a 'non-benzenoid dehydrobenzene' (see Section 16.1). The [18]annulene synthesis was concluded by the partial hydrogenation of **153** over 10% Pd/C; the yield was 32%.[176] Hydrocarbon **154**, a compound crystallizing as reddish-brown needles, is the first annulene—after benzene—in which hydrogen-hydrogen interactions are not extreme (see above for, *e.g.*, [10]annulene) and consequently the deviation of the carbon skeleton from planarity is small.[177]

What looks rather straightforward in Scheme 29 is actually considerably more complicated, since the Eglinton coupling of **74** not only yielded **149** but also an acyclic dimer, 1,5,7,11-dodecatetrayne, formed in 9 % yield, and the cyclic tetra- (**155**, 6%), penta- **156** (6%), and hexamer **157** (2%); in one experiment even a **74**-heptamer was produced (Scheme 30).

The cyclopolyacetylenes could be separated by painstaking chromatography (providing 700 fractions!), isomerized by treatment with potassium *tert*-butoxide; the resulting dehydroaromatics—except **160**, which could not be obtained in pure form—could be partially hydrogenated to [24]annulene (**158**) and even [30]annulene (**159**). Changing the substrate but keeping the general strategy—oxidative cyclization of an appropriate acetylenic precursor, base-catalyzed isomerization, partial hydrogenation over a Pd/C or Lindlar catalyst—Sondheimer and co–workers were able to synthesize many other members of the [*n*]annulene family, including [14]annulene,[178] [16]annulene (**146**), which was first obtained by this route,[179] [20]annulene,[180] and [22]annulene.[181] For **146** we have already encountered the preparatively superior synthesis from the cyclooctatetraene dimer **140** (see above, Scheme 28). With the vinylogous series of the [*n*]annulenes the predictions of the Hückel rule could be most convincingly validated. Expressed in the language of the NMR criterion of aromaticity[182] the (4*n*+2) π-systems benzene (**3**), [14]annulene, [18]annulene (**154**) and [22]annulene are *diatropic*, *i.e.* they give low-field signals for their outer, and high-field signals for their inner protons (which are, of course, absent in **3**). The 4*n* π-systems cyclobutadiene (**2**, see Section 10.1), [12]annulene (**144**, see above),

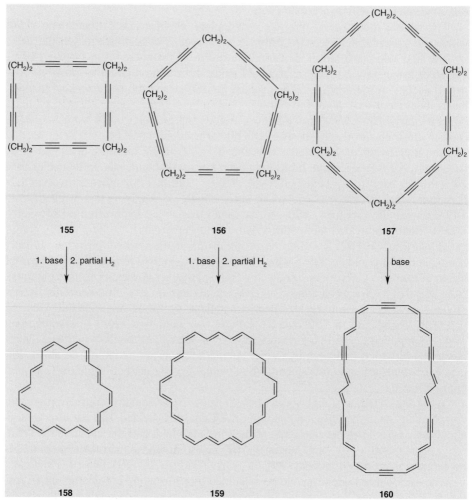

Scheme 30. Additional annulenes from 1,5-hexadiyne (**74**).

[16]annulene (**146**, see above), [20]annulene, and [24]annulene (**158**, see above), on the other hand, are *paratropic*, *i.e.* in these annulenes the inner protons (if present) absorb at low, and the outer protons at high fields. The two annulenes which do not fall into these categories, cyclooctatetraene (**4**, see Section 10.3) and [10]annulene (**5**, **116**, **117**), are called *atropic*; and the reason for their atropicity can be found in their non-planar structures.

Dehydroannulenes were encountered as intermediates in several of the above annulene syntheses. Since their additional π-electrons are in orbitals orthogonal to the conjugated π-perimeter the dehydroannulenes should show the same type of magnetic behavior as their annulenic counterparts. Numerous of these still higher unsaturated hydrocarbons have been prepared,[183] and we will discuss here only two examples, synthesized by routes different from those which we have encountered so far in the annulene field.

1,5,9-Trisdehydro[12]annulene (**165**), first prepared by base-catalyzed isomerization of 1,3,7,9-cyclododecatetrayne,[184] can also be synthesized from 1,5,9-cyclododecatriene (**161**), a large-ring triene, which we have come across already in Section 7.3. Subjecting **161** to a bromination-dehydrobromination sequence led to the trisvinylbromide **162**, which was converted to the hexabromide **163** by treatment with NBS. When this intermediate was dehydrobrominated with four equivalents of sodium ethoxide in ethanol, 5-bromo-1,9-bisdehydro[12]annulene (**164**) was formed, whereas excess base led directly to the triyne **165** (Scheme 31).[185]

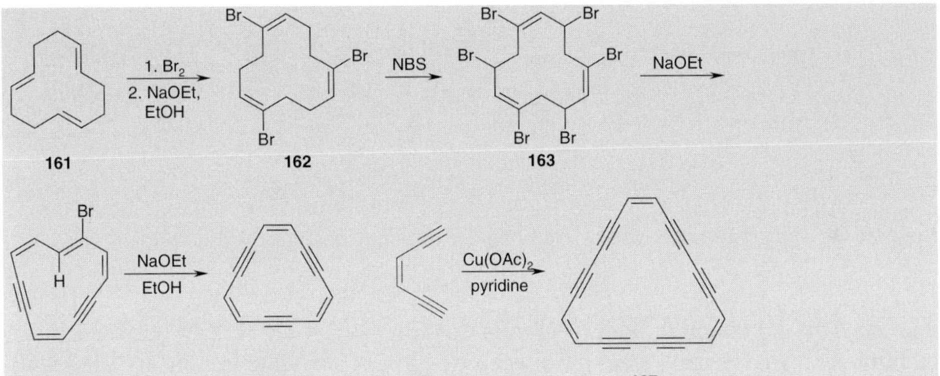

Scheme 31. The synthesis of 1,5,9-trisdehydro[12]annulene (**165**) and 1,3,7,9,13,15-hexakisdehydro[18]annulene (**167**).

A 'stretched version' of **165**, 1,3,7,9,13,15-hexakisdehydro[18]annulene (**167**), in which the triple bonds of the former molecule have been replaced by butadiynyl groups, is accessible by Eglinton coupling of *cis*-3-hexen-1,5-diyne (**166**, see Section 16.1).[186] Because **165** is a [4n] and **167** a [4n+2] π-electron system (neglecting the out-of-plane π-electrons), it is of interest to compare the chemical shifts of the olefinic protons of these geometrically very similar molecules. Whereas **165** gives a singlet at δ = 4.45, for **167** a singlet is registered at δ = 7.02, indicating the (expected) paramagnetic ring current in the dehydro[12]annulene and the diamagnetic ring current in the dehydro[18]annulene. The ^{1}H NMR spectrum of the precursor bromide **164** is even more revealing, because it contains an internal hydrogen substituent as a reporter atom for electronic effects. At room temperature this inner proton absorbs at δ = 16.4, *i.e.* it is experiencing very strong deshielding because of the paramagnetic ring current induced in the nearly planar 12π-electron system.[187]

Very large-ring dehydroannulenes, as exemplified by the hydrocarbon **173**, and many of their derivatives, have been prepared and studied extensively by Nakagawa and co-workers. A typical synthetic protocol is summarized in Scheme 32.[183, 188]

This building-block synthesis, which makes use of several synthetic transformations, which we have already encountered in earlier chapters of this book, begins with the condensation of the aldehyde **168** with a ketone to yield the chain-extended

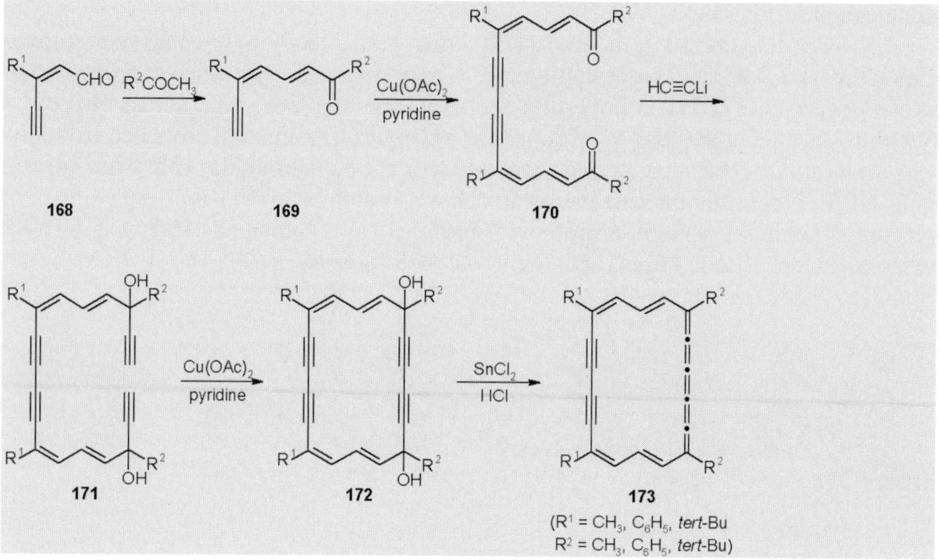

Scheme 32. The preparation of very large-ring dehydroannulenes according to Nakagawa.

dienyne **169**. Its coupling under Eglinton conditions then led to **170** into which two additional ethynyl groups were introduced by lithium acetylide treatment. The resulting bis propargyl alcohol **171** was cyclized by another, now intramolecular, Eglinton step, and the produced **172** was reduced to the final product **173** by tin(II) chloride reduction as discussed in Section 9.2. As a tetradehydro[18]annulene **173** should be diatropic and, indeed, in the ^{1}H NMR spectrum of the tetramethyl derivative (**173**, R^1 = R^2 = CH$_3$) the two inner protons absorb as a triplet at δ = –5.24, whereas the four outer protons are registered as a doublet at δ = 9.66. Modification of this approach enabled the preparation of numerous other [4n+2]- and [4n]-dehydroannulenes of the acetylene/cumulene type, and they were shown to be either dia- or paratropic by NMR spectroscopy.[188]

As for the annulenes (see above) benzannelation has also been performed for the dehydroannulenes and leads to an increase in stablity. A particular interesting example is the tribenzo derivative of **165**, 5,11,17-tris-dehydro-tribenzo-[a,e,i]-[12]annulene (**178**), which was prepared nearly simultaneously by Staab and by Eglinton, Raphael and co–workers by the reaction sequences shown in Scheme 33. The former workers[189] coupled phthalaldehyde (**174**) with the (preformed) bis Wittig reagent **175** to the dehydro[12]annulene **176**, obtained as a mixture of its *trans, trans*- and *trans,cis* isomers. This was subsequently brominated, and the resulting tetrabromide—only the olefinic double bonds of **176** are attacked—dehydrobrominated with potassium *tert*-butoxide in boiling tetrahydrofuran.

Although the desired **178** was produced, the main elimination product was the monoolefin **177**. However, repetitive bromine addition and dehydrobromination of the initially isolated product mixture eventually caused the complete conversion of **177** into **178**.[189, 190] In the Eglinton–Raphael route to **178**,[191] copper *ortho*-iodophenylacetylide (**179**) was trimerized under Stephens–Castro conditions.[192] Extended hydrocarbon structures composed of benzene rings and triple bonds, and in which **178** is contained as a substructure, will be discussed in greater detail in Section 15.2.

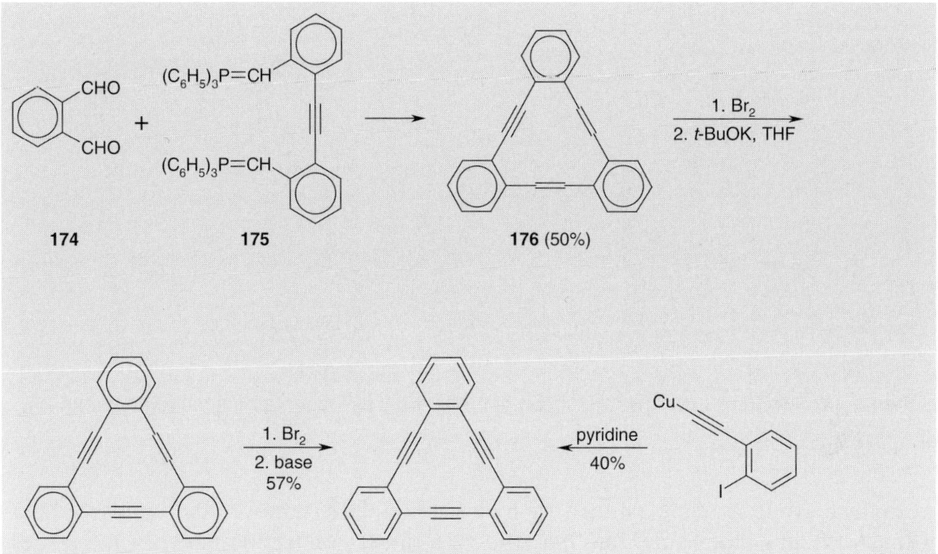

Scheme 33. The synthesis of a tribenzo derivative of **165**.

10.5 Bridged annulenes

As discussed above (Section 10.4), the instability of [10]annulene can be traced back to bond angle deformations and intraannular hydrogen-hydrogen repulsion. In particular we noted that in the di-*trans*, tri-*cis*-configured hydrocarbon **5** the two internal hydrogen atoms must—if the molecule is planar—occupy the same space—a physical impossibility. If preparable or even isolable at all, **5** must be non-planar. If, on the other hand, it is this particular steric interaction which destabilizes the annulene, should then its complete or partial removal not lead to stabilization of the corresponding hydrocarbon? In Scheme 34 I have listed several possible means whereby this steric relief might be accomplished.

Removal of the two internal hydrogen atoms and concomitant bond formation between the transannular carbon atoms leads to naphthalene (**180**)—certainly an aromatic hydrocarbon, but not of the monocyclic type to which the original Hückel rules refers. If one *trans*-configured double bond in **5** is replaced by a carbon-carbon triple bond, as in **181**, only one internal hydrogen atom is left (again assuming a planar conformation), and if both *trans*-linkages are exchanged for triple bonds, as in **183**, the resulting hydrocarbon no longer has any internal substituents. Finally, replacement of the two hydrogen atoms by a molecular bridge, in the simplest case a

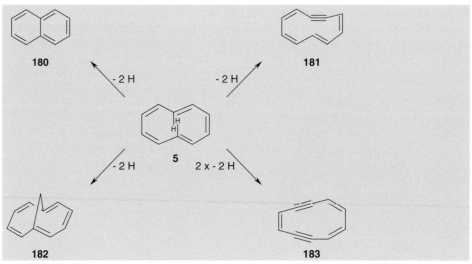

Scheme 34. How to remove the interaction of the internal hydrogen atoms in [10]annulene (**5**).

methylene group as demonstrated by hydrocarbon **182**, removes the repulsive inter-action without at the same time introducing a direct connection between the respective carbon atoms. Possibility **182** is formally related to **180** because one is dealing with bicyclic systems in both cases, the latter having a 'zero-bridge'. Provided, how-ever, that the bridging unit in **182** does not significantly change the steric and/or electronic properties of the π-perimeter, *i.e.* that it is non-interactive, one can still re-gard the hydrocarbon as of the Hückel-type, *i.e.* as a monocyclic and fully conjugated molecule. In other words, the purpose of the molecular bridge is to remove an intoler-able steric situation. Of the three approaches symbolyzed by **181**–**183** only one, **182**, could be realized—and most successfuly, as we shall see below. However, whereas there have apparently been no attempts to verify possibility **181**, an interesting ex-periment was conducted by Masamune and co–workers to test alternative **183**[193, 194] (Scheme 35).

Coupling of the enediyne **184** with the bis tosylate **185**, followed by functional group manipulation provided the ten-membered ring compound **186**, with two mesyl-ate groups in correct orientation to enable base-induced β-elimination. Although the intended derivative of **183**, hydrocarbon **187**, could not be detected, the isolated re-action products **190** and **191** (formed in 30–40% yield) are highly suggestive of the intermediate generation of the dehydro[10]annulene. Using knowledge not available to the original workers we would explain the formation of the benzenoid products by postulating a Bergman cyclization[195] of **187** to the diradical **188** which could either abstract hydrogen atoms from the solvent and form the tetrahydroanthracene **190** or experience ring-opening *via* **189** to yield the ten-membered ring compound **191**. We shall return to similar hydrocarbons and the diradicals derived from them in Chapter 16.1.

Two tetradehydrocyclodecabiphenylenes, which incorporate **183** as a subsystem, have been prepared by reacting *ortho*-phthalaldeyde (**174**) with appropriate bis Wittig

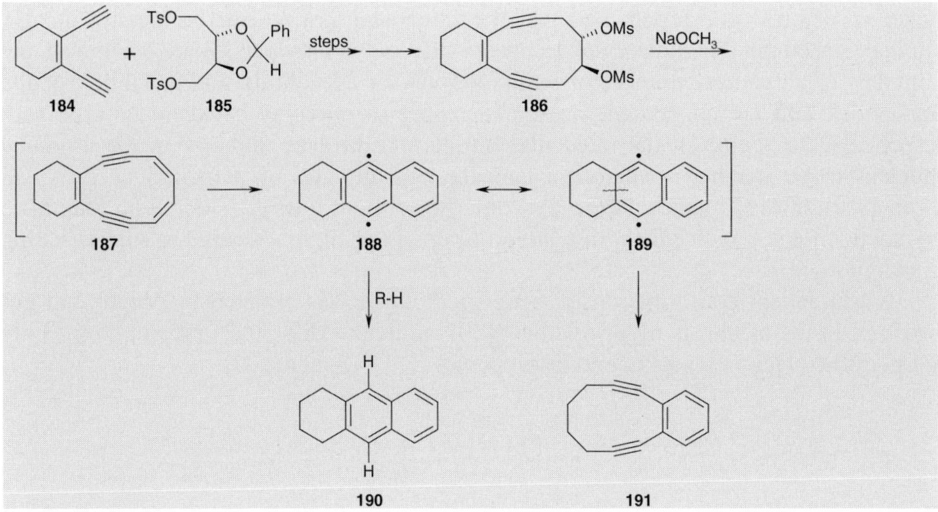

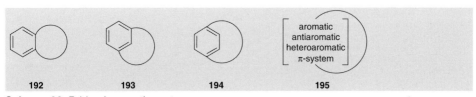

Scheme 35. An attempt to prepare a tetradehydro[10]annulene—an early example of a Bergman cyclization.

reagents as described for the synthesis of **178**. Because these (stable) hydrocarbons do no longer have hydrogen atoms on their π-perimeter, they are not amenable to ¹H NMR studies.[196]

Before turning to the bridged [10]annulene **182** some general remarks on bridged aromatic compounds are in order. In an aromatic molecule, benzene in the simplest case, the molecular bridge can be anchored in three different ways (Scheme 36)—in the *ortho* (**192**), *meta* (**193**) or *para* positions (**194**). The *ortho*-bridged systems **192** are commonly treated as 'normal benzenoid aromatics', excluding the *ortho*-methano- and -ethano-bridged compounds which will be discussed Section 16.3.

Scheme 36. Bridged aromatic systems.

That systems **193** and **194**, the so-called meta- and paracyclophanes, will also be treated in a separate chapter (see Section 12.3) is justified by the particular geometry of these hydrocarbons. In contrast to the [*n*]annulenes, for which planarity *prima facie* is a prerequisite (see below, however), **193** and **194** very often have non-planar benzene rings, particulary when the molecular bridge is short. Generalizing the bridged-aromatic concept, as illustrated in **195**, we see that that the role of the aromatic subsystem can be played by any aromatic, antiaromatic or heteroaromatic nucleus, whether neutral or charged, and that the molecular bridge can principally display similar structural and functional richness—the polymethylene chain being just the simplest of an endless number of variations. As far as stability and isolability of the molecules **195** are concerned, we might expect an interplay between the type and, especially, the length of the molecular bridge and the type and size of the aromatic nucleus to be spanned—whereas a benzene ring bridged by a methylene group in *para* position (a [1]paracyclophane) would not currently be considered an attainable synthetic target, a molecule with a larger π-perimeter might succumb to such bridging much more readily.

A triumphant conformation of these predictions was realized by Vogel and co-workers in the synthesis of 1,6-methano[10]annulene (**182**), the first and so far only carbocyclic [1]paracyclophane to be prepared[197, 198] (Scheme 37).

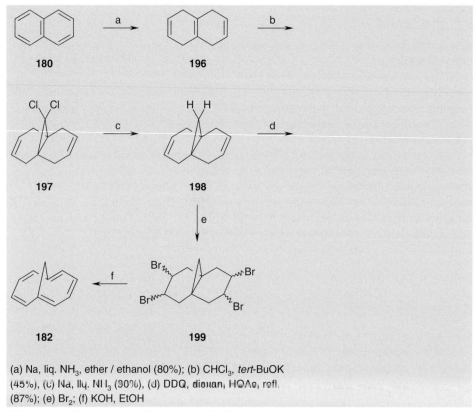

(a) Na, liq. NH$_3$, ether / ethanol (80%); (b) CHCl$_3$, *tert*-BuOK (45%); (c) Na, liq. NH$_3$ (90%); (d) DDQ, dioxan, HOAc, refl. (87%); (e) Br$_2$; (f) KOH, EtOH

Scheme 37. The synthesis of 1,6-methano[10]annulene (**182**) by Vogel.

The starting material of the synthesis, naphthalene (**180**), illustrates the task to be accomplished—its central bond must be replaced by a methylene group. This type of ring enlargement is often performed by carbene reactions, and we can thus expect the use of a carbene or carbenoid at some time during the course of the synthesis. Be-

cause methylenation of **180** by *e.g.* diiodomethane and zinc-copper couple is unspe-
cific and takes place preferentially at the peripheral double bonds,[198] naphthalene
was first converted to the triene **196** (1,4,5,8-tetrahydronaphthalene, isotetralin) by
Birch reduction. This hydrocarbon offers an attacking carbene two sites of addition—
dichlorocarbene, generated from chloroform and potassium *tert*-butoxide, prefers the
(electron-rich) internal double bond with high selectivity. The produced *gem*-
dichloride **197** was dehalogenated by treatment with sodium in liquid ammonia, and
the resulting propellane **198** (see Section 6.1) was oxidized to the bridged annulene
182 either by 2,3-dichloro-5,6-dicyano-1,4-benzoquinone (DDQ) in dioxan or *via*
bromination to **199** followed by dehydrobromination, the intermediate norcaradiene
derivative evidently undergoing electrocyclic ring opening under the reaction conditions. Although **182** was not considered an aromatic compound in the original publication,[197] later investigations revealed it to be a true (4n+2)-hydrocarbon with n = 2.
Its ¹H NMR spectrum shows it to be a diatropic hydrocarbon, the ring protons ap-
pearing as an AA'BB' multiplet at δ = 7.27 and 6.95, and the methylene protons, positioned above the π-plane, as a singlet at δ = –0.52.[199] According to X-ray structural
analysis the periphery of **182** is not planar—the bridgehead carbon atoms C1 and C6
lie out of the plane that passes through the other olefinic carbon atoms by *ca* 0.4 Å.
The average bond lengths in the perimeter (1.373–1.419 Å) are of benzenoid type (the
C–C bond length in **3** is 1.398 Å) and indicate significant delocalization of the π-
system despite the lack of overall planarity. The distance between C1 and C6 is 2.235
Å, suggesting that there is no bonding between these atoms, *i.e.* that **182** does not
exist as a norcaradiene valence isomer.[199] Chemically methano[10]annulene is stable
towards oxygen and thermally to at least 220 °C—a dramatic stability increase when
compared with the non-bridged [10]annulenes **116** and **117** (see above). Like a benzenoid aromatic, **182** undergoes substitution with bromine, copper nitrate, and acetyl
chloride (preferentially in the α-position as in **180**). The role of the molecular bridge
in **182** can also be played by oxygen and an imino group as shown soon afterwards
by the preparation of 1,6-oxido[10]annulene[200] and 1,6-imino[10]annulene,[201]
bridged annulenes that also qualify as aromatic compounds in both physical (spectroscopic) and chemical respects.

The successful preparation of **182** was not only a milestone in annulene, or more
generally, aromatic chemistry—it also marked the beginning of a synthetic undertaking with the goal of preparing a homologous series of bridged [4n+2]annulenes corresponding to the classical aromatic hydrocarbons, the acenes or linearly annelated
aromatics, naphthalene, anthracene, tetracene *etc.* (see Section 14.1). Although a
strategy for preparing these higher-bridged annulenes by a route similar to the one
which yielded **182**, *i.e.* starting from the appropriate Birch reduced acene, immediately suggests itself, it met with many practical problems, and was finally abandoned.

To begin with, in the multiply methano-bridged [4n+2]annulenes the problem of
geometrical isomers arises (Scheme 38).[202]

Thus carbene addition to the central double bonds of hexahydroanthracene (**200**)
can in principle yield an *anti* or a *syn* adduct. Experimentally all carbenes and carbenoids tried, yielded the *anti* adduct, shown in **201** for dichlorocarbene addition.[203]
The transformation of **201**, *via* **202**, to the [14]annulene **203** proceeded as expected.
This hydrocarbon, however, has a puckered annulene ring with torsional angles of the
2_{pz}-orbitals at neighboring carbon atoms of up to 70°, and shows pronounced bond

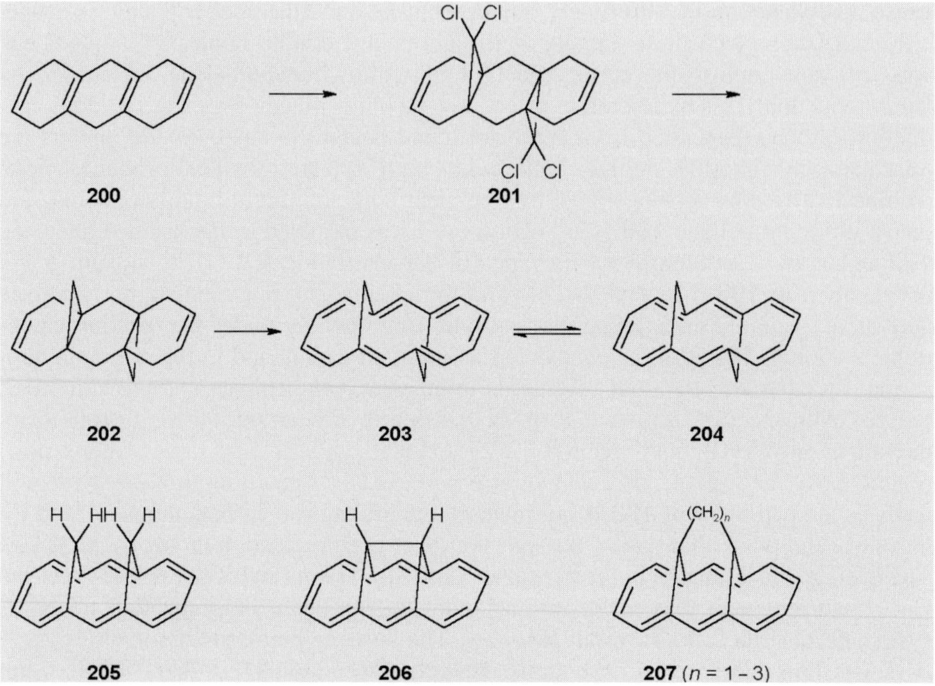

Scheme 38. The preparation of bismethano[14]annulenes.

alternation and typical olefin behavior. Interestingly, **203** is also a dynamic annulene, experiencing a rapid π-bond shift which converts it into its valence isomer **204**. Again, this is a degenerate process since **203** and **204** are identical as long as their carbon atoms are not marked.

That the *syn* isomers are not found in the cyclopropanation experiments is a result of the very unfavorable transannular hydrogen-hydrogen interaction, as illustrated in **205**. Considering the success of the amputation of this interaction in the [10]annulene series (see above), it was obvious to exploit this concept for **205** also. Although syntheses were completed not only for **206** from **200** but also for the hydrocarbons **207** in general, in which the critical hydrogen atoms are replaced by oligomethylene bridges of different lengths,[204] the preparative expense was large and no generalisation of this approach was attempted.

Both **206** and the homologs **207** behave as typical aromatic hydrocarbons. This is surprising because these molecules become progressively bent when the length of the 'secondary bridge' is increased. This is true even for the last member of the series **207** with $n = 3$, in which the profile of the π-perimeter approaches that of a semicircle and the torsional angles of the 2_{pz}-orbitals at neighboring carbon atoms reach values of up to 40°. Clearly, the steric condition—*i.e.* planarity—for aromaticity demanded by Hückel's rule is far less stringent than assumed earlier, an observation which will be corroborated later in Section 12.3 on the cyclophanes.

To prepare *syn*-1,6:8,13-bismethano[14]annulene (**205**) and its higher homologs, an entirely new synthetic strategy had to be developed; this begins with cycloheptatriene-1,6-dicarboxaldehyde (**212**) (Scheme 39), and then proceeds in a modular fashion (see below).[205]

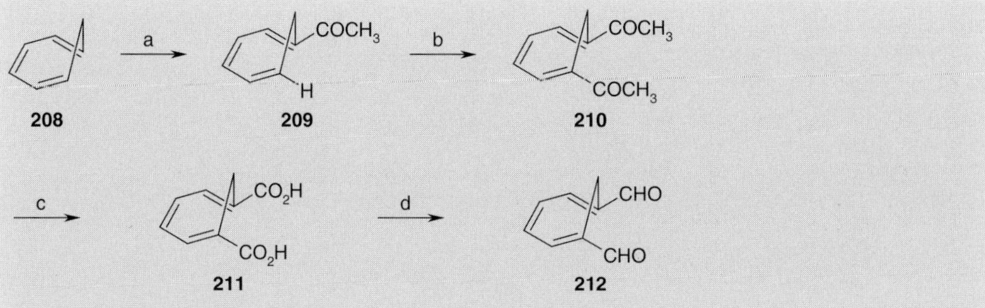

(a) CH₃COCl, ZnCl₂, HOAc / CH₂Cl₂ (45-50%); (b) CH₃COCl, AlCl₃, CH₂Cl₂ (50%); (c) NaOH, Br₂ (66%); (d) CH₃COCl, then Li[AlH(*tert*-BuO)₃] (61%)

Scheme 39. Cycloheptatriene-1,6-dicarboxaldeyhde (**212**)—the basic starting material for multi-bridged [4*n*+2]annulenes.

When 1,3,5-cycloheptatriene (**208**) was acetylated with acetylchloride in the presence of the relatively mild Lewis acid zinc chloride, the monoketone **209** was produced; this, on further acetylation with the more agressive aluminum trichloride, yielded the diketone **210**. Haloform reaction subsequently provided the diacid **211** which, after conversion to its dianhydride with acetylchloride, was reduced to the dialdehyde **212**.

Having multigram quantities of the 'homophthalaldehyde' **212** in hand, it could be employed in a homologation sequence with the double Wittig–Horner reagent **213**, itself easily available *via* Michaelis–Arbuzov reaction from diethyl α,α'-dibromoglutarate and triethyl phosphite, yielding the diester **214**, which was converted by standard methodology to the dialdehyde **215** (Scheme 40).

Various routes were developed to construct the missing six-membered ring in **215**;[206] the most useful in practice[207] was successive treatment of **215** with methylene(triphenyl)phosphorane and chloromethylene(triphenyl)phosphorane and heating the resulting vinylchloride **216** in DMF at 150 °C. As expected, ring closure in a 14π-electrocyclic process to **217** took place, which spontaneously lost hydrogenchloride to produce **205**. This doubly-bridged [14]annulene is again not planar, but according to its ¹H NMR spectrum it must certainly be classified as aromatic. Among the structural features of **205** the extremely short distance between the inner methano hydrogen atoms of only 1.78 Å is particularly noteworthy. This is one of the shortest distances ever reported between non-bonded hydrogen atoms.[207] Having accomplished the synthesis of **205** from **215**, everything was ready to extend this *aufbau* sequence. As summarized in Scheme 41, which also includes the lower homolog **182**, the whole series of multi-bridged [4*n*+2]annulenes could be obtained—a most remarkable achievement both from the viewpoints of synthetic design and preparative economy!

Although the hydrocarbons in the series **182**→**205**→**220**→**221** become increasingly unstable, they are all aromatic by the NMR criterion. It should be remem-

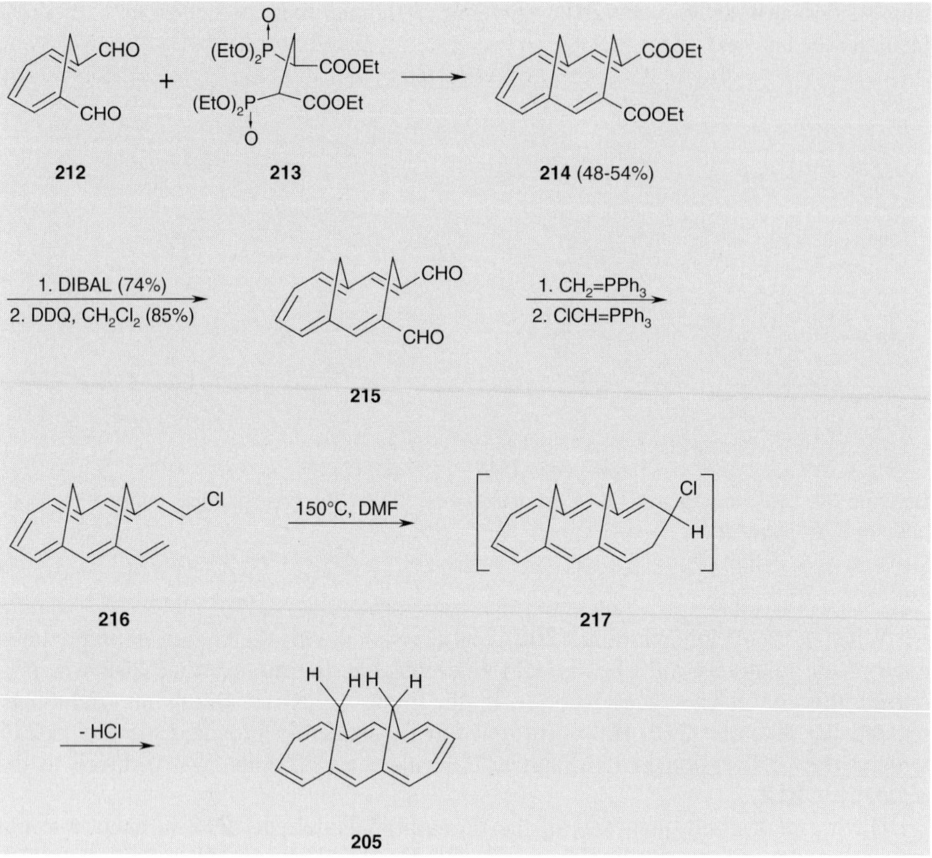

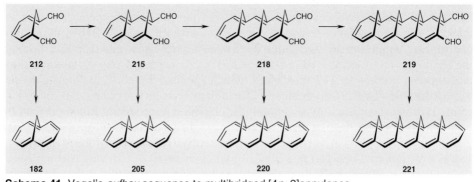

Scheme 40. The synthesis of *syn*-1,6:8,13-bismethano[14]annulene (**205**).

Scheme 41. Vogel's *aufbau* sequence to multibridged [4*n*+2]annulenes.

bered that the stability of the classical (benzenoid) acenes also decreases with increasing annelation (see Section 14.1).

The doubly-bridged hydrocarbon **205** was actually not the first bridged [14]annulene to be synthesized; this credit goes to the 15,16-dihydropyrene derivative **226**, prepared by Boekelheide and co-workers in 1963 by the route summarized in Scheme 42.[208]

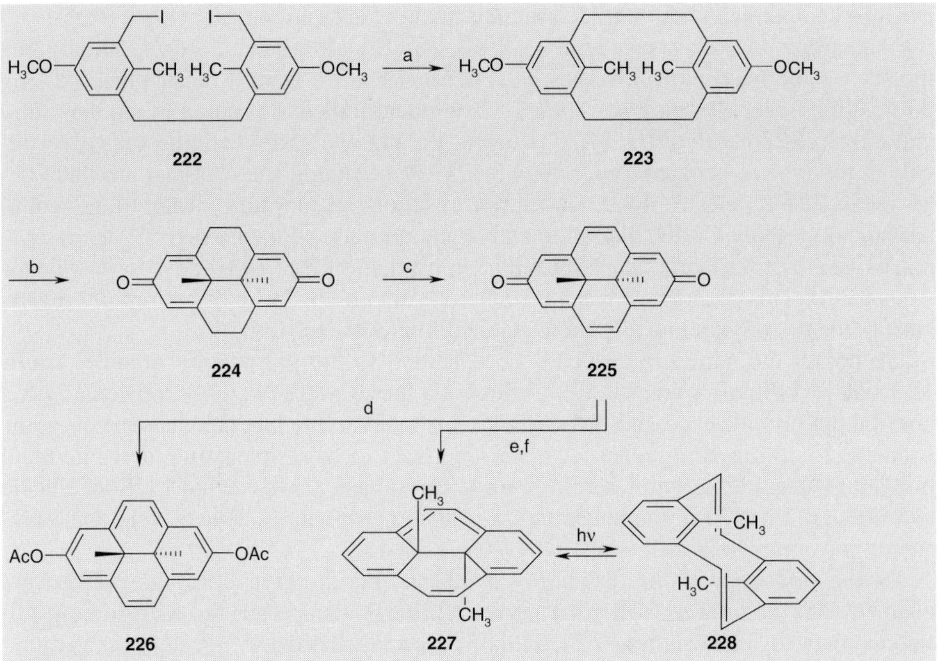

(a) Na, Ph₂C=CPh₂ (56%); (b) FeCl₃, CHCl₃ (92%); (c) NBS (76%); (d) Zn / Ac₂O, Et₃N (92%); (e) LiAlH₄, AlCl₃, -80°C; (f) Pd / C (comb. yield 75%)

Scheme 42. The first synthesis of a bridged [14]annulene by Boekelheide.

Wurtz coupling of the diiodode **222** led to the [2.2]metacyclophane **223** which was oxidized in almost quantitative yield to the bis dienone **224**. Unsaturation was subsequently introduced into the bridges by oxidation with NBS, and when the resulting **225** was reduced with zinc in acetic anhydride in the presence of triethylamine, 2,7-diacetoxy-*trans*-15,16-dimethyl-15,16-dihydropyrene (**226**) was formed. In a later investigation the parent hydrocarbon **227** could be prepared from **225** by first reducing its keto functions to methylene groups and then dehydrogenating the resulting dihydro intermediate[209] (Scheme 42).

The transformations shown in the scheme are, in fact, only the last four of a synthetic sequence involving 17 steps. These last steps are more than routine, though, and illustrate several important points. First of all, they demonstrate the decisive concept in this area of bridged annulene chemistry—the conversion of a [2.2]metacyclophane precursor (see Section 12.3) into the 15,16-dihydropyrene nucleus. This process, in which a three-dimensional, layered structure is 'flattened', was maintained in all later synthetic work. Secondly, the question might be asked why oxygen-carrying substituents were introduced into the 2 and 7 positions of the future [14]annulene, the electronic structure of which they certainly would influence. The answer—phenol-phenol coupling is a very well known process in natural products and synthetic organic chemistry, and the **223**→**224** interconversion exploits the driving force of this C–C bond-forming reaction. Only later it became clear, that **227** and other dihydropyrenes can also be obtained without this particular mode of activation (see below).

According to X-ray structural analysis, the dihydropyrenes are essentially flat molecules, the maximum deviation of a perimeter atom from a mean plane passing through the molecule being *ca* 3 pm.[210] The internal methyl protons of the dark, emerald-green **227** aborb at δ = -4.25, whereas the peripheral aromatic protons give signals in the typical benzene range (δ = 8.67–7.98). Among the classical aromatic behavior of **227** its electrophilic substitution reactions of nitration, deuteration, bromination, and Friedel-Crafts alkylation and acylation must be mentioned;[211] the overall reactivity is greater than that of benzene. On irradiation **227** ring-opens to its valence isomer **228**. This photoisomer is not stable on standing in the dark but rapidly reverts back to the thermodynamically more stable dihydropyrene form.[212]

Although the above route could be extended to the preparation of other *trans*-15,16-dialkyl-dihydropyrenes,[213, 214] these syntheses were not only too lengthy but also did not provide access to the various *cis* isomers or the parent substances, *e.g.* the isomeric 15,16-dihydropyrenes. A novel approach to these interesting and important bridged [14]annulenes had, therefore, to be developed. It was found by Boekelheide and Mitchell in 1970 in the construction of the appropriate [2.2]metacyclophane precursor in a completely different way[215] (Scheme 43).

In the first step of this far shorter synthesis, the *m*-xylyl dibromides **229** were coupled with the dithiols **230** (alternatively **229** can also be reacted with sodium sulfide in ethanol). The resulting 2,11-dithia[3.3]metacyclophanes[216] can occur in either the *anti* (**231**) or *syn* form (**232**). Methylation of **231** with dimethoxycarbonium tetrafluoroborate gave the corresponding *anti*-bis sulfonium salt **234** in near quantitative yield. Subjecting this to treatment with base (sodium hydride in THF under heterogeneous conditions), resulted in a Stevens rearrangement and formation of the sulfides **233** as mixture of isomers. A second methylation step followed, providing the bis sulfonium salts **235**, which, on Hofmann elimination with potassium *tert*-butoxide in THF, yielded the *trans*-dimethyl derivative **227** directly. When the *syn* isomer **232** (R = CH₃) was taken through the same series of transformations *cis*-15,16-dimethyldihydropyrene (**237**) was produced. This latter hydrocarbon is distinctly less stable than its *trans* isomer **227**; it reacted slowly with air and must be stored in the dark. The increased reactivity is probably due to **237** having one sterically unprotected side, which reagents can approach more readily than in **227** where attack from either side of the molecular plane is protected by one methyl substituent. This rationalization is supported by the observation that the parent hydrocarbon **236**, which was prepared by the same synthetic strategy, is also light- and oxygen-sensitive. In fact, in the unsubstituted case, the dihydropyrene **236** is not formed directly, but its metacyclophane diene precursor (**228**, without the methyl groups) could be isolated. For all dialkylated dihydropyrenes this valence isomer is unstable under the reaction conditions (see above). Only on irradiation with 254-nm light is *trans*-[2.2]metacylophane-1,9-diene converted into **236** (with pyrene always being produced as a byproduct). In

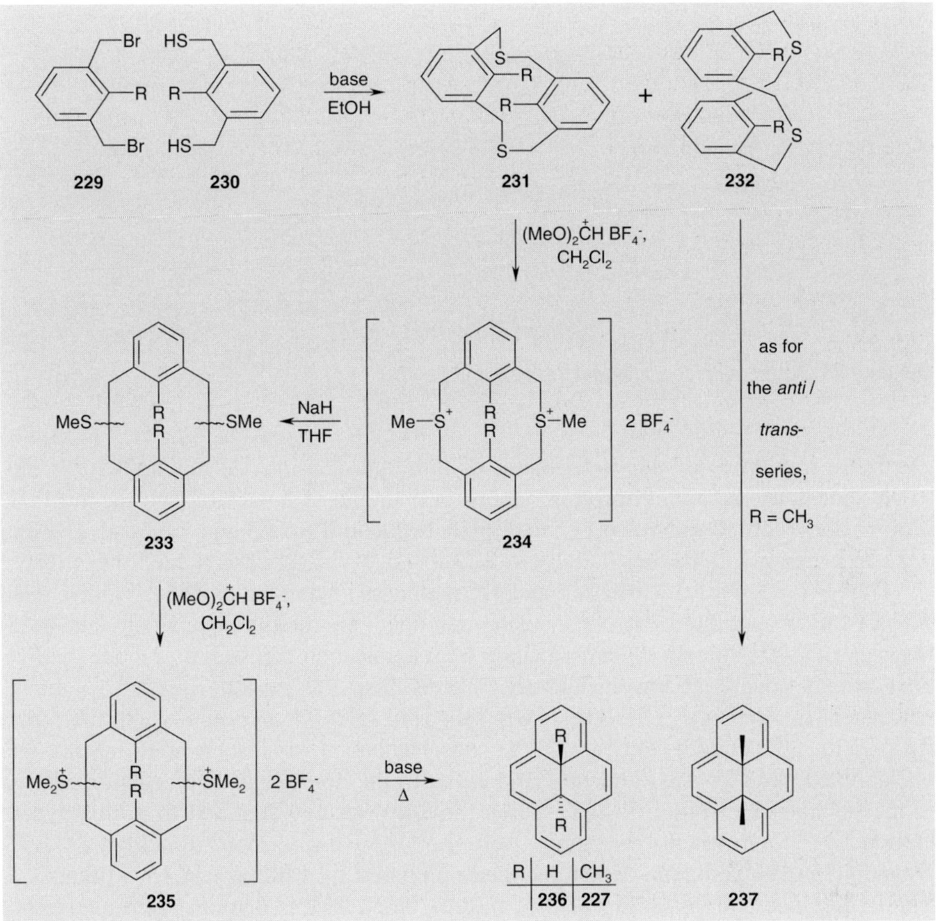

Scheme 43. The synthesis of 15,16-dihydropyrenes by transformation of sulfide linkages to carbon-carbon double bonds.

the ¹H NMR spectrum the π-perimeter protons of **236** appear as multiplets between δ = 8.58 and 7.89, and the bridge protons as a singlet far beyond the TMS reference signal at δ = –5.49, a convincing proof of the diatropicity of this ethanobridged [14]annulene. Comparison with *cis*-15,16-dihydropyrene would be very instructive; this hydrocarbon has not yet been prepared, however.

Although 1,6-methano[10]annulene (**182**) and the 15,16-dihydropyrenes (especially **227**) stand out in this area of aromatic chemistry, they are by no means the only bridged [4*n*+2]annulenes. To begin with, both Masamune[217] and Scott and co–workers[218] have described syntheses for 1,5-methano[10]annulene (**238**; Scheme 44), an isomer of **182** which has strong structural similarity with azulene (see Section 11.4).

According to NMR spectroscopy this hydrocarbon, which is stable up to 300 °C, qualifies as an aromatic compound, and so does hydrocarbon **239**, a triply short-circuited derivative of tri-*trans*-di-*cis*-[10]annulene.[219] With its internal methyl substituent **239** once again has a reporter group which can provide information on the

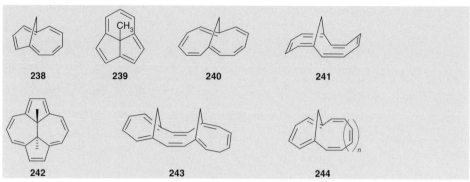

Scheme 44. A selection of additional bridged annulenes.

electronic properties of the π-perimeter of this hydrocarbon. As required for a dia-tropic molecule, the methyl protons absorb as a singlet at δ = –1.67 and the aromatic protons are recorded as two sets of multiplets between δ = 7.53 and 7.83 and 7.89 and 7.92. In keeping with the aromaticity of **239** its reactivity towards dienophiles is low.

The [4n]systems 1,7- (**240**)[220] and 1,6-methano[12]annulene (**241**)[221] have been prepared by Vogel and co–workers using annulenic precursors once again. The para-tropicity of **240** is nicely illustrated again by 'chemical shift reversal', *i.e.* the methyl-ene protons absorb at lower field (δ = 6.16) than the protons bonded to the π-perimeter (δ = 5.5 and 5.2). The NMR spectrum of this compound is, furthermore, temperature dependent indicating the occurrence of a fast π-bond shift. The [12]annulene **241** is also paratropic and exists in the non-planar conformation shown in the scheme. An isomer of **227**, which formally can be generated from this hydro-carbon by rotating its bridging unit by 90°, *trans*-15,16-dimethyl-1,4:8,11-ethandiyliden[14]annulene (**242**), has been prepared by Müllen and co–workers.[222] The importance and versatility of the building block **212** is demonstrated once more by its reductive dimerization (McMurry coupling) to **243**,[223] and that the homotro-pylidene subunit can finally also be incorporated into very-large-ring [4n]- and [4n+2]annulenes is shown by the 1,6-methanoannulene series **244**. Whereas the 18- and 22-membered ring systems are diatropic, the [12]-, [20]-, and [24]annulenes are paratropic. When, however, a certain ring size is exceeded, the difference between diatropicity ('aromatic character') and paratropicity ('antiaromatic character') van-ishes, and the 28-, 30-, 32-, 34- and 38-membered ring systems are atropic.[224]

References

1.   E. Clar, *Polycyclic Hydrocarbons*, Academic Press, London, **1964**, Vol. I and II.
2.   G. Schröder, *Cyclooctatetraen*, Verlag Chemie, Weinheim, **1965**.
3.   American Chemical Society, *Kekulé Centennial*, Washington, D. C. **1966**, Advances in Chemistry Series, Vol. 61.
4.   The literature in the cyclobutadiene field up to *ca* 1965 has been summarized by M. P.

Cava, *Cyclobutadiene and Related Compounds*, Academic Press, New York, **1967**.

5. G. M. Badger, *Aromatic Character and Aromaticity*, Cambridge University Press, Cambridge, **1969**.
6. P. J. Garratt, *Aromaticity*, McGraw Hill, London, **1971**.
7. E. Clar, *The Aromatic Sextet*, John Wiley and Sons, London, **1972**.
8. G. I. Fray, R. G. Saxton, *The Chemistry of Cyclooctatetraene and its Derivatives*, Cambridge University Press, Cambridge, **1978**.
9. The Chemical Society, Specialist Periodical Reports, *Aromatic and Heteroaromatic Chemistry*, London, **1973**–**1979**, Volume 1–7.
10. J. D. Memory, N. K. Wilson, *NMR of Aromatic Compounds*, J. Wiley and Sons, New York, **1982**.
11. D. Lloyd, *Non-benzenoid Conjugated Carbocyclic Compounds*, Elsevier, Amsterdam, **1984**. For reviews on annulenes see P. Skrabal in *MTP International Reviews of Sciences. Organic Chemistry*, Series 1, Vol. 3, *Aromatic Compounds,* H. Zollinger, (*Ed.*), Butterworths, London, **1973**, 237–269 and Series 2, London, **1976**, 229–257.
12. American Chemical Society, *Polycyclic Hydrocarbons and Carcinogenisis*, R. G. Harvey (*Ed.*), Washington, **1985**, ACS Symposium Series, No. 283.
13. Houben-Weyl, *Methoden der Organischen Chemie, Arene und Arine* Chr. Grundmann, (*Ed.*), Georg Thieme Verlag, Stuttgart, **1981**, Vol. V/2b; *cf.* Houben-Weyl, *Methoden der Organischen Chemie, Carbocyclische π-Systeme*, Thieme Verlag, Stuttgart, **1985**, Vol. V/2c.
14. P. J. Garratt, *Aromaticity,* J. Wiley and Sons, New York, **1986**.
15. H.-G. Franck, J. W. Stadelhofer, *Industrial Aromatic Chemistry*, Springer-Verlag, Berlin, **1988**.
16. J. R. Dias, *Handbook of Polycyclic Hydrocarbons*, Elsevier, Amsterdam, **1988**.
17. D. Lloyd, *The Chemistry of Conjugated Cyclic Compounds. To Be Or Not To Be Like Benzene?*, John Wiley and Sons, Chichester, **1989**.
18. R. Taylor, *Electrophilic Aromatic Substitution*, John Wiley and Sons, Chichester, **1990**.
19. M. K. Shepherd, *Cyclobutarenes*, Elsevier, Amsterdam, **1991**.
20. R. G. Harvey, *Polycyclic aromatic hydrocarbons–chemistry and carcinogenicity*, Cambridge University Press, Cambridge, **1991**.
21. V. I. Minkin, M. N. Glukhovtsev, B. Ya. Simikin, *Aromaticity and Antiaromaticity*, J. Wiley and Sons, New York, **1994**.
22. H. Hellmann, *Umweltanalytik von Kohlenwasserstoffen*, VCH Verlagsgesellschaft, Weinheim, **1995**.
23. M. Zander, *Polycyclische Aromten*, Teubner Studienbücher Chemie, Stuttgart, **1995**.
24. R. G. Harvey, *Polycyclic Aromatic Compounds*, Wiley-VCH, New York, N. Y., **1997**.
25. Although so far no volume dedicated exclusively to aromatic compounds has appeared in *The Chemistry of Functional Groups* S. Patai, Z. Rappoport (*Eds*.), J. Wiley and Sons, Chichester, many volumes of this important series contain a wealth of important information on aromatic hydrocarbons.
26. W. H. Perkin, Jr, *Ber. Dtsch. Chem. Ges.*, **1893**, *26*, 2243–2246; *cf.* W. H. Perkin, Jr, *J. Chem. Soc.*, **1894**, *65*, 950–975.
27. A. Kekulé, *Liebigs Ann. Chem.*, **1872**, *162*, 77–124.
28. R. Willstätter, W. v. Schmaeel, *Ber. Dtsch. Chem. Ges.*, **1905**, *38*, 1992–1999.
29. R. Malachowski, Ph. d. dissertation, ETH Zürich, **1911**.
30. J. Thiele, *Justus Liebigs Ann.*, **1899**, *306*, 87–142.
31. H. C. Longuet-Higgins, L. Orgel, *J. Chem. Soc.*, **1956**, 1969–1972.
32. R. Criegee, G. Schröder, *Liebigs Ann. Chem.*, **1959**, *623*, 1–8; *cf.* R. Criegee, G. Schröder, *Angew. Chem.*, **1959**, *71*, 70–71.

33. For modern reviews on cyclobutadiene and related compounds see G. Maier, *Angew. Chem.*, **1974**, *86*, 491–505; *Angew. Chem. Int. Ed. Engl.* **1974**, *13*, 425–439; T. Bally, S. Masamune, *Tetrahedron*, **1980**, *36*, 343–370; D. Lloyd, *Non-Benzenoid Conjugated Carbocyclic Compounds*, Elsevier, Amsterdam, **1984**, 197–244; G. Maier, *Angew. Chem.*, **1988**, *100*, 317–341; *Angew. Chem. Int. Ed. Engl.*, **1988**, *27*, 309–333; M. Regitz, W. Heydt, U. Bergsträsser in *Adv. in Strain in Organic Chemistry*, B. Halton, (*Ed.*), JAI Press, Greenwich, **1996**, Vol. 5, 161–243; S. Haber, M. Regitz in Houben-Weyl, *Methods of Organic Chemistry, Carbocyclic Three- and Four-membered Ring Compounds*, A. de Meijere (*Ed.*), Thieme Verlag, Stuttgart, **1997**, Vol E17f, 1051–1120.
34. G. F. Emerson, L. Watts, R. Pettit, *J. Am. Chem. Soc.*, **1965**, *87*, 131–133; *cf.* J. D. Fitzpatrick, L. Watts, G. F. Emerson, R. Pettit, *J. Am. Chem. Soc.*, **1965**, *87*, 3254–3255.
35. L. Watts, J. D. Fitzpatrick, R. Pettit, *J. Am. Chem. Soc.*, **1965**, *87*, 3253–3254.
36. R. H. Grubbs, D. A. Gray, *J. Am. Chem. Soc.*, **1973**, *95*, 5765–5767; *cf.* R. H. Grubbs, T. Λ. Pancoast, *J. Am. Chem. Soc.*, **1977**, *99*, 2382–2383.
37. E. K. G. Schmidt, *Angew. Chem.*, **1973**, 85, 820–821; *Angew. Chem. Int. Ed. Engl.*, **1973**, *12*, 777–778.
38. J. Rebek, Jr, F. Gaviña, *J. Am. Chem. Soc.*, **1975**, *97*, 3453–3456.
39. R. Gompper, F. Holsboer, W. Schmidt, G. Seybold, *J. Am. Chem. Soc.*, **1973**, *95*, 8479–8481 and lit. cited.
40. O. L. Chapman, C. L. McIntosh, J. Pacansky, *J. Am. Chem. Soc.*, **1973**, *95*, 614–617.
41. O. L. Chapman, D. De La Cruz, R. Roth, J. Pacansky, *J. Am. Chem. Soc.*, **1973**, *95*, 1337–1338.
42. A. Krantz, C. Y. Lin, M. D. Newton, *J. Am. Chem. Soc.*, **1973**, *95*, 2744–2746.
43. The lactone **13** was first prepared by E. J. Corey and J. Streith, *J. Am. Chem. Soc.*, **1964**, *86*, 950–951 who also proposed to use it as a precursor for the generation of cyclobutadiene.
44. G. Maier, B. Hoppe, *Tetrahedron Lett.*, **1973**, 861–864. These early photofragmentations were carried out in an organic matrix (ether/THF/isopentane) at -196 °C. In later studies argon matrices at 10 K were commonly employed: G. Maier, H.-G. Hartan, T. Sayrac, *Angew. Chem.* **1976**, *88*, 252–253; *Angew. Chem. Int. Ed. Engl.*, **1976**, *15*, 226–227.
45. G. Maier, G. Fritschi, B. Hoppe, *Tetrahedron Lett.*, **1971**, 1463–1468.
46. M. J. Gerace, D. M. Lemal, H. Ertl, *J. Am. Chem. Soc.*, **1975**, *97*, 5584–5586.
47. G. Maier, M. Hoppe, H. P. Reisenauer, *Angew. Chem.*, **1983**, *95*, 1009–1010; *Angew. Chem. Int. Ed. Engl.*, **1983**, *22*, 990.
48. S. Masamune, N. Nakamura, J. Sapadaro, *J. Am. Chem. Soc.*, **1975**, *97*, 918–919.
49. D. W. Whitman, B. K. Carpenter, *J. Am. Chem. Soc.*, **1980**, *102*, 4272–4274; *cf.* D. W. Whitman, B. K. Carpenter, *J. Am. Chem. Soc.*, **1982**, *104*, 6473–6474.
50. S. Masamune, M. Suda, H. Ona, L. M. Leichter, *J. Chem. Soc. Chem. Commun.*, **1972**, 1268–1269.
51. R. D. Miller, E. Hedaya, *J. Am. Chem. Soc.*, **1969**, *91*, 5401–5402.
52. A. M. Orendt, B. R. Arnold, J. G. Radziszewski, J. C. Facelli, K. D. Malsch, H. Strub, D. M. Grant, J. Michl, *J. Am. Chem. Soc.*, **1988**, *110*, 2648–2650.
53. D. J. Cram, M. E. Tanner, R. Thomas, *Angew. Chem.*, **1991**, *103*, 1048–1051; *Angew. Chem. Int. Ed. Engl.*, **1991**, *30*, 1024–1027; *cf.* D. J. Cram, J. M. Cram, *Container Molecules and Their Guests*; *Monographs in Supramolecular Chemistry*, Royal Soc. of Chemistry, Cambridge, **1994**.
54. G. Maier, F. Fleischer, *Tetrahedron Lett.*, **1991**, *32*, 57–60; *cf.* G. Maier, M. Hoppe, K. Lanz, H. P. Reisenauer, *Tetrahedron Lett.*, **1984**, *25*, 5645–5648.
55. S. Masamune, N. Nakamura, M. Suda, H. Ona, *J. Am. Chem. Soc.*, **1973**, *95*, 8481–8483.
56. P. Eisenbarth, M. Regitz, *Angew. Chem.*, **1982**, *94*, 935–936; *Angew. Chem. Int. Ed.*

Engl., **1982**, *21*, 913–914; *cf.* J. Fink, H. Gümbel, P. Eisenbarth, M. Regitz, *Chem. Ber.*, **1987**, *120*, 1027–1037.

57. G. Maier, R. Wolf, H.-O. Kalinowski, *Angew. Chem.*, **1992**, *104*, 764–766; *Angew. Chem. Int. Ed. Engl.*, **1992**, *31*, 738–740; *cf.* G. Maier, D. Born, *Angew. Chem.*, **1989**, *101*,1085–1087; *Angew. Chem. Int. Ed. Engl.*, **1989**, *28*, 1050–1051.

58. J. B. Koster, G. J. Timmermans, H. van Bekkum, *Synthesis,* **1971**, 139–140. Interestingly, irradiation of the so-called T-shaped acetylene dimer in a xenon matrix with a KrF laser (λ = 248 nm) yields - besides other products - cyclobutadiene (**2**): G. Maier, C. Lautz, *Eur. J. Org. Chem.*, **1998**, 796–776.

59. H. Hogeveen, R. F. Kingma, D. M. Kok, *J. Org. Chem.*, **1982**, *47*, 989–997 and refs. cited.

60. B. Witulski, L. Ernst, P. G. Jones, H. Hopf, *Angew. Chem.*, **1989**, *101*, 1290–1291; *Angew. Chem. Int. Ed. Engl.*, **1989**, *28*, 1279–1280.

61. H. Kimling, A. Krebs, *Angew. Chem.*, **1972**, *84*, 952–953; *Angew. Chem. Int. Ed. Engl.*, **1972**, *11*, 932–933.

62. J. Pocklinton, Ph. d. dissertation, University of Hamburg, **1979**.

63. E. Hückel, *Zeitschr. für Physik*, **1931**, *70*, 204; *cf.* J. A. Berson, *Angew. Chem.*, **1996**, *108*, 2922–2937; *Angew. Chem. Int. Ed. Engl.*, **1996**, *35*, 2750–2764.

64. This prediction is borne out as the rich chemistry of cyclobutadiene and its derivatives shows; for a summary see ref.[33]

65. B. A. Hess, Jr, P. Cársky, L. J. Schaad, *J. Am. Chem. Soc.,* **1993**, *105*, 695–701.

66. G. Maier, *Angew. Chem.*, **1988**, *100*, 317–341; *Angew. Chem. Int. Ed. Engl.*, **1988**, *27*, 309–333.

67. H. Irngartinger, M. Nixdorf, N. H. Riegler, A. Krebs, H. Kimling, J. Pocklington, G. Maier, K.-D. Malsch, K.-A. Schneider, *Chem. Ber.*, **1989**, *121*, 673–677; *cf.* H. Irngartinger, M. Nixdorf, *Chem. Ber.*, **1988**, *121*, 679–683.

68. H. Irngartinger, H. Rodewald, *Angew. Chem.*, **1974**, *86*, 783–784; *Angew. Chem. Int. Ed. Engl.*, **1974**, *13*, 740–741.

69. F. Fratev, V. Monev, R. Janoschek, *Tetrahedron*, **1982**, *38*, 2929–2932; *cf.* W. T. Borden, E. R. Davidson, P. Hart, *J. Am. Chem. Soc.,* **1978**, *100*, 388–392 and J. A. Jafri, M. D. Newton, *J. Am. Chem. Soc.*, **1978**, *100*, 5012–5017.

70. R. Janoschek, J. Kalcher, *Int. J. Quant. Chem.,* **1990**, *38*, 653–664; *cf.* R. Janoschek, *Chem. in uns. Zeit*, **1991**, *25*, 59–66.

71. B. K. Carpenter, *J. Am. Chem. Soc.,* **1983**, *105*, 1700–1701; *cf.* G. Maier, H.-O. Kalinowski, K. Euler, *Angew. Chem.*, **1982**, *94*, 706–707; *Angew. Chem. Int. Ed. Engl.*, **1982**, *21*, 693–694.

72. G. Maier, R. Wolf, H.-O. Kalinowski, R. Boese, *Chem. Ber.,* **1994**, *127*, 191–200 and lit. cited.

73. The total number of C_6H_6 isomers was first evaluated by H. Hopf, *Habilitationsschrift*, Karlsruhe, **1972**, applying graph theoretical arguments. Nowadays computer programs are available to determine the number of possible constitutional isomers of organic compounds, *e.g.* Program RAIN 2.0 by E. Fontain, Munich.

74. A. Ihde, *The Development of Modern Chemistry*, Dover Publications, paperback edn., New York, N. Y., **1984**.

75. J. R. Partington, *A Short History of Chemistry*, Dover Publications, New York, N. Y., 3rd ed. (paperback), **1989**; *cf.* J. R. Partington, *Origins and Development of Applied Chemistry*, Ayer Comp. Publ., **1975** (out of print).

76. G. Schultz, *Bericht über die Feier der Deutschen Chemischen Gesellschaft zu Ehren August Kekulés*, *Ber. Dt. Chem. Ges.*, **1890**, *23*, 1265–1312.

77. For a compact summary of the numerous methods to prepare benzene and its derivatives

see Houben-Weyl, *Methoden der Organischen Chemie*, Thieme Verlag, Stuttgart, **1981**, Vol. 5/2b, 24–495.

78. E. E. van Tamelen, S. P. Pappas, *J. Am. Chem. Soc.*, **1963**, *85*, 3297–3298. As far as the name Dewar benzene is concerned see J. Dewar, *Proc. Roy. Soc. Edinburgh*, **1866/67**, 84 and C. K. Ingold, *J. Chem Soc.*, **1922**, *121*, 1133–1143; *cf.* A. H. Schmidt, *Chem. in uns. Zeit*, **1977**, *11*, 118–128. For more recent syntheses of Dewar benzene (**35**) see: N. J. Turro, C. A. Renner, W. H. Waddell, T. J. Katz, *J. Am. Chem. Soc.*, **1976**, *98*, 4320–4322; N. J. Turro, V. Ramamurthy, *Recl. Trac. Chim. Pays-Bas*, **1974**, *98*, 173–176; M. Christl, B. Mattauch, *Chem. Ber.*, **1985**, *118*, 4203–4223.

79. E. E. van Tamelen, S. P. Pappas, *J. Am. Chem. Soc.*, **1962**, *84*, 3789–3791; Review: E. E. van Tamelen, *Acc. Chem. Res.*, **1972**, *5*, 186–192.

80. W. Schäfer, *Angew. Chem.*, **1966**, *78*, 716; *Angew. Chem. Int. Ed. Engl.*, **1966**, *5*, 669. For reviews see: W. Schäfer, H. Hellmann, *Angew. Chem.*, **1967**, *79*, 566–573; *Angew. Chem. Int. Ed. Engl.*, **1967**, *6*, 518–525; H. Hogeveen, D. M. Kok in *The Chemistry of Triple-Bonded Functional Groups* S. Patai, Z. Rappopport (*Eds.*), J. Wiley and Sons, New York, N. Y., **1983**, Suppl. C, Part 2, Chapt. 23, 981–1013. The structure of the intermediate **50** has been elucidated by X-ray crystal structure determination: C. Krüger, P. J. Roberts, Y. H. Tsay, J. B. Koster, *J. Organomet. Chem.*, **1974**, *78*, 69–74.

81. N. J. Turro, V. Ramamurthy, T. J. Katz, *Nouv. J. Chem.*, **1977**, *1*, 363–365.

82. H. G. Viehe, R. Merényi, J. F. M. Oth, J. R. Senders, P. Valange, *Angew. Chem.*, **1964**, *76*, 922; *Angew. Chem. Int. Ed. Engl.*, **1964**, *3*, 755.

83. K. E. Wilzbach, L. Kaplan, *J. Am. Chem. Soc.*, **1965**, *87*, 4004–4006.

84. I. E. Den Besten, L. Kaplan, K. E. Wilzbach, *J. Am. Chem. Soc.*, **1968**, *90*, 5868–5872. For the preparation of benzvalene derivatives carrying bulky groups, the photoisomerization of the corresponding benzene isomers is the method of choice: R. West, M. Furue, V. Mallikarjuna Rao, *Tetrahedron Lett.*, **1973**, 911–914; M. G. Barlow, R. N. Haszeldine, R. Hubbard, *J. Chem. Soc. Chem. Commun.*, **1969**, 202–203; D. M. Lemal, J. V. Staros, V. Austel, *J. Am. Chem. Soc.*, **1969**, *91*, 3373–3374.

85. K. E. Wilzbach, J. S. Ritscher, L. Kaplan, *J. Am. Chem. Soc.*, **1967**, *89*, 1031–1032; *cf.* L. Kaplan, K. E. Wilzbach, *J. Am. Chem. Soc.*, **1967**, *89*, 1030–1031.

86. T. J. Katz, E. Jang Wang, N. Acton, *J. Am. Chem. Soc.*, **1971**, *93*, 3782–3783; *cf.* T. J. Katz, R. J. Roth, N. Acton, E. J. Carnahan, *Org. Synth.*, **1973**, *53*, 157.

87. For a review on the chemistry of benzvalene (**37**) see M. Christl, *Angew. Chem.*, **1981**, *93*, 515–531; *Angew. Chem. Int. Ed. Engl.*, **1981**, *20*, 529–546.

88. M. Christl, G. Freitag, G. Brüntrup, *Chem. Ber.*, **1978**, *111*, 2307–2319.

89. G. W. Klumpp. J. J. Vrielink, *Tetrahedron Lett.*, **1972**, 539–542; *cf.* M. Christl, C. Herzog, D. Brückner, R. Lang, *Chem. Ber.*, **1986**, *119*, 141–155 and lit. cited; R. T. Taylor, L. A. Paquette, *Tetrahedron Lett.*, **1976**, 2741–2744.

90. M. Christl, R. Lang, *J. Am. Chem. Soc.*, **1982**, *104*, 4494–4496; *cf.* M. Christl, C. Herzog, P. Kemmer, *Chem. Ber.*, **1986**, *119*, 3045–3058 and M. Christl, R. Lang, C. Herzog, *Tetrahedron*, **1986**, *42*, 1585–1596.

91. M. Christl, R. Lang, M. Lechner, *Liebigs Ann. Chem.*, **1980**, 980–996; *cf.* M. Christl, M. Lechner, *Angew. Chem.*, **1975**, *87*, 815–816; *Angew. Chem. Int. Ed. Engl.*, **1975**, *14*, 765.

92. M. Christl, G. Brüntrup, *Chem. Ber.*, **1974**, *107*, 3908–3914.

93. U. Szeimies-Seebach, J. Harnisch, G. Szeimies, M. Van Meerssche, G. Germain, J.-P. Declerq, *Angew. Chem.*, **1978**, *90*, 904–905; *Angew. Chem. Int. Ed. Engl.*, **1978**, *17*, 848–849.

94. W. E. Billups, M. M. Haley, *Angew. Chem.*, **1989**, *101*, 1735–1737; *Angew. Chem. Int. Ed. Engl.*, **1989**, *28*, 1711–1713; *cf.* W. E. Billups, D. J. McCord, *Angew. Chem.*, **1994**, *106*, 1394–1406; *Angew. Chem. Int. Ed. Engl.*, **1994**, *33*, 1332–1343.

95. R. Breslow, P. Gal, H. W. Chang, L. J. Altman, *J. Am. Chem. Soc.*, **1965**, *87*, 5139–5144.

96. W. H. de Wolf, W. Stol, I. J. Landheer, F. Bickelhaupt, *Recl. Trav. Chim. Pays-Bas*, **1971**, *90*, 405–409; *cf.* W. H. de Wolf, J. W. van Straten, F. Bickelhaupt, *Tetrahedron Lett.*, **1972**, 3509–3510; W. H. de Wolf, I. J. Landheer, F. Bickelhaupt, *Tetrahedron Lett.*, **1975**, 179–182.

97. R. Weiss, St. Andrae, *Angew. Chem.*, **1973**, *85*, 145–147; *Angew. Chem. Int. Ed. Engl.*, **1973**, *12*, 150–151; R. Weiss, St. Andrae, *Angew. Chem.*, **1973**, *85*, 147–148; *Angew. Chem. Int. Ed. Engl.*, **1973**, *12*, 152–153; R. Weiss, H. Kölbl, *J. Am. Chem. Soc.*, **1975**, *97*, 3222–3224; R. Weiss, H. Kölbl, *J. Am. Chem. Soc.*, **1975**, *97*, 3224–3225.

98. J. H. Davis, K. J. Shea, R. G. Bergman, *J. Am. Chem. Soc.*, **1977**, *99*, 1499–1507; *cf.* J. H. Davies, K. J. Shea, R. G. Bergman, *Angew. Chem.*, **1976**, *88*, 254–255; *Angew. Chem. Int. Ed. Engl.*, **1976**, *15*, 232–233.

99. R. Breslow, P. Gal, *J. Am. Chem Soc.*, **1959**, *81*, 4747–4748; R. Breslow, P. Gal, H. W. Chang, L. J. Altman, *J. Am. Chem. Soc.*, **1965**, *87*, 5139–5144; G. Maier, A. Schick, I. Bauer, R. Boese, M. Nussbaumer, *Chem. Ber.*, **1992**, *125*, 2111–2117.

100. M. W. Grayston, D. M. Lemal, *J. Am. Chem. Soc.*, **1976**, *98*, 1278–1280.

101. For a summary of the photochemical behavior of benzene see J. Malkin, *Photophysical and Photochemical Properties of Aromatic Compounds*, CRC Press, Boca Rotan, FL, **1992**, 63–65; J. Kopecky, *Photochemistry: A Visual Approach*, VCH Publishers, Inc., New York, N. Y., **1992**, Chapter 7, 97–115. The preparation and interconversion of valence-bond isomers of aromatic and heteroaromatic compounds has been reviewed: Y. Kobayashi, I. Kumadaki, *Top. Curr. Chem.*, **1984**, *123*, 103–150.

102. For a review of photochemical benzene isomerizations see D. Bryce-Smith, A. Gilbert in *Rearrangements in Ground and Excited States* P. DeMayo, (*Ed.*), Academic Press, New York, N. Y., **1980**, Vol. 3, 349.

103. G. Zimmermann, M. Nüchter, H. Hopf, K. Ibrom, L. Ernst, *Liebigs Ann. Chem.*, **1996**, 1407–1412; U. Nüchter, H. Hopf, G. Zimmermann, V. Francke, *Liebigs Ann. Chem.*, **1997**, 1505–1515.

104. W. D. Huntsman, H. J. Wristers, *J. Am. Chem. Soc.*, **1963**, *85*, 3308–3309; B. A. W. Coller, M. L. Heffernan, A. J. Jones, *Austr. J. Chem.*, **1968**, *21*, 1807–1826; J. E. Kent, A. J. Jones, *Austr. J. Chem.*, **1970**, *23*, 1059–1062; H. Hopf, *Angew. Chem.*, **1970**, *82*, 703; *Angew. Chem. Int. Ed. Engl.*, **1970**, *9*, 732; Review: F. Théron, M. Verny, R. Vessière in *The Chemistry of Functional Groups, The Chemistry of the Carbon-Carbon Triple Bond* S. Patai, (*Ed.*), J. Wiley and Sons, Chichester, **1978**, 381–445.

105. G. Zimmermann, M. Nüchter, M. Remmler, H. Hopf, L. Ernst, C. Mlynek, *Chem. Ber.*, **1994**, *127*, 1747–1753.

106. I am grateful to *L. Ernst* for performing these calculations with the RAIN program.[73]

107. For a compilation of $(CH)_{2n}$-hydrocarbons with $n \geq 2$ see A. T. Balaban, *Rev. Roum. Chim.*, **1966**, *11*, 1097; *cf.* A. T. Balaban, M. Banciu, *J. Chem. Educ.*, **1984**, *61*, 766–770; L. R. Smith, *J. Chem. Educ.*, **1978**, *55*, 569–570; A. T. Balaban, M. Banciu, V. Ciorba, *Annulenes, Benz-, Hetero-, Homo-Derivatives and their Valence Isomers*, CRC Press, Boca Raton, FL, **1987** (3 volumes).

108. For a summary of rearrangements and interconversions of the $(CH)_{2n}$-systems see L. T. Scott, M. Jones, Jr, *Chem. Rev.*, **1972**, *72*, 181–202.

109. For a review of the influence of strain on the spectroscopic and chemical behavior of the $(CH)_8$ hydrocarbons see K. Hassenrück, H.-D. Martin, R. Walsh, *Chem. Rev.*, **1989**, *89*, 1125–1146.

110. H. Röttele in Houben-Weyl, *Methoden der Organischen Chemie*, Thieme Verlag, Stuttgart, **1972**, Vol. V/1d, 417–525. For a review on the conformational and π-electronic dynamics within polyolefinic [8]annulene frameworks see: L. A. Paquette in ref.[139]

111. At the BASF company at Ludwigshafen the monthly production of cyclooctatetraene amounted to 10 tons for a long time, see preface of ref.[2]

112. R. Willstätter, E. Waser, *Ber. Dtsch. Chem. Ges.*, **1911**, *44*, 3423–3445; *cf.* R. Willstätter, M. Heidelberger, *Ber. Dtsch. Chem. Ges.*, **1913**, *46*, 517–527.

113. A. C. Cope, C. G. Overberger, *J. Am. Chem. Soc.*, **1948**, *70*, 1433–1437.

114. Because of the Second World War this work was only published much later: W. Reppe, O. Schlichting, K. Klager, T. Toepel, *Liebigs Ann. Chem.*, **1948**, *560*, 1–92; *cf.* W. Reppe, O. Schlichting, H. Meister, *Liebigs Ann. Chem.*, **1948**, *560*, 93–104.

115. M. Avram, I. G. Dinulescu, E. Marica, G. Mateescu, E. Sliam, C. D. Nenitzescu, *Chem. Ber.*, **1964**, *97*, 382–389.

116. W. Merk, R. Pettit, *J. Am. Chem. Soc.*, **1967**, *89*, 4787–4788.

117. L. Watts, J. R. Fitzpatrick, R. Pettit, *J. Am. Chem. Soc.*, **1966**, *88*, 623–624.

118. A. C. Cope, W. J. Bailey, *J. Am. Chem. Soc.*, **1948**, *70*, 2305–2309; *cf.* R. E. Foster, R. S. Schreiber, *J. Am. Chem. Soc.*, **1948**, *70*, 2303–2305. A more recent and efficient alternative for the conversion of **93** to **4** has been described by W. Gansing, G. Wilke, *Angew. Chem.*, **1978**, *90*, 380; *Angew. Chem. Int. Ed. Engl.*, **1978**, *17*, 371.

119. H-D. Martin, P. Pföhler, T. Urbanek, R. Walsh, *Chem. Ber.*, **1983**, *116*, 1415–1421.

120. H.-D. Martin, T. Urbanek, P. Pföhler, R. Walsh, *J. Chem. Soc. Chem. Commun.*, **1985**, 964–965.

121. K. Hassenrück, H.-D. Martin, R. Walsh, *Chem. Ber.*, **1988**, *121*, 369–372.

122. J. Meinwald, D. Schmidt, *J. Am. Chem. Soc.*, **1969**, *91*, 5877; *cf.* J. Meinwald, H. Tsuruta, *J. Am. Chem. Soc.*, **1969**, *91*, 5877–5878.

123. H. E. Zimmerman, J. D. Robbins, J. Schantl, *J. Am. Chem. Soc.*, **1969**, *91*, 5878–5879; *cf.* H. Iwamura, H. Kihara, *Chem. Lett.*, **1973**, 71.

124. H.-D. Martin, T. Urbanek, R. Walsh, *J. Am. Chem. Soc.*, **1985**, *107*, 5532–5534 and refs. cited.

125. R. Huisgen, C. F. Mietzsch, *Angew. Chem.*, **1964**, *76*, 36–38; *Angew. Chem. Int. Ed. Engl.*, **1964**, *3*, 83–85; *cf.* E. Vogel, H. Kiefer, W. R. Roth, *Angew. Chem.*, **1964**, *76*, 432–433; *Angew. Chem. Int. Ed. Engl.*, **1964**, *3*, 442.

126. H. E. Zimmerman, R. W. Binkley, R. S. Givens, G. L. Grunewald, M. A. Sherwin, *J. Am. Chem. Soc.*, **1969**, *91*, 3316–3323; *cf.* H. E. Zimmerman, P. S. Mariano, *J. Am. Chem. Soc.*, **1969**, *91*, 1718–1727.

127. S. S. Hixson, P. S. Mariano, H. E. Zimmerman, *Chem. Rev.*, **1973**, *73*, 531–551; *cf.* D. Döpp, H. E. Zimmerman in Houben-Weyl, *Methoden der Organischen Chemie*, Thieme Verlag, Stuttgart, **1975**, Photochemie, Vol. I, 413–432; H. E. Zimmerman, D. Armesto, *Chem. Rev.*, **1996**, *96*, 3065–3112.

128. J. Meinwald, H. Tsuruta, *J. Am. Chem. Soc.*, **1970**, *92*, 2579–2580.

129. D. Dudek, K. Glänzer, J. Troe, *Ber. Bunsenges. Phys. Chem.*, **1979**, *83*, 789–797.

130. H. E. Zimmerman, H. Iwamura, *J. Am. Chem. Soc.*, **1970**, *92*, 2015–2022.

131. H.-D. Martin, T. Urbanek, R. Walsh, K. Hassenrück, B. Mayer, unpublished results.

132. E. Osawa, K. Aigami, Y. Inamoto, *J. Org. Chem.*, **1977**, *42*, 2621–2626 and refs. cited.

133. L. Cassar, P. E. Eaton, J. Halpern, *J. Am. Chem. Soc.*, **1970**, *92*, 6366–6368.

134. J. E. Byrd, L. Cassar, P. E. Eaton, J. Halpern, *J. Chem. Soc. Chem. Commun.*, **1971**, 40–41.

135. L. Cassar, P. E. Eaton, J. Halpern, *J. Am. Chem. Soc.*, **1970**, *92*, 3515–3518.

136. W. Merk, R. Pettit, *J. Am. Chem. Soc.*, **1967**, *89*, 4788–4789.

137. C. Rücker, B. Trupp, *J. Am. Chem. Soc.*, **1988**, *110*, 4828–4829.

138. For the synthesis of octabisvalene derivatives see C. Rücker, H. Prinzbach, *Angew. Chem.*, **1985**, *97*, 426–427; *Angew. Chem. Int. Ed. Engl.*, **1985**, *24*, 411–412; C. Rücker, H. Prinzbach, H. Irngartinger, R. Jahn, H. Rodewald, *Tetrahedron Lett.*, **1986**, *27*, 1565–

1568; C. Rücker, *Chem. Ber.*, **1987**, *120*, 1629–1644; B. Trupp, D.-R. Handreck, H.-P. Böhm. L. Knothe, H. Fritz, H. Prinzbach, *Chem. Ber.*, **1991**, *124*, 1757–1775.

139. For a review see L. A. Paquette in *Advances in Theoretically Interesting Molecules* R.P. Thummel, (*Ed.*), The JAI Press, Greenwich, CT, **1992**, Vol. 2, 1–77.

140. For a recent success to prepare transient planar cyclooctatetraene see P. G. Wenthold, D. A. Hrovat, W. T. Borden, W. C. Lineberger, *Science*, **1996**, *272*, 1456–1459; *cf.* W. T. Borden, H. Iwamura, J. A. Berson, *Acc. Chem. Res.*, **1994**, *27*, 109–116. The conformational behavior of cyclooctatetraene (**4**) is determined by two processes: ring inversion and double bond shifts. The dependence of both of these pathways on the type and degree of subsition has been studied in classical investigations by Paquette and co-workers; for a summary see ref.[139]

141. A. Krebs, *Angew. Chem.*, **1965**, *77*, 966; *Angew. Chem. Int. Ed. Engl.*, **1965**, *4*, 954; *cf.* A. Krebs, D. Byrd, *Liebigs Ann. Chem.*, **1967**, *707*, 66–74.

142. H. N. C. Wong, P. J. Garratt, F. Sondheimer, *J. Am. Chem. Soc.*, **1974**, *96*, 5604–5605; *cf.* H. N. C. Wong, F. Sondheimer, *Tetrahedron*, **1981**, *37*, *R. B. Woodward Memorial Issue,* 99–109.

143. R. A. G. de Graaff, S. Gorter, C. Romers, H. N. C. Wong, F. Sondheimer, *J. Chem. Soc. Perkin Trans* 2, **1981**, 478–480.

144. R. Destro, T. Pilati, M. Simonetta, *J. Am. Chem. Soc.*, **1975**, *97*, 658–659 and ref. cited; *cf.* R. Destro, T. Pilati, M. Simonetta, *Acta Cryst. Sec.*, **1977**, *B33*, 447–456.

145. H. N. C. Wong, F. Sondheimer, *Angew. Chem.*, **1976**, *88*, 126–127; *Angew. Chem. Int. Ed. Engl.*, **1976**, *15*, 117–118. For a review on planar dehydro[8]annulenes see H. N. C. Wong in *Adcances in Theoretically Interesting Molecules*, R.P. Thummel (*Ed*), JAI Press, Greenwich, CT, **1995**, Vol. 3, 109–146 and lit. mentioned.

146. D. Loos, J. Leska, *Coll. Czech. Chem. Comm.*, **1980**, *45*, 187–200.

147. L. Farnell, J. Kao, L. Radom, H. F. Schaefer III, *J. Am. Chem. Soc.*, **1981**, *103*, 2142–2151.

148. E. E. van Tamelen, T. L. Burkoth, *J. Am. Chem. Soc.*, **1967**, *89*, 151–152.

149. S. Masamune, K. Hojo, G. Bigam, D. L. Rabenstein, *J. Am. Chem. Soc.*, **1971**, *93*, 4966–4968; *cf.* S. Masamune, N. Darby, *Acc. Chem. Res.*, **1972**, *5*, 272–281.

150. If stereoisomers are included the number of possible $(CH)_{10}$ hydrocarbons increases to approximately 100. Altogether there are 369067 possible $C_{10}H_{10}$ isomers, including, for example, 1283 acyclic combinations, 44 with a ten-, 187 with a nine-, and 466 with an eight-membered ring; *cf.* G. Maier, N. H. Wiegand, St. Baum, R. Wüllner, *Chem. Ber.*, **1989**, *122*, 781–794; G. Maier, N. H. Wiegand, St. Baum, R. Wüllner, W. Mayer, R. Boese, *Chem. Ber.*, **1989**, *122*, 767–779.

151. M. Avram, E. Sliam, C. D. Nenitzescu, *Liebigs Ann Chem..*, **1960**, *636*, 184–189.

152. G. Schröder, *Angew. Chem.*, **1963**, *75*, 722.

153. R. B. Woodward, T. Fukunaga, R. C. Kelly, *J. Am. Chem. Soc.*, **1964**, *86*, 3162–3164.

154. S. Masamune, H. Cuts, M. G. Hogben, *Tetrahedron Lett.*, **1966**, 1017–1021.

155. M. Jones, Jr, L. T. Scott, *J. Am. Chem. Soc.*, **1967**, *89*, 150–151.

156. S. Masamune, C. G. Chin, K. Hojo, R. T. Seidner, *J. Am. Chem. Soc.*, **1967**, *89*, 4804–4805.

157. E. Hedaya, D. W. McNeil, P. Schissel, D. J. McAdoo, *J. Am. Chem. Soc.*, **1968**, *90*, 5284–5286.

158. K. Hojo, R. T. Seidner, S. Masamune, *J. Am. Chem. Soc.*, **1970**, *92*, 6641–6644; *cf.* T. J. Katz, J. J. Cheung, N. Acton, *J. Am. Chem. Soc.*, **1970**, *92*, 6643–6644.

159. L. A. Paquette, J. C. Stowell, *J. Am. Chem. Soc.*, **1970**, *92*, 2584–2586; *cf.* L. A. Paquette, J. C. Stowell, *J. Am. Chem. Soc.*, **1971**, *93*, 2459–2463.

160. J. S. McKennis, L. Brener, J. S. Ward, R. Pettit, *J. Am. Chem. Soc.*, **1971**, *93*, 4957–4958.

161. A. de Meijere, D. Kaufmann, O. Schallner, *Angew. Chem.*, **1971**, *83*, 404–405; *Angew. Chem. Int. Ed. Engl.*, **1971**, *10*, 417; H. Prinzbach, D. Stusche, *Helv. Chim. Acta*, **1971**, *54*, 755–759.

162. D. Bosse, A. de Meijere, *Angew. Chem.*, **1974**, *86*, 706–707; *Angew. Chem. Int. Ed. Engl.*, **1974**, *13*, 663–664.

163. P. E. Eaton, Y. S. Or, S. J. Branca, *J. Am. Chem. Soc.*, **1981**, *103*, 2134–2136.

164. H.-D. Martin, B. Mayer, M. Pütter, H. Höchstetter, *Angew. Chem.*, **1981**, *93*, 695–696; *Angew. Chem. Int. Ed. Engl.*, **1981**, *20*, 677–678.

165. G. Schröder, W. Martin, *Angew. Chem.*, **1966**, *78*, 117; *Angew. Chem. Int. Ed. Engl.*, **1966**, *5*, 130.

166. H. Röttele, W. Martin, J. F. M. Oth, G. Schröder, *Chem. Ber.*, **1969**, *102*, 3985–3995; *cf.* J. F. M. Oth, H. Röttele, G. Schröder, *Tetrahedron Lett.*, **1970**, 61–66.

167. G. Schröder, W. Martin, H. Röttele, *Angew. Chem.*, **1969**, *81*, 33; *Angew. Chem. Int. Ed. Engl.*, **1969**, *8*, 69.

168. J. F. M. Oth, J.-M. Gilles, G. Schröder, *Tetrahedron Lett.*, **1970**, 67–72; *cf.* J. F. M. Oth, J.-M. Gilles, *Tetrahedron Lett.*, **1968**, 6259–6264; J. F. M. Oth, G. Anthoine, J.-M. Gilles, *Tetrahedron Lett.*, **1968**, 6265–6270.

169. H. A. Staab, F. Graf, K. Doerner, A. Nissen, *Chem. Ber.*, **1971**, *104*, 1159–1169; *cf.* H. A. Staab, U. E. Meissner, A. Gensler, *Chem. Ber.*, **1979**, *112*, 3907–3913 for the synthesis of benzo[18]annulenes.

170. H. Brunner, K. H. Hausser, M. Rawitscher, H. A. Staab, *Tetrahedron Lett.*, **1966**, 2775–2779; *cf.* H. A. Staab, F. Graf, B. Junge, *Tetrahedron Lett.*, **1966**, 743–749.

171. F. Sondheimer, R. Wolovsky, *J. Am. Chem. Soc.*, **1962**, *84*, 260–269 and lit. cited; *cf.* K. Stöckel, F. Sondheimer, *Org. Synth.*, **1974**, *54*, 1–10. For a discussion of [18]annulene in its historical context see H. Baumann, J. F. M. Oth, *Helv. Chim. Acta*, **1982**, *65*, 1885–1893.

172. F. Sondheimer, *Proc. Roy. Soc.*, **1967**, *A297*, 173; F. Sondheimer, I. C. Calder, J. A. Elix, Y. Gaoni, P. J. Garratt, K. Grohmann, G. Di Maio, J. Mayer, M. V. Sargent, R. Wolovsky in *Aromaticity*, Special Publication, No. 21, The Chemical Society, London, **1967**, 75–107; F. Sondheimer, *Pure Appl. Chem.*, **1971**, *23*, 331–353; F. Sondheimer, *Acc. Chem. Res.*, **1972**, *5*, 81–91.

173. Besides Eglinton oxidation the Glaser coupling has been used for the ring building step. However, this alternative also yields substantuial amounts of linear coupling products: F. Sondheimer, Y. Amiel, R. Wolovsky, *J. Am. Chem. Soc.,* **1959**, *81*, 4600–4606.

174. F. Sondheimer, D. A. Ben-Efraim, Y. Gaoni, *J. Am. Chem. Soc.*, **1961**, *83*, 1682–1685.

175. H. Hopf, *Chem. Ber.*, **1971**, *104*, 3087–3095; H. Hopf, *Tetrahedron Lett.*, **1970**, 1107–1110.

176. F. Sondheimer, R. Wolovsky, Y. Amiel, *J. Am. Chem. Soc.*, **1962**, *84*, 274–284.

177. J. Bregman, F. L. Hirshfeld, D. Rabinovich, G. M. J. Schmidt, *Acta Crystallogr.*, **1965**, *19*, 227, 235.

178. F. Sondheimer, Y. Gaoni, *J. Am. Chem. Soc.*, **1960**, *82*, 5765–5766.

179. I. C. Calder, Y. Gaoni, F. Sondheimer, *J. Am. Chem. Soc.*, **1968**, *90*, 4946–4954.

180. B. W. Metcalf, F. Sondheimer, *J. Am. Chem. Soc.*, **1971**, *93*, 6675–6677 and refs. cited.

181. R. M. McQuilkin, B. W. Metcalf, F. Sondheimer, *J. Chem. Soc. Chem. Commun.*, **1971**, 338–339.

182. P. J. Garratt, *Aromaticity*, J. Wiley & Sons, New York, N. Y., **1986**, 30–34.

183. M. Nakagawa, in *The Chemistry of the Carbon-Carbon Triple Bond*, S. Patai, (*Ed.*), John Wiley & Sons, Chichester, **1978**, Chapter 15, 635–712.

184. F. Sondheimer, R. Wolowsky, P. J. Garratt, I. C. Calder, *J. Am. Chem. Soc.,* **1966**, *88*, 2610; *cf.* R. Wolovsky, F. Sondheimer, *J. Am. Chem. Soc.*, **1965**, *87*, 5720–5727.

185. K. G. Untch, D. C. Wysocki, *J. Am. Chem. Soc.*, **1967**, *89*, 6386–6387; *cf.* K. G. Untch, D. C. Wysocki, *J. Am. Chem. Soc.*, **1966**, 88, 2608–2610.

186. W. H. Okamura, F. Sondheimer, *J. Am. Chem. Soc.*, **1967**, *89*, 5991–5992.

187. J. A. Pople, K. G. Untch, *J. Am. Chem. Soc.*, **1966**, *88*, 4811–4815.

188. M. Nakagawa in *Topics in Nonbenzenoid Aromatic Chemistry*, T. Nozoe, R. Breslow, K. Hafner, S. Ito, I. Murata (*Eds.*) Hirokawa, Tokyo, **1973**, Vol. I, 191.

189. H. A. Staab, F. Graf, *Tetrahedron Lett.*, **1966**, 751–757.

190. H. A. Staab, F. Graf, *Chem. Ber.*, **1970**, *103*, 1107–1118; *cf.* H. A. Staab, R. Bader, *Chem. Ber.*, **1970**, *103*, 1157–1167.

191. J. D. Campbell, G. Eglinton, W. Henderson, R. A. Raphael, *J. Chem. Soc. Chem. Commun.*, **1966**, 87–89. This process also leads to a tetrameric hydrocarbon, an oktakis-dehydro[16]annulene. For still other syntheses of benzannelated dehydroannulenes see N. Darby, T. M. Cresp, F. Sondheimer, *J. Org. Chem.*, **1977**, *42*, 1960–1967; C. Huynh, G. Linstrumelle, *Tetrahedron Lett.*, **1988**, *44*, 6337–6374.

192. R. D. Stephens, C. E. Castro, *J. Org. Chem.*, **1963**, *28*, 3313–3315.

193. N. Darby, C. U. Kim, J. A. Salaün, K. W. Shelton, S. Takada, S. Masamune, *J. Chem. Soc. Chem. Commun.*, **1971**, 1516–1517.

194. If the two internal hydrogen atoms of the butadienyl unit of **183** are removed, a hexakis-dehydro[10]annulene of the Nakagawa type results. All attempts to generate this highly unsaturated [10]annulene lead only to azulene and naphthalene: R. W. Alder, D. T. Edley, *J. Chem. Soc (C)*, **1971**, 3485–3487. For an attempt to synthesize a dibenzo derivative of this hydrocarbon see H. W. Whilock, Jr, J. K. Reed, *J. Org. Chem.*, **1969**, *34*, 874–878. For the synthesis of an isomer of **183**, 1,6-didehydro[10]annulene see A. G. Myers, N. S. Finney, *J. Am. Chem. Soc.*, **1992**, *114*, 10986–10987.

195. For recent reviews see: K. C. Nicolaou, A. L. Smith in *Modern Acetylene Chemistry*, P. J. Stang, F. Diederich, (*Eds.*), VCH Verlagsgesellschaft, Weinheim, **1995**, 203–283; K. K. Wang, *Chem. Rev.*, **1996**, *96*, 207–222.

196. C. F. Wilcox, Jr, K. A. Weber, *J. Org. Chem.*, **1986**, *51*, 1088–1094.

197. E. Vogel, H. D. Roth, *Angew. Chem.*, **1964**, *76*, 145; *Angew. Chem. Int. Ed. Engl.*, **1964**, *3*, 228; *cf.* E. Vogel, W. Klug, A. Breuer, *Org. Synth.*, **1974**, *54*, 11–18.

198. E. Müller, H. Fricke, H. Kessler, *Tetrahedron Lett.*, **1964**, 1525–1530.

199. For leading references to extensive studies on the NMR properties of **182** and its derivatives see H. C. Dorn, C. S. Yannoni, H.-H. Limbach, E. Vogel, *J. Phys. Chem.*, **1994**, *98*, 11628–11629. For the X-ray structure of **182** see R. Bianchi, T. Pilati, M. Simonetta, *Acta Cryst.*, **1980**, *B36*, 3146–3148. For reviews on the generation of long carbon-carbon bands in strained molecules see E. Osawa, K. Kanematsu in *Molecular Structure and Energetics*, J. F. Liebman, A. Greenberg (*Eds.*), VCH Publishers, Deerfield Beach, **1986**, vol. 3, Chapter 7, 329–369; K. P. Lipkowitz, M.A. Peterson, *Chem. Rev.*, **1993**, *93*, 2463–2486; F. Toda, K. Tanaka, Z. Stein, J. Goldberg, *Acta Cryst.*, **1996**, *C52*, 177–180; G. Kaupp, J. Boy, *Angew. Chem.*, **1997**, *109*, 48–50; *Angew. Chem. Int. Ed. Engl.*, **1977**, *36*, 48–49.

200. E. Vogel, M. Biskup, W. Prater, W. A. Böll, *Angew. Chem.*, **1964**, *76*, 785; *Angew. Chem. Int. Ed. Engl.*, **1964**, *3*, 642; *cf.* F. Sondheimer, A. Shani, *J. Am. Chem. Soc.*, **1964**, *86*, 3168–3169.

201. E. Vogel, W. Pretzer, W. A. Böll, *Tetrahedron Lett.*, **1965**, 3613–3617. How 1,6-imino[10]annulene and its 14π-homolog, *syn*-1,6:8,13-diimino[14]annulene served as stepping stones to novel porphyrins is described by E. Vogel in *Pure Appl. Chem.*, **1993**, *65*, 143–152.

202. For a summary of work on annulenes beyond 1,6-methano[10]annulene (**182**) see E. Vogel, *Pure Appl. Chem.*, **1971**, *28*, 355–377; E. Vogel, *Israel J. Chem.*, **1980**, *20*, 215–224;

E. Vogel, *Pure Appl. Chem.*, **1982**, *54*, 1015–1039; E. Vogel in *Current Trends in Organic Synthesis*, H. Nozaki, (*Ed.*), Pergamon Press, Oxford, **1983**, 379–400; *cf.* W. Wagemann, M. Iyoda, H. M. Deger, J. Sombroek, E. Vogel, *Angew. Chem.*, **1978**, *90*, 988–990; *Angew. Chem. Int. Ed. Engl.*, **1978**, *17*, 956–957.

203. E. Vogel, U. Haberland, H. Günther, *Angew. Chem.*, **1970**, *82*, 510–512; *Angew. Chem. Int. Ed. Engl.*, **1970**, *9*, 513–514; *cf.* C. M. Gramaccioli, A. S. Mimun, A. Mugnoli, M. Simonetta, *J. Am. Chem. Soc.*, **1973**, *95*, 3148–3154.

204. E. Vogel, H. Reel, *J. Am. Chem. Soc.*, **1972**, *94*, 4388–4389; *cf.* E. Vogel, A. Vogel, H.-K. Kübbeler, W. Sturm, *Angew. Chem.*, **1970**, *82*, 512–513; *Angew. Chem. Int. Ed. Engl.*, **1970**, *9*, 515–516 and E. Vogel., W. Sturm, H.-D. Cremer, *Angew. Chem.*, **1970**, *82*, 513–514; *Angew. Chem. Int. Ed. Engl.,* **1970**, *9*, 516.

205. E. Vogel, H. M. Deger, J. Sombroek, J. Palm, A. Wagner, J. Lex, *Angew. Chem.*, **1980**, *92*, 43–45; *Angew. Chem. Int. Ed. Engl.*, **1980**, *19*, 41–43.

206. E. Vogel., E. Schmidbaucr, H.-J. Altenbach, *Angew. Chem.*, **1974**, *86*, 818–819; *Angew. Chem. Int. Ed. Engl.*, **1974**, *13*, 736–737.

207. E. Vogel, J. Sombroek, W. Wagemann, *Angew. Chem.*, **1975**, *87*, 591–592; *Angew. Chem. Int. Ed. Engl.*, **1975**, *14*, 564–565.

208. V. Boekelheide, J. B. Phillips, *J. Am. Chem. Soc.*, **1963**, *85*, 1545–1546. For a comprehensive review on the chemistry of dihydropyrenes see R. H. Mitchell in *Advances in Theoretically Interesting Molecules*, R. P. Thummel, (*Ed.*), JAI Press, Inc. Greenwich, CT, Vol. I, **1989**, 35–199; *cf.* Y.-H. Lai, P. Chen, T.-G. Peck, *Pure Appl. Chem.,* **1993**, *65*, 81–87.

209. V. Boekelheide, J. B. Phillips, *J. Am. Chem. Soc.,* **1967**, *89*, 1695–1704.

210. A. W. Hanson, *Acta Cryst.*, **1965**, *18*, 599; *cf.* A. W. Hanson, *Acta Cryst.*, **1967**, *23*, 476. For more recent studies on the X-ray structure of dimethyldihydropyrene see R. V. Williams, W. D. Edwards, A. Vij, R. W. Tolbert, R. H. Mitchell, *J. Org. Chem.*, **1998**, *63*, 3125–3127; *cf.* R. H. Mitchell, V. S. Iyer, N. Khalifa, R. Mahadevan, S. Venugopalan, S. A. Weerawarna, P. Zhou, *J. Am. Chem. Soc.*, **1995**, *117*, 1514–1532.

211. J. B. Phillips, R. J. Molyneux, E. Sturm, V. Boekelheide, *J. Am. Chem. Soc.*, **1967**, *89*, 1704–1709.

212. C. E. Ramey, V. Boekelheide, *J. Am. Chem. Soc.*, **1970**, *92*, 3681–3684.

213. V. Boekelheide, T. Miyasaka, *J. Am. Chem. Soc.*, **1967**, *89*, 1709–1714.

214. V. Boekelheide, T. A. Hylton, *J. Am. Chem. Soc.*, **1970**, *92*, 3669–3675; *cf.* H. Blaschke, C. E. Ramey, I. Calder, V. Boekelheide, *J. Am. Chem. Soc.*, **1970**, *92*, 3675–3681.

215. R. H. Mitchell, V. Boekelheide, *J. Am. Chem. Soc.*, **1970**, *92*, 3510–3512; *cf.* V. Boekelheide, R. H. Mitchell, *J. Am. Chem. Soc.*, **1974**, *96*, 1547–1557.

216. 2,11-Dithia[3.3]metacyclophanes were first described by T. Sato, M. Wakabayashi, M. Kainosho, K. Hata, *Tetrahedron Lett.*, **1968**, 4185–4189 and F. Vögtle, L. Schunder, *Chem. Ber.*, **1969**, *102*, 2677–2683.

217. S. Masamune, D. W. Brooks, K. Morio, R. L. Sobczak, *J. Am. Chem. Soc.*, **1976**, *98*, 8277–8279; *cf.* S. Masamune, D. W. Brooks, *Tetrahedron Lett.,* **1977**, 3239–3240.

218. L. T. Scott, W. R. Brunsvold, *J. Am. Chem. Soc.*, **1978**, *100*, 4320–4321; *cf.* L. T. Scott, W. R. Brunsvold, M. A. Kirms, I. Erden, *J. Am. Chem. Soc.,* **1981**, *103*, 5216–5220.

219. T. L. Gilchrist, C. W. Rees, D. Tuddenham, D. J. Williams, *J. Chem. Soc. Chem. Commun.*, **1980**, 691–692; *cf.* Z. Lidert, C. W. Rees, *J. Chem. Soc. Chem. Commun.*, **1982**, 499–500.

220. E. Vogel, H. Königshafen, K. Müllen, J. F. M. Oth, *Angew. Chem.*, **1974**, *86*, 229–231; *Angew. Chem. Int. Ed. Engl.*, **1974**, *13*, 281–283; *cf.* A. Mugnoli, M. Simonetta, *Acta Cryst.*, **1975**, *A31*, 121.

221. E. Vogel, M. Mann, Y. Sakata, K. Müllen, J. F. M. Oth, *Angew. Chem.*, **1974**, *86*, 231–232; *Angew. Chem. Int. Ed. Engl.*, **1974**, *13*, 283–284.

2; *Angew. Chem. Int. Ed. Engl.*, **1974**, *13*, 283–284.

222. W. Huber, J. Lex, T. Meul, K. Müllen, *Angew. Chem.*, **1981**, *93*, 401–402; *Angew. Chem. Int. Ed. Engl.,* **1981**, *20*, 391–392.

223. D. Tanner, O. Wennerström, E. Vogel, *Tetrahedron Lett.*, **1982**, *23*, 1221–1224; *cf.* E. Vogel, U. Kürschner, H. Schmickler, J. Lex, O. Wennerström, D. Tanner, U. Norinder, C. Krüger, *Tetrahedron Lett.*, **1985**, *26*, 3087–3090.

224. K. Yamamoto, S. Kuroda, Y. Nozawa, S. Fujita, J. Ojima, *J. Chem. Soc. Chem. Commun.*, **1987**, 199–200; for the preparation of the corresponding 1,6-methano tetradehydro annulenes see J. Ojima, S. Fujita, M. Masamoto, E. Ejiri, T. Kato, S. Kuroda, Y. Nozawa, H. Tatemitsu, *J. Chem. Soc. Chem. Commun.*, **1987**, 534–536.

11 Cross-Conjugated and Related Hydrocarbons

Besides linear and cyclic conjugation, which were covered in Chapters 7 to 10, there is a third major type of interaction between unsaturated groups in organic molecules—cross-conjugation. Cross-conjugated compounds have been defined as systems 'possessing three unsaturated groups, two of which, although conjugated to a third unsaturated center, are not conjugated to each other'.[1] If the word 'conjugated' is employed in the classical sense of describing a system of alternating single and double bonds, the two centers separately conjugated to the third are themselves separated by two single bonds. The simplest representative of this bonding situation is hence provided by 3-methylene-1,4-pentadiene (**1**) (Scheme 1).

Each unsaturated center possesses $2n$ π-electrons where n is an integer. The lone pair of electrons of a singly bonded nitrogen or oxygen atom is considered an unsaturated center isoelectronic with a vinyl group. According to this definition compounds such as divinylether (**2**) or benzamide (**3**) can also be regarded as cross-conjugated systems. Indeed, the first organic compound to be prepared by total synthesis, urea (**4**), is a legitimate example of a cross-conjugated molecule. Returning to hydrocarbons, the replacement of one or both of the vinyl substituents in **1** by triple bonds leads to the simplest cross-conjugated alkynes—3-methylene-1-penten-4-yne (**5**, 2-ethynyl-1,3-butadiene) and 3-methylene-1,4-pentadiyne (**6**). The term cross-conjugtion is, however, commonly used in a broader sense than defined above. For example, quinones and derived systems are usually also referred to as cross-conjugated. The parent molecule of these compounds is p-xylylene (**7**) or p-quinodimethane, the all-carbon analog of p-benzoquinone. Finally, molecules of the type represented by the general structure **8**, in which the sp^2-hybridized carbon centers can be charged, can carry unpaired electrons, can be part of a longer π-system, or can simply be short-circuited have often been called cross-conjugated (or having Y-conjugation), because two types of conjugated subsystems share a common unsaturated unit. In this sense even a molecule which contains only two double bonds such as triafulvene (**9**) can—and has!—also been called a cross-conjugated hydrocarbon, and so are all compounds derived from it by the vinylog principle, e.g. pentafulvene (**10**) would be the next cross-conjugated hydrocarbon in the series. Obviously, pentafulvene itself can occur as a subsystem in more complex molecules, 1,2-dihydropentalene (**11**), pentalene (**12**) and azulene (**13**) being representative examples. In pentafulvalene (**14**) two fulvene molecules formally share the central double bond. Note that molecules such as **12** or **13** can also be regarded as [8]- and [10]annulenes (see Sections 10.3 and 10.4) with 'zero-bridges' ('short-circuited annulenes'). Formal consideraton of this type can sometimes be quite useful because

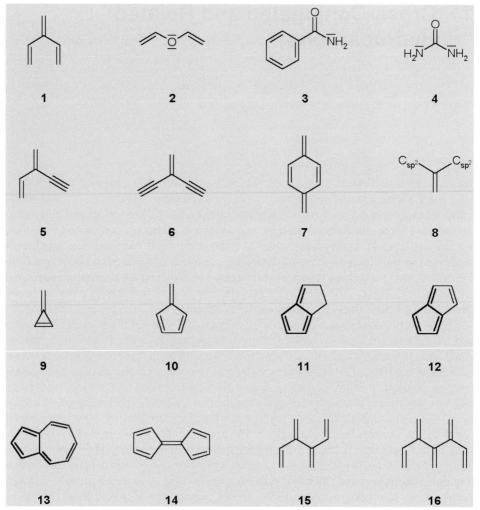

Scheme 1. A representative selection of cross-conjugated molecules.

they enable the 'transfer' of synthesis design and planning from one area of hydrocarbon chemistry to another.

Besides the parent systems such as **1**, **5** and **6** in this chapter we will also describe the preparation of molecules such as **9–14**. In the acyclic series the formulas shown in Scheme 1 provide no information about the actual conformation of the different molecules. These are always drawn so as to guarantee maximum overlap (maximum cross-conjugation). This might not be so for an individual hydrocarbon, and indeed 3-methylene-1,4-pentadiene (**1**) is not at all the planar molecule that might be expected from casual inspection of formula **1** (see below). Extensive MO calculations have been performed for cross-conjugated systems; for the prototype hydrocarbon **1** they suggest that the double-bond character for the central double bond (C3–C6) is less than that of the double bonds in 1,3-butadiene, whereas the C1–C2 double bond of **1** has more double-bond character than that of 1,3-butadiene. As a phenomenon cross-conjugation occurs widely in organic chemistry, as illustrated by numerous natural and non-natural quinones, many dyestuffs (including indigo and the triphenylmethane dyes), fused-ring aromatics, and cross-conjugated polymers. To understand these rather complex π-systems, thorough knowledge of the underlying hydrocarbon parent systems cannot be dispensed with.

11.1 The dendralenes

The simplest acyclic cross-conjugated hydrocarbon, 3-methylene-1,4-pentadiene (**1**) is the parent compound of a class of polyolefins known as *dendralenes*, from the Greek word for tree.[2] In trivial nomenclature **1** is [3]dendralene, and the higher members of the vinylogous series are obtained by replacing non-terminal hydrogen atoms by vinyl substituents thus generating additional cross-conjugated structures— [4]dendralene (**15**, 3,4-dimethylene-1,5-hexadiene), [5]dendralene (**16**, 3,4,5-trimethylene-1,6-heptadiene) *etc.* (Scheme 1). As we shall see, no acyclic dendralenes beyond **15** are known; higher dendralenes have, however, been incorporated into cyclic structures.[3]

Although polymethyl derivatives of [3]dendralene had already been prepared around the turn of the 20th century,[4] the parent hydrocarbon **1** was only obtained 50 years later by Blomquist who subjected the diacetate **17** to ester pyrolysis (Scheme 2).[5]

Almost simultaneously Bailey reported that the triacetate **18** can also be used as the starting material (flash vacuum pyrolysis at 540 °C, 43% yield),[6] and for a long time this approach has been the best practical way of preparing **1**.[7] Instead of the ester pyrolysis Hofmann elimination of **19** can also be employed to prepare **1**.[8] The Lindlar hydrogenation of 2-ethynyl-1,3-butadiene (**5**) is interesting because it involves as the starting material the simplest cross-conjugated acetylene,[9, 10] itself obtained by dehydration of the propargyl alcohol **21** over molecular sieves 5 Å at 300 °C.

Although the electrocyclic ring-opening of 1-vinyl-cyclobutene (**20**) took place quantitatively, this route to **1** suffers from the fact that the precursor hydrocarbon **20** had to be prepared by a multi-step sequence from methylenecyclobutane (**26**, not a commercial product), its Prins reaction with formaldehyde to **23**, and ring-opening and dehydration of the latter.[11] The most recent and efficient route (overall yield 20%) to [3]dendralene (**1**) starts from the butadiene-synthon 2,5-dihydro-thiophen-1,1-dioxide (**29**, 3-sulfolene), which was converted by the standard steps **29**→**28**→**27**→**24** into the acetate **24**. Although this intermediate could in principle fragment directly to **1** under vapor-phase pyrolysis conditions, it first eliminated acetic acid on heating to yield the non-conjugated sulfone **25**. After this had been isomerized into **22** by treatment with 1,8-diazabicyclo[5.4.0]undec-7-ene (DBU), cheletropic ring-opening ultimately provided **1**.[12] There are preparative and structural reasons for interest in this cross-conjugated triene. For synthetic applications

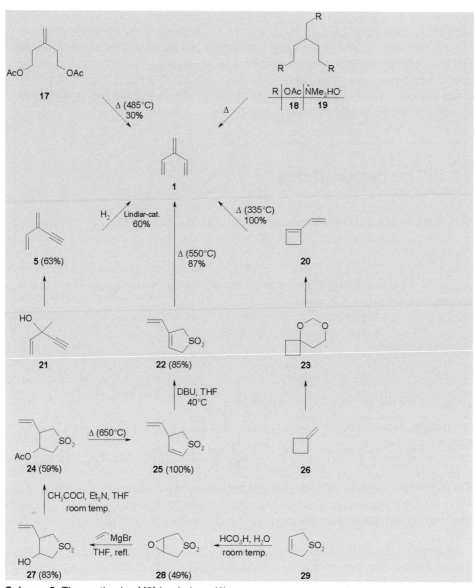

Scheme 2. The synthesis of [3]dendralene (**1**).

[4+2]cycloadditions to **1** (and its vinylogs, see below) are particularly valuable, because these molecules can function as tandem-annelating reagents[12] or dienes in diene-transmissive Diels–Alder additions (Scheme 3).[13]

   Thus cycloaddition of a double-bond dienophile **30** first produces the 1:1 adduct **31**. Although this reaction 'destroys' one conjugated diene unit, it simultaneously 'generates' a new one, because the newly formed double bond is in conjugation with the so far 'unused' vinyl substituent of **1**. Consequently a second Diels–Alder addition

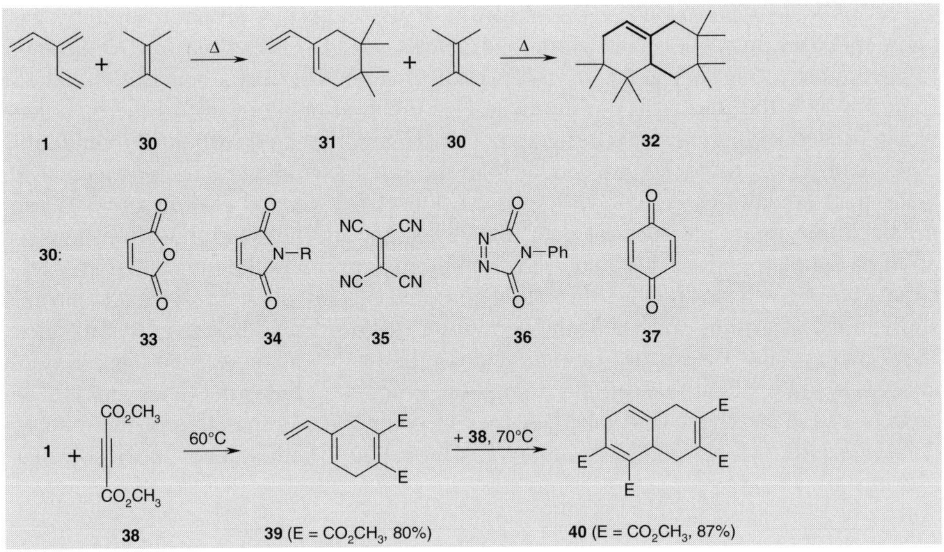

Scheme 3. [3]Dendralene (**1**) as a tandem-annelating reagent.

can take place, and it yields the 2:1 adduct **32**, in which the branched central core of the product has been provided by the dendralene. Numerous dienophiles **30** have been added to **1**,[5, 6, 12] ranging from maleic anhydride (**33**) through **34**, **35**, and **36** to p-benzoquinone (**37**). Since the addition process can be interrupted at the 1:1 stage, mixed additions, using different dienophiles **30**, are possible. With a typical triple-bond dienophile such as dimethyl acetylenedicarboxylate (**38**) mono, **39**, and bis adducts, **40**, can be prepared, useful precursors for aromatic compounds carrying versatile functional groups in predetermined positions.[12]

Obviously, every additional vinyl group in the starting dendralene enables the construction (annelation) of an additional six-membered ring.

That **1** must possess a unique, non-planar structure is suggested by its electronic spectrum[5, 6]—with an absorption maximum of 224 nm (log ε = 4.14) it hardly differs from the UV spectrum of 1,3-butadiene (λ_{max} = 217 nm, log ε = 4.32).[14] Evidently only two of the three double bonds participate in conjugation, thus excluding any of the planar structures **41**–**43** as a possibilty for its actual conformation. In fact, as shown in Scheme 4 all these alternatives involve severe intramolecular hydrogen-hydrogen repulsions, which can only be relieved by rotating one of the vinyl groups out of the plane passing through the other two double bonds.

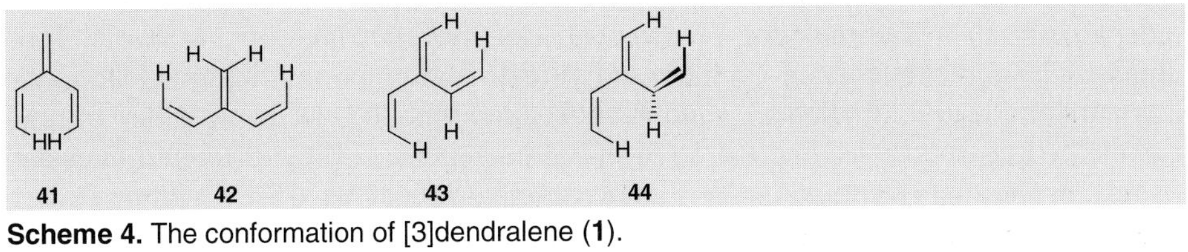

Scheme 4. The conformation of [3]dendralene (**1**).

As proven by an extensive investigation of the molecular structure of **1** by gas-phase electron diffraction, and vibrational, NMR, and UV spectroscopy,[15] the most stable conformation of **1** is the *anti-skew* conformation **44**, with a bent *anti*-butadiene fragment with the third vinyl group making a dihedral angle of 40° to the butadiene plane, rotated from the *cis* conformation. The cross-conjugated carbon–carbon double bond is observed to be 7.5 pm longer than the other C=C-bonds, in agreement with theoretical predictions (see above). The existence of a second conformation is suggested from vibrational spectral data. Planarization, and hence conjugation between all three double bonds of **1**, can be achieved by incorporating the molecule into cyclic structures. As will be shown below, this enforced planarization causes bathochromic shifts in the electronic spectra of the resulting [3]dendralene derivatives. Many alkyl, vinyl, and aryl derivatives of [3]dendralene are known.[3] As far as their syntheses are concerned they often rely on β-elimination processes, but sometimes unique approaches have also been developed. A case in point is the preparation of the tetraaryl derivatives **48**, synthesized by Staudinger, who called [3]dendralenes 'open fulvenes' (Scheme 5).[16]

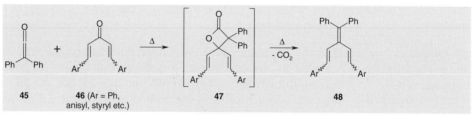

45   **46** (Ar = Ph,   **47**   **48**
             anisyl, styryl etc.)

Scheme 5. Dendralenes from cross-conjugated precursors.

Addition of diphenylketene (**45**) to the dibenzylideneacetone derivatives **46**—themselves cross-conjugated systems!—initially yielded the β-lactones **47**, which were unstable under the reaction conditions and eliminated carbon dioxide. The scope and limitations of this remarkable route to [3]dendralenes remain to be established.

The next higher vinylog, [4]dendralene (**15**) has been obtained by a variety of transformations; because none of these is preparatively satisfactory, not much is known about the chemical behavior of this cross-conjugated tetraene.

As shown in Scheme 6, **15** has been prepared by oxidative dimerization of the Grignard reagent prepared from 1-chloro-2,3-butadiene (**49**),[17] Hofmann elimination of the tetra quaternary salt **50a**[18]—this being the first synthesis of the hydrocarbon—and ester pyrolysis of the tetraacetate **50b**.[19]

The simplicity of these reactions does not, however, compensate for the preparative effort required to prepare the penultimate precursors. In the pyrolysis of the monoacetate **51** a β-elimination precedes an electrocyclic ring-opening step.[20] The diradical 2,3-dimethylene-1,4-cyclohexandiyl (**52**) is an important intermediate in the thermal dimerization of allenes[21] and in connection with the electronic nature of non-Kekulé hydrocarbons (see Section 16.2). It has been generated by different routes and readily opens to [4]dendralene by homolytic cleavage of its C5-C6 carbon-carbon bond. The precursors for **52** include its bisallene isomer **53** (see Section 9.1),[22, 23]

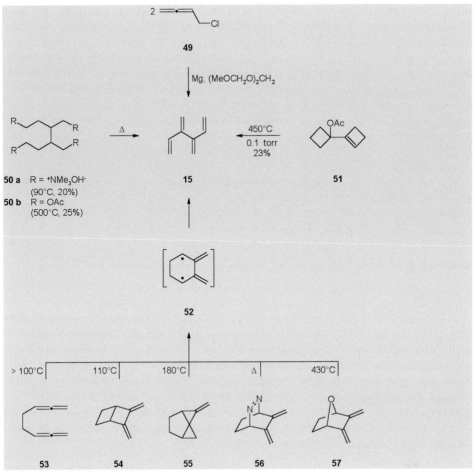

Scheme 6. The synthesis of [4]dendralene (**15**).

the bi- and tricyclic hydrocarbons **54**[24] and **55**,[25] the azo compound **56**,[24] and the cyclic ether **57**.[23] According to its UV spectrum (λ_{max} = 216.5 nm, log ε = 4.56),[18, 19, 22] which again is remarkably similar to that of 1,3-butadiene, this dendralene also exists in a non-planar conformation.[26]

The parent hydrocarbons [3]- (**1**) and [4]dendralene (**15**) have been incorporated in numerous ways into more or less complex molecular frameworks, including those of many dyestuffs.[3] Because the amount of rotational freedom are lower in these molecules than in the acyclic models, increased planarity and hence electronic interaction result. Among the fixed [3]dendralenes, 3-methylene-1,4-cyclohexadiene (**60**) deserves special note. The first derivatives of this hydrocarbon, which is also known as *p*-isotoluene, carrying 'protecting' *gem* dimethyl groups in the 6 position, were prepared by v. Auwers and co-workers at the beginning of the 20th century.[27, 28] The parent system **60** was synthesized sixty years later by Plieninger et al. from 1,4-dihydrobenzoic acid chloride (**58**) which was first converted to the amide **59**. Hydride reduction subsequently afforded a tertiary amine, which was subjected to a Hofmann elimination to yield the hydrocarbon in small yield (Scheme 7).[29]

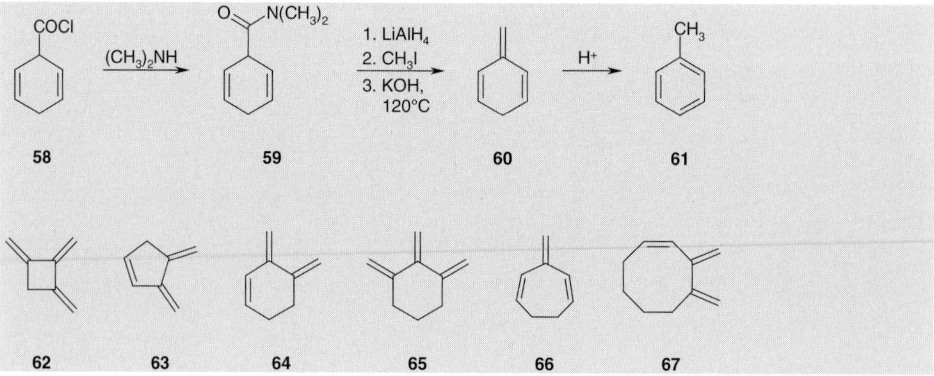

Scheme 7. A selection of [3]dendralenes sterically fixed by incorporation into a ring system.

With λ_{max} of 242 nm (shoulder at 247 nm) **60** showed the expected bathochromic shift; this is particularly obvious when compared with the model compound 3-methylene-cyclohexene which absorbs at 231 nm.[29] Traces of acid converted **60** into its stabilomer toluene (**61**). This process, which is known as the semibenzene rearrangement, can also be induced thermally as was also shown with numerous derivatives of the hydrocarbon.[30, 31] According to MNDO calculations *p*-isotoluene is a planar compound,[32] a prediction later confirmed by microwave spectroscopy.[33] An interesting application of the thermal **60→61**-isomerization will presented later in our discusssion on [*n*]paracyclophanes (see Section 12.3).

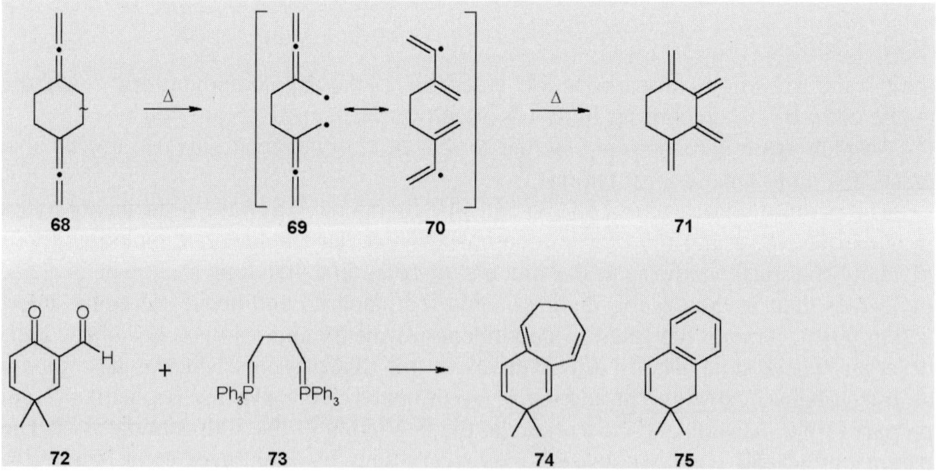

Scheme 8. A small selection of sterically fixed [4]dendralenes.

Many other planarized [3]dendralenes have been reported;[3] a small selection, hydrocarbons **62** to **67** is included in Scheme 7.[34] Whenever the electronic spectra of these compounds have been recorded they have contained absorption bands at longer wavelengths than those of the parent hydrocarbon **1**.

The high tendency of allenes to rearrange to conjugated dienes (see Section 9.1) has been exploited for the preparation of the sterically fixed [4]dendralene 1,2,3,4-tetramethylenecyclohexane (**71**) (Scheme 8),[35] a hydrocarbon showing strong resemblance to the [*n*]radialenes discussed below in Section 11.4.

Gas-phase pyrolysis (500 °C, 0.1 torr) of the bis-allenic substrate **68** cleaved one of the cyclohexane ring bonds to yield the diradical **69**, which—by way of the resonance structure **70**—recyclized to **71** with the surprisingly high yield of *ca* 80%. The UV spectrum of **71** in cyclohexane shows several absorption maxima between 228 and 280 nm.

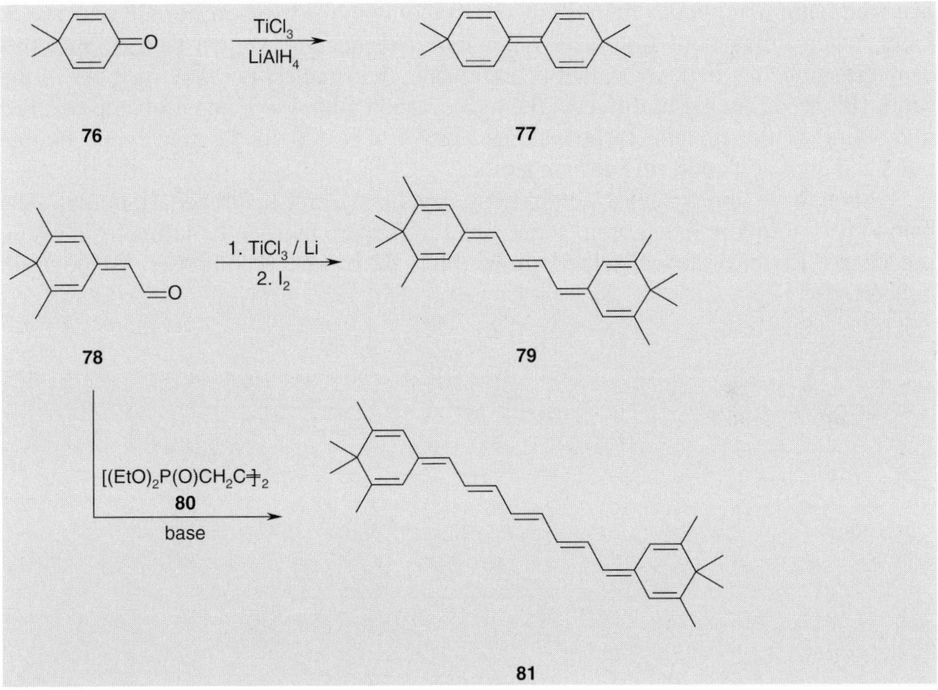

Scheme 9. The synthesis of extended dendralenes.

Another planarized [4]dendralene, the bicyclic hydrocarbon **74**, was accessible from 2-formyl-4,4-dimethylcyclohexadienone (**72**) by Wittig reaction with the double ylid **73**. Employing the lower homolog of **73** led to the aromatic product **75**—evidently the intermediate [4]dendralene derivative with its (unprotected) isotoluene subunit underwent the semibenzene rearrangement (see above) under the reaction conditions.[36] According to its spectroscopic data **74** is a coplanar [4]dendralene (λ_{max} = 266 nm, log ε = 4.06, cyclohexane). That the other important olefination process, the McMurry coupling, has also been used for the synthesis of dendralenes should come as no surprise. As illustrated in Scheme 9, the doubly cross-conjugated hydrocarbons **77**[37] and **79**[38] have been obtained from the cross-conjugated carbonyl precursors **76** and **78**, respectively, by this reductive dimerization. For the preparation of still higher vinylogs of this type, such as **81**, double Wittig–Horner condensation—with, *e.g.*, **80**–has been employed again.[38]

11.2 The fulvenes

Fulvenes, monocyclic hydrocarbons of the general structure **82**, formally consist of an odd-membered ring in which all carbon atoms are sp^2-hybridized, and a semicyclic double bond. Originally the name, proposed by Thiele (see below), which is derived from the Latin word *fulvus* for yellow, referred only to the hydrocarbon **82** with $n = 2$. Later, when fulvenes of ring sizes other than five became known (see below), this deep orange compound was called pentafulvene, the smallest possible member of the series (**82**, $n = 1$) being triafulvene, the hydrocarbon with $n = 3$, heptafulvene *etc*. The numbering of the fulvene carbon atoms always begins α to the semicyclic double bond and ends at its non-ring carbon atom.

   Although fulvene syntheses employing acyclic starting materials are known (see below) it is normally much more economical to prepare these polyolefins from cyclic precursors. Retrosynthetically the hydrocarbons **82** can be decomposed as shown in Scheme 10.

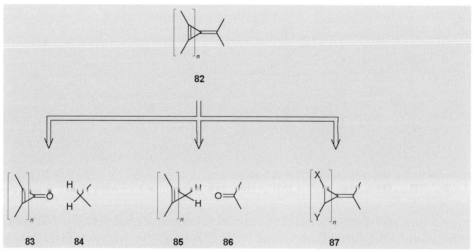

Scheme 10. Retrosynthesis of fulvenes.

   Beginning with the semicyclic double bond, this can be cleaved either to yield an unsaturated cyclic ketone **83** and a methylene component **84**, or a cyclic oligoolefin

85 and a carbonyl fragment **86**. Which combination is selected for a particular synthesis depends on the ring size of the cyclic precursor molecules. When n is even in **83**, the ketone has strong anti-aromatic character because of the polarity of the C–O double bond. Hence this condensation route is recommended for $n = 1$ (**83** = cyclopropenone) and $n = 3$ (**83** = tropone, cycloheptatrienone). On the other hand, for the **85/86**-combination the situation is reversed—base-induced condensation will now work well with $n = 2$ because the aromatic cyclopentadienyl anion is involved, whereas both the cyclopropenyl anion and the cycloheptatrienyl anion are antiaromatic $4n$ π-systems. Rather than generating the semicyclic double bond by an intermolecular reaction, it can also be present in the starting material already, as shown in **87**. In this case the ring double bonds must be generated by elimination reactions. Clearly, this approach will be problematic for the higher fulvenes, because multiple eliminations would have to be performed. It is, however, a viable possibility for $n = 1$ in **87**, as we shall see below, and it could also be useful if the carbocyclic ring already contains C–C double bonds in addition to the groups to be eliminated. The polarity of the synthetic approaches **83+84** and **85+86**, respectively, carries over to the product fulvenes, which—in contrast to many other unsaturated hydrocarbons—have an inherent dipole moment. Again depending on the ring size the negative end of the dipole will be at the non-ring carbon atom (for $n = 1$ and 3 in **82**) or at the ring (for $n = 2$ in **82**). Because of their polar nature most fulvenes react readily both with nucleophiles and electrophiles. Whereas an electrophile will attack the non-ring carbon atom of triafulvene, a nucleophile will prefer the endocyclic double bond of this smallest fulvene. In both cases, however, resonance stabilized intermediates result. For the next higher vinylog, pentafulvene, the situation is reversed—nucleophilic attack at C6 generates an (aromatic) cyclopentadienyl anion, electrophilic attack at C1 produces a 1,3-pentadienyl cation.

Condensation reactions with cyclopropenones led to the first stable triafulvene derivatives, which were usually fully substituted or functionalized. For example, alkyl- or aryl-substituted cyclopropenones **88** react with active methylene compounds such as malononitrile (**89**) under acidic conditions in a Knoevenagel-type condensation to the triafulvenes **90**. Occasionaly β-alanine is added as a catalyst[39, 40] (Scheme 11).

In a related, but more general approach, the cyclopropenones **88** were first O-alkylated or acylated to give alkoxy-, **91**, or acyloxycyclopropenylium salts. When these were reacted with active methylene components **92** in the presence of a base such as triethylamine, the triafulvenes **93** resulted, normally in good yield.[41] Applying this approach to the parent compound cyclopropenone (**94**) met with failure, however. Low-temperature acylation with acetyl tetrafluoroborate (**95**) produced the desired cyclopropenylium salt **96**, but this was unstable even at -80 °C and decomposed to the BF_3 complex of the starting ketone, **97**, and acetylfluoride (**98**).[42]

Success was finally achieved in 1984 by three different groups (Scheme 12). Whereas Billups and co–workers first prepared 1-chloro-2-methylene-cyclopropane (**100**) by chlorocarbene addition to allene (**99**), and then dehydrochlorinated this adduct by a vacuum-gas-solid reaction over potassium *tert*-butoxide impregnated Chromsorb W,[43] Staley et al. prepared the bromide **101** from **99** and then performed the dehydrobromination under similar conditions.[44]

The third route, developed by Maier and co-workers[45] subjected the diazoketone

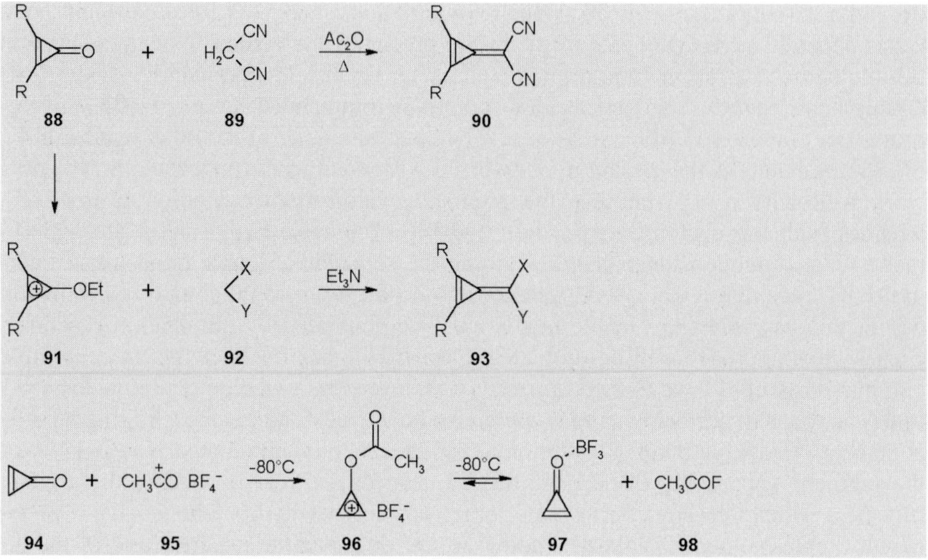

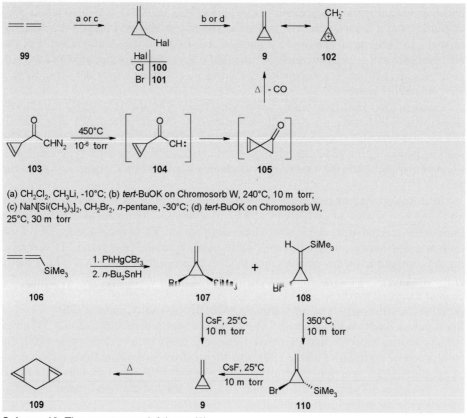

Scheme 11. The preparation of triafulvenes by condensation reactions.

(a) CH₂Cl₂, CH₃Li, -10°C; (b) *tert*-BuOK on Chromosorb W, 240°C, 10 m torr;
(c) NaN[Si(CH₃)₃]₂, CH₂Br₂, *n*-pentane, -30°C; (d) *tert*-BuOK on Chromosorb W,
25°C, 30 m torr

Scheme 12. Three routes to triafulvene (**9**).

103 to flash vacuum pyrolysis and condensed the fragmentation and rearrangement products on a window cooled to 10 K to enable spectroscopic investigation. Besides other C$_4$H$_4$ hydrocarbons—including cyclobutadiene (see Section 10.1)—triafulvene (**9**) could be identified by its IR spectrum. Presumably the cross-conjugated hydrocarbon was formed by α-insertion of the initially generated ketocarbene **104** to the spiro compound **105**, which subsequently eliminated carbon monoxide.

The unsubstituted triafulvene is a highly reactive hydrocarbon which can be stored and studied only at *ca* –100 °C. Its ^{1}H NMR spectrum, measured in [D$_8$]-THF at –96 °C contains two triplets at δ = 3.47 and 8.61, indicating that the polar structure **102** is an important resonance contributor.[43] Despite the high instability of **9**, its microwave spectrum could be measured; a dipole moment of 1.9 D units was determined.[46] In a more recent improvement and simplification Billups and co–workers,[47] first synthesized the trimethylsilyl derivatives **107** and **108** from trimethylsilyl allene (**106**) and then performed the vacuum-gas-solid process with solid cesium fluoride, this time supported on glass helices. Whereas **107** could be converted into **9** directly, **108** was first isomerized to the *trans* isomer of **107**, the β-bromocyclopropylsilane **110**, before elimination could be achieved. Above –75 °C **9** underwent a remarkable head-to-tail dimerization to the presumably very highly strained tricyclic diene **109**.

These syntheses of triafulvene were preceded by an interesting trapping experiment, due to Neuenschwander and co-workers,[48] which made the formation of **9** very likely but did not enable its isolation and spectroscopic characterization. As shown in Scheme 13, this route to **9** again relies heavily on β-elimination reactions.

Metalation of the *gem* dibromide **111** with *n*-butyllithium at low temperature provided the organolithium intermediate **112**, which, on quenching with methyl iodide,

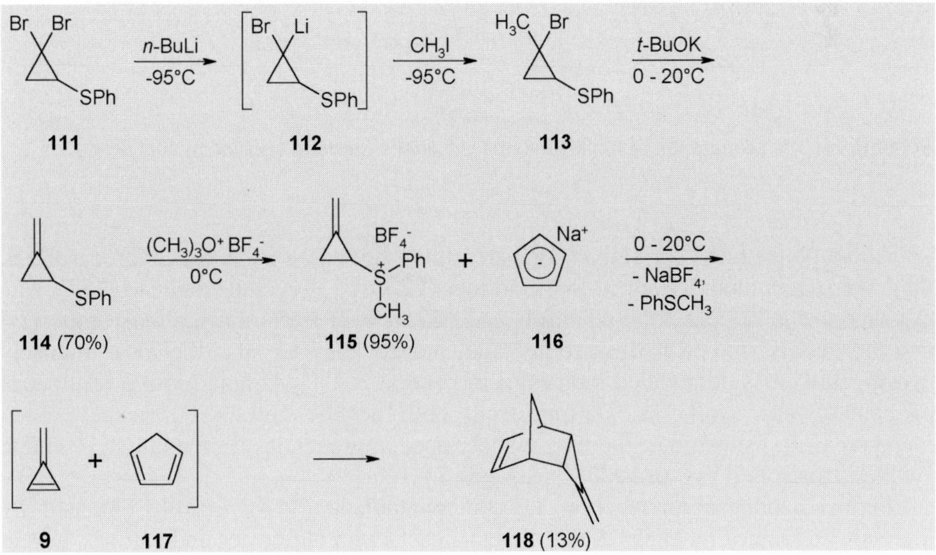

Scheme 13. Trapping of triafulvene (**9**) by 1,3-cyclopentadiene (**117**).

was transformed into the cyclopropane derivative **113**. This intermediate has the ultimately required number of carbon atoms and two leaving groups at positions which enable the introduction of two double bonds. In actual fact the elimination was performed stepwise—initial reaction with potassium *tert*-butoxide to the methylenecyclopropane derivative **114** then methylation with Meerwein's reagent to the sulfonium salt **115**. Rather than using the typical bases normally employed in Hofmann eliminations, cyclopentadienyl sodium (**116**) was used: it not only caused the formation of the desired **9** but was also simultaneously converted to the trapping reagent 1,3-cyclopentadiene (**117**). Because the latter is in close proximity to triafulvene the formation of the [4+2]cycloaddition product **118** is more or less inevitable. Diels–Alder additions are often reversible, and products such as **118** suggest that **9** and its derivatives might also be generated by a retro-Diels–Alder process, and this is indeed so.[49]

For a triafulvene synthesis not beginning with a three-membered ring compound, the addition of carbenes to acetylenes, often used for the synthesis of cyclopropenes, is another alternative. It was realized by Stang and co–workers who obtained the tetraalkyl triafulvenes **122** (R = CH$_3$, C$_2$H$_5$), when isobutylidene (**120**) was generated by α-elimination of the vinyl triflate **119** in the presence of various alkynes **121**[50] (Scheme 14).

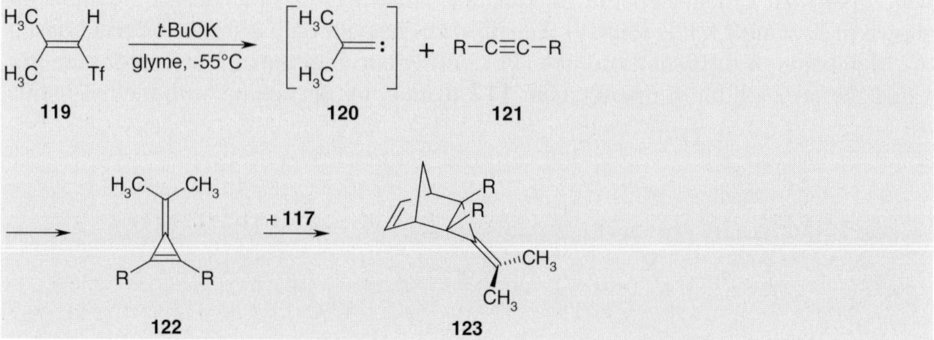

Scheme 14. The preparation of triafulvenes by vinylidene carbene addition to acetylenes.

Although the alkyl derivatives **122** are more stable than the parent hydrocarbon **9**, they were again readily trapped to the adducts **123** by 1,3-cyclopentadiene (**117**).

Among the fulvenes, the pentafulvenes (**82**, *n* = 2) are by far the most important and thoroughly studied hydrocarbons, and, indeed, they are also the most intensely investigated cross-conjugated molecules in general.[39–41,51] Although there are dozens of pentafulvene syntheses, starting from both acyclic and cyclic (usually five-membered ring) substrates, the most widely used approach is still the classical Thiele synthesis described first in 1900[52] (Scheme 15).

In this condensation reaction 1,3-cyclopentadiene (**117**)—readily obtained by thermal cracking of its Diels–Alder dimer—reacts with aldehydes and ketones **124** in the presence of sodium ethoxide, sodium or potassium hydroxide in ethanol. The base

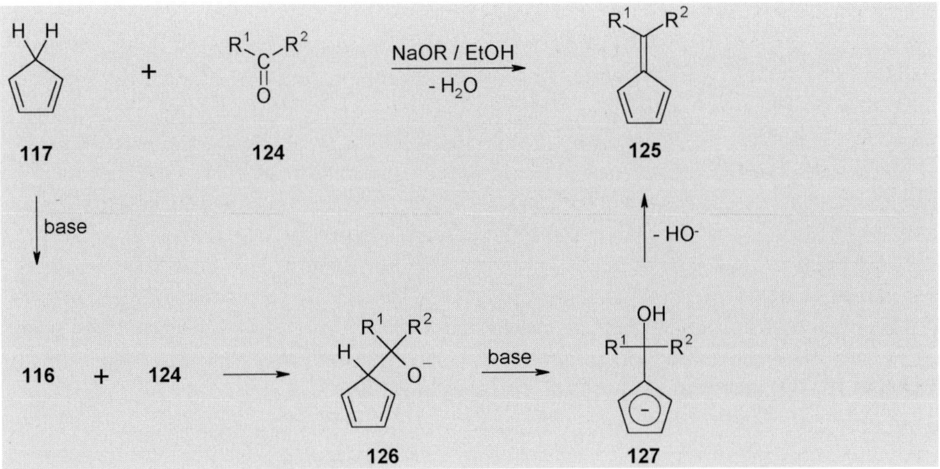

Scheme 15. The Thiele fulvene synthesis.

not only deprotonates **117** (pk$_a$ *ca* 15) to **116**, but also catalyzes the dehydration of the primary adduct **126** *via* the substituted cyclopentadienide **127**. The fulvenes **125** are obtained in good yields with aliphatic and alicyclic ketones and in medium yields with simple aromatic ketones—both 6,6-dimethyl- (**125**, R^1 = R^2 = CH$_3$) and 6,6-diphenylfulvene (**125**, R^1 = R^2 = Ph) are commercial products. With sterically hindered diarylketones the yields are mediocre to poor and with simple aliphatic aldehydes normally very low. In fact, with formaldehyde (**124**, R^1 = R^2 = H), Thiele could obtain only what he called "pentafulvene in polymeric form". That this simplest possible condensation works at all was only shown more than half a century later.[53] A yet more recent reinvestigation of the reaction showed that pentafulvene is produced in 0.6% yield only (referred to **117**), enough, however, to record its spectroscopic data for the first time.[54]

For synthesis of preparative amounts of pentafulvene other methods had to be developed. Although they still rely on **117** as the source of the five-membered ring, the semicyclic double bond is introduced differently.

Thus Neuenschwander and co–workers first alkylated **116** with 1-acetoxy-1-bromomethane (**128**, easily prepared from Lewis-acid-catalyzed reaction of formaldehyde with acetylbromide) to provide the acetoxymethylcyclopentadienes **129** (mixture of isomers), which were subsequently treated with triethylamine. Starting from formaldehyde, pentafulvene (**130**) was produced in 38% overall yield (Scheme 16).[55]

Numerous fulvene derivatives—including the 6-vinyl- and 6-ethynyl derivatives—have been prepared from the corresponding aldehydes by the same procedure.[56]

Preceding this work, Hafner and co-workers had already described another efficient fulvene synthesis in which **116** was first converted into a functionalized fulvene, 6-dimethylamino-pentafulvene (**132**), which was subsequently reduced to the parent hydrocarbon. The **116**→**132** conversion can be accomplished by various reagents,[41] the *N,N*'-dimethylformamide-dimethyl sulfate complex **131** being particularly convenient.[57] Lithium aluminum hydride reduction of **132** followed by aque-

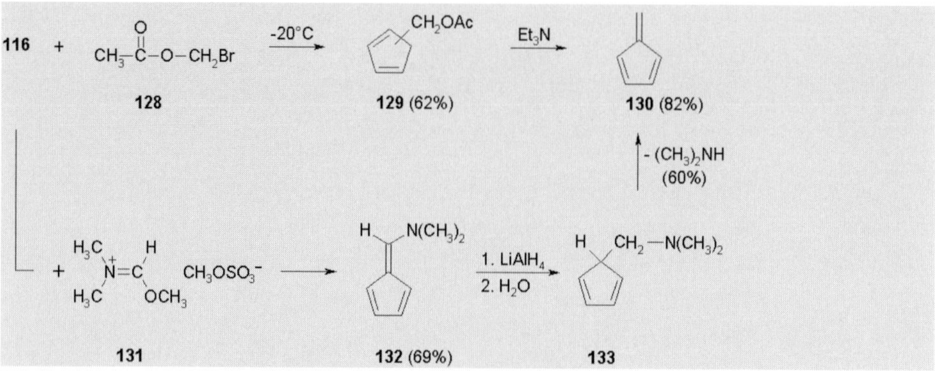

Scheme 16. Two improved syntheses of pentafulvene (**130**).

ous work-up at 0 °C yielded the Mannich base **133** which lost dimethylamine above 100 °C under high vacuum conditions or during chromatography on neutral alumina, furnishing **130** in 60% yield.[58] Since **132** reacts analogously with various organolithium compounds, 6-substituted fulvenes such as 6-methylfulvene (81%) and numerous 6-arylfulvenes are readily prepared by the same approach.

Among the 'acyclic routes' to pentafulvene (**130**) the pyrolysis of 1,5-hexadiyne, *i.e.* an open-chain isomer of fulvene, has already been discussed[59] (see Section 10.2). Even ethyne, a substrate which can hardly be surpassed in structural simplicity, has been trimerized to **130** (30% yield) in the presence of a palladium catalyst.[60]

As we advance through the fulvene series **82**, we should expect another 'reversal of polarity' (see above) in progressing from pentafulvene (**130**) to the next higher vinylog heptafulvene (**82**, $n = 3$). In other words, the routes leading to this hydrocarbon and its derivatives, and their physical and chemical properties, should resemble those for triafulvene (**9**). This is indeed the case, and instead of cyclopropenones and cyclopropenylium ions cycloheptatrienone (**134**, tropone), cycloheptatriene derivatives, and tropylium cations are now used as starting materials.

Thus the dinitrile **136** was formed when **134** and malononitrile were heated under reflux in acetic anhydride (Scheme 17).[61]

That the acetoxytropylium acetate **135** is probably formed as an intermediate in this reaction, is supported by the observation that **134** could be alkylated with triethyloxonium tetrafluoroborate to the corresponding tropylium ion salt, and that this subsequently furnished **136** (65%) with malononitrile.[62]

Although unsuccessful for the synthesis of triafulvene (see above), the acetylation of **134** with acetyltetrafluoroborate at low temperature to **137**, interception of this salt with methyllithium to the cycloheptatriene derivative **138** (mixture of isomers), and gas-phase pyrolysis of the latter in the presence of base yielded the parent hydrocarbon, heptafulvene (**139**).[63]

The first synthesis of **139** was accomplished in 1955 by Doering and co-workers[64] who first prepared a suitable cycloheptatriene derivative, the ester **143**, by methoxycarbonyl carbene addition (generated by decomposition of methyl diazoacetate **141**) to benzene (**140**). As expected, the initially produced norcaradiene derivative **142** opened to **143**, which was subsequently transformed into the quaternary

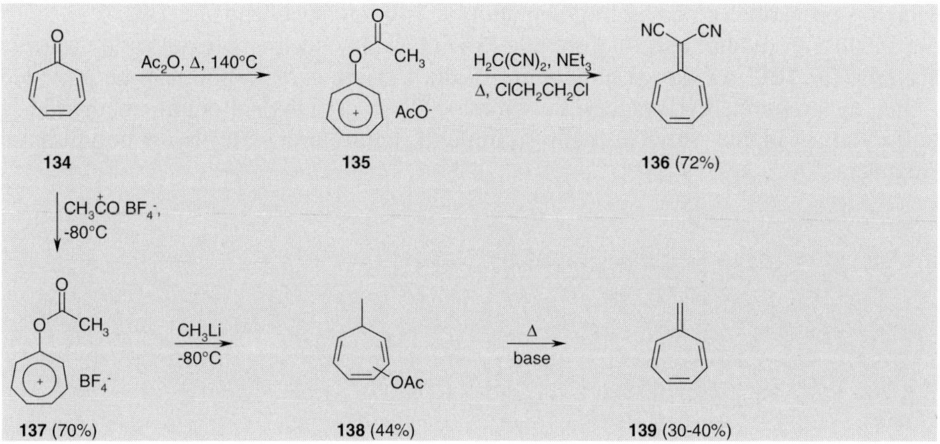

Scheme 17. The formation of heptafulvenes from tropone (**134**).

ammonium salt **144** by conventional steps. Hofmann elimination to **139** concluded the synthesis (Scheme 18),[64] which, however, provided the hydrocarbon in poor yields only.

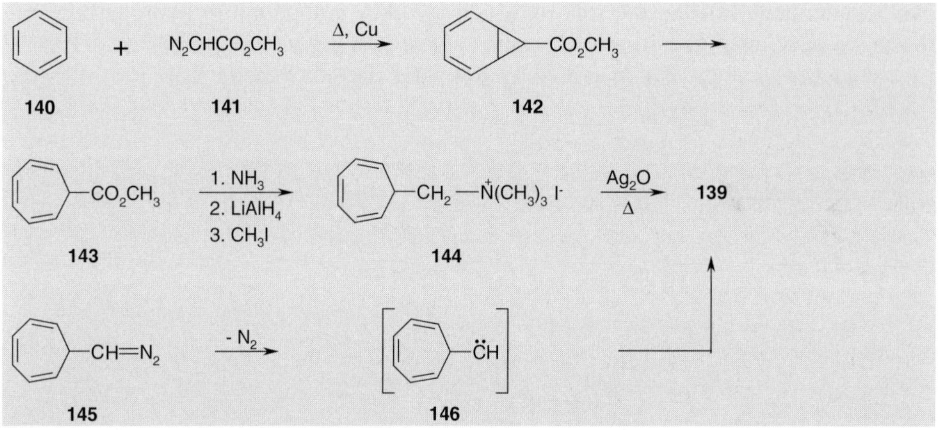

Scheme 18. The first synthesis of heptafulvene (**139**) by Doering.

Because NMR spectroscopy was not available as a method for structure determination at the time, UV spectroscopy and chemical behavior were employed to establish the structure of **139**.

Although access to heptafulvene by α-insertion of the carbene **146** is an obvious possibility, its practical realization met with difficulties. Zimmerman and co–workers prepared the diazo compound **145** from the ethyl ester analog of **143**, but when they decomposed it in cyclopentane at 50 °C **139** was only produced in 37% yield. The emainder of the product mixture being cyclooctatetraene (38%) and benzene (19%), which is probably formed by fragmentation of **146** into acetylene and **140**.[65]

Finally, tropylium tetrafluoroborate (**147**) has also been employed as the starting material for **139**. As shown in Scheme 19 the missing carbon atom can be provided either by bromomethyl magnesium bromide[66] or dimethylsulfonium methylide, a sulfurylid available from trimethylsulfonium tetrafluoroborate by *n*-butyllithium treatment.[67]

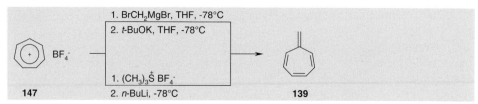

Scheme 19. From tropylium tetrafluoroborate (**147**) to heptafulvene (**139**).

The highest unsubstituted monocyclic fulvene so far reported is nonafulvene (**153**); it was prepared by Neuenschwander and co–workers by analogy to their penta-fulvene (**130**) synthesis discussed above.[68] Just as cyclopentadienide salts provided the carbocyclic ring in the construction of **130**, lithium cyclononatetraenide (**151**) was needed and used for the synthesis of **153**. This building block is available from cyclooctatetraene (**148**)—the role of which as a key compound in hydrocarbon syn-thesis we have encountered already many times (see, *e.g.*, Sections 10.3 and 10.4)—by a sequence of steps first described by Katz and Garratt (Scheme 20).[69]

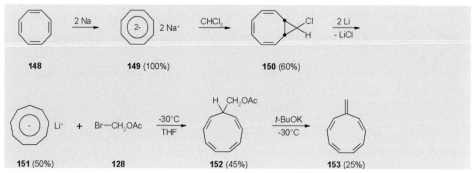

Scheme 20. Neuenschwander's synthesis of nonafulvene (**153**) from cyclooctatetraene (**148**).

Metalation of **148** with sodium first led to the aromatic bis anion **149** in quantita-tive yield. When this was intercepted with chloroform (*cf.* the synthesis of benzvalene discussed in Section 10.2) the bicyclo[6.1.0]nonatriene derivative **150** resulted; on treatment with lithium this opened to **151**. The alkylation with 1-acetoxy-1-bromomethane (**128**) in tetrahydrofuran was carried out just as in fulvene synthesis (see above) and furnished 9-acetoxymethylcyclononatetraene (**152**). Although this is

a very reactive compound which, *e.g.*, readily underwent disrotatory electrocyclization,[68] it could be subjected to β-elimination with potassium *tert*-butoxide at low
temperature. The structure of the thermally extremely unstable **153** was established
by chemical and spectroscopic means,[70] and it was shown that nonafulvene is a non-
planar hydrocarbon[70, 71]—in pronounced contrast to all its lower vinylogs which
have co-planar structures.

11.3 The fulvalenes

Having prepared the parent hydrocarbons tria- (**9**), penta- (**130**), hepta- (**139**), and
nonafulvene (**153**) we can now begin to incorporate these structures into more complex frameworks, *i.e.* again apply our *aufbau* principle.

One group of cross-conjugated hydrocarbons that consists of fulvene subunits is
the fulvalenes, molecules of the general structure **154** in which the fulvene building
blocks share the central (connecting) double bond. As shown by the compound matrix
in Scheme 21 combinations in which both rings are of even or odd size are possible
('symmetrical and unsymmetrical fulvalenes').

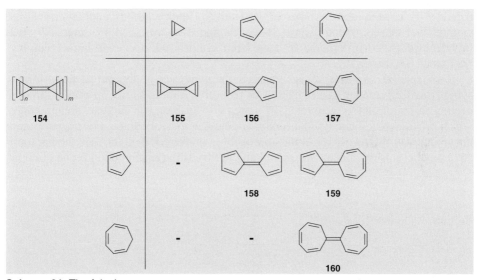

Scheme 21. The fulvalenes.

The symmetrical hydrocarbons—**155**, **158**, **160**—are called tria-, penta- and
heptafulvalene, whereas for some of the unsymmetrical structures both trivial, calicene and sesquifulvalene, and systematic names, pentatriafulvalene and heptapentafulvalene for **156** and **159** are customary. There are several reasons for interest in the
study of these hydrocarbons. As illustrated in Scheme 22 for calicene (**156**) the unsymmetrical fulvalenes can have a dipole moment (see resonance structure **161**), because both halves of the molecule would profit from reduction of the double bond

character of the central double bond, because both would thus acquire aromatic character. The barrier to rotation around this central double bond would consequently expected to be significantly lower than the barrier to isomerization for conventional olefins. Furthermore, heterocyclic variants of the fulvalenes have become important components of novel charge-transfer (CT) complexes, which have high electric conductivity. A particularly important class of compounds are CT complexes in which tetrathiafulvalenes such as **162** (TTF) function as π-electron donors[72] (Scheme 22). Knowledge gained during study of the parent hydrocarbons might hence be applied to the heteroorganic variants.

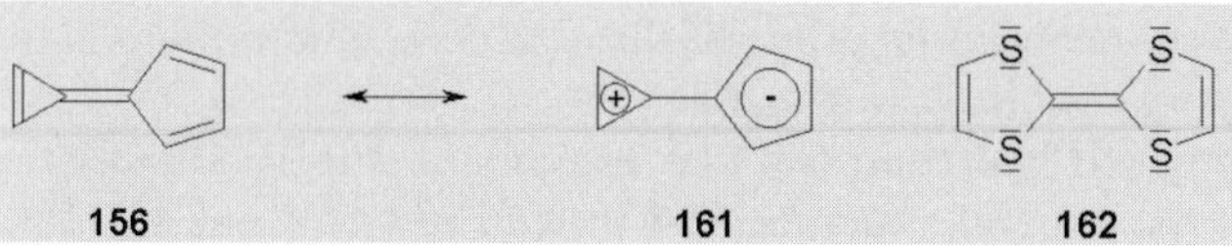

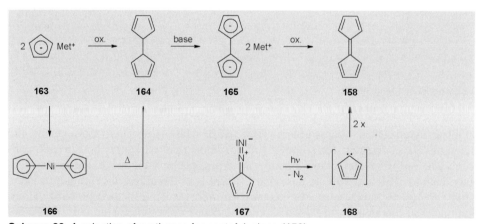

Scheme 22. Why fulvalenes are interesting objects of study.

Considering the low stability of triafulvene (see above), it is not very surprising that none of the triafulvalene parent hydrocarbons **155–157** has so far been prepared.[73] A considerable number of derivatives—hydrocarbons and functionalized systems—of these compounds are, however, known as we shall see later. All other fulvalenes depicted in Scheme 21 have been synthesized, and even higher vinylogs are known (see below).

Two general strategies have emerged for synthesis of the fulvalenes—for the symmetrical representatives oxidative or reductive dimerizations, in which the connecting double bond is generated in a stepwise fashion, or carbene dimerizations, which provide the double bond directly, are the methods of choice. For preparation of unsymmetrical fulvalenes an 'ionic approach' is preferred; it either uses the appropriate aromatic ions or salts as precursors and produces the double bond by an elimina-

Scheme 23. A selection of syntheses for pentafulvalene (**158**).

tion process after the charges have 'annihilated' each other by formation of a carbon-carbon single bond, or it exploits condensation reactions which can ultimately be traced back to the Thiele fulvene process.

The syntheses for pentafulvalene (**158**), collected in Scheme 23, are typical of routes towards the symmetrical hydrocarbons.

The first synthesis of **158**, performed by Doering and co-workers in 1959, generated the dimer **164** (9,10-dihydropentafulvalene) by reaction of the cyclopentadienide **163** with iodine.[74] Deprotonation subsequently led to the bis anion **165** which could be oxidized to **158**. Initially pentafulvalene was only obtained in trace amounts—the typical hydrocarbon saga!—and its full spectral characterization, in particular, had to wait for a more efficient access to the compound. This was found by Neuenschwander and co-workers, who performed the coupling of **163** to **164** with anhydrous cupric chloride (84%), metalated the latter hydrocarbon with *n*-butyllithium in tetrahydrofuran to **165** in quantitative yield, and used cupric chloride again for the terminating oxidation step (73%). The same approach had also been employed to prepare 1,2:5,6-dibenzo-pentafulvalene.[75]

9,10-Dihydropentafulvalene (**164**), a $C_{10}H_{10}$ hydrocarbon (see Section 10.4) and important 'double diene', which we encountered already in Paquette's dodecahedrane synthesis (see Section 5.3), had previously been prepared by an indirect route from **163** *via* nickelocene (**166**). When Hedaya et al. pyrolyzed this sandwich compound at 950 °C and 0.08 torr they obtained a mixture of hydrocarbons, from which **164** could be separated in 15% yield.[76]

The carbene route started from 1-diazo-2,4-cyclopentadiene (**167**), which can be deazotized to **168**, the 'monomer' of **158**. Dimerization took place under a variety of conditions, including irradiation of **167** in an inert gas matrix at low temperatures, the first studies again going back to the late 1950s already.[77-80]

Although heptafulvalene (**160**) was first prepared from tropylium bromide—now, of course, a reductive dimerization constitutes the first step—the experimental procedure is not readily available,[81] and recourse to the carbene route has often been taken (Scheme 24).

Thus photolysis of the tosylhydrazone salt **169** in tetrahydrofuran provided heptafulvalene (**160**) in yields up to 50%.[82] Alternatively, the corresponding diazocycloheptatriene could be used as a precursor for **171**.[83] Another substrate for **160** is a mixture of 1-, 2-, and 3-chlorocycloheptatriene (**170**), which could be converted into the fulvalene in almost quantitative yield.[84] When trimethylsilyltropylium tetrafluoroborate (**172**) was desilylated by treatment with tetrabutyl ammonium fluoride in dichloromethane, **160** was produced in *ca* 70% yield; because of the instability of the hydrocarbon, however, only 20% survived purification by column chromatography on basic alumina.[85]

In an interesting rearrangement and elimination route, cycloheptatrienylidene (**171**) was generated by gas-phase pyrolysis of 7-acetoxynorbornadiene (**173**) and heptafulvalene obtained in 58% yield.[86] Although all these syntheses can be rationalized by postulating the generation of **171** as a reactive intermediate, competing reaction channels involving its resonance structure 1,2,4,6-cycloheptatetraene (**174**), a highly strained and unsaturated cycloallene, which we have already encoutered in Section 9.3, are also possible.[82]

Because of steric hindrance between its inner hydrogen atoms, heptafulvalene is a

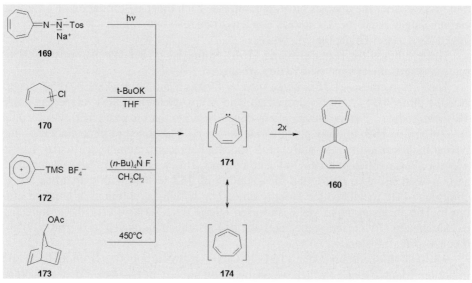

Scheme 24. Heptafulvalene (**160**) by dimerization of cycloheptatrienylidene (**171**).

non-planar ('S-shaped') hydrocarbon in which the largest deviations from the best molecular plane amount to 35 pm.[87] Today **160** is a thoroughly studied cross-conjugated hydrocarbon, both from the reactivity[88] and the structural and theoretical viewpoints.[89-91]

By following the general strategy presented in Scheme 23 for the synthesis of symmetrical fulvalenes, nonafulvalene (bicyclononylidene-2,4,6,8,2',4',6',8'-octaene) has finally been prepared from cyclononatetraenide. The hydrocarbon, which is the next entry on extending the diagonal of the compound matrix in Scheme 21, is extremely reactive because of its very easy valence isomerization to tetrahydro-dibenzopentafulvalene (mixture of isomers).[92]

Turning to the fulvalenes with rings of uneven size, the focus will be on sesquifulvalene (**159**, heptapentafulvalene) because the most important synthetic principles for the preparation of the unsymmetrical fulvalenes can be demonstrated particularly well with this molecule.

The first attempts to prepare **159** were undertaken in the early 1960s when Prinzbach et al. obtained a deep red solution on treating tropone (**134**) with sodium cyclopentadienide (**163**, Met$^+$ = Na$^+$). From the UV spectrum, which is similar to the spectra of several substituted sesquifulvalenes, they assigned structure **159** to the formed hydrocarbon.[93] Full spectroscopic and chemical characterization of sesquifulvalene was, however, only accomplished a decade later by Neuenschwander and co-workers who developed the pathway summarized in Scheme 25 for its preparation,[94] a series of steps which, incidentally, could only by followed through by the application of highly sophisticated laboratory techniques.[95]

Reaction of acetoxy tropylium tetrafluoroborate (**137**) with sodium cyclopentadienide (**163**, Met$^+$ = Na$^+$) furnished acetoxycyclopentadienyl cycloheptatriene (**175**) as a mixture of isomers in almost quantitative yield. When these acetates were sub-

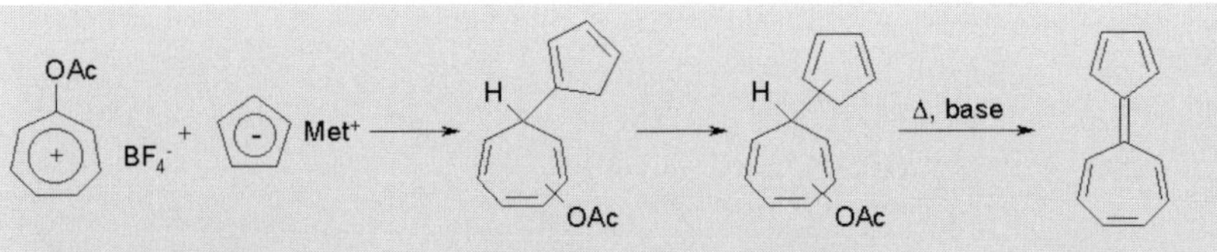

Scheme 25. The preparation of sesquifulvalene (159) by Neuenschwander.

jected to flash vacuum pyrolysis at 360 °C (contact time 20 s) in the presence of triethylamine, sigmatropic hydrogen shifts first converted them to the isomeric esters 176, from which 159 was produced in excellent yield (total yield over all steps: 53%).

Alternate routes to 159 have been published,[96, 97] and one will be presented here, because it involves transformations which we have come across previously in another context. When the tricyclic valence isomer of tropone, 177,[98] an important intermediate in the preparation of 'stretched versions' of the fulvalenes also (see below), was subjected to Thiele conditions (cyclopentadiene, 5% methanolic potassium hydroxide, 0 °C), the fulvene derivative 178 was obtained in 40-45% yield. Both silver-ion catalyzed (see Section 10.3) and photochemical isomerization converted this precursor in excellent yield (ca 90%) into sesquifulvalene (Scheme 26).

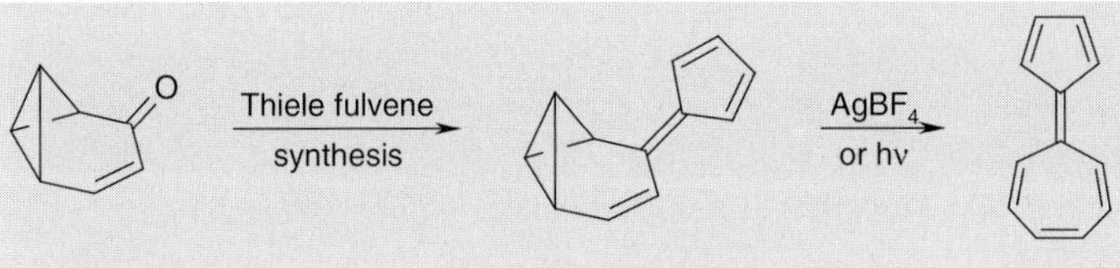

Scheme 26. An alternative route to sesquifulvalene (159).

Despite their increasing instability still higher fulvalenes have been prepared. For the synthesis of nonapentafulvalene the coupling approach described in Scheme 23 was again successful; sodium cyclopentadienide (163, Met$^+$ = Na$^+$) and sodium cyclononatetraenide (151, Na$^+$ instead of Li$^+$) were used as the aromatic building blocks.[99]

Eleven-membered ring systems are so far the largest subsystems which have been incorporated into fulvalenes. They are, however, no longer monocyclic but have a methano bridge, which results in a drastic increase in stability, i.e. the very large fulvalenes contain methanoannulene subunits (see Section 10.5).

To prepare the simplest of these hydrocarbons, pentahendecafulvalene or fidecene (181)—from the Latin word fides for violin—Prinzbach and co-workers[100] coupled the cation 179[101] with lithium cyclopentadienide (163, Met$^+$ = Li$^+$) (Scheme 27), and then dehydrogenated the resulting dihydro product 180 (mixture of isomers) with chloranil.

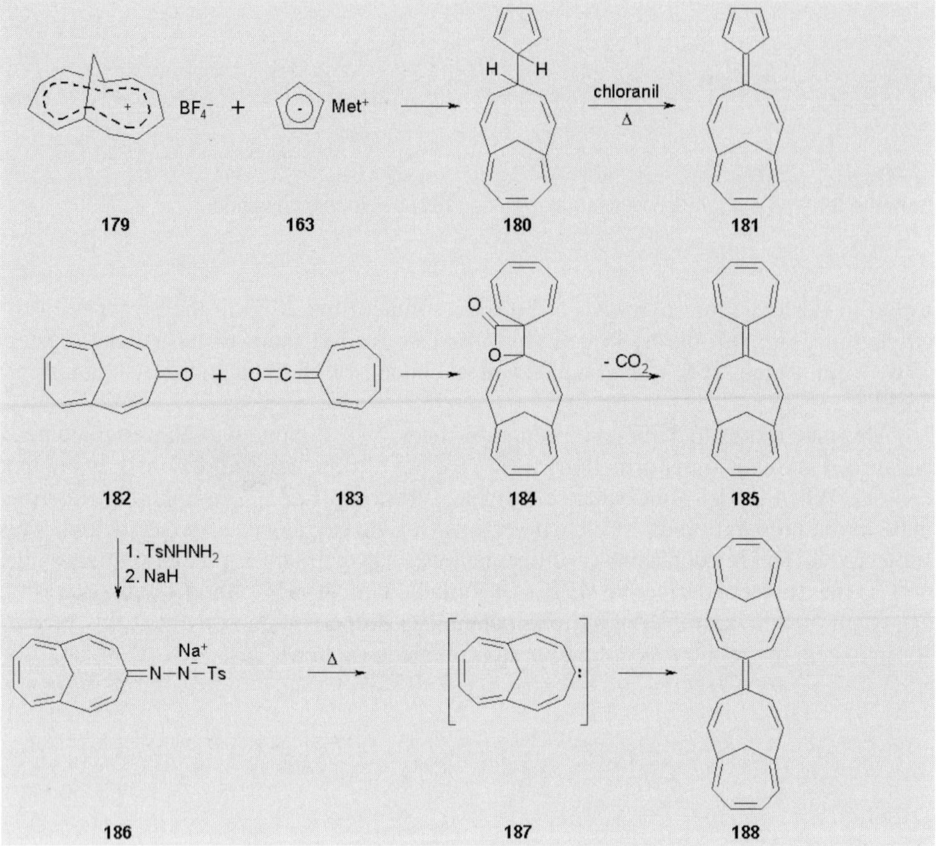

Scheme 27: The preparation of fidecene (**181**) and related methano-bridged fulvalenes.

Whereas the unstable **181** could only be obtained in solution, its derivatives are crystalline compounds the spectroscopic and chemical properties of which have been investigated.[100] The heptahendecafulvalene **185**, formed in good yield (80–85%) by addition of cycloheptatrienylidene ketene (**183**) to 4,9-methano[11]annulenone (**182**) and decarboxylation of the β-lactone intermediate **184**, is, on the other hand, surprisingly stable.[102] Its X-ray structural analysis reveals S-shaped geometry, reminiscent of that of heptafulvalene (see above). The ketone **182**[103] also served as the starting material for the 22-π-electron fulvalene **188**, which was produced in 67% yield when the sodium salt of the tosylhydrazone of **182**, **186**, was thermally decomposed at 130 °C.[104] Just as for heptafulvalene (**160**, see above) trapping experiments with various alkenes, alkynes, *etc.* established the intermediacy of a carbene (**187**)intermediate.

To control the high reactivity of the fulvalenes, extensive use has been made of the introduction of bulky (mostly *tert*-butyl) substituents, and of benzoannelation. This is most obvious for the triafulvalenes, for which none of the parent molecules shown in Scheme 21 has yet been synthesized or even generated only temporarily. In the calicene series, *i.e.* compounds derived from **156**, the least substituted hydrocarbon is 1,3-di-*tert*-butyl-5,6-dimethylcalicene (**193**), prepared by Prinzbach and co-workers from the two aromatic salts **189** and **190** as shown in Scheme 28.[105]

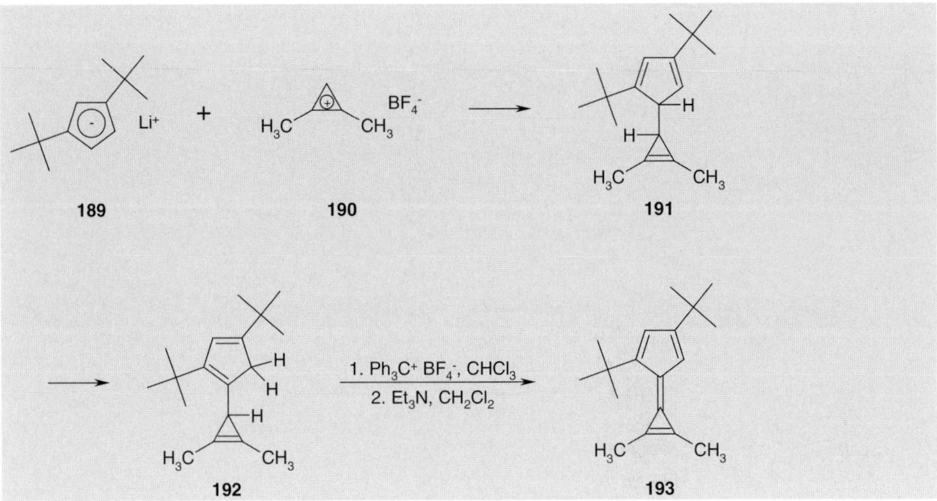

Scheme 28. Prinzbach's preparation of a sterically shielded calicene.

Coupling of the two subunits first led to the dihydro derivative **191** which isomerized to the isomer **192** by 1,5-hydrogen migration. Oxidation was subsequently achieved by a stepwise protocol: hydride abstraction with trityl tetrafluoroborate in chloroform first provided the corresponding cationic salt which was then deprotonated to **193** by treatment with triethylamine in dichloromethane. Although the hydrocarbon is sufficiently stable to be characterized by, *e.g.*, ^{1}H NMR spectroscopy, it is rapidly attacked by oxygen. Interestingly the proton spectra are temperature dependent, and this dependence might be a result of rotation around the connecting double bond, involving a (perpendicular) transition state with dipolar character, as illustrated by structure **161**, above. Introduction of the aforementioned stabilizing groups into several of the other fulvalenes collected in Scheme 21 has also been accomplished.[106–108]

So far we have varied only the sizes of the rings and the number and types of the substituents of the fulvalenes. There is, however, a third parameter which can be modified: the connecting π-bond system. If the double bond in **154** is exchanged for a 1,3-diene unit the fulvadienes **194** result; these can, in principle, be extended to the higher vinylogs **195**, o ≥ 1 (Scheme 29).

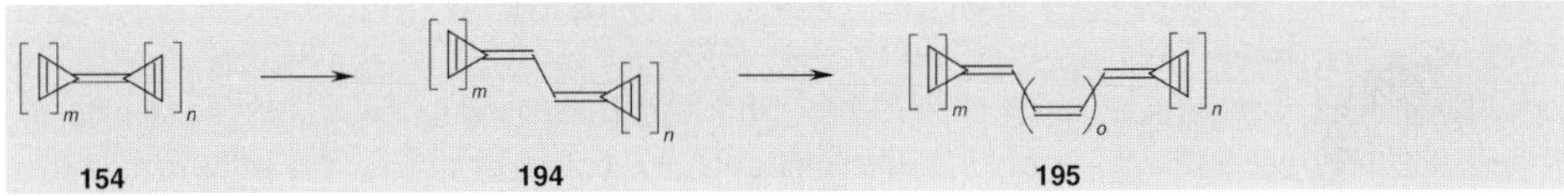

Scheme 29. How to extend the fulvalene system (**154**).

These stretched fulvalenes are of interest because of their dipolar character[109] and as substrates in multi-electron pericyclic processes.[110] A sizeable number has been prepared, and a small selection is shown in Scheme 30 to demonstrate some of the methods employed in their synthesis and the structural diversity which has been achieved.

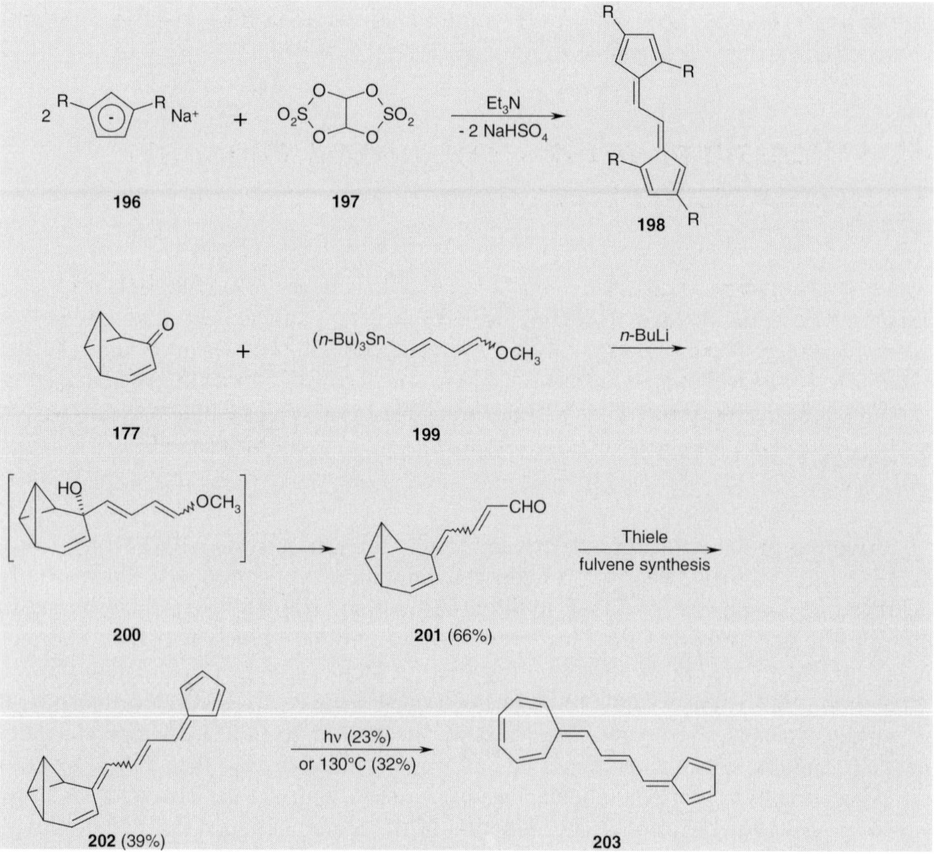

Scheme 30. The preparation of extended fulvenes.

The parent pentafulvadiene (**198**, R = H) was first prepared in small amounts (3–5%) by reacting sodium cyclopentadienide (**196**, R = H) with the double electrophile **197** ('glyoxal sulfate'), which functions as a substitute for glyoxal which itself does not undergo the Thiele condensation. Hydrocarbon **198** (R = H) is significantly more stable than pentafulvalene (**158**), and can be stored for extended periods of time at –20 °C under an inert gas atmosphere.[111] By using the di-*tert*-butyl derivative **196** (R = *tert*-butyl) Hafner and co–workers prepared the tetrasubstituted hydrocarbon **198** (R = *tert*-butyl) in 79 % yield as air- and light-stable crystals.[107] Highervi-nylogs of **198** are known—up to four consecutive double bonds—and in synthesizing these stretched fulvalenes both unsaturated dialdehydes and certain equivalents of them—*e.g.* 2,7-dibromo-2,7-dihydrooxepin for Z,Z-mucodialdehyde—have been used to form the ring-connecting parts of the different target molecules.[112] The synthesis of the extended sesquifulvalene **203** is modeled on the preparation of the parent compound **159**. Again it starts with the tropone isomer, **177**, which was first chain-elongated by treatment with 4-methoxy-1-tributylstannyl-1,3-butadiene (**199**) in the presence of *n*-butyllithium. The alcohol intermediate formed, **200**, was not isolated but provided the aldehyde **201** directly as a mixture of isomers. Thiele condensation and photochemical or thermal isomerization of the resulting **202** concluded the synthesis of this cross-conjugated 16π-electron system.[113]

11.4 Pentalene, azulene, heptalene and other zero-bridged annulenes

Many of the hydrocarbons discussed in Section 11.3, *e.g.* **158**, **159**, **181**, **198** (R = H), and **203**, contain a pentafulvene (**130**) unit as a subsystem, and we have seen how the methodology developed for fulvene synthesis (see Section 11.2) can often be employed to prepare the fulvalenes also. There are, however, other ways to construct more complex and extended π-sytems from **130**, and in this section we shall largely deal with two of them—pentalene (**204**) and azulene (**205**) (Scheme 31).

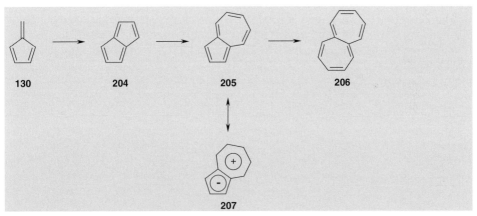

Scheme 31. From fulvene (**130**) to heptalene (**206**).

Formally pentalene can be derived from fulvene by connecting its C1 and C6 positions by an etheno bridge, and azulene is generated if a 1,3-butadieno unit is used for that purpose. Of course, **205** can also be regarded as the next higher vinylog of **204**. If we continue the vinylogization by also enlarging the five-membered ring of **205** by an additional double bond we arrive at heptalene (**206**), a bicyclic polyolefin which will be discussed towards the end of this section.

Clearly, this building process can be extended further, and numerous other com-

binations and connections of endocyclic double bonds arise when we proceed to tri-
or even polycyclic hydrocarbon frameworks as will be demonstrated below. That we
here restrict ourselves largely to **204–206** is because of the importance of these compounds in synthetic, structural, and theoretical chemistry. Pentalene (**204**) has attracted the efforts of many chemists because it can be regarded as a short-circuited
cyclooctatetraene. Because the latter avoids antiaromaticity by assuming a non-planar
(tub-like) structure (see Section 10.3), connecting its C1 and C5 carbon atoms by a
'zero-bridge' should result in planarization and hence antiaromatic character. This, in
turn, should lead to increased reactivity ('instability') of **204**. In general, we would
expect the reactivity of pentalene to be similar to that of cyclobutadiene, the smallest
of the uncharged antiaromatic hydrocarbons (see Section 10.1). Indeed, many attempts to prepare **204** failed, and only recently has this highly reactive species been
isolated under matrix conditions (see below). This fleeting existence contrasts with
the stability of its vinylog, azulene (**205**). The history of this beautifully deep-blue
compound—*azur* means blue in French—can be traced as far back as the 15th century. After an extensive and extended effort to establish the structure of certain naturally occurring derivatives, the parent hydrocarbon **205** was finally synthesized in
1937 (see below).[114] In contrast to **204**, **205** is a short-circuited 4n+2-hydrocarbon
(a [10]annulene, see Section 10.4); its dipole moment (μ = 1.0 D) indicates, however,
that the polar structure **207** must be an important resonance contributor. As formula
207 shows, azulene can be regarded as a fusion of a tropylium cation and a cyclopentadienyl anion, just as its isomer naphthalene consists of two fused benzene rings.

That pentalene (**204**) and cyclobutadiene are of comparable reactivity is supported by many observations. Both hydrocarbons are stabilized by benzoannelation—
indeno[2.1-*a*]indene (**208**) and several of its derivatives are isolable compounds[115–
117]—and by substituents that modify their electronic and/or steric character, as illustrated by the two pentalene derivatives **209**[118] and **210**[119] (Scheme 32)—just as for
cyclobutadiene (see Section 10.1).

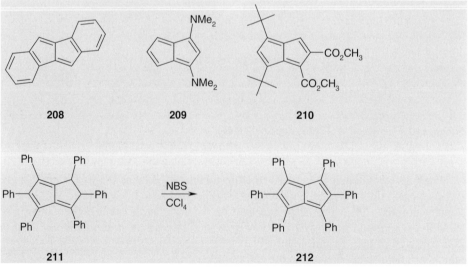

Scheme 32. A selection of pentalenes stabilized by benzoannelation and substitution.

Among the pure hydrocarbons perphenylpentalene (**212**), prepared in 77% yield by LeGoff in 1962[120] by oxidation of 1,2,3,4,5,6-hexaphenyl-1,2-dihydropentalene (**211**) with N-bromosuccinimid in oxygen-free tetrachloromethane, remained the only non-annelated representative for a long time. However, complete substitution as in this example not only blocks further reactions very effectively, it also prevents the measurement of *e.g.* NMR data for pentalenic hydrogen atoms. And, of course, substituents with delocalizable π-electrons change the electronic properties of the core of the molecule.

This is not so for 1,3,5-tri-*tert*-butyl-pentalene (**217**), the first stable, not fully substituted pentalene hydrocarbon, prepared by Hafner and co–workers in 1973 (Scheme 33).[121]

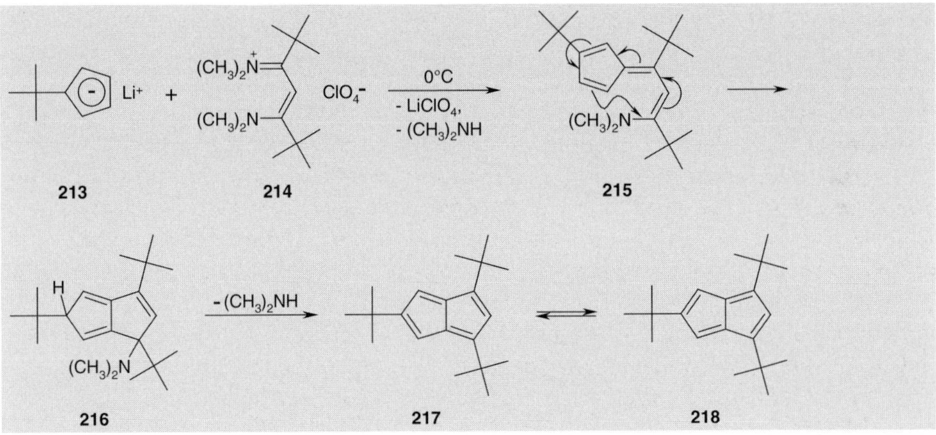

Scheme 33. Hafner's synthesis of 1,3,5-tri-*tert*-butyl-pentalene (**217**).

In a remarkable one-pot process lithium *tert*-butylcyclopentadienide (**213**) reacts with the vinylogous amidinium perchlorate **214** to yield **217** in 86% yield. This reaction cascade is initiated by formation of the fulvene derivative **215**, which is converted into the 1,5-dihydropentalene derivative **216** by 8π-electrocyclization followed by a formal 1,3-hydrogen shift. The final stabilization to **217** is achieved by 1,4-elimination of dimethylamine. Although this pentalene is air-sensitive it is surprisingly stable thermally (no dimerization up to 300 °C, see below). The chemical shifts of the ring protons (δ = 4.72 and 5.08) speak for a paramagnetic ring current in **217**. In accordance with the D_{2h} symmetry of the hydrocarbon its ^{13}C NMR spectrum contains only five lines for the eight ring carbon atoms at room temperature. However, when the temperature is reduced to –140 °C signal broadening begins, and in the range of –160 to –180 °C the carbon signals for C1/C3, C3a/C6a and C4/C6 split, indicating valence isomerisation between the two isomers **217** and **218**. Indeed, the activation barrier for this isodynamic double-bond shift process has been determined as 4 kcal mol^{-1} by line-shape analysis.[122] This experimental proof of bond alternation—as required by theory—is also suggested by the X-ray structure determination of **217**.[123]

How critical the stability of the hydrocarbon depends on the *tert*-butyl substituents—and how similar the structural and chemical properties of **217** are to those of cyclobutadiene (see Section 10.1)—is underlined by the observation that exchange of one of the protective groups (the 5-*tert*-butyl substituent) for an isopropyl substituent resulted in dimerization of the corresponding pentalene derivative above -20 °C in solution. When a substituent is absent from this position, *i.e.* in an attempt to prepare 1,3-di-*tert*-butyl-pentalene, only the dimer is obtained.[124]

Before discussing these (substituted) dimeric structures it is useful to present an approach to the parent molecule **204**, also due to Hafner and co–workers,[125] exploiting the 8π-electrocyclization process of 6-[(dialkylamino)vinyl]fulvene derivatives described above (Scheme 34).

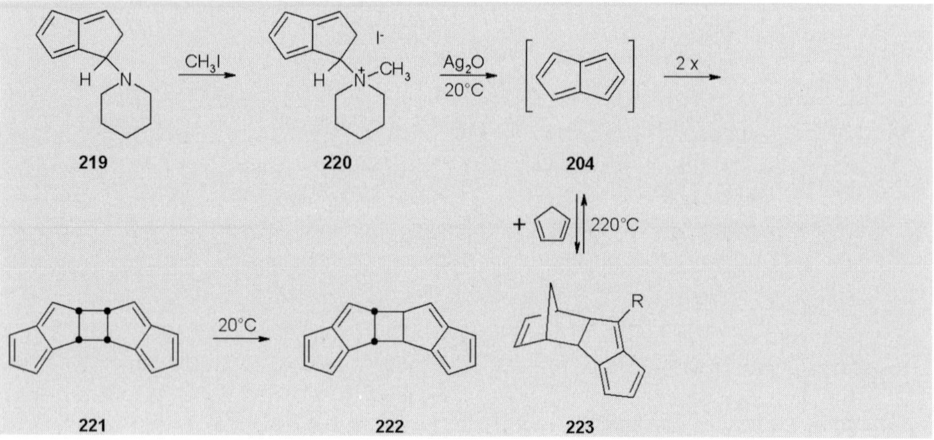

Scheme 34. The first generation of pentalene (**204**) as a reactive intermediate by Hafner and co–workers.

1-Piperidino-1,2-dihydropentalene (**219**) was first prepared by analogy with the synthesis of **216**. When this was methylated to **220** and this methoiodide subjected to Hofmann elimination by treatment with silver oxide at room temperature the *syn-cis*-dimer **221** was formed in 22% yield. That pentalene (**204**) served as the immediate precursor for **221** is supported by the observation that Hofmann elimination in the presence of excess cyclopentadiene led to the trapping product **223** (R = H). This Diels–Alder addition is reversible, as is shown by the thermal cleavage of **223** (R = H) back to its precursor dienophile **204**. At room temperature **221** isomerized in poor yield into its *syn-trans* dimer **222**. Comparable eliminations, performed with different methyl derivatives of either dialkylamino-1,5- or -1,2-dihydropentalenes, *i.e.* derivatives of type **216** and **219**, respectively, produced mono- and multiply-methylated pentalenes, which always selftrapped to the corresponding dimers.[126] Although these dimers could be subjected to cycloreversion by either flash-vacuum pyrolysis or low-temperature photolysis, enabling UV- and IR-spectroscopic identification of the appropriate methylpentalene,[124, 126] and could also be employed as substrates for the preparation of pentacarbonyl iron complexes of different methylpentalenes,[127] for a long time the parent hydrocarbon **204** was not accessible by these routes or modifications thereof. And again the reason was simple and hydrocarbon-like—lack of material prevented the use of **221** or **222** for the preparation of pentalene by either photochemical [2+2]-or thermal [2+8]cycloreversion. This only changed very recently when 1,5-dihydropentalene (**225**) became available (in *ca* 60% yield) by gas-phase pyrolysis of cyclooctatetraene (**224**) (Scheme 35, see Section 10.3).[128]

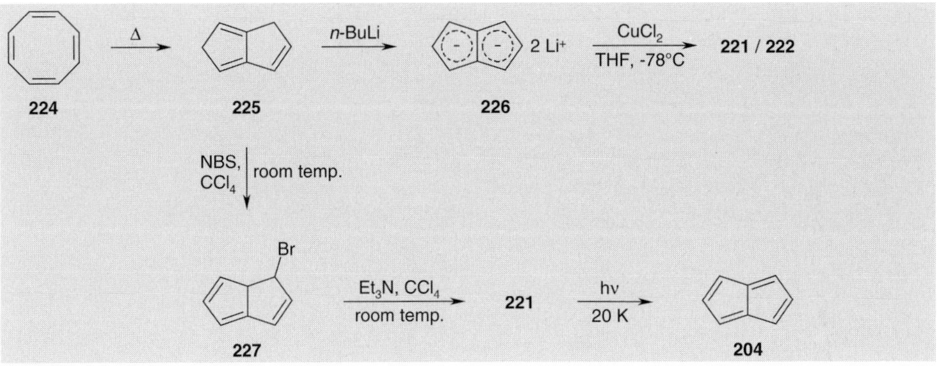

Scheme 35. Preparation of pentalene (**204**) by cleavage of the dimers **221** and **222**.

Metalation with *n*-butyl lithium then provided dilithium pentalenediide (**226**), a bis anion previously available only with difficulty.[129] When this was subjected to the CuCl$_2$-induced oxidative coupling already used so successfully for the synthesis of fulvalenes (see above) Neuenschwander and co–workers obtained a 1:1-mixture of **221** and **222** in 12% yield—low, but superior to the previous yields of these hydrocarbons.[130] Furthermore, NBS bromination of **225** enabled the preparation of the (unstable) allyl bromide **227**, which—on dehydrobromination with triethylamine— yielded the *cis* dimer **221** as the sole product. As in Hafner's synthesis (see above) the most likely precursor to this product is again pentalene (**204**). That this assumption is indeed correct was finally proven by photocleavage of **221** in an argon matrix.[131] Pentalene, generated from the *cis* dimer in a two-step process, was characterized by means of its electronic and vibrational absorption spectra, which could be analyzed and interpreted with reference to various quantum chemical calculations. The single and double bonds of **204** are localized (C_{2h} symmetry), in accordance with the antiaromatic character of the hydrocarbon—again comparable with what we have learned about cyclobutadiene (see Section 10.1).

The next higher vinylog of pentalene, azulene (**205**), has attracted the interest of chemists for a very long time.[114] Derivatives of azulene such as chamazulene (**228**), guaiazulene (**229**) and vetivazulene (**230**) (Scheme 36) are responsible for the blue color of certain fractions of distilled essential oils from chamomille, yarrow, worm-wood, and other sources.

The correct structures for **229** and **230** were established by Pfau and Plattner in the 1930s; these workers also developed the first synthesis of the parent hydrocarbon[132] (Scheme 37).

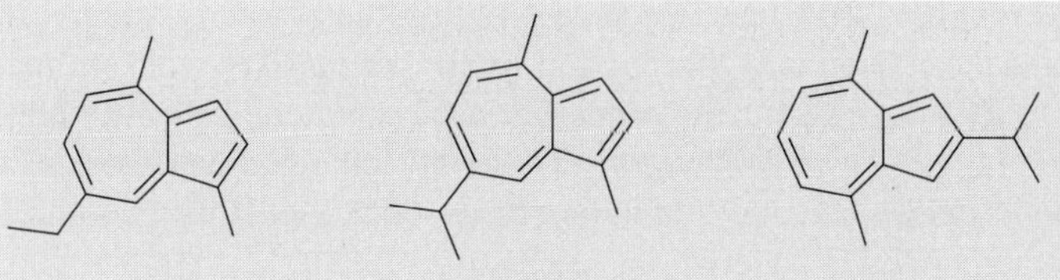

Scheme 36. A selection of naturally ocurring azulene hydrocarbons.

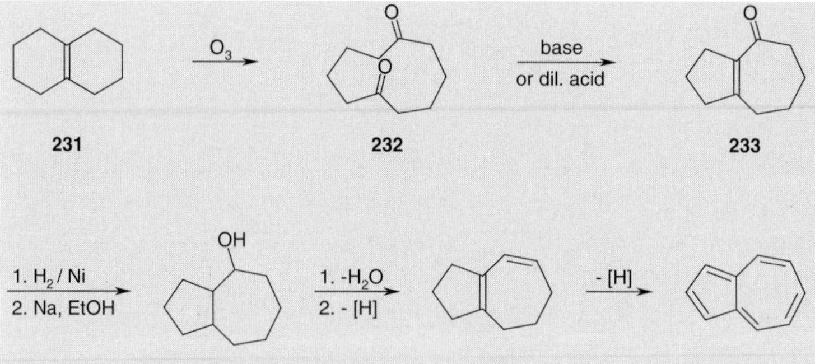

Scheme 37. The first synthesis of azulene (**205**) by Pfau and Plattner.

In this classical route, Δ⁹-octalin (**231**) was first cleaved by ozonolysis to the diketone **232** which on treatment with either dilute acetic acid or alkali recyclized to the so-called Hückel ketone **233**[133] by an intramolecular aldol reaction. After hydrogenation and reduction of the carbonyl group, the secondary alcohol **234** was obtained; this after dehydration and subsequent dehydrogenation over palladium on charcoal furnished **235**. The total yield was not given in the original synthesis, but subsequent improvement—use of lithium aluminum hydride for the ketone reduction, dehydrogenation over a special Pd/C/MnO₂ catalyst—led to yields in the 20% range for the **233**→**205** conversion.[134]

Numerous alternatives to the Plattner–Pfau synthesis appeared in subsequent years; these differ primarily in the steps used to construct the bicyclo[5.3.0]decane skeleton. Many of these approaches are mere modes of formation, not preparatively useful syntheses, and certainly not classics.[135] Among the more general routes ring-enlargements of indane derivatives by carbene addition followed by oxidation, annelation of a seven-membered ring to a cyclopentane ring and the reverse strategy must be mentioned, as must the [2+8]cycloaddition of, *e.g.*, triple bond dienophiles to heptafulvene and the widely used copper-catalyzed decomposition of 4-aryl-1-diaza-oxobutanes.[135]

A common feature of these azulene syntheses is that they make use of a terminating, often forceful and consequently yield-reducing dehydrogenation step. This is not so for the other classical route to azulene and numerous of its derivatives—the Hafner–Ziegler synthesis. In an early retrosynthesis—a *gedankenexperiment*—azulene is hydrolyzed to 5-cyclopentadienyl-2,4-pentadienal (**236**) which, by tautomerization, can be converted into the fulvene derivative **237**. A second hydrolysis step converts this intermediate into cyclopentadiene (**117**) and glutacondialdehyde (**238**) (Scheme 38).[136]

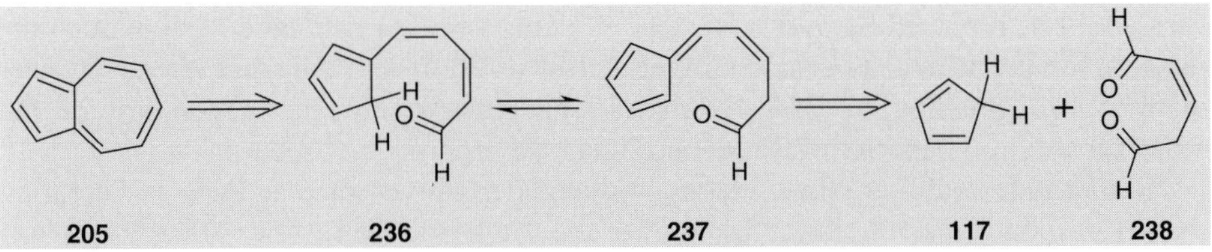

Scheme 38. The Hafner–Ziegler synthesis of azulene (**205**)—the retrosynthesis.

The realization of this elegant concept failed, since it was impossible to condense the unstable aldehyde, its sodium salt, or an acyl derivative of the enol form of **238** with **117**. However, when the so-called Zincke-aldehyde **243** was condensed with **117** the bisvinylogous aminofulvene **244** was produced; this, on heating to 290–300 °C or treatment with superheated water vapor, split off *N*-methylaniline and provided azulene (**205**) in yields up to 60% (Scheme 39).[136,137]

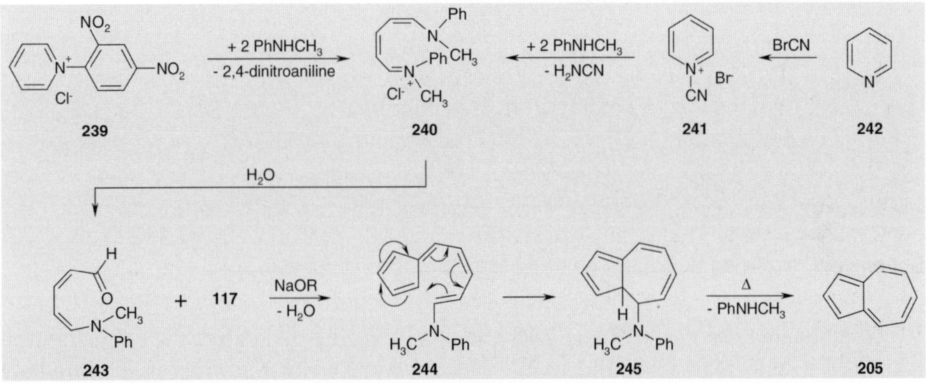

Scheme 39. The Hafner–Ziegler azulene synthesis—the practical realization.

The required aldehyde **243** can be prepared by cleavage of the pyridinium chloride **239** with *N*-methylaniline, and hydrolysis of the resulting iminium salt **240**. The same intermediate was also accessible from pyridine (**242**) after this had been converted to **241** by cyanation with BrCN and the pyridinium bromide been cleaved by reaction with *N*-methylaniline. Mechanistically **244** first undergoes an electrocyclic 10π-process to **245**, which subsequently looses the amine in a β-elimination step. The scope of the method is very large because numerous desired substituents—alkyl, aryl, functional groups—can be introduced by way of both components **117** and **238**. Indeed, even dimeric hydrocarbons such as 6,6'-biazulenyl could be prepared from 4,4'-bipyridyl.[138]

Formally the Hafner–Ziegler synthesis has been regarded as a condensation of pyridine (**242**) with cyclopentadiene (**117**) and loss of ammonia[136]—and, indeed, the concept has been realized in practice by treating quaternary pyridinium salts with cyclopentadienides and heating the reaction mixture to 250 °C in appropriate solvents.[139] Moreover, in an excellent application of heuristic reasoning, the pyridinium salts could be replaced by pyrylium salts.[140] This has the practical advantage that the reaction temperature can occasionally be as low as 20 °C and the isolation of intermediates—corresponding to **244**—becomes unnecessary. For the generation of the 4,6,8-substitution pattern in **205**, in particular, this approach is unsurpassed.[137]

Rather than cyclizing bisvinylogous fulvene derivatives such as **244**, and creating the seven-membered ring in the last part of the synthesis, the reverse strategy can also be achieved with various vinylogous heptafulvenes which form the missing five-memberd ring by thermal isomerization in the last step.[141]

The generic relationships between fulvene (**130**), pentalene (**204**), azulene (**205**), and heptalene (**206**), referred to in Scheme 31, are more than a mere formalism, and numerous interconversions between these hydrocarbons and their derivatives have been performed in practice. The following examples are a small selection from a vast literature and reflect personal predilection more than anything else.

[6+4]Cycloaddition of thiophene-1,1-dioxides such as **246** to the versatile 6-*N,N*-dimethylaminofulvene[57] (**132**, see above) provided the adduct **247** which lost sulfur dioxide and dimethylamine even at room temperature and yielded (25%) 6-methylazulene (**248**) (Scheme 40).[142, 143]

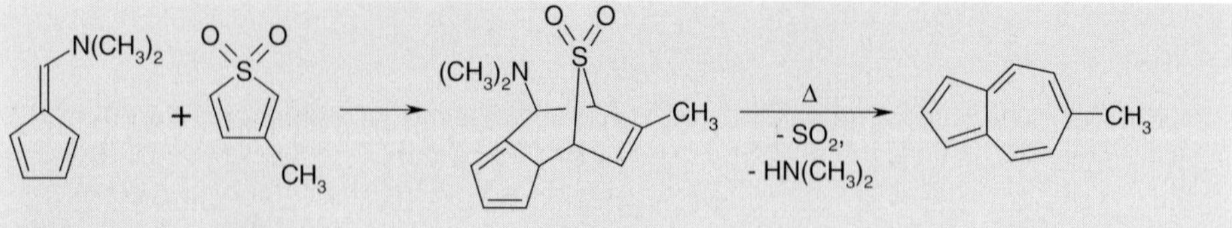

Scheme 40. Azulenes from fulvenes by addition of a 4π-component.

Components such as **132** and **246** are usually readily available, the cycloaddition takes place with high regioselectivity, and the work-up of the process is extremely simple.

Heptafulvene derivatives as substrates for an azulene synthesis require only the addition of a 2π-component (to produce the missing five-membered ring). For example, 2-oxo-2*H*-(cyclohept[*b*]furan) (**249**) reacted with the *in situ*-prepared enamine 250 to the [8+2]cycloadduct 251, which was converted to azulene in 60% yield by decarboxylation to the dihydroazulene **252** followed by loss of dimethylamine (Scheme 41).[144]

A remarkable azulene synthesis has been accomplished with two reactive hydrocarbons encountered earlier in this book—1,3,5-tri-*tert*-butylpentalene (**217**, see above) and cyclooctyne (**253**, see Section 8.2). Heating these two 'borderline' hydrocarbons to 270 °C led to the bridged azulene **255** in 70% yield, presumably *via* the initially produced cyclobutene derivative **254** which underwent electrocyclic ring-opening under the reaction conditions (Scheme 42).[145]

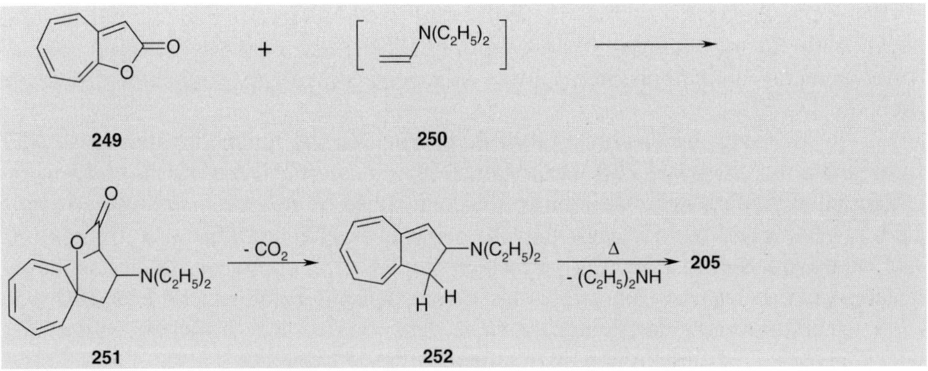

249   **250**

251   **252**

Scheme 41. Azulenes from heptafulvenes by addition of a 2π-component.

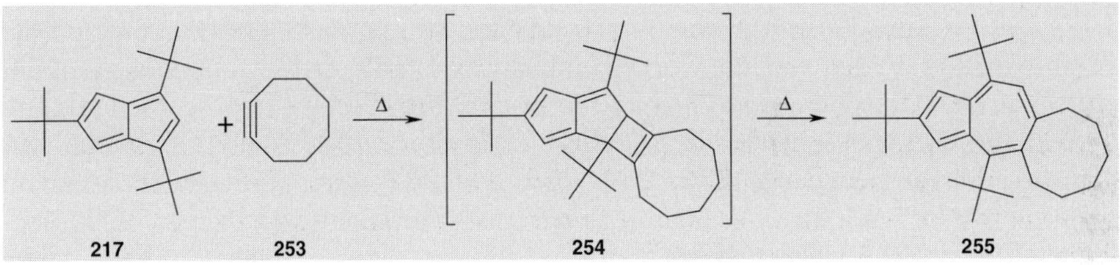

217   **253**   **254**   **255**

Scheme 42. Azulenes from pentalenes by addition of a reactive alkyne.

Although heptalene derivatives have been prepared by a very similar approach—cycloaddition of a reactive alkyne such as dimethyl acetylenedicarboxylate to azulene,[145] the parent compound **206** was prepared differently. And it is again very instructive to compare the different syntheses leading to this 'short-circuited' [12]annulene.

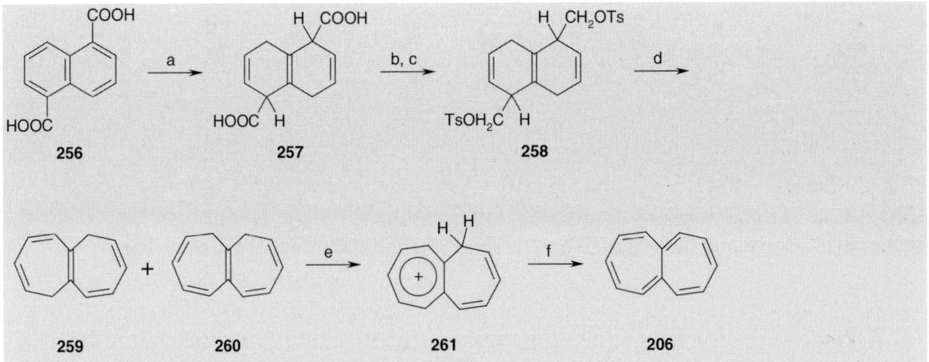

256   **257**   **258**

259   **260**   **261**   **206**

(a) Na, NH₃, C₂H₅OH (55-60%); (b) LiAlH₄, ether; (c) TsCl, pyridine (comb. yield 59%);
(d) NaH₂PO₄ / CH₃CO₂H, 90°C (77%); (e) Ph₃C⁺BF₄⁻, CH₂Cl₂ (88%); (f) (C₂H₅)₃N, CHCl₃ (41%)

Scheme 43. The first synthesis of heptalene (**206**) by Dauben and Bertelli.

The first synthesis of heptalene (**206**), described by Dauben and Bertelli in 1961 began with 1,5-naphthalenedicarboxylic acid (**256**), *i.e.* a starting material that already contains all carbon atoms ultimately required in the product hydrocarbon (Scheme 43).[146]

In the first step Birch reduction of **256** provided the tetrahydro derivative **257** from which the ditosylate **258** was obtained by reduction and esterification. Ring enlargement to the heptalene skeleton was accomplished by solvolysis in buffered acetic acid, providing the two isomeric dihydro derivatives **259** and **260** as a 3:2 mixture. The final oxidation to **206** was performed stepwise, an approach which has subsequently often been used for similar tasks (see below). Treatment of **259/260** with trityl tetrafluoroborate in dichloromethane first yielded the tropylium cation **261** which gave up its excess proton on treatment with triethylamine.

All other classic heptalene syntheses start from 1,4,5,8-tetrahydronaphthalene (**262**), a hydrocarbon substrate the great preparative worth of which we already encountered in annulene chemistry, when 1,6-methano[10]annulene (**263**, see Section 10.5) was prepared from it in three steps. In fact, by cuprous chloride-promoted cyclopropanation Vogel and co–workers transformed **263** to the norcaradiene **264**, which, at 140 °C, isomerized to its bicyclic isomer **265**. When a hydrocarbon mixture of **264** and **265** was subjected to gas-phase pyrolysis at 400 °C and 0.2 torr the four isomeric dehydroheptalenes **259**, **260**, **266**, and **267** were produced in excellent yield (above 90%). Indirect oxidation as described in Scheme 43 then afforded heptalene (**206**) in 60% yield (Scheme 44).[148]

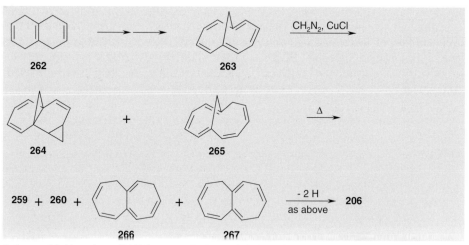

Scheme 44. Heptalene (**206**) from tetrahydronaphthalene (**262**): Vogel's first synthesis.

A route delivering 1,6-dihydroheptalene (**259**) exclusively was developed by Paquette and co–workers, who added appropriate amounts of dibromocarbene to **262** to furnish the bis adduct **268** in 45–50% yield (Scheme 45).[149]

When **268** was debrominated with methyllithium, regioselective twofold intramolecular cyclopropylidene carbene insertion took place to furnish the highly

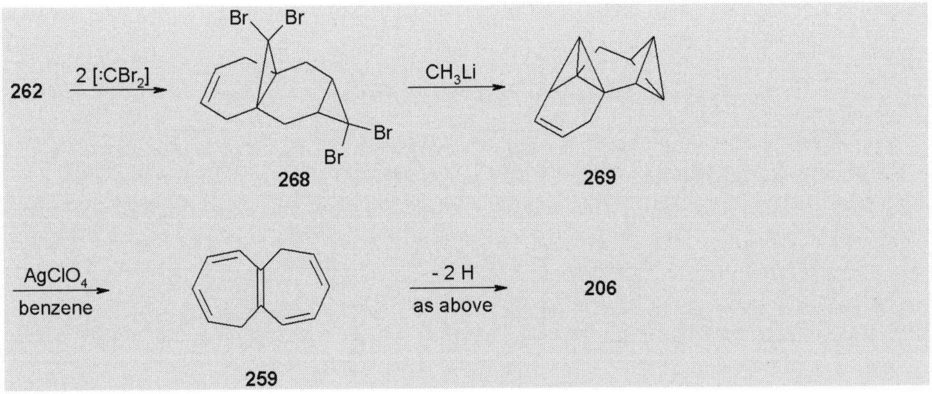

Scheme 45. Heptalene (**206**) from tetrahydronaphthalene (**262**): Paquette's synthesis.

strained bis-bicyclobutane **269** in 59% yield. Dissolving this in a benzene solution of silver perchlorate caused quantitative isomerization to **259**, exploiting the already described ability of silver ions to induce the isomerization of bicyclo[1.1.0]butanes to conjugated dienes (see Section 10.3).[150] Stepwise dehydrogenation of **259** again concluded this heptalene synthesis.

In a later, second synthesis of **206**,[151] which can also be employed to prepare selected heptalene derivatives,[152] Vogel and co–workers first blocked the central double bond of **262** by epoxidation (Scheme 46).

Subsequent dibromocarbene addition to **270** provided the *anti*-bis adduct **271**, which was debrominated by treatment with lithium in *tert*-butanol/tetrahydrofuran. NBS bromination of the resulting hydrocarbon **272** then furnished a mixture of tetrabromides of general structure **273** which—without purification—was reduced with zinc in THF to 3,8-dihydroheptalene (**274**). Dauben–Bertelli dehydrogenation again completed the synthesis. Although heptalene (**206**) can be crystallized from ethanol, it is an extremely reactive, highly oxygen-sensitive hydrocarbon. As indicated by the first ¹H NMR spectra, **206** must be a nonplanar, chiral cyclopolyolefin.[147] Its temperature-dependent NMR spectra indicate dynamic interconversions between different conformers comparable with the bond shifting and ring inversion processes observed for cyclooctatetraene (see Section 10.3).[153] The nonplanarity of the heptalene skeleton has been confirmed by X-ray structural analysis of various stable and crystalline derivatives such as dimethyl 1,2-heptalenedicarboxylate, the two seven-membered rings of which both prefer a boat structure in the crystalline state.[154] Not surprisingly, heptalene can also be stabilized by poly (methyl) substitution,[155] incorporation into larger (condensed) π-systems,[156] and metal complexation.[151, 157]

Numerous other 'short-circuited' annulenes containing pentalene, azulene, and heptalene as substructures are known,[146] and I will conclude this section by presenting the synthesis of just two interesting representatives which have recently been reported.

The kinetically stabilized 6-chloropentafulvene **276** is accessible on a 100-g scale by reaction of the easily available formylcyclopentadiene **275** with phosgene or oxalyl chloride (Scheme 47).[156, 158]

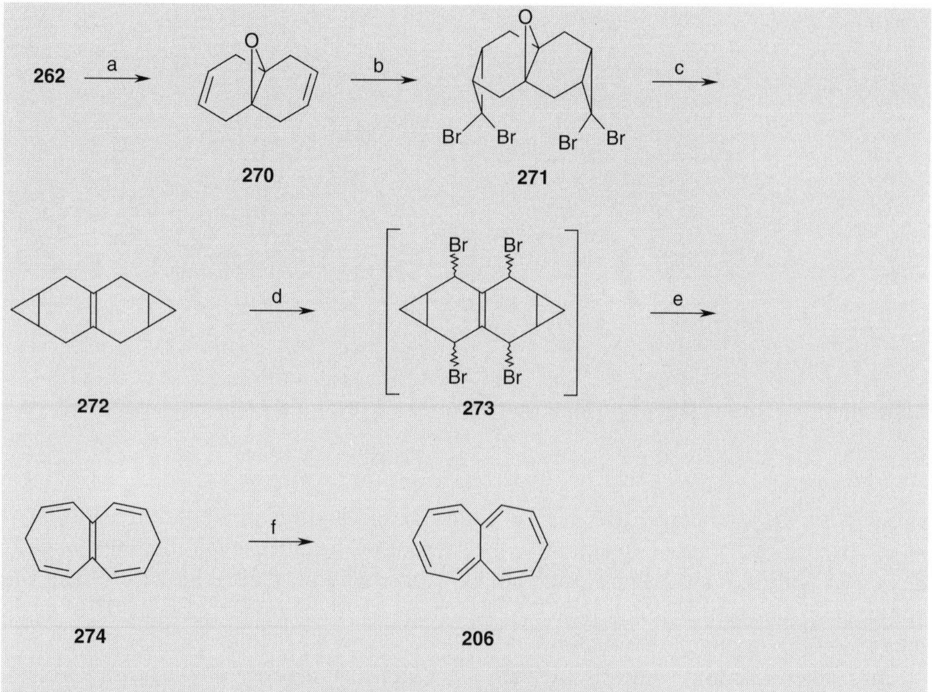

(a) CH$_3$CO$_3$H, CH$_3$CO$_2$Na, CH$_2$Cl$_2$ (81%); (b) CHBr$_3$, NaOH, phase transfer catalysis (35-70%); (c) Li, (CH$_3$)$_3$COH, THF (50-55%); (d) NBS, CCl$_4$; (e) Zn, THF (comb. yield 35-40%); (f) Ph$_3$C$^+$ BF$_4^-$, then (C$_2$H$_5$)$_3$N, (55-60%)

Scheme 46. Heptalene (**206**) from tetrahydronaphthalene (**262**): Vogel's second synthesis.

When **276** was metalated with lithium dialkylamides at –78 °C the fulvene carbenoid **277** was generated with mainly retention of configuration. Quenching this intermediate with methyl triflate produced the 6-chloro-6-methylfulvene **278**, which on dehydrochlorination provided the crystalline, yellow 5-vinylidene cyclopentadiene **279**,[159] a fulvenallene[160] with most remarkable and surprising properties. Between 0 and 25 °C it dimerized in a solid-state, topochemically controlled reaction[161] to the tetrahydrodicyclopenta[a,e]pentalene **280**, a linearly annelated tetraquinane hydrocarbon. Proton-catalyzed tautomerization of **280** yielded the isomeric hydrocarbon **281** with a 6-butadienylpentafulvene subsystem, and which could easily be oxidized with 2,3-dichloro-5,6-dicyano-1,4-benzoquinone (DDQ) to the fully conjugated tetracycle **282**, an extended stable pentalene derivative containing [4n] partial systems within a [4n+2]perimeter.

The versatility of **276** as a substrate for the preparation of short-circuited annulenes is underlined by the observation that on heating this pentafulvene derivative in methanol in the presence of traces of acid 1,3,5,7-tetrakis-*tert*-butyl-*s*-indacene (**285**) was produced in excellent yield (90%) (Scheme 48).[162]

Presumably, the reaction is triggered by (reversible) protonation of **276** to the resonance-stabilized cation **283**. Electrophilic addition of this intermediate to unreacted substrate then could lead to the pentafulvenyl pentafulvalene **284**, which by

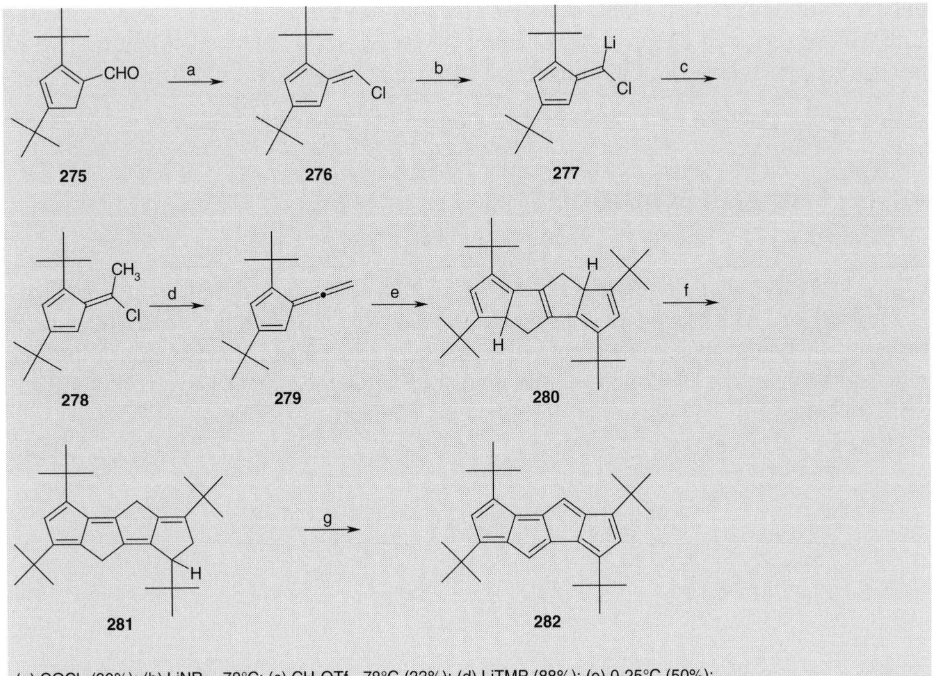

(a) COCl$_2$ (90%); (b) LiNR$_2$, -78°C; (c) CH$_3$OTf, -78°C (33%); (d) LiTMP (88%); (e) 0-25°C (50%); (f) H$^+$ (75%); (g) DDQ, Δ (72%)

Scheme 47. The synthesis of the first hydrocarbon with a [4*n*+2]perimeter having [4*n*] partial systems.

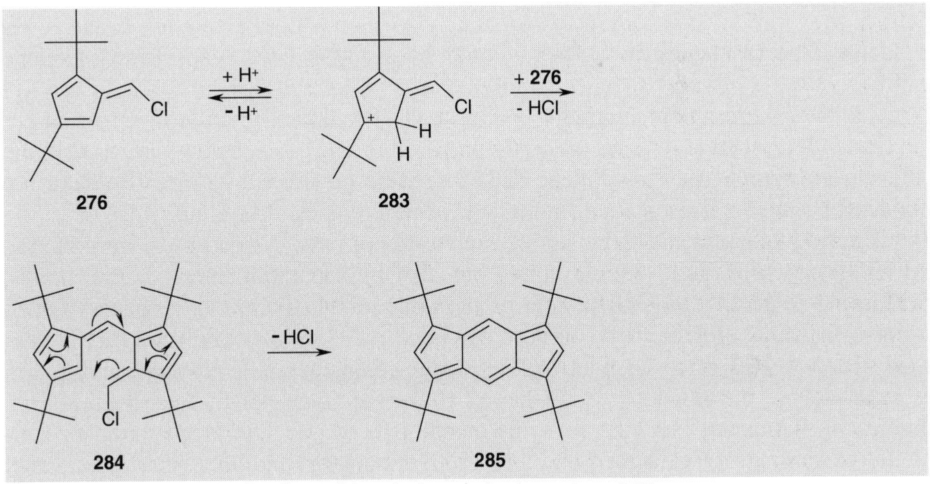

Scheme 48. The synthesis of a stable *s*-indacene hydrocarbon, **285**.

12π-electrocylization with subsequent loss of hydrogen chloride could provide **285**. The crystalline, brownish hydrocarbon is thermally stable, even on exposure to air, but its solutions are extremely sensitive towards oxygen or traces of acid. Formally **285** can be regarded either as a [12]annulene (see Section 10.4) disturbed by two σ-bonds or a *p*-xylylene (*p*-quinodimethane) derivative (see Section 12.3).

11.5 The [*n*]radialenes

Radialenes are alicyclic hydrocarbons in which all ring carbon atoms are sp^2-hybridized and which carry as many semicyclic double bonds as possible. The general term for the parent molecules is [*n*]radialenes where $n \geq 3$ and stands both for the ring size and the number of double bonds involved. Thus the parent hydrocarbons **286–289** are called [3]-, [4]-, [5]-, and [6]radialene, respectively (Scheme 49).[163]

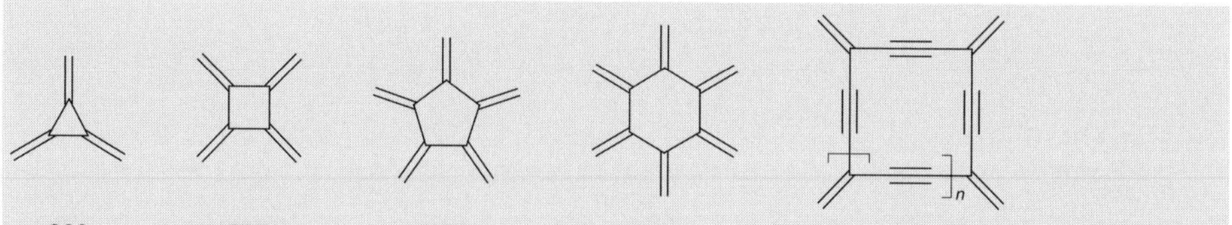

                286   287   288   289   290

Scheme 49: [*n*]Radialenes—the parent systems.

The radialenes have an uninterrupted cyclic arrangement of cross-conjugated π-systems and can formally be related to the dendralenes (see above, Section 11.1) of which they are a short-circuited version, *e.g.* connecting positions 2 and 4 in [3]dendralene (3-methylene-1,4-pentadiene) by a single bond leads to [3]radialene (**286**).

Radialenes are a comparatively young class of hydrocarbon; the first member, a hexamethyl derivative of **289**, was only prepared in 1961 (see below) and among the still elusive radialenes, [5]radialene (**288**) must be mentioned before all others, not the least because it is the formal 'monomer' of fullerene C_{60} (see Chapter 2).[164]

The ring-size of a radialene can be increased not only by successive introduction of additional semicyclic double bonds but also by separating these double bonds, within the respective ring, by additional unsaturated substructures ('π-spacers') such as benzene rings or triple bonds. These extended radialenes, symbolyzed by the general structure **290**, have been termed 'exploded radialenes'.[165] Although not as important as, *e.g.*, the annulenes (see Chapter 10) or the linear polyolefins (Section 7.1), interest in radialenes has been growing steadily for years. This is not only for synthetic reasons (development of new preparative methods) but also because of structural and theoretical interest, because radialenes are important reference structures in the development of computational methods. Radialenes carrying polarizing substituents are, furthermore, becoming interesting substrates in material science.[164]

Of the three conceivable synthetic strategies leading to the radialenes: (i) generation of the ring system first, followed by formation of the double bonds; (ii) preparation of the double bonds first, followed by ring closure; and (iii) generation of the desired ring system plus double bonds in one synthetic transformation, the first is by far the most important; the second has not yet been applied. It is not surprising that the first approach relies strongly on the common routes of alicyclic chemistry, *e.g.* carbene addition to appropriate double bonds systems (including cumulenes) to prepare appropriately substituted cyclopropanes, [2+2]cycloadditions of double bonds systems to furnisch cyclobutane derivatives, *etc.* There are, however, alternatives, as we shall see below, and pathways involving carbenes and carbenoids are not only of importance for the preparation of [3]radialenes.

Rather than following historical development I will begin our discussion of this class of hydrocarbon with the simplest representative, [3]radialene (**286**) and then proceed stepwise to the higher vinylogs.

The parent hydrocarbon **286** has been generated by classical β-elimination reactions from the dibromide **291** by Dorko (Scheme 50),[166] from the tris Hofmann base **292** and the triiodide **293** by Griffin and co-workers,[167] and from the tribromide **294** by Bally, Haselbach et al.[168]

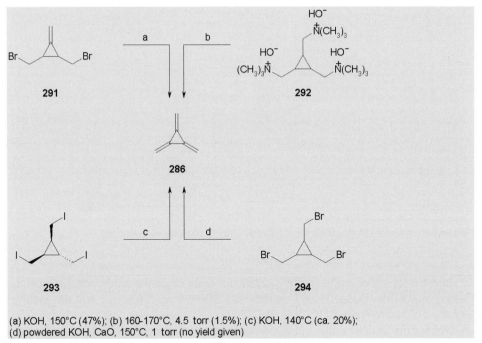

(a) KOH, 150°C (47%); (b) 160-170°C, 4.5 torr (1.5%); (c) KOH, 140°C (ca. 20%);
(d) powdered KOH, CaO, 150°C, 1 torr (no yield given)

Scheme 50. Various elimination routes to [3]radialene (**286**).

All these reactions have been performed as gas-phase processes, and the radialene formed has been collected at –63 °C or below. At –78 °C liquid **286** is stable for several days, but polymerization occurs readily when the vapor is exposed to room temperature, especially when it comes into contact with air.

Various alkyl- and aryl-substituted [3]radialenes could be prepared from 1,1-

dihalo-1-alkenes by use of organometallic pathways. In fact, the first [3]radialene to be synthesized, hexamethyl[3]radialene (**298**), was obtained by Köbrich and co–workers in very low yield by treatment of 1,1-dibromo-2-methyl-1-propene (**295**) with *n*-butyllithium in tetrahydrofuran at low temperatures. The lithium carbenoid **296** and permethylbutatriene (**297**) are probably intermediates in this transformation (Scheme 51).[169]

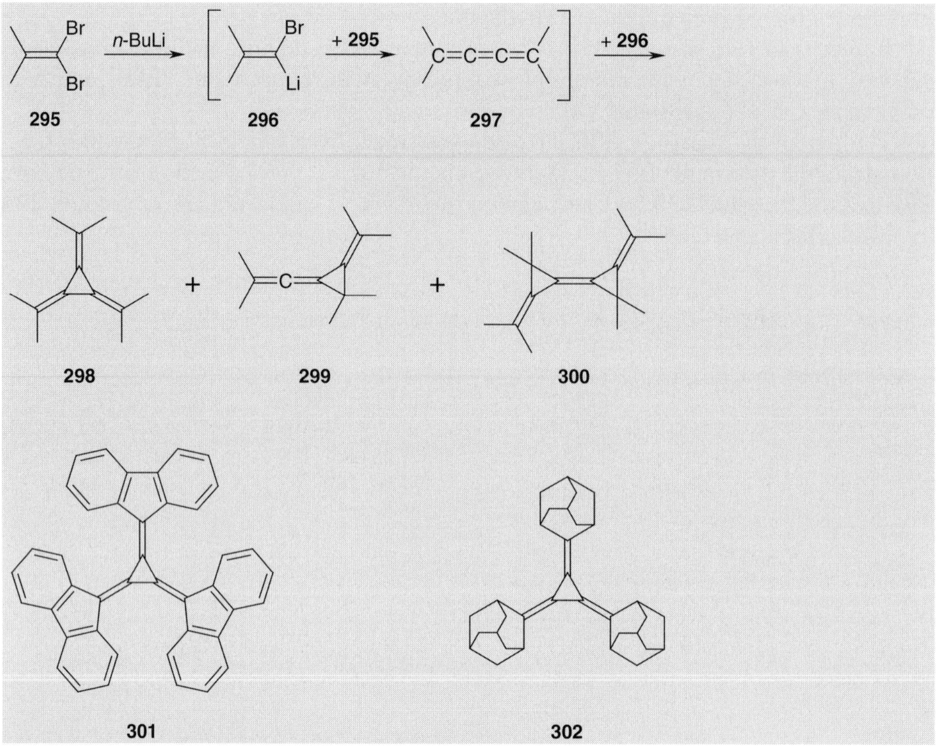

Scheme 51. Hexamethyl[3]radialene (**298**), the first [3]radialene hydrocarbon.

Because the outer double bonds of **297** are more readily cyclopropanated than the central one, other products such as **299** and **300** are also formed, and the yield of [3]radialene is correspondingly low (0.2–2%).

Neither the lithium carbenoid pathway nor the cyclopropanation of butatrienes is a general route to [3]radialenes. More successful is the cyclotrimerization of 1,1-dihaloalkenes *via* copper or nickel carbenoids, provided the substituents at the other end of the C-C double bond are not too small. Thus tris-(fluoren-9-ylidene)-cyclopropane (**301**)[170] and tris-(2-adamantylidene)cyclopropane (**302**)[171, 172] were synthesized from the corresponding unsaturated *gem* dihalides *via* copper and nickel organic intermediates by Iyoda and Oda[171] and Komatsu and co-workers,[172] respectively.

Following the classical investigations of West and Zecher on the preparation of

[3]radialenes with quinoid substructures[164, 173] tetrachlorocyclopropane and tetra-chlorocyclopropene (**303**) have often been used as building blocks for the preparation of functionalized, sometimes highly polarized [3]radialene derivatives. In hydrocarbon chemistry, the reaction of **303** with excess 1,5-bis(trimethylsilyl)1,4-pentadiyne in the presence of *tert*-butyllithium in THF at –78 °C caused a condensation/substitution reaction providing an intermediate which could be oxidized with DDQ to the hexaethynyl[3]radialene derivative **304** (Scheme 52).[174]

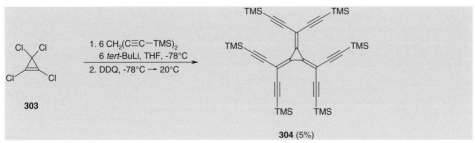

Scheme 52. Hexaethynyl[3]radialenes—precursors for novel carbon networks?

It was intended to use **304** as a building block for novel carbon networks; however, desilylation under very mild conditions led to an unstable product of so far unknown nature. The extended conjugation present in **304** manifests itself in the deep-red color of its crystals.

[4]Radialenes represent the biggest and best investigated subset of the radialene family, and more methods exist for preparing these hydrocarbons than for any other class of radialene. The major routes towards **287** and its derivatives make use of appropriately functionalized cyclobutanes and cyclobutenes, the thermal and Ni(0)-catalyzed cyclodimerization of butatrienes or higher cumulenes, and the cyclotetramerization of (1-bromo-1-alkenyl)cuprates.

Cyclobutane derivatives are involved in most known syntheses of the parent hydrocarbon **287**. It was synthesized first by Griffin and Peterson in 1962 by subjecting the tetra(aminoxide) **305a** to Cope elimination or the tetra(ammonium) salt **305b** to Hofmann degradation; in both instances the yields were poor and the product contaminated (Scheme 53).[175]

Later the tetrabromide **306** was dehydrobrominated with either sodium hydroxide in ethanol at 0 °C (yield < 50%)[176] or by adding it slowly to a molten KOH–water eutectic mixture at 150 °C while passing a stream of helium through the reaction vessel.[177] A novel approach by Hopf and Trabert used the well known thermal cyclization of 1,5-hexadiyne and its derivatives to 3,4-bismethylenecyclobutenes (see Section 10.2).[178] When this electrocyclization was applied to the bisprotected diol **307** the bismethylenecyclobutene **308** was formed in good yield. This intermediate, formed by a tandem process ([3.3]sigmatropic rearrangement, then electrocyclization of the resulting bisallene intermediate) contains not only the four-membered ring and two of the ultimately required double bonds, but it is also set up for the final transformations—hydrolysis of **308** led to the corresponding diol which was converted to

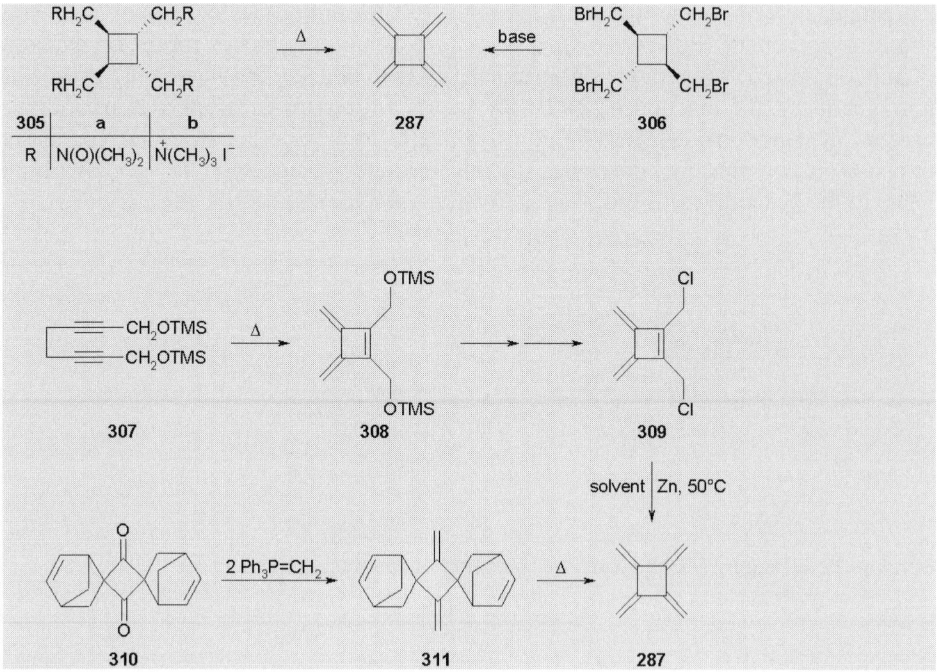

Scheme 53. The known routes to [4]radialene (**287**).

the dichloride **309** by treatment with thionyl chloride. Reductive 1,4-dechlorination of the latter with zinc in a high boiling solvent such as ethyleneglycol di-*n*-butyl ether and rapid removal of **287** from the reaction mixture, as it is formed, by vacuum transfer then yielded the hydrocarbon in 77% yield[179]—for the first time in amounts large enough to enable comprehensive investigation of the chemical behavior of [4]radialene.[180]

Finally, Ripoll and co-workers have olefinated the 1,3-cyclobutandione derivative **310** by a Wittig reaction to the spiro hydrocarbon **311**, which, when heated to 680 °C at 5×10^{-5} torr, underwent double retro-Diels–Alder cleavage to **287** with loss of cyclopentadiene.[181]

The thermally and photochemically induced [2+2]cyclodimerization of butatrienes across the central C–C double bond or of a higher cumulene at an inner double bond, seems on first sight as a reasonable route to [4]radialenes. In practice, however, this process fails as often as it is successful. Butatriene itself, as we have seen in Section 9.2, yields 1,5-cyclooctadiyne and other products, but no **287** on warming. In contrast to earlier assumptions, the photochemical dimerization of tetraphenyl- and tetrakis(4-methoxyphenyl)butatriene does not yield the respective [4]radialene derivatives, but occurs at one of the terminal C–C double bonds to give a head-to-tail dimer.[182] Among the acyclic butatrienes, apparently only those carrying electron-withdrawing substituents are able to dimerize to [4]radialenes, the high yield formation of octachloro[4]radialene from perchlorobutatriene being a particularly important example.[183]

Higher cumulenes, on the other hand, do dimerize thermally to derivatives of **287**.

Thus **312** was formed from the corresponding pentatetraene[184] and **313** from a hexapentaene 'half'.[185] That the dimerization of a cumulene to a [4]radialene is a complex process which is not only influenced by steric and electronic effects of substituents but also by the presence of certain metal catalysts is demonstrated *inter alia* by 1,2,3-cycloheptatriene (**315**), one of the smallest cyclocumulenes ever prepared. As was mentioned in Section 9.4, **315** was released when the bridged bicyclo[1.1.0]butane derivative **314** was treated with cesium fluoride. This highly reactive intermediate could be trapped by various dienes, but it did not dimerize. In the presence of Ni(PPh₃)₄, however, the doubly bridged [4]radialene **316** was formed (Scheme 54).[186]

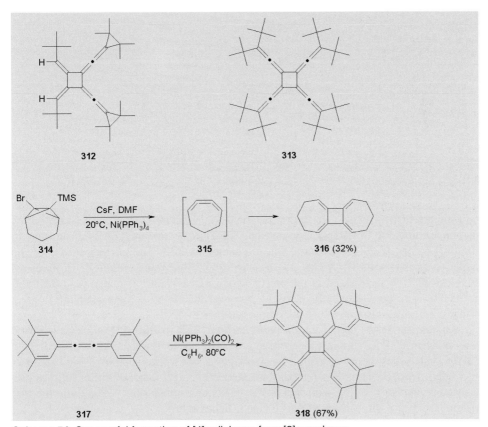

Scheme 54. Successful formation of [4]radialenes from [3]cumulenes.

Other Ni(0)-mediated syntheses of [4]radialenes from [*n*]cumulenes are known, such as the dimerization of the cross-conjugated butatriene **317** to the radialene **318**, which also took place in good yield,[187] and of permethylbutatriene to permethy[4]radialene.[188, 189] On the other hand, tetraphenylbutatriene could not be cyclodimerized in the presence of catalytic amounts of Ni(PPh₃)₄ in benzene.[189]

Instead of employing butatrienes as substrates in these reactions, 2,3-dihalo-1,3-

butadienes can also be used as radialene precursors. This approach has proven its preparative worth especially for the synthesis of peralkylated[189] and various functionalized [4]radialenes.[190]

The elusive octaphenyl[4]radialene (**323**) was finally prepared by Iyoda and co-workers from 1,1-dibromo-2,2-diphenylethene (**319**) *via* organocuprate intermediates (Scheme 55).[191]

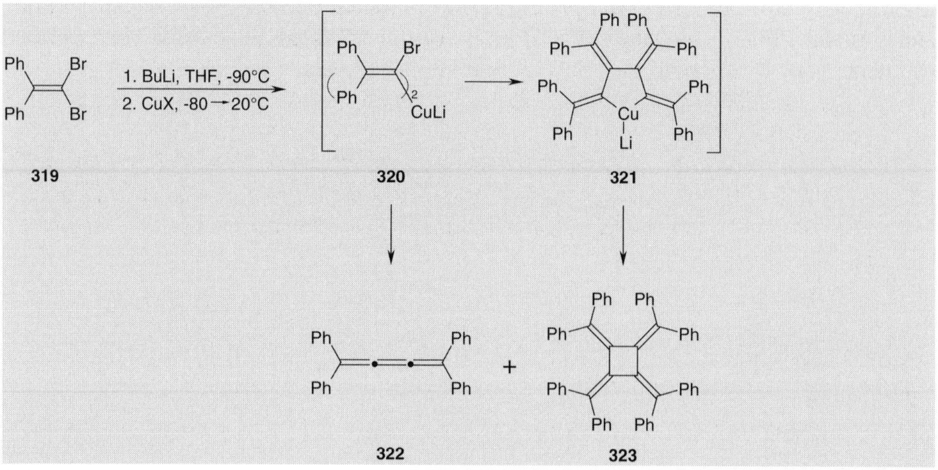

Scheme 55. The preparation of octaphenyl[4]radialene (**323**) by Iyoda and co-workers.

The initial intermediate in the 'tetramerization' of **319** is the cuprate **320**, which can either dimerize to tetraphenylbutatriene (**322**) or cyclize under loss of bromine to the cyclic cuprate **321**, the immediate precursor of **323**. The correct choice of the added copper(I) salts is crucial for the success of this transformation. Best results (41% of **323**) were realized in the presence of CuCN. With this copper salt **322** was formed in trace quantities only; it might, however, become the main coupling product when other salts, *e.g.* CuI-PBu$_3$, are employed.

Among the series of the parent hydrocarbons **286** to **289**, [5]radialene (**288**) is still unknown. The simplest derivative reported so far is decamethyl[5]radialene (**325**) which has been prepared by Iyoda and co-workers from 1,1-dibromo-2-methylpropene (**324**) by low-temperature metalation with *n*-butyllithium followed by a metal exchange reaction with nickel[192] or, better, copper salts,[193] and the thermal decomposition of the carbenoid thus formed (Scheme 56).

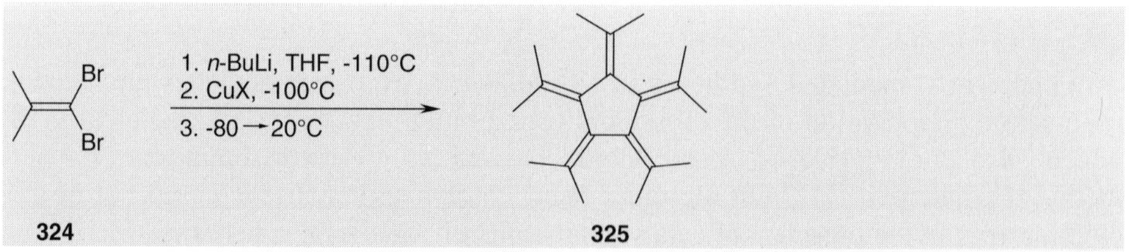

Scheme 56. Iyoda's synthesis of the first [5]radialene hydrocarbon, **325**.

The yield of **325** varies—it was only 14% with CuBr·SMe₂, but it more than doubled to 32% when CuI·P(n-Bu)₃ was employed. The formation of the [5]radialene was accompanied by di-, tri-, and tetramerization of the dimethylvinylidene intermediate produced from **324**, leading to permethylbutatriene, and the respective permethylated [3]- and [4]radialenes. It is probable, however, that this important, formally so simple reaction, will remain a singular event. Whenever the bulkiness of the substituents in **324** was increased the amount of [3]- and [4]radialenes produced increased drastically. Several derivatives of **288** have been described with sulfur-containing substituents around the perimeter of the [5]radialene nucleus;[194] these compounds are of interest as electron-rich π-donors in material science.[195]

The preparation of the 'totally benzoannelated' derivative of [5]radialene, the bowl-shaped hydrocarbon corannulene, will be described in Section 12.2.

The first [n]radialene ever to be prepared was all-(E)-7,8,9,10,11,12-hexamethyl[6]-radialene (**327**) obtained in 1961 by Hopff and Wick from either the hexachloride or hexabromide **326** by treatment with magnesium in methanol (Scheme 57).[196]

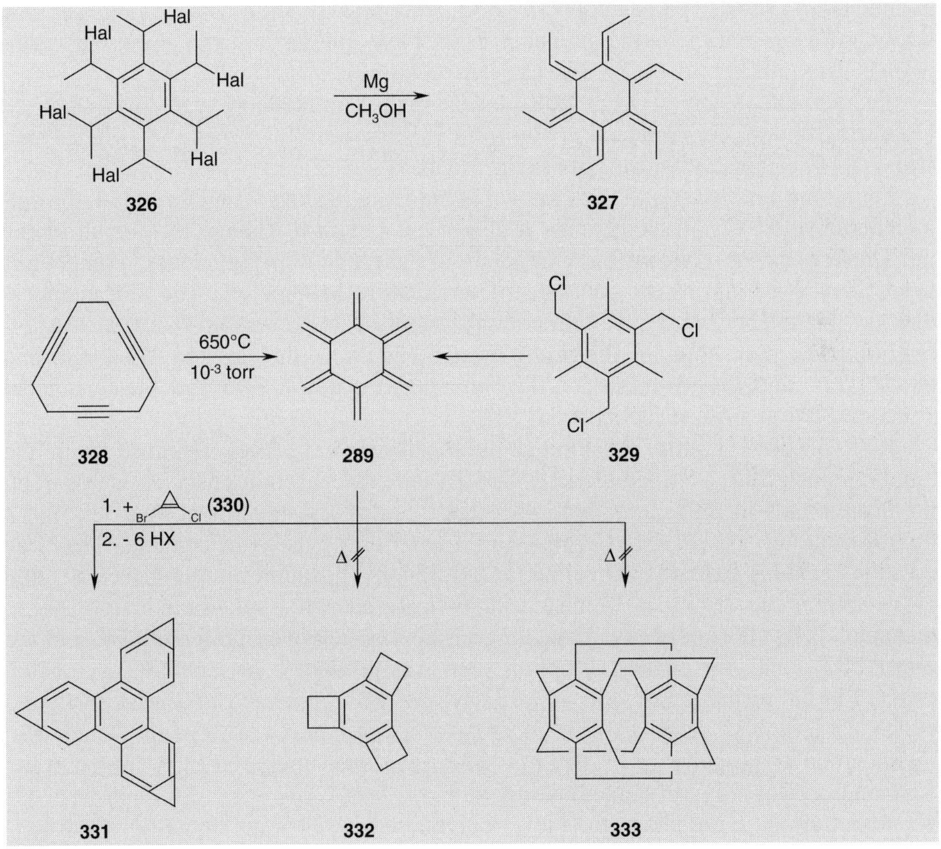

Scheme 57. The first [n]radialene—all-(E)-hexamethyl[6]radialene (**327**) and different routes to [6]radialene (**289**).

Considering that the generation of **327** requires the cleavage of six carbon-halogen bonds, the isolated [6]radialene yield of *ca* 30% is certainly acceptable. When instead of the hexahalide **326** the corresponding hexabromohexa-*n*-propylbenzene derivative was dehalogenated, the expected hexaethyl[6]radialene (hexapropylidenecyclohexane) was produced.[197] The isomer shown (**327**), with its characteristic paddle-wheel configuration—proven by X-ray structural analysis[198]—is the main elimination product, but not the only diastereomer, as assumed by the original authors.[199]

The hexaalkyl[6]radialenes are stable solids, which can be sublimed and handled readily in air. This is in pronounced contrast to the parent hydrocarbon [6]radialene (**289**), obtained nearly simultaneously by three different research groups.

Whereas the thermal isomerization of 1,5,9-cyclododecatriyne (**328**) to **289** is of no practical importance,[200] the thermal dehydrochlorination of 2,4,6-tris(chloromethyl)mesitylene (**329**) is a preparatively useful procedure (isolated yield *ca* 50%),[201, 202] making **289** available for further transformations, *e.g.* Diels–Alder additions.[201] Among these the threefold addition of 1-bromo-2-chloro-cyclopropene (**330**) followed by dehydrohalogenation led to tricycloproparene (**331**) (see Section 16.3).[203] In contrast, neither the electrocyclization of **289** to the trisbutarene **332** nor its dimerization to [2₆]paracyclophane (**333**, superphane) could be accomplished. We shall, however, return to the preparation of these interesting and important compounds later (see Sections 12.3 and 16.3, respectively).

The permethyl derivative of **289**, dodecamethyl[6]radialene, was obtained when tetramethylbutatriene, produced *in situ* from 3,4-dibromo-2,5-dimethyl-2,4-hexadiene with a Ni(0) catalyst, was trimerized with Ni(0).[204]

For a long time very little was known about [*n*]radialenes with larger than the six-membered rings. Only recently this situation has begun to change as a result of the introduction of the concept of 'extended' or 'exploded' [*n*]radialenes, compounds symbolized *inter alia* by the general structure **290** in Scheme 49. The first members of these interesting π-systems, with butadiyne units as spacer groups between the sp²-hybridized carbon atoms of the original radialene, were prepared by Diederich and co–workers, and the syntheses of the square hydrocarbon **336** and the hexagonal structure **337** are typical (Scheme 58).[205, 206]

Both of these highly unsaturated hydrocarbons have been obtained from the building block **334**, a tetraethynylethene derivative (see Section 15.2). As is apparent, the triple bonds in **334** carry three different types of substituents at their termini: a hydrogen atom, a triethylsilyl group, and a tris-isopropylsilyl substituent. The 'free position' enables oxidative dimerization of **334** by Eglinton coupling (see Section 10.4). When the resulting 'dimer' was hydrolyzed with potassium carbonate in methanol-THF the triethylsilyl protecting group was removed preferentially, and the dimer **335**, ready for another coupling step, was produced. Its Eglinton cyclization resulted in the formation of **336** and **337** which were generated in remarkably good yields (15 and 20%, respectively). Other stretched radialenes such as the pentagonal hydrocarbon corresponding to **336** and **337** have been prepared by a similar route from appropriate polyacetylenic building blocks.

Although not a radialene, the bicyclic hydrocarbon **344**, colloquially called heri-cene from the Latin word *hericeus* for hedgehog, shows some family relationship to the hydrocarbons discussed so far in this section, because it is also composed of as

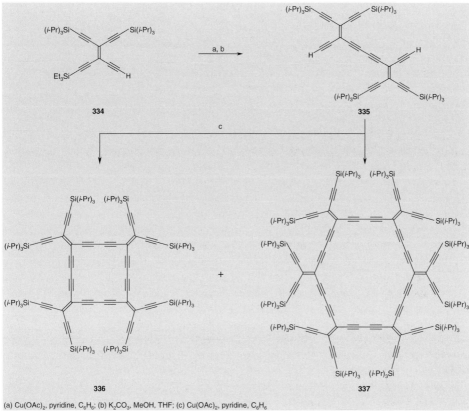

(a) Cu(OAc)$_2$, pyridine, C$_6$H$_6$; (b) K$_2$CO$_3$, MeOH, THF; (c) Cu(OAc)$_2$, pyridine, C$_6$H$_6$

Scheme 58. The preparation of the extended [*n*]radialenes **336** and **337** by Diederich and co–workers.

many semicyclic double bonds as possible. The hexaene may be regarded as a 'three-dimensional version' of a radialene and could equally well be discussed in Chapter 13 in which three-dimensional oligoolefins will be presented in a broader context.

[2.2.2]Hericene (**344**) has been synthesized by P. Vogel and co–workers. by the route summarized in Scheme 59.[207, 208]

Thus double Diels–Alder addition of excess maleic anhydride to coumalic acid (**338**) at 200 °C furnished the bis anhydride **339** which was esterified to the pen-taester **340** with methanol–sulfuric acid. Note that **338** is a derivative of α-pyrone, which we have already encountered in Section 10.1 as a precursor for cyclobutadiene. On heating **340** in the presence of potassium cyanide, acetone cyanohydrin, and tet-rabutylammonium cyanide in anhydrous acetonitrile at 80 °C Michael addition took place and the cyano pentaester **341** was obtained in good yield. Under these condi-tions base-catalyzed epimerization occured simultaneously and provided **342** as the major product. Hydrolysis of the latter, then esterification with methanol led to the all-*trans* hexaester **343**, which was converted into the hydrocarbon **344** by routine steps (reduction with lithium aluminum hydride, chlorination with thionyl chloride, base-induced dehydro-chlorination). By modification of the reaction conditions the

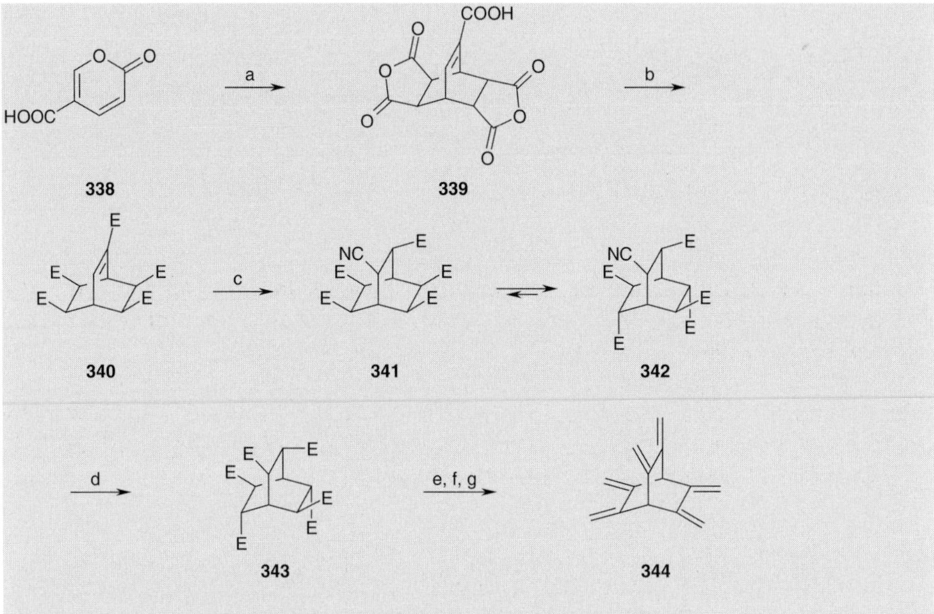

(a) 3 equiv. maleic anhydride, 200°C; (b) MeOH, H+ (87%); (c) KCN, Me$_2$C(OH)CN, n-Bu$_4$NCN, CH$_3$CN (97%); (d) CH$_3$COOH, H$_2$SO$_4$, then MeOH, H+ (85%); (e) LiAlH$_4$, THF (45%); (f) SOCl$_2$, pyridine (45%); (g) tert-BuOK, THF, 3 d (total yield from **340**: 7%)

Scheme 59. The synthesis of [2.2.2]hericene (**344**) by P. Vogel et al. (E in all structures = COOCH$_3$).

overall yield of the **340**→**344** transformation could be increased from the 7% given in the Scheme to nearly 30 %, enabling comprehensive investigation of the chemical and spectroscopic properties of this prototype hericene.[208]

11.6 The [*n*]rotanes

[*m.n*]Rotanes are a special class of polyspiranes (polyspiro compounds) composed of a *m*-membered central ring (*m* = 3, 4, 5...) and *n*-membered peripheral rings (*n* = 3, 4, 5...) such that each atom of the central ring is at the same time part of a peripheral ring.[209] Originally these compounds were just called [*n*]rotanes and restricted to polyspiranes of three-membered rings,[210] and it is this class of rotanes on which this section will concentrate. Thus the hydrocarbons in the series **345** to **348** have traditionally been called [3]- (**345**), [4]- (**346**), [5]- (**347**), and [6]rotane (**348**), whereas in the more comprehensive nomenclature **345** would be called [3.3]rotane, **346** [4.3]rotane *etc.* (Scheme 60).

That the chemical properties of the C–C double bond and the cyclopropane ring are similar, has often been noted, and it hence appears logical to let a section on

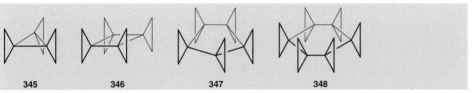

Scheme 60. The [*n*]rotane family.

[*n*]rotanes follow a discussion of the [*n*]radialene family. Actually the radialene/rotane relationship is more than a formal one, because the former have been converted to the latter by exhaustive cyclopropanation (see below).

The smallest and most highly strained[211] member of the rotane family, [3]rotane (**345**), was first prepared from bicylopropylidene (**349**, see Section 7.2 and below) by Conia and Fitjer in 1973 by the addition of cyclopropylidene generated from 1-cyclopropyl-1-nitroso-urea (**350**) by treatment with base (Scheme 61).[212]

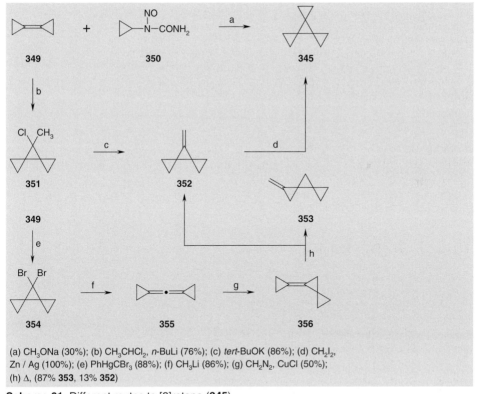

(a) CH₃ONa (30%); (b) CH₃CHCl₂, *n*-BuLi (76%); (c) *tert*-BuOK (86%); (d) CH₂I₂, Zn / Ag (100%); (e) PhHgCBr₃ (88%); (f) CH₃Li (86%); (g) CH₂N₂, CuCl (50%); (h) Δ, (87% **353**, 13% **352**)

Scheme 61. Different routes to [3]rotane (**345**).

Most other reported [3]rotane syntheses also start from **349**. Thus Erden performed the spiroalkylation step by treating **349** with chloromethylcarbene, produced by α-elimination of 1,1-dichloroethane with *n*-butyllithium. The resulting chloride **351** was converted to the spiroolefin **352** by base-induced dehydrochlorination, and subsequent Simmons–Smith cyclopropanation then gave the desired **345** in high yield.[213]

In a second route to **345** Conia and Fitjer added dibromocarbene (produced by decomposition of tribromomethyl phenylmercury) to **349** and opened the resulting dibromo derivative **354** with methyllithium to the allene **355** (DMS allene synthesis, see Section 9.1). When the latter hydrocarbon was cyclopropanated with di-azomethane-CuCl a hydrocarbon mixture resulted, which contained **356** in *ca* 50% yield. Thermolysis of **356** (200 °C, 150–200 torr) subsequently provided a 87:13 mixture of **352** and **353**, and methylenation of the former by a modified Simmons–Smith reaction finally led to **345**.[214]

As can be seen in Scheme 61 the crucial substrates and intermediates in these routes to **345** are the two hydrocarbons **349** (bicyclopropylidene) and **352**, and because these are important hydrocarbons in their own right[215] two directed approaches to these cyclopropane derivatives are summarized in Scheme 62.

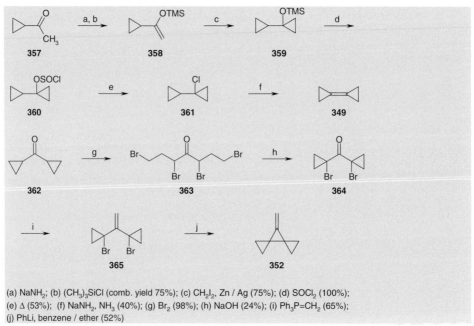

(a) NaNH₂; (b) (CH₃)₃SiCl (comb. yield 75%); (c) CH₂I₂, Zn / Ag (75%); (d) SOCl₂ (100%);
(e) Δ (53%); (f) NaNH₂, NH₃ (40%); (g) Br₂ (98%); (h) NaOH (24%); (i) Ph₃P=CH₂ (65%);
(j) PhLi, benzene / ether (52%)

Scheme 62. The preparation of the rotane precursors bicyclopropylidene (**349**) and 7-methylene-dispiro[2.0.2.1]heptane (**352**).

Cyclopropylmethylketone (**357**), a commercially available starting material, was first converted into its silyl enolether **358** by conventional methodology. Simmons–Smith cyclopropanation to **359** followed by replacement of the trimethylsiloxy group by chloride, by use of thionylchloride, and heating of the intermediately produced **360** gave the chloride **361**. When this was dehydrochlorinated with sodium amide in liquid ammonia 1-cyclopropyl-cyclopropene was produced which isomerized to **349** under the reaction conditions.[216] To prepare 7-methylene-dispiro[2.0.2.1]heptane (**352**) in a preparatively effective way the two three-membered rings of dicyclopropylketone (**362**) were cleaved by bromine treatment to provide the acyclic tetrabromoketone **363**. Recyclization was accomplished by dehydrobromination with sodium hydroxide, and when the resulting 1,1'-dibromodicyclopropylketone (**364**) was olefinated, **365**, as the immediate precursor for **352**, was obtained. To induce the intramolecular coupling **365** was subjected to the action of phenyllithium in benzene-ether.[217, 218]

The next higher homolog in the series, [4]rotane (**346**), was also first synthesized by the Conia group (Scheme 63).[219]

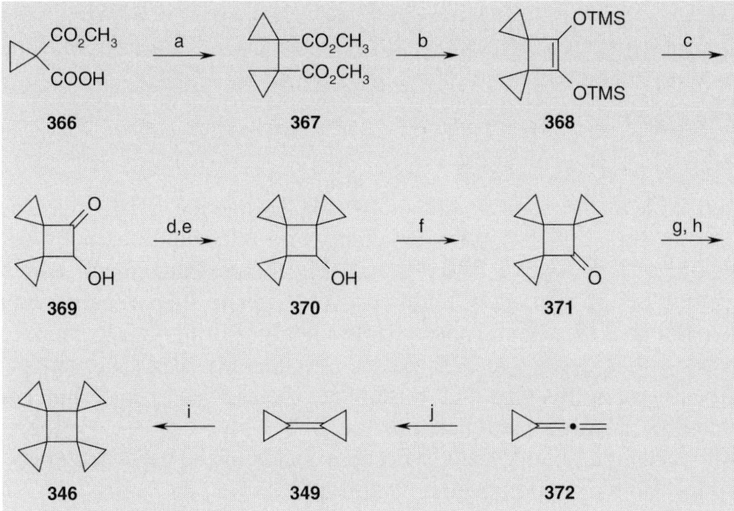

(a) electrolysis, Pt electrode (30%); (b) Na/(CH₃)₃SiCl, toluene (42%); (c) H₂O, acetone (70%); (d) Wittig (60%); (e) Simmons-Smith (50%); (f) CrO₃, pyridine (100%); (g) Wittig (100%); (h) Simmons-Smith (80%); (i) 210°C (15%); (j) Simmons-Smith (33%)

Scheme 63. Two routes to [4]rotane (**346**)—by Conia and co–workers...

The synthesis began with the electrochemical dimerization of the monoester of cyclopropane-1,1-dicarboxylic acid, **366**; this provided the bicylopropyl derivative **367** in fair yield. The cyclobutane nucleus was then formed by an acyloin condensation (Rühlmann variant) and the resulting bis-silyl enolether **368** was converted to the free acyloin **369** by treatment with aqueous acetone. The missing three-membered rings were next introduced by standard organic methods—Wittig reaction of **369** followed by Simmons–Smith cyclopropanation transformed the ketol **369** into the trispiro derivative **370**, which after Sarett oxidation to the ketone **371**, yielded [4]rotane (**346**) by another Wittig and then Simmons–Smith sequence. Subjecting **346** to a retro[2+2]cycloaddition led to bicyclopropylidene (**349**) again. And indeed heating this reactive olefin—prepared this time by methylenation of the allene **372**—

at 210 °C furnished **346** in 15% yield, accompanied (40%) by methylenespiro-[2.2]pentane formed by thermal isomerization of **349**.[220] Trispiro[2.0.2.0.2.1]decan-10-one (**371**)[217] also served as a precursor for [4]rotane in a synthesis developed by de Meijere et al. (Scheme 64).[221]

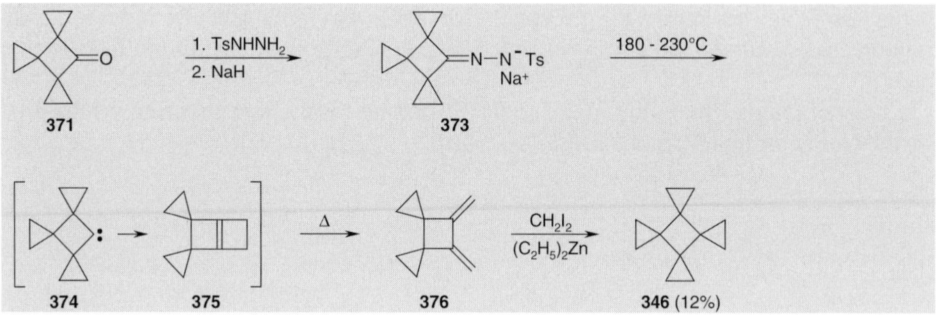

Scheme 64. ...and by de Meijere.

Having converted **371** to the sodium salt of its tosylhydrazone, **373**, this was thermally decomposed at 180–230 °C to provide–along with other hydrocarbons–the 1,2-bismethylenecyclobutane derivative **376**. Presumably the rearrangement, which temporarily destroys one of the ultimately required cyclopropane rings, begins with the formation of the carbene **374**, which subsequently undergoes ring-enlargement to the bicyclo[2.2.0]hex-1(4)-ene derivative **375**. As we have already seen (Section 7.4) highly strained hydrocarbons of this type tend to stabilize themselves by cyclobutene-to-butadiene ring-opening. Double methylenation of **376** concluded the synthesis which yielded **346** in varying amounts, mainly because yields of the carbene generation step depend strongly on the decomposition conditions.

The formal relationship between the [n]radialenes and the [n]rotanes has already been discussed (see Section 11.5), and, indeed, carbene addition—methylene itself, dichloro- and dibromocarbene—to [4]radialene (**287**) led to various 'rotaradialenes' and finally to the fully cyclopropanated starting material. The best results were obtained with dichlorocarbene which yielded the bis adducts **377** and **378**, and—by a third dichlorocarbene addition—the tris adduct **379** and finally octachloro[4]rotane (**380**) (Scheme 65).[180]

The structures of several of these adducts (**377**, **380**) were proven by X-ray structural analysis.

The first rotane to be prepared was [5]rotane (**047**), synthesized by Conia and Ripoll in 1969 from the cyclohexene derivative **381**, as shown in Scheme 66.[210]

The first part of the synthesis is concerned with the construction of the five-membered core of the target molecule, and this goal was accomplished by ozonolysis (oxidative work-up) of **381**, treating the resulting diacid **382** with acetic anhydride, and subsequent thermolysis at 250 °C. Two further cyclopropane rings were introduced into the ring-contracted ketone obtained, **383**, by a four-step sequence involving condensation with four equivalents of formaldehyde, transformation of the resulting tetraalcohol *via* its tetratosylate to the corresponding tetrabromide, and finally

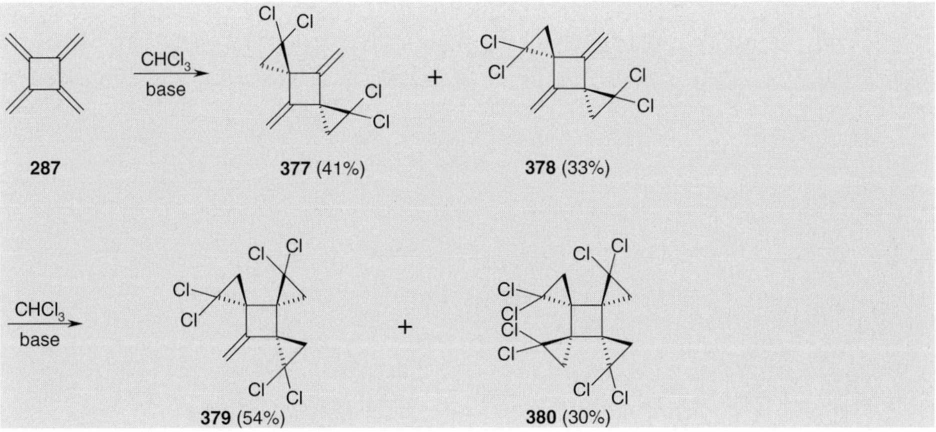

CHCl3
base
Cl
Cl
+
287
377 (41%)
378 (33%)
379 (54%)
380 (30%)

Scheme 65. From radialenes to rotanes.

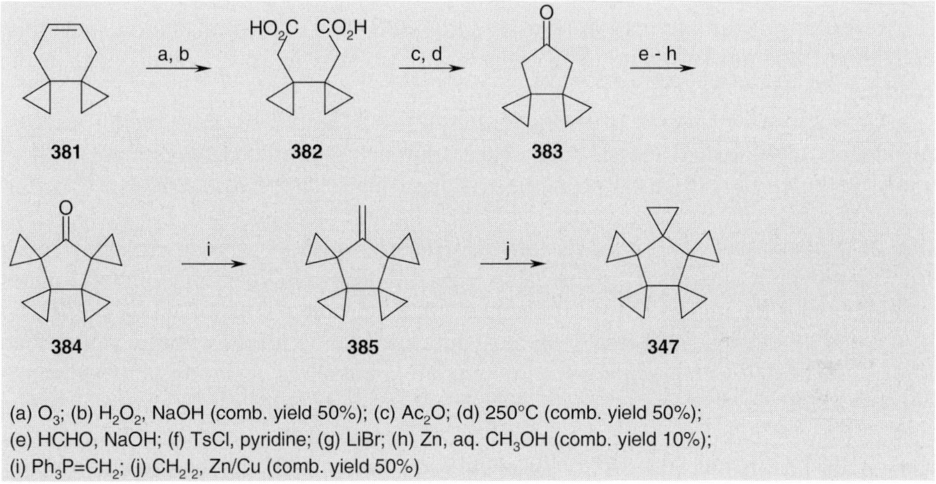

a, b
HO2C CO2H
c, d
e - h
O
i
j
381
382
383
384
385
347
(a) O3; (b) H2O2, NaOH (comb. yield 50%); (c) Ac2O; (d) 250°C (comb. yield 50%);
(e) HCHO, NaOH; (f) TsCl, pyridine; (g) LiBr; (h) Zn, aq. CH3OH (comb. yield 10%);
(i) Ph3P=CH2; (j) CH2I2, Zn/Cu (comb. yield 50%)

Scheme 66: [5]Rotane (**347**)—the first rotane to be synthesized.

reductive ring-closure with zinc to the ketone **384**—an admittedly conventional synthesis signalizing that there was yet no means of convenient introduction of cyclopropane rings into the 'growing' rotane system. Wittig olefination, followed by Simmons–Smith cyclopropanation concluded the preparation of [5]rotane (**347**), which was obtained in 1.25% overall yield.

One synthesis of [6]rotane (**348**), that of de Meijere and Proksch,[222] exploits the observation that cyclopropyl hydrocarbons are selectively oxidized by ozone in the α-position to give cyclopropylketones in high yield.[223] When this reaction was applied to the tetraspirane **388**—readily obtained from the diketone **386** *via* **387** by Wittig olefination and Simmons–Smith cyclopropanation, the bis ketone **389** was obtained (Scheme 67).[222]

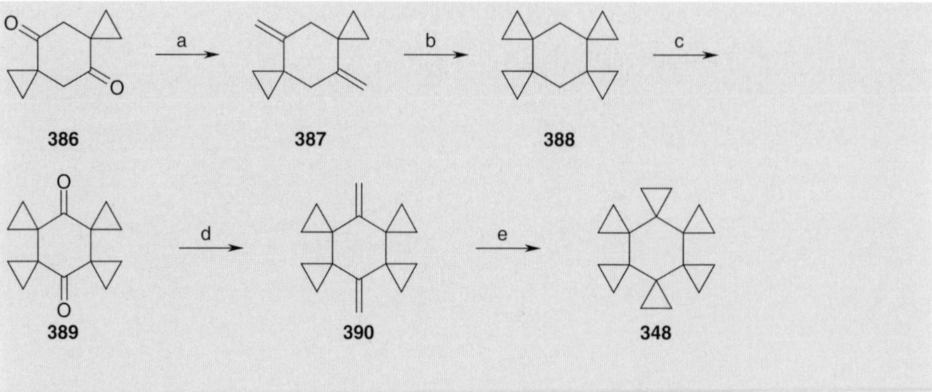

386   **387**   **388**

389   **390**   **348**

(a) Ph$_3$P=CH$_2$ (76%); (b) CH$_2$I$_2$, Zn(C$_2$H$_5$)$_2$ (70%); (c) O$_3$, SiO$_2$ (61%); (d) Ph$_3$P=CH$_2$ (67%); (e) CH$_2$I$_2$, Zn(C$_2$H$_5$)$_2$ (72%)

Scheme 67. The synthesis of [6]rotane (**348**) by de Meijere and Proksch.

Repetition of the olefination (formation of **390**) and cyclopropanation steps then converted **389** into [6]rotane (**348**) which was isolated in 16% overall yield in the five-step sequence shown.

Although the syntheses of the parent systems **345** to **348** are certainly accomplishments to be called classic, and these approaches provided enough material to study *inter alia* the structural properties of these interesting hydrocarbons,[224] an inherent disadvantage of these syntheses is their mutual independence. For the preparation of this series of 'cyclopropylogous' hydrocarbons an iterative approach, starting from a common precursor and leading successively to higher and higher rotanes would be most desirable. This formidable synthetic problem was solved by Fitjer who developed an elegant 'universal' rotane synthesis yielding all the rotanes higher than three.[225] The decisive precursor for this unified approach is cyclopropylidenedispiroheptane (**393**), which was prepared from the dibromide **364**, a substrate which had proven its worth already for the synthesis of [3]rotane (**345**, see above). Wittig reaction of **364** with the ylid **391** yielded the dibromide **392** which was cyclized intramolecularly by treatment with phenyllithium (Scheme 68).[225]

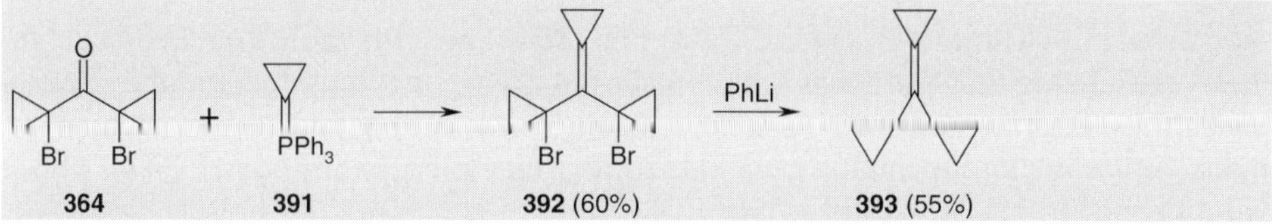

364   **391**   **392** (60%)   **393** (55%)

Scheme 68. Preparation of cyclopropylidenedispiroheptane (**393**), the starting material for Fitjer's universal rotane synthesis.

Hydrocarbon **393** and structurally similar compounds, **394**, underwent a highly regioselective [3+2]cycloaddition with *p*-nitrobenzenesulfonylazide to yield the unstable triazoline adducts **395**. By nitrogen extrusion accompanied by ring expansion these were transformed into the imines **397**, which on hydrolysis with potassium hydroxide in methanol furnished the (central) ring-expanded ketones **396**. At this point the synthesis can follow either of two courses—methylenation with methylenetriphenylphosphorane gave the olefins **398** which were converted into the rotanes **399** by cyclopropanation, whereas Wittig reaction between the ketones **396** and the ylid **391** led back to the (now ring-enlarged) cyclopropenylidenes **394**—and the cycle can start all over again (Scheme 69). The method has been used to prepare [4]- (**346**), [5]-(**347**) and—for the first time—[6]rotane (**348**), and it has been extended to the preparation of precursors for [7]- and [8]rotane.[226]

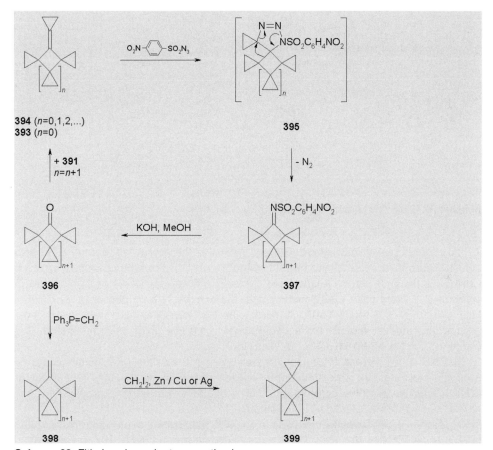

Scheme 69. Fitjer's universal rotane synthesis.

With the chemistry of the rotanes composed exclusively out of three-membered rings solidly established, more recent work in this area has branched out into different directions. de Meijere and co–workers have prepared expanded rotanes such as **400** and **401**,[227] the former prepared from **393** by a three-step sequence involving dibromocyclopropanation, methyllithium-induced dehalogenation of the formed dibromo-cyclopropane adduct, and cyclopropylidenation of the resulting allene intermediate. When **400** was cyclopropanated with ethoxycarbonyl carbene, the resulting ester was converted into a N-nitrosourea derivative, and the carbene generated from it by treatment with base was added to bicyclopropylidene (**349**), hydrocarbon **401** was produced. This unique rotane formally consists of three [3]rotane subunits which share a common three-membered ring (Scheme 70). How even more extended hydrocarbon structures can be constructed from cyclopropane rings alone will be discussed in Section 15.3.

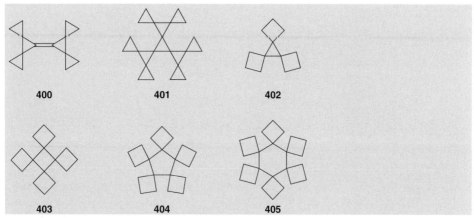

Scheme 70. Expanded [3]rotanes and the [n.4]rotane family.

Fitjer and co–workers have developed routes to the new rotane family **402–405** constructed out of cyclobutane building blocks only.[228] For a long time [6.3]rotane (**348**) was the first per(cyclo)alkylated cyclohexane existing at room temperature in solution in a fixed chair conformation with the extremely high barrier of 21.4 ± 0.2 kcal mol^{-1}.[229]. [6.4]rotane (**405**) surpasses the free energy of activation by far, possessing the highest barrier for a cyclohexane chair to chair interconversion (see Chapter 3.3) ever reported: 37.5 ± 0.2 kcal mol^{-1}.

Several of the ketone intermediates passed *en route* to the [n.4]rotanes could be used as substrates for the preparation of new types of propellanes (see Chapter 6.1)[230] and the so-called coronanes[231], a particularly instructive example being the preparation of [6.5]coronane (**411**) (Scheme 71)[232].

Addition of allylmagnesiumbromide to **406** led to two conformationally stable homoallylic alcohols, the one with an axial hydroxyl group (OH$_a$), **407**, being formed in 60% whereas its equatorial isomer was produced in 17% yield. When **407** was treated with thionylchloride in pyridine a remarkable cascade process was triggered in which five 1,2-carbon shifts took place and the rotane precursor was isomerized to the hexacyclic diene **408** in excellent yield. When this diene was treated with zircono-cenechlorohydride and N-bromosuccinimide the bromide **409** was formed. For the closure of the last five-membered ring the radical **410** was generated by refluxing a solution of **409** in benzene in the presence of tri(n-butyl)tin hydride/azobisisobuty-

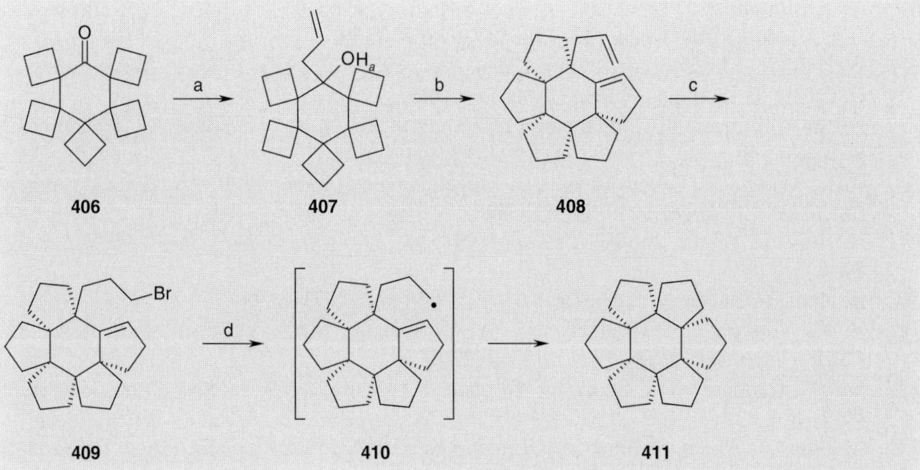

Scheme 71. From the polyspiroketone **406** to [6.5]coronane (**411**) by a cascade rearrangement.

rylnitrile (AIBN) (ratio 10:1). Unfortunately, the selectivity of this process was poor: the hydrocarbon mixture produced contained at least eight compounds, and [6.5]coronane (**411**) was only a minor one (8% yield). Still, an X-ray structure determination and a ^{13}C DNMR investigation could be performed: they showed that **411** is all-*cis*-configurated, adopts a (flat) chair conformation in the crystalline state and in solution and exhibits an extremely low barrier of inversion ($\Delta G^{\neq} < 8.6$ kcal mol^{-1} at 137 K) - in marked contrast to its isomer **405**.[233]

All rotanes discussed in this section are saturated hydrocarbons. When unsaturated spacer elements, such as the butadiyne unit, are inserted between the spiro carbon atoms, the so-called 'exploded' rotanes result[234], a class of hydrocarbons already discussed in the chapter on cyclic acetylenes (see Chapter 8.3).

References

1. N. F. Phalan, M. Orchin, *J. Chem. Educ.*, **1968**, *45*, 633–637; *cf.* J. March, *Advanced Organic Chemistry*, John Wiley & Sons, New York, N. Y., **1992**, 4th ed., 33.
2. This trivial name was first proposed by P. G. Farrell, G. Grinter, S. F. Mason in *Optische Anregung organischer Systeme* W. Foerst (*Ed.*), Verlag Chemie, **1966**, 88–99.
3. For a review on the preparation of dendralenes see H. Hopf, *Angew. Chem.*, **1984**, *96*, 947–959; *Angew. Chem. Int. Ed. Engl.*, **1984**, *23*, 948–960; for an early approach to calculate π-resonance energies of cross-conjugated hydrocarbons see B. A. Hess, Jr, L. J. Schaad, *J. Am. Chem. Soc.*, **1971**, *93*, 305–310.
4. Th. v. Fellenberg, *Ber. Dtsch. Chem. Ges.*, **1904**, *37*, 3578–3581.

5. A. T. Blomquist, J. A. Verdol, *J. Am. Chem. Soc.*, **1955**, *77*, 81–83; *cf.* A. T. Blomquist, J. A. Verdol, *J. Am. Chem. Soc.*, **1955**, *77*, 1806–1809.
6. W. J. Bailey, J. Economy, *J. Am. Chem. Soc.*, **1955**, *77*, 1133–1136; *cf.* W. J. Bailey, C. H. Cunov, L. Nicholas, *J. Am. Chem. Soc.*, **1955**, *77*, 2787–2790.
7. For an optimization of this method providing yields up to 76% see W. S. Trahanovsky, K. A. Koeplinger, *J. Org. Chem.*, **1992**, *57*, 4711–4716.
8. H.-D. Martin, M. Eckert-Maksic, B. Mayer, *Angew. Chem.*, **1980**, *92*, 833–835; *Angew. Chem. Int. Ed. Engl.*, **1980**, *19*, 807–808.
9. H. Hopf, H. Priebe, *Angew. Chemie*, **1984**, *94*, 299–300; *Angew. Chem. Int. Ed. Engl.*, **1984**, *21*, 286.
10. H. Hopf, H. Bader, H. Jäger, *Chem. Ber.*, **1989**, *122*, 1193–1198.
11. V. M. Vdovin, E. Sh. Finkelshtein, A. V. Shelkov, M. S. Yatsenko, *Bull. Acad. Sci. USSR, Chem. Ser.*, **1986**, *35* (11), 2364–2366.
12. J. I. G. Cadogan, S. Cradock, S. Gillman, I. Gosney, *J. Chem. Soc. Chem. Commun.*, **1991**, 114–115.
13. O. Tsuge, E. Wada, S. Kanemasa, *Chem. Lett.*, **1983**, 239–242; *cf.* O. Tsuge, E. Wada, S. Kanemasa, *Chem. Lett.*, **1983**, 1525–1528.
14. A. I. Scott, *Ultraviolet Spectra of Natural Products*, Pergamon Press, Oxford, **1964**, 35.
15. A. Almenningen, A. Gatail, D. S. B. Grace, H. Hopf, P. Klaeboe, F. Lehrich, C. J. Nielsen, D. L. Powell, M. Traetteberg, *Acta Chem. Scand.*, **1988**, *A42*, 634–650.
16. H. Staudinger, *Ber. Dtsch. Chem. Ges.*, **1908**, *41*, 1493–1500; *cf.* H. Staudinger, N. Kon, *Liebigs Ann. Chem.*, **1911**, *384*, 38–135 and W. Schlenk, E. D. Bergmann, *Liebigs Ann. Chem.*, **1928**, *463*, 2–97.
17. C. A. Aufdermarsh, U. S. Patent 3 264 366; *Chem. Abstr.*, **1966**, *65*, 20 003 g; *cf.* C. A. Aufdermarsh, *J. Org. Chem.*, **1964**, *29*, 1994–1996.
18. K. Greiner, Ph. d. dissertation, University of Erlangen, **1960**.
19. W. J. Bailey, N. A. Nielson, *J. Org. Chem.*, **1962**, *27*, 3088–3091.
20. L. K. Bee, J. W. Everett, P. J. Garratt, *Tetrahedron*, **1977**, *33*, 2143–2150.
21. H. Hopf in *The Chemistry of the Allenes* S. R. Landor, *(Ed.)*, Academic Press, London, **1982**, Vol. 2, Chapt. 5, 525–562.
22. L. Skattebøl, S. Solomon, *J. Am. Chem. Soc.*, **1965**, *87*, 4506–4513; *cf.* W. R. Roth, M. Heiber, G. Erker, *Angew. Chem.*, **1973**, *85*, 511–512; *Angew. Chem. Int. Ed. Engl.*, **1973**, *12*, 504.
23. W. Grimme, H.-J. Rother, *Angew. Chem.*, **1973**, *85*, 512–514; *Angew. Chem. Int. Ed. Engl.*, **1973**, *12*, 505–507; *cf.* W. R. Roth, B. P. Scholz, R. Breuckmann, K. Jelich, H.-W. Lennartz, *Chem. Ber.*, **1982**, *115*, 1934–1946.
24. W. R. Roth, G. Erker, *Angew. Chem.*, **1973**, *85*, 510–511; *Angew. Chem. Int. Ed. Engl.*, **1973**, *12*, 505–506.
25. W. R. Roth, G. Erker, *Angew. Chem.*, **1973**, *85*, 512; *Angew. Chem. Int. Ed. Engl.*, **1973**, *12*, 505.
26. Formally **15** may be regarded as a subsystem of biphenyl, and it could well be that derivatives of this dendralene have stereochemical properties, i.e. show atropisomerism, as interesting and important as those of biphenyl.
27. K. v. Auwers, K. Müller, *Ber. Dtsch. Chem. Ges.*, **1911**, *44*, 1595–1608 and literature cited.
28. K. v. Auwers, *Ber. Dtsch. Chem. Ges.*, **1911**, *44*, 788–809; *cf.* K. v. Auwers, K. Ziegler, *Liebigs Ann. Chem.*, **1921**, *425*, 217–313.
29. H. Plieninger, W. Maier-Borst, *Angew. Chem.*, **1963**, *75*, 1177; *cf.* H. Plieninger, W. Maier-Borst, *Chem. Ber.*, **1965**, *98*, 2504–2508.
30. B. Miller, K.-H. Lai, *J. Am. Chem. Soc.*, **1972**, *94*, 3472–3481 and literature cited; *cf.* H.

Hart, J.-D. DeVrieze, *J. Chem. Soc. Chem. Commun.*, **1968**, 1651–1652.

31.   J. J. Gajewski, A. M. Gortva, *J. Org. Chem.*, **1989**, *54*, 373–378 and refs. cited; *cf.* J. E. Bartmess, *J. Am. Chem. Soc.*, **1982**, *104*, 335–337.

32.   J. Janssen, W. Lüttke, *J. Mol. Struct.*, **1982**, *81*, 73–86.

33.   W. Hutter, H.-K. Bodenseh, A. Koch, *J. Mol. Struct.*, **1994**, *319*, 73–83.

34.   **62**: H.-D. Martin, B. Mayer, *Tetrahedron Lett.*, **1979**, 2351–2352; *cf.* J. K. Williams, W. H. Sharkey, *J. Am. Chem. Soc.*, **1959**, *81*, 4269–4272; **63**: J. M. E. Krekels, J. W. de Haan, H. Kloosterziel, *Tetrahedron Lett.*, **1970**, 2751–2754; *cf.* R. A. Clark, W. J. Hayles, D. S. Young, *J. Am. Chem. Soc.*, **1975**, *97*, 1966–1977; **64**: see ref.[23]; **65**: see ref.[7]; **66**: W. E. Billups, W. Y. Chow, K. H. Leavell, E. S. Lewis, *J. Org. Chem.*, **1974**, *39*, 274–275; W. E. Billups, B. A. Baker, W. Y. Chow, K. H. Leavell, E. S. Lewis, *J. Org. Chem.*, **1975**, *40*, 1702–1704; P. Radlick, G. Fenical, G. Alford, *Tetrahedron Lett.*, **1970**, 2707–2710; **67**: J. J. Gajewski, *Hydrocarbon Thermal Isomerizations*, Academic Press, New York, N. Y., **1981**, 382; *cf.* H. Hopf, R. Kirsch, *Tetrahedron Lett.*, **1985**, *26*, 3327–3330.

35.   H. Hopf, W. Lenk, *Tetrahedron Lett.*, **1982**, *23*, 4073–4076.

36.   A. Cassens, Ph. d. dissertation, University of Göttingen, **1979**.

37.   J. Janssen, W. Lüttke, *Chem. Ber.*, **1982**, *115*, 1234–1243; *cf.* T. Bally, L. Neuhaus, S. Nitsche, E. Haselbach, J. Janssen, W. Lüttke, *Helv. Chim. Acta*, **1983**, *66*, 1288–1295 and literature cited.

38.   S. Hünig, B. Hagenbruch, *Liebigs Ann. Chem.*, **1984**, 340–353; *cf.* H. Berneth, B. Hagenbruch, S. Hünig, B. Ort, *Liebigs Ann. Chem.*, **1984**, 354–369.

39.   E. D. Bergman, I. Agranat, *J. Am. Chem. Soc.,* **1964**, *86*, 3587; *cf.* A. S. Kende, P. T. Izzo, *J. Am. Chem. Soc.*, **1964**, *86*, 3587–3589. In the same vein the Wittig (see T. Eicher, E. v. Angerer, A.-M. Hansen, *Liebigs Ann. Chem.,* **1971**, *746*, 102–119) and the Peterson olefination (see H. H. Schubert, P. J. Stang, *J. Org. Chem.*, **1984**, *49*, 5087–5090) have been used to prepare triafulvene derivatives carrying stabilizing (polarizing) functional groups. For the generation of the first silatriafulvene derivative by a condensation reaction see K. Sakamoto, J. Ogasawara, H. Sakurai, M. Kira, *J. Am. Chem. Soc.*, **1997**, *119*, 3405–3406.

40.   For reviews of the older fulvene literature see E. D. Bergman, *Chem. Rev.*, **1968**, *68*, 41–84; T. Eicher, J. L. Weber, *Top. Curr. Chem.*, **1975**, *57*, 1–109.

41.   For summaries see M. Neuenschwander in *The Chemistry of Double-bonded Functional Groups* S. Patai, (*Ed.*), J. Wiley and Sons, Chichester, **1989**, Supplement A, Vol. 2, Part 2, 1131–1268; M. Oda in Houben-Weyl, *Methods of Organic Chemistry, Carbocyclic Three- and Four-membered Ring Compounds* A. de Meijere, (*Ed.*), Thieme Verlag, Stuttgart, **1997**, Vol. E17d, 2955–2969; K. P. Zeller in Houben-Weyl, *Methoden der Organischen Chemie*, Thieme Verlag, Stuttgart, **1985**, Vol.V/2c, 504 -684.

42.   M. Neuenschwander, W. K. Schenck, *Chimia*, **1975**, *29*, 215–217.

43.   W. E. Billups, L.-J. Liu, E. W. Casserly, *J. Am. Chem. Soc.*, **1984**, *106*, 3698–3699.

44.   S. W. Staley, T. D. Norden, *J. Am. Chem. Soc.*, **1984**, *106*, 3699–3700.

45.   G. Maier, M. Hoppe, K. Lanz, H. P. Reisenauer, *Tetrahedron Lett.*, **1984**, *25*, 5645–5648.

46.   T. D. Norden, S. W. Staley, W. H. Taylor, M. D. Harmony, *J. Am. Chem. Soc.*, **1986**, *108*, 7912–7918.

47.   W. E. Billups, Ch. Gesenberg, R. Cole, *Tetrahedron Lett.*, **1997**, *38*, 1115–1116.

48.   A. Weber, M. Neuenschwander, *Angew. Chem.*, **1981**, *93*, 788–791; *Angew. Chem. Int. Ed. Engl.*, **1981**, *20*, 774–775.

49.   J. Krebs, D. Guggisberg, U. Stämpfli, M. Neuenschwander, *Helv. Chim. Acta*, **1986**, *69*, 835–848.

50.   P. J. Stang, M. G. Mangum, *J. Am. Chem. Soc.*, **1975**, *97*, 3854–3856; *cf.* W. M. Jones, J.

M. Denham, *J. Am. Chem. Soc.*, **1964**, *86*, 944–945; *cf.* M. A. Battiste, *J. Am. Chem. Soc.*, **1964**, *86*, 942–944.

51. For a review covering the literature up to 1967 see P. Yates in *Advances in Alicyclic Chemistry*, *Vol. 2*, H. Hart, (*Ed.*), Academic Press, New York, N. Y., **1968**, 59–184; *cf.* K. P. Zeller in Houben-Weyl, *Methoden der Organischen Chemie*, Thieme Verlag, Stuttgart, **1985**, Vol 5/2c, 504–684.
52. J. Thiele, *Ber. Dtsch. Chem. Ges.*, **1900**, *33*, 666–673; *cf.* J. Thiele, H. Balhorn, *Liebigs Ann. Chem.*, **1906**, *348*, 1–15.
53. J. Thiec, J. Wiemann, *Bull. Soc. Chim. France*, **1956**, *23*, 177; *cf. Bull. Soc. Chim. France*, **1960**, *27*, 1066.
54. D. Meuche, M. Neuenschwander, H. Schaltegger, H. U. Schlunegger, *Helv. Chim. Acta,* **1964**, *47*, 1211–1215; *cf.* M. Neuenschwander, D. Meuche, H. Schaltegger, *Helv. Chim. Acta*, **1964**, *47*, 1022–1032.
55. M. Neuenschwander, R. Iseli, *Helv. Chim. Acta*, **1977**, *60*, 1061–1072.
56. R. Kyburz, H. Schaltegger, M. Neuenschwander, *Helv. Chim. Acta*, **1971**, *54*, 1037–1046; *cf.* H. Schaltegger, H. Brändli, M. Neuenschwander, *Chimia*, **1966**, *20*, 246–248 and M. Neuenschwander, P. Kyburz, R. Iseli, *Chimia*, **1970**, *24*, 342–344.
57. K. Hafner, K. H. Vöpel, G. Ploss, C. König, *Org. Synth. Coll. Vol. 5*, **1973**, 431–433.
58. E. Sturm, K. Hafner, *Angew. Chem.*, **1964**, *76*, 862–863; *Angew. Chem. Int. Ed. Engl.*, **1964**, *3*, 749; *cf.* B. M. Trost, R. M. Cory, *J. Org. Chem.*, **1972**, *37*, 1106–1110 for some improvements on the original Hafner fulvene synthesis.
59. J. E. Kent, A. J. Jones, *Austr. J. Chem.*, **1970**, *23*, 1059–1062; *cf.* B. A. W. Coller, M. L. Heffernan, A. J. Jones, *Austr. J. Chem.*, **1968**, *21*, 1807–1826; G. Zimmermann, M. Nüchter, H. Hopf, K. Ibrom, L. Ernst, *Liebigs Ann. Chem.*, **1996**, 1407–1412; U. Nüchter, H. Hopf, G. Zimmermann, V. Francke, *Liebigs Ann. Chem.*, **1997**, 1505–1515.
60. G. A. Chukhadzhyan, Zh. I. Abramyan, G. M. Tonyan, L. I. Sagradyan, T. S. Elbakyan, *J. Org. Chem. USSR*, **1981**, *17*, 1636–1640.
61. M. Oda, M. Funamizu, Y. Kitahara, *Bull. Chem. Soc. Jpn.*, **1969**, *42*, 2386–2387.
62. K. Hafner, H. W. Riedel, M. Danielisz, *Angew. Chem.*, **1963**, *75*, 344–345; *Angew. Chem. Int. Ed. Engl.*, **1963**, *2*, 215.
63. M. Neuenschwander, W. K. Schenk, *Chimia*, **1972**, *26*, 194–197; *cf.* W. K. Schenk, R. Kyburz, M. Neuenschwander, *Helv. Chim. Acta*, **1975**, *58*, 1099–1119; R. Hollenstein, A. Mooser, M. Neuenschwander, W. v. Philipsborn, *Angew. Chem.*, **1974**, *86*, 595–596; *Angew. Chem. Int. Ed. Engl.*, **1974**, *13*, 551–552.
64. W. v. E. Doering, D. H. Wiley, *Angew. Chem.*, **1955**, *67*, 429; W. v. E. Doering, D. H. Wiley, *Tetrahedron*, **1960**, *11*, 183–198.
65. H. E. Zimmerman, L. R. Sousa, *J. Am. Chem. Soc.*, **1972**, *94*, 834–842.
66. P. Bönzli, M. Neuenschwander, P. Engel, *Helv. Chim. Acta*, **1990**, *73*, 1685–1699.
67. B. M. Trost, R. C. Atkins, L. Hoffman, *J. Am. Chem. Soc.*, **1973**, *95*, 1285–1295.
68. M. Neuenschwander, A. Frey, *Chimia*, **1974**, *28*, 119–120; *cf.* M. Neuenschwander, A. Frey, *Chimia*, **1974**, *28*, 117–118. Prior to the synthesis of the parent hydrocarbon only 10,10-bis(dimethylamino)nonafulvene (K. Hafner, H. Tappe, *Angew. Chem.,* **1969**, *81*, 564–565; *Angew. Chem. Int. Ed. Engl.*, **1969**, *8*, 593–594) and several substituted benzannelated derivatives of **153** were known (M. Rabinovitz, A. Gazit, *Tetrahedron Lett.*, **1972**, 721–724).
69. T. J. Katz, P. J. Garratt, *J. Am. Chem. Soc.*, **1963**, *85*, 2852–2853; *cf.* E. A. La Lancette, R. E. Benson, *J. Am. Chem. Soc.*, **1963**, *85*, 2853. The situation is actually more complex than presented in Scheme 20 since **151** may not only exist in the all-*cis*-configuration shown but also as a tri-*cis*, mono-*trans*-structure, see *e.g.* G. Boche, D. Martens, W. Danzer, *Angew. Chem.*, **1969**, *81*, 1003–1004; *Angew. Chem. Int. Ed. Engl.*, **1969**, *9*, 984; *cf.*

G. Boche, H. Weber, D. Martens, A. Bieberbach, *Chem. Ber.*, **1978**, *111*, 2480–2496.

70. G. Sabbioni, M. Neuenschwander, *Helv. Chim. Acta*, **1986**, *68*, 623–634; *cf.* J. Furrer, P. Bönzli, A. Frey, M. Neuenschwander, P. Engel, *Helv. Chim. Acta*, **1987**, *70*, 862–880.

71. P. Bönzli, M. Neuenschwander, *Helv. Chim. Acta*, **1991**, *74*, 255–274.

72. J. Ferraris, D. O. Cowan, V. Walatka, Jr, J. H. Perlstein, *J. Am. Chem. Soc.*, **1973**, *95*, 948–949; *cf.* L. B. Coleman, M. J. Cohen, D. J. Sandman, F. G. Yamagishi, A. F. Garito, A. J. Heeger, *Solid State Commun.*, **1973**, *12*, 1125; R. V. Gemmer, D. O. Cowan, T. O. Poehler, A. N. Bloch, R. E. Pyle, R. H. Banks, *J. Org. Chem.*, **1975**, *40*, 3544–3547 and references cited therein.

73. MO calculations on triafulvalenes have been reported, though, see B. A. Hess, Jr, L. J. Schaad, *J. Am. Chem. Soc.*, **1971**, *93*, 305–310; *cf.* E. J. J. Groenen, *Mol. Phys.*, **1978**, *36*, 1555–1564; N. Tyutyulkov, D. Bontschev, *Monatshef. Chem.*, **1969**, *100*, 1941–1947.

74. W. v. E. Doering, *The Kekulé-Symposium*, London, **1958**, Butterworths and Co., London, **1959**, 45.

75. A. Escher, W. Rutsch, M. Neuenschwander, *Helv. Chim. Acta*, **1986**, *69*, 1644–1654; *cf.* W. Rutsch, A. Escher, M. Neuenschwander, *Chimia*, **1983**, *37*, 160–162. For the preparation of other stable pentafulvalenes see R. Brand, H.-P. Krimmer, H.-J. Lindner, V. Sturm, K. Hafner, *Tetrahedron Lett.*, **1982**, 5131–5134.

76. E. Hedaya, D. W. McNeil, P. Schissel, D. J. McAdoo, *J. Am. Chem. Soc.*, **1968**, *90*, 5284–5286.

77. E. A. Matzner, Ph. d. thesis, Yale University, **1958**.

78. M. S. Baird, I. R. Dunkin, M. Poliakoff, *J. Chem. Soc. Chem. Commun.*, **1974**, 704–705; *cf.* M. S. Baird, I. R. Dunkin, N. Hacker, M. Poliakoff, J. J. Turner, *J. Am. Chem. Soc.*, **1981**, *103*, 5190–5195.

79. W. B. DeMore, H. O. Pritchard, N. Davidson, *J. Am. Chem. Soc.*, **1959**, *81*, 5874–5879; *cf.* C. LeVanda, K. Bechgaard, D. O. Cowan, U. T. Mueller-Westerhoff, P. Eilbracht, G. A. Candela, R. L. Collins, *J. Am. Chem. Soc.*, **1976**, *98*, 3181–3187.

80. O. M. Nefedov, P. S. Zuev, A. K. Maltsev, Y. V. Tomilov, *Tetrahedron Lett.*, **1989**, *30*, 763–764.

81. W. v. E. Doering, U. S. Dep. Comm. Off. Tech. Serv. Rep., **1958**, *147 858*; *cf.* J. R. Mayer, Ph. d. thesis, Yale University, **1955**.

82. W. M. Jones, C. L. Ennis, *J. Am. Chem. Soc.*, **1967**, *89*, 3069–3071; *cf.* W. M. Jones, C. L. Ennis, *J. Am. Chem. Soc.*, **1969**, *91*, 6391–6397.

83. R. J. McMahon, O. L. Chapman, *J. Am. Chem. Soc.*, **1986**, *108*, 1713–1714; *cf.* R. J. McMahon, C. J. Abelt, O. L. Chapman, J. W. Johnson, C. L. Kreil, J.-P. LeRoux, A. M. Mooring, P. R. West, *J. Am. Chem. Soc.*, **1987**, *109*, 2456–2469.

84. K. Untch, *Int. Symp. On the Chemistry of Non-benzenoid Aromatic Compounds*, Sendai, Japan, **1970**; *cf.* B. L. Duell, W. M. Jones, *J. Org. Chem.*, **1978**, *43*, 4901–4903.

85. M. Reiffen, R. W. Hoffmann, *Tetrahedron Lett.*, **1978**, 1107–1110; *cf.* R. W. Hoffmann, M. Lotze, M. Reiffen, K. Steinbach, *Liebigs Ann. Chem.*, **1981**, 581–590.

86. R. W. Hoffmann, I. H. Loof, C. Wentrup, *Liebigs Ann. Chem.*, **1980**, 1198–1206.

87. R. Thomas, Ph. Coppens, *Acta Cryst.*, **1972**, *B28*, 1800–1806.

88. I. Erden, D. Kaufmann, *Tetrahedron Lett.*, **1981**, *22*, 215–218; *cf.* H. Volz, M. Volz-deLecea, *Liebigs Ann. Chem.*, **1971**, *750*, 136–148.

89. A. Riemann, R. W. Hoffmann, J. Spanget-Larsen, R. Gleiter, *Chem. Ber.*, **1985**, *118*, 1000–1007.

90. G. Binsch, E. Heilbronner, *Tetrahedron*, **1968**, *24*, 1215–1223.

91. D. J. Bertelli, C. Galino, D. L. Dreyer, *J. Am. Chem. Soc.*, **1964**, *86*, 3329–3334.

92. A. Escher, M. Neuenschwander, P. Engel, *Helv. Chim. Acta*, **1987**, *70*, 1623–1637.

93. H. Prinzbach, W. Rosswog, *Angew. Chem.*, **1961**, *73*, 543; *cf.* H. Prinzbach, W. Rosswog,

Tetrahedron Lett., **1963**, 1217–1221.

94. M. Neuenschwander, W. K. Schenk, *Chimia*, **1972**, *26*, 194–197.

95. W. K. Schenk, R. Kyburz, M. Neuenschwander, *Helv. Chim. Acta*, **1975**, *58*, 1099–1119. For a route starting from thallium cyclopentadienide (**163**, Met$^+$ = Tl$^+$) and tropylium tetrafluoroborate see J. D. White, T. Furuta, *Synth. Commun.*, **1973**, *3*, 459–463.

96. R. W. Hoffmann, A. Riemann, B. Mayer, *Chem. Ber.*, **1985**, *118*, 2493–2513.

97. H. Babsch, H. Prinzbach, *Tetrahedron Lett.*, **1978**, 645–648.

98. H. Prinzbach, H. Babsch, H. Fritz, *Tetrahedron Lett.*, **1976**, 2129–2132.

99. A. Escher, M. Neuenschwander, *Helv. Chim. Acta*, **1987**, *70*, 49–58; *cf.* A. Escher, M. Neuenschwander, *Angew. Chem.*, **1984**, *96*, 983–984; *Angew. Chem. Int. Ed. Engl.*, **1984**, *23*, 973–974.

100. L. Knothe, H. Prinzbach, E. Hädicke, *Chem. Ber.*, **1981**, *114*, 1656–1678; *cf.* H. Prinzbach, L. Knothe, *Angew. Chem.*, **1968**, *80*, 698–699; *Angew. Chem. Int Ed. Engl.*, **1968**, *7*, 729–730.

101. W. Grimme, H. Hoffmann, E. Vogel, *Angew. Chem.*, **1965**, *77*, 348–349; *Angew. Chem. Int. Ed. Engl.*, **1965**, *4*, 354; *cf.* R. H. A. M. Brounts, P. Schipper, H. M. Buck, *J. Chem. Soc. Chem. Commun.*, **1980**, 522–524.

102. A. Beck, L. Knothe, D. Hunkler, H. Prinzbach, *Tetrahedron Lett.*, **1986**, *27*, 485–488.

103. W. Grimme, J. Reisdorff, W. Jünemann, E. Vogel, *J. Am. Chem. Soc.*, **1970**, *92*, 6335–6337.

104. W. M. Jones, R. A. LaBar, U. Brinker, P. H. Gebert, *J. Am. Chem. Soc.*, **1977**, *99*, 6379–6391.

105. For a review of these cyclic cross-conjugated π-systems see H. Prinzbach, *Pure Appl. Chem.*, **1971**, *28*, 281–329. Calicenes stabilized by aryl substituents, benzannelation and functional groups are discussed in a summary by G. Becker in Houben-Weyl, *Methoden der Organischen Chemie, Carbocyclische π-Elektronensysteme*, Thieme Verlag, Stuttgart, **1985**, Vol. 5/2c, 479–485.

106. H. Prinzbach, H. Knöfel, E. Woischnik, *The Jerusalem Symposium on Quantum Chemistry and Biochemistry*, III, Jerusalem, **1971**, 296–283; *cf.* H. Prinzbach, H. Knöfel, *Angew. Chem.*, **1969**, *81*, 900–901; *Angew. Chem. Int. Ed. Engl.*, **1969**, *8*, 881–882.

107. R. Brand, H.-P. Krimmer, H.-J. Lindner, V. Sturm, K. Hafner, *Tetrahedron Lett.*, **1982**, *23*, 5131–5134.

108. G. Becker in Houben-Weyl, *Methoden der Organischen Chemie, Carbocyclische π-Elektronensysteme*, Thieme Verlag, Stuttgart, **1985**, Vol. V/2c, 685–690. For a review on the synthesis of several heteroanalogous sesquifulvenes see G. Seitz, *Angew. Chem.*, **1969**, *81*, 518–522; *Angew. Chem. Int. Ed. Engl.*, **1969**, *8*, 478–482.

109. H.-J. Lindner, K. Hafner, K. Römer, B. v. Gross, *Liebigs Ann. Chem.*, **1975**, 731–742.

110. H. Prinzbach, L. Knothe, *Pure Appl. Chem.*, **1986**, *58*, 25–37.

111. H. Sauter, H. Prinzbach, *Angew. Chem.*, **1972**, *84*, 297–299; *Angew. Chem. Int. Ed. Engl.*, **1972**, *11*, 296–298; *cf.* H. Sauter, B. Gallenkamp, H. Prinzbach, *Chem. Ber.*, **1972**, *110*, 1382–1402.

112. O. Schweikert, Th. Netscher, G. L. McMullen, L. Knothe, H. Prinzbach, *Chem. Ber.*, **1984**, *117*, 2006–2026. For the preparation of vinylogous 6-(cyclopentadienyl)-pentafulvenes, fulvenes which carry a polyolefinic side chain to which a chargeable cyclopentadienyl end group has been attached see M. Eiermann, B. Stowasser, K. Hafner, K. Bierwirth, A. Frank, A. Lerch, J. Reußwig, *Chem. Ber.*, **1990**, *123*, 1421–1431.

113. O. Schweikert, Th. Netscher, L. Knothe, H. Prinzbach, *Chem. Ber.*, **1984**, *117*, 2027–2044.

114. For a thorough and highly readable account of the history of the azulene problem see H.-J. Hansen, *Chimia*, **1996**, *50*, 489–496 and *Chimia*, **1997**, *51*, 147–159.

115. K. Brand, *Ber. Dtsch. Chem. Ges.*, **1912**, *45*, 3071–3077; *cf.* K. Brand, F. W. Hoffmann, *Ber. Dtsch. Chem. Ges.*, **1920**, *53*, 815–821.

116. C. T. Blood, R. P. Linstead, *J. Chem. Soc.*, **1952**, 2263–2268; *cf.* E. LeGoff, *J. Am. Chem. Soc.*, **1962**, *84*, 1505–1506; W. Ried, D. Freitag, *Chem. Ber.*, **1968**, *101*, 756–762.

117. G. Schaden, *Angew. Chem.*, **1977**, *89*, 50–51; *Angew. Chem. Int. Ed. Engl.*, **1977**, *16*, 50–51; *cf.* C. C. Chuen, S. W. Fenton, *J. Org. Chem.*, **1958**, *23*, 1538–1539.

118. K. Hafner, K. F. Bangert, V. Orfanos, *Angew. Chem.*, **1967**, *79*, 414–415; *Angew. Chem. Int. Ed. Engl,* **1967**, *6*, 451–452. For a review of this early work on functionalized pentalenes see K. Hafner, *Pure Appl. Chem.*, **1971**, *28*, 153–180.

119. K. Hafner, M. Suda. *Angew. Chem.*, **1976**, *88*, 341–342; *Angew. Chem. Int. Ed. Engl.*, **1976**, *15*, 314.

120. E. LeGoff, *J. Am. Chem. Soc.*, **1962**, *84*, 3975–3976.

121. K. Hafner, H. U. Süss, *Angew. Chem.*, **1973**, *85*, 626–628; *Angew. Chem. Int. Ed. Engl.*, **1973**, *12*, 575–577.

122. K. Hafner, *Nachr. Chem. Techn. Lab.*, **1980**, *28*, 222–226.

123. B. Kitschke, H. J. Lindner, *Tetrahedron Lett.*, **1977**, 2511–2514; *cf.* A. Falchi, C. Gellini, P. R. Salvi, K. Hafner, *J. Phys. Chem.*, **1995**, *99*, 14659–14666.

124. H. U. Süss, Ph. d. dissertation, Technische Hochschule Darmstadt, **1977**.

125. K. Hafner, R. Dönges, E. Goedecke, R. Kaiser, *Angew. Chem.*, **1973**, *85*, 362–364; *Angew. Chem. Int. Ed. Engl.*, **1973**, *12*, 337–339; *cf.* R. Dönges, K. Hafner, H. J. Lindner, *Tetrahedron Lett.*, **1976**, 1345–1348.

126. Depending on degree and position of the methyl substitution the structure of the dimers can be quite complex, even more so since not only *syn-* but also *anti-*isomers may be formed. For a summary see H. J. Lindner in Houben-Weyl, *Methoden der Organischen Chemie, Carbocyclische π-Systeme*, Thieme Verlag, Stuttgart, **1985**, Vol. V/2c, 103–122. A retro-Diels-Alder cleavage of the methyl derivative **223** (R = CH$_3$) was used for the first generation and spectroscopic characterization of 1-methylpentalene: R. Bloch, R. A. Marty, P. de Mayo, *J. Am. Chem. Soc.*, **1971**, *93*, 3071–3072; *cf.* R. Bloch, R. A. Marty, P. de Mayo, *Bull Soc. Chim. France,* **1972**, 2031–2037; *cf.* N. C. Baird, R. M. West, *J. Am. Chem. Soc.*, **1971**, *93*, 3072–3073.

127. Various routes to pentalenes *via* metal complexes have been attempted; they have been summarized by S. A. R. Knox, F. G. A. Stone, *Acc. Chem. Res.*, **1974**, *7*, 321–328. Note again the similarity to the generation of cyclobutadiene from different metal complexes (see Chapter 10.1).

128. H. Meier, A. Pauli, P. Kochhan, *Synthesis*, **1987**, 573–574; *cf.* H. Meier, A. Pauli, H. Kolshorn, P. Kochhan, *Chem. Ber.*, **1987**, *120*, 1607–1610.

129. T. J. Katz, M. Rosenberger, *J. Am. Chem. Soc.*, **1962**, *84*, 865–866; *cf.* T. J. Katz, M. Rosenberger, R. K. O'Hara, *J. Am. Chem. Soc.*, **1964**, *86*, 249–252.

130. S. You, M. Neuenschwander, *Chimia*, **1996**, *50*, 24–26.

131. T. Bally, S. Chai, M. Neuenschwander, Z. Zhu, *J. Am. Chem. Soc.*, **1997**, *119*, 1869–1875.

132. P. A. Plattner, A. S. Pfau, *Helv. Chim. Acta*, **1937**, *20*, 224–232.

133. W. Hückel, A. Gercke, A. Groß, *Ber. Dtsch. Chem. Ges.*, **1933**, *66*, 563–567; *cf.* W. Hückel, L. Schnitzspahn, *Liebigs Ann. Chem.*, **1933**, *505*, 274–282.

134. H. O. House, J. H. C. Lee, D. VanDerveer, J. E. Wissinger, *J. Org. Chem.,* **1983**, *48*, 5285–5288 and refs. cited; *cf.* A. G. Anderson, J. P. Nelson, J. J. Tazuma, *J. Am. Chem. Soc.*, **1953**, *75*, 4980–4989.

135. For a very comprehensive summary of these azulene syntheses see K.-P. Zeller in Houben-Weyl, *Methoden der Organischen Chemie, Carbocyclische π-Elektronensysteme*, Thieme Verlag, Stuttgart, **1985**, Vol. V/2c, 127–418.

136. K. Ziegler, K. Hafner, *Angew. Chem.*, **1955**, *67*, 301; *cf.* K. Ziegler, *Angew. Chem.*, **1955**, *67*, 301; K. Hafner, *Angew. Chem.*, **1955**, *67*, 301–302. Practically simultaneously H. Rösler and W. König published the same synthesis of azulene: *Naturwiss.*, **1955**, *42*, 211.

137. K. Hafner, *Liebigs Ann. Chem.*, **1957**, *606*, 79–89; K. Hafner, H. Weldes, *Liebigs Ann. Chem.*, **1957**, *606*, 90–99. For the synthesis of azulene (**205**) in kg-amounts see K. Hafner, K.-P. Meinhardt, *Org. Synth.*, **1984**, *62*, 134–139. For the large scale preparation of 4,6,8-trimethylazulene see K. Hafner, H. Kaiser, *Org. Synth.*, **1973**, Coll. Vol. V, 1088–1091.

138. M. Hanke, C. Jutz, *Synthesis*, **1980**, 31–32; *cf.* M. Hanke, C. Jutz, *Angew. Chem.*, **1979**, *91*, 227; *Angew. Chem. Int. Ed. Engl.*, **1979**, *18*, 214 for the preparation of 5,5′-biazulenyl.

139. K. Hafner, *Angew. Chem.*, **1958**, *70*, 419–430.

140. K. Hafner, *Angew. Chem.*, **1957**, *69*, 393.

141. H. Prinzbach, H.-J. Herr, *Angew. Chem.*, **1972**, *84*, 117–118; *Angew. Chem. Int. Ed. Engl.*, **1972**, *11*, 135–136.

142. D. Copland, D. Leaver, W. B. Menzies, *Tetrahedron Lett.*, **1977**, 639–640.

143. S. E. Reiter, L. C. Dunn, K. N. Houk, *J. Am. Chem. Soc.*, **1977**, *99*, 4199–4201.

144. M. Yasunami, S. Miyoshi, N. Kanegae, K. Takase, *Bull. Chem. Soc. Jpn.*, **1993**, *66*, 892–899.

145. K. Hafner, H. Diehl, H. U. Süss, *Angew. Chem.*, **1976**, *88*, 121–123; *Angew. Chem. Int. Ed. Engl.*, **1976**, *15*, 104–105. Review: K. Hafner, G. L. Knaup, H. J. Lindner, *Bull. Chem. Soc. Jpn.*, **1988**, *61*, 155–163.

146. For a review of heptalene chemistry see L. A. Paquette, *Israel J. Chem.*, **1980**, *20*, 233–239; *cf.* G. Becker, H. Kolshorn, Houben-Weyl, *Methoden der Organischen Chemie, Carbocyclische π-Elektronensysteme*, Vol.5/2c, Thieme Verlag, Stuttgart, **1985**, 418–461.

147. H. J. Dauben, D. J. Bertelli, *J. Am. Chem. Soc.*, **1961**, *83*, 4657–4659; *cf.* H. J. Dauben, D. J. Bertelli, *J. Am. Chem. Soc.*, **1961**, *83*, 4659–4660.

148. E. Vogel, H. Königshofen, J. Wassen, K. Müllen, J. F. M. Oth, *Angew. Chem.*, **1974**, *86*, 777–778; *Angew. Chem. Int. Ed. Engl.*, **1974**, *13*, 732–733.

149. L. A. Paquette, A. R. Browne, E. Chamot, *Angew. Chem.*, **1979**, *91*, 581–582; *Angew. Chem. Int. Ed. Engl.*, **1979**, *18*, 546–547.

150. Reviews: L. A. Paquette, *Synthesis*, **1975**, 347–357; K. C. Bishop III, *Chem. Rev.*, **1976**, *76*, 461–486.

151. E. Vogel, D. Kerimis, N. T. Allsion, R. Zellerhoff, J. Wassen, *Angew. Chem.*, **1979**, *91*, 579–581; *Angew. Chem. Int. Ed. Engl.*, **1979**, *18*, 545–555.

152. E. Vogel, F. Hogrefe, *Angew. Chem.*, **1974**, *86*, 779–780; *Angew. Chem. Int. Ed. Engl.*, **1974**, *13*, 735–736. For the synthesis of stable heptalene derivatives see E. Vogel, J. Ippen, *Angew. Chem.*, **1974**, *86*, 778–779; *Angew. Chem. Int. Ed. Engl.*, **1974**, *13*, 734–735.

153. J. F. M. Oth, K. Müllen, H. Königshofen, J. Wassen, E. Vogel, *Helv. Chim. Acta*, **1974**, *57*, 2387–2398; *cf.* K. Müllen, *Helv. Chim. Acta*, **1974**, *57*, 2399–2406.

154. H. J. Lindner, B. Kitschke, *Angew. Chem.*, **1976**, *88*, 123–124; *Angew. Chem. Int. Ed. Engl.*, **1976**, *15*, 106; *cf.* K. Hafner, G. L. Knaup, H. J. Lindner, *Bull. Chem. Soc. Jpn.*, **1988**, *61*, 155–163.

155. K. Hafner, N. Hock, G. L. Knaup, K.-P. Meinhardt, *Tetrahedron Lett.*, **1986**, *27*, 1669–1672.

156. Ch. Jutz, E. Schweiger, *Angew. Chem.*, **1971**, *83*, 886; *Angew. Chem. Int. Ed. Engl.*, **1971**, *10*, 808; *cf.* K. Hafner, *Pure Appl. Chem.*, **1971**, *28*, 153–180.

157. F. W. Grimm, K. Hafner, H. J. Lindner, *Chem. Ber.*, **1996**, *129*, 1569–1572.

158. H.-P. Krimmer, B. Stowasser, K. Hafner, *Tetrahedron Lett.*, **1982**, *23*, 5135–5138.

159. B. Stowasser, K. Hafner, *Angew. Chem.*, **1986**, *98*, 477–479; *Angew. Chem. Int. Ed.*

Engl., **1986**, *25*, 466–467.

160. Fulvenallenes, including the parent hydrocarbon, have long been known in hydrocarbon chemistry; for a review see C. Wentrup, *Top. Curr. Chem.*, **1976**, *62*, 173–251; *cf.* C. Wentrup, *Chimia*, **1977**, *31*, 258–262.

161. H. J. Lindner, K. Hafner, *Mol. Cryst. Liq. Cryst.*, **1996**, *276*, 339–348.

162. K. Hafner, B. Stowasser, H.-P. Krimmer, St. Fischer, M. C. Böhm, H. J. Lindner, *Angew. Chem.*, **1986**, *98*, 646–648; *Angew. Chem. Int. Ed. Engl.*, **1986**, *25*, 630–631; *cf.* K. Hafner, H.-P. Krimmer, *Angew. Chem.*, **1980**, *92*, 202–204; *Angew. Chem. Int. Ed. Engl.*, **1980**, *19*, 199–200 and T. S. Balaban, St. Schardt, V. Sturm, K. Hafner, *Angew. Chem.*, **1995**, *107*, 360–363; *Angew. Chem. Int. Ed. Engl.*, **1995**, *34*, 330–332.

163. The term 'radialene', obviously referring to the 'radiant structures' of these hydrocarbons, was introduced by J. R. Platt and E. Heilbronner; *cf.* E. Waltin, F. Gerson, J. N. Murrell, E. Heilbronner, *Helv. Chim. Acta*, **1961**, *44*, 1400–1413. The term was originally only applied to [6]radialene (**289**) and its derivatives but soon used for the whole vinylogous series.

164. So far only two reviews on radialenes have appeared: H. Hopf, G. Maas, *Angew. Chem.*, **1992**, *104*, 953–977; *Angew. Chem. Int. Ed. Engl.*, **1992**, *31*, 931–955; G. Maas, H. Hopf, *The Chemistry of Functional Groups*, S. Patai, Z. Rappoport, (*Eds.*), *The Chemistry of Dienes and Polyenes*, Vol. 1, J. Wiley & Sons, Chichester, **1997**, Chapter 21, 927–977. For a review on the use of radialenes as possible substrates for molecular magnets see J. S. Miller, A. J. Epstein, W. M. Reiff, *Acc. Chem. Res.*, **1988**, *21*, 114–120.

165. Review: F. Diederich in *Modern Acetylene Chemistry* P. J. Stang, F. Diederich, (*Eds.*), VCH Verlagsgesellschaft, Weinheim, **1995**, 443-471.

166. E. A. Dorko, *J. Am. Chem. Soc.*, **1965**, *87*, 5518–5520.

167. P. A. Waitkus, L. I. Peterson, G. W. Griffin, *J. Am. Chem. Soc.*, **1966**, *88*, 181–183; *cf.* P. A. Waitkus, E. B. Sanders, L. I. Peterson, G. W. Griffin, *J. Am. Chem. Soc.*, **1967**, *89*, 6318–6327.

168. T. Bally, H. Baumgärtel, U. Büchler, E. Haselbach, W. Lohr, J. P. Maier, J. Vogt, *Helv. Chim. Acta*, **1978**, *61*, 741–753; *cf.* T. Bally, E. Haselbach, *Helv. Chim. Acta*, **1978**, *61*, 754–761.

169. G. Köbrich, H. Heinemann, *Angew. Chem.*, **1965**, *77*, 590; *Angew. Chem. Int. Ed. Engl.*, **1965**, *4*, 594; *cf.* G. Köbrich, H. Heinemann, W. Zündorf, *Tetrahedron*, **1967**, *23*, 565–584.

170. M. Iyoda, H. Otani, M. Oda, *Angew. Chem.*, **1988**, *100*, 1131–1132; *Angew. Chem. Int. Ed. Engl.*, **1988**, *27*, 1080–1081.

171. M. Iyoda, A. Mizusuna, H. Kurata, M. Oda, *J. Chem. Soc. Chem. Commun.*, **1989**, 1690–1692.

172. K. Komatsu, H. Kamo, R. Tsuji, K. Takeuchi, *J. Org. Chem.*, **1993**, *58*, 3219–3211.

173. R. West, D. Zecher, *J. Am. Chem. Soc.*, **1967**, *89*, 152–153; *cf.* R. West, D. Zecher, *J. Am. Chem. Soc.*, **1970**, *92*, 155–161.

174. T. Lange, V. Gramlich, W. Amrein, F. Diederich, M. Gross, C. Boudon, J.-P. Gisselbrecht, *Angew. Chem.*, **1995**, *107*, 898–901; *Angew. Chem. Int. Ed. Engl.*, **1995**, *34*, 805–808.

175. G. W. Griffin, L. I. Peterson, *J. Am. Chem. Soc.*, **1962**, *84*, 3398–3400; *cf.* G. W. Griffin, L. I. Peterson, *J. Am. Chem. Soc.*, **1963**, *85*, 2268–2273.

176. T. Bally, U. Buser, E. Haselbach, *Helv. Chim. Acta*, **1978**, *61*, 38–45.

177. F. A. Miller, F. R. Brown, K. H. Rhee, *Spectrochim. Acta, Part A*, **1972**, *28*, 1467–1478.

178. W. D. Huntsman, H. J. Wristers, *J. Am. Chem. Soc.*, **1963**, *85*, 3308–3309; *cf.* W. D. Huntsman, H. J. Wristers, *J. Am. Chem. Soc.*, **1967**, *89*, 342–347.

179. H. Hopf, L. Trabert, *Liebigs Ann. Chem.*, **1980**, 1786–1800.

180. H. Hopf, L. Trabert, D. Schomburg, *Chem. Ber.*, **1981**, *114*, 2405–2414.

181. M. C. Lasne, J. L. Ripoll, J. M. Denis, *Tetrahedron*, **1981**, *37*, 503–508.

182. Z. Berkovitch-Yellin, M. Laha, L. Leiserowitz, *J. Am. Chem. Soc.*, **1974**, *96*, 918–920.

183. B. Heinrich, A. Roedig, *Angew. Chem.*, **1968**, *80*, 367–368; *Angew. Chem. Int. Ed. Engl.*, **1968**, *7*, 375.

184. P. J. Stang, A. E. Learned, *J. Chem. Soc. Chem. Commun.*, **1988**, 301–302; *cf.* A. E. Learned, A. M. Arif, P. J. Stang, *J. Org. Chem.*, **1988**, *53*, 3122–3123.

185. H. D. Hartzler, *J. Am. Chem. Soc.*, **1971**, *93*, 4527–4531; *cf.* M. Iyoda, M. Oda, Y. Kai, N. Kanehisa, N. Kasai, *Chem. Lett.*, **1990**, 2149–2152.

186. H.-G. Zoch, G. Szeimies, R. Römer, G. Germain, J.-P. Declercq, *Chem. Ber.*, **1983**, *116*, 2285–2310; *cf.* S. Hashmi, K. Polborn, G. Szeimies, *Chem. Ber.*, **1989**, *122*, 2399–2401.

187. B. Hagenbruch, K. Hesse, S. Hünig, G. Klug, *Liebigs Ann. Chem.*, **1981**, 256–263.

188. M. Iyoda, S. Tanaka, M. Nose, M. Oda, *J. Chem. Soc. Chem. Commun.*, **1983**, 1058–1059.

189. M. Iyoda, S. Tanaka, H. Otani, M. Nose, M. Oda, *J. Am. Chem. Soc.*, **1988**, *110*, 8494–8500.

190. T. Sugimoto, H. Awaji, Y. Misaki, Z. Yoshida, Y. Kai, H. Nakagawa, N. Kasei, *J. Am. Chem. Soc.*, **1985**, *107*, 5792–5793; *cf.* Y. Misaki, Y. Matsumura, T. Sugimoto, Z. Yoshida, *Tetrahedron Lett.*, **1989**, *30*, 5289–5292.

191. M. Iyoda, H. Otani, M. Oda, Y. Kai, Y. Baba, N. Kasai, *J. Am. Chem. Soc.*, **1986**, *108*, 5371–5372.

192. M. Iyoda, A. Mizusuna, H. Kurata, M. Oda, *J. Chem. Soc. Chem. Commun.*, **1989**, 1690–1692.

193. M. Iyoda, H. Otani, M. Oda, Y. Kai, Y. Baba, N. Kasai, *J. Chem. Soc. Chem. Commun.*, **1986**, 1794–1796.

194. Z. Yoshida, T. Sugimoto, *Angew. Chem. Adv. Mat.*, **1988**, *100*, 1633–1637; *Angew. Chem. Int. Ed. Engl. Adv. Mat.*, **1988**, *1*, 1573–1577; *cf.* T. Sugimoto, Y. Misaki, Y. Arai, Y. Yamamoto, Z. Yoshida, Y. Kai, N. Kasai, *J. Am. Chem. Soc.*, **1988**, *110*, 628–629.

195. T. Sugimoto, Y. Misaki, Z. Yoshida, J. Yamauchi, *Mol. Cryst. Liq. Cryst.*, **1989**, *176*, 259–270; *cf.* K. Kano, T. Sugimoto, Y. Misaki, T. Enoki, H. Hatakeyama, H. Oka, Y. Hosotani, Z. Yoshida, *J. Phys. Chem.*, **1994**, *98*, 252–258.

196. H. Hopff, A. K. Wick, *Helv. Chim. Acta*, **1961**, *44*, 19–24; *cf.* H. Hopff, A. K. Wick, *Helv. Chim. Acta*, **1961**, *46*, 380–386.

197. H. Hopff, A. Gati, *Helv. Chim. Acta*, **1965**, *48*, 1289–1296; H. Hopff, G. Kormany, *Helv. Chim. Acta*, **1965**, *48*, 437–443.

198. W. Marsh, J. D. Dunitz, *Helv. Chim. Acta*, **1975**, *58*, 707–712.

199. Th. Höpfner, Ph. d. dissertation, Braunschweig, **1997**.

200. A. J. Barkovich, E. S. Strauss, K. P. C. Vollhardt, *J. Am. Chem. Soc.*, **1977**, *99*, 8321–8322; W. V. Dower, K. P. C. Vollhardt, *Tetrahedron*, **1986**, *42*, 1873–1881.

201. P. Schiess, M. Heitzmann, *Helv. Chim. Acta*, **1978**, *61*, 844–847; *cf.* P. Schiess, M. Heitzmann, S. Rutschmann, R. Stäheli, *Tetrahedron Lett.*, **1978**, 4569–4572.

202. L. G. Harruff, M. Brown, V. Boekelheide, *J. Am. Chem. Soc.*, **1978**, *100*, 2893–2894; *cf.* R. Gray, L. G. Harruff, J. Krymowski, J. Peterson, V. Boekelheide, *J. Am. Chem. Soc.*, **1978**, *100*, 2892–2893.

203. W. E. Billups, D. J. McCord, B. R. Maughon, *J. Am. Chem. Soc.*, **1994**, *116*, 8831–8832.

204. G. Wilke, *Angew. Chem.*, **1988**, *100*, 190–211; *Angew. Chem. Int. Ed. Engl.*, **1988**, *27*, 185–206.

205. A. M. Boldi, F. Diederich, *Angew. Chem.*, **1994**, *106*, 482–485; *Angew. Chem. Int. Ed. Engl.*, **1994**, *33*, 468–470.

206. J. Anthony, A. M. Boldi, C. Boudon, J.-P. Gisselbrecht, M. Gross, P. Seiler, C. B. Kno-

bler, F. Diederich, *Helv. Chim. Acta.*, **1995**, *78*, 797–817.

207. O. Pilet, P. Vogel, *Angew. Chem.*, **1980**, *92*, 1036–1037; *Angew. Chem. Int. Ed. Engl.*, **1980**, *19*, 1003–1004; *cf.* O. Pilet, J. L. Birbaum, P. Vogel, *Helv. Chim. Acta*, **1983**, *66*, 19–34. For a recent improvement of this synthesis of hericene see F. H. Köhler, A. Steck, *J. Organomet. Chem.*, **1993**, *444*, 165–177.

208. For a review of [*l.m.n*]hericenes and related semicyclic hydrocarbons see P. Vogel in *Advances in Theoretically Interesting Molecules* R. P. Thummel, (*Ed.*), JAI Press, Inc., Greenwich, CT, **1989**, Vol. 1, 201–355.

209. M. Giersig, D. Wehle, L. Fitjer, N. Schormann, W. Clegg, *Chem. Ber.*, **1988**, *121*, 525–531.

210. J. L. Ripoll, J. M. Conia, *Tetrahedron Lett.*, **1969**, 979–980; *cf.* J. L. Ripoll, J. C. Limasset, J. M. Conia, *Tetrahedron,* **1971**, *27*, 2431–2452.

211. For the experimental determination of the strain energies of various polyspiranes, including **345** and **346**, see H.-D. Beckhaus, C. Rüchardt, S. I. Kozhushkov, V. N. Belov, S. P. Verevkin, A. de Meijere, *J. Am. Chem. Soc.*, **1995**, *117*, 11854–11860.

212. L. Fitjer, J. M. Conia, *Angew. Chem.*, **1973**, *85*, 349–350; *Angew. Chem. Int. Ed. Engl.*, **1973**, *12*, 334–335.

213. I. Erden, *Synth. Commun.*, **1986**, *16*, 117–121. A nearly identical approach to [3]rotane (**345**) has been described by N. S. Zefirov, K. A. Lukin, S. I. Kozhushkov, T. S. Kuznetsova, A. M. Domarev, I. M. Sosonkin, *J. Org. Chem. USSR*, **1989**, *25*, 278–284.

214. L. Fitjer, J. M. Conia, *Angew. Chem.*, **1973**, *85*, 832–833; *Angew. Chem. Int. Ed. Engl.*, **1973**, *12*, 761–762.

215. A. de Meijere, I. Erden, W. Weber, D. Kaufmann, *J. Org. Chem.*, **1988**, *53*, 152–161.

216. L. Fitjer, J. M. Conia, *Angew. Chem.*, **1973**, *85*, 347–349; *Angew. Chem. Int. Ed. Engl.*, **1973**, *12*, 332–333. For the preparation of **349** from methylenecyclopropane see P. LePerchec, J. M. Conia, *Tetrahedron Lett.*, **1970**, 1587–1588.

217. L. Fitjer, *Angew. Chem.*, **1976**, *88*, 803–804; *Angew. Chem. Int. Ed. Engl.*, **1976**, *15*, 762.

218. L. Fitjer, *Chem. Ber.*, **1982**, *115*, 1035–1046.

219. J. M. Conia, J. M. Denis, *Tetrahedron Lett.*, **1969**, 3545–3546.

220. P. LePerchec, J. M. Conia, *Tetrahedron Lett.*, **1970**, 1587–1588.

221. A. de Meijere, H. Wenck, J. Kopf, *Tetrahedron*, **1988**, *44*, 2427–2438.

222. E. Proksch, A. de Meijere, *Tetrahedron Lett.*, **1976**, 4851–4854,

223. E. Proksch, A. de Meijere, *Angew. Chem.*, **1976**, *88*, 802–803; *Angew. Chem. Int. Ed. Engl.*, **1976**, *15*, 761–762.

224. For the X-ray structural analyses of **345–348** see C. Pascard, Th. Prangé, A. de Meijere, W. Weber, J.-P. Barnier, J.-M. Conia, *J. Chem. Soc. Chem. Commun.*, **1979**, 425–426; Th. Prangé, C. Pascard, A. de Meijere, U. Behrens, J.-P. Barnier, J.-M. Conia, *Nouv. J. Chim.*, **1980**, *4*, 321–327; R. Boese, Th. Miebach, A. de Meijere, *J. Am. Chem. Soc.*, **1991**, *113*, 1743–1748; P. Aped, N. L. Allinger, *J. Am. Chem. Soc.*, **1992**, *114*, 1–16. For the photoelectron spectra of rotanes and related polyspiranes see R. Gleiter, R. Haider, J.-M. Conia, J.-P. Barnier, A. de Meijere, W. Weber, *J. Chem. Soc. Chem. Commun.*, **1979**, 130–132.

225. L. Fitjer, *Angew. Chem.*, **1976**, *88*, 804–805; *Angew. Chem. Int. Ed. Engl.*, **1976**, *15*, 763–764.

226. L. Fitjer, *Chem. Ber.*, **1982**, *115*, 1047–1060.

227. S. Zöllner, H. Buchholz, R. Boese, R. Gleiter, A. de Meijere, *Angew. Chem.*, **1991**, *103*, 1544–1546; *Angew. Chem. Int. Ed. Engl.*, **1991**, *30*, 1518–1520.

228. L. Fitjer, Chr. Steeneck, S. Gaini-Rahimi, U. Schröder, K. Justus, P. Puder, M. Dittmer, C. Hassler, J. Weiser, M. Noltemeyer, M. Teichert, *J. Am. Chem. Soc.*, **1998**, *120*, 317-328. 'Mixed' rotanes containing both three- and four-membered rings have also been

synthesized, see ref.[229]

229. L. Fitjer, U. Klages, W. Kühn, D. S. Stephenson, G. Binsch, M. Noltemeyer, E. Egert, G. M. Sheldrick, *Tetrahedron*, **1984**, *40*, 4337–4349; *cf.* L. Fitjer, M. Giersig, D. Wehle, M. Dittmer, G.-W. Koltermann, N. Schormann, E. Egert, *Tetrahedron*, **1988**, *44*, 393–404.

230. L. Fitjer, U. Quabeck, *Angew. Chem.*, **1989**, *101*, 55–57; *Angew. Chem. Int. Ed. Engl.*, **1989**, *28*, 94–95.

231. L. Fitjer, U. Quabeck, *Angew. Chem.*, **1987**, *99*, 1054–1056; *Angew. Chem. Int. Ed. Engl.*, **1987**, *26*, 1023–1025.

232. D. Wehle, L. Fitjer, *Angew. Chem.*, **1987**, *99*, 135–137; *Angew. Chem. Int. Ed. Engl.*, **1987**, *26*, 130–132.

233. D. Wehle, N. Schormann, L. Fitjer, *Chem. Ber.*, **1988**, *121*, 2171–2177.

234. A. de Meijere, S. Kozhushkov, C. Puls, T. Haumann, R. Boese, M. J. Cooney, L. T. Scott, *Angew. Chem.*, **1994**, *106*, 934–936; *Angew. Chem. Int. Engl.*, **1994**, *33*, 869–871; *cf.* A. de Meijere, S. Kozhushkov, T. Haumann, R. Boese, C. Puls, M. J. Cooney, L.T. Scott, *Chem. Eur. J.*, **1995**, *1*, 124–131.

12 Leaving the π-Plane—Non-Planar Aromatic Compounds

Many of the π-systems discussed so far have planar structures: the small-ring cycloolefins (Chapter 8), many of the fulvenes and fulvalenes and hydrocarbons derived therefrom (Chapter 11), the small-ring radialenes (Chapter 11) and in particular the annulenes (Chapter 10). In fact, in the Hückel definition of aromaticity co-planarity of the annulene ring is a prerequisite. However, as we asked to what extent the tetrahedral carbon atom (Chapter 6), the carbon-carbon double (Chapter 8) and the carbon-carbon triple bond (Chapter 9) may be deformed from their normal geometry, the question may and must be raised how far aromatic and especially benzenoid hydrocarbons can deviate from their 'classical', planar geometry. There are many ways of inducing or enforcing this geometric change, and I shall discuss three major approaches in this chapter.

In the first approach to non-planar aromatic compounds we will follow the geometric changes which take place on angular annelation of benzene rings (linear annelation which ultimately leads to partial structures of graphite will be discussed in Section 14.4). Thus starting from phenanthrene (**1**) by successive benzoannelation benzo[c]phenanthrene (**2**) is formed first, followed by dibenzo[c,g]phenanthrene (**3**). Because the internal hydrogen atoms H_i in these molecules come into increasing contact which each other with increasing degree of annelation, we would expect deviations from planarity in these hydrocarbons from a certain number of benzene rings onwards. This is indeed so, and because helical structures are formed by this building process, this class of non-planar aromatic compounds is called helicenes. Interest in these hydrocarbons is largely caused by their inherent chirality.

In the second route to non-planar aromatics we replace the central six-membered ring of (the planar) hydrocarbon coronene (**4**, see Section 14.2) by rings of different ring sizes and adjust the number of surrounding benzene rings accordingly. A particularly important representative of this group of aromatic compounds is corannulene (**5**) with its five-membered core. As already mentioned in Chapter 2, corannulene has a bowl-shaped structure and can be regarded as a subsystem of fullerene C_{60}. Other central rings are possible, however, and the resulting circulenes—which in principle could also be used to construct other forms of carbon—are also non-planar.

Finally, we introduce non-planarity in aromatic hydrocarbons[1] by molecular bridging. Taking benzene (**7**) as the simplest example we can bridge it in either *meta* (**8**) or *para* position (**9**) (Scheme 1).[2]

The resulting system are called cyclophanes and we expect their planarity to depend primarily on the length of the bridging unit—the shorter it is the more the

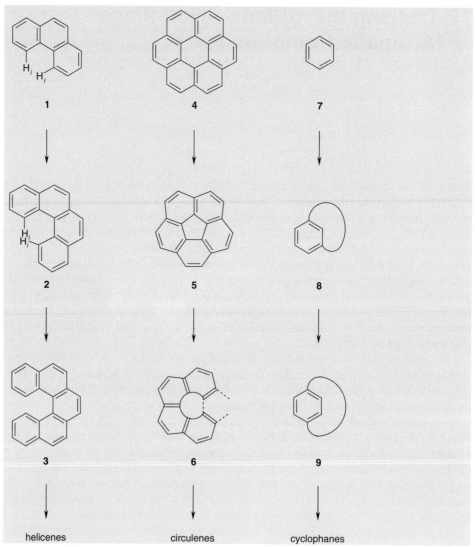

Scheme 1. Three routes to generate non-planar aromatic hydrocarbons.

benzene ring should be bent. The variability of the molecular bridge is enormous—beginning with a simple polymethylene chain, certain of its links can be replaced by all types of heteroatoms, complete functional groups or other more or less complex hydrocarbon systems, including, of course, benzene rings. The resulting [*m.n*]cyclophanes have been studied extensively as we shall see below. There can, furthermore, be more than one bridging unit—giving rise to the so-called multibridged cyclophanes, which include the so-called superphanes in which all aromatic hydrogen atoms have been replaced by bridges. And if a hydrocarbon system contains more than one cyclophane unit we speak of multi-layered cyclophanes. As the benzene rings in these molecules are increasingly deformed, the question of whether from a certain point onwards the 'aromatic character' of a benzene ring will be replaced by 'cyclohexatriene character' of the deformed six-membered ring has strongly motivated research in this area of hydrocarbon chemistry. Another approach to benzenoid hydrocarbons showing bond alternation will be discussed in Section 16.3.

12.1 The helicenes

The first work on helicenes dates back to the turn of the 20th century, although neither the helical nor the chiral nature of these angularly annelated aromatic compounds was recognized at that time and many of the early preparations of these hydrocarbons turned out to be mixtures of isomers when subjected to the increasingly more powerful separation and structure elucidation techniques.

For example, benzo[c]phenanthrene (**2**) was only obtained in pure form in 1931[3] after having been described in the chemical literature much earlier.[4] Directed syntheses—applying methods which will be presented in the discussion of the higher helicenes (see below)—were published in the early 1950s.[5-7] According to X-ray structural analysis **2** is a helical hydrocarbon;[8] that it can be resolved into its enantiomers seems unlikely from its calculated racemization barrier (6 kcal mol^{-1} by AM1 and 10 kcal mol^{-1} by PM3[9]). Not surprisingly, phenanthrene (**1**), with a distance between the hydrogen atoms H$_i$ larger than in **2**, is a planar hydrocarbon. Introduction of methyl substituents into these positions results, however, in non-planarity and, hence, chirality. The angle between the mean planes of the outer rings of 4,5-dimethylphenanthrene amounts to 27.9° and the racemization barrier is 16 ± 2 kcal mol^{-1} in hexane.[10]

Returning to the higher helicenes, dibenzo[c,g]phenanthrene (**3**, pentahelicene) is the first helicene for whose preparation the whole range of methods customary in this area of hydrocarbon chemistry, has been applied. Originally the carbon skeleton of **3** was prepared from the diamino dicarboxylic acid **10** by the Pschorr reaction, an early and quite powerful route to condensed aromatic systems (Scheme 2).[11]

The resulting dicarboxylic acid **11** was subsequently thermally decarboxylated to the parent hydrocarbon **3**. In another conventional route, the central ring of **3** was formed by a protocol very often used to prepare benzene rings—the Diels–Alder addition of maleic anhydride to appropriate dienes followed by decarboxylation/aromatization, i.e. a process in which maleic anhydride is employed as an acetylene equivalent. In this instance on addition of maleic anhydride the diene **12** yielded the adduct **13**, which was converted to **3** via its tetrahydro derivative **14**.[12]

Other derivatives of 1,1'-binaphthyl have also been used to prepare **3** by non-photochemical routes.[13] Of particular importance for the preparation of optically pure **3** was the oxidative cyclization of the enantiomerically pure bis-phosphonium salt **15** (yield 85%)[14] and the Stevens rearrangement of (S)-(+)-**16** (100%).[15] As shown by these experiments, the dextrorotatory enantiomer has the P configuration.[15-17] The resolution of **3**, first accomplished by use of 2-(2,4,5,7-

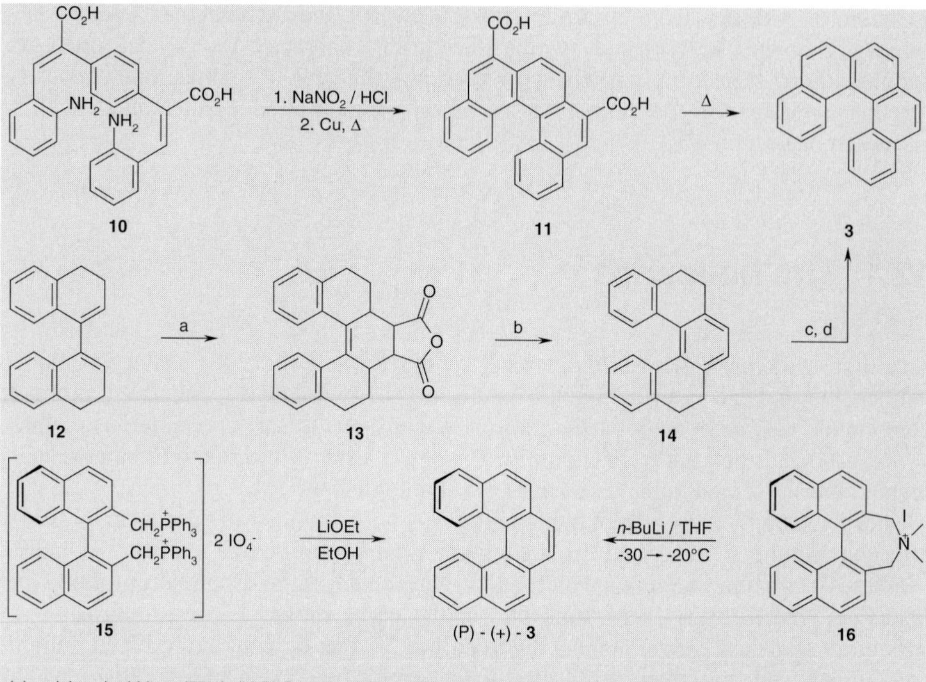

(a) maleic anhydride, Δ (75%); (b) PbO₂, Δ (41%); (c) NBS, dibenzoylperoxide; (d) NaOAc, HOAc (comb. yield 92%)

Scheme 2. Non-photochemical routes to dibenzo[c,g]phenanthrene (**3**, [5]-helicene).

tetranitro-9-fluorenylideneaminooxy)propionic acid (TAPA)[18] (see below), was, of course, the prerequisite to determination of the racemization barrier ($\Delta G^{\ddagger}$) of optically active [5]-helicene. With 24.2 kcal mol^{-1} (in chloroform)—corresponding to a half-life of *ca* 14 h at room temperature—this is significantly higher than the value measured for 4,5-dimethylphenanthrene (see above).

The most general route to helicenes makes use of the photochemical conversion of stilbenes into phenanthrenes in the presence of an oxidizing agent (I_2, O_2, tetracyanoethylene (TCNE)) (Scheme 3).[19,20] For the synthesis of the higher helicenes in particular (see below) there is practically no alternative to this universal approach.

When this photocylodehydrogenation was applied to either the dinaphthylethene **17** or the distyrylbenzene derivative **18**, however, hydrocarbon **3** was evidently formed only as an intermediate, because the oxidation did not stop at this stage but proceeded all the way to benzo[g,h]perylene (**10**).[21,22] A way out of this dilemma, which also solves another problem often encountered in the stilbene-phenanthrene photocyclization, namely that of the formation of regioisomeric cyclization products, was provided by Katz and co–workers,[23] who found that bromine substituents on benzene rings direct photocylizations away from the *ortho* position. Thus irradiation of **20**, which in principle could yield two regioisomers, led solely to the tribromide **21**; its debromination by treatment with butyllithium followed by hydrolysis furnished the desired **3**, both steps occurring in excellent yield. Besides the problems of overoxidation and regioselectivity the photocyclization route suffers from a third

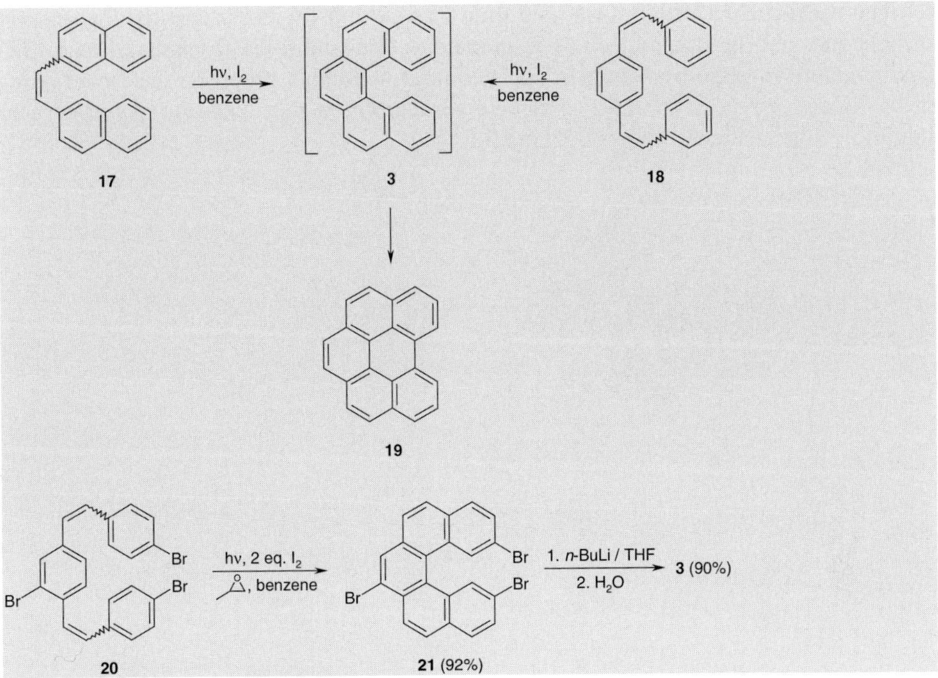

Scheme 3. Photochemical routes to [5]-helicenes.

drawback—at higher substrate concentrations stilbenes tend to undergo photodimerizations to cyclobutane derivatives. The preparation of amounts large enough to use helicenes for further studies, *e.g.* for the preparation of chiral catalysts or in material science, can therefore become difficult. It is against this background that the Diels–Alder route to helicenes has recently been rediscovered.[24, 25] For example, when 1,4-divinylbenzene (**22**) was heated under reflux in toluene with excess *p*-benzoquinone (**23**) in the presence of trichloroacetic acid and 4-*tert*-butylcatechol the bis quinone derivative **24** of [5]-helicene was formed (Scheme 4);[26] likewise 2,7-divinylnaphthalene furnished the corresponding hexahelicene bis-quinone.

Although the yields of these double Diels–Alder additions are low (< 20%) this disadvantage is offset by the ready availability of the starting materials and the simplicity of the method.

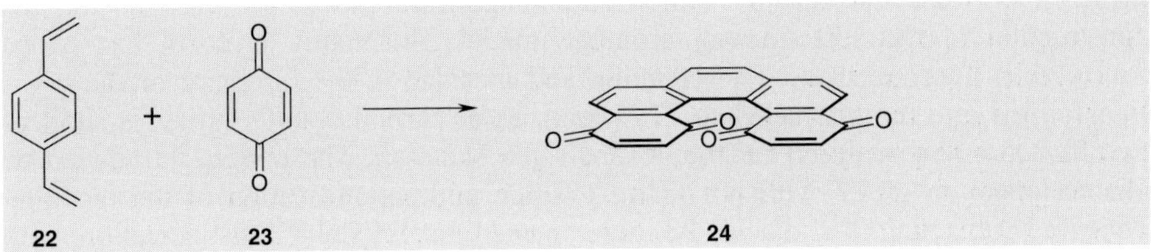

Scheme 4. A simple route to helicene quinones.

The hydrocarbon which for a long time was the most widely accepted incarnation of helicenes and helical molecules in general is phenanthro[3,4-*c*]phenanthrene (**32**) first prepared in 1956 by Newman and Lednicer, who also proposed the trivial name hexahelicene for it, a "purely aromatic hydrocarbon which owes its asymmetry to intramolecular overcrowding" (Scheme 5).[27]

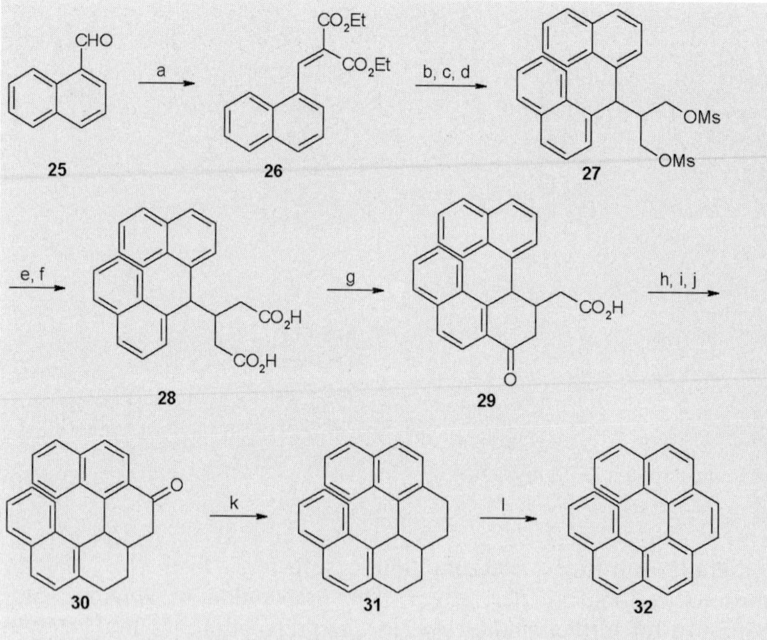

(a) diethyl malonate, base (70%); (b) 1-naphthyl magnesium bromide (30-50%); (c) LiAlH₄ (90%); (d) mesyl chloride, Et₃N (86%); (e) KCN; (f) NaOH (comb. yield 55%); (g) HF (60-65%); (h) N₂H₄, NaOH (87%); (i) PCl₅, o-dichloro benzene; (j) SnCl₄ (comb. yield 56%); (k) N₂H₄, NaOH (89%); (l) Rh / Al₂O₃, 300°C (63%)

Scheme 5. The synthesis of hexahelicene (**32**) by Newman and Lednicer.

Progress in providing the six benzene rings required in the target molecule was fast in the initial steps of the synthesis. In just two steps—condensation of α-naphthaldehyde (**25**) with diethyl malonate followed by Grignard addition of α-naphthyl magnesium bromide to the produced α,β-unsaturated diester **26**—four of these rings were introduced, while—simultaneously—the stage was set for the construction of the still missing aromatic nuclei. A lengthy protocol had to be followed to generate these. By reduction and mesylation the Grignard product was transformed into the bis mesylate **27**, which, as a 'carbon count' shows, is short of two carbon atoms required for the hexahelicene skeleton. These were introduced by chain elongation of **27** with potassium cyanide and saponification of the resulting dinitrile. Subjecting the diacid **28** thus formed to Friedel–Crafts acylation with anhydrous hydrogen fluoride furnished the keto acid **29** which was transformed by a

Haworth cyclization to the ketone **30**. After its Wolff–Kishner reduction to tetrahydrohexahelicene **31**, the aromatization of this precursor to **32** required substantial optimization. Best results were finally achieved with rhodium on alumina at 300 °C in benzene in an autoclave.

As for so many other hydrocarbon syntheses discussed in this book, a study of the chemical, structural, and—especially—chiroptical properties of **32** became possible only with the development of more efficient syntheses. Among these, photocyclizations of the stilbene-phenanthrene type (see above) are by far the most important,[28] and we will take the preparation of [6]-helicene by this route as the prototype for the synthesis of the higher helicenes (see below) also.

Photodehydrocyclization of the precursor hydrocarbons **33**,[29] **35**,[29] and **37**[30] in benzene, using iodine as the oxidant, led to [6]-helicene in good yields (Scheme 6).

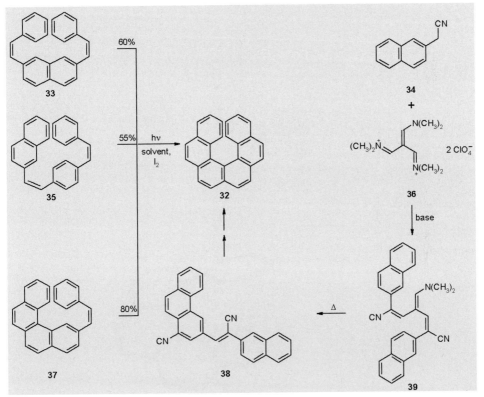

Scheme 6. Photochemical preparation of hexahelicene (**32**).

The precursor olefins for these cyclizations were obtained from simple starting materials by the Siegrist[31] or the Wittig reaction. Alternatives to these conventional olefination reactions are, however, known. For example the condensation of β-naphthylacetonitrile (**34**) with the diperchlorate **36** in the presence of sodium methoxide furnished the cross-conjugated (see Section 11.1) dinitrile **39** in 78%

yield. When this was heated to 160–170 °C electrocyclic ring-closure followed by elimination of dimethylamine took place and the diarylethene derivative **38** was produced in quantitative yield. The next step, photocyclization in the presence of iodine, was also nearly quantitative (96%), and the parent hydrocarbon **32** was obtained after alkaline hydrolysis of the dinitrile and copper-mediated decarboxylation in quinoline. With *ca* 50% overall yield, this sequence to hexahelicene is more than satisfactory.[32]

From [7]-helicene on, which incidentally was the first carbohelicene to be synthesized by dehydrophotocyclization,[33] all higher benzologs, up to [14]-helicene,[34, 35] have been prepared by this approach from appropriately substituted bis-

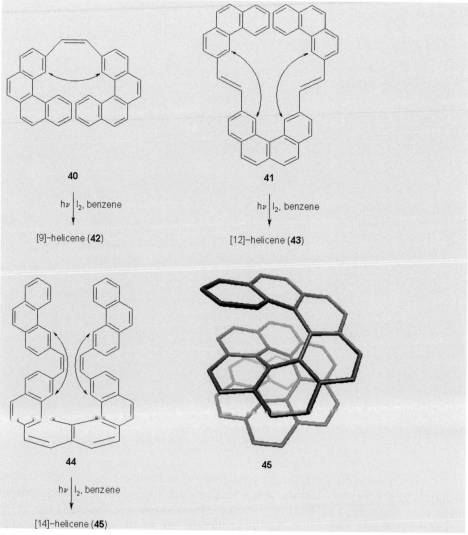

Scheme 7. The synthesis of several higher [*n*]-helicenes by dehydrophotocylization.

arylethenes. For example [9]- (**42**),[30], [12]- (**43**),[36], and [14]-helicene (**45**)[35, 36] were obtained from the symmetrical precursors **40**, **41**, and **44** (Scheme 7).

The helical nature of the hydrocarbons becomes evident by a three-dimensional representation, *e.g.* for [14]-helicene (**45**) in Scheme 8—but how have the structures and configurations of the helicenes been proven? By resolution of these molecules and/or by X-ray structural analysis. The first resolution of hexahelicene (**32**) was described by Newman and Lednicer already in their original synthesis.[27] They prepared diasteromeric charge-transfer (CT) complexes of the hydrocarbon with the chiral complexing agent (*R*)- or (*S*)-2-(2,4,5,7-tetranitro-9-fluorenylideneaminooxy) propionic acid (TAPA, **46** in Scheme 8 is (*R*)-TAPA). Chromatography on chiral columns has been applied widely for resolution; the stationary phase was coated with TAPA,[37] riboflavin,[38] or other reagents.[35] The extraordinarily high specific rotation ($[\alpha]^{25}_{579}$) of these hydrocarbons, ranging from 3640° for [6]-helicene (**32**)[27, 39] to 9620° for [13]-helicene,[40] makes these compounds ideal targets for asymmetric syntheses because small optical inductions can be detected reliably. Indeed, asymmetric photocyclizations have been performed under a variety of chiral conditions, among them asymmetric photosynthesis with circularly polarized light (CPL),[41] asymmetric synthesis in chiral solvents,[42] and in cholesteric liquid crystals.[43] Optical yields were generally low (in rare cases betwen 2 and 3%, usually in the 0.1 to 0.5% range), but when the 1,2-diarylethene derivative already carries a chiral substituent high diastereoselection may be achieved, as was shown by Martin and co-workers (Scheme 8).[44]

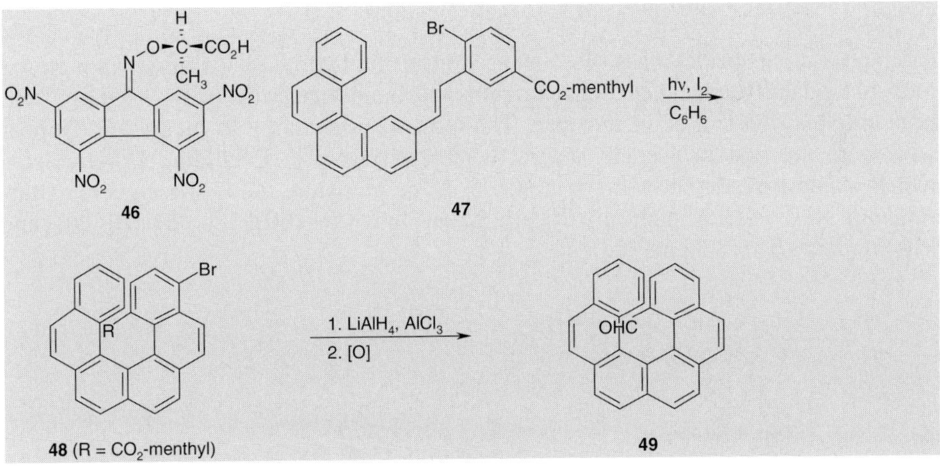

Scheme 8. The preparation of optically active helicenes.

For example, irradiation of the menthyl ester **47** at −78 °C in benzene in the presence of iodine yielded the (+) diastereomer of the hexahelicene derivative **48** in 96% diastereomeric excess. When **48** was reduced with lithium aluminum hydride/ aluminum trichloride a debrominated 1-hydroxymethyl-[6]-helicene was formed, which on oxidation provided the optically active aldehyde **49**. This, in turn, could be used to synthesize optically active [8]-, [9]-, [10]-, [11]-, and [13]-helicene.[40]

The ultimate proof of the helical structure of the [n]-helicenes was provided by X-ray structure determinations; this has been conducted for a sizeable number of the parent systems, their derivatives and CT-complexes.[35] For example, in [6]-helicene (**32**) the angle between the terminal rings is 58.5°[45] and for [7]-helicene it is 30.7°.[46] The structural parameters of [10]-[47] and [11]-helicene[48] have also been reported.

Among the reactions of the helicenes[49] the question of the configurational stability of the optically active hydrocarbons, *i.e.* their (thermal) racemization, has been of prime importance ever since these chiral molecules became synthetically available. As already mentioned, the free enthalphy of racemization of [5]-helicene (**3**) is 24.9 kcal mol^{-1} (at 196 °C);[18] for [6]-helicene this barrier increases significantly to 36.9 kcal mol^{-1} (196 °C).[50] On going to the next benzologs, however, this increase levels off and [7]-, [8]-, and [9]-helicene (**42**) show $\Delta G^{\neq}$ values of 42.6 (269 °C), 43.5 (270 °C), and 45.0 kcal mol^{-1} (270 °C).[51] The comparative ease with which these hydrocarbons lose their optical activity has been rationalized by a 'direct inversion' process, in which the terminal rings have to pass each other *via* a transition state in which they are parallel. For such a process to occur, molecular deformation of the whole hydrocarbon structure is necessary, spread over a large number of bonds and angles;[51] this hypothesis is supported by force-field calculations.[52]

12.2 The circulenes

Circulenes are a special class of condensed aromatic hydrocarbons in which a closed loop of angularly annelated benzene rings surrounds an inner ring (a 'cavity'). In principle this ring can be of any size. The term was coined by Wynberg and Dopper who were the first to prepare several heterocirculenes.[53] Excluding [3]-circulene, which would probably be too strained ever to be prepared, the first members of this series of polycyclic aromatic hydrocarbons are thus [4]- (**50**), [5]- (**5**), [6]- (**4**), and [7]-circulene (**51**) (Scheme 9).

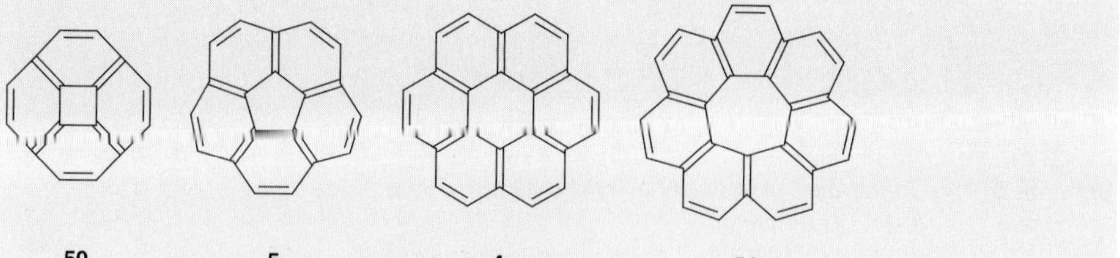

| 50 | 5 | 4 | 51 |

Scheme 9. The first members of the circulene family.

Two of these compounds traditionally have different names—**5** is commonly called corannulene,[54] and **4** coronene.[55] Coronene is a planar molecule with which we will deal with in a later chapter (see Section 14.2), just as with another famous 'circulene', kekulene (see Section 14.2). Whereas [4]-circulene (**50**) has not been prepared, the higher and the lower benzologs of coronene, [5]- (**5**) and [7]-circulene (**51**) are well known. For a long time these hydrocarbons have attracted the chemists' attention because of their non-planar structures (see below).[56] Recently this interest has increased sharply because corannulene (**5**) can be regarded as a subsystem (the 'cap' of) of C_{60} fullerene (see Chapter 2) and **51** can, in principle, also be incorporated into novel forms of carbon.

Corannulene ([5]-circulene, **5**) was first synthesized in 1966 by Lawton and Barth[54, 57] in a 17-step synthesis, classical in every meaning of the word. From the beginning the authors noted the "unusual strain resulting from the geometrical requirement that the bond angles deviate appreciably from the normal values found for benzenoid compounds".[54] To overcome this strain they decided to prepare less strained, *e.g.* saturated precursor molecules and intermediates and introduce the unsaturation—and hence the expected strain—as late as possible, defer it even to the last step if required.

The synthesis begins with acenaphthene (**52**) which was first converted to methyl 4,5-methylenephenanthrene-3-carboxylate (**53**) by a known procedure[58] (Scheme 10).

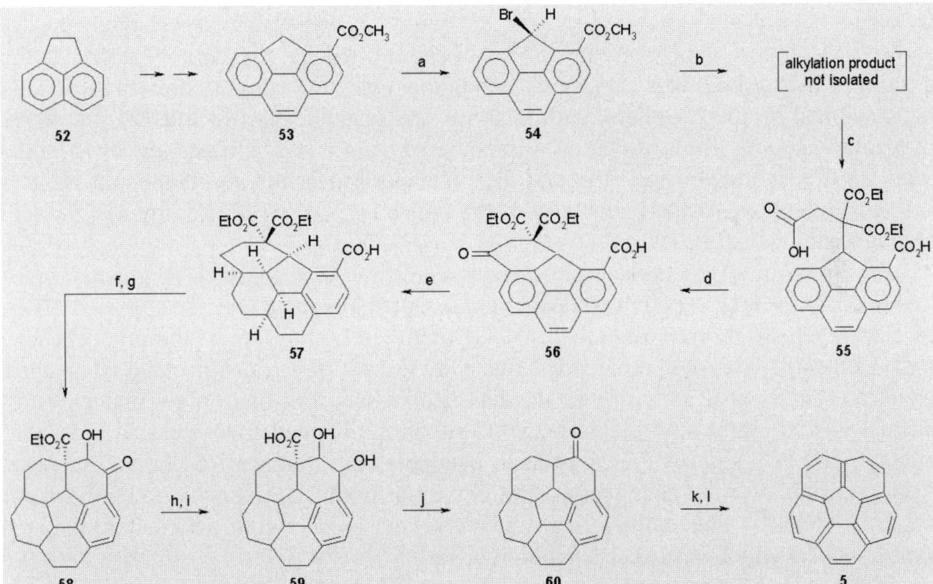

(a) NBS (92%); (b) $_{\text{EtO}_2C}\diagdown^{\text{CO}_2\text{Et}}$, *t*-BuOK / *t*-BuOH; (c) aq. KOH (comb. yield 97%); (d) PPA, 80°C (70%); (e) Pd / C, H$_2$, 5 atm, 20°C (100%); (f) CH$_2$N$_2$ (93%); (g) Na / liq. NH$_3$ (48%); (h) NaBH$_4$; (i) OH$^-$; (j) 225°C, - CO$_2$, -H$_2$O (comb. yield 92%); (k) NaBH$_4$ (94%); (l) Pd / C, 270°C (16%)

Scheme 10. The first synthesis of corannulene ([5]-circulene, **5**) by Lawton and Barth.

Derivative **53** contains three of the five six-membered rings, the cyclopentanoid core, and an ester function to be used in later ring-building processes. Counting the number of carbon atoms reveals that **53** is four carbon atoms short of those needed in the target **5**. The missing atoms were introduced by an alkylation reaction with triethyl 1,1,2-ethanetricarboxylate after **53** had been converted into the electrophile **54** by NBS bromination. When the resulting alkylation product was hydrolyzed with aqueous potassium hydroxide solution, the diacid diester **55** was formed in near quantitative yield. Note that this intermediate contains not only all required carbon atoms but also the functionality necessary for creating the missing six-membered rings. Friedel–Crafts cyclization of **55** proceeded well furnishing the keto acid diester **56** in good yield again. When an attempt was made to reduce the keto group of **56** by catalytic hydrogenation over Pd/C this aim was accomplished, but two of the three aromatic rings were hydrogenated also. Fortunately this process yielded one stereoisomer only—the dodecahydrobenzo[*g,h,i*]fluoranthene diester acid **57** in which all the hydrogen atoms have been added from one side only, thus bringing one of the ester groups (on the concave side of the molecule) close to the acid function on the aromatic ring. Diazomethane esterification of **57**, then acyloin condensation with excess sodium in liquid ammonia-ether afforded the acyloin ester **58** with the complete carbon skeleton of corannulene. Reduction of **58** with sodium borohydride gave a diol ester which was hydrolyzed, without isolation, to the diol acid **59**. When this intermediate was heated to 225 °C it lost carbon dioxide and water and gave the ketone **60**. After reduction 4-hydroxytetradecahydrocorannulene was produced; dehydrogenation and dehydration of this over 5% Pd/C finally produced corannulene (**5**). Although the authors in their original publication[54] hesitated to predict whether **5** would be a planar or bowl-shaped molecule, X-ray structural anaylsis quickly showed the latter to be correct.[59] The plane formed by the carbon atoms at the termini of the radial bonds (*i.e.* those 'extending' from the central core) and a second plane formed by the peripheral carbon atoms are, respectively, 87 and 121 pm above the plane passing through the five-membered ring. The average dihedral angle between the five-membered ring and the inner portion of the six-membered rings is 26.8°, and the bowl-shaped geometry of the molecule is also reflected by bond-length and bond-angle distortions.

As shown by many other hydrocarbon syntheses—*e.g.* those of adamantane (see Section 3.1), [1.1.1]propellane (see Section 6.1), cyclooctatetraene (see Section 10.3), [2.2]paracyclophane (see Section 12.3)—it seems to be the fate of the pioneers who first synthesize 'signpost molecules' that they do not have enough material in their hands to systematically investigate the chemical behavior of their target molecules. In all these—and many more—instances the situation changed 'overnight' when short, elegant, and efficient syntheses became available. Corannulene (**5**) is no exception from this rule, and the change was brought about by Scott and co–workers by work begun in 1991[60] and culminating six years later in a highly efficient, three-step synthesis of **5** which is summarized in Scheme 11.[61]

In the first step acenaphthenequinone (**61**), 2,4,6-heptantrione (**62**), and norbornadiene (**63**) were heated under reflux in toluene in the presence of glycine to produce 7,10-diacetylfluoranthene (**66**) in 70–75% yield. This transformation is initiated by a Knoevenagel condensation between **61** and **62** to give the cyclopentadienone derivative **65**. When **63** adds to this reactive intermediate the Diels–Alder adduct **64** is formed which, under the reaction conditions used, eliminates carbon monoxide and cyclopentadiene. Conversion of the acetyl groups of **66** to 1-chlorovinyl side chains with PCl₅ followed, and the resulting **67**—which carries

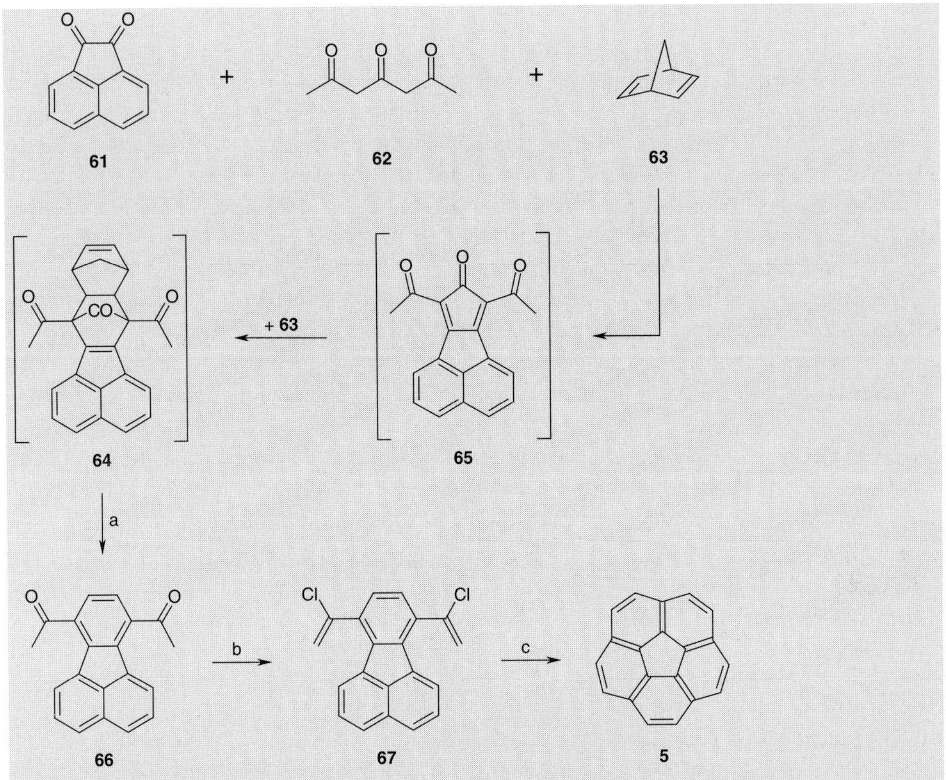

(a) glycine, toluene, reflux (70-75%); (b) PCl₅, toluene, reflux (85%); (c) flash pyrolysis, 1100°C (35-40%)

Scheme 11. The simplest version of Scott's synthesis of corannulene (**5**).

two latent ethynyl substituents—on high temperature pyrolysis cyclized to **5** with loss of hydrogen chloride. One mechanistic proposal for this last step suggests that the (known)[62] rearrangement of terminal acetylenes into vinylidene carbenes followed by C,H-insertion is involved in this cyclization step.

The Scott synthesis of **5** is not only remarkable because of its brevity but also because the immediate precursor of **5** contains already all ultimately needed π-electrons. In other words, this high-temperature approach is diametrically opposed to Lawton's (low-temperature) strategy, in which unsaturation was deliberately avoided until the very last step, the argument being that 'early' unsaturation would cause the strain to rise too soon in the course of the synthesis.

Scott's success precipitated numerous other high-energy approaches to corannulene, some of which are summarized in Scheme 12 along with the original thermal isomerization of 7,10-diethynylfluoranthene (**68**) to **5**.[60]

Thus the tetrabromide **69**,[61] the bis enol ether **70**,[63] the bis sulfone **71**,[64] the cross-conjugated enediyne derivative **72**,[65] the tetrabromide **73**,[64] and the two [4]-helicene derivatives **74** and **75**[66] could all be converted into corannulene (**5**) by sequences involving at least one high-temperature pyrolysis step.

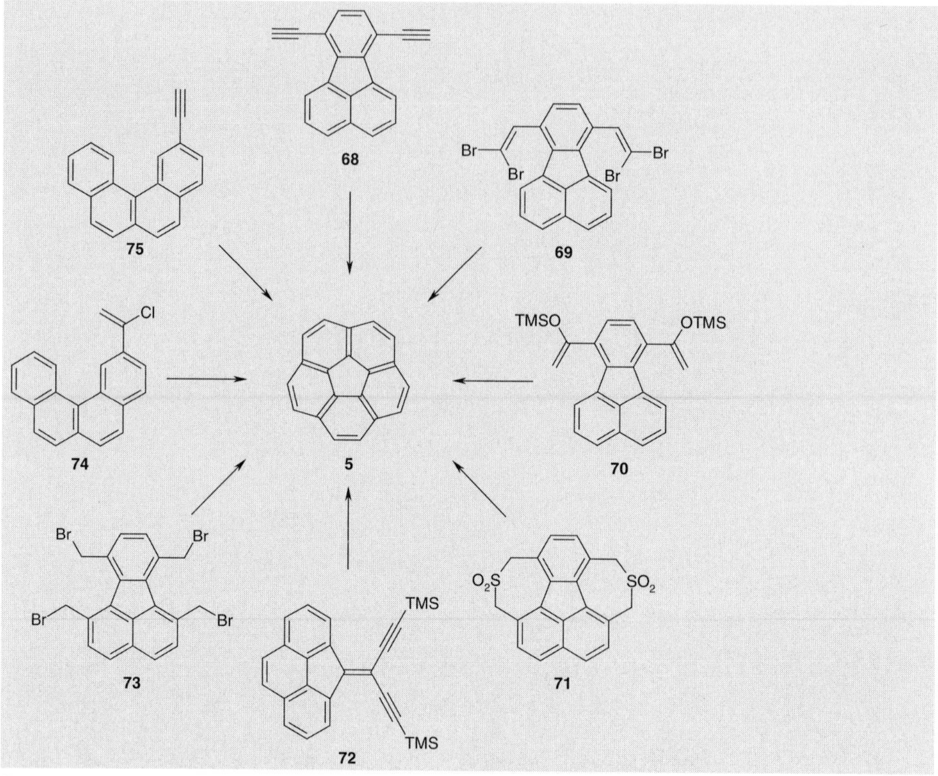

Scheme 12. Further high-temperature routes to corannulene (**5**).

When corannulene became available in preparative amounts reports on its chemical behavior appeared quickly;[67] these included its use as starting material for large condensed aromatic compounds (see Chapter 14) and its reduction to the mono-to tetraanion,[68] the latter forming stable high-order molecular sandwich structures with lithium ions.[69, 70]

The configurational stability of the corannulene-bowl has been determined by low-temperature NMR measurements on simple derivatives.[71] The low bowl-to-bowl inversion barrier of only 10.2 kcal mol^{-1} is surprising on first sight, but this at least partially reflects the rigidity of our thinking about the flexibility of organic compounds (*cf.* the racemization of the helicenes discussed above).

The next higher benzolog in the circulene series, [7]-circulene (**51**) has been of interest for many years, because according to molecular models it should have a unique saddle-shaped molecular structure with C$_2$ symmetry. After several unsuccessful attempts to prepare this hydrocarbon,[53, 72] it was finally obtained by Yamamoto and co–workers in a multi-step sequence (Scheme 13) largely applying steps which had proven their preparative worth in cyclophane (see below) and helicene chemistry (see above).[73, 74]

Thus double dehydrocyclization of the diene **76**, which had been assembled from smaller building blocks, led to 1,16-dehydro-2,15-dibromohexahelicene (**77**) in which the two bromine substituents served as anchor points for the missing last etheno

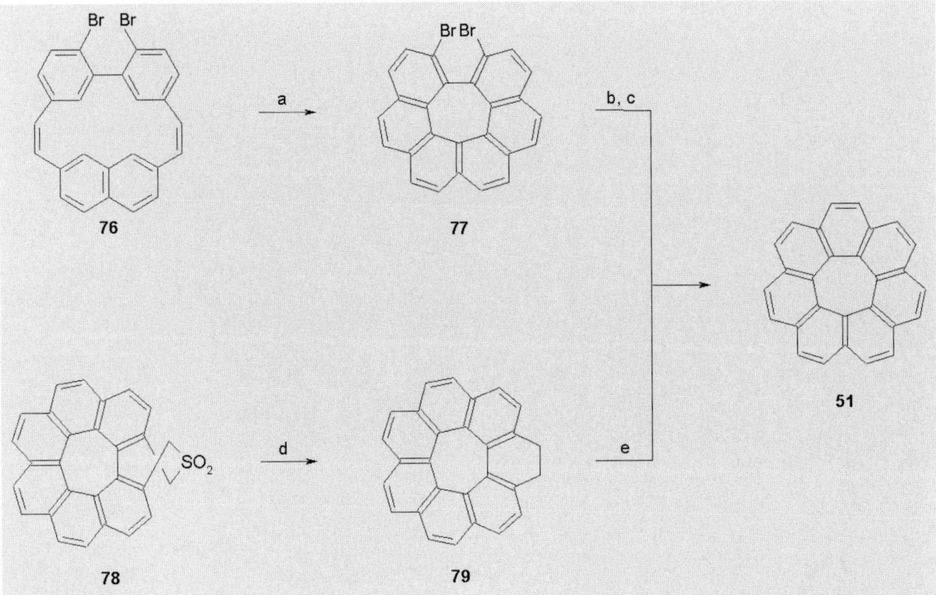

(a) hν, cyclohexane, I₂ (47%); (b) n-BuLi/THF, DMF (35%); (c) TiCl₃/LiAlH₄ (35%); (d) 550°C, 0.01 torr (24%); (e) Pd/C, 280°C (85%)

Scheme 13. The synthesis of [7]-circulene (**51**) from hexahelicene precursors by Yamamoto and co–workers.

bridge. This was constructed by replacing the halogen atoms by formyl groups (lithiation of **77** followed by a DMF quench) and subjecting the resulting dialdehyde to a McMurry coupling process. In an alternative route, also beginning with a 'short-circuited' hexahelicene precursor, the sulfone **78** was pyrolyzed at 550 °C and the resulting dihydro derivative **79** subsequently dehydrogenated over Pd/C.[75] As shown by low-temperature single crystal X-ray diffraction [7]-circulene is indeed non-planar. Its central seven-membered ring is boat-configurated, and the seven annelated benzene rings form the saddle with twofold symmetry suggested by model building.[73]

The problem of the archetypical bowl-shaped hydrocarbon corannulene (**5**) having been solved, the obvious next step in the area of concave hydrocarbons was the creation of bigger curved structures with the eventual goal of synthesizing C_{60} and other fullerenes themselves by a directed and controlled route. After all, in the language of synthetic organic chemistry the formation of C_{60} by vaporization of carbon rods[76] is a mode of formation, not a synthesis.

Although the (preparative) road to the fullerenes will probably be a stony one, the first successful steps have been completed.[77] As shown by the selection in Scheme 14, the so-called C_{2v} semibuckminsterfullerene **81** has been prepared from the tetrakis-chlorovinyl derivative **80** in a synthesis patterned after the Scott route to corannulene by Rabideau and co–workers.[78] Alternatively, Hagen, Zimmermann and Scott have arrived at this $C_{30}H_{12}$ bowl-shaped polycyclic hydrocarbon by flash vacuum pyrolysis (FVP) from the spiro compound **82** (Scheme 14).[79]

Another $C_{30}H_{12}$ hydrocarbon, C_3 semibuckminsterfullerene **85** was prepared by

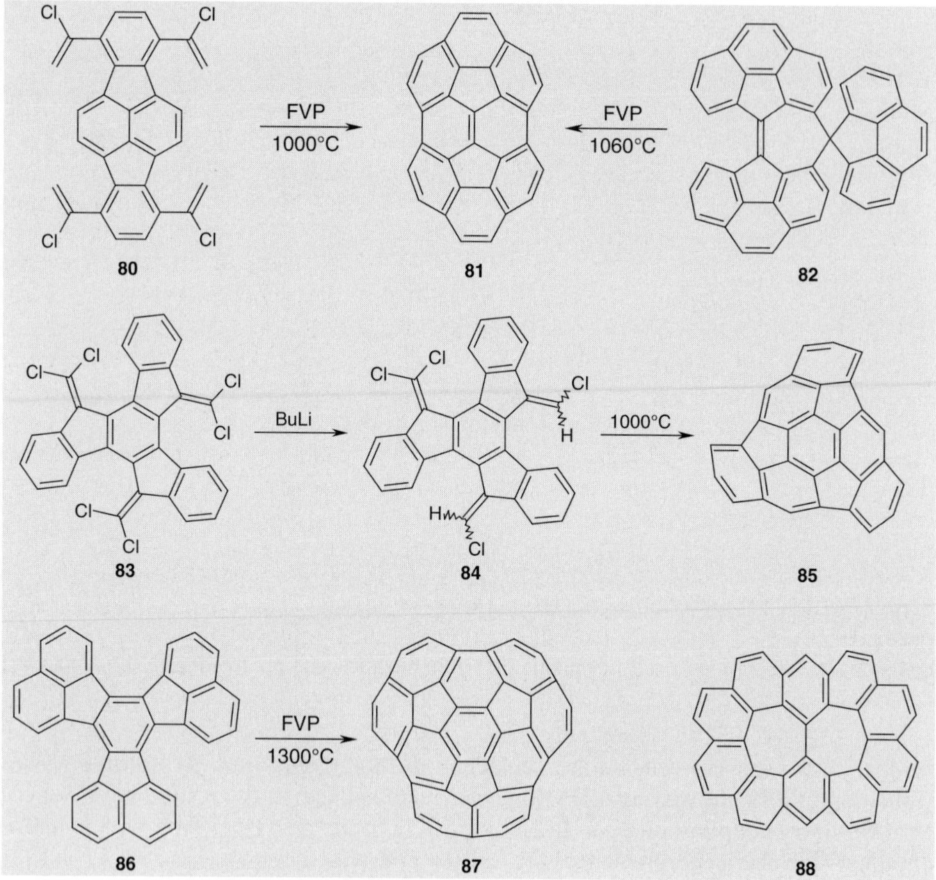

Scheme 14. *En route* to fullerenes? The synthesis of various fullerene subunits.

Rabideau and co-workers from the hexachloride **83**. After this had been partially dechlorinated by treatment with *n*-butyllithium followed by hydrolysis, the resulting isomeric tetrachlorides **84** on high-temperature pyrolysis yielded the first C$_{30}$ hydrocarbon representing exactly half of the C$_{60}$ carbon framework.[80] An even bigger fullerene subunit, the C$_{36}$H$_{12}$ hydrocarbon triacenaphthotriphenylene (**87**), which comprises 60% of the C$_{60}$ sphere has been obtained by Scott and co-workers[81] in just one step by FVP from decacyclene (**86**), a hydrocarbon known since the turn of the last century[82] and today available commercially in kilogram quantities. The enthusiasm raised by these synthetic successes is damped by their currently very low yields. For example, the **86**→**87** dehydrocyclization took place in 0.2% yield only; furthermore, many of these steps towards fullerenes furnish product mixtures requiring extensive separation work.

In closing this section on circulenes we note that the preparation of extended curved polycyclic hydrocarbons is not restricted to C$_{60}$ substructures. A case in point is provided by the so-called [7.7]-circulene **88**, in which two conjoined seven-membered rings are surrounded by a wreath of benzene rings.[74]

12.3 The cyclophanes

Although the beginning of cyclophane chemistry can be traced back to the late 19th century, and bridged aromatic compounds of the general type **8** and **9**, whether of synthetic or natural origin, were occasionally discussed in the chemical literature, it is probably no slanting of history to state that cyclophane chemistry as a separate field of aromatic chemistry was only established with Cram's epoch-making studies on [*m.n*]paracyclophanes in the early 1950s,[83] which culminated about 15 years later (see below). Cram also coined the term for this class of bridged aromatic compounds which expresses their *cyc*lic nature and their two constituent parts—a *ph*enyl and a bridging alk*ane* subunit—in an euphonious manner. Beginning with these groundbreaking investigations no other group of aromatic compounds has been studied as extensively as the cyclophanes during the last quarter of a century. Even if only pure hydrocarbon systems are considered, *i.e.* heteroatom-containing bridged aromatics are explicitly excluded—and this means, *e.g.*, exclusion of Lüttringhaus' famous ansa compounds[84] and many crown ethers and structures derived therefrom—the structural diversity is breathtaking. As already mentioned (see above and Section 10.5) the part of the 'aromatic subunit' can be played not only by any type of aromatic system, whether benzenoid or not, uncharged or charged, in principle [4*n*] π-systems ('antiaromatic' subunits) can also be bridged. As far as the molecular bridge is concerned the polymethylene chain is only the simplest variant. The bridge can contain unsaturation of any type—double and triple bonds, cumulenic units, a second aromatic or antiaromatic ring system *etc.* In the latter instance, the cyclophanes become *binuclear* or consist of two 'decks'. Additional bridges give rise to *multibridged* cyclophanes and additional aromatic units to *multilayered* cyclophanes, respectively (see below). Clearly, a text of this size and with its intended purpose can only try to describe the most important developments in this area. Fortunately, a rich review literature enables rapid entry into the vast literature.[85-97]

The reason for the unabated fascination of cyclophane chemistry is to be found in the unusual structural features of many of these compounds, especially the non-planarity of their benzene rings, their anomalous bond lengths and bond angles, their novel electronic properties, especially electronic interactions between parallel benzene nuclei, their stereochemistry—most cyclophanes possess a plane of chirality—and their use as building blocks for the construction of more extended three-dimensional π-systems. Many of these properties can be found even in the simplest representative cyclophanes assembled in Scheme 15.

Thus both [*n*]para- (**89**) and [*n*]metacyclophanes (**90**) are very well suited to study the problem of benzene ring deformation. It is intuitively obvious that the amount of ring bending will depend on the length of the polymethylene bridge—the shorter it is, the less planar should be the six-membered ring. In fact, the question has often been asked whether by the introduction of a short polymethylene bridging unit[94] the benzene ring cannot be made to forfeit its aromatic character and acquire the electronic properties of a Kekulé benzene, *i.e.* a cyclohexatriene (see Section 16.3). Among the binuclear cyclophanes [2.2]paracyclophane (**91**) is by far the most thoroughly studied hydrocarbon. Not only is the deformation of the benzene rings of importance in this sandwich-structured aromatic molecule (see below), but also if,

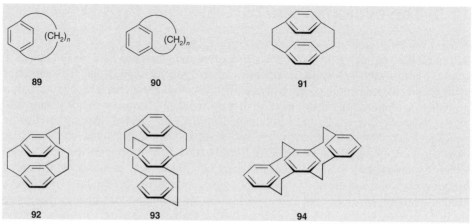

Scheme 15. Some fundamental cyclophane hydrocarbons.

and to what extent, there is electronic interaction between its aromatic subunits. Note that the bridges in a [2.2]cyclophane need not only be anchored in the *para* positions of the benzene rings.

When the two benzene rings in **91** are bridged by one additional ethano bridge the multibridged cyclophane [1.2.4]paracyclophane (**92**) results;[98, 99] this is one of three possible triply-bridged cyclophanes with parallel ethano bridges. Clearly, the ultimate goal in this series of hydrocarbons is the fully bridged molecule—superphane (see below). Among the extended three-dimensional aromatics the multilayered cyclophanes exemplified by the triple-layered hydrocarbon **93** and multistepped systems such as **94** are prominent. Obviously, in these cases the question of increasing the number of decks or steps still further—how far?—is a preparative challenge.

Before discussing the preparation of the most important [*n*]paracyclophanes (**89**) some general remarks on the (presumed) geometry of these hydrocarbons are in order. If we take a deformation angle α (formed by the plane passing through the non-bridged aromatic carbon atoms and the plane defined by a bridgehead and its two neighboring carbon atoms as shown in Scheme 16) as a measure of the non-planarity and hence strain in these compounds, the assumption that both α and the heats of formation of the hydrocarbons **89** should increase with shortening of the bridge is supported by the AM1 calculations[9, 100, 101] summarized in Scheme 16.

Note that bending out of the aromatic plane becomes apparent even with bridges as long as pentadecamethylene. It remains low, however, until the nona- and octamethylene bridged compounds, *i.e.* up to a point at which the calculated heat of formation is also negative. Beginning with [8]paracyclophane α and ΔH_f grow quickly, reaching limits at *n* = 4 or 5 which to overcome synthetically will require creative solutions indeed.[100, 101]

As suggested by these calculations, [*n*]paracyclophanes with *n* decreasing from 15 to approximately 9 should pose no particular synthetic difficulties. And, indeed, in these cases the bridge can be closed over a preexisting aromatic nucleus by classical methods such as acyloin condensation (*n* = 9, 10, 12, 14), Friedel–Crafts acylation

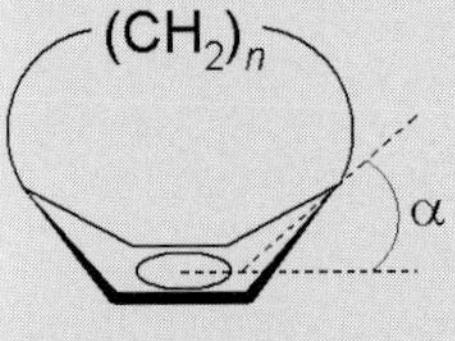

n	α [°]	ΔH_f [kcal mol^{-1}]
15	0.6	-68.2
14	0.5	-62.3
13	1.3	-54.9
12	1.9	-48.9
11	2.2	-40.5
10	3.4	-29.5
9	9.6	-18.9
8	13.4	-8.7
7	16.4	5.1
6	22.3	26.2
5	28.6	53.8
4	35.6	87.8
3	79.4	117.8

Scheme 16. AM1-calculated deformation angles (α) and heats of formation (ΔH_f) of the [n]paracyclophanes (**89**).

under high dilution conditions (n = 9–14, 16) or the intramolecular Eglinton coupling of p-bis-ethynylbenzenes (n = 10–12, 14) as shown by Cram,[102] Huisgen,[103] and Misumi[104] and their co-workers, respectively.

For the lower homologs new approaches had to be developed,[105] because for these strained systems intermolecular coupling of their precursor molecules dominates over intramolecular ring closure. In principle the increasing strain can be overcome either by the generation of high-energy intermediates (carbenes, radicals) or by the preparation of energy-rich starting materials; both of these general strategies have been used extensively. To exploit the release of the benzene resonance energy as a driving force has been particularly valuable.

The preparation of [8]paracyclophane (**98**) by Cram and Knox is a good example for the reactive-intermediate approach (Scheme 17).[106]

Bis vinylogous (1,6-) Hofmann elimination of a mixture of the two Hofmann bases **95** and **99** resulted in the formation of p-xylylene (**96**, p-quinodimethane) and its furanoid analog **100**. In both intermediates the tendency to regain their resonance energy is large, and this can be accomplished by a [6+6]cycloaddition to the benzofuranophane **97**. Of course, in this cross-coupling experiment the symmetrical dimers of **96** and **100** were also formed. In fact, the dimerization of **96** is one of the most important methods of making [2.2]phanes, whether binuclear ([2.2]para-cyclophane, **91**) or layered (*e.g.* **93**), as we shall see later in this chapter. In the present case Clemmensen reduction of **97** yielded [8]paracyclophane (**98**) accompanied by its 3,8-diketo derivative. Addition of bromine to **97** provided the enediones **101** and **102**, respectively, with the configuration of the ring-opened products depending on the reaction conditions. Both diastereomers could be reduced to **98**, which, after optimization of this procedure, could be obtained in gram quantities.[107, 108]

Two other interesting [8]paracyclophane syntheses proceeding *via* reactive intermediates employ the 'bishomo' p-xylylene **103**[109] and the cross-conjugated (see Chapter 11) spirodienone **108** (Scheme 18).[110]

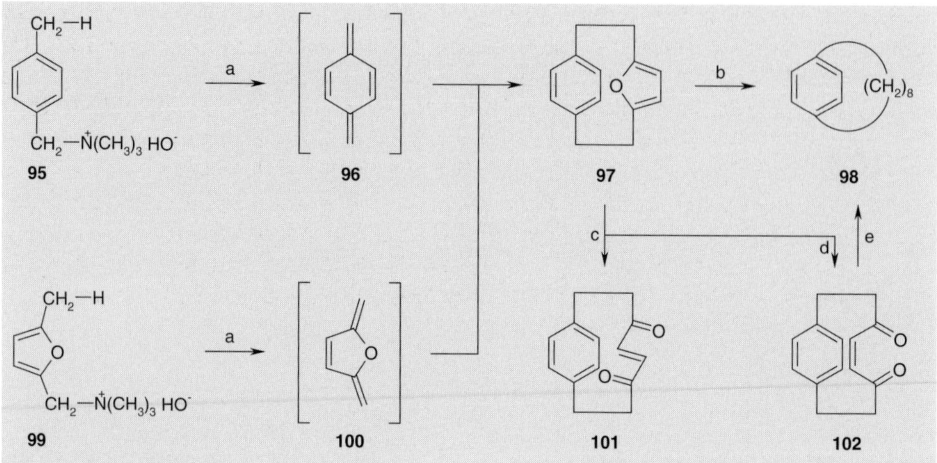

(a) Δ (23%); (b) Zn / Hg, HOAc, HCl (54%); (c) Br$_2$, MeOH, 5% H$_2$SO$_4$, -30°C (47%); (d) Br$_2$, MeOH, -30°C, H$_2$O (47%); (e) LiAlH$_4$ / AlCl$_3$ (65%)

Scheme 17. The preparation of [8]paracyclophane (**98**) by Cram and Knox.

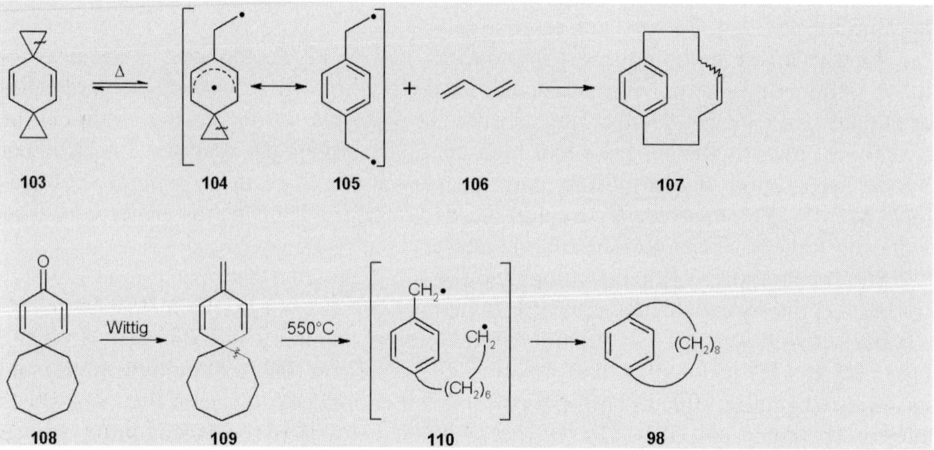

Scheme 18. Additional routes to [8]paracyclophane (**98**) involving reactive intermediates ...

As shown by Tsuji and Nishida, pyrolysis of **103** in the presence of various conjugated dienes, including 1,3-butadiene (**106**), afforded the isomeric [8]para-cyclophanes **107** in over 65% yield. CIDNP measurements showed[111, 112] that the reaction begins with the homolytic cleavage of one cyclopropane ring, thus generating the diradical **104**. This is further stabilized by destruction of the second three-membered ring, leading to the aromatic 1,8-diradical **105**. Both the formation of **104** and **105** are reversible. In the presence of **106** the diradical **105** was trapped and the paracyclophanes **107** were produced. Clearly, the process derives its driving force from the reduction of strain on the one hand, and on the gain of the benzene resonance energy on the other. High-temperature pyrolysis of another spirocompound, the triene

109, available from the ketone **108** by Wittig olefination (70%) was used by Jenneskens, de Wolf, and Bickelhaupt in their approch to [8]paracyclophane (**98**). On heating of **109** to 550 °C at 0.4 torr a semibenzene rearrangement was triggered which caused the formation of the diradical **110**, the reclosure of which yielded **98** in a remarkable 70% isolated yield.[110] As we shall see, this type of isomerization is very useful for the preparation of several lower homologs of **98** also.

In the approach entailing slow build up of strain relatively strain-free paracyclophanes are constructed first and subjected to ring-contraction processes at the end of the synthetic sequence. This is well illustrated, for example, by Allinger's preparation of 4-carboxy-[8]paracyclophane (**114**), the first crystalline derivative of a [n]paracyclophane (Scheme 19).[113]

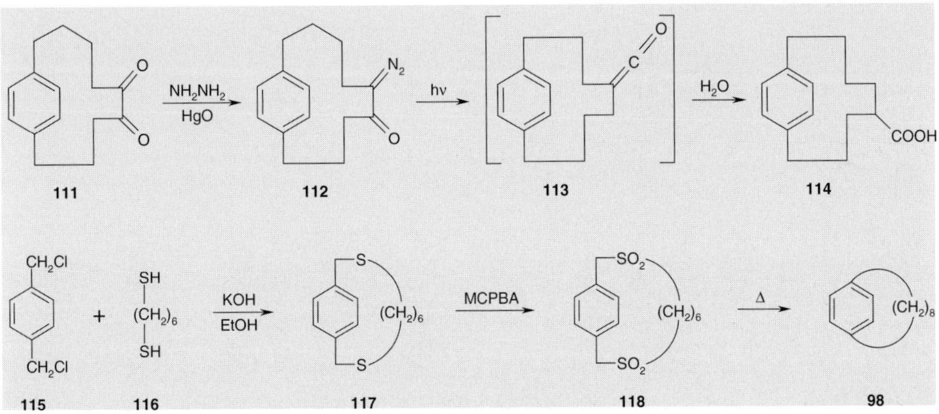

Scheme 19. ... and ring contraction of larger phane systems.

The relatively strain-free (see Scheme 16 above) [9]paracyclophane derivative **111** was first prepared by an acyloin condensation/oxidation sequence from appropriate precursors. Condensation with hydrazine then oxidation with HgO transformed **111** into the α-diazoketone **112** (mixture of regioisomers) which on photolysis underwent Wolff rearrangement to the ketene intermediate **113**, trapped by water to the acid **114**. Not only could the structure of this derivative be determined by X-ray structural analysis but the compound also served as a precursor for the preparation of the first derivative of [7]paracyclophane, 3-carboxy-[7]paracyclophane, prepared by converting **114** into 4-keto-[8]paracyclophane and then repeating the whole diazoketone ring contraction sequence.[114, 115] Although this stepwise protocol could, in principle, be continued, pratical considerations—numerous steps, low overall yield—made the development of novel alternatives for the preparation of still lower [n]paracyclophane homologs a *conditio sine qua non* (see below).

Currently the simplest route to [8]paracyclophane (**98**) is that of Otsubo and Misumi who obtained the hydrocarbon in just three steps from the commercially available starting materials α,α'-dichloro-*p*-xylene (**115**) and 1,6-hexandithiol (**116**).[116] Coupling of these components furnished the bis-thia-[10]paracyclophane **117** and after this had been oxidized with *m*-chloroperbenzoic acid (MCPBA) to the bis sulfone **118** the strained product **98** was generated in the last, strain-producing step applying one of the most powerful C–C-bond forming processes in cyclophane chemistry—the sulfone pyrolysis.[117] The overall yield of this [8]paracyclophane synthesis was 40%.

Unsubstituted [7]paracyclophane (**126**) was first prepared by M. Jones, Jr and co-workers in 1973 by an imaginative new approach which again exploits the driving force of a rearomatization process—the isomerization of the spirocarbene **125** to the target hydrocarbon **126** (Scheme 20).[118]

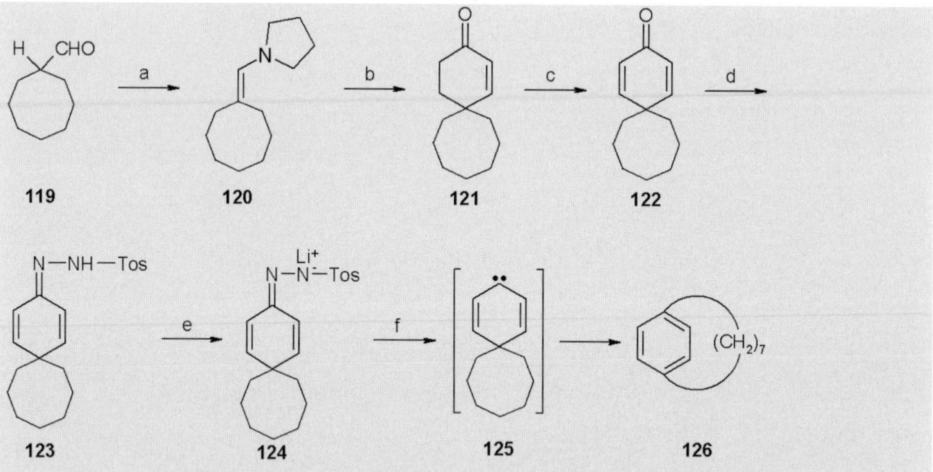

(a) pyrrol, toluene, H+; (b) methyl vinyl ketone, EtOH, NaOAc, HOAc, NaOH, H₂O; (c) DDQ, dioxan; (d) TsNHNH₂, EtOH; (e) n-BuLi, THF; (f) 360-380°C, 1 torr

Scheme 20. The synthesis of [7]paracyclophane (**126**) by M. Jones, Jr and co-workers.

The precursor for **125**, the lithium salt **124**, was prepared by a five-step sequence, beginning with the aldehyde **119**. After its conversion to the enamine **120**, a Robinson-type annelation with methyl vinyl ketone yielded the spiroketone **121**; this was dehydrogenated to the cross-conjugated cyclohexadienone **122** (see Chapter 11). Reaction with tosylhydrazine subsequently furnished the hydrazide **123** from which **124** was prepared by treatment with butyllithium. Flash vacuum pyrolysis of the latter then led to **126**, which was isolated in 7–10% yield, the overall yield of the whole sequence amounting to 2–3%. Because starting aldehydes with ring-sizes between 4 and 12 could be converted to the corresponding ketones **122** a first general synthesis of [n]paracyclophanes seems to be evolving.[119] We shall see soon where it meets its limits!

The most convenient route to [7]paracyclophane (**126**) currently involves the thermal isomerization of the lower homolog of **109**. By this route **126** could be prepared from 3-methylenespiro[5.6]dodeca-1,4-diene in 19% yield.[110, 120]

Although the Jones spirocarbene method was successful,[121] in the first preparation of [6]paracyclophane (**128**) from the salt **127**, yields were low (2%), and the cyclophane had to be separated by gas chromatography from other aromatic compounds also produced. The need to develop alternatives was furthermore stressed by the observation that the scope of this approach cannot be extended beyond **128**—an attempt to convert **129** into [5]paracyclophane (**130**) by flash vacuum pyrolysis (FVP) failed (Scheme 21).[122]

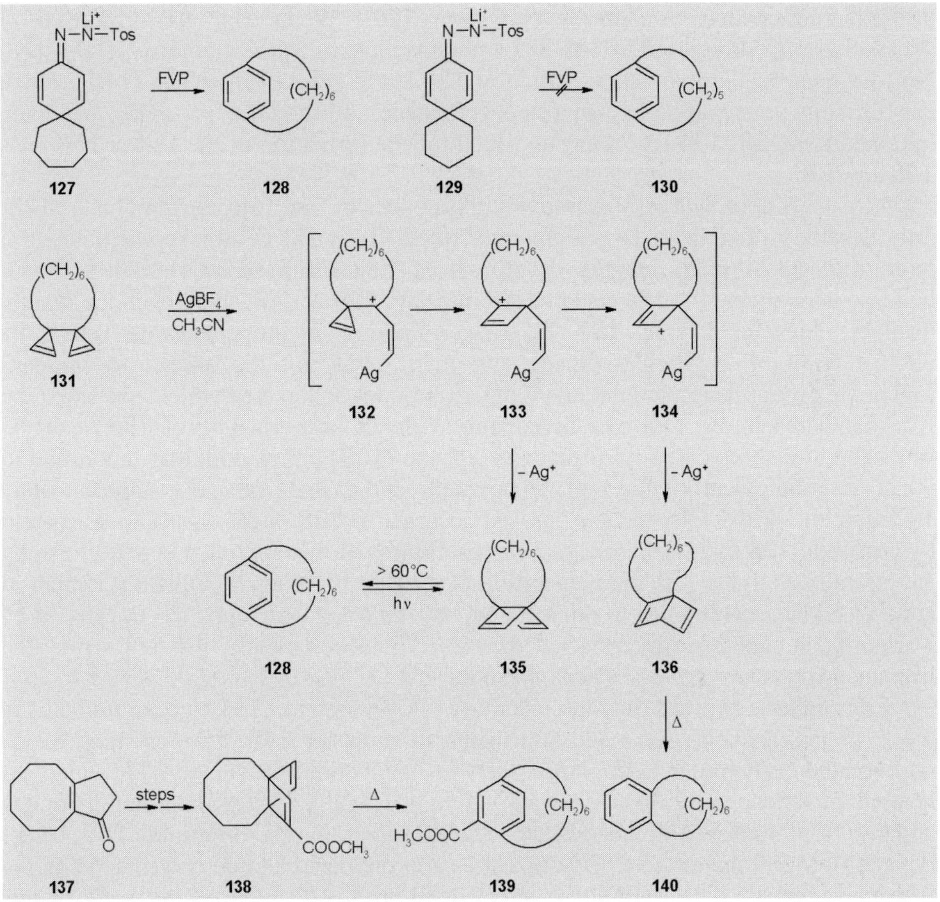

Scheme 21. Various routes to [6]paracyclophane (**128**).

Because with decreasing bridge length the distance between the *para* positions of a [*n*]paracyclophane (**89**) becomes increasingly shorter, one could expect that from a certain point onwards the aromatic nucleus could collapse to a Dewar benzene derivative (see Section 10.2). Conversely, a 1,4-bridged Dewar benzene might, in principle, open to its [*n*]paracyclophane valence isomer. That this is indeed so was shown by extensive investigations by the groups of Bickelhaupt, Jones, Tobe and others.

The Dewar benzene derivatives required for this methodology have been synthesized by different routes. By a series of rearrangements reminiscent of the

(CH)$_6$-valence isomerizations discussed in Section 10.3 the bridged bicyclopropenyl **131** was first isomerized to a mixture of the Dewar benzenes **135** and **136** by silver tetrafluoroborate treatment in anhydrous acetonitrile (see Section 10.3 for Ag$^+$-catalyzed rearrangements of hydrocarbons).[123-125] Although the 1,2-bridged isomer **136** was produced in vast excess (**135**:**136** = 1:9), the required **135** could be separated by gas chromatography. On heating this propellane (see Section 6.1) to 60–90°C it ring-opened to [6]paracyclophane (**128**). When this was irradiated in cyclohexane-d_{12} it reclosed to **135** and a photostationary equilibrium was established. Not surprisingly, thermal isomerization of **136** led to benzocyclooctene (**140**), which can formally be regarded as an orthocyclophane. A sequence involving the ionic intermediates **132**–**134** has been suggested for the formation of the Dewar benzenes **135** and **136**.

Again this approach is preparatively unsatisfactory and [6]paracyclophane (**128**) only became available in larger amounts when Tobe and co–workers developed a more directed—though lengthy—synthesis of functionalized [1,4]bridged Dewar benzene derivatives.[126] The decisive intermediate **138** was obtained from the readily available bicyclic ketone **137** by first preparing a propellane derivative by photoaddition of 1,2-dichloroethene and then subjecting the adduct to standard functional group manipulations involving a ring-contraction step which converted the five-membered into a four-membered ring. Valence isomerization of **138** at 60 °C furnished 8-methoxycarbonyl-[6]paracyclophane (**139**) which could be saponified to the corresponding carboxylic acid. The overall yield of the synthesis is approximately 10% and one of the intermediates passed *en route* to **138** opens access to the parent hydrocarbon **128** by ester pyrolysis. The usefulness of this approach is underlined by successfully applying it to the preparation of 8-methoxycarbonyl-[6]paracyclophan-3-ene.[127] A remarkable route to other functionalized [6]paracyclophanes, developed by Tochtermann and co-workers (Scheme 22),[128] is completely different from the [n]paracyclophane syntheses discussed so far.

The synthesis exploits the high reactivity of cyclooctyne (**141**, see Section 8.2) in the first step, a Diels–Alder addition to the furandiester **142**. The resulting 7-oxa-norbornadiene derivative **143** photocyclized to the oxaquadricyclane **144**, which, on heating in xylene, ring-opened again but to the bridged oxepin diester **145**. On addition of bromine 1,4-homoaddition occurred, providing the dibromide **146**. Under typical McMurry conditions this adduct was debrominated and deoxygenated and the [6]paracyclophane diester **147** was obtained. With 15–20% the overall yield of this tandem process is more than satisfactory, considering what has been accomplished. Although conceptionally very attractive—*i.e.* by beginning with cycloheptyne (see Section 8.2) instead of **141**—this route could not be extended to the preparation of a [5]paracyclophane derivative.[128]

When the bridge is reduced to five or even less methylene units, preparative difficulties in synthesizing, or even of obtaining only hints of the intermediacy of the appropriate [n]paracyclophane, increase even more sharply, because according to calculations [5]- to [3]paracyclophane are thermodynamically and kinetically much less stable then [6]paracyclophane (**128**).[129]

Despite this, Bickelhaupt, Tobe and co-workers were able to generate [5]paracyclophane by irradiating 1,4-pentamethylene Dewar benzene—the lower homolog of **135**—in THF-d_8 at –60 °C, the precursor having been prepared from the

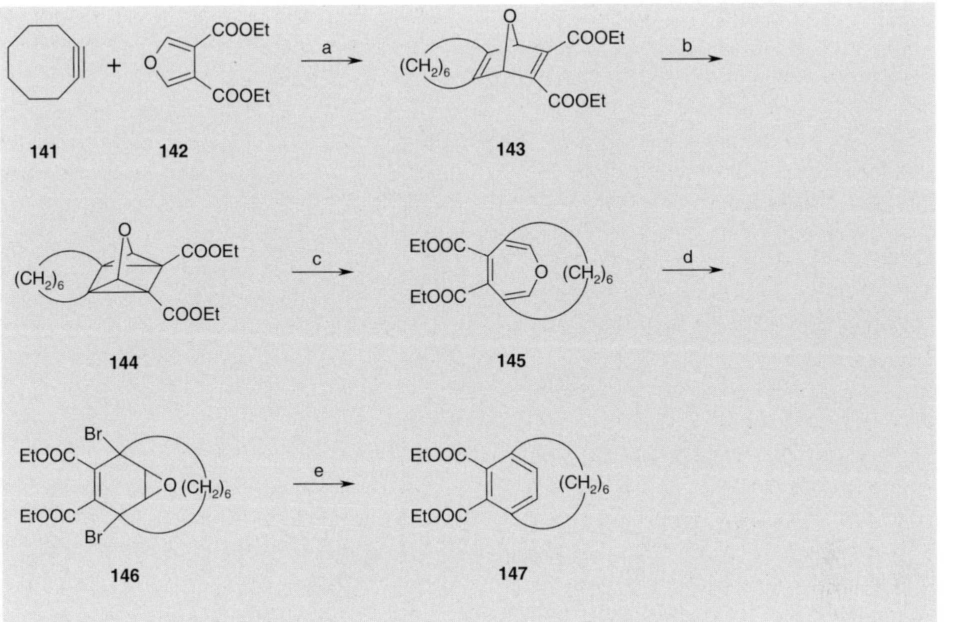

(a) 120°C, 4 h (78%); (b) hv, Et₂O, -20°C (74%); (c) xylene, reflux (62%); (d) Br₂, CH₂Cl₂
(55%); (e) TiCl₃ / LiAlH₄, THF (40%)

Scheme 22. Tochtermann's route to the [6]paracyclophane diester **147**.

corresponding bridged bicyclopropenyl by an analogous route. The extremely reactive hydrocarbon, formed in 6–7% yield, could not be isolated, but was identified by its UV and ^{1}H NMR spectra.[130] Whereas the unsubstituted [5]paracyclophane decomposed rapidly above 0 °C, two of its derivatives, 7-methoxycarbonyl-[5]paracyclophane[131] and 7,8-bis-methoxycarbonyl-10,11-dimethyl-[5]paracyclophane[132]—also obtainable from their Dewar benzene valence isomers by photochemical conversion—have half-lifes of several hours at room temperature.

Currently synthetic efforts to fathom the [*n*]paracyclophane concept, *i.e.* to test the deformability of the benzene ring, culminate and end at [4]paracyclophane (**149**). That the existence of this extremely distorted molecule (see angle α in Scheme 16) could only be inferred from trapping or matrix isolation experiments should be of no surprise after experiences with its higher homolog. All attempts to synthesize **149** use the bridged Dewar benzene route, which thus has turned out to be the method of choice for the preparation of the lowest [*n*]paracyclophane homologs.

When Bickelhaupt and co–workers irradiated 1,4-tetramethylene Dewar benzene (**148**) at -20 °C in tetrahydrofuran in the presence of excess trifluoroacetic acid, they isolated the 1,4-addition products **151** and **152** (Scheme 23).[133]

Replacement of the tetrahydrofuran by methanol resulted in the formation of the ether **153**. To rationalize these observations it has been proposed that after the photochemical ring-opening of **148** to the desired [4]paracyclophane (**149**) this is instantaneously protonated to the bicyclic σ-complex **150**. Whereas the trapping products **151** and **153** are easily explained, the formation of **152** evidently involved

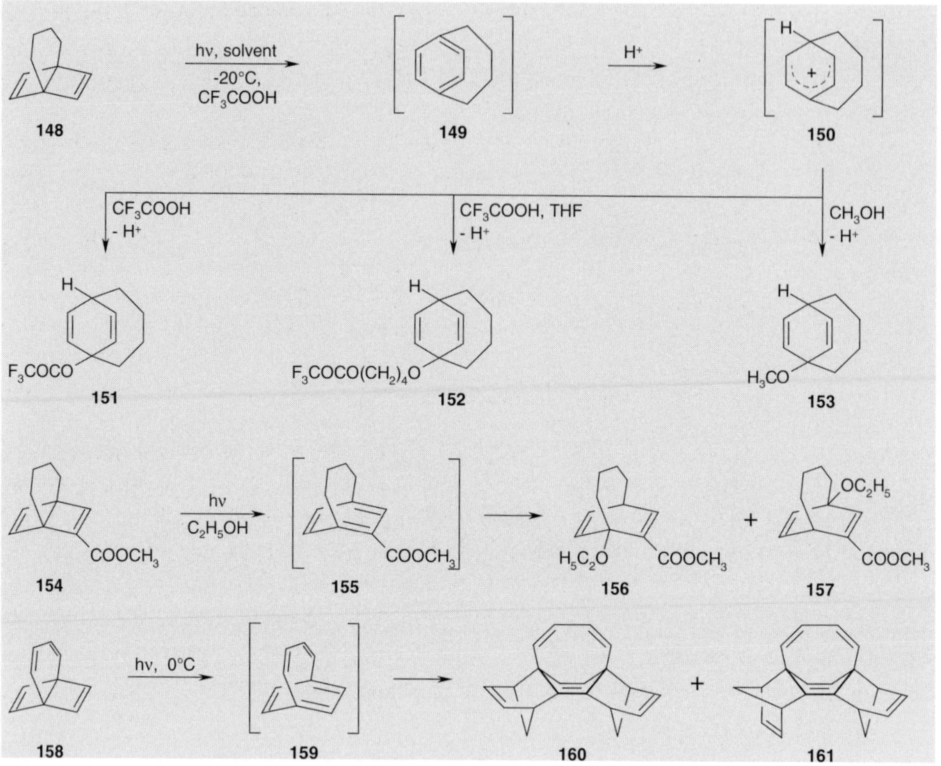

Scheme 23. The generation of [4]paracyclophane (**149**) by Bickelhaupt and by Tsuji and Nishida.

one molecule of the solvent which was ring-opened by **150** before the resulting carbocation was trapped by the acid. Repetition of the experiment under neutral conditions (irradiation of **148** in THF-d_8 at -50 °C) gave no hint on the formation of [4]paracyclophane. Tsuji and Nishida prepared not only **148** and **149**, which they intercepted with various alcohols to yield adducts of type **153**,[134] but also studied derivatives of Dewar benzenes such as **154**. When this was irradiated in the presence of ethanol two regioisomeric 1,4-addition products **156** and **157** were produced, clearly as a result of addition to the unsymmetrical [4]paracyclophane ester **155**. Furthermore, irradiation of **148** in a matrix at 77 K led to an absorption maximum at 340 nm with a shoulder at 370 nm, UV bands in the range expected for **149**. Nothing can better illustrate the success of the Dewar benzene approach to the very strained [n]paracyclophanes than the photochemical ring-opening of the propellane **158**, viz. a hydrocarbon which contains two additional, strain-enhancing double bonds in its four-membered bridge! When this hydrocarbon was photoisomerized in hexane at 0 °C in the presence of excess 1,3-cyclopentadiene, by means of a high-pressure mercury lamp the two 2:1 adducts **160** and **161** were formed in 63% yield in a ratio of 2:1. Obviously the [4]paracyclophane diene **159** was generated as an intermediate and subsequently trapped in a Diels–Alder addition in which a double bond of a 'benzene ring' functions as a dienophile.[135]

As remarked above, irradiation of [6]paracyclophane (**128**) caused reclosure to the Dewar benzene isomer **135**. That strained [*n*]paracyclophanes can also be transformed thermally into their Dewar forms has recently been demonstrated for the first time on a [4]paracyclophane derivative.[136]

Compared to the [*n*]paracyclophanes (**89**) far less is known about their *meta* isomers **90**. According to molecular models and spectroscopic data, boat-type deformation of the benzene rings of these hydrocarbons begins at a bridge length of approximately seven. A considerable number of [*n*]metacyclophanes with larger bridges is known,[105, 137-140] and for the syntheses of the hydrocarbons with $n \leq 8$ many of the methods we have encountered in [*n*]paracyclophane chemistry already (*e.g.* as sulfone pyrolysis[139, 140] or intramolecular Eglinton coupling,[141] see above) have been employed.

There are, however, distinct differences (although it is not known with certainty whether these approaches just have not been tried for the synthesis of the *para*-substituted hydrocarbons **89**). For example, in what must be one of the most direct routes to any cyclophane Kumada and co–workers connected 1,3-dichlorobenzene (**162**) with the bis Grignard compound **163** in the presence of dichloro[1,3-bis-(diphenylphosphino)-propane]nickel(II) ([Ni(dppp)Cl$_2$], **164**) by a coupling process that later should bear its senior author's name (Scheme 24).[142]

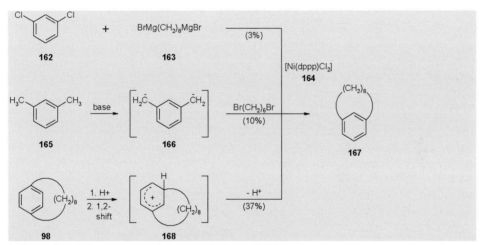

Scheme 24. Various routes to [8]metacyclophane (**167**).

Another direct route to **167**, described by Bates and co–workers, involves the dianion, **166**, of *m*-xylene (**165**). The former has been generated from **165** by treatment with the Lochmann–Schlosser base (*n*-butyllithium–potassium *tert*-butoxide). The low yield of the bis-alkylation is outweighed by the simplicity and availability of the starting materials.[143]

That small [*n*]paracyclophanes can serve as precursors for their [*n*]*meta* isomers was shown by Hopf and co–workers who isomerized [8]paracyclophane (**98**) to **167** by treatment with hydrogenchloride-aluminum trichloride at −10 °C in dichloro-methane; the process is assumed to occur *via* the σ-complex **168**.[107, 108, 144]

Most of these methods have also been used to prepare [7]metacyclophane (**174**),[145] but unprecedented routes to this hydrocarbon have also been reported (Scheme 25).[146]

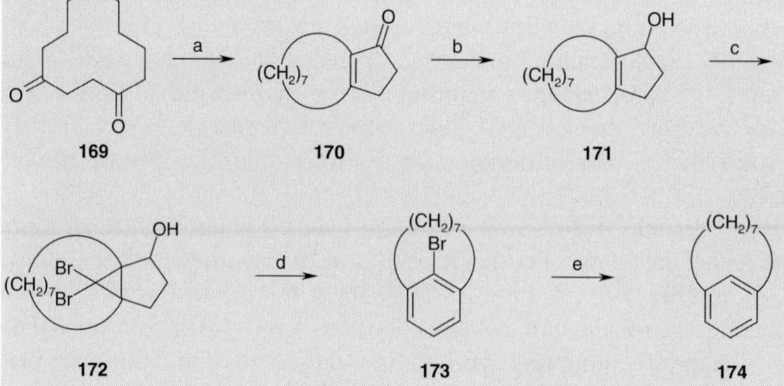

(a) NaOH / CH₃OH (87%); (b) LiAlH₄, ether (86%); (c) HCBr₃ / tert-BuOK; (d) 135°C (comb. yield 32%); (e) n-BuLi, n-hexane, then H₂O (70%)

Scheme 25. [7]Metacyclophane (**174**) by contraction of a medium-sized ring.

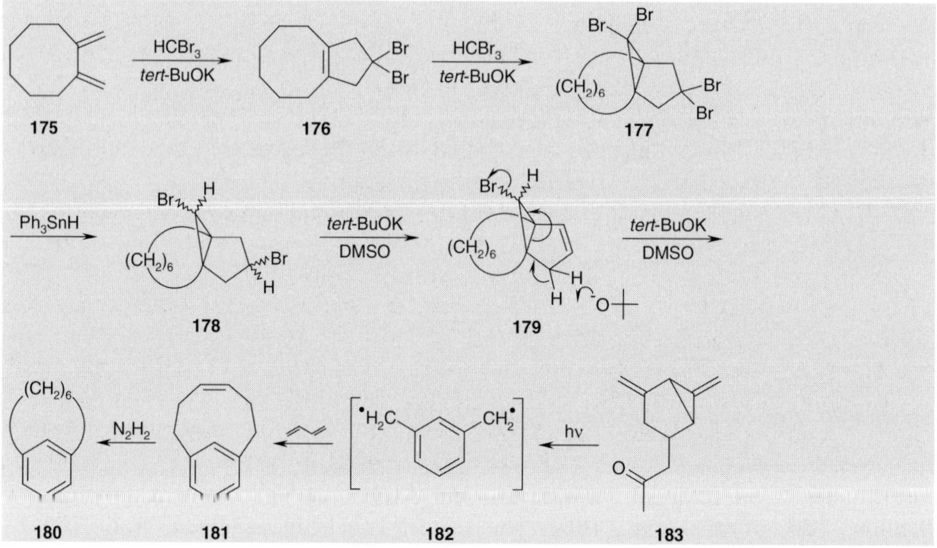

Scheme 26. Two mechanistically interesting routes to [6]metacyclophane (**180**).

As discovered by Nozaki, Hirano and co-workers, intramolecular aldol condensation of cyclododeca-1,4-dione (**169**) provided the bicyclic enone **170**, which on reduction with lithium aluminum hydride was converted to the allylic alcohol **171**. When this was dibromocyclopropanated and the resulting propellane **172** (see Section

6.1) subjected to flash vacuum pyrolysis (FVP) the bromo-[7]metacyclophane **173** was formed; this was reduced to the parent compound **174** by metalation/hydrolysis. Because this methodology could also be employed for the synthesis of the next lower homolog in the [*n*]metacyclophane series, [6]metacyclophane (**180**), it constitutes a useful addition to the preparative arsenal of cyclophane chemistry.[146, 147]

Two mechanistically interesting routes to **180** have been reported by Bickelhaupt[148] and by Berson[149] and their co-workers (Scheme 26).

Addition of dibromocarbene to 1,2-bismethylenecyclooctane (**175**) furnished the 1:1 adduct **176** in a rare example of a 1,4-addition of a carbene to a conjugated diene. When **176** was dibromocyclopropanated the expected tetrabromide **177** was isolated; this, on partial reduction with triphenyltinhydride, was converted to the dibromide **178** as a mixture of isomers. On treatment with base (potassium *tert*-butoxide in DMSO) normal 1,2-elimination yielded the tricyclic monobromide **179**. Further exposure to base caused 1,4-homoelimination and direct formation of [6]metacyclophane (**180**). Alternatively this hydrocarbon was obtained by photolysis of the ketone **183** (Norrish type II cleavage) and trapping the cleavage product, the non-Kekulé hydrocarbon **182** (see Section 16.2), with 1,3-butadiene. The [6]metacyclophan-3-ene (**181**) thus generated could then be reduced with diimine to **180**. As in the examples discussed above, acid-catalyzed isomerization of a [6]paracyclophane precursor constitutes a useful entry into the *meta* series also. Thus both the hydrocarbon **180**[150] and the olefin **181** have been prepared from their *para* isomers by trifluoracetic acid-catalyzed isomerizations.[151]

Just as [6]metacyclophane (**180**) could be synthesized from 1,2-bismethylene-cyclooctane (**175**) the still more strained [5]metacyclophane is accessible from the lower homolog of **175**, 1,2-bismethylenecycloheptane; the main difference from the sequence summarized in Scheme 26 is that dibromocarbene was replaced by its dichloro analog, and the dichloride corresponding to **176** was prepared in a stepwise fashion (1,2-addition of one equivalent of dichlorocarbene to the diene and thermal isomerization of the resulting bridged vinylcyclopropane).[152]

The ultimate test of this approach involved propellanes **184** and **185** (Hal = Cl or Br) (Scheme 27).[153] When these were dehydrohalogenated with potassium *tert*-butoxide in DMSO it was not [4]metacyclophane (**189**) that was isolated (*cf.* the elimination of **178** to **180** above), but a complex mixture of eliminations products, including the three hydrocarbons **186**-**188**.

The dienes **187** and **188** are probably isomerization products of **186**, which could be isolated in 30% yield. Thermal ring-opening of **186** caused the formation of dimers, with [4]metacyclophane (**189**) very likely a highly reactive intermediate. Attempted photoconversion of **186** led to the bridged prismane derivative **190** (see Section 4.1). Clearly, the 'benzene ring' in **189** has lost all its classical stability and become very fragile indeed.

No other bridged aromatic hydrocarbon has fascinated chemists more than [2.2]paracyclophane (**91**), a molecule as archetypical of a layered organic system as ferrocene of a layered 'inorganic' structure. Interesting aspects of **91** and its derivatives were recognized almost immediately after these substances were first synthesized.

Structurally these molecules are attractive because of their unusual bond lengths and bond angles, and —in particular—their short intraannular distance, *ca* 3 Å, which

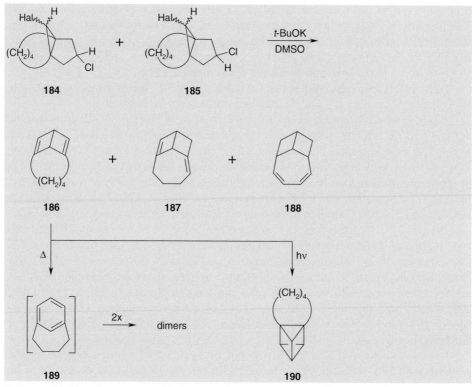

Scheme 27. The close relationship between [4]metacyclophane (**189**) and its Dewar benzene (**186**) and prismane isomers (**190**).

is distinctly shorter than the distance between the layers of graphite (3.40 Å). Because of the proximity of the benzene rings one might expect electronic interactions between them, which, width differently substituted (*i.e.* polarized) aromatic subsystems, would lead to intramolecular charge transfer. Compared with benzene one would expect effects on reaction rates and unusual substituent-directing effects because of the layered structure. Because any (mono) substituted [2.2]paracyclophane is chiral—having a plane of chirality—these compounds would be very useful for stereochemical studies also. Planar chiral compounds are rare in organic chemistry as we have seen in the discussion on *trans* cycloolefins in Section 7.4. Finally, [2.2]paracyclophane, a 'brick of π-electron density', can be used as a building block in the construction of more extended three dimensional aromatic compounds (see below).

Before presenting the most important preparative developments during the half-century-old history of **91**, I would like to deconstruct this molecule retrosynthetically, because this enables us to summarize a large amount of synthetic work in a single–and simple—scheme (Scheme 28).

Homolytic cleavage of one of the ethano bridges of **91** results in the diradical **191**, for which either another paracyclophane containing a nominal fracture point, X in **196**, or, *e.g.*, the dihalide **198**(X=Hal)—or any other similarly functionalized 1,2-diphenylethane derivative—could serve as a precursor. Normally [*n*.2]phanes such as

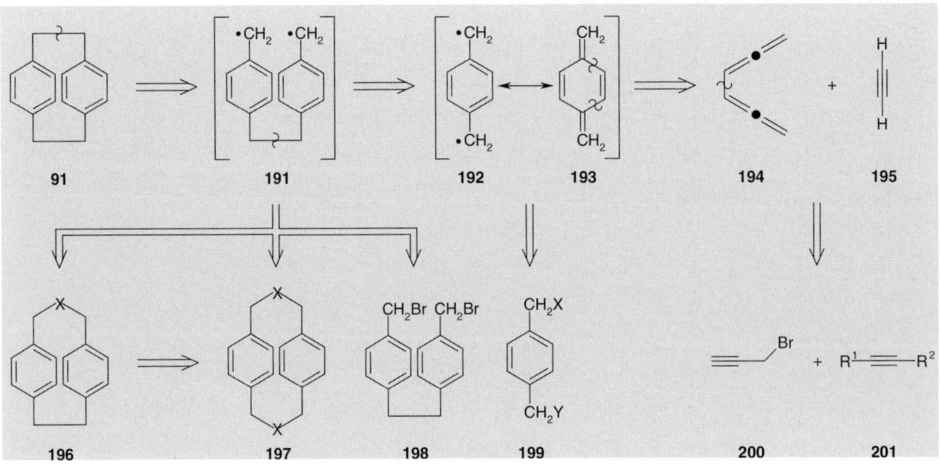

Scheme 28. [2.2]Paracyclophane (**91**)–retrosynthesis.

196 are only generated as intermediates, the doubly ring-expanded phane derivatives **197** serving as their precursors (see below). Continuing our analysis we next break the second ethano bridge and arrive at the diradical **192** which can also be written in its resonance form **193**, a *p*-xylylene (or *p*-quinodimethane). 'Natural' precursors of **192** are *p*-xylene derivatives **199**, whether X is different from Y or not. In fact, X and Y might even be hydrogen substituents, *e.g.* **199** might be *p*-xylene as we shall see. As a six-membered ring system **193** immediately offers itself for a retro-Diels–Alder cleavage; by this process it is decomposed into 1,2,4,5-hexatetraene (biallenyl, **194**, see Section 9.1) as the diene component and acetylene (**195**) as the dienophile. Cleaving the central C–C single bond of **194** transforms the hydrocarbon into a propargyl derivative, *e.g.* propargyl bromide (**200**), and because **195** is such a poor dienophile we should attach two activating groups (R^1, R^2) to it to generate a reactive dienophile, **201**. Both **200** and **201** can be prepared from acetylene (**195**), and in ending our retrosynthesis we note that [2.2]paracyclophane (**91**) is a formal octamer of **195**.

The first description and structure determination of **91** is remarkably similar in some ways with the early history of adamantane (see Section 3.1)—both compounds were obtained initially in minute amounts only, and their structures were established by X-ray structural analysis, a physical method by no means routine at the time of discovery of these hydrocarbons. Cyclophane **91** was, furthermore, originally prepared by the least likely method of our retrosynthesis—the high-temperature pyrolysis of *p*-xylene (**202**). On the basis of earlier studies of Scwarc,[154] in 1949 Brown and Farthing extracted traces of **91** from a complex pyrolysis mixture formed from **202** at temperatures above 750 °C (Scheme 29).[155]

It was suggested by the original authors that **91** is produced by dimerization of either **192** or **193**, available by cleavage of a benzylic bond in **202**. Alternatively, however, a step-wise process is conceivable in which only one C–H bond is broken initially. The thus generated monoradical **203** could subsequently dimerize to 4,4'-dimethylbibenzyl (**204**), which is—in fact—the main product of the low-molecular

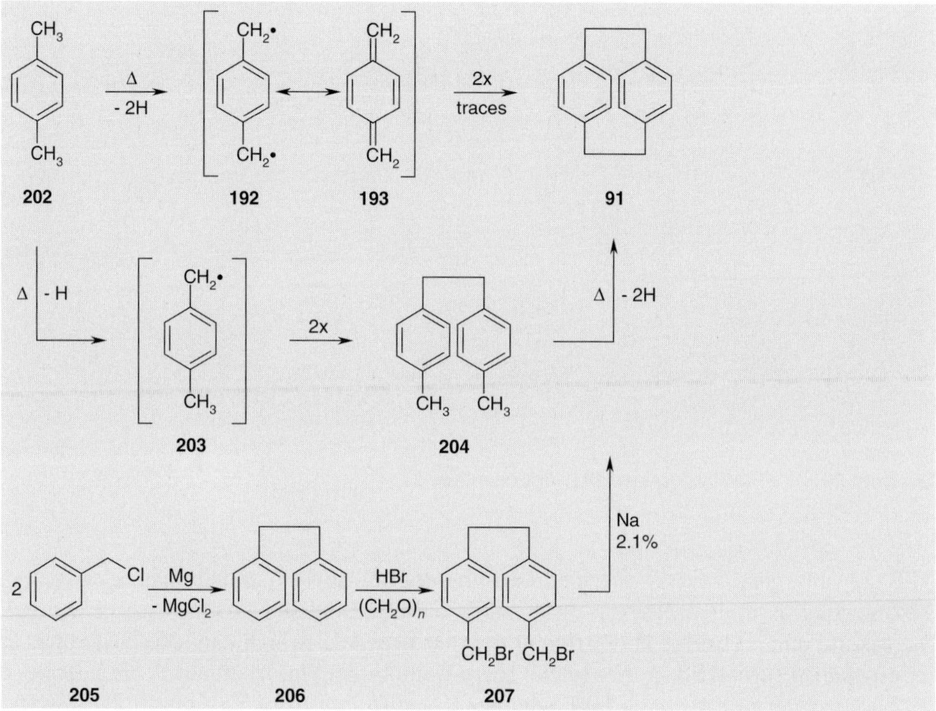

Scheme 29. The first syntheses of [2.2]paracyclophane (**91**) by Brown and Farthing and by Cram.

weight pyrolysis products isolated. Dehydrogenation of **204** by a similar mechanism could then lead to **91**.

The first directed route to [2.2]paracyclophane (**91**) was described only two years after these pyrolysis studies. In 1951 Cram and Steinberg subjected the dibromide **207**—prepared by conventional steps from benzylchloride (**205**) *via* bibenzyl (**206**) and its bromomethylation—to Wurtz coupling with sodium and obtained **91** in 2.1 % yield.[156] The low yield is undoubtedly a result of the unfavorable conformation which **207** must adopt for ring closure and—especially—the high strain energy of **91**; this was later determined to be 29.6 kcal mol^{-1}.[157] The desired intramolecular process can, therefore, hardly compete with intermolecular polymerization reactions.

Separation of these yield-reducing effects could be achieved by the so-called dithiacyclophane route to **91**, a protocol developed by Vögtle,[158] Boekelheide,[159] Staab,[160] and many others[85–97, 117] and applied successfully for the synthesis of numerous derivatives of **91** also (Scheme 30).

In the first step of this approach two aromatic building blocks, **208** and **209**, are combined to the bis-thia-[3.3]paracyclophane **210**. In this intermediate the two benzene rings are already arranged in the parallel orientation ultimately required—but the molecule is far less strained than **91**. In the second step the 'strain barrier' must be overcome, and this can be accomplished directly, *i.e.* by photolytic desulfurization in the presence of triethylphosphite or indirectly *via* the sulfone **212**, easily prepared by

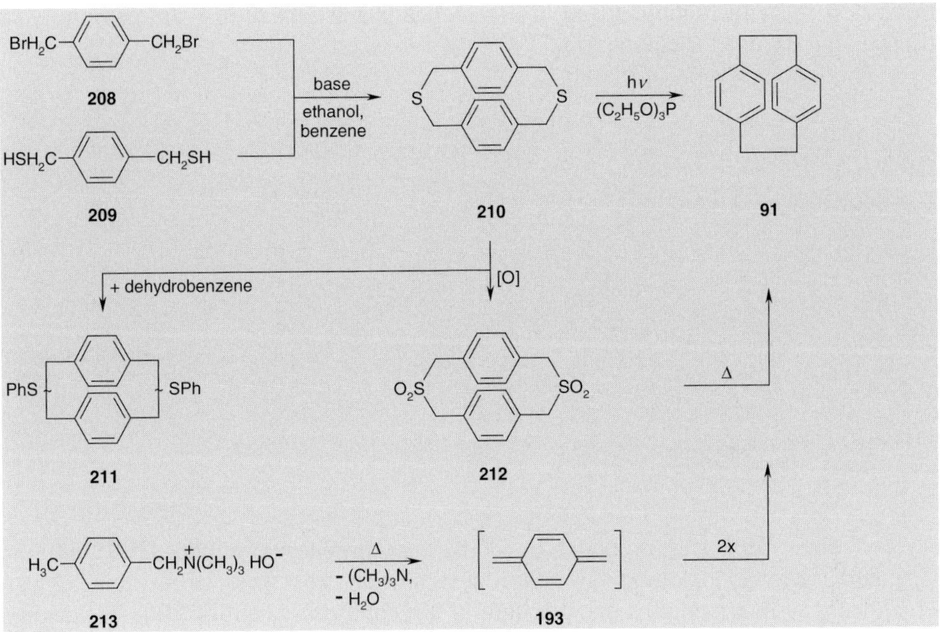

Scheme 30. Preparation of [2.2]paracyclophane (**91**) by the dithiacyclophane and the Hofmann elimination routes.

oxidation of **210** with, *e.g.*, hydrogenperoxide. Sulfone pyrolysis, a most useful and important route to strained macrocycles of all types,[117] then provided **91** in yields as high as 80%. This double extrusion process probably occurs stepwise, but singly ring-contracted intermediates of type **196** are usually not isolated.[158] Other ways of ring-contraction are known—*e.g.* the reaction of **210** with dehydrobenzene (see Section 16.1) in a Stevens-type process led to the phenylsulfides **211**, which, after oxidation and pyrolytic elimination of phenylsulfinic acid, yielded [2.2]paracyclophanediene, easily hydrogenated to **91**. Because both **208** and **209** can be varied over a considerable range, this route to [2.2]paracyclophanes, especially in the form of the sulfone pyrolysis, is probably the most general,[85–90] providing numerous functionalized phanes also. Despite this, it is not the method of choice for the preparation of the parent hydrocarbon. The standard route to **91**, discovered by Winberg and Fawcett in 1960—and also used commercially—submits *p*-methylbenzyltrimethylammonium hydroxide (**213**; X = H, Y = N(CH₃)₃OH in **199**) to a 1,6-elimination process, a bis-vinylogous Hofmann elimination. This leads to the highly reactive *p*-xylylene (**193**), which subsequently dimerizes to **91** in yields up to 20%.[161, 162] Modern variants of this process, which is also applicable to the synthesis of phanes with condensed aromatic subunits and heterophanes, give considerably higher yields.[93] The Hofmann elimination route is of particular value for the preparation of multilayered cyclophanes, as we shall see below.

Proceeding in the retrosynthesis scheme (Scheme 28) we finally arrive at the Diels–Alder approach between 1,2,4,5-hexatetraene (**194**) and different triple-bond dienophiles **201**. The route was developed by Hopf and co–workers and constitutes the fastest route to functionalized [2.2]paracyclophanes, excluding the possibility of derivatizing **91** itself (Scheme 31).[163–165]

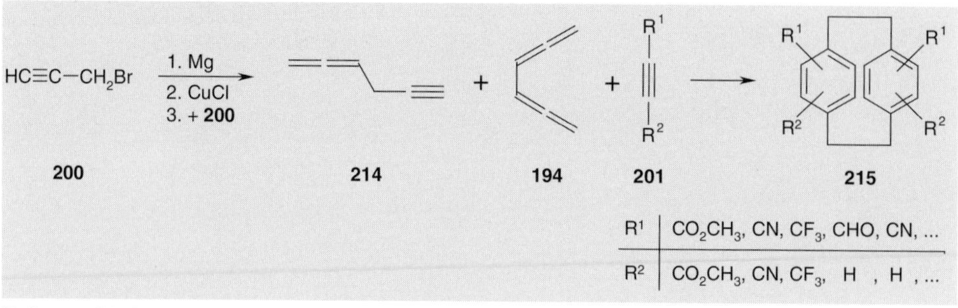

Scheme 31. Functionalized [2.2]paracyclophanes **215** by Hopf's method.

The diene **194** was prepared by converting propargylbromide (**200**) into its Grignard reagent, addition of cuprous chloride (possible formation of an organocuprate intermediate), and addition of a second equivalent of **200**. Besides the hexatetraene **194** its propargylallene isomer **214** was produced (**214**:**194** *ca* 3:2); this, as expected, did not participate in the [2+4]cycloaddition with **201**. The triple-bond dienophiles have to be activated by electron-withdrawing groups; hydrocarbons such as 2-butyne, tolane, or even cyclooctyne (see Section 8.2) do not react. Symmetrical dienophiles (**201**, $R^1 = R^2$) lead to tetrasubstituted [2.2]paracyclophane derivatives with the substituents in the two benzene rings in *anti* orientation. With monosubstituted dienophiles **201** (usually $R^1 = H$, R^2 = electron-withdrawing group) mixtures of all possible four regioisomers are obtained; these can, however, be separated by crystallization or chromatography and are useful precursors for the synthesis of multi-bridged cyclophanes (see below).

As a result of being clamped together by ethano bridges anchored in *para* positions, the two benzene rings are not only deformed into boat configuration, but the planes passing through the non-bridged positions of **91** are only separated by 3.09 Å, *i.e.* they are more than 10% closer than the planes of graphite, which are 3.40 Å apart.[166] Because the intraannular distance should decrease even further—and the electronic interaction between the benzene rings increase correspondingly—when additional ethano bridges are introduced into **91**, the question arises how these bridges can be constructed and how many of them can be introduced. As long as the bridges are anchored in identical positions of the two aromatic halves ('parallel phanes') the number of cyclophane isomers corresponds to the number of isomeric methylbenzenes, *i.e.* there are three [2.2]- (or [2₂]-) cyclophanes, three [2₃]cyclophanes, three [2₄]cyclophanes, one [2₅]cyclophane, and one [2₆]cyclophane; the last has been nicknamed 'superphane'. All these hydrocarbons are known, having been prepared largely by the groups of Cram, Boekelheide, and Hopf.[167, 168] Most of these syntheses provided sufficient amounts of material to enable comprehensive studies of structural, spectroscopic, and chemical properties of this unique series of hydrocarbons.[169, 170] Because this work has been reviewed several times[167, 168] I shall concentrate myself here on the most important synthetic aspects only, especially on those steps which are of general significance to the preparation of multi-bridged cyclophanes. These reactions can be recognized most clearly in the syntheses of two triply-bridged cyclophanes and of superphane.

The first multi-ethano-bridged cyclophane to be prepared was $[2_3](1,3,5)$cyclophane (**223**), synthesized by Boekelheide and Hollins in 1970 (Scheme 32).[171]

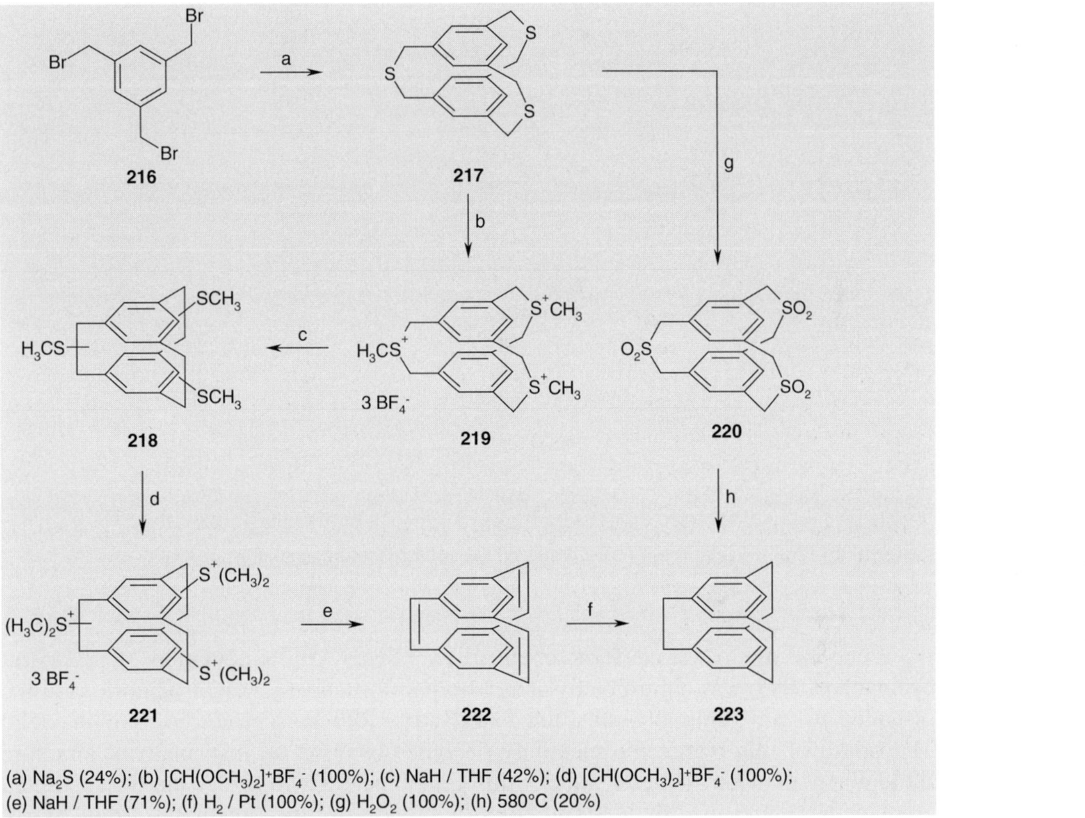
Br
Br
Br
a
S
S
S
216
217
g
b
SCH3
S+CH3
SO2
H3CS
c
H3CS+
O2S
SCH3
S+CH3
SO2
3 BF4-
218
219
220
d
h
S+(CH3)2
(H3C)2S+
e
f
S+(CH3)2
3 BF4-
221
222
223
(a) Na2S (24%); (b) [CH(OCH3)2]+BF4- (100%); (c) NaH / THF (42%); (d) [CH(OCH3)2]+BF4- (100%);
(e) NaH / THF (71%); (f) H2 / Pt (100%); (g) H2O2 (100%); (h) 580°C (20%)

Scheme 32. The first threefold ethanobridged cyclophane—$[2_3](1,3,5)$cyclophane (**223**).

The triply-bridged cyclophane **217** was first prepared by the thiaphane route (see above)—the reaction between 1,3,5-tris-bromomethylbenzene (**216**) and sodium sulfide. For the ring-contraction the S-analogous Stevens rearrangement was used again, converting the tris sulfonium salt **219** (from **217** by Meerwein-type alkylation) to the tris-thioether **218**. Alkylation followed by Hofmann elimination then furnished the cyclophanetriene **222**, the salt **221** being passed *en route*. Saturation of the bridges was accomplished quantitatively by catalytic hydrogenation. Although the synthesis is straightforward, the number of steps is too high. Significant abbreviation was achieved by oxidizing **217** to the trissulfone **220** which lost three equivalents of sulfur dioxide on flash vacuum pyrolysis at 580 °C. In **223** the distance between the bridged carbon atoms of the aromatic nuclei has been reduced to 2.75 Å, whereas the non-bridged atoms are 2.83 Å apart.[172]

Other multi-bridged cyclophanes are also available by sulfone pyrolysis, attesting to the preparative importance and generality of this method.[173–175]

The introduction of an additional ethano bridge into an already existing cyclophane, *viz.* **91**, was first realized by Cram and Truesdale (Scheme 33).[98]

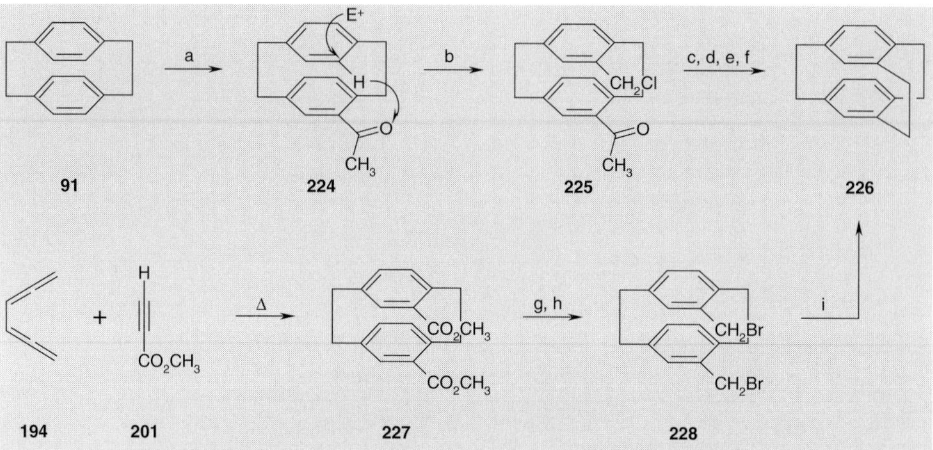

(a) CH₃COCl, AlCl₃ (74%); (b) (CH₂O)ₙ / HCl (55%); (c) KOBr, KOH, dioxan; (d) LiAlH₄, THF (comb. yield 72%); (e) PBr₃, CH₂Cl₂ (81%); (f) *n*-BuLi, THF (68%); (g) LiAlH₄, THF (58%); (h) PBr₃, CH₂Cl₂ (77%); (i) Zn, DMSO (89%)

Scheme 33. The introduction of an additional bridge into [2.2]paracyclophane (**91**).

When the methylketone **224**, obtained by Friedel–Crafts acylation of [2.2]para-cyclophane (**91**), was chloromethylated with paraformaldehyde-hydrochloric acid the so-called *pseudo*-geminally-substituted derivative **225** was produced in high yield. The origin of this high regioselectivity becomes obvious on inspection of structure **224**. When an electrophile E⁺ attacks this cyclophane it will prefer the unsubstituted aromatic ring, and in it the carbon atom directly across the functional group. When this carbon atom is rehybridized on formation of the σ-complex, its hydrogen atom will be pushed towards the oxygen atom of the acetyl function, *i.e.* towards an internal nucleophile ideally positioned to pick it up after intraannular migration. By proton transfer the originally unsubstituted ring is rearomatized, the proton being finally lost from the (protonated) carbonyl group to surrounding solvent molecules. With the carbon atoms of the future bridge properly positioned all that remained to be done was C–C bond formation. This was accomplished by a series of classical steps: haloform oxidation, hydrolysis, lithium aluminum hydride reduction, conversion of the resulting diol to the dibromide **228**, and Wurtz-type coupling employing *n*-butyllithium. As we shall see below this bridging protocol can be employed universally for the preparation of multi-bridged phanes. In a later preparation of **226** Hopf and co-workers exploited their Diels–Alder route (see above); the diester **227** was produced (accompanied by its regioisomers from which it was separated) by [2+4]cycloaddition of methyl propiolate (**201**, R^1 = H, R^2 = CO_2CH_3) to 1,2,4,5-hexatetraene (**194**) and after its conversion to **228** bridging was readily achieved by reduction with zinc dust in DMSO.[99] Use of regioisomers of **227** in similar sequences provided all three isomeric [2$_4$]cyclophanes.[176–178]

The name speaking for itself, the ultimate phane is superphane (**238**, [2$_6$]cyclophane), the fully ethano-bridged hydrocarbon. Two syntheses for this fascinating compound have been published, and, in keeping with the philosophy of this book, it is interesting to compare these two approaches.

An extraordinary feature of the first synthesis of **238**, performed by Sekine, Brown and Boekelheide in 1979 is the consistent application of a single concept throughout the entire reaction sequence (Scheme 34).[167, 179]

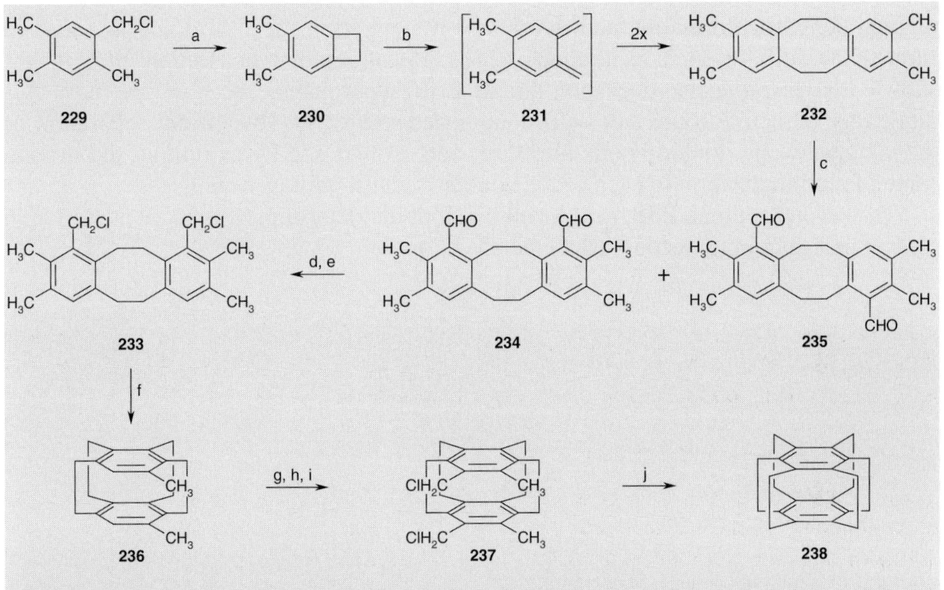

(a) 750°C, 10^{-3} torr (53%); (b) 400°C, N$_2$, 1 atm (63%); (c) Cl$_2$CHOCH$_3$, SnCl$_4$ (60% **234**); (d) NaBH$_4$, CH$_3$OH (100%); (e) SOCl$_2$, benzene (100%); (f) 700°C, 10^{-2} torr (40%); (g) Cl$_2$CHOCH$_3$, SnCl$_4$ (98%); (h) LiAlH$_4$, ether (96%); (i) SOCl$_2$, pyridine (93%); (j) 650°C, 10^{-2} torr (57%).

Scheme 34. The first synthesis of superphane (**238**) by Boekelheide et al. ...

Often in hydrocarbon chemistry, and especially when highly symmetric target molecules are to be prepared, cycloadditions play an important and even critical role in the initial stages of the synthesis (see, *e.g.*, tetraasterane (Section 3.4), cubane (Section 5.2), dodecahedrane (Section 5.3), pagodane (Section 5.3), and many more). Later steps often suffer from the need for several synthetic transformations to produce just one single C–C bond, making the whole scheme very lengthy, and yields often minuscule. The Boekelheide superphane synthesis avoids these pitfalls because it consists basically of a threefold repetition of the *o*-xylylene dimerization. That it furthermore begins with an inexpensive commercial product, chlorodurene (**229**), makes it even more attractive.

When **229** was heated to 750 °C under high-vacuum, dehydrochlorination took place, and the benzocyclobutene **230** (see Section 16.3) was produced. On heating, this highly strained hydrocarbon opened to the *o*-xylylene intermediate **231** which, as its *para* isomer **193**, readily dimerized—this time to an 'orthocyclophane', the hydrocarbon **232**, with two completed ethano bridges and two 'halfbridges', the two methyl substituents. By Rieche formylation **232** was converted to a mixture of the regioisomeric dialdehydes **234** and **235**, the former being used to proceed with the bridge-building process. Borohydride reduction then reaction with thionyl chloride transformed **234** into **233**, an intermediate with the structural and functional prerequisites for another thermally induced *o*-xylylene dimerization. This step of the synthesis is, in fact, the most crucial one, because the *o*-xylylene subunits generated must be able to approach each other close enough to enable intramolecular bridge-closure. Fortunately this was so. For the resulting **236** the subsequent steps were predetermined—formylation, reduction, chlorination yielded **237**, which underwent thermal dehydrochlorination and *o*-xylylene dimerization for a third and last time to furnish the fully-bridged superphane (**238**). Recalling that in Section 11.5 it was shown that [6]radialene does not dimerize to superphane, we now see how this intuitively attractive route can be accomplished indirectly—by formal separation of the six semicyclic double bonds into three sets of two and by exploiting the driving force of rearomatization in each of the double-bridge-building steps.

The second route to **238**, developed by Hopf and El-Tamany, is more attractive in its first half than in its second (Scheme 35).[180]

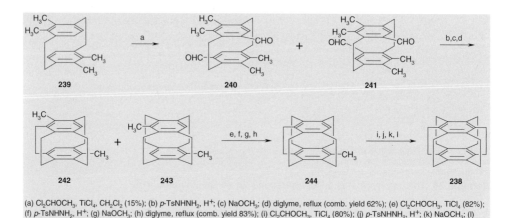

(a) Cl$_2$CHOCH$_3$, TiCl$_4$, CH$_2$Cl$_2$ (15%); (b) *p*-TsNHNH$_2$, H⁺; (c) NaOCH$_3$; (d) diglyme, reflux (comb. yield 62%); (e) Cl$_2$CHOCH$_3$, TiCl$_4$ (82%); (f) *p*-TsNHNH$_2$, H⁺; (g) NaOCH$_3$; (h) diglyme, reflux (comb. yield 83%); (i) Cl$_2$CHOCH$_3$, TiCl$_4$ (80%); (j) *p*-TsNHNH$_2$, H⁺; (k) NaOCH$_3$; (l) diglyme, reflux (comb. yield 74%)

Scheme 35 ...and a second one by Hopf and El-Tamany.

The starting material, the tetramethyl-[2.2]paracyclophane **239**, is available in multigram quantities by the Diels–Alder approach presented in Scheme 31, using dimethyl acetylenedicarboxylate as the dienophile and conversion of the ester functions of the paracyclophane product into methyl groups. The closure of the missing bridges was, however, more demanding than in the Boekelheide route. Formylation of **239** led to a mixture of the dialdehydes **240** and **241**, and the production of the monoformyl derivative. The bridges were subsequently formed by converting the aldehydes into the corresponding tosylhydrazones, preparation of the sodium salts of these intermediates, and a final pyrolysis step. The bridge was generated in this reaction by intramolecular carbene insertion. When the resulting hydrocarbon mixture of **242** and **243** was again put through the same sequence, the penta-bridged hydrocarbon **244** was obtained; this was subjected once again to the same bridging routine to yield superphane (**238**).

The most remarkable structural feature of **238** is the extremely short distance between its (planar) benzene 'decks'—at 2.624 Å this is *ca* 80 pm shorter than the graphite inter-layer distance.[179]

The pronounced 'face-to-face' interaction between the layered rings in [2.2]paracyclophane (**91**), which manifests itself in numerous chemical and physical properties of this sandwich compound,[85–97] soon led to studies of the preparation of multilayered cyclophanes, a development *inter alia* influenced by the synthesis of triple- and quadrupel-layered transition metal complexes. For layered compounds of this type there is the intriguing possibility that electronic effects could be transmitted through the whole columnar structure. To prepare multilayered cyclophanes, the Hofmann elimination of substrates containing functionalized *p*-xylene subunits and the generation of '*p*-xylylenophane intermediates' therefrom, is by far the most general approach. The ground-breaking experiment was performed by Longone and Chow in 1964 (Scheme 36).[181]

1,6-Elimination of the Hofmann base **245**, prepared in three steps from durene (NBS-bromination, substitution with trimethylamine, replacement of bromide by hydroxide) yielded 2,5-dimethyl-*p*-xylylene (**246**) which, as expected, dimerized to the tetramethyl[2.2]paracyclophane derivative **247**, containing two sets of (crossed) *p*-xylene subsystems. Consequently, the sequence which led to **245** could be repeated,

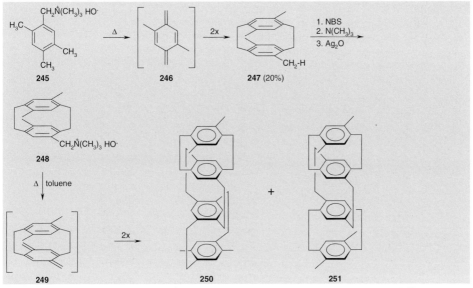

Scheme 36. The preparation of the first multilayered cyclophanes by Longone and Chow.

and the Hofmann base **248** was obtained. This, on heating, lost water and trimethylamine and yielded a *p*-xylylene, **249**, which is already a phane. As intended, dimerization occurred and provided the quadruple-layered [2.2]cyclophanes **250** and **251**, initially isolated as a mixture in low yield only (<10%).[182] Later Misumi and co–workers in very extensive studies not only optimized these dimerization reactions—their yields depend strongly on reactions conditions, even on the shape of the reaction flask—but also developed cross-breeding pyrolysis of two different Hofmann bases as a means of preparing multi-layered cyclophanes with an uneven number of decks. Eventually all parent systems and many (methyl) derivatives with layers up to six could be prepared and investigated chemically and structurally; yields up to 40% were achieved. Typical examples are summarized in Scheme 37.[183]

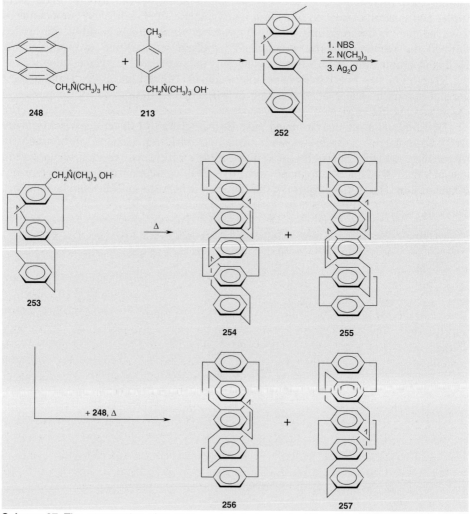

Scheme 37. The construction of multi-layered cyclophanes by Misumi and co–workers.

The reactions in this scheme are mostly self-explanatory. Co-thermolysis of the ammonium bases **248** and **213** led to the triply-layered cyclophane **252**, as a result of the reaction between the *p*-quinodimethanes **249** and **193**. Having two *para*-positioned methyl substituents left, **252** could be converted to the Hofmann base **253**, again by the standard methodology; 1,6-elimination then provided either a hexa-layered cyclophane as a mixture of the two isomers **254** (D_{2h} symmetry, 4.7%) and **255** (C_{2h} symmetry, 2.4%), or—if the process was conducted in the presence of an equimolar amount of **248**—a mixture of the two penta-layered hydrocarbons **256** and **257**.[184, 185]

That intraannular π–π interaction along the column axis is indeed present in these multi-layered compounds is borne out by their electronic and NMR spectra.[183] Particularly revealing are the UV/Vis spectra of the charge-transfer complexes prepared from various multi-layered cyclophanes (donors) and tetracyanoethylene (acceptor). Electron release to the complexed ring from the remaining benzene decks in the donor column is expected, and is, indeed, observed. Increasing the number of layers led to a progressive red shift of the long-wavelength maxima of the TCNE complexes. Even in the penta-layered hydrocarbon **256** such an electron release is clearly discernible.

Needless to say, the ethano bridges in a [2.2]cyclophane must not be anchored in the *para* positions. In fact, all isomeric [2.2]cyclophanes have been synthesized.[186] Apart from the *ortho* isomer (= dibenzocyclooctadiene), [2.2]metacyclophane (**259**) has been studied most thoroughly among this group of isomeric hydrocarbons. [2.2]Metacyclophane is of interest not only for structural reasons (deformation of benzene rings, steric interference between substituents) and because of the electronic interaction between its benzene subsystems, but also because of its dynamic behavior—in contrast to the rigid [2.2]paracyclophane (**91**), **259** can exist in two readily interconvertible forms, an 'open' *anti* and a 'closed' *syn* isomer, as will be discussed below. The hydrocarbon also bears the distinction of being the first cyclophane to be prepared. It was obtained in variable yields (up to 12%) by Pellegrin in 1899 from 1,3-bis(bromomethyl)benzene (**258**) by Wurtz coupling with sodium in ether (Scheme 38).[187]

Subsequent systematic studies led to a drastic increase in yield, making the hydrocarbon readily available in gram quantities. Applying the so-called Müller–Röscheisen conditions[188] (sodium in THF, low temperature, tetraphenylethene (TPE) as a catalyst) **259** has been synthesized in 77% yield starting from the 'half-closed dimer' 3,3'-bis-(bromomethy)bibenzyl.[189] Reinvestigation of the original Wurtz coupling of **258** by Jenny and co-workers[190, 191] under Müller–Röscheisen conditions showed that not only **259** is produced but a whole homologous series of *m*-xylylene oligomers **260**, with *n* increasing from 2 to 10. Note that the largest macrocycle contains a fifty-membered ring! Because most of the cyclophane-producing reactions discussed already for the *para* series (*inter alia* sulfone pyrolysis,[117] ring-contraction by Stevens rearrangement[192] of dithia[2.2]metacyclophanes or of diseleno[2.2] metacyclophanes[193]—see above) have also been applied successfully to the synthesis of the *meta* isomers, numerous [2.2]metacyclophanes, including a rich variety of compounds carrying a wide selection of functional groups, are available today on a routine basis.[83, 93]

[2.2]Metacyclophane (**259**) prepared by these routes exists in the stepwise, *anti*

260

n	2	3	4	5	6	7	8	9	10
yield, %	35	8	2	5	4	0.5	0.6	0.5	0.3

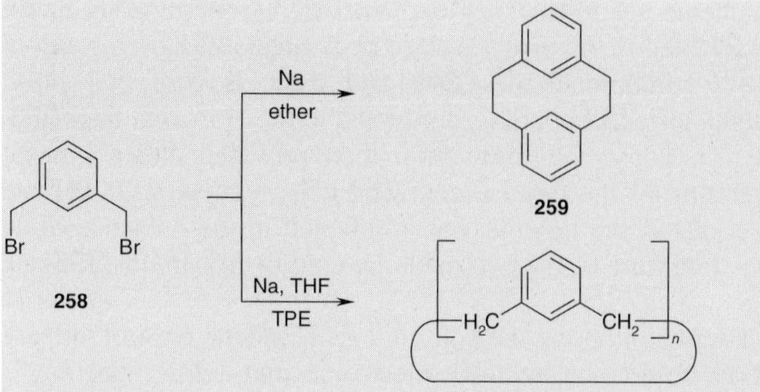

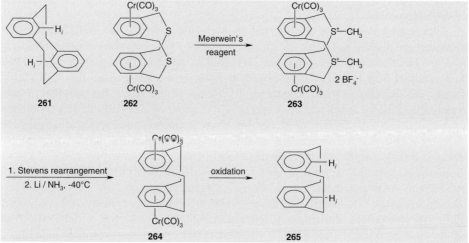

Scheme 38. [2.2]Metacyclophane (**259**), the first cyclophane to be prepared.

structure depicted by **261** (C_{2h} symmetry) both in the crystalline state and in solution.[194] After *syn–anti* conformational isomerism was observed for [2.2]metacyclophane derivatives, however,[195] the question arose whether the *syn* isomer of **259** could not be prepared. The problem was eventually solved by Mitchell and co–workers by an approach, summarized in Scheme 39,[196] exploiting the observation that electron-withdrawing substituents favor the *syn* configuration in 2,11-dithia[3.3]metacyclophanes.

By use of the strongly electron-accepting $Cr(CO)_3$ fragment, the conformationally fixed bis-chromium complex **262** was first prepared by heating the conformationally

264 265

Scheme 39. The two conformers of [2.2]metacyclophane (**259**)—*anti-* (**261**) and *syn*-[2.2]metacyclophane (**265**).

flexible *syn*-2,11-dithia[3.3]metacyclophane under reflux with excess chromium-hexacarbonyl in di-*n*-butylether (62% yield). After methylation of **262** with Meerwein's reagent the resulting **263** was subjected to Stevens rearrangement then reduction with lithium in liquid ammonia at –40 °C. This led to a mixture of the mono-chromium complex of **265** and the bis complex **264** shown in the scheme. Demetalation of the latter was accomplished by oxidation with either *m*-chloroperbenzoic acid or cerium(IV) in acetonitrile at –45 °C. *syn*-[2.2]Metacyclophane (**265**) is stable only at low temperatures (–40 °C); at 0 °C it rapidly isomerized to the *anti* isomer **261**. The intraannular hydrogen atoms H$_i$ of **265**, which do not face an opposing benzene ring, absorb at δ = 6.58, as compared with a chemical shift of 4.25 for these hydrogen atoms in the *anti* conformer **261**.

Just as appropriately functionalized [2.2]paracyclophanes have been used as building blocks for the construction of multi-layered hydrocarbons (see above), derivatives of [2.2]metacyclophanes can be employed for the preparation of multi-stepped cyclophanes.[197]

This is illustrated in Scheme 40 for the synthesis of various triply- and quadruply-stepped metacyclophanes, for which the bis-(bromomethyl)derivative **266** served as a common precursor.[198]

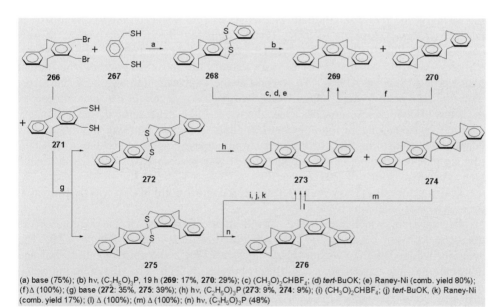

(a) base (75%); (b) hv, (C$_2$H$_5$O)$_3$P, 19 h (**269**: 17%, **270**: 29%); (c) (CH$_3$O)$_2$CHBF$_4$; (d) *tert*-BuOK; (e) Raney-Ni (comb. yield 80%); (f) Δ (100%); (g) base (**272**: 35%, **275**: 39%); (h) hv, (C$_2$H$_5$O)$_3$P (**273**: 9%, **274**: 9%); (i) (CH$_3$O)$_2$CHBF$_4$; (j) *tert*-BuOK; (k) Raney-Ni (comb. yield 17%); (l) Δ (100%); (m) Δ (100%); (n) hv, (C$_2$H$_5$O)$_3$P (48%)

Scheme 40. The preparation of multi-stepped [2.2]metacyclophanes by Misumi and co–workers.

To obtain three-stepped hydrocarbons, **266** was first coupled with 1,3-bis-(mercaptomethyl)benzene (**267**) to furnish the dithiaphane **268**, which was isolated in good yield. Desulfurization was subsequently performed by irradiating **268** in triethylphosphite at room temperature and provided the *up,down* isomer (*u,d* isomer) **269** in 17% yield, and the *u,u* isomer **270** in 29% yield. Alternatively **268** was converted into **269** exclusively by the S-methylation—Stevens rearrangement—Raney nickel desulfurization route. That the *u,d* isomer is thermodynamically favored is also apparent from the thermal isomerization of the *u,u* hydrocarbon **270**. Replacement of **267** by the bis-thiol **271**, which is already a [2.2]metacyclophane, led to approximately equal amounts of the quadruply-stepped dithiaphanes **272** and **275**, from which the hydrocarbons **273**, **274**, and **276** were obtained by the methods just described. Again, the hydrocarbon with alternating steps, **273** (*u,d,u*) is the most stable isomer.

In this section we have so far discussed only cyclophanes constructed from two basic building elements, benzene rings and ethano bridges. By variation of the number of these units and the position of the bridges in the rings we have created an astonishing variety of structures already. The extraordinary structural richness of the [*m.n*]cyclophane family becomes, however, obvious only, when we begin to vary these construction elements also. For example, the length of the (saturated) bridging unit has been varied over a considerable range. Intuitively we would expect that the molecular strain of an [*m.n*]cyclophane and the deformation of its benzene rings would decrease with increasing bridge length. At the same time the flexibility (conformational mobility) of the corresponding hydrocarbons and the intraannular distances should increase. These predictions have generally turned out to be correct.[87, 93] All these trends should, however, be reversed should the bridge become increasingly shorter. The ultimate challenge in [*m.n*]paracyclophane chemistry is hence [1.1]paracyclophane (**287**). Having prepared the first derivative of this compound, 1,10-bis(methoxycarbonyl)[1.1]paracyclophane in 1993,[199] Tsuji and co–workers met this challenge four years later with the synthesis of the parent hydrocarbon (Scheme 41).[200] To overcome the expected high strain of the target molecule (see below), the strategy which had been so successful for the synthesis of the short-bridged [*n*]cyclophanes (see above), *i.e.* the generation of the ultimately required benzene ring by photoisomerization of a Dewar benzene precursor, was employed here also. In **287** the difficulty of the task is heightened by the presence of two six-membered rings in the product, *i.e.* it was necessary to synthesize a double Dewar benzene isomer of **287**.

Tsuji's synthesis is risky in the sense that the all-or-nothing step was performed at the very end of a lengthy route.

The problem which had to be solved in the first part of the synthesis was obtaining convenient access to the functionalized tricyclic bisketone **279**. The solution involved the conversion of the diester **277** *via* the corresponding diol and dibromide to the chain-elongated di-*tert*-butylester **278**, and the acid-promoted intramolecular cyclization of the diacid derived therefrom. The over-all yield of these first five steps was 20%. In the central part of the synthesis the Dewar benzene 'halves' had to be constructed. This was achieved by photocycloaddition of two equivalents of acetylene to **279**. A process leading only to the *anti* isomer **281**, presumably because this is a stepwise process and the initially generated *mono* adduct **280** allows further acetylene addition from its 'open' side only. A double ring-contraction involving the bis-diazoketone **282**, performed photochemically in methanol, then led to the diester **283** as a mixture of three stereoisomers. The diesters were hydrolyzed to a mixture of dicarboxylic acids, which were converted *via* their isocyanates into the corresponding carbamates by treatment with 2-(methylthio)

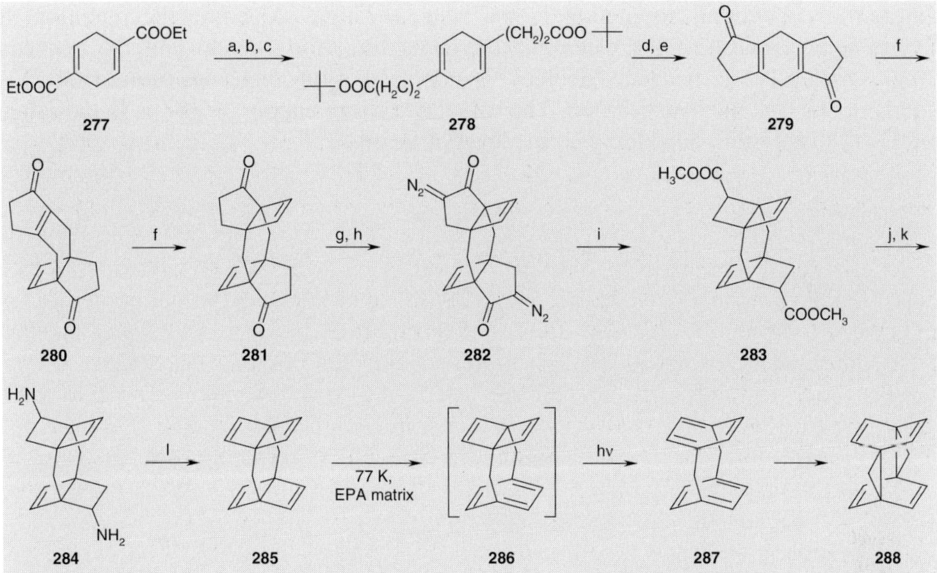

(a) LiAlH$_4$; (b) PBr$_3$; (c) CH$_3$COO-*tert*-Bu, LDA; (d) CF$_3$COOH; (e) (CF$_3$CO)$_2$O, BF$_3$·Et$_2$O; (f) hν, HC≡CH , acetone, CH$_2$Cl$_2$; (g) HCOOEt, NaOEt; (h) TsN$_3$, Et$_3$N; (i) hν, CH$_3$OH; (j) NaOH, H$_2$O, H$^+$; (PhO)$_2$P(O)N$_3$, Et$_3$N, H$_3$CSCH$_2$CH$_2$OH; (k) (CH$_3$)$_2$SO$_4$, NaOH, CH$_3$OH; H$_2$O, H$^+$; NaOH; (l) CH$_3$I, NaHCO$_3$, *tert*-BuOK, DMF

Scheme 41. The synthesis of [1.1]paracyclophane (**287**) by Tsuji and co–workers.

ethanol. When these were reacted with dimethylsulfate and with aqueous sodium hydroxide, carbamate salts resulted, from which diamine **284** (mixture of isomers) was liberated by acidification and decarboxylation. After quaternization with methyl iodide, Hofmann elimination with potassium *tert*-butoxide in DMF-pentane finally furnished the desired bis-Dewar benzene **285**, the yield of the **282**→**285** interconversion being 15%. In the third, and all-decisive, part of the synthesis a glassy mixture of **285** in EPA (diethyl ether-isopentane-ethanol, 5:5:2) was irradiated with 254-nm light at 77 K. As indicated by UV/Vis and NMR monitoring the intended ring-opening did indeed occur. Initially one Dewar benzene subsystem isomerized— formation of **286**—and then [1.1]paracyclophane (**287**) was obtained. Although the spectra of **287** remained unchanged for a few hours at –20 °C (while complete destruction was observed after 4 h at room temperature), a secondary photoisomeri- zation occurred on further irradiation, leading to the interesting hydrocarbon **288** in which two 1,3-cyclohexadiene units are clamped together by cyclopropane rings. To gain insight into the thermodynamic stability and strain energy of **287** thermochemical calculations were performed by *ab initio* and DFT quantum- mechanical methods. These calculations revealed that the extent of distortion of the benzene rings is comparable with that in [5]paracyclophane (see above), the bending angle by which the bridgehead carbon atoms are 'lifted' out of the plane passing through the four unbridged carbon atoms being *ca* 23° (compared with the (experimental) angle of 12.6° for [2.2]paracyclophane (**91**) and 23.5° (calculated) for [5]paracyclophane (**130**)). At 2.36–2.40 Å the shortest transannular inter-atom distance is *ca* 1 Å shorter than the inter-layer distance in graphite. There is, furthermore, no bond alternation in the benzene rings, indicating the retention of cyclic delocalization of the electrons despite of the strong bending of the benzene ring. Obviously, a benzene nucleus is far less rigid than our notion of this fundamental organic hydrocarbon. The calculated strain energy of **287** is in the range of 93–128 kcal mol^{-1} depending on the level of theory.[201]

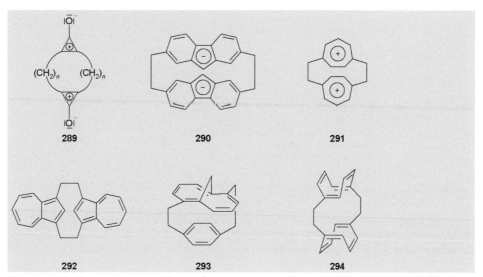

Scheme 42. Various types of bridged aromatics—a glimpse of the structural richness of cyclophane chemistry.

Whereas in the initial phase of cyclophane chemistry only saturated (polymethylene) bridges were employed, later studies showed that the bridging elements tolerate various amounts of unsaturation, *e.g.* double[202] and triple bonds[203] and allene groups.[108] Recently even complex polycyclic bridging units such as the adamantane skeleton (see Section 3.1) have been incorporated into phane systems.[204]

The extraordinary variability of the cyclophane motif, however, reveals itself only when we consider the near endless possibilities of exchange and permutation of the aromatic subunit—it can be carbo- or heterocyclic, benzenoid or non-benzenoid, charged or neutral. Scheme 42 presents only a very small selection of this huge variety of phane hydrocarbons and whose syntheses—although clearly deserving of discussion in a text on classics in systems chemistry—are not described here in detail only because of space limitations.

Thus cyclophanes containing both positively and negatively polarized or charged decks have been synthesized, as is demonstrated by the bridged bis-cyclopropenones **289**,[205] the dianionic dibenzocyclopentadienophane **290**,[206] and the dicationic tropyliophane **291**.[207] Cyclophanes incorporating nonbenzenoid, aromatic (4n+2) or 'antiaromatic' (4n) subsystems include the [2.2]azulenophane **292**,[208, 209] the methano[10]annulenophane **293**,[210] and the [2.2]cyclooctatetraenophane **294**.[211]

Several concepts originally developed for cyclophane hydrocarbon systems have been extended to other classes of compound, a case in point being the superphanes, *i.e.* polycyclic systems which can all be traced back to [2₆](1,2,3,4,5,6)cyclophane (**238**) (see above). For example Shinmyozu and co–workers were recently able to synthesize a 'hexahomo' superphane, [3₆](1,2,3,4,5,6)cyclophane (**295**), a compound which resembles a pinwheel with six blades and forms a TCNE complex with the highest CT band (λ = 728 nm) ever reported for a charge-transfer complex of an [*m.n*]paracyclophane or multibridged benzenophane (Scheme 43).[212]

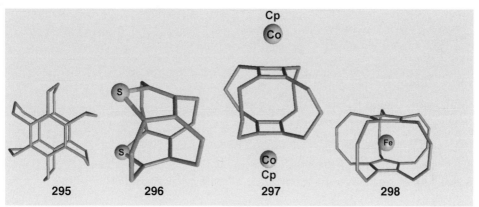

Scheme 43. Extending the superphane concept.

A heteroaromatic superphane, [2₄](1,2,3,4)thiophenophane (**296**) has been obtained by Tashiro and co–workers[213] and a superphane containing two cyclobutadiene subunits and capped (and therefore stabilized) by cyclopentadienyl-cobalt fragments by Gleiter et al. in an elegant one-step reaction starting from 1,6-cyclodecadiyne (see Section 8.3) and different cyclopentadienyl cobalt derivatives.[214] Interestingly, metals cannot only be bound to the exterior surfaces of superphanes but have also been encapsulated by these 'hollow' molecules, as was shown by Hisatome and co–workers with the preparation of [4₅](1,2,3,4,5)ferrocenophane.[215] These novel metal organic compounds are not only of interest in their own right but also in connection with endohedral fullerenes, in which various metal atoms are incarcerated by an all-carbon molecular container.[216]

We started this chapter with general considerations on how to prepare non-planar aromatic compounds. The three approaches subsequently discussed—the helicenes, circulenes and cyclophanes—by no means represent the only ways of leaving the π-plane. As already shown for the preparation of Dewar benzenes (Section 10.2) the introduction of sterically demanding substituents into a benzene nucleus can also lead to more or less severe distortion of its π-plane. A remarkable non-planar hydrocarbon, 9,10,11,12,13,14,15,16-octaphenyldibenzo[*a,c*]naphthacene (**302**), which owes its highly distorted π-system to multisubstitution, has recently been synthesized by Pascal and co–workers (Scheme 44).[217]

The carbon framework of **302** was produced by addition of the highly substituted dehydrobenzene (see Section 16.1) generated by diazotization and fragmentation of the anthranilic acid **299** to hexaphenylisobenzofuran (**300**). Reduction of the Diels–Alder adduct **301** by zinc in acetic acid then furnished the hydrocarbon **302**. According to

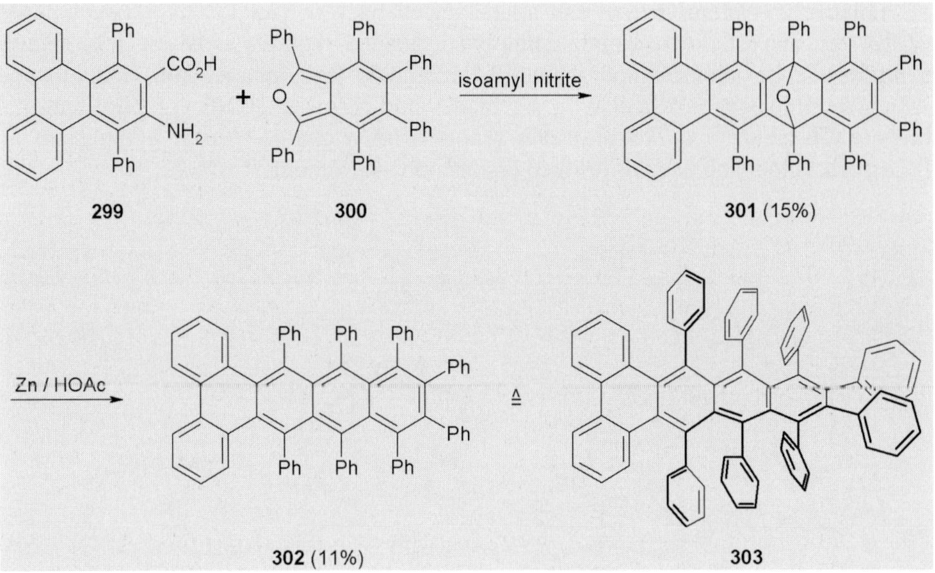

Scheme 44. An extremely twisted polycyclyclic aromatic hydrocarbon (PAH).

crystal structure analysis the molecule has the twist configuration shown by formula **303** with an end-to-end-twist angle of 105°—the highest value observed for a distorted PAH. Yet the molecule, an orange-colored solid which is soluble in most organic solvents, still has a conjugated π-system. The severe distortion has no effect on the thermal stability of the hydrocarbon, which even at 400 °C shows no sign of decomposition!

We will return to yet another type of three-dimensional aromatic hydrocarbon— the iptycenes—after discussing, in Chapter 13, the olefinic parent system from which they are derived.

References

1.   For a theoretical treatment of three-dimensional π-electron systems see R. C. Haddon, *Acc. Chem. Res.*, **1988**, *21*, 243–249 and W. C. Herndon, P. C. Nowak in *Advances in Theoretically Interesting Molecules* R. P. Thummel, (*Ed.*), JAI Press, **1992**, Vol. 2, 113–141. The proportion of highly non-planar hydrocarbons increases rapidly with the growing number of rings, *e.g.* it reaches 79% for the condensed polycyclic aromatic hydrocarbons (PAHs) with ten rings. The great majority of the PAHs capable of existence will exist as chiral and nonchiral highly non-planar molecules.

2.   The *ortho*-bridged benzene derivatives such as benzocyclopropene, -cyclobutene (see Chapter 16.3), indane, tetralin *etc.* are not regarded as cyclophanes. Depending on the length of the bridge they are usually planar compounds.

3.   J. W. Cook, *J. Chem. Soc.*, **1931**, 2524–2528; J. W. Cook, *J. Chem. Soc.*, **1931**, 2529–

2532.

4. R. Weitzenböck, H. Lieb, *Monatsh.*, **1912**, *33*, 549–565; *cf.* F. Mayer, T. Oppenheimer, *Ber. Dtsch. Chem. Ges.*, **1918**, *51*, 510–516.
5. A. L. Wilds, R. G. Werth, *J. Org. Chem.*, **1952**, *17*, 1154–1161.
6. J. Smuszkovicz, E. J. Modest, *J. Am. Chem. Soc.*, **1952**, *70*, 2542–2543.
7. M. S. Newman, H. V. Anderson, K. H. Takemura, *J. Am. Chem. Soc.*, **1953**, *75*, 347–349; by a modern route **2** is obtained from 1-chloro-2-(2'-styryl)-naphthalene by dehydrochlorination under flash vacuum pyrolysis conditions: M. J. Plater, *Tetrahedron Lett.*, **1994**, *35*, 801–802.
8. F. L. Hirshfeld, S. Sandler, G. M. J. Schmidt, *J. Chem. Soc.*, **1963**, 2108–2125; *cf.* F. L. Hirshfeld, *J. Chem. Soc.*, **1963**, 2126–2135. For the molecular structure of the benzo[c]phenanthrene/2,3-dichloro-5,6-dicyanobenzoquinone complex see J. Bernstein, H. Regev, F. H. Herbstein, *Acta Crystallogr. Sect. B*, **1977**, *33*, 1716–1724.
9. I thank Dr. Norman Goldberg (Braunschweig) for these calculations.
10. R. N. Armstrong, H. L. Ammon, J. N. Darnow, *J. Am. Chem. Soc.*, **1987**, *109*, 2077–2082; *cf.* F. Imashiro, A. Saika, Z. Taira, *J. Org. Chem.*, **1987**, *52*, 5727–5729 and R. Fritsch, E. Hartmann, D. Andert, A. Mannschreck, *Chem. Ber.*, **1992**, *125*, 849–855 and previous publications in this series.
11. R. Weitzenböck, A. Klingler, *Monatsh.*, **1918**, *39*, 315–323. As shown later this synthesis neither yields the pure diacid **11** nor the pure hydrocarbon **3**, see. J. W. Cook, *J. Chem. Soc.*, **1933**, 1592–1597.
12. The original synthesis is due to H. A. Weidlich, *Ber. Dtsch. Chem. Ges.*, **1938**, *71*, 1203–1204, an improvement shown in Scheme 2 has been worked out by Y. Altman, D. Ginsburg, *J. Chem. Soc.*, **1959**, 466–468.
13. E. D. Bergman, J. Szmuszkovicz, *J. Am. Chem. Soc.*, **1951**, *73*, 5153–5155.
14. H.-J. Bestmann, R. Armsen, H. Wagner, *Chem. Ber.*, **1969**, *102*, 2259–2269.
15. I. G. Stará, I. Starý, M. Tichý, J. Závada, V. Hamus, *J. Am. Chem. Soc.*, **1994**, *116*, 5084–5088. For the application of the McMurry and carbenoid couplings to prepare **3** see F. Dubois, M. Gingras, *Tetrahedron Lett.*, **1998**, *39*, 5039–5040.
16. H.-J. Bestmann, W. Both, *Angew. Chem.*, **1972**, *84*, 293; *Angew. Chem. Int. Ed. Engl.*, **1972**, *11*, 296; *cf.* H.-J. Bestmann, W. Both, *Chem. Ber.*, **1974**, *107*, 2923–2925.
17. For the X-ray structure determination of **3** see A. O. McIntosh, J. M. Robertson, V. Vand, *J. Chem. Soc.*, **1954**, 1661–1668; R. Kuroda, *J. Chem. Soc. Perkin II*, **1982**, 789–794; D. Fabbri, A. Dore, S. Gladiali, O. De Lucchi, G. Valle, *Gazz. Chim. Ital.*, **1996**, *126*, 11–18.
18. Ch. Goedicke, H. Stegemeyer, *Tetrahedron Lett.*, **1970**, 937–940.
19. A. Smakula, *Z. Phys. Chem. Abt. B*, **1934**, *25*, 90–98; C. O. Parker, P. E. Spoerri, *Nature*, **1950**, *166*, 603; R. E. Buckles, *J. Am. Chem. Soc.*, **1955**, *77*, 1040–1041; P. Hugelshofer, J. Kalvoda, K. Schaffner, *Helv. Chim. Acta*, **1960**, *43*, 1322–1332.
20. Reviews on this oxidative photocylization: F. B. Mallory, C. W. Mallory, *Org. Reactions*, **1983**, *30*, 1–456; W. H. Laarhoven, W. J. C. Prinsen, *Top. Curr. Chem.*, **1984**, *125*, 63–130; W. H. Laarhoven, *Org. Photochem.*, **1987**, *9*, 129; H. Meier, *Angew. Chem.*, **1992**, *104*, 1425–1446; *Angew. Chem. Int. Ed. Engl.*, **1992**, *31*, 1399–1420.
21. M. Scholz, M. Mühlstädt, F. Dietz, *Tetrahedron Lett.*, **1967**, 665–668.
22. F. Dietz, M. Scholz, *Tetrahedron*, **1968**, *24*, 6845–6849; *cf.* W. H. Laarhoven, T. J. H. M. Cuppen, R. J. F. Nivard, *Tetrahedron*, **1970**, *26*, 1069–1083.
23. L. Liu, Th. J. Katz, *Tetrahedron Lett.*, **1991**, *32*, 6831–6834.
24. L. Liu, Th. J. Katz, *Tetrahedron Lett.*, **1990**, *31*, 3983–3986; *cf.* N. D. Willmore, D. A. Hoic, Th. J. Katz, *J. Org. Chem.*, **1994**, *59*, 1889–1891.
25. L. Minuti, A. Taticchi, A. Marrocchi, E. Gacs-Baitz, *Tetrahedron*, **1997**, *53*, 6873–6878.
26. N. D. Willmore, L. Liu, Th. J. Katz, *Angew. Chem.*, **1992**, *104*, 1081–1082; *Angew.*

Chem. Int. Ed. Engl., **1992**, *31*, 1093–1094. More recent studies show that the yield is increased when the cycloadditions are performed in the presence of basic alumina as a catalyst. Furthermore, the yield drastically rises when the aromatic diene is activated by electron-releasing substituents, such as alkoxy groups: Th. J. Katz, L. Liu, N. D. Willmore, J. M. Fox, A. L. Rheingold, S. Shi, C. Nuckolls, B. H. Rickman, *J. Am. Chem. Soc.*, **1997**, *119*, 10054–10063.

27.   M. S. Newman, D. Lednicer, *J. Am. Chem. Soc.*, **1956**, *78*, 4765–4770.

28.   For the preparation of hexahelicene-7-carboxylic acid by a non-photochemical route see D. Bogaert-Verhoogen, R. H. Martin, *Tetrahedron Lett.*, **1967**, 3045–3048.

29.   R. H. Martin, M.-J. Marchant, M. Baes, *Helv. Chim. Acta*, **1971**, *54*, 358–360.

30.   R. H. Martin, M. Flammang-Barbieux, J. P. Cosyn, M. Gelbcke, *Tetrahedron Lett.*, **1968**, 3507–3510. The mechanism of these helicene-forming processes can be quite complex; it has been investigated particularly carefully for 2-styryl-benzo[*c*]phenanthrene, see W. J. C. Prinsen, W. H. Laarhoven, *J. Org. Chem.*, **1989**, *54*, 3689–3694; *cf.* W. H. Laarhoven, Th. J. H. M. Cuppen, H. H. Brinkhof, *Tetrahedron*, **1982**, *38*, 3179–3182.

31.   A. E. Siegrist, *Helv. Chim. Acta*, **1967**, *50*, 906–957; *cf.* A. E. Siegrist, P. Liechti, H. R. Meyer, K. Weber, *Helv. Chim. Acta*, **1969**, *52*, 2521–2554.

32.   Chr. Jutz, H.-G. Löbering, *Angew. Chem.*, **1975**, *87*, 415–416; *Angew. Chem. Int. Ed. Engl.*, **1975**, *14*, 418.

33.   M. Flammang-Barbieux, J. Nasielski, R. H. Martin, *Tetrahedron Lett.*, **1967**, 743–744.

34.   R. H. Martin, *Angew. Chem.*, **1974**, *86*, 727–738; *Angew. Chem. Int. Ed. Engl.*, **1974**, *13*, 649–660.

35.   Review: W. H. Laarhoven, W. J. C. Prinsen, *Top. Curr. Chemistry*, **1984**, *125*, 63–130. This review also summarizes synthetic work aimed at the preparation of substituted carbohelicenes, benzannelated and double helicenes as well as heterohelicenes.

36.   R. H. Martin, M. Baes, *Tetrahedron*, **1975**, *31*, 2135–2137.

37.   W. H. Laarhoven, Th. J. H. M. Cuppen, R. J. F. Nivard, *Tetrahedron*, **1974**, *30*, 3343–3347; *cf.* L. H. Klemm, D. Reed, *J. Chromatogr.*, **1960**, *3*, 364; L. H. Klemm, K. B. Desai, J. R. Spooner, *J. Chromatogr.*, **1964**, *14*, 300.

38.   Y. H. Kim, A. Tishbee, E. Gil-Av, *J. Am. Chem. Soc.*, **1980**, *102*, 5915–5917.

39.   The dextrorotatory enantiomer of **32** possesses the right-handed (*P*-) helicity.

40.   R. H. Martin, V. Libert, *J. Chem. Res. (S)*, **1980**, 130.

41.   A. Moradpour, J. F. Nicoud, G. Balavoine, H. Kagan, G. Tsoucaris, *J. Am. Chem. Soc.*, **1971**, *93*, 2353–2354; W. J. Bernstein, M. Calvin, O. Buchardt, *J. Am. Chem. Soc.*, **1973**, *95*, 527–532; H. Kagan, A. Moradpour, J. F. Nicod, G. Balavoine, R. H. Martin, J. P. Cosyn, *Tetrahedron Lett.*, **1971**, 2479–2482; O. Buchardt, *Angew. Chem.*, **1974**, *86*, 222–228; *Angew. Chem. Int. Ed. Engl.*, **1974**, *13*, 179–185. In all cases right-handed CPL forms *M*- and left-handed CPL forms *P*-helicenes.

42.   W. H. Laarhoven, Th. J. H. M. Cuppen, *J. Chem. Soc. Chem. Commun.*, **1977**, 47; *cf.* W. J. C. Prinsen, W. H. Laarhoven, *Recl. Trav. Chim. Pays-Bas*, **1995**, *114*, 470–475.

43.   M. Nakazaki, K. Yamamoto, K. Fujiwara, *Chem. Lett.*, **1978**, 863; M. Hibert, G. Solladie, *J. Org. Chem.*, **1980**, *45*, 5393–5394.

44.   J.-M. Vanest, R. H. Martin, *Recl. Trav. Chim. Pays-Bas*, **1979**, *98*, 113. For more recent applications of the chiral auxiliary approach see A. Sudhakar, Th. J. Katz, *J. Am. Chem. Soc.*, **1986**, *108*, 179–181.

45.   C. de Rango, G. Tsoucaris, J. P. Declerq, G. Germain, J. P. Putzeys, *Cryst. Struct. Commun.*, **1973**, *2*, 189.

46.   P. T. Beurkens, G. Beurkens, G. T. E. M. van der Hark, *Cryst. Struct. Commun.*, **1976**, *5*, 240.

47.   G. Le Bas, A. Navaza, Y. Mauguen, C. de Rango, *Cryst. Struct. Commun.*, **1976**, *5*, 357;

cf. S. Ramdos, J. M. Thomas, M. E. Jordan, C. J. Eckhardt, *J. Phys. Chem.*, **1981**, *85*, 2421–2425.

48. G. Le Bas, A. Navaza, M. Knossow, C. de Rango, *Cryst. Struct. Commun.*, **1976**, *5*, 713.
49. For electrophilic substitution and substitution reactions of [6]-helicene see P. M. op den Brouw, W. H. Laarhoven, *Recl. Trav. Chim. Pays-Bas*, **1978**, *97*, 265–268; for the oxidation of [6]-helicene see J. W. Diesveld, J. H. Borkent, W. H. Laarhoven, *Recl. Trav. Chim. Pays-Bas*, **1980**, *99*, 391–394.
50. R. H. Martin, M-J. Marchant, *Tetrahedron Lett.*, **1972**, 3707–3708; *cf.* R. H. Martin, N. Defay, H. P. Figeys, M. Flammang-Barbieux, J. P. Cosyn, M. Gelbke, J. J. Schurter, *Tetrahedron*, **1969**, *25*, 4981–4998.
51. R. H. Martin, M.-J. Marchant, *Tetrahedron*, **1974**, *30*, 347–349. For the thermal racemization of methyl-substituted hexahelicenes see J. H. Borkent, W. H. Laarhoven, *Tetrahedron*, **1978**, *34*, 2565–2567.
52. H.-J. Lindner, *Tetrahedron*, **1975**, *31*, 281–284.
53. J. H. Dopper, H. Wynberg, *Tetrahedron Lett.*, **1972**, 763–766; *cf.* J. H. Dopper, D. Oudman, H. Wynberg, *J. Am. Chem. Soc.*, **1973**, *95*, 3692–3698; J. H. Dopper, H. Wynberg, *J. Org. Chem.*, **1975**, *40*, 1957–1966.
54. W. E. Barth, R. G. Lawton, *J. Am. Chem. Soc.*, **1971**, *93*, 1730–1745.
55. E. Clar, M. Zander, *J. Chem. Soc.*, **1957**, 4616–4619. The first synthesis of coronene (**4**) has been described by R. Scholl, K. Meyer, *Ber. Dtsch. Chem. Ges.*, **1932**, *65*, 902–915. For the X-ray structure see J. M. Robertson, J. G. White, *J. Chem. Soc.*, **1945**, 607–617.
56. According to molecular mechanics calculations all circulenes up to [12]-circulene are non-planar, with the only exception of [6]circulene (**4**, coronene); I thank Norman Goldberg for these calculations.
57. The name corannulene is derived from the Latin: *cor* means heart or within and *annulus* means ring. The name was also chosen so as to connote its relationship to coronene (**4**), its parent.
58. A. Sieglitz, W. Schidlo, *Chem. Ber.*, **1963**, *96*, 1098–1108.
59. J. C. Hanson, C. E. Nordman, *Acta Crystallogr. Ser. B*, **1976**, *32*, 1147–1153.
60. L. T. Scott, M. M. Hashemi, D. T. Meyer, H. B. Warren, *J. Am. Chem. Soc.*, **1991**, *113*, 7082–7084. In this initial study corannulene (**5**) was prepared by flash vacuum pyrolysis at 1000 °C of 7,10-diethynylfluoranthene in ca. 10% yield (see also Scheme 12)
61. L. T. Scott, P.-C. Cheng, M. M. Hashemi, M. S. Bratcher, D. T. Meyer, H. B. Warren, *J. Am. Chem. Soc.*, **1997**, *119*, 10963–10968.
62. For the first (unsuccessful) attempt to use 7,10-disubstituted fluoranthene derivatives as precursors for corannulenes see J. T. Chang, M. D. W. Robins, *Austr. J. Chem.*, **1968**, *21*, 2237–2245; *cf.* J. R. Davy, M. N. Iskander, J. A. Reiss, *Austr. J. Chem.*, **1979**, *32*, 1067–1078.
63. C. Z. Liu, P. W. Rabideau, *Tetrahedron Lett.*, **1996**, *37*, 3437–3440.
64. A. Borchardt, A. Fuchicello, K. V. Kilway, K. K. Baldrige, J. S. Siegel, *J. Am. Chem. Soc.*, **1992**, *114*, 1921–1923; *cf.* T. J. Seiders, K. K. Baldrige, J. S. Siegel, *J. Am. Chem. Soc.*, **1996**, *118*, 2754–2755.
65. G. Zimermann, U. Nüchter, S. Hagen, M. Nüchter, *Tetrahedron Lett.*, **1994**, *35*, 4747–4750.
66. G. Mehta, G. Panda, *Tetrahedron Lett.*, **1997**, *38*, 2145–2148.
67. For a summary see L. T. Scott, *Pure Appl. Chem.*, **1996**, *68*, 291–300.
68. A. Ayalon, M. Rabinovitz, P.-C. Cheng, L. T. Scott, *Angew. Chem.*, **1992**, *104*, 1691–1692; *Angew. Chem. Int. Ed. Engl.*, **1992**, *31*, 1636–1637.
69. A. Ayalon, A. Sygula, P.-C. Cheng, M. Rabinovitz, P. W. Rabideau, L. T. Scott, *Science*, **1994**, *265*, 1065–1067.

70. M. Baumgarten, L. Gherghel, M. Wagner, A. Weitz, M. Rabinovitz, P.-C. Cheng, L. T. Scott, *J. Am. Chem. Soc.*, **1995**, *117*, 6254–6257.

71. L. T. Scott, M. M. Hashemi, M. S. Bratcher, *J. Am. Chem. Soc.*, **1992**, *114*, 1920–1921; *cf.* P. U. Biedermann, S. Pogodin, I. Agranat, *J. Org. Chem.*, **1999**, *64*, 3655–3662 for recent theoretical calculations on the inversion barriers of bowl-shaped aromatic hydrocarbons.

72. P. J. Jessup, J. A. Reiss, *Austr. J. Chem.*, **1977**, *30*, 851–857.

73. K. Yamamoto, T. Harada, Y. Okamoto, H. Chikamatsu, M. Nakazaki, Y. Kai, T. Nakao, M. Tanaka, S. Harada, N. Kasai, *J. Am. Chem. Soc.*, **1988**, *110*, 3578–3584.

74. K. Yamamoto, *Pure Appl. Chem.*, **1993**, *65*, 157–163.

75. K. Yamamoto, H. Sonobe, H. Matsubara, M. Sato, S. Okamoto, K. Kitaura, *Angew. Chem.*, **1996**, *108*, 69–70; *Angew. Chem. Int. Ed. Engl.*, **1996**, *35*, 69–70

76. W. Krästchmer, L. D. Lamb, K. Fostiropoulos, D. R. Huffmann, *Nature*, **1990**, *347*, 354.

77. Reviews: P. W. Rabideau, A. Sygula in *Advances in Theoretically Interesting Molecules* R. P. Thummel, (*Ed.*), JAI Press, Inc., Greenwich, CT, **1995**, Vol. 3, 1–36; P. W. Rabideau, A. Sygula, *Accounts Chem. Res.*, **1996**, *29*, 235–242; G. Mehta, H. Surya Prakash Rao, *Tetrahedron*, **1998**, *54*, 13325–13370.

78. A. H. Abdourazak, Z. Marcinow, H. E. Folsom, F. R. Fronczek, R. Sygula, A. Sygula, P. W. Rabideau, *Tetrahedron Lett.*, **1994**, *35*, 3856–3860; *cf.* P. W. Rabideau, A. H. Abdourazak, H. E. Folsom, Z. Marcinow, R. Sygula, A. Sygula, *J. Am. Chem. Soc.*, **1994**, *116*, 7891–7892; *cf.* M. D. Clayton, Z. Marcinow, P. W. Rabideau, *J. Org. Chem.*, **1996**, *61*, 6052–6054.

79. S. Hagen, M. S. Bratcher, M. S. Erickson, G. Zimmermann, L. T. Scott, *Angew. Chem.*, **1997**, *109*, 407–409; *Angew. Chem. Int. Ed. Engl.*, **1997**, *36*, 406–408.

80. A. H. Abdourazak, Z. Marcinow, A. Sygula, R. Sygula, P. W. Rabideau, *J. Am. Chem. Soc.*, **1995**, *117*, 6410–6411; A. Sygula, P. W. Rabidean, *J. Am. Chem. Soc.*, **1978**, *120*, 12666–12667.

81. L. T. Scott, M. S. Bratcher, S. Hagen, *J. Am. Chem. Soc.*, **1996**, *118*, 8743–8744.

82. P. Rehländer, *Ber. Dtsch. Chem. Ges.*, **1903**, *36*, 1583–1587; *cf.* K. Dziewonski, *Ber. Dtsch. Chem. Ges.*, **1903**, *36*, 962–971.

83. For a look at these early years of cyclophane chemistry see D. J. Cram, J. M. Cram, *Acc. Chem. Res.*, **1971**, *4*, 204–213.

84. A. Lüttringhaus, *Liebigs Ann. Chem.*, **1937**, *528*, 181–210; *cf.* A. Lüttringhaus, *Liebigs Ann. Chem.*, **1937**, *528*, 211–222; A. Lüttringhaus, H. Gralheer, *Liebigs Ann. Chem.*, **1942**, *550*, 67–98; A. Lüttringhaus, G. Eyring, *Angew. Chem.*, **1957**, *69*, 137.

85. B. H. Smith, *Bridged Aromatic Compounds*, Academic Press, New York, N.Y., **1964**.

86. F. Vögtle, P. Neumann, *Top. Curr. Chem.*, **1974**, *48*, 67–128.

87. P. M. Keehn, S. M. Rosenfeld (*Eds.*), *Cyclophanes I , II*, Academic Press, New York, N.Y., **1983**.

88. F. Vögtle (*Ed.*), *Cyclophanes I, Top. Curr. Chem.*, **1983**, *113*.

89. F. Vögtle (*Ed.*), *Cyclophanes II, Top. Curr. Chem.*, **1983**, *115*.

90. H. Hopf, *Naturwiss.*, **1903**, *70*, 349 358.

91. F. Diederich, *Cyclophanes*, Royal Society of Chemistry, London, **1991**.

92. D. J. Cram, J. M. Cram, *Container Molecules and Their Guests*, Royal Society of Chemistry, London, **1992**.

93. F. Vögtle, *Cyclophane Chemistry*, J. Wiley & Sons, Chichester, **1993**.

94. F. Vögtle (*Ed.*), *Cyclophanes, Top. Curr. Chem.*, **1994**, *172*.

95. G. J. Bodwell, *Angew. Chem.*, **1996**, *108*, 2221–2224; *Angew. Chem. Int. Ed. Engl.*, **1996**, *35*, 2085–2088.

96. A. de Meijere, B. König, *Synlett*, **1997**, 1221–1232.

97. B. König, *Top. Curr. Chem.*, **1998**, *196*, 89–133.

98. D. J. Cram, E. A. Truesdale, *J. Am. Chem. Soc.*, **1973**, *95*, 5825–5827; *cf.* D. J. Cram, E. A. Truesdale, *J. Org. Chem.*, **1980**, *45*, 3974–3981; D. J. Cram, R. B. Hornby, E. A. Truesdale, H. J. Reich, M. H. Delton, J. M. Cram, *Tetrahedron*, **1974**, *30*, 1757–1768.

99. H. Hopf, S. Trampe, K. Menke, *Chem. Ber.*, **1977**, *110*, 371–372; *cf.* A. E. Murad, H. Hopf, *Chem. Ber.*, **1980**, *113*, 2358–2371.

100. High level *ab initio* calculations have also been performed on selected [n]paracyclophanes: J. E. Rice, T. J. Lee, R. B. Remington, W. D. Allen, D. A. Clabo, Jr, H. F. Schaefer III, *J. Am. Chem. Soc.*, **1987**, *109*, 2902–2909; *cf.* T. J. Lee, J. E. Rice, W. D. Allen, R. B. Remington, H. F. Schaefer III, *Chem. Phys.*, **1988**, *123*, 1–25.

101. St. Grimme, *J. Am. Chem. Soc.*, **1992**, *114*, 10542–10547

102. S. M. Rosenfeld, K. A. Choe in *Cyclophanes* P. M. Keehn, S. M. Rosenfeld, (*Eds.*), Academic Press, New York, N. Y., **1983**, Vol. I, 311–357. [9]Paracyclophane was first prepared by a Diels-Alder route involving cycloaddition of maleic anhydride to cyclotrideca-1,3-diene: M. F. Bartlett, S. K. Figdor, K. Wiesner, *Can. J. Chem.*, **1952**, *30*, 291–294 and lit. cited. For a later application of this route to [n]paracyclophanes (n = 8, 7) see P. G. Gassman, T. F. Bailey, R. C. Hoye, *J. Org. Chem.*, **1980**, *45*, 2923–2924; *cf.* P. G. Gassman, S. R. Korn, T. F. Bailey, T. H. Johnson, J. Finer, J. Clardy, *Tetrahedron Lett.*, **1979**, 3401–3404; P. G. Gassman, R. P. Thummel, *J. Am. Chem. Soc.*, **1972**, *94*, 7183–7184.

103. R. Huisgen, *Angew. Chem.*, **1957**, *69*, 341–359; *cf.* R. Huisgen, W. Rapp, I. Ugi, H. Walz, I. Glogger, *Liebigs Ann. Chem.*, **1954**, *586*, 52–69.

104. T. Inoue, T. Kaneda, S. Misumi, *Tetrahedron Lett.*, **1974**, 2969–2972.

105. Reviews: V. V. Kane, W. H. de Wolf, F. Bickelhaupt, *Tetrahedron,* **1994**, *50*, 4575–4622; Y. Tobe, *Top. Curr. Chem.,* **1994**, *172*, 1–40.

106. D. J. Cram, G. R. Knox, *J. Am. Chem. Soc.*, **1961**, *83*, 2204–2205; D. J. Cram, C. S. Montgomery, G. R. Knox, *J. Am. Chem. Soc.*, **1966**, *88*, 515–525.

107. K. L. Noble, H. Hopf, L. Ernst, *Chem. Ber.*, **1984**, *117*, 455–473.

108. K. L. Noble, H. Hopf, L. Ernst, *Chem. Ber.*, **1984**, *117*, 474–488.

109. T. Tsuji, T. Shibata, Y. Hienuki, S. Nishida, *J. Am. Chem. Soc.*, **1978**, *100*, 1806–1814; T. Tsuji, S. Nishida, *J. Am. Chem. Soc.*, **1974**, *96*, 3649–3650.

110. L. W. Jenneskens, W. H. de Wolf, F. Bickelhaupt, *Tetrahedron*, **1986**, *42*, 1571–1574.

111. T. Tsuji, S. Nishida, *Acc. Chem. Res.*, **1984**, *17*, 56–61.

112. G. L. Closs, M. S. Czeropski, *Chem. Phys. Lett.*, **1977**, *45*, 115–116.

113. N. L. Allinger, L. A. Freiberg, R. B. Hermann, M. A. Miller, *J. Am. Chem. Soc.,* **1963**, *85*, 1171–1176; *cf.* M. G. Newton, T. J. Walter, N. L. Allinger, *J. Am. Chem. Soc.*, **1973**, *95*, 5652–5658; D. J. Cram, M. F. Antar, *J. Am. Chem. Soc.*, **1958**, *80*, 3103–3109; D. J. Cram, M. F. Antar, *J. Am. Chem. Soc.*, **1958**, *80*, 3109–3114.

114. N. L. Allinger, T. J. Walter, *J. Am. Chem. Soc.*, **1972**, *94*, 9267–9268.

115. N. L. Allinger, T. J. Walter, M. G. Newton, *J. Am. Chem. Soc.*, **1974**, *96*, 4588–4597.

116. T. Otsubo, S. Misumi, *Synth. Comm.*, **1978**, *8*, 285–289; *cf.* H. Higuchi, E. Kobayashi, Y. Sakata, S. Misumi, *Tetrahedron* **1986**, *42*, 1731–1739.

117. Reviews: J. Dohm, F. Vögtle, *Top. Curr. Chem.*, **1992**, *161*, 69–106; *cf.* F. Vögtle, L. Rossa, *Angew. Chem.*, **1979**, *91*, 534–549; *Angew. Chem. Int. Ed. Engl.*, **1979**, *18*, 274–289.

118. A. D. Wolf, V. V. Kane, R. H. Levin, M. Jones, Jr, *J. Am. Chem. Soc.*, **1973**, *95*, 1680; *cf.* T. E. Berdick, R. H. Levin, A. D. Wolf, M. Jones, Jr, *J. Am. Chem. Soc.*, **1973**, *95*, 5087–5088.

119. V. V. Kane, M. Jones, Jr, *Org. Synth.*, **1981**, *61*, 129–133.

120. J. W. van Straten, W. H. de Wolf, F. Bickelhaupt, *Recl. Trav. Chim. Pays-Bas,* **1977**, *96*,

88.

121. V. V. Kane, A. D. Wolf, M. Jones, Jr. *J. Am. Chem. Soc.*, **1974**, *96*, 2643–2644.

122. R. H. Levin, *Ph. d. thesis*, Princeton University, **1970**.

123. J. W. van Straten, I. J. Landheer, W. H. de Wolf, F. Bickelhaupt, *Tetrahedron Lett.*, **1975**, 4499–4502.

124. S. L. Kammula, L. D. Iroff, M. Jones, Jr., J. W. van Straten, W. H. de Wolf, F. Bickelhaupt, *J. Am. Chem. Soc.*, **1977**, *99*, 5815; *cf.* F. Bokisch, H. Dreeskamp, T. v. Haugwitz, W. Tochtermann, *Chem. Ber.*, **1991**, *124*, 1831–1835.

125. R. Weiss, C. Schlierf, *Angew. Chem.*, **1971**, *83*, 887–888; *Angew. Chem. Int. Ed. Engl.*, **1971**, *10*, 811; *cf.* R. Weiss, S. Andrae, *Angew. Chem.*, **1973**, *85*, 145–147; *Angew. Chem. Int. Ed. Engl.*, **1973**, *12*, 150–152.

126. Y. Tobe, K. Ueda, K. Kakiuchi, Y. Odaira, Y. Kai, N. Kasai, *Tetrahedron*, **1986**, *42*, 1851–1858 and refs. cited.

127. Y. Tobe, K. Ueda, T. Kaneda, K. Kakiuchi, Y. Odaira, Y. Kai, N. Kasai, *J. Am. Chem. Soc.*, **1987**, *109*, 1136–1144.

128. W. Tochtermann, P. Rösner, *Tetrahedron Lett.*, **1980**, *21*, 4905–4908; W. Tochtermann, P. Rösner, *Chem. Ber.*, **1981**, *114*, 3725–3736; J. Liebe, Ch. Wolff, W. Tochtermann, *Tetrahedron Lett.*, **1982**, *23*, 171–172; W. Tochtermann, M. Haase, *Chem. Ber.*, **1984**, *117*, 2293–2299; C. Wolff, C. Krieger, J. Weiss, W. Tochtermann, *Chem. Ber.*, **1985**, *118*, 4144–4178; J. Hünger, C. Wolff, W. Tochtermann, E. M. Peters, K. Peters, H. G. von Schnering, *Chem. Ber.*, **1986**, *119*, 2698–2722; *cf.* J. L. Jessen, C. Wolff, W. Tochtermann, *Chem. Ber.*, **1986**, *119*, 297–312.

129. L. W. Jenneskens, J. N. Louwen, W. H. de Wolf, F. Bickelhaupt, *J. Phys. Org. Chem.*, **1990**, *3*, 295.

130. L. W. Jenneskens, F. J. J. de Kanter, P. A. Kraakman, L. A. M. Turkenburg, W. E. Koolhaas, W. H. de Wolf, F. Bickelhaupt, Y. Tobe, K. Kakiuchi, Y. Odaira, *J. Am. Chem. Soc.*, **1985**, *107*, 3716–3717.

131. Y. Tobe, T. Kaneda, K. Kakiuchi, Y. Odeira, *Chem. Lett.*, **1985**, 1301–1304.

132. G. B. M. Kostermans, W. H. de Wolf, F. Bickelhaupt, *Tetrahedron Lett.*, **1986**, *27*, 1095–1098; *cf.* G. B. M. Kostermans, W. H. de Wolf, F. Bickelhaupt, *Tetrahedron*, **1987**, *43*, 2955–2966.

133. G. B. M. Kostermans, M. Bobeldijk, W. H. de Wolf, F. Bickelhaupt, *J. Am. Chem. Soc.*, **1987**, *109*, 2471–2475.

134. T. Tsuji, S. Nishida, *J. Am. Chem. Soc.*, **1988**, *110*, 2157–2164; *cf.* T. Tsuji, S. Nishida, *J. Chem. Soc. Chem. Commun.*, **1987**, 1189–1190.

135. T. Tsuji, S. Nishida, *J. Am. Chem. Soc.*, **1989**, *111*, 368–369.

136. M. Okuyama, M. Ohkita, T. Tsuji, *J. Chem. Soc. Chem. Commun.*, **1997**, 1277–1278.

137. J. v. Braun, K. Heider, W. Wyczatkowska, *Ber. Dtsch. Chem. Ges.*, **1918**, *51*, 1215–1227; *cf.* J. v. Braun, L. Neumann, *Ber. Dtsch. Chem. Ges.*, **1919**, *52*, 2015–2019.

138. A. Lorenzi-Riatsch, H. Wälchli, M. Hesse, *Helv. Chim. Acta*, **1985**, *68*, 2177–2181 and refs. cited therein; *cf.* P. H. Nelson, J. T. Nelson, *Synthesis*, **1991**, 192–194.

139. F. Vögtle, P. Koo Tze Mew, *Angew. Chem.* **1978**, *90*, 58–60; *Angew. Chem. Int. Ed. Engl.*, **1978**, *17*, 60–62; *cf.* F. Vögtle, L. Rossa, *Angew. Chem.*, **1979**, *91*, 534–549; *Angew. Chem. Int. Ed. Engl.*, **1979**, *18*, 514–529.

140. H. Higuchi, K. Tani, T. Otsubo, Y. Sakata, S. Misumi, *Bull. Chem. Soc. Jpn.*, **1987**, *60*, 4027–4036; *cf.* H. Higuchi, M. Kugimiya, T. Otsubo, Y. Sakata, S. Misumi, *Tetrahedron Lett.*, **1983**, *24*, 2593–2594.

141. A. J. Hubert, J. Dale, *J. Chem. Soc.*, **1963**, 86–93.

142. K. Tamao, S. Kodama, T. Nakatsuka, Y. Kiso, M. Kumada, *J. Am. Chem. Soc.*, **1975**, *97*, 4405–4406.

143. R. B. Bates, F. A. Camou, V. V. Kane, P. K. Mishra, K. Suvannachut, J. J. White, *J. Org. Chem.*, **1989**, *54*, 311–317.

144. [10]Paracyclophane derivatives have previously been rearranged to their *meta*-isomers: A. T. Blomquist, R. E. Stahl, Y. C. Meinwald, B. H. Smith, *J. Org. Chem.*, **1961**, *26*, 1687–1691.

145. [7]metacyclophane (**174**) by bis-alkylation of **162**: R. B. Bates, C. A. Ogle, *J. Org. Chem.*, **1982**, *47*, 3949–3952; by isomerization of [7]paracyclophane (**126**): K. L. Noble, H. Hopf, M. Jones, Jr, S. L. Kammula, *Angew. Chem.*, **1978**, *90*, 629–630; *Angew. Chem. Int. Ed. Engl.*, **1978**, *17*, 602.

146. S. Hirano, H. Hara, T. Hiyama, S. Fujita, H. Nozaki, *Tetrahedron*, **1975**, *31*, 2219–2227.

147. S. Hirano, T. Hiyama, S. Fujita, H. Nozaki, *Chem. Lett.*, **1972**, 707; *cf.* S. Fujita, S. Hirano, H. Nozaki, *Tetrahedron Lett.*, **1972**, 403–406.

148. J. W. van Straten, W. H. de Wolf, F. Bickelhaupt, *Tetrahedron Lett.*, **1977**, 4667–4670; *cf.* N. A. Le, M. Jones, Jr, F. Bickelhaupt, W. H. de Wolf, *J. Am. Chem. Soc.*, **1989**, *111*, 8491–8493; L. W. Jenneskens, W. H. de Wolf, F. Bickelhaupt, *Angew. Chem.*, **1985**, *97*, 568–569; *Angew. Chem. Int. Ed. Engl.*, **1985**, *24*, 585–586.

149. J. L. Goodman, J. A. Berson, *J. Am. Chem. Soc.*, **1985**, *107*, 5409–5424.

150. Y. Tobe, T. Sorori, K. Kobiro, K. Kakiuchi, Y. Odaira, *Tetrahedron Lett.*, **1987**, *28*, 2861–2862.

151. Y. Tobe, K. Ueda, K. Kakiuchi, Y. Odaira, *Chem. Lett.*, **1983**, 1645–1646; *cf.* Y. Tobe, A. Takemura, M. Jimbo, T. Takahashi, K. Kobiro, K. Kakiuchi, *J. Am. Chem. Soc.*, **1992**, *114*, 3479–3491.

152. L. A. M. Turkenburg, P. M. L. Blok, W. H. de Wolf, F. Bickelhaupt, *Tetrahedron Lett.*, **1981**, *22*, 3317–3320; L. W. Jenneskens, F. I. J. de Kanter, L. A. M. Turkenburg, H. J. R. de Boer, W. H. de Wolf, F. Bickelhaupt, *Tetrahedron*, **1984**, *40*, 4401–4413.

153. G. B. M. Kostermans, Ph. d. thesis, Vrije Univ., Amsterdam, **1989**; *cf.* G. B. M. Kostermans, M. Hogenbirk, L. A. M. Turkenburg, W. H. de Wolf, F. Bickelhaupt, *J. Am. Chem. Soc.*, **1987**, *109*, 2855–2857; G. B. M. Kostermans, P. van Dansik, W. H. de Wolf, F. Bickelhaupt, *J. Org. Chem.*, **1988**, *53*, 4531–4534.

154. M. Scwarc, *J. Chem. Phys.*, **1948**, *16*, 128–136.

155. C. J. Brown, A. C. Farthing, *Nature*, **1949**, *164*, 915–916; *cf.* C. J. Brown, *J. Chem. Soc.*, **1953**, 3265–3270.

156. D. J. Cram, H. Steinberg, *J. Am. Chem. Soc.*, **1951**, *73*, 5691–5704.

157. K. Nishiyama, M. Sakiyama, S. Seki, H. Horita, T. Otsubo, S. Misumi, *Tetrahedron Lett.*, **1977**, 3739–3740; *cf.* K. Nishiyama, M. Sakiyama, S. Seki, H. Horita, T. Otsubo, S. Misumi, *Bull. Chem. Soc. Jpn.*, **1980**, *53*, 869–877 and refs. cited therein.

158. F. Vögtle, *Angew. Chem.*, **1969**, *81*, 258–259; *Angew. Chem. Int. Ed. Engl.*, **1969**, *8*, 274.

159. V. Boekelheide, *Topics Curr. Chem.*, **1983**, *113*, 87–143 and refs. cited therein.

160. H. A. Staab, M. Haenel, *Tetrahedron Lett.*, **1970**, 3585–3588.

161. H. E. Winberg, F. S. Fawcett, W. E. Mochel, C. W. Theobald, *J. Am. Chem. Soc.*, **1960**, *82*, 1428–1435.

162. H. E. Winberg, F. S. Fawcett, *Org. Synth. Coll. Vol. V*, J. Wiley and Sons, New York, N. Y., **1973**, 883–886. Formally the dimerization may be regarded either as a [6+6]- or a [8+8]cycloaddition. In both cases it is forbidden on orbital-symmetry grounds to occur in a concerted manner. The process is thus expected to take place step-wise, *i.e.* by formation of one ethano bridge after the other. For the application of [2.2]paracyclophane (**91**) in polymer chemistry see: W. F. Gorham, *J. Polym. Sci. Part A-1*, **1966**, *4*, 3027–3039. For structural characterization of isolable *p*-quinodimethanes see W. J. Y. Cheng, N. R. Janosy, J. M. C. Nadeau, S. Rossenfeld, M. Rushing, J. P. Jasinski, V. Rotello, *J. Org. Chem.*, **1998**, *63*, 379–382.

163. H. Hopf, *Angew. Chem.*, **1972**, *84*, 471–472; *Angew. Chem. Int. Ed. Engl.*, **1972**, *11*, 419; *cf.* J. Kleinschroth, H. Hopf, I. Böhm, *Org. Synth.*, **1981**, *60*, 41–48.
164. H. Hopf, F. T. Lenich, *Chem. Ber.*, **1974**, *107*, 1891–1902.
165. H. Hopf, F. W. Raulfs, D. Schomburg, *Tetrahedron*, **1986**, *42*, 1655–1663.
166. H. Hope, J. Bernstein, K. N. Trueblood, *Acta Crystallogr. Sect. B*, **1972**, *B28*, 1733–1743.
167. V. Boekelheide in *Strategies and Tactics in Organic Synthesis* Th. Lindberg, (*Ed.*), Academic Press Inc., New York, N. Y., **1984**, 1–19.
168. H. Hopf in *Cyclophanes II*, P. M. Keehn, S. R. Rosenfeld, (*Eds.*), Academic Press, New York, N. Y., **1983**, 521–572.
169. Z. Yang, B. Kovac, E. Heilbronner, J. Lecoultre, C. W. Chan, H. N. C. Wong, H. Hopf, F. Vögtle, *Helv. Chim. Acta*, **1987**, *70*, 299–307; E. Heilbronner, Z. Yang, *Top. Curr. Chem.*, **1983**, *115*, 1–55; F. Gerson, *Top. Curr. Chem.*, **1983**, *115*, 57–105.
170. L. Ernst, V. Bockelheide, H. Hopf, *Magn. Reson. Chem.*, **1993**, *31*, 669–676.
171. V. Boekelheide, R. A. Hollins, *J. Am. Chem. Soc.*, **1970**, *92*, 3512–3513; *cf.* V. Boekelheide, R. A. Hollins, *J. Am. Chem. Soc.*, **1973**, *95*, 3201–3208.
172. A. W. Hanson, *Cryst. Struct. Commun.*, **1980**, 1243–1247.
173. [2₃](1,2,4)(1,3,5)Cyclophane: N. Nakazaki, K. Yamamoto, Y. Miura, *J. Chem. Soc. Chem. Commun.*, **1977**, 206–207.
174. [2₄](1,2,4,5)cyclophane: R. Gray, V. Boekelheide, *Angew. Chem.*, **1975**, *87*, 138; *Angew. Chem. Int. Ed. Engl.*, **1975**, *14*, 107; *cf.* R. Gray, V. Boekelheide, *J. Am. Chem. Soc.*, **1979**, *101*, 2128–2136.
175. [2₄](1,2,4,5)cyclophane quinhydrone: H. A. Staab, V. M. Schwendemann, *Angew. Chem.*, **1978**, *90*, 805–807; *Angew. Chem. Int. Ed. Engl.*, **1978**, *17*, 756–757; H. A. Staab, V. M. Schwendemann, *Liebigs Ann. Chem.*, **1979**, 1258–1269.
176. J. Kleinschroth, Ph. d. thesis, University of Würzburg, **1980**.
177. H. Hopf, W. Gilb, K. Menke, *Angew. Chem.*, **1977**, *89*, 177–178; *Angew. Chem. Int. Ed. Engl.*, **1977**, *16*, 191.
178. J. Kleinschroth, H. Hopf, *Angew. Chem.*, **1979**, *91*, 336–337; *Angew. Chem. Int. Ed. Engl.*, **1979**, *18*, 329.
179. Y. Sekine, M. Brown, V. Boekelheide, *J. Am. Chem. Soc.*, **1979**, *101*, 3126–3127; Y. Sekine, M. Brown, V. Boekelheide, *J. Am. Chem. Soc.*, **1981**, *103*, 1777–1785.
180. H. Hopf, S. El-Tamany, *Chem. Ber.*, **1983**, *116*, 1682–1685.
181. D. T. Longone, H. S. Chow, *J. Am. Chem. Soc.*, **1964**, *86*, 3898–3899; *cf.* D. T. Longone, H. S. Chow, *J. Am. Chem. Soc.*, **1970**, *92*, 994–998.
182. For the isolation of the pure isomer **251** and the determination of its structure by X-ray structural analysis see H. Mizuno, K. Nishiguchi, T. Otsubo, S. Misumi, N. Morimoto, *Tetrahedron Lett.*, **1972**, *13*, 4981–4984; *cf.* H. Mizuno, K. Nishiguchi, T. Toyoda, T. Otsubo, S. Misumi, N. Morimoto, *Acta Crystallogr. Sect. B*, **1977**, *B33*, 329–334.
183. Reviews: S. Misumi, T. Otsubo, *Acc. Chem. Res.*, **1978**, *11*, 251–256; S. Misumi in *Cyclophanes II* P. M. Keehn, S. M. Rosenfeld, (*Eds.*) Academic Press, New York, N. Y., **1983**, 573–628. For a review on layered [3.3]orthocyclophanes see S. Mataka, T. Thiemann, M. Taniguchi, T. Sawada, *Synlett*, **1999**, in press.
184. T. Otsubo, S. Mizogami, I. Otsubo, Z. Tozuka, A. Sakagami, Y. Sakata, S. Misumi, *Bull. Chem. Soc. Jpn.*, **1973**, *46*, 3519–3530; *cf.* T. Otsubo, S. Mizogami, Y. Sakata, S. Misumi, *Tetrahedron Lett.*, **1973**, *14*, 2457–2460.
185. T. Otsubo, Z. Tozuka, S. Mizogami, Y. Sakata, S. Misumi, *Tetrahedron Lett.*, **1972**, *13*, 2927–2930.
186. [2.2]Metaparacyclophane: D. J. Cram, R. C. Helgeson, D. Lock, L. A. Singer, *J. Am. Chem. Soc.*, **1966**, *88*, 1324–1325; *cf.* F. Vögtle, *Chem. Ber.*, **1969**, *102*, 3077–3081;

[2.2]Orthometacyclophane: G. J. Bodwell, L. Ernst, M. W. Haenel, H. Hopf, *Angew. Chem.*, **1989**, *101*, 509–510; *Angew. Chem. Int. Ed. Engl.*, **1989**, *28*, 455–456; [2.2]Orthoparacyclophane: Y. Tobe, M. Kawaguchi, K. Kakiuchi, K. Naemura, *J. Am. Chem. Soc.*, **1993**, *115*, 1173–1174.

187. M. Pellegrin, *Recl. Trav. Chim. Pays-Bas*, **1899**, *18*, 457–465; *cf.* W. Baker, J. F. W. McOmie, J. M. Norman, *Chem. Ind.*, London, **1950**, 77.

188. E. Müller, G. Röscheisen, *Chem. Ber.*, **1957**, *90*, 543–553 (20% yield of **259**); N. L. Allinger, M. A. DaRooge, R. B. Herrmann, *J. Am. Chem. Soc.*, **1961**, *83*, 1974–1978 (39% yield of **259**).

189. W. S. Lindsay, P. Stokes, L. G. Humber, V. Boekelheide, *J. Am. Chem. Soc.*, **1961**, *83*, 943–949.

190. W. Jenny, R. Paioni, *Chimia*, **1968**, *22*, 142–143.

191. R. Paioni, W. Jenny, *Helv. Chim. Acta*, **1969**, *52*, 2041–2054.

192. R. H. Mitchell, V. Boekelheide, *J. Am. Chem. Soc.*, **1974**, *96*, 1547–1557; *cf.* R. H. Mitchell, T. Otsubo, V. Boekelheide, *Tetrahedron Lett.*, **1975**, 219–222.

193. H. Higuchi, K. Tani, T. Otsubo, Y. Sakata, S. Misumi, *Bull. Chem. Soc. Jpn.*, **1987**, *60*, 4027–4036; *cf.* M. Hojjatie, S. Muralidharan, H. Freiser, *Tetrahedron*, **1989**, *45*, 1611–1622.

194. For the most comprehensive compilation of X-ray structural data of the cyclophanes see P. M. Keehn in *Cyclophanes I*, P. M. Keehn, S. M. Rosenfeld (*Eds.*), Academic Press, New York, N. Y., **1983**, 69–238.

195. F. Vögtle, L. Schunder, *Chem. Ber.*, **1969**, *102*, 2677–2683.

196. R. H. Mitchell, T. K. Vinod, G. W. Bushnell, *J. Am. Chem. Soc.*, **1985**, *107*, 3340–3341.

197. Reviews: F. Vögtle, G. Hohner, *Top. Curr. Chem.*, **1978**, *74*, 1–29; S. Misumi in *Cyclophanes II*, P. M. Keehn, S. M. Rosenfeld, (*Eds.*), Academic Press, New York, N. Y., **1983**, 573–628.

198. T. Umemoto, T. Otsubo, S. Misumi, *Tetrahedron Lett.*, **1974**, 1573–1576; *cf.* T. Umemoto, T. Otsubo, Y. Sakata, S. Misumi, *Tetrahedron Lett.*, **1973**, 593–596.

199. T. Tsuji, M. Ohkita, S. Nishida, *J. Am. Chem. Soc.*, **1993**, *115*, 5284–5285.

200. T. Tsuji, M. Ohkita, T. Konno, S. Nishida, *J. Am. Chem. Soc.*, **1997**, *119*, 8425–8431.

201. Meanwhile stable derivatives of the **286–288** family have been prepared and were subjected to X-ray structural analysis: H. Kawai, T. Suzuki, M. Ohkita, T. Tsuji, 9th International Symposium of Novel Aromatic Compounds (ISNA-9), Hong Kong, **1998**, Poster 84.

202. K. C. Dewhirst, D. J. Cram, *J. Am. Chem. Soc.*, **1958**, *80*, 3115–3125; *cf.* M. Stöbbe, O. Reiser, R. Näder, A. de Meijere, *Chem. Ber.*, **1987**, *120*, 1667–1674.

203. H. Hopf, M. Psiorz, *Angew. Chem.*, **1982**, *94*, 639–640; *Angew. Chem. Int. Ed. Engl.*, **1982**, *21*, 623. For the use of [2.2]paracyclophyne intermediates for the construction of extended paracyclophane hydrocarbon systems see the summaries by B. König, A. de Meijere, *Synlett*, **1997**, 1221–1232 and B. König, *Top. Curr. Chem.*, **1998**, *196*, 89–133. Triple bonds have also been included in the bridges of cyclophanes with longer bridges as well as the the aromatic nucleus, see: T. Matsuoka, T. Negi, T. Otsubo, Y. Sakata, S. Misumi, *Bull. Chem. Soc. Jpn.*, **1972**, *45*, 1825–1833; T. Kaneda, T. Ogawa, S. Misumi, *Tetrahedron Lett.*, **1973**, 3373–3376; T. Inoue, T. Kaneda, S. Misumi, *Tetrahedron Lett.*, **1974**, 2969–2972; T. Kaneda, T. Inoue, Y. Yasufuku, S. Misumi, *Tetrahedron Lett.*, **1975**, 1543–1544; D. T. Longone, J. A. Gladysz, *Tetrahedron Lett.*, **1976**, 4559–4562.

204. J. Dohm, M. Nieger, K. Rissanen, F. Vögtle, *Chem. Ber.*, **1991**, *124*, 915–922.

205. R. Gleiter, M. Merger, A. Altreuther, H. Irngartinger, *J. Org. Chem.*, **1996**, *61*, 1946–1953.

206. M. W. Haenel, *Tetrahedron Lett.*, **1976**, 3121–3124; *cf.* M. W. Haenel, *Tetrahedron Lett.*,

1977, 1273–1276.

207. J. G. O´Conner, P. M. Keehn, *Tetrahedron Lett.*, **1977**, 3711–3714; *cf.* H. Horita, T. Otsubo, S. Misumi, *Chem. Lett.*, **1978**, *7*, 807–812.

208. Y. Fukazawa, M. Aoyagi, S. Ito, *Tetrahedron Lett.,* **1979**, 1055–1058. For the preparation of related azulenophanes see R. Luhowy, P. M. Keehn, *Tetrahedron Lett.*, **1976**, 1043–1046; N. Kato, Y. Fukazawa, S. Ito, *Tetrahedron Lett.*, **1976**, 2045–2048.

209. Isomers of **292** have also been synthesized: S. Ito, N. Kato, H. Matsunaga, S. Oeda, Y. Fukazawa, *Tetrahedron Lett.*, **1979**, 2419–2422; K. Rudolf, T. Koenig, *Tetrahedron Lett.*, **1985**, *26*, 4835–4838; *cf.* T. Kawashima, T. Otsubo, Y. Sakata, S. Misumi, *Tetrahedron Lett.*, **1978**, 1063–1066.

210. M. Matsumoto, T. Otsubo, Y. Sakata, S. Misumi, *Tetrahedron Lett.*, **1977**, 4425–4428.

211. L. A. Paquette, M. A. Kesselmayer, *J. Am. Chem. Soc.*, **1990**, *112*, 1258–1259; *cf.* J. E. Garbe, V. Boekelheide, *J. Am. Chem. Soc.*, **1983**, *105*, 7384–7388. For the synthesis of [*n*]cyclooctatetraeneophanes see L. A. Paquette, M. P. Trova, *Tetrahedron Lett.,* **1987**, *28*, 2795–2798; *cf.* T.-Z. Wang, L. A. Paquette, *Tetrahedron Lett.*, **1988**, *29*, 41–44. For a recent review see L. A. Paquette in *Advances in Theoretically Interesting Molecules*, R. P. Thummel, (*Ed.*), JAI Press, Inc. Greenwich, CT, **1992**, Vol. *2*, 1–77.

212. Y. Sakamoto, N. Miyoshi, T. Shinmyozu, *Angew. Chem.*, **1996**, *108*, 585–586; *Angew. Chem. Int. Ed. Engl.*, **1996**, *35*, 549–550; *cf.* Y. Sakamoto, N. Miyoshi, M. Hirakida, S. Kusumoto, H. Kawase, J. M. Rudzinski, T. Shinmyozu, *J. Am. Chem. Soc.*, **1996**, *118*, 12267–12275; K. Hori, W. Sentou, T. Shinmyozu, *Tetrahedron Lett.*, **1997**, *38*, 8955–8958.

213. M. Takeshita, M. Koike, H. Tsuzuki, M. Tashiro, *J. Org. Chem.*, **1992**, *57*, 4654–4658.

214. R. Gleiter, M. Karcher, M. L. Ziegler, B. Nuber, *Tetrahedron Lett.*, **1987**, *28*, 195–198. For a review on superphanes see R. Gleiter, M. Karcher, *Acc. Chem. Res.*, **1993**, *26*, 311–318; *cf.* R. Gleiter, M. Karcher, *Angew. Chem.*, **1988**, *100*, 851–852; *Angew. Chem. Int. Ed. Engl.*, **1988**, *27*, 840–841 for the demetalation of this complex.

215. M. Hisatome, J. Watanabe, K. Yamakawa, Y. Iitaka, *J. Am. Chem. Soc.*, **1986**, *108*, 1333–1334.

216. Review: H. Schwarz, *Angew. Chem.*, **1992**, *104*, 301–305; *Angew. Chem. Int. Ed. Engl.*, **1992**, *31*, 293–297.

217. X. Qiao, D. M. Ho, R. A. Pascal Jr, *Angew. Chem.*, **1997**, *109*, 1588–1589; *Angew. Chem. Int. Ed. Engl.*, **1997**, *36*, 1531–1532.

13 Three-dimensional Oligoolefins

The multi-layered cyclophanes presented in Section 12.3 are important model compounds for the investigation of electronic interactions over comparatively large distances in hydrocarbons in which the π-systems are not—as in conjugated molecules such as the polyolefins (Section 7.1) or polyacetylenes (Section 8.1)— directly connected to each other by a single bond. As demonstrated by, *e.g.*, the UV/Vis spectra of the tetracyanoethylene (TCNE) complexes of the multi-layered cyclophanes, there is indeed overlap between all benzene decks, such that an electron deficiency in one aromatic subunit (the one which is complexed) is compensated by the facing benzene nucleus, which, in turn, receives electron density from the next benzene ring in line—all the way through the molecular stack.

In principle the benzene rings in these layered hydrocarbons can be replaced in part or completely by other π-systems, in particular by (isolated) double bonds. And, in fact, we have encountered a representative of such a layered alkene already with the diene **1**, the preparation and properties of which were discussed in Section 7.4.[1] In this molecule the two double bonds are precisely oriented in parallel arrangement by the four ethano bridges (Scheme 1).

Clearly, other such distinct arrangements of double bonds in three-dimensional space are possible—for example, an orthogonal arrangement as illustated in **2** or a

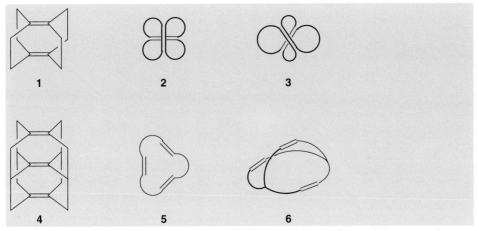

Scheme 1. Three-dimensional oligoenes—a selection of representative model compounds.

crossed structure as in **3**. Of course, all kinds of orientation are conceivable; we would, however, expect certain structures to have more interesting and unique chemical and structural properties than others. In all these cases the precise nature of the σ-skeleton, the scaffold which 'holds' the double bonds, is obviously of utmost importance.

As always when we think and plan in terms of building blocks, the complexity of these three-dimensional di- and polyolefins also increases rapidly with the number of the repeated unit. Three double bonds, for example, can be oriented in a layered fashion again as in (the unknown) hydrocarbon **4**, or—simpler—in the mono- and bicyclic arrangements symbolized by **5** and **6**, respectively; the former has the more flexible arrangement of double bonds. Although we would not expect electronic interactions of any type in hydrocarbons such as **4** to **6** if the bridging elements are too long, the situation will probably change when the sites of unsaturation are separated by two or even one saturated carbon atom only. The latter situation was termed *homoconjugation* in a seminal paper by Hine et al. in 1955, in which the electronic properties of one of the most important three-dimensional trienes, bicyclo[2.2.2]octa-2,5,7-triene (**6**, with CH bridgeheads) were discussed for the first time;[2] a hydrocarbon which later became known as barrelene.

13.1 Barrelene

The first synthesis of bicyclo[2.2.2]octa-2,5,7-triene was accomplished in 1960 by Zimmerman and Paufler,[3] who also proposed the trivial name barrelene (**13**), because of the barrel-shaped array of its molecular orbitals (see below). A prerequisite for the synthesis was an efficient means of preparing α-pyrone (**8**) (*cf.* Section 10.1), an unsaturated lactone previously available only on a small scale (Scheme 2).

This was accomplished by passing coumalic acid (**7**, see Section 11.5) *in vacuo*

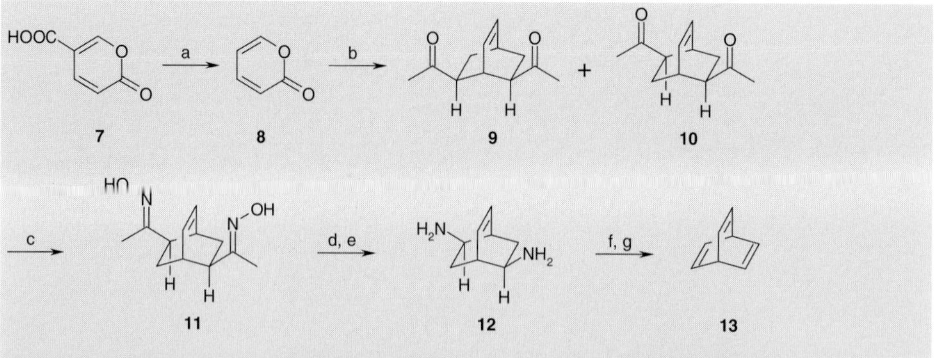

(a) Cu, 650°C (60-65%); (b) methyl vinyl ketone, 160°C (**10**: 18%); (c) NH₂OH, C₂H₅OH / H₂O (98%); (d) p-TsCl, dioxan, then NaHCO₃ / H₂O (28%); (e) NaOH, H₂O (97%); (f) CH₃I, CH₃OH, NaOH (100%); (g) Ag₂O, then Δ (76%)

Scheme 2. The first synthesis of barrelene (**13**) according to Zimmerman and Paufler.

over fine copper turnings at 650 °C. When the diene **8** was heated under reflux with methyl vinyl ketone in a decarboxylative domino Diels–Alder addition, a mixture of two diketones, **9** and **10**, was obtained. 5,7-Diacetyl-bicyclo[2.2.2]oct-2-ene (**10**) was isolated in 18 % yield. Treatment of **10** with hydroxylamine provided the bis oxime **11** from which the ditosylate was prepared; the latter was subsequently subjected to a solvolytic Beckmann rearrangement under alkaline conditions. The resulting bis amide was hydrolyzed to the diamine **12** which was converted to barrelene (**13**) in a typical twofold Hofmann elimination sequence. Although the yield of the overall sequence remained low even after optimization (in the few percent range), enough **13** could be prepared to study its structural and chemical properties

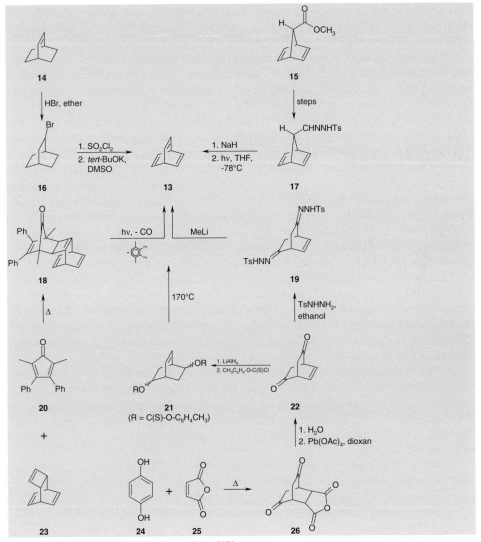

Scheme 3. Alternative routes to barrelene (**13**).

(bromine addition, hydrogenation, and, in particular, its photochemical behavior—see below).

Because this synthesis was rather lengthy, alternative approaches to this unique hydrocarbon were developed, and many shorter routes appeared in the chemical literature; a selection is summarized for comparison in Scheme 3.

Thus ionic addition of hydrogen bromide to bicyclo[2.2.2]oct-2-ene (**14**) led to 2-bromo bicyclo[2.2.2]octane (**16**, 94%), from which a complex mixtures of halides was obtained by radical chlorination with sulfuryl chloride-dibenzoylperoxide. Low-temperature dehydrohalogenation with potassium *tert*-butoxide then provided a hydrocarbon mixture from which **13** could be separated in approximately 2% overall yield.[4] Although this route is short, the brevity is paid for by very poor selectivity and hence yield. The second route—conversion of 7-methoxycarbonylnorbornadiene (**15**) into the tosylhydrazone of norbornadiene-7-carboxaldehyde and low-temperature photolysis of its sodium salt—furnished barrelene (**13**) in 3–4% yield, but also five other hydrocarbons, and **13** had to be separated by gas chromatography.[5]

A real improvement in yield was realized in the third route, in which Nenitzescu's hydrocarbon **23**—prepared from cyclooctatetraene (see Section 10.3)—was first reacted with 2,5-dimethyl-3,4-diphenylcyclopentadienone (**20**) to give a mixture of Diels–Alder adducts from which the polycyclic ketone **18** could be separated and submitted to ultraviolet irradiation in tetrahydrofuran at room temperature.[6] The yield of barrelene in this photodecarbonylation and photofragmentation (removal of 1,4-dimethyl-2,3-diphenylbenzene) was 50%, which translates into an overall yield of 24% from cyclooctatetraene. Because the latter has become rare and costly, however, alternatives were clearly desirable. One involves melting together hydroquinone (**24**) and maleic anhydride (**25**) at 200 °C for 2 h. This led to the Diels–Alder adduct **26**, which—after hydrolysis to the corresponding diacid—was decarboxylated by treatment with lead tetraacetate to give bicyclo[2.2.2]oct-2-ene-5,7-dione (**22**). To produce the two missing C–C double bonds, **22** was first transformed to the bis tosylhydrazone **19** which on treatment with methyllithium furnished a 9:1 mixture of benzene (resulting from the [4+2]cycloreversion of barrenyl lithium) and barrelene.[7] Although the overall yield of the whole sequence is low again (*ca* 1%), none of the compounds involved is expensive or difficult to obtain. Alternatively the diketone **22** has been reduced to its diol and the latter been converted to the thionocarbonate **21**. When this was thermolyzed at 170 °C barrelene (**13**) was generated in approximetely 20% yield.[8]

All of these methods still have more or less serious drawbacks, including low yields, difficult separations (often by gas chromatography), and the need for special substrates and experimental techniques *etc*. A straightforward, simple synthesis was hence mandatory, should barrelene ever become a 'readily available hydrocarbon'. The solution to the problem, provided by de Lucchi and co–workers in 1997, is shown in Scheme 4.[9]

In this synthesis the complete carbon framework of **13** was produced in the first step. It makes use of a Diels–Alder addition between oxepin (**27**), which is in equilibrium with its valence isomer benzene oxide (**28**) as a diene component and 1,2-bis(phenylsulfonyl)ethene (**29**) as a dienophile. In other words, in this [2+4]cyclo-addition **28** serves as an equivalent for an 'activated benzene', whereas **29** is an activated acetylene equivalent, thus formally reducing the barrelene synthesis to a

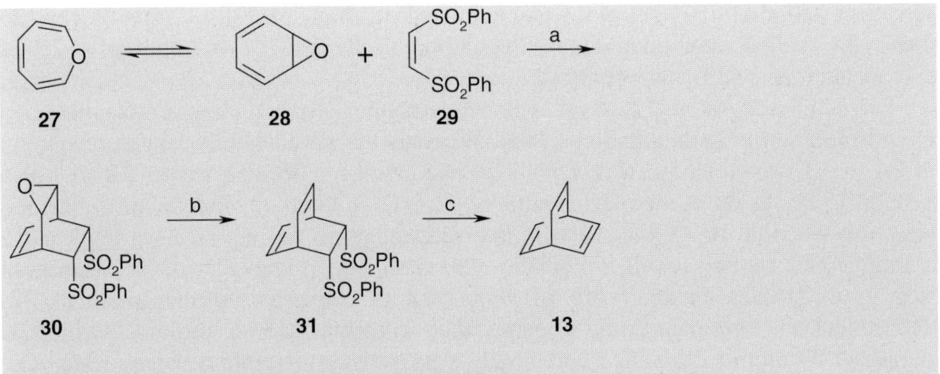

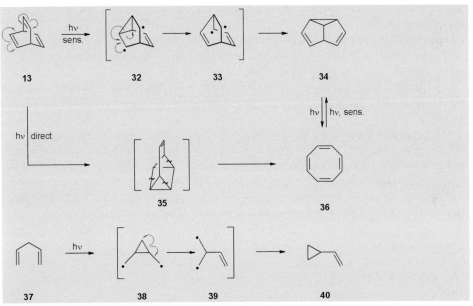

(a) benzene, 48 h, 80°C (96%); (b) WCl₆, *n*-BuLi, THF, 24 h, room temp. (90%); (c) Na / Hg, buffered CH₃OH (90%)

Scheme 4. De Lucchi's synthesis of barrelene (**13**).

Scheme 5. The first preparation of semibullvalene (**34**) from barrelene by Zimmerman and Grunewald.

1,4-addition of acetylene to benzene. The obtained Diels–Alder adduct, **30**, was subsequently deoxygenated by treatment with tungsten hexachloride and butyllithium, and the resulting **31** cleaved to barrelene with sodium amalgam. The overall yield of this approach, which only uses rapidly prepared starting materials and intermediates, is a stunning 75%!

Among the reactions of barrelene (**13**), its photochemical isomerization had one of the largest impacts on hydrocarbon chemistry and on photochemistry in general. Photolysis of 1–2% solutions of **13** in isopentane containing 3-8% of acetone as a sensitizer afforded 25–40% of an isomeric hydrocarbon, tricyclo[3.3.0.0^{2,8}]octa-3,6-diene (**34**), called semibullvalene by the authors (Scheme 5).[10] In addition, 1–2% of cyclooctatetraene (**36**) was produced.

Direct photolysis of **13** gave cyclooctatetraene (**36**), which was also obtained upon irradiation of semibullvalene (**34**). Whereas the photochemical interconversion of **34** to **13** was not observed, sensitized irradiation of **36** also led to **34** (see also Section 13.3). As far as the mechanisms of these C$_8$H$_8$ interconversions are concerned (see also Section 10.3), the **13**→**34** rearrangement is the classical example of a rearrangement termed the di-π-methane rearrangement. It begins with a 1,3-bridging step in the triplet excited state in which **13** is converted to the diradical **32**. Subsequent cyclopropane ring opening then furnishes **33**, a diradical which is stabilized by allylic resonance and finally 1,3-cyclizes to semibullvalene (**34**). The direct conversion of **13** to cyclooctatetraene probably involves the intramolecular [2+2]cycloadduct **35**, which on ring-opening (wavy lines) furnishes cyclooctatetraene (**36**). The di-π-methane (or divinylmethane) rearrangement, in which a 'skipped diene', **37**, is photochemically isomerized to a vinylcyclopropane **40**, *via* the intermediate diradicals **38** and **39**, is one of the most important and general of all photoprocesses.[11, 13]

Photoisomerizations of the barrelene-to-semibullvalene type are quite general; three examples from the area of pure hydrocarbons involve benzobarrelene,[14] naphthobarrelene,[15] and anthrobarrelene.[16]

The most distinct property of semibullvalene is its 'fluxional behavior'[10] which will be discussed—together with the presentation of more recent routes to **34**—in Section 13.3.

Theoretically barrelene (**13**) has been discussed extensively as an example of a Möbius-like molecule.[2, 17] As shown in Scheme 6, in which for reasons of clarity only one half of **13** is drawn, homoconjugative overlap between the p-orbitals of the double bonds leads to a situation in which the requirement of like algebraic signs of overlapping lobes of p-orbitals cannot wholly be satisfied. The resulting sign inversion in a cyclic array of p-orbitals is typical of Möbius-type hydrocarbons.[17]

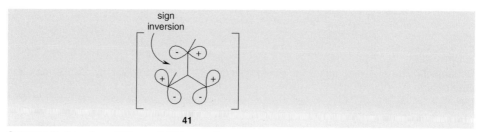

Scheme 6. Barrelene (**13**) as a Möbius-system.

It has been concluded from measurements of heats of hydrogenation that although the obtained data do not rigorously exclude the possibility of ground-state delocalization in barrelene, this effect, if present at all, is swamped by strain effects. The three-dimensional 6π-system **13** can hence not be compared with the planar and conjugated 6π system benzene.[18] Although quantum-mechanical computations show that the three double bonds in **13** do interact with typical Möbius splitting (two MOs decreasing in energy and one increasing), because the decrease in energy of the lowest two bonding MOs is exactly counterbalanced by an increase of the third bonding MO, there is no homoaromaticity in barrelene. In fact, the hydrocarbon has been described as 'non-aromatic'.[19]

13.2 Triptycene and the iptycenes

Replacement of all three double bonds in barrelene (**13**) by benzene rings leads to a hydrocarbon which has been called triptycene (**50**), the trivial name for 9,10-[1',2']benzeno-9,10-dihydroanthracene or tribenzobicyclo[2.2.2]octatriene.[20]

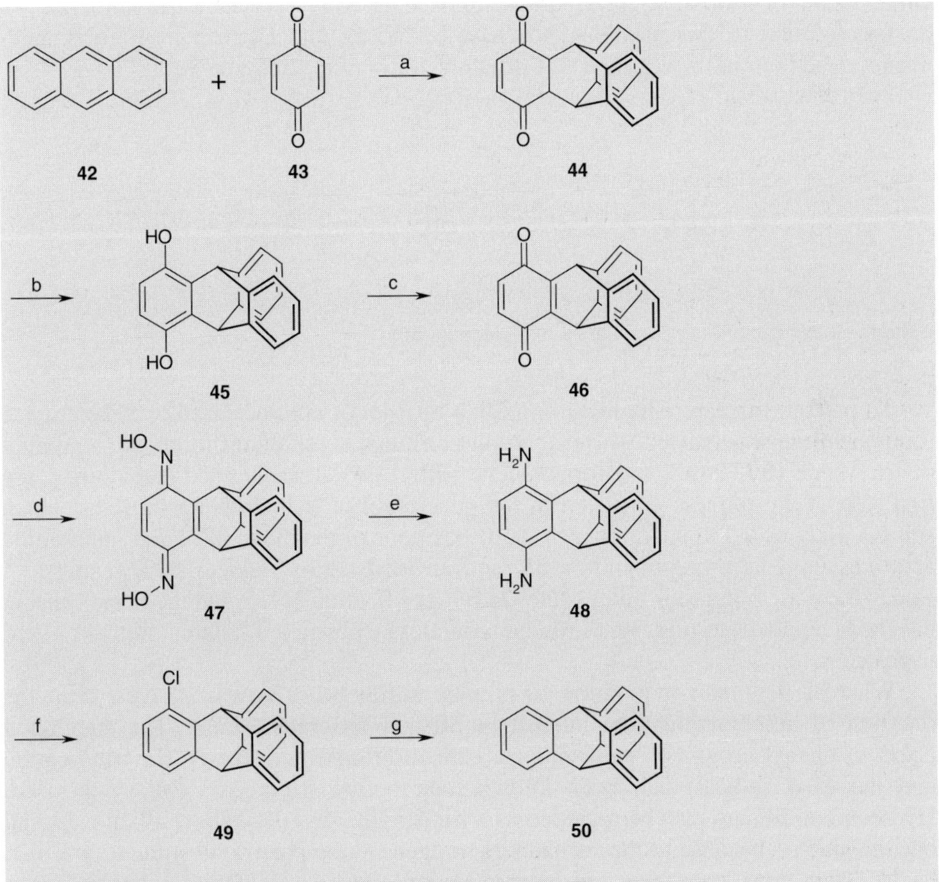

(a) xylene, refl. (83%); (b) HOAc, HBr (90%); (c) KBrO$_3$, HOAc (93%);
(d) NH$_2$OH, EtOH (78%); (e) SnCl$_2$, HCl, EtOH (86%); (f) NaNO$_2$, HCl / H$_2$SO$_4$,
then NaHPO$_2$, HCl; (g) Pd / C, H$_2$

Scheme 7. The first synthesis of triptycene (**50**) according to Bartlett and co–workers

On the basis of earlier studies by Clar, who was the first to prepare several derivatives of triptycene,[21] the parent hydrocarbon **50** was synthesized by Bartlett and co–workers in 1942 as shown in Scheme 7.[20]

Cycloaddition of *p*-benzoquinone (**43**) to anthracene (**42**) in xylene provided the 9,10-adduct **44**, which is already a tautomer of a triptycene derivative—the hydroquinone **45**, formed from **44** by acid-catalyzed isomerization as already observed by Clar.[21] Oxidation to **46** with potassium bromate, formation of the bis oxime **47** with hydroxylamine and stannous chloride reduction then led to the bis amine **48**. The final reduction to triptycene (**50**) turned out to be difficult, but was finally accomplished by catalytic hydrogenation of the monochloride **49**, which was formed in admixture with its dichloride when **48** was diazotized and reduced.[22] Although exploratory reactions could be undertaken with **50**, a general investigation of the chemical behavior of the triptycenes had to await a far simpler synthesis—the usual 'hydrocarbon story' which we have come across so many times in this volume already (compare adamantane (Section 3.1), cyclooctatetraene (Section 10.3), corannulene (Section 12.2), [2.2]paracyclophane (Section 12.4) and many others).

The solution to this problem, provided by Wittig and Ludwig in 1956 remains unsurpassed both in its elegance and structural variability (Scheme 8).[23, 24]

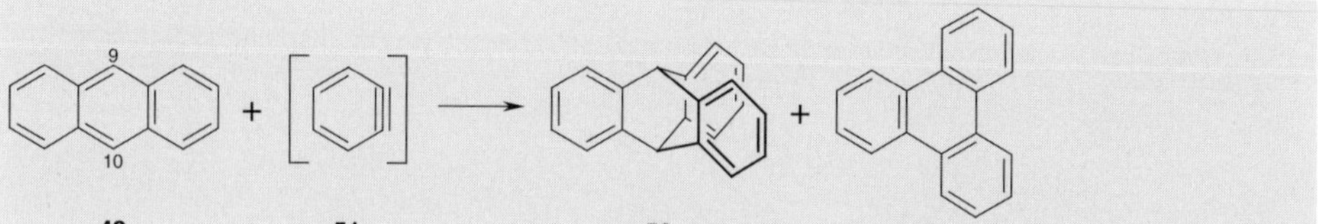

Scheme 8. Wittig's one-step synthesis of triptycene (**50**).

To prepare the parent hydrocarbon **50**, a mixture of anthracene (**42**) and *o*-fluoro-bromobenzene was reacted with magnesium turnings in tetrahydrofuran and a mixture of triptycene (**50**, 28%) and triphenylene (**52**, 11%) was obtained! Obviously, the first step of the process consisted in the generation of dehydrobenzene (**51**), which subsequently added to the (activated) 9,10-positions of the diene **42**. Later, anthranilic acid was used as a precursor for **51**, and an increase in yield to 60% realized.[25] Today there are numerous other routes to **51** (see Section 16.1), and the basic concept has been applied to the synthesis of countless substituted and/or functionalized triptycenes.

Whereas the main motivation to prepare barrelene (**13**) was derived from the question of homoconjugation, interest in **50** had different origins. The stabilizing effect of phenyl groups on carbocations, carbanions and radicals (*cf.* the trityl cation and the trityl radical) had been known long before triptycene was synthesized. Hydrocarbon **50** can also be regarded as a triphenylmethyl derivative, albeit a special one, in which—because of the second CH-bridgehead the phenyl substituents are 'tied back'. Thus their π-systems are kept in an orthogonal orientation relative to any developing p-orbital on one of the bridgeheads, and no overlap is possible. The derived reactive intermediate, once generated, will thus have very different properties from those observed for the unbridged variants. This has been validated by experimental observation. Furthermore, a triptycene carrying an appropriate leaving group in the 9 position (the bridgehead) will show neither typical S_N1 nor S_N2-behavior—a bridgehead carbocation cannot flatten, and the usual backside approach of a nucleophile is blocked in a triptycene.[26, 27] In addition, the rigid structure of the triptycene nucleus with its three *ortho* hydrogen atoms pointing exactly in the direction of any bridgehead C–X bond, has made derivatives of **50** ideal model compounds for the study of phenomena of hindered rotation around C–C single bonds.[28]

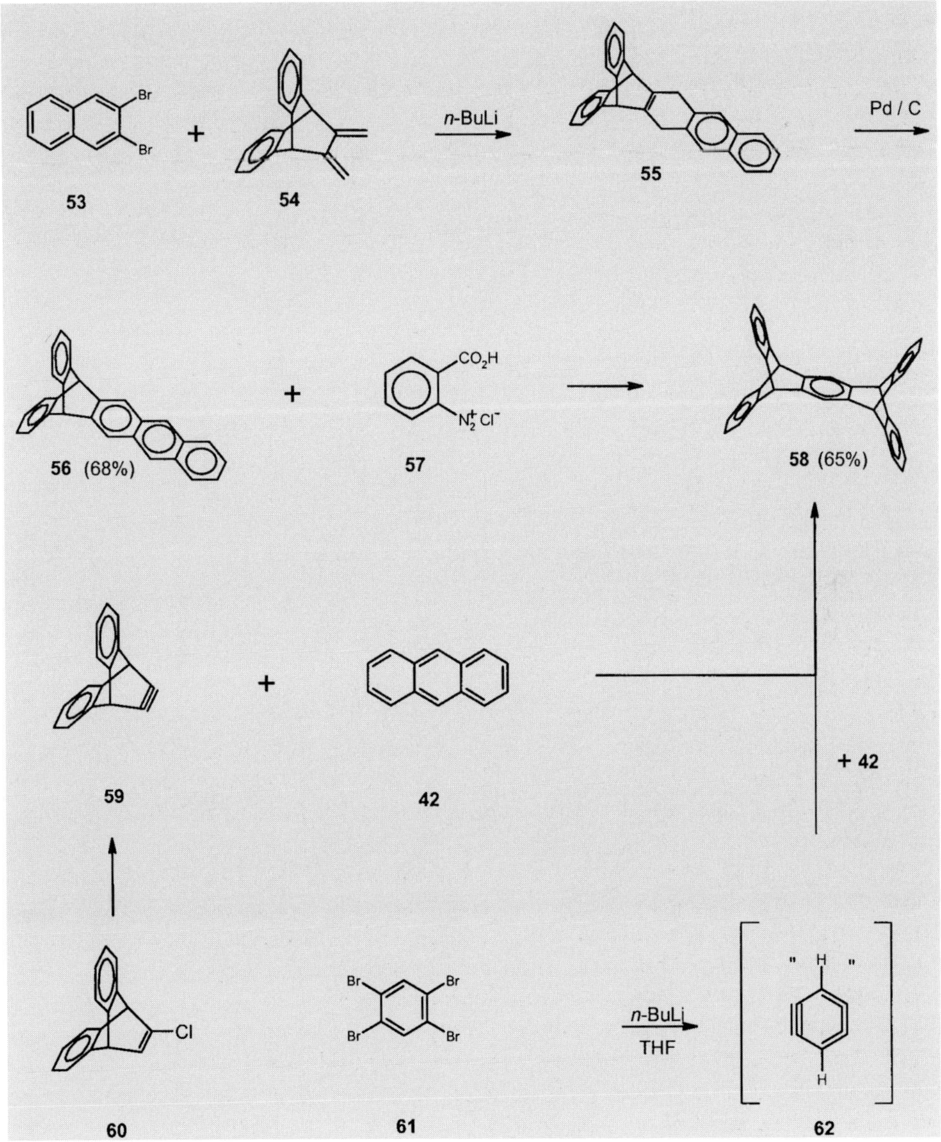

Scheme 9. Dehydrobenzene intermediates in the synthesis of pentiptycene (**58**).

Just as layered paracyclophanes have been constructed from the
[2.2]paracyclophane building block, extended three-dimensional benzannelated
hydrocarbons incorporating several triptycene units, the so-called iptycenes, a term
coined by Hart,[29] have been synthesized from derivatives of **50**. A typical example is

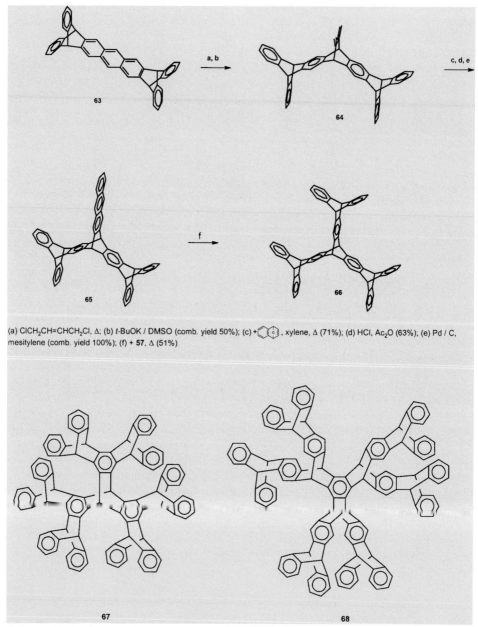

(a) ClCH$_2$CH=CHCH$_2$Cl, Δ; (b) *t*-BuOK / DMSO (comb. yield 50%); (c) + ⌬ , xylene, Δ (71%); (d) HCl, Ac$_2$O (63%); (e) Pd / C,
mesitylene (comb. yield 100%); (f) + **57**, Δ (51%)

Scheme 10. A selection of higher iptycenes—hydrocarbons with a dendritic structure.

the synthesis of pentiptycene (**58**), a hydrocarbon in which two triptycene units share a common (central) benzene ring (Scheme 9).[30]

Diels–Alder addition of 2,3-dehydronaphthalene, generated from the dibromide **53** by treatment with butyllithium, to the anthracene-derived diene **54** provided the adduct **55**, which was dehydrogenated in good yield to the naphthannelated triptycene **56**. To prepare the target molecule **58**, a second aryne addition was employed, the [2+4]cycloaddition of dehydrobenzene, generated from its 'storage form' **57**. This route to **58** resulted in good overall yield again and compares favorably with several alternatives. Previously **58** had been prepared in poorer yield by the cycloaddition of the strained and hence highly reactive acetylene **59** (see Section 8.2)—generated *in situ* from the vinyl chloride **60**—to anthracene (**42**),[31] and by a much shorter, higher yielding route from 1,2,4,5-tetrabromobenzene (**61**), **42**, and butyllithium in tetrahydrofuran.[29] Although the last transformation certainly takes place stepwise, the tetrabromide **61** formally behaves as a diaryne equivalent (symbolized by **62** in Scheme 9).

By judicious selection of various dehydroaromatic and diene intermediates the above approach has been extended to furnish still higher branched iptycenes, hydrocarbons with growing dendritic character.[32] Thus tritriptycene (**66**), a nonaiptycene is available by first converting the pentiptycene **63** to the diene **64**, generation of a new anthracene unit in one of the growing arms of the space-station type molecular system, and finally addition of dehydrobenzene to the most reactive part of **65** (Scheme 10).[33]

Still more branching is present in supertriptycene (**67**), a $C_{104}H_{62}$ hydrocarbon with D_{3h} symmetry[34] and the nonadecaiptycene **68**, the largest ($C_{132}H_{78}$) monomeric iptycene synthesized to date.[35] The graphical representation of these complicated hydrocarbons becomes increasingly difficult. Structures such as **67** and **68** which certainly are graphically unsatisfactory, still show the connectivity of the carbon atoms very well. If an attempt is made to draw these structures in similar geometric quality as, *e.g.*, **65** and **66** the corresponding representations become very difficult to decipher.

Among the interesting properties of the iptycenes their extremely high melting points and thermal stability must be mentioned; supertriptycene (**67**), for example, loses weight only on thermogravimetric analysis above 580 °C. Another interesting feature of these hydrocarbons is their molecular cavities; **67** has three large cavities, lined by benzene rings, and large enough to enclose solvent molecules. Thus the iptycenes could be useful in the formation of novel host–guest complexes. Finally, the chiroptical properties of these star-shaped, non-planar aromatic compounds warrant closer scrutiny; some chiral iptycenes have already been prepared.[30]

13.3 Bullvalene and semibullvalene

In Chapter 2 we established generic relationships between hydrocarbon reactions, and demonstrated how the Cope rearrangement can be 'incorporated' into numerous hydrocarbon systems more complex than 1,5-hexadiene, the simplest molecule capable of this valence isomerization. It was demonstrated that Cope processes can also be 'built' into three-dimensional olefinic structures, a particular important representative being tricyclo[3.3.2.0^{4,6}]deca-2,7,9-triene or bullvalene (**71**). The synthesis of this unique $C_{10}H_{10}$ hydrocarbon and, especially, its thermal isomerization has made an impact on organic chemistry which more than many other experiments deserves to be called a 'classic'.

Bullvalene (**71**) was prepared in what has been called a "characteristically spectacular stroke"[36] from cyclooctatetraene (**69**, COT, see Section 10.3) in a two-step process *via* the COT-dimer **70** by Schröder in 1963 (Scheme 11).[37, 38]

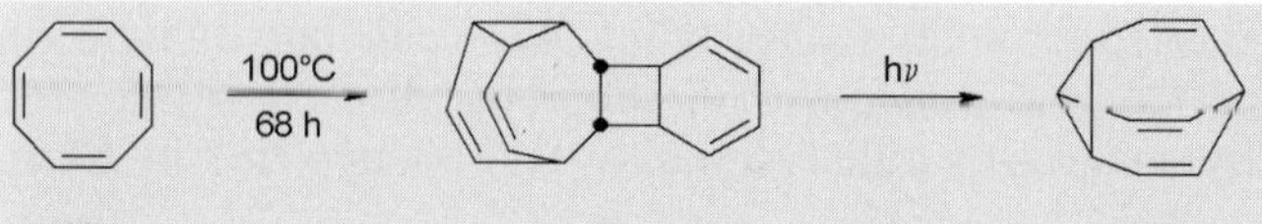

Scheme 11. The first synthesis of bullvalene (**71**) by Schröder...

Actually, when treated thermally COT dimerizes to at least four dimers, the exact structures of which were, for a long time, a subject of controversy.[39] Hydrocarbon **70** is the '76°-dimer' (its melting point is 76 °C) which is available on a multi-gram scale from **69** in *ca* 20 % yield (relative to reacted **69**).[40] UV-irradiation of **70** in ether then led to bullvalene, produced in *ca* 75 % yield. In a typical run 10–12 g of **71** can be prepared from 100 g of **69**. Schröder's synthesis stands out not only for its 'brilliance'[41] but also because it remains one of the very few first syntheses of an

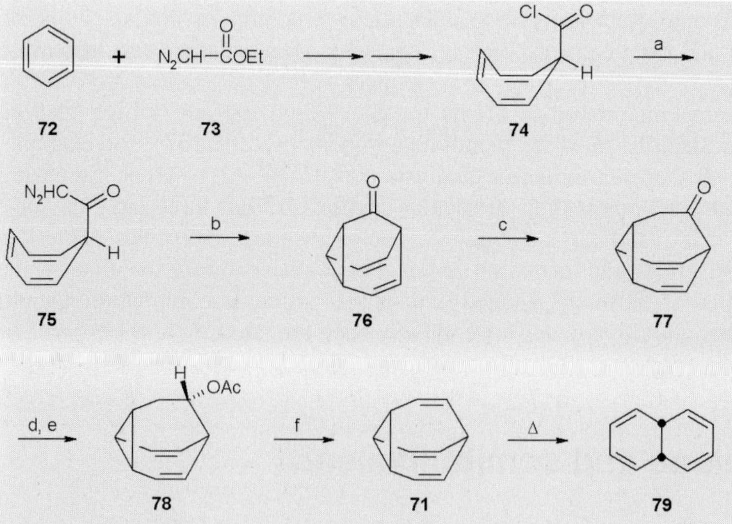

(a) CH$_2$N$_2$, 0°C; (b) CuSO$_4$, benzene, hexane, reflux; (c) CH$_2$N$_2$, ether (25%); (d) NaBH$_4$, ethanol (60%); (e) Ac$_2$O, pyridine (80%); (f) 345°C

Scheme 12. ...and Doering's rational synthesis of **71**.

important hydrocarbon which has not subsequently been surpassed.

This is demonstrated, for example, by a second, a 'rational route' to **71** published by Doering and co–workers in 1967 (Scheme 12).[36]

The sequence first proceeded from benzene (**72**) and ethyl diazoacetate (**73**)[42] to cycloheptatrien-7-ylcarbonyl chloride (**74**) as the decisive synthetic intermediate. Conversion to the diazomethyl ketone **75** was accomplished next and **75**, on reaction with copper powder or cuprous ion provided tricyclo[3.3.1.0^{2,8}]nona-3,6-dien-9-one (**76**, barbaralone), a compound to which I shall return below in the context of 'fluxional molecules'. Classical ring enlargement with diazomethane transformed **76** into bullvalone (**77**), which on hydride reduction and esterification furnished the acetate **78**. Ester pyrolysis in a flow system at 345 °C then concluded this synthesis of bullvalene. Note that the elimination step is accompanied by the generation of another $(CH)_{10}$ hydrocarbon, *cis*-9,10-dihydronaphthalene (**79**)[43] (see below).

The last directed bullvalene synthesis to be presented here, that of Serratosa and co–workers,[44] follows a completely different strategy, and although it is no competitor whatsoever to the Schröder route, it is worthy of evaluation (Scheme 13).[44]

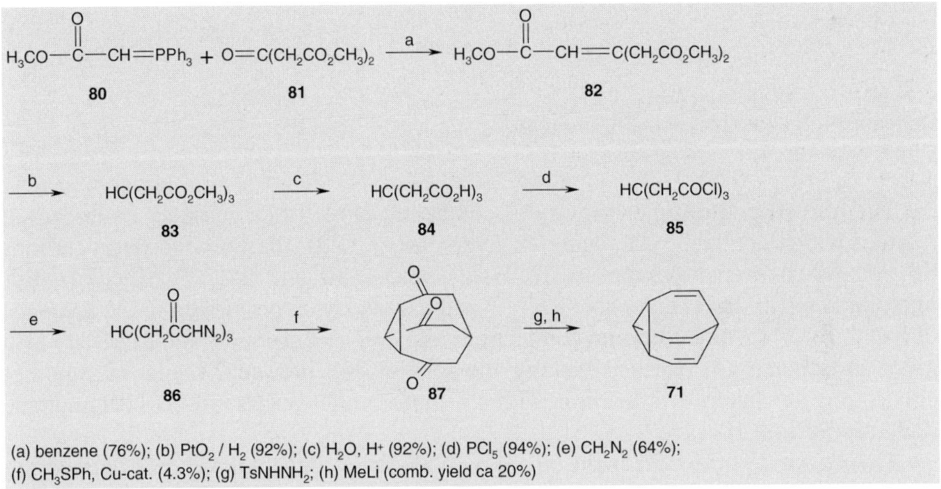

H$_3$CO—C(=O)—CH=PPh$_3$ + O=C(CH$_2$CO$_2$CH$_3$)$_2$ —a→ H$_3$CO—C(=O)—CH=C(CH$_2$CO$_2$CH$_3$)$_2$

80   **81**   **82**

—b→ HC(CH$_2$CO$_2$CH$_3$)$_3$ —c→ HC(CH$_2$CO$_2$H)$_3$ —d→ HC(CH$_2$COCl)$_3$

83   **84**   **85**

—e→ HC(CH$_2$C(=O)CHN$_2$)$_3$ —f→ **87** —g, h→ **71**

86

(a) benzene (76%); (b) PtO$_2$ / H$_2$ (92%); (c) H$_2$O, H$^+$ (92%); (d) PCl$_5$ (94%); (e) CH$_2$N$_2$ (64%); (f) CH$_3$SPh, Cu-cat. (4.3%); (g) TsNHNH$_2$; (h) MeLi (comb. yield ca 20%)

Scheme 13. Serratosa's route to bullvalene (**71**).

Its main difference from the other two approaches lies in the very late generation of the cyclopropane ring. Although the first steps, Wittig reaction between **80** and **81** to provide **82**, catalytic hydrogenation of the latter to **83**, which was subsequently converted to the tris-diazoketone **86** *via* **84** and **85** by routine steps, took place in good to excellent yields, the bottleneck of this alternative was the copper-catalyzed thermal decomposition of **86** to the triketone **87**. Although high yields would not be expected for this step (probably a two-step process involving a monocyclic enedione intermediate) the observed 4% is disappointing. Triple 'dehydration' *via* the tristosylhydrazone of **87** and its treatment with methyllithium then yielded **71** in approximately 20% yield.

It has already been pointed out that during its thermal generation from **78** bullvalene evidently isomerized to the $C_{10}H_{10}$ hydrocarbon 9,10-dihydronaphthalene (**79**). In fact, it is within the context of thermal and photochemical isomerizations of $(CH)_{10}$ isomers (and reactive intermediates such as carbenes[45] of the same composition) that one comes across the production of bullvalene (**71**) most often.[41] In Section 10.4 a part of a very extensive and complex 'map' of '$(CH)_{10}$ land' has already been discussed. Here this map is enlarged, yet still without attempting to cover all its aspects (Scheme 14) (see Section 13.4).

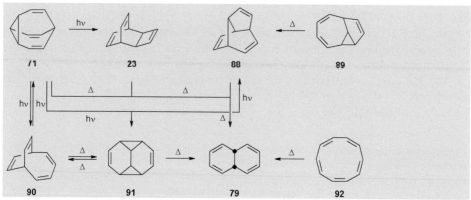

Scheme 14. Another part of the $(CH)_{10}$ map.

The (direct) photoisomerization of bullvalene (**71**) led to a complex hydrocarbon mixture which contained Nenitzescu's hydrocarbon (**23**), the tricyclic hydrocarbons **88** and **89**, and two symmetrical $(CH)_{10}$ isomers, **90** and **91**.[41,46] The **71**→**90**-photointerconversion is reversible[45, 47, 48] and has also been performed in the presence of metal salts.[49] Some 'thermal connections' as well as a second route to **88** are also given in Scheme 14, one establishing the relationship betwen **71** and **79**, another connecting the latter hydrocarbon with its monocyclic isomer all-*cis*-[10]annulene (**92**) (see Section 10.4).[50]

The *raison d'être* for **71** is the so-called bullvalene concept developed by Doering in the early 1960s[51, 52] which deals with fast and reversible valence isomerizations and for which bullvalene was considered to be the ideal model and test compound. We have seen already in the introductory Chapter 2 how a generic relationship can be established between the Cope rearrangement of 1,5-hexadiene and **71**, with *cis*-1,2-divinylcyclopropane, cycloheptatriene (tropylidene), and homotropylidene serving as intermediate 'generations'. As illustrated in Scheme 15, subjecting bullvalene to a series of [3,3]sigmatropic rearrangements results in the complete interconversion of all three types of carbon and hydrogen atoms—saturated, cyclopropane and olefinic. For example, the seven steps (**93** to **100**) highlighted in the Scheme exchange the positions 6, 7, and 8.

In other words, if the Cope processes in **71** occur, no two carbon atoms of the molecule will remain bonded to each other for an extended period of time. The ten carbon atoms relentlessly change positions, but remain bonded to each other in a bullvalene skeleton—a situation unprecedented in organic chemistry! Altogether

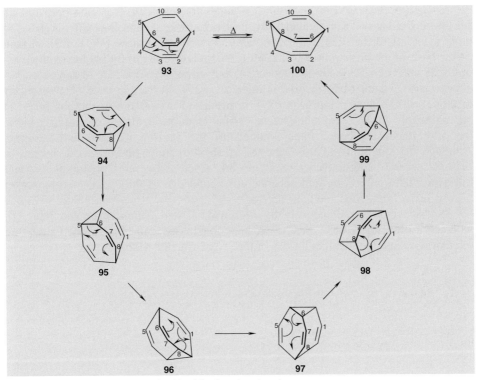

Scheme 15. Bullvalene (**71**)—the ideal fluxional molecule.

10!/3 = 1209600 structurally identical isomers exist. Immediately after bullvalene became available, Doering's predictions were borne out by experiment[37, 38]—the low-temperature (−85 °C) ^{1}H NMR spectrum of **71** contained two complex signal groups centered at δ = 6.65 (olefinic protons) and 2.08 (cyclopropyl and bridgehead protons). When the temperature was increased, coalescence was observed at 13 °C and at 100 °C one sharp singlet at δ = 4.22 was visible, confirming the predicted fluxional nature of **71**.[53] The rate constants and activation parameters for this unique process were determined as early as 1963 by ^{1}H NMR spectroscopy[54] and later by ^{13}C NMR analysis also.[55] The most often quoted activation parameters for this automerization are those of Oth and co-workers; in solution an Arrhenius activation energy E_a = 13.18 ± 0.07 kcal mol^{-1} and log A = 13.37 have been determined.[56] In the solid state, where the equilibration process can be investigated by ^{13}C MAS NMR measurements, these values are slightly higher (E_a = 14.5 kcal mol^{-1}, log A = 14.14).[57, 58] How these values compare with those obtained for other degenerate Cope processes will be seen upon discussion of the lower vinylog of bullvalene, semibullvalene.

Today, bullvalene (**71**) is not only a structurally thoroughly studied hydrocarbon,[59] it has become the starting material for numerous further transformations. In pure hydrocarbon chemistry its cyclopropanation has enabled the synthesis of many new cage hydrocarbons with interesting reactivity and stereochemical features.[60]

As already discussed in Section 13.1 semibullvalene (**34**) was first prepared by sensitized irradiation of barrelene (**13**); it also became evident how **34** is related to other members of the (CH)$_8$ family, notably to cyclooctatetraene (COT, **69**). In fact, on the basis of the observation[61] that **69** can be reconverted to **34** photochemically, a practical approach to semibullvalene (**34**) has been developed.[62] Whereas earlier experiments[61] were performed in solution at -60 °C in the presence of a sensitizer (acetone), in the newer approach COT is irradiated at 300 nm in the gas phase at approximately 70 °C.[62] Semibullvalene (**34**) is produced in almost quantitative yield, the only impurities being trace amounts of the starting material and benzene. Although the simplicity of this synthesis probably cannot be surpassed, I want to present several alternative routes to **34** since they are preparatively and mechanistically interesting and enhance our knowledge of the chemical behavior of polycyclic hydrocarbons and their derivatives.

As discovered simultaneously by Meinwald[63] and Zimmerman and their co-workers[64] tricyclo[3.3.0.0^{2,6}]octane (**101**) upon photochlorination afforded a mixture of chlorides, among them dichlorides of the gross structure **102** (Scheme 16).

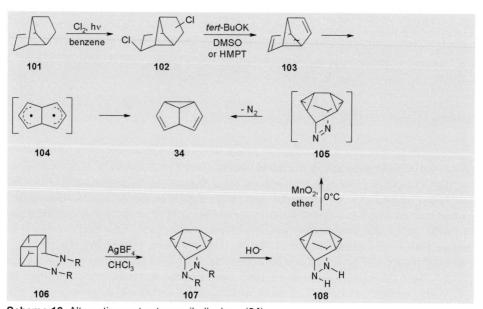

Scheme 16. Alternative routes to semibullvalene (**34**).

Dehydrochlorination of **102** under carefully controlled conditions at room temperature enabled the isolation and characterization of tricyclo[3.3.0.0^{2,6}]octa-3,7-diene (**103**), another (CH)$_8$ hydrocarbon,[64, 65] (see Section 10.3) which, however, at slightly elevated temperatures isomerized rapidly to **34** (estimated half-life of 1 h at 20 °C), most probably *via* the bis-allylic diradical **104** as a reaction intermediate. In another simultaneous discovery Askani[66] and Paquette[67] effected the COT→semibullvalene interconversion in a stepwise fashion, by metal-catalyzed isomerization. Diazabasketanes, **106**, readily available from cyclooctatetraene (**69**) and various azadienophiles (diethyl and dimethyl azodicarboxylate, *N*-phenyltriazolindione; R in **106**: CO$_2$CH$_3$, CO$_2$Et, -CO-N(Ph)-CO-), by silver(I)-ion-catalyzed rearrangement (see Section 10.3) provided the azasnoutanes **107** in good yields. These could be converted to semibullvalene (**34**) either by hydrolysis to the hydrazo derivative **108** and its subsequent oxidation with manganese dioxide (for R = CO$_2$CH$_3$ and CO$_2$Et)[66] or directly by saponification-decomposition of the urazole **107** (R–R = –CO–N(Ph)–CO) with potassium hydroxide in aqueous ethylene glycol at 100 °C.[67] In both reactions 9,10-diazasnoutene (**105**), a very unstable azo compound, is passe *en route*; it could, however, be isolated as a CuCl complex.[68] Although this route is longer than the direct **69**→**34**-photoisomerization (see above) it enables the preparation of gram amounts of semibullvalene.

Without any doubt the most interesting property of semibullvalene (**34**) is its fluxional behavior.[69] Like bullvalene (**71**) it can undergo a thermal isomerization,

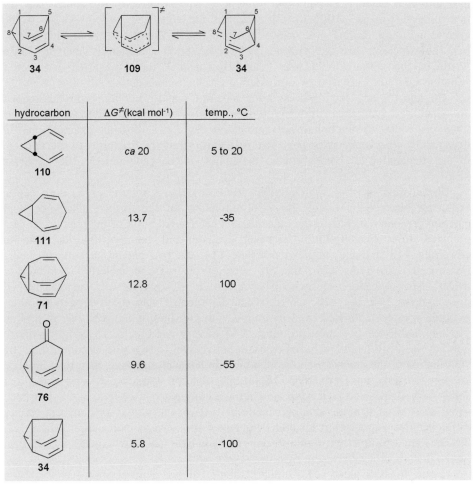

hydrocarbon	$\Delta G^{\neq}$(kcal mol^{-1})	temp., °C
110	*ca* 20	5 to 20
111	13.7	-35
71	12.8	100
76	9.6	-55
34	5.8	-100

Scheme 17. The automerization of semibullvalene (**34**) and other Cope systems.

which in the absence of a label is degenerate, *i.e.* it can automerize (Scheme 17). The free enthalpy of activation for this process, **34** ⇌ **34**, which occurs *via* the transition structure **109**, has been determined several times,[70] the value given in Scheme 17 is from a study by dynamic ^{13}C NMR spectroscopy.[71]

Comparison of the extremely low value of less than 6 kcal mol^{-1} with the $\Delta G^{\ddagger}$ values of Cope rearrangements in other hydrocarbons containing a *cis*-1,2-divinylcyclopropane moiety reveals an interesting trend. Whereas the parent hydrocarbon, *cis*-1,2-divinylcyclopropane (**110**), isomerizes (irreversibly to 1,4-cycloheptadiene) with a free enthalpy of activation of *ca* 20 kcal mol^{-1},[72] connecting its vinyl ends by a methylene group to homotropylidene (**111**), one of the simplest fluxional molecules, results in a drastic reduction of the activation energy.[73] Further bridging of the methylene groups in **111** by a double bond (bullvalene, **71**), a keto function (barbaralone, **76**),[74] and finally a 'zero-bridge' (semibullvalene, **34**) causes the shown reduction of $\Delta G^{\ddagger}$ and raises the question whether a still further decrease is possible—even to the extent that $\Delta G^{\ddagger}$ would eventually become negative. In such a case the two valence isomers **34** in Scheme 17 would be less stable than the delocalized structure **109**. In other words the localized structures **34** would correspond to the Kekulé structures of benzene, and **109** to the 'real', *i.e.* delocalized benzene structure. Just as benzene is aromatic, such a semibullvalene derivative would be 'homoaromatic'. Dewar[75] and Hoffmann[76] have suggested that such a situation might be achieved by introduction of electron-accepting substituents at C2, C4, C6, and C8 and electron-donating groups at C1 and C5. This proposal triggered an—ongoing—synthetic effort to produce this pattern of substituents. From the numerous routes to functionalized semibullvalenes[67, 77–83] just one example, that of Quast and co-workers, can be presented here to illustrate the feasibility of this concept in the preparation of homoaromatic molecules of the general type **109** (Scheme 18).[84, 85]

Bromination of *cis*-1,5-dimethylbicyclo[3.3.0]octan-2,6-dione (**112**) with cupric bromide, followed by double dehydrobromination of the formed dibromide with calcium carbonate furnished the dienedione **113**. 1,4-Addition of the cuprate reagent prepared from phenyllithium, cuprous cyanide and boron trifluoride etherate converted **113** into the saturated diketone **114**, the two phenyl substituents being introduced exclusively from the *exo* face of the starting material. Addition of trimethylsilyl cyanide subsequently yielded a 2:3-mixture of the *O*-silylated cyanohydrins **115** and **116**, which, when deprotected with the hydrogen fluoride-pyridine complex followed by dehydration with phosphorous oxychloride in boiling pyridine, led to the unsaturated dinitrile **117** in very good yield. The ring closure to 2,6-dicyano-1,5-dimethyl-4,8-diphenylsemibullvalene (**118**) was accomplished by chlorination of the anion formed from **117** with hexachloroethane, then base-induced dehydrohalogenation. Derivative **118** strongly absorbs visible light—it is a deep red solid—although it does not have an extended conjugated π-system, and it has been suggested[85] that it is the semibullvalene derivative with the as yet smallest energy difference between a localized and a delocalized structure, as indicated in Scheme 17. 2,6-Dicyano-1,5-dimethyl-semibullvalene has the extremely low barrier of 3.45 kcal mol^{-1} ($\Delta G^{\ddagger}$; –73 °C) for valence isomerization.[85]

Just as for bullvalene (**71**), semibullvalene valence isomerization has also been studied in the solid state.[86, 87]

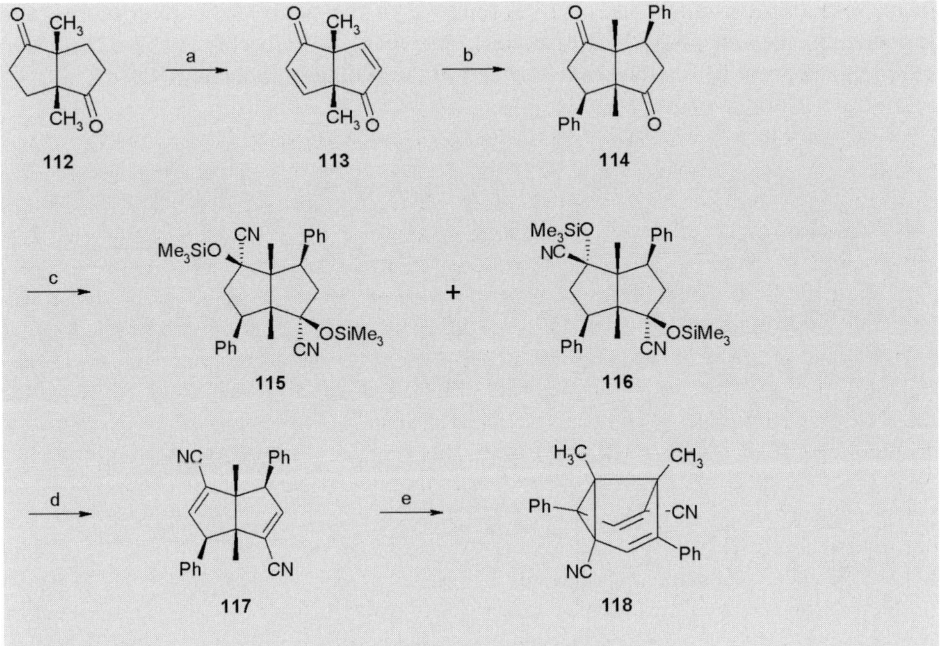

Compounds 112, 113, 114, 115, 116, 117, 118 with reagents:

(a) CuBr$_2$, CHCl$_3$ / EtOAc; CaCO$_3$, Δ (60-70%); (b) CuCN, PhLi, THF, BF$_3$·Et$_2$O, -78°C (70%); (c) KCN, 18-crown-6, TMSCl, 20°C (64%); (d) HF, pyridine, POCl$_3$, Δ (78%); (e) NaOH, CH$_2$Cl$_2$ / C$_2$Cl$_6$, Bu$_4$N$^+$ HO$^-$ (51%)

Scheme 18. Towards homoaromatic semibullvalenes.

13.4 Triquinacene

When Woodward and co–workers prepared triquinacene (**130**) for the first time in 1964, they stated several reasons why this hydrocarbon should be an interesting object of study. Its three double bonds, positioned in a rigid skeleton, should provide valuable information about the phenomenon of homoaromaticity; a study of the capacity of **130** to form metal complexes would be of interest—the hydrocarbon has a curved surface and should thus in principle be capable of binding metal or metal containing fragments endo- or exohedrally—and, finally, triquinacene could be an interesting precursor for other hydrocarbons, notably acepentalene (its tetradehydro derivative, see below) and dodecahedrane, its dimer. These suggestions turned out to be very stimulating and have, in fact, with the exception of the conversion of **130** to dodecahedrane (see Section 5.3) become a reality in the decades since the original publication.

Woodward's multi-step synthesis began with the alcohol **119**. This starting material already contained all carbon atoms of the final product, so no addition of any carbon-containing reagent was required. In fact, **119** has two more carbon atoms than the number eventually needed and thus the main objective of the synthesis consisted in reconnecting certain bonds of **119**, removal of the two extra carbon atoms, and introducing the missing double bonds. As seen in Scheme 19, the last two requirements were fulfilled towards the end of the synthesis (Scheme 19).[88]

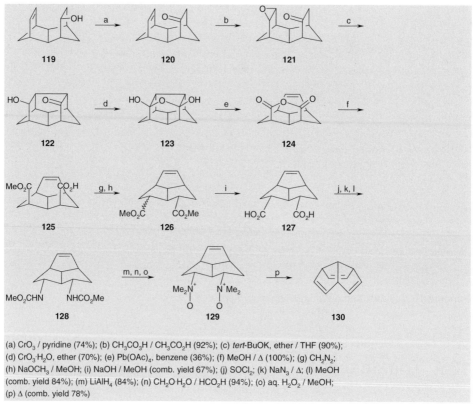

(a) CrO$_3$ / pyridine (74%); (b) CH$_3$CO$_3$H / CH$_3$CO$_2$H (92%); (c) *tert*-BuOK, ether / THF (90%); (d) CrO$_3$·H$_2$O, ether (70%); (e) Pb(OAc)$_4$, benzene (36%); (f) MeOH / Δ (100%); (g) CH$_2$N$_2$; (h) NaOCH$_3$ / MeOH; (i) NaOH / MeOH (comb. yield 67%); (j) SOCl$_2$; (k) NaN$_3$ / Δ; (l) MeOH (comb. yield 84%); (m) LiAlH$_4$ (84%); (n) CH$_2$O·H$_2$O / HCO$_2$H (94%); (o) aq. H$_2$O$_2$ / MeOH; (p) Δ (comb. yield 78%)

Scheme 19. The first synthesis of triquinacene (**130**) by Woodward and co–workers.

Oxidation of **119**, first with chromium trioxide and then by peracetic acid, transformed the alcohol, *via* the ketone **120**, into the epoxide **121**. This intermediate is set up for an intramolecular ring-closing/ring-opening process exploiting the acidity of its α-methylene group. The ketoalcohol **122**, resulting from treatment of **121** with base, contains a fifth five membered ring, the newly formed ring being that which will eventually be part of the triquinane skeleton. Having fulfilled their roles, the two C$_2$ bridges in **119-121** can now be dismantled. This was accomplished by oxidation of **122** to the dihydroxyether **123**, which, as a result of a fourth oxidation, suffered partial destruction of its σ-framework and conversion into the unsaturated anhydride **124**, in which the structure of the target molecule was beginning to emerge. The degradation of the anhydride ring—although simple in principle—turned out to be cumbersome in practice, because a stereochemical problem had to be solved. The carboxyl functions in **125**, obtained by methanolysis of **124**, are not in the proper orientation for further manipulation, which required the *exo,exo* orientation rather than the *endo,endo* configuration shown in **125**. Esterification of the latter and subsequent base-catalyzed epimerization provided a mixture of the *exo,exo*- and the *endo, exo* diesters **126**, from which the diacid **127** was prepared by hydrolysis with aqueous methanolic sodium hydroxide; the overall yield of the **125**→**127**-conversion was a comfortable 67%. The two missing double bonds were finally generated by a series of conventional steps. Curtius degradation of **127** led to the bis urethane **128**, which—after hydride reduction, Eschweiler-Clarke methylation of the resulting secondary amine, and oxidation—furnished the bis aminoxide **129**. Tricyclo[5.2.1.0^{4,10}] deca-2,5,8-triene (triquinacene, **130**) was finally obtained by a twofold Cope elimination.

Although other triquinacene syntheses were developed after Woodward's pioneering work[89–93] none of these was sufficiently effective preparatively to enable the chemistry of **130** to be investigated on a broader scale. This only changed when two routes were reported which not only start from readily available materials but were also significantly shorter than the Woodward synthesis and could be scaled-up to the multi-gram level.

The first of theses syntheses was accomplished by Deslongchamps and co–workers and is summarized in Scheme 20.[94, 95]

Starting from the so-called Thiele acid, **131**, easily prepared by carboxylation of sodium cyclopentadienide and again containing two extra carbon atoms and two of

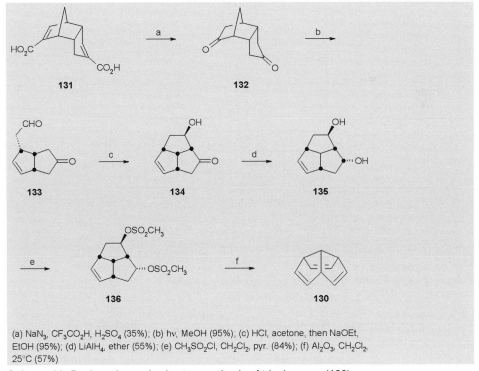

(a) NaN_3, CF_3CO_2H, H_2SO_4 (35%); (b) hv, MeOH (95%); (c) HCl, acetone, then NaOEt, EtOH (95%); (d) $LiAlH_4$, ether (55%); (e) CH_3SO_2Cl, CH_2Cl_2, pyr. (84%); (f) Al_2O_3, CH_2Cl_2, 25°C (57%)

Scheme 20. Deslongchamps's six-step synthesis of triquinacene (**130**).

the finally needed five-membered rings in the correct arrangement, the diketone **132** was first prepared by treatment of **131** with sodium azide in a mixture of sulfuric and trifluoroacetic acid. The next step, a photocleavage, is critical—as shown in Scheme 20 irradiation with ultraviolet light resulted in chemoselective cleavage and led solely to the ketone aldehyde **133**. Liberated from its fixed position in the tricyclic framework of **132** the aldehyde function is free to rotate into position for an intramolecular aldol reaction with the cyclopentanone ring. Indeed, the aldol cyclization step took place as expected and with excellent yield to provide the keto alcohol **134**, with the complete σ-skeleton of **130**. The remaining steps were more or less predetermined and involved reduction of **134** to **135**, conversion of the diol to the bis mesylate **136**, and finally twofold β-elimination to triquinacene (**130**).

The second preparatively high-yielding route to **130**, that of Cook and co–workers (Scheme 21),[96,97] also impresses with its elegance. Its first step makes use of the Weiss reaction,[98] a means of assembling the bicyclo[3.3.0]octane skeleton which wehave already encountered in the section on fenestranes (Section 6.2), and, currently, probably the best way of preparing oligoquinanes in numerous modifications.[99]

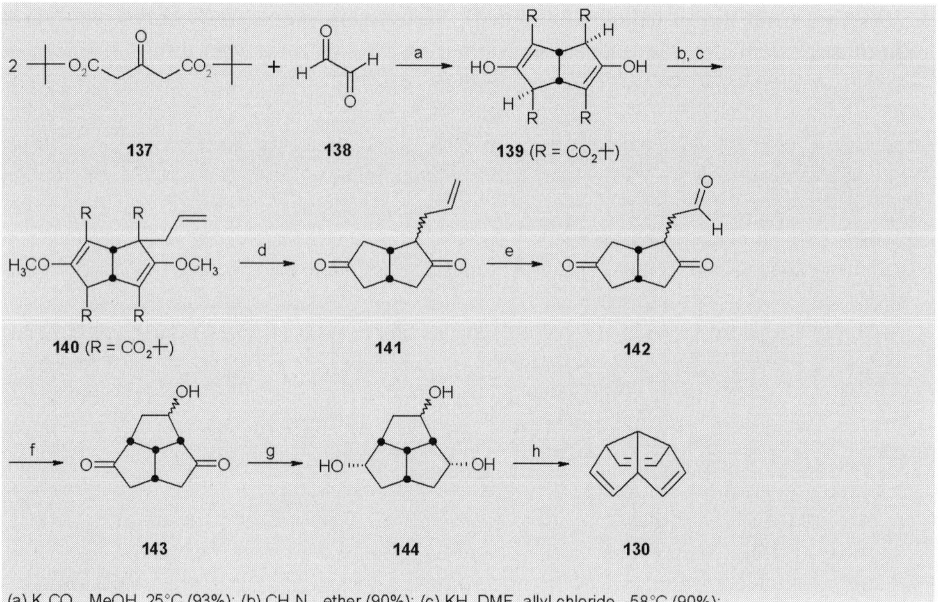

(a) K$_2$CO$_3$, MeOH, 25°C (93%); (b) CH$_2$N$_2$, ether (90%); (c) KH, DMF, allyl chloride, -58°C (90%); (d) HCl, HOAc, 90°C (90%); (e) O$_3$, EtOAc, -60°C (81%); (f) THF, HCl, 25°C (85%); (g) BH$_3$ / THF (93%); (h) HMPA, refl. (80%)

Scheme 21. Cook's eight-step synthesis of triquinacene (**130**).

To prepare a suitable '[3.3.0]-precursor' for triquinacene (**130**) two equivalents of di-*tert*-butyl-3-oxoglutarate (**137**) were condensed with glyoxal (**138**) in alkaline solution to provide the tetraester **139** in better than 90% yield. After this had been transformed into a bis enol ether with diazomethane, monoalkylation with allyl chloride provided **140** in excellent yield with high regioselectivity. On acid-catalyzed hydrolysis **140**, as expected, also underwent decarboxylation and furnished the diketone **141** as a mixture of two diastereoisomers. The selectivity of this transformation is underlined by the observation that only 2% of a dialkylated 3,7-dione was formed. Ozonolysis of **141** then led to the aldehyde **142**, *i.e.* the allyl group was being employed as a masked acetaldehyde equivalent. With ten carbon atoms, **142** has reached the number needed to synthesize **130**, and the last third of the synthesis consisted in formation of the triquinacene σ-skeleton and the introduction of the double bonds. These tasks were accomplished by first preparing the tricyclic diketo alcohol **143** (mixture of stereoisomers) from **142** by acid-catalyzed intramolecular aldol reaction, subsequent Lewis-acid-mediated reduction of **143** to **144** with borane-THF, and final dehydration of this triol by heating it under reflux in HMPA. The threefold elimination was not completely regioselective and the triquinacene was accompanied by some isotriquinacene (8%),[100] the amount of which could be reduced (to < 2%) by first converting **144** to the trimesylate and treating this with aluminum oxide as decribed by Deslongchamps et al. (see above). The whole sequence could easily be scaled up, and has also been used to prepare derivatives of **130**, such as its 1,10-dimethyl or 1,10-cyclohexano-derivative.[97] The latter is a bridged triquinacene in which a six-membered ring shields the convex face of the triquinacene skeleton and a molecule which incorporates a [4.3.3]propellane subsystem (see Section 6.1).

Concerning the homoaromaticity of triquinacene (**130**) recent[101] thermochemical and theoretical studies show that the hydrocarbon is definetely[102] not homoaromatic, in line with previous interpretations of spectroscopic[103] and structural data[104] from the hydrocarbon and from its calculated overlap integral of non-bonded interactions.[105]

The usefulness of **130** as a precursor for the synthesis of other hydrocarbons with interesting chemical and structural properties has been amply demonstrated by de Meijere and co-workers (Scheme 22).

Thus treatment of **130** with the superbasic Lochmann–Schlosser base mixture (*n*-BuLi, *tert*-BuOK, TMEDA, hexane) converted the hydrocarbon in practically quantitative yield into dipotassium acepentalenediide (**145**).[106, 107] Trapping of this salt with a variety of electrophiles led to 4,7-disubtituted dihydroacepentalene derivatives, for example quenching with trimethylstannyl chloride provided **146**.[108] Quenching with water led to the cross-conjugated (see Section 11.1), highly reactive tetraene **148** which dimerized in a [4+2]cycloaddition mode to the heptacyclic hydrocarbon **147**. When **148** was set free from **145** in the presence of trapping reagents such as anthracene (**150**) or cyclopentadiene (**151**) the Diels–Alder adducts **153** and **154**, respectively, were formed.[108] To generate the elusive acepentalene (**152**), a $C_{10}H_6$ hydrocarbon which, according to calculations, has a curved surface, **146** was first converted in the mass spectrometer by chemical ionization with N_2O to the anion radical **149**; a subsequent neutralization-reionization sequence (a NRMS experiment) of the mass-selected ion **149** furnished the acepentalene cation radical **155**, thereby proving the intermediate existence of the neutral acepentalene (**152**).[109]

4,7-bis(Trimethylstannyl)dihydroacepentalene (**146**) again proved to be the starting material of choice for obtaining metal complexes of triquinacene-derived hydrocarbons. When de Meijere and co-workers transmetalated **146** with salt-free

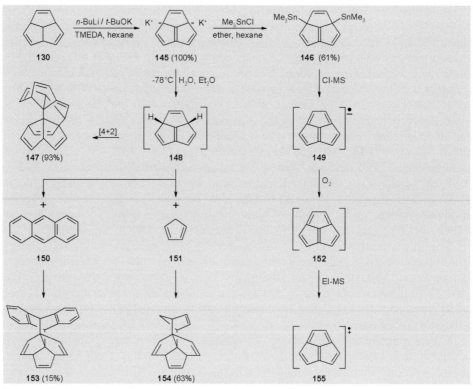

Scheme 22. From triquinacene (**130**) to acepentalene (**152**).

methyllithium in dimethoxyethane, they isolated a crystalline compound which according to X-ray structural analysis was a dimeric aggregate of two ion triplets held together by solvating dimethoxyethane bridges between the lithium ions. As shown in **156** two of the lithium ions are sandwiched between the convex surfaces of the acepentalene units and two are located on the outer surface (Scheme 23).[110] Dilithium acepentalenediide (**156**) also provided a route to other metal and transition metal complexes as illustrated by the zirconium, hafnium, and uranium complexes **157–159**.[106, 111]

As a tricyclic unsaturated hydrocarbon we would expect **130** to undergo isomerization reactions which would not only enable us to enlarge our (CH)$_{10}$ map (see Section 10.4 and above) but could also lead to new members of this hydrocarbon family. This, indeed, is observed experimentally. Low temperature photolysis of **100** in a dilute pentane solution with a medium pressure mercury lamp provided the known (CH)$_{10}$ hydrocarbons **90** (see Section 10.4), Nenitzescu's hydrocarbon (**23**, see above and Section 10.4) and bullvalene (**71**, see Section 13.3) in small amounts, the dehydrogenation products naphthalene (**160**) and azulene (**161**, see Section 11.4) in traces, and as main products the new isomers **162** and **163**, the latter a saturated (CH)$_{10}$ isomer with the trivial name barretane (Scheme 24).[112]

With the exception of its high thermal stability, which makes **130** one of the thermally most stable (CH)$_{10}$ isomers, the pyrolysis behavior of triquinacene is

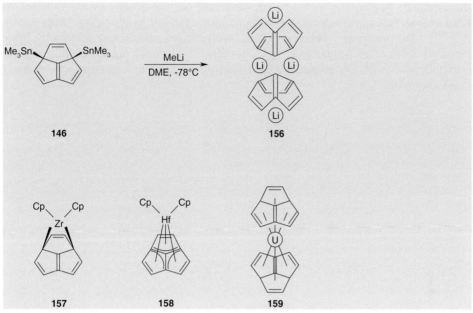

Scheme 23. Dihydroacepentalene derivatives as precursors for acepentalene metal complexes.

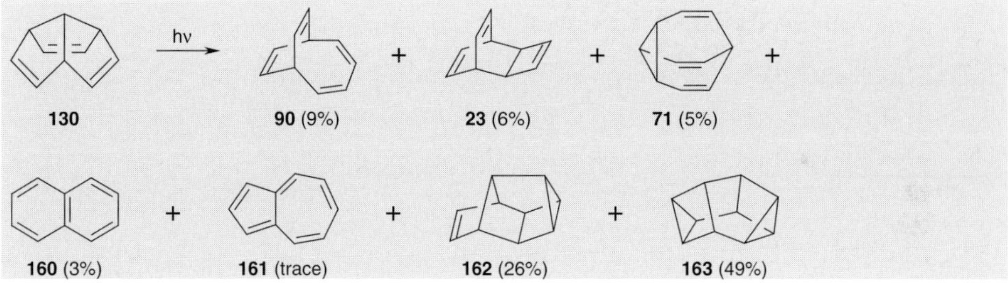

Scheme 24. Photochemical transformations of triquinacene (**130**).

unexciting. The hydrocarbon survived passage through a quartz flow system at 500 °C, and even at 900 °C it remained incompletely consumed. In the range covered by these temperatures naphthalene (**160**), azulene (**161**), 1,2-dihydronaphthalene, and indene are among the pyrolysis products of **130**.[113]

Returning to the photochemistry of triquinacene (**130**) it is at first sight surprising that its valence isomer hexacyclo[4.4.0.0^{2,10}.0^{3,5}.0^{4,8}.0^{7,9}]decane (**165**), colloquially called diademane, because of its crown-like structure[90, 114] is not among the photoproducts of **130**. After all, because of the skipped arrangement of the double bonds in the hydrocarbon it should readily undergo a di-π-methane photore-arrangement (see above). This process could initially lead to the diradical **164** which in principle could undergo a twofold ring closure to **165**. Although **164** has been invoked to explain the formation of several of the actually isolated photoproducts[112] it evidently cannot close to **165**, a hydrocarbon much more strained than **130**. That the (thermal) equilibrium between **165** and **130** lies completely on the side of the latter was shown when diademane (**165**) became available synthetically (Scheme 25)[114, 115] from snoutene (**168**), a $(CH)_{10}$ hydrocarbon which had previously been prepared from cyclooctatetraene (COT, **69**),[116, 117] once again underlining the outstanding role of this compound as a starting material in hydrocarbon chemistry (see Chapter 10).

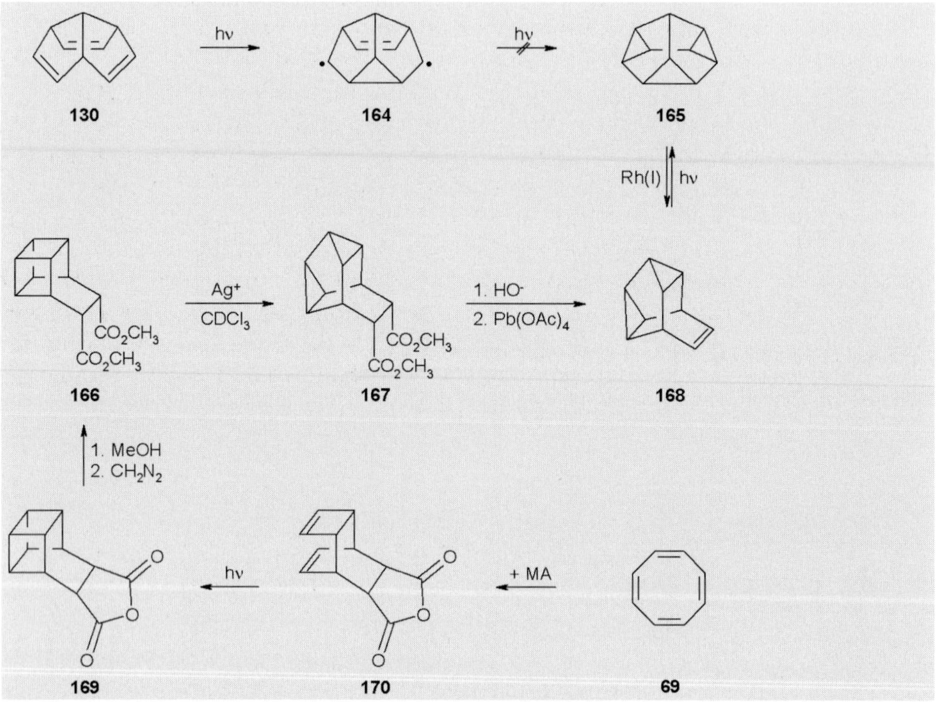

Scheme 25. The synthesis of diademane (**165**) and its relationship to triquinacene (**130**).

To prepare snoutene (**168**) maleic anhydride (MA) was added to COT (**69**) and the resulting Diels–Alder adduct **170** closed photochemically to the basketane derivative **169**. After this anhydride had been converted to the diester **166**, Ag^+-catalyzed isomerization (see Section 10.3) provided the isomeric diester **167** which, by way of saponification and decarboxylation, yielded the desired **168**. To obtain diademane (**165**) from **169** an intramolecular [2+2]cycloaddition of a double bond to a three-membered ring has to take place.[118] This was accomplished by low-temperature (–40 to –70 °C) photolysis in pentane, and **165** was eventually isolated in *ca* 15% yield.[115] Whereas rhodium(I) complexes induce the back-reaction **165**→**168**, simple heating of diademane or treatment with various other metal ions[119] bring about its rearrangement to triquinacene (**130**). In fact, the thermal lability even resulted in a reduction in yield during the isolation of **165** from the photolysate of **168**. With $E_a = 31.6 \pm 0.7$ kcal mol^{-1} and log A = 14.64 ± 0.3 ($E_a = 28.4 \pm 0.2$ kcal mol^{-1} by differential scanning calorimetry[101]) the Arrhenius activation energy for the ring-opening of **165** is extraordinarily low. This indicates that a concerted process is taking place, in which all three cyclopropyl σ-bonds open and all three π-bonds form simultaneously. According to simple MINDO/3[102] and more recent high-level *ab initio* and density functional theory (DFT)[101] calculations, the transition state of this rearrangement has the same C_{3v} symmetry as the starting material **165** and the product **130**, as expected for a cycloreversion of the $[_\sigma 2_s + _\sigma 2_s + _\sigma 2_s]$-type.

Diademane (**165**) is also of importance in relation to another classical problem in hydrocarbon chemistry. Formal removal of the central bridgehead in **165** leads to *cis*-tris-σ-homobenzene (**171**), a hydrocarbon which so far has resisted preparation—despite many efforts. The similarity between the C–C double bond and the cyclopropane ring has often been noted[120] and just as benzene is the archetype of an aromatic molecule (see Section 10.2) the tris-homo analog **171** should be an ideal homoaromatic hydrocarbon.[121] Homoaromaticity should *inter alia* manifest itself in the transition state **172** of the thermal ring-opening of **171** to all-*cis*-1,4,7-cyclononatriene (**173**), the 'π-isomer' of **171**, and a skipped monocyclic triene, in which electronic interaction between the double bonds has been proven to occur.[122] Because of the bonding situation prevailing in cyclopropane rings, the *cis*-orientation of the three-membered rings in **171** should ensure optimum overlap of the bonding orbitals involved in **172**. The activation barrier for the **171**→**173**-interconversion would thus be expected to be low, even more so because there is severe crowding of the interior hydrogen atoms of the substrate (Scheme 26).

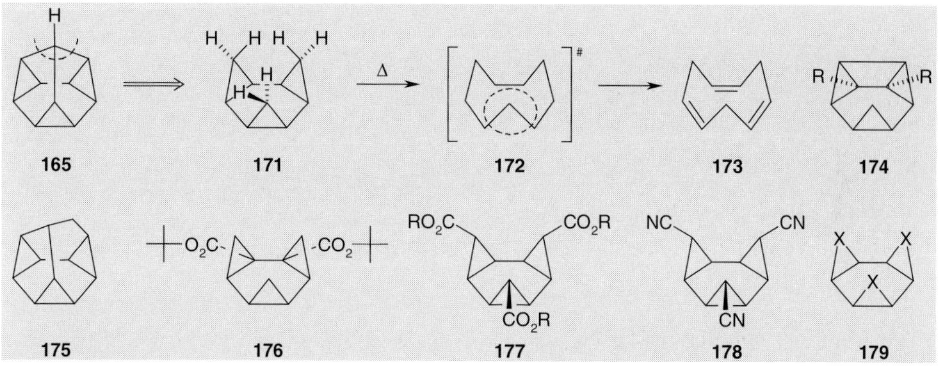

Scheme 26. The stability of *cis*-tris-σ-homobenzene (**171**) and some of its derivatives.

We would expect stabilization of a *cis*-tris-σ-homobenzene derivative if the unfavorable internal steric interactions were removed by a molecular bridge—just as 1,3,7-tri-*cis*-5,9-di-*trans*-1,3,5,7,9-cyclodecapentaene is stabilized by its formal transformation into methano[10]annulene (replacement of the inner hydrogen atoms by a methano bridge, see Section 10.5). Indeed, besides **165**, the singly bridged *cis*-tris-σ-homobenzene **174** (R = CO$_2$CH$_3$)[123] and the triply bridged 'homodiademane' **175**[124] have been prepared.

Removal of the bridging element resulted in a drastic increase in reactivity, as demonstrated by the di-*tert*-butyl ester **176**, which with an (estimated) half-life of 15 min at 8 °C could not be isolated, but rapidly isomerized to di-*tert*-butyl 1,4,7-

cyclononatriene-1,2-dicarboxylate.[125] On the other hand, as shown by Prinzbach et al. introduction of electron-withdrawing substituents at the methylene groups of **171** also caused stabilization of the *cis*-tris-σ-homobenzene framework, and the triester **177** and the trinitrile **178** are isolable compounds.[126] Furthermore, as shown by comprehensive studies of Prinzbach[127, 128] and Vogel[129] and their respective co-workers, complete replacement of the methylene groups in **171** by heteroatoms as in **179** (X = O) or heteroatom containing fragments (**179**, X = NR) led to the corresponding 'heterotrishomoaromatics' as isolable and stable compounds with a rich chemistry.

If the all-*cis*-orientation of the three-membered rings in a tris-σ-homobenzene is really responsible for the ease of isomerization to the π-isomer, then repositioning of one of the cyclopropane rings as in *cis,cis,trans*-tetracyclo[6.1.0.0^{2,4}.0^{5,7}]nonane (**183** *trans*-tris-σ-homobenzene) should result in a pronounced increase in the activation energy for the isomerization. Hydrocarbon **183** was first prepared by Lüttke and Engelhard in 1972 as shown in Scheme 27.[130]

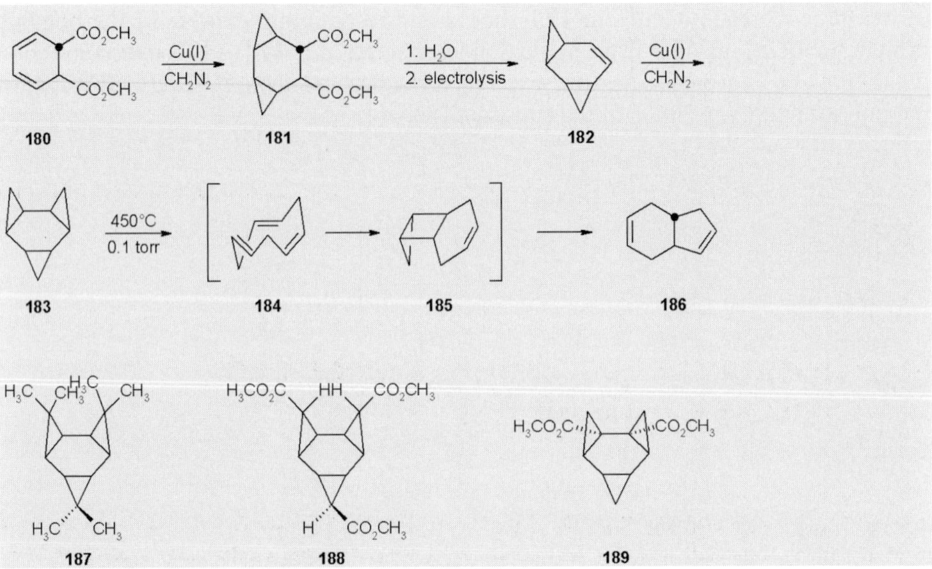

Scheme 27. Preparation and thermal isomerization of *trans*-tris-σ-homobenzene (**183**).

Cyclopropanation of dimethyl *trans*-1,2-dihydrophthalate (**180**) with diazomethane-CuCl first yielded the bis adduct **181** from which the previously reported[131] *trans*-tricyclo[5.1.0.0^{2,4}]oct-5-ene (**182**) was prepared by saponification then Kolbe electrolysis. A second cyclopropanation step then converted **182** into the *trans*-tris-homobenzene **183**, a hydrocarbon which, indeed, could be subjected to much higher thermal stress than the all-*cis*-configurated trishomobenzenes discussed earlier. Flow-pyrolysis at 380–400 °C (contact time 10-75 s) provided *trans*-bicyclo[4.3.0]nona-3,7-diene (**186**) in *ca* 75% yield, presumably *via* cis,trans,trans-1,4,7-cyclononatriene (**184**) and the tricyclic isomer **185**—a [π2$_s$ + π2$_a$]cycloadduct of the strained **184**—as reaction intermediates. For the overall process activation parameters $E_a = 42.0 \pm 1$ kcal mol^{-1} and log A = 13.39 have been determined[115, 130] by de Meijere et al. Several derivatives of **183**, for example the hexamethyl compound **187**[132] (see Section 7.2 for its preparation), the triester **188**,[133] and the diester **189**[134] have also been prepared, illustrating the relative ease of access to the *trans*-tris-σ-homobenzenes.

The high tendency of **171** to ring-open carries over to its lower homolog *cis*-bis-σ-homobenzene (**190**) which also has not yet been synthesized. Again, introduction of electron-withdrawing groups such as in **191**[135] and **192**[136] resulted in a pronounced increase in thermal stability, especially when, as in **192**, molecular bridges are incorporated as additional rate-retarding elements (see above) (Scheme 28).

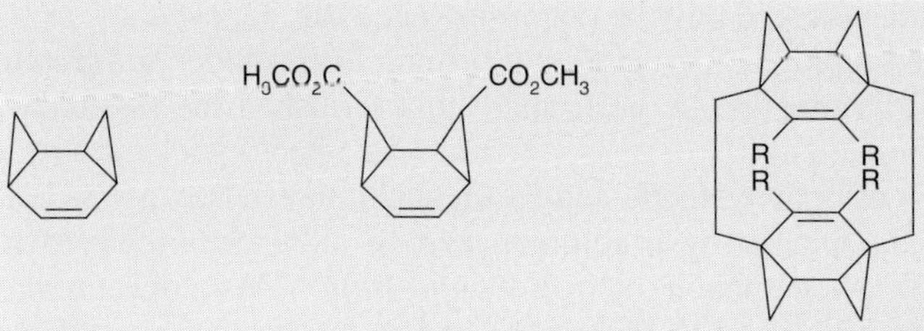

H$_3$CO$_2$C   CO$_2$CH$_3$

190   191   192 (R = CO$_2$CH$_3$)

Scheme 28. Stabilizing *cis*-bis-σ-homobenzenes with electron-withdrawing groups and molecular bridges.

13.5 Spiropolyenes, stellapolyenes and related hydrocarbons

The orthogonal arrangement of double bonds symbolized by the general structure **2** (see above) can be realized in many different ways. A structurally simple arrangement is encountered in spiropentadiene (**196**, 'bowtiediene'), the simplest small-ring, spiro-connected cycloalkene. This highly strained hydrocarbon was prepared by Billups and Haley in 1991,[137] by means of the vacuum gas phase elimination technique of β-halocyclopropylsilanes with solid fluoride which had already been so successful in the preparation of triafulvene (see Section 11.2). The synthesis, the decisive steps of which are summarized in Scheme 29, transforms the central carbon atom of an allene intermediate, **193**, into the spiro-center of the target molecule.

Treatment of **193** with chlorocarbene, generated from dichloromethane and methyllithium, yielded the monoadduct **194**, which was converted to the spiropentane derivative **195** by repetition of this step. The double dehydrohalogenation was subsequently accomplished by evaporating **195** through a 'fluoride column',[137] and the volatile **196** was condensed into a liquid nitrogen trap. The NMR spectroscopic

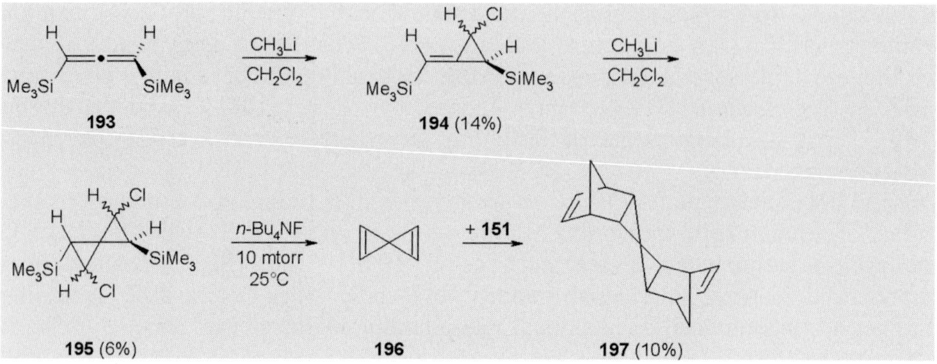

Scheme 29. The synthesis of spiropentadiene (**196**) by Billups and Haley.

properties of the spiropentadiene could only be determined at –105 °C; even at –78 °C the hydrocarbon polymerized rapidly. A structure-confirming Diels–Alder addition to cyclopentadiene (**151**) could, however, be performed; this furnished the bis adduct **197**.

Spiropentadiene (**196**) is of both synthetic and theoretical interest, because two π-systems held in perpendicular planes by a common carbon atom are predicted to interact and thus show the phenomenon of *spiroconjugation*. This concept was introduced in 1967 by Simmons[138] and Hoffmann[139] and co-workers who predicted that spiroconjugation would produce characteristic effects in the electronic spectra and on the chemical reactivity of the π-systems involved. The strongest interaction of this type is expected in molecules with D_{2d} symmetry, for which the overlap integral of the orbitals in different planes is estimated to reach *ca* 20% of the value for adjacent p-orbitals in planar π-sytems. Quantum-chemical calculations[138–140] and experimental[141, 142] studies involving (especially) UV- and photoelectron spectroscopy (PES) support the notion of this kind of through-space interaction. Interest in the synthesis of spiroalkaoligoenes hence grew quickly; the preparation of several of the most important reference compounds will be presented below,[143] proceeding stepwise from **196** to the higher vinylogs, just as for fulvenes (see Section 11.2).

Fluoride-induced β-elimination was the method of choice also for the preparation of spiro[2.2]hepta-1,4,6-triene (**202**) (Scheme 30).[144]

Photolysis of diazocyclopentadiene (**198**) in the presence of (2-bromovinyl)trimethylsilane (**200**) furnished the cycloadduct **201**, presumably *via* the cyclopentadienylidene intermediate **199**. Subsequent treatment of **201** with cesium fluoride in DMSO produced the spiroheptatriene **202**, a hydrocarbon significantly more stable than the spiropentadiene **196**. Carbene addition of this type had previously been employed by Dürr and co–workers for direct preparation of simple derivatives of **202** by photodecomposition of **198** in the presence of different alkynes as trapping reagents; 2-butyne (**203**), for example, yielded the dimethyl derivative **204**.[145]

The most thoroughly studied spiropolyene is spiro[4.4]nonatetraene (**208**), prepared by Semmelhack and co-workers in 1972 by the route shown in Scheme 31.[146]

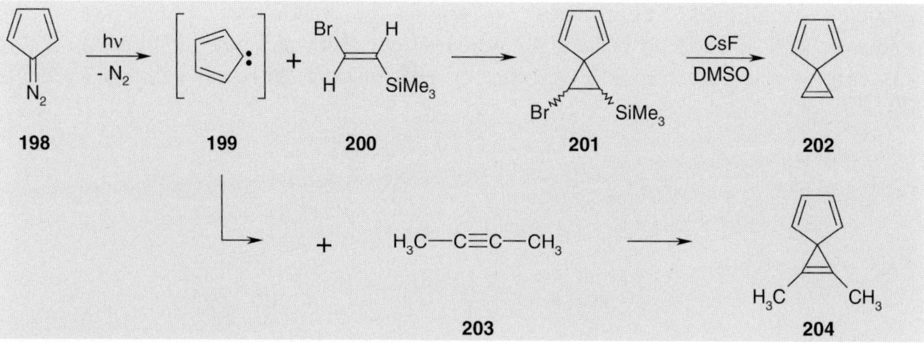

Scheme 30. The preparation of spiro[2.4]hepta-1,4,6-triene (**202**) and some of its derivatives.

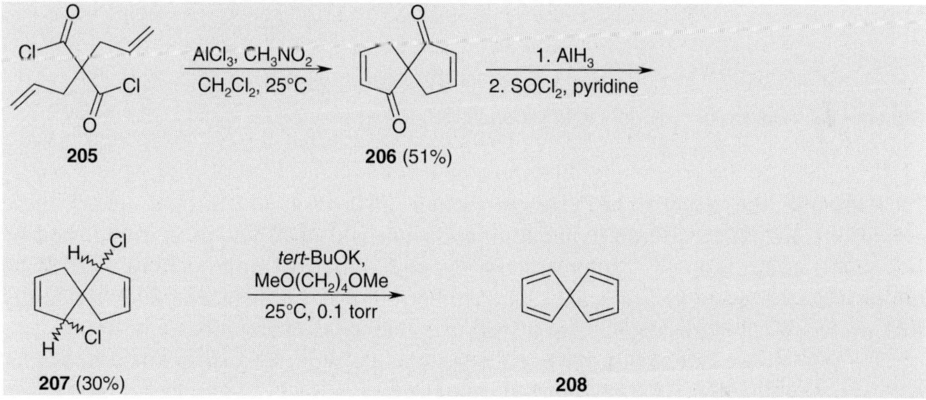

Scheme 31. The preparation of spiro[4.4]nonatetraene (**208**) by Semmelhack and co–workers.

Friedel–Crafts-type cyclization of the diacid dichloride **205** (readily prepared from allyl bromide and diethyl malonate) afforded spiro[4.4]nona-2,6-diene-1,5-dione (**206**) in acceptable yield. Its reduction with aluminum hydride then led to a diol which was reacted with thionyl chloride to produce the dichloride **207** as a mixture of isomers. Potassium *tert*-butoxide-induced dehydrochlorination of **207** finally provided the spirononatetraene **208**. That this is indeed a spiroconjugated hydrocarbon was demonstrated by spectroscopic and structural studies. The electronic spectrum of **208**, for example, shows a red shift of 22 nm and a reduced extinction coefficient compared with that of the (less unsaturated) reference compounds spiro[4.4]nona-1,3,7-triene and spiro[4.4]nona-1,3-diene, in accordance with theoretical predictions.[138, 139] The pronounced 'spiro-splitting' observed in the photoelectron spectrum of **208**,[142] the slight shortening of the double bonds, and small elongation of the single bonds connecting the spiro atom to its four neighbors, as determined by X-ray diffraction,[147] point in the same direction. Among the chemical reactions of spiro[4.4]nonatetraene its thermal isomerization to indene, involving an unusually rapid 1,5-vinyl migration, is noteworthy.[148]

The addition of cyclopentadienylidene (**199**) to unsaturated systems is not restricted to alkenes and alkynes. As shown by Schönleber, photolysis of the precursor **198** in benzene yielded the norcaradiene **209**, and when cyclooctatetraene was employed as the trapping reagent for **199**, adduct **211** was isolated (Scheme 32).[149]

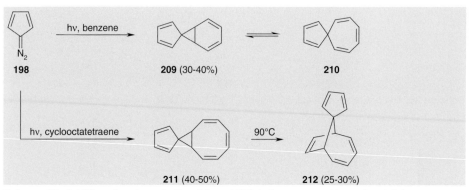

Scheme 32. Addition of cyclopentadienylidene (**199**) to benzene and to cyclooctatetraene.

Although the tropylidene valence isomer **210** was not formed under these conditions, its existence in a dynamic equilibrium with **209** has been established by ^{13}C NMR spectroscopy.[150] Interestingly, attempted thermal ring-opening of **211** did not provide the spiro[4.8]tridecahexaene isomer but the spiropentaene **212**. Evidently **211** prefers a 1,5-sigmatropic carbon shift to a valence isomerization step.[149]

Through-space interaction between π-systems is by no means confined to homo- and spiroconjugation. As recognized by Hoffmann and co–workers, interaction between π-units can also be mediated by the σ-skeleton of a whole molecule,[151, 152] irrespective of whether this is a four-[153] or higher-membered ring. A particularly interesting class of hydrocarbons in which the π-units are separated by a six-membered ring are the stellapolyenes, star-shaped hydrocarbons prepared and studied extensively by Gleiter and co–workers.[154]

For the preparation of 2,6-dimethylenetricyclo[3.3.0.0^{3,7}]octane (**218**, stelladiene) the key intermediate, the acid **216**, was obtained by converting the *endo* carboxylic acid **213** *via* the ketene intermediate **214** to the cyclobutanone **215** and opening of the latter by treatment with base, by following established methodology[155, 156] for the synthesis of substituted tricyclo[3.3.0.0^{3,7}]octane derivatives (overall yield for these steps 10%) (Scheme 33).[157]

To transform the carboxylate function of **216** into the second semicyclic double bond in **218**, the acid was converted to the corresponding *N,N*-dimethylamide and this was reduced to the primary amine **217**. After oxidation with hydrogen peroxide the resulting amine oxide was subjected to Cope elimination, furnishing **218**—which incidentally is an isomer of twistadiene discussed in Section 3.3—in 10% yield. Strong interaction between the central six-membered ring and the two double bonds is indicated by the 0.9-eV split of the first two bands in the PE spectrum of **218**.[158]

For the synthesis of 2,4,6-trimethylenetricyclo[3.3.0.0^{3,7}]octane (**224**, stellatriene) this methodology was modified (Scheme 34).[159]

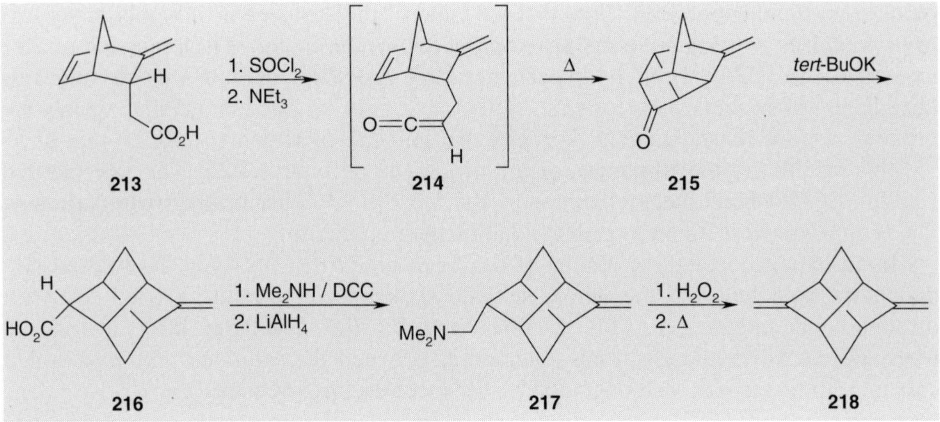

Scheme 33. The synthesis of stelladiene (**218**) ...

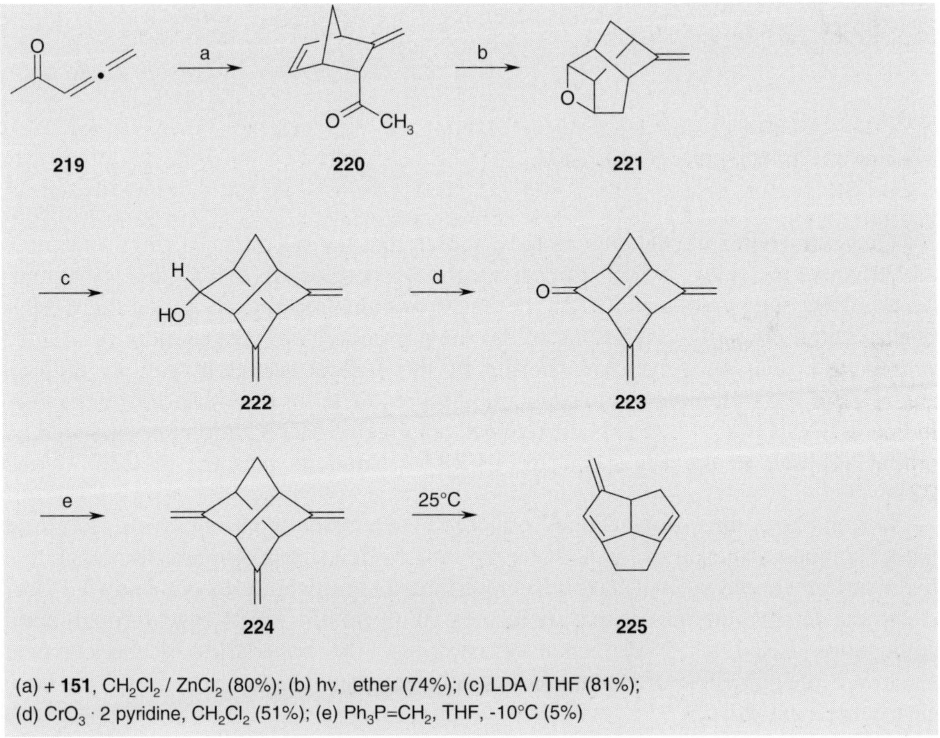

(a) + **151**, CH_2Cl_2 / $ZnCl_2$ (80%); (b) hν, ether (74%); (c) LDA / THF (81%);
(d) $CrO_3 \cdot 2$ pyridine, CH_2Cl_2 (51%); (e) $Ph_3P=CH_2$, THF, -10°C (5%)

Scheme 34. ... and of stellatriene (**224**) by Gleiter and co–workers.

Lewis-acid-catalyzed Diels–Alder addition of acetylallene (**219**) to cyclopentadiene (**151**) gave the expected *endo* methyl ketone **220**, which is ideally set-up for an intramolecular Paterno–Büchi reaction. Like the ketone **215** the oxetane **221** could be ring-opened with base (lithium diisopropylamide, LDA) and the resulting alcohol **222** oxidized to the dienone **223**. Note that the yield of this last step is only moderate. It is even poorer in the very last transformation of the synthesis, the Wittig reaction of **223** to stellatriene (**224**). Both hydrocarbons, **223** and **224**, contain one or two 1,5-hexadiene subsystems, respectively, within a highly strained molecule, which are prone to a Cope rearrangement. For **224**, the half-life of which is only 30 min at 30 °C, this resulted in the formation of the triquinane derivative **225**. The low thermal stability of **224** hints that stellatetraene, the most highly unsaturated hydrocarbon of this series, will be difficult to isolate at ambient temperature.

In an extension of these studies it has been noted that not only the twisted six-membered ring but also the whole stellane skeleton can provide a good relay for through-bond interaction. This is illustrated by the distellene **226**, a rod-like molecule, in which the electronic interaction between the saturated and unsaturated subunits could again be demonstrated by PE spectroscopy (Scheme 35).[160]

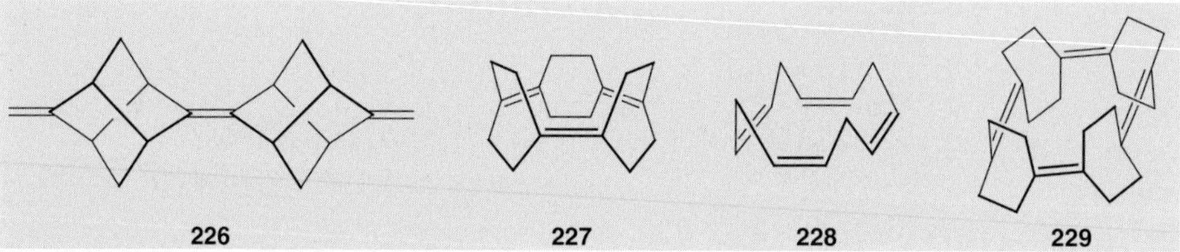

Scheme 35. The limits of σ/π-conjugation.

Quantum-chemical calculations have shown that the σ-system of rings and cages are effective as relays when the basis orbital energies of the π- and σ-systems involved are comparable, and when the linear combinations of the π- and the σ-MOs of the central coupling unit belong to the same irreducible representation. A strained central unit such as a cyclobutane ring or the stellane skeleton with its high p-character of the C–C bonds is hence a good prerequisite for effective σ/π-interaction. In line with this interpretation is the reduced conjugative stabilization (as measured by orbital splitting in the PE spectrum) in **227**[161] and its absence in **228**[162] and **229**.[143]

In summary, 'three-dimensional' oligoenes such as the representative selection of model compounds discussed here, have not only been a strong stimulus to preparative hydrocarbon chemistry, but have also contributed significantly to our knowledge of electronic interactions in organic molecules. With homo-, spiro-, and through-bond conjugation based on experimental observations, the boundaries of the classical electronic phenomenon of conjugation solely between neighboring p-orbitals have clearly been extended.[163]

References

1.   K. B. Wiberg, M. G. Matturro, P. J. Okarma, M. E. Jason, *J. Am. Chem. Soc.*, **1984**, *106*, 2194–2200; *cf.* K. B. Wiberg, M. G. Matturro, R. D. Adams, *J. Am. Chem. Soc.*, **1981**,

103, 1600–1602.

2.   J. Hine, J. A. S. Brown, L. H. Zalkow, W. R. Gardner, M. Hine, *J. Am. Chem. Soc.*, **1955**, *77*, 594–598.

3.   H. E. Zimmerman, R. M. Paufler, *J. Am. Chem. Soc.*, **1960**, *82*, 1514–1515; *cf.* H. E. Zimmerman, G. L. Grunewald, R. M. Paufler, M. A. Sherwin, *J. Am. Chem. Soc.*, **1969**, *91*, 2330–2338.

4.   G. N. Taylor, *J. Org. Chem.*, **1972**, *37*, 2904–2905.

5.   J. Stapersma, I. D. C. Rood, G. W. Klumpp, *Tetrahedron*, **1982**, *38*, 191–199.

6.   W. G. Dauben, G. T. Rivers, R. J. Twieg, W. T. Zimmerman, *J. Org. Chem.*, **1976**, *41*, 887–889.

7.   C. W. Jefford, T. W. Wallace, M. Acar, *J. Org. Chem.*, **1977**, *42*, 1654–1655.

8.   C. Weitemeyer, A. de Meijere, *Angew. Chem.*, **1976**, *88*, 721–722; *Angew. Chem. Int. Ed. Engl.*, **1976**, *15*, 686–687; A. de Meijere, O. Schallner, C. Weitemeywer, U. Spielmann, *Chem. Ber.*, **1979**, *112*, 908–935.

9.   S. Cossu, S. Battaggia, O. de Lucchi, *J. Org. Chem.*, **1997**, *62*, 4162–4163.

10.   H. E. Zimmerman, G. L. Grunewald, *J. Am. Chem. Soc.*, **1966**, *88*, 183–184; *cf.* H. E. Zimmerman, R. W. Binkley, R. S. Givens, M. A. Sherwin, *J. Am. Chem. Soc.*, **1967**, *89*, 3932–3933; H. E. Zimmerman, R. W. Binkley, R. S. Givens, G. L. Grunewald, M. A. Sherwin, *J. Am. Chem. Soc.*, **1969**, *91*, 3316–3323.

11.   H. E. Zimmerman, P. S. Mariano, *J. Am. Chem. Soc.*, **1969**, *91*, 1718–1727.

12.   D. O. Cowan, R. L. Drisko, *Elements of Organic Photochemistry*, Plenum Press, New York, N. Y., **1976**, Chapter 8, 337–365.

13.   D. Döpp, H. E. Zimmerman, in Houben-Weyl, *Methoden der Organischen Chemie*, Thieme Verlag, Stuttgart, **1975**, Vol. IV/5b, *Photochemie*, part 1, 413–432.

14.   H. E. Zimmerman, R. S. Givens, R. M. Pagni, *J. Am. Chem. Soc.*, **1968**, *90*, 6096–6108.

15.   H. E. Zimmerman, C. O. Bender, *J. Am. Chem. Soc.*, **1970**, *92*, 4366–4376.

16.   H. E. Zimmerman, D. R. Amick, *J. Am. Chem. Soc.*, **1973**, *95*, 3977–3982.

17.   E. Heilbronner, *Tetrahedron Lett.*, **1964**, 1923–1928; *cf.* H. E. Zimmerman, *Acc. Chem. Res.*, **1971**, *4*, 272–280.

18.   R. B. Turner, *J. Am. Chem. Soc.*, **1964**, *86*, 3586–3587.

19.   H. E. Zimmerman, private communication (September 9, 1998).

20.   P. D. Bartlett, M. J. Ryan, S. G. Cohen, *J. Am. Chem. Soc.*, **1942**, *64*, 2649–2653. The trivial name refers to the triptych of antiquity, a book with three leaves hinged on a common axis; *cf.* P. D. Bartlett, F. D. Greene, *J. Am. Chem. Soc.*, **1954**, *76*, 1088–1096.

21.   E. Clar, *Ber. Dtsch. Chem. Ges.*, **1931**, *64*, 1676–1688.

22.   A simplification of the Bartlett synthesis of **50** was achieved by hydride reduction of **44** and subsequent dehydration of the resulting diol: A. C. Craig, C. F. Wilcox, *Jr. J. Org. Chem.*, **1959**, *24*, 1619; *cf.* W. Theilacker, U. Berger-Brose, K. H. Beyer, *Chem. Ber.*, **1960**, *93*, 1659–1681.

23.   G. Wittig, R. Ludwig, *Angew. Chem.*, **1956**, *68*, 40

24.   G. Wittig, *Org. Synth. Coll. Vol. IV*, 964–966.

25.   M. Stiles, R. G. Miller, U. Burckhardt, *J. Am. Chem. Soc.*, **1963**, *85*, 1792–1797; *cf.* L. Friedman, F. M. Logullo, *J. Org. Chem.*, **1969**, *34*, 3089–3092.

26.   P. D. Bartlett, S. G. Cohen, J. D. Cotman, Jr, N. Kornblum, J. R. Landry, E. S. Lewis, *J. Am. Chem. Soc.*, **1950**, *72*, 1003–1004.

27.   P. D. Bartlett, E. S. Lewis, *J. Am. Chem. Soc.*, **1950**, *72*, 1005–1009; *cf.* R. Huisgen, C. Rüchardt, *Liebigs Ann. Chem.*, **1956**, *601*, 1–21.

28.   M. Oki, *Applications of Dynamic NMR Spectroscopy to Organic Chemistry*, VCH-Verlagsgesellschaft, Weinheim, **1985**; *cf.* M. Oki, *Topics in Stereochemistry* N. L. Allinger, E. L. Eliel, S. H. Wilen, (*Eds.*), J. Wiley & Sons, New York, N. Y., **1983**, Vol.

14, 1–81. One of the highest barriers of rotation around a C–C single bond ever observed involves di(2-methyl-9-triptycyl) for which a value of 54.1 kcal mol^{-1} has been determined: L. H. Schwartz, C. Koukotas, C.-S. Yu, *J. Am. Chem. Soc.*, **1977**, *99*, 7710–7711; for a review see L. Ernst, *Chem. in uns. Zeit.*, **1983**, *17*, 21–30.

29. H. Hart, S. Shamouilian, Y. Takehira, *J. Org. Chem.*, **1987**, *46*, 4427–4432.
30. H. Hart, A. Bashir-Hashemi, J. Luo, M. A. Meador, *Tetrahedron,* **1986**, *42*, 1641–1654.
31. V. R. Skvarchenko, V. K. Shalaev, E. I. Klabunovskii, *Russ. Chem. Rev.*, **1974**, *43*, 951–966; *cf.* V. R. Skvarchenko, V. K. Shalaev, *Proc. Acad. Sci. USSR, Chem. Sect.,* **1974**, *216*, 307.
32. G. R. Newkome, C. N. Moorfield, F. Vögtle, *Dendritic Molecules: Concepts, Syntheses, Perspectives*, VCH, Weinheim, **1996**; *cf.* M. Fischer, F. Vögtle, *Angew. Chem.*, **1999**, *111*, 934–935; *Angew. Chem. Int. Ed.*, **1999**, *38*, 884–905 and references cited therein.
33. A. Bashir-Hashemi, H. Hart, D. L. Ward, *J. Am. Chem. Soc.*, **1986**, *108*, 6675–6679.
34. H. Hart, *Pure Appl. Chem.*, **1993**, *65*, 27–34 and refs. cited therein.
35. S. B. Singh, H. Hart, *J. Org. Chem.*, **1990**, *55*, 3412–3415.
36. W. v. E. Doering, B. M. Ferrier, E. T. Fossel, J. H. Hartenstein, M. Jones, Jr, G. Klumpp, R. M. Rubin, M. Saunders, *Tetrahedron*, **1967**, *23*, 3943–3963.
37. G. Schröder, *Angew. Chem.*, **1963**, *75*, 722–723.
38. G. Schröder, *Chem. Ber.*, **1964**, *97*, 3140–3149; *cf.* R. Merényi, J. F. M. Oth, G. Schröder, *Chem. Ber.*, **1964**, *97*, 3150–3161.
39. For a summary see G. I. Fray, R. G. Saxton, *The Chemistry of Cyclooctatetraene and its Derivatives*, Cambridge University Press, Cambridge, **1978**; *cf.* G. Schröder, *Cyclooctatetraen*, Verlag Chemie, Weinheim, **1965**.
40. G. Schröder, *Chem. Ber.*, **1964**, *97*, 3131–3139; *cf.* G. Schröder, G. Kirsch, J. F. M. Oth, *Chem. Ber.*, **1974**, *107*, 460–476.
41. L. T. Scott, M. Jones, Jr, *Chem. Rev.*, **1972**, *72*, 181–202.
42. W. v. E. Doering, D. W. Wiley, *Tetrahedron*, **1960**, *11*, 183–189; *cf.* M. J. S. Dewar, R. Pettit, *J. Chem. Soc.*, **1956**, 2021–2025.
43. E. E. v. Tamelen, B. Pappas, *J. Am. Chem. Soc.*, **1963**, *85*, 3296–3297.
44. J. Font, F. López, F. Serratosa, *Tetrahedron Lett.*, **1972**, 2589–2590; *cf.* C. Almansa, M. L. Garcia, C. Jaime, A. Moyano, M. A. Pericás, F. Serratosa in *Strain and its Implications in Organic Chemistry*, A. de Meijere, S. Blechert, (*Eds.*), NATO ASI Series, Kluwer Academic Publishers, Dordrecht, **1989**, 447–450.
45. M. Jones, Jr, S. D. Reich, L. T. Scott, *J. Am. Chem. Soc.*, **1970**, *92*, 3118–3126; *cf.* S. Masamune, H. Zenda, M. Wiesel, N. Nakatsuka, G. Bigam, *J. Am. Chem. Soc.*, **1968**, *90*, 2727–2728.
46. M. Jones, Jr, *J. Am. Chem. Soc.*, **1967**, *89*, 4236–4238.
47. M. Jones, Jr, L. T. Scott, *J. Am. Chem. Soc.*, **1967**, *89*, 150–151.
48. W. v. E. Doering, J. W. Rosenthal, *Tetrahedron Lett.*, **1967**, 349–350.
49. H.-P. Löffler, G. Schröder, *Angew. Chem.*, **1968**, *80*, 758–759; *Angew. Chem. Int. Ed. Engl.*, **1968**, *7*, 736–737.
50. E. E. van Tamelen, T. L. Burkoth, R. H. Greely, *J. Am. Chem. Soc.*, **1971**, *93*, 6120–6129, *cf.* E. E. van Tamelen, T. L. Burkoth, *J. Am. Chem. Soc.*, **1967**, *89*, 151–152.
51. W. v. E. Doering, W. R. Roth, *Angew. Chem.*, **1963**, *75*, 27–35.
52. W. v. E. Doering, W. R. Roth, *Tetrahedron*, **1963**, *19*, 715–737.
53. Review of molecules with fluxional behavior: G. Schröder, J. F. M. Oth, R. Merényi, *Angew. Chem.*, **1965**, *77*, 774–784; *Angew. Chem. Int. Ed. Engl.*, **1965**, *4*, 752–762.
54. M. Saunders, *Tetrahedron Lett.*, **1963**, 1699–1702.
55. H. Günther, J. Ulmen, *Tetrahedron*, **1974**, *30*, 3781–3786.
56. J. F. M. Oth, K. Müllen, J.-M. Gilles, G. Schröder, *Helv. Chim. Acta*, **1974**, *57*, 1415–

1433.

57. J. J. Titman, Z. Luz, H. W. Spiess, *J. Am. Chem. Soc.*, **1992**, *114*, 3765–3771.

58. Z. Luz, R. Poupko, S. Alexander, *J. Chem. Phys.*, **1993**, 99, 7544–7553.

59. First X-ray structure determination: A. Amit, R. Huber, W. Hoppe, *Acta Cryst.*, **1968**, *B24*, 865–869; low-temperature, high-resolution X-ray structure: T. Koritsanszky, J. Buschmann, P. Luger, *J. Phys. Chem.*, **1996**, *100*, 10547–10553; low-temperature neutron diffraction: P. Luger, J. Buschmann, R. K. McMullan, J. R. Ruble, P. Matias, G. A. Jeffrey, *J. Am. Chem. Soc.*, **1986**, *108*, 7825–7827. Review of the chemical behavior: G. Schröder, J. F. M. Oth, *Angew. Chem.*, **1967**, *79*, 458–467; *Angew. Chem. Int. Ed. Engl.*, **1967**, *6*, 414–423.

60. For summaries see A. de Meijere in *Cage Hydrocarbons*, G. A. Olah, (*Ed.*), J. Wiley & Sons, New York, N. Y., **1990**, chapter 8, 261–311; A. de Meijere, *Angew. Chem.*, **1979**, *91*, 867–884; *Angew. Chem. Int. Ed. Engl.*, **1979**, *18*, 809–926.

61. H. E. Zimmerman, H. Iwamura, *J. Am. Chem. Soc.*, **1970**, *92*, 2015–2022; *cf.* H. E. Zimmerman, H. Iwamura, *J. Am. Chem. Soc.*, **1968**, *90*, 4763–4764. The first isomerization of a cyclooctatetraene to a semibullvalene involved octamethyl-cyclooctatetraene which thermally rearranges at elevated temperatures to permethylsemibullvalene: R. Criegee, R. Askani, *Angew. Chem.*, **1968**, *80*, 531–532; *Angew. Chem. Int. Ed. Engl.*, **1968**, *7*, 537. Whereas this process cannot be accomplished for the parent hydrocarbon **36**, 1,3,5,7-tetramethylcyclooctatetraene also undergoes this thermal isomerization[61]; *cf.* H. Iwamura, *Tetrahedron Lett.*, **1973**, 369–372.

62. N. J. Turro, J.-M. Liu, H. E. Zimmerman, R. E. Factor, *J. Org. Chem.*, **1980**, *45*, 3511–3512; *cf.* D. Dudek, H. Glänzer, J. Troe, *Ber. Bunsenges. Phys. Chem.*, **1978**, *82*, 1243.

63. J. Meinwald, D. Schmidt, *J. Am. Chem. Soc.*, **1969**, *91*, 5877.

64. H. E. Zimmerman, J. D. Robbins, J. Schantl, *J. Am. Chem. Soc.,* **1969**, *91*, 5878–5879.

65. J. Meinwald, H. Tsurata, *J. Am. Chem. Soc.*, **1969**, *91*, 5877–5878.

66. R. Askani, *Tetrahedron Lett.*, **1970**, 3349–3350; *cf.* R. Askani, *Chem. Ber.*, **1969**, *102*, 3304–3309; R. Askani, I. Gurang, W. Schwertfeger, *Tetrahedron Lett.*, **1975**, 1315–1318; R. Askani, Th. Hornykiewytsch, W. Schwertfeger, M. Jansen, *Chem. Ber.*, **1980**, *113*, 2154–2174 and refs. cited.

67. L. A. Paquette, *J. Am. Chem. Soc.*, **1970**, *92*, 5765–5767; *cf.* D. R. James, G. H. Birnberg, L. A. Paquette, *J. Am. Chem. Soc.*, **1974**, *96*, 7465–7477 and refs. cited therein; R. K. Russell, R. E. Wingrad, Jr, L. A. Paquette, *J. Am. Chem. Soc.*, **1974**, *96*, 7483–7491.

68. R. M. Moriarty, C.-L. Yeh, N. Ishibi, *J. Am. Chem. Soc.*, **1971**, *93*, 3085–3086.

69. The high interest in semibullvalene (**34**) is underlined by still more syntheses than can be discussed here; *cf.* E. W. Turnblum, T. J. Katz, *J. Am. Chem. Soc.*, **1973**, *95*, 4292–4311; R. Malherbe, *Helv. Chim. Acta*, **1973**, *56*, 2845–2846; P. K. Freeman, K. E. Swenson, *J. Org. Chem.*, **1982**, *47*, 2033–2039; J. Stapersma, G. W. Klumpp, *Recl. Trav. Chim. Pays-Bas*, **1982**, *101*, 274–275.

70. A. K. Cheng, F. A. L. Anet, J. Mioduski, J. Weinwald, *J. Am. Chem. Soc.*, **1974**, *96*, 2887–2891.

71. D. Moskau, R. Aydin, W. Leber, H. Günther, H. Quast, H.-D. Martin, K. Hassenrück, L. S. Miller, K. Grohmann, *Chem. Ber.*, **1989**, *122*, 925–931.

72. J. M. Brown, B. T. Golding, J. J. Stofko, Jr, *J. Chem. Soc. Chem. Commun.*, **1973**, 319–320.

73. H. Günther, J. B. Pawliczek, J. Ulmen, W. Grimme, *Angew. Chem.*, **1972**, *84*, 539–540; *Angew. Chem. Int. Ed. Engl.*, **1972**, *11*, 517–518.

74. J. B. Lambert, *Tetrahedron Lett.*, **1963**, 1901–1906. Just as the derivatives of semibullvalene, those of barbaralane have been extensively investigated for their fluxional behavior; for leading references see: H. Quast, E. Geisser, A. Mayer, L. M.

Jackman, K. L. Colson, *Tetrahedron*, **1986**, *42*, 1805–1813; H. Quast, E. Geißler, Th. Heckert, K. Knoll, E.-M. Peters, K. Peters, H. G. von Schnering, *Chem. Ber.*, **1993**, *126*, 1465–1475; H. Quast, K. Knoll, E.-M. Peters, K. Peters, H. G. von Schnering, *Chem. Ber.*, **1993**, *126*, 1047–1060; H. Quast, M. Witzel, E.-M. Peters, K. Peters, H. G. von Schnering, *Liebigs Ann. Chem.*, **1995**, 725–738; H. Quast, Chr. Becker, M. Witzel, E.-M. Peters, K. Peters, H. G. von Schnering, *Liebigs Ann. Chem.*, **1996**, 985–992.

75.   M. J. S. Dewar, D. H. Lo, *J. Am. Chem. Soc.*, **1971**, *93*, 7201–7207; *cf.* M. J. S. Dewar, Z. Náhlovská, B. D. Náhlovský, *J. Chem. Soc. Chem. Commun.*, **1971**, 1377–1378.

76.   R. Hoffmann, W. D. Stohrer, *J. Am. Chem. Soc.*, **1971**, *93*, 6941–6948.

77.   L. S. Miller, K. Grohmann, J. J. Dannenberg, L. Todaro, *J. Am. Chem. Soc.*, **1981**, *103*, 6249–6251.

78.   R. Gompper, M.-L. Schwarzensteiner, *Angew. Chem.*, **1982**, *94*, 447–448; *Angew. Chem. Int. Ed. Engl.*, **1982**, *21*, 438–439; *Angew. Chem. Suppl.*, **1982**, 1028–1035; *cf.* R. Gompper, M.-L. Schwarzensteiner, H.-U. Wagner, *Tetrahedron Lett.*, **1985**, *26*, 611–614.

79.   R. Askani, M. Littmann, *Tetrahedron Lett.*, **1982**, *23*, 3651–3652; *cf.* R. Askani, *Tetrahedron Lett.*, **1971**, 447–450; R. Askani, R. Kirsten, B. Dugall, *Tetrahedron*, **1981**, *37*, 4437–4444. For an improvement of the Askani synthesis of 1,5-dialkylsemibullvalenes see H. Quast, T. Dietz, E.-M. Peters, K. Peters, H. G. von Schnering, *Liebigs Ann. Chem.*, **1995**, 1159–1168. See also L. A. Paquette, M. A. Kesselmayer, G. E. Underiner, S. D. House, R. D. Rogers, K. Meerholz, J. Heinze, *J. Am. Chem. Soc.*, **1992**, *114*, 2644–2652 and references cited therein.

80.   G. Mehta, C. Ravikrishna, *Tetrahedron Lett.*, **1998**, *39*, 4899–4900.

81.   D. Paske, R. Ringshandl, I. Sellner, H. Sichert, J. Sauer, *Angew. Chem.*, **1980**, *92*, 464–465; *Angew. Chem. Int. Ed. Engl.*, **1980**, *19*, 456–457; *cf.* C. Schnieders, K. Müllen, C. Braig, H. Schuster, J. Sauer, *Tetrahedron Lett.*, **1984**, *25*, 749–752.

82.   H. Quast, C. Becker, E.-M. Peters, K. Peters, H. G. von Schnering, *Liebigs Ann. Chem./Recueil*, **1997**, 685–698.

83.   R. V. Williams, V. R. Gadgil, K. Chanhan, D. van der Helm, M. B. Hossain, L. M. Jackman, E. Fernandes, *J. Am. Chem. Soc.*, **1996**, *118*, 4208–4209.

84.   H. Quast, Th. Herkert, A. Witzel, E.-M. Peters, K. Peters, H. G. von Schnering, *Chem. Ber.*, **1994**, *127*, 921–932; *cf.* H. Quast, J. Christ, *Liebigs Ann. Chem.*, **1984**, 1180–1192.

85.   L. M. Jackman, E. Fernandes, M. Heubes, H. Quast, *Eur. J. Org. Chem.*, **1998**, 2209–2227. Note added in proof: In the meantime the equilibrium between localized (structure type **34**) and delocalized (structure type **109**) states of (thermochromic) semibullvalenes and barbaralanes could be observed directly by Quast and co-workers; H. Quast, M. Seefelder, *Angew. Chem.* **1999**, *111*, 1132–1136; *Angew. Chem. Int. Ed. Engl.*, **1999**, *38*, 1064–1071; *cf.* M. Seefelder, H. Quast, *Angew. Chem.*, **1999**, *111*, 1136–1139; *Angew. Chem.*, **1999**, *38*, 1068–1071.

86.   R. D. Miller, C. S. Yannoni, *J. Am. Chem. Soc.*, **1980**, *102*, 7396–7397.

87.   L. M. Jackman, A. Benesi, A. Mayer, H. Quast, E.-M. Peters, K. Peters, H. G. von Schnering, *J. Am. Chem. Soc.*, **1989**, *111*, 1512–1513.

88.   R. B. Woodward, T. Fukunaga, R. C. Kelly, *J. Am. Chem. Soc.*, **1964**, *86*, 3162–3164.

89.   L. T. Jacobson, *Acta Chem. Scand.*, **1967**, *21*, 2235–2246.

90.   A. de Meijere, D. Kaufmann, O. Schallner, *Angew. Chem.*, **1971**, *83*, 404–405; *Angew. Chem. Int. Ed. Engl.*, **1971**, *10*, 417–418.

91.   M. J. Wyvratt, L. A. Paquette, *Tetrahedron Lett.*, **1974**, 2433–2436; *cf.* L. A. Paquette, S. V. Ley, W. B. Farnham, *J. Am. Chem. Soc.*, **1974**, *96*, 312–313; L. A. Paquette, W. B. Farnham, S. V. Ley, *J. Am. Chem. Soc.*, **1975**, *97*, 7273–7279.

92.   L. A. Paquette, P. B. Lavrik, R. H. Summerville, *J. Org. Chem.*, **1977**, *42*, 2659–2665.

93.   E. Carceller, M. L. García, A. Moyano, F. Serratosa, *J. Chem. Soc. Chem. Commun.*,

1984, 825–826; *cf.* E. Carceller, M. L. García, A. Moyano, M. A. Pericàs, F. Serratosa, *Tetrahedron,* **1986**, *42*, 1831–1839.

94. R. Russo, Y. Lambert, P. Deslongchamps, *Can. J. Chem.,* **1971**, *49*, 531–533.

95. J.-C. Mercier, P. Soucy, W. Rosen, P. Deslong champs, *Synth. Comm.,* **1973**, *3*, 161–164; full paper: P. Deslongchamps, U. O. Cheryan, Y. Lambert, J.-C. Mercier, L. Ruest, R. Russo, P. Soucy, *Can. J. Chem.,* **1978**, *56*, 1687–1704.

96. St. H. Bertz, G. Lannoye, J. M. Cook, *Tetrahedron Lett.,* **1985**, *26*, 4695–4698.

97. A. K. Gupta, G. S. Lannoye, G. Kubiak, J. Schkeryantz, S. Wehrli, J. M. Cook, *J. Am. Chem. Soc.,* **1989**, *111*, 2169–2179.

98. U. Weiss, J. M. Edwards, *Tetrahedron Lett.,* **1968**, 4885–4887.

99. Review: A. K. Gupta, X. Fu, J. B. Snyder, J. M. Cook, *Tetrahedron,* **1991**, *47*, 3665–3710; *cf.* H. Quast, *Janssen Chim. Acta,* **1986**, *4*, 26.

100. L. A. Paquette, J. D. Kramer, *J. Org. Chem.,* **1984**, *49*, 1445–1446.

101. S. P. Verevkin, H.-D. Beckhaus, C. Rüchardt, R. Haag, S. I. Kozhushkov, T. Zywietz, A. de Meijere, H. Jiao, P. v. R. Schleyer, *J. Am. Chem. Soc.,* **1998**, *120*, 11130–11135.

102. Based on heats of hydrogenation measurements, a ground state stabilization of **130** by 4.5 kcal mol^{-1} has previously been claimed and assigned to homoaromaticity; *cf.* J. F. Liebman, L. A. Paquette, J. R. Peterson, D. W. Rogers, *J. Am. Chem. Soc.,* **1986**, *108*, 8267–8268. See, however, for criticism of this interpretation: M. A. Miller, J. M. Schulman, R. L. Disch, *J. Am. Chem. Soc.,* **1988**, *110*, 7681–7684; *cf.* J. Spanget-Larsen, R. Gleiter, *Angew. Chem.,* **1978**, *90*, 471–472; *Angew. Chem. Int. Ed. Engl.,* **1978**, *17*, 441–442; M. J. S. Dewar, A. J. Holder, *J. Am. Chem. Soc.,* **1989**, *111*, 5384–5387; A. J. Holder, *J. Comput. Chem.,* **1993**, *14*, 251.

103. IR, UV/Vis: R. F. Childs, *Acc. Chem. Res.,* **1984**, *17*, 347–352; CD: L. A. Paquette, F. R. Kearney, A. F. Drake, S. F. Mason, *J. Am. Chem. Soc.,* **1981**, *103*, 5064–5069; PES: G. G. Christoph, J. L. Muthard, L. A. Paquette, M. C. Böhm, R. Gleiter, *J. Am. Chem. Soc.,* **1978**, *100*, 7782–7784; *cf.* J. C. Bünzli, D. C. Frost, L. Weiler, *Tetrahedron Lett.,* **1973**, 1159–1162.

104. E. D. Stevens, J. D. Kramer, L. A. Paquette, *J. Org. Chem.,* **1976**, *41*, 2266–2269.

105. L. A. Paquette, R. A. Snow, J. L. Muthard, T. Cynkowski, *J. Am. Chem. Soc.,* **1979**, *101*, 6991–6996.

106. Review: R. Haag, A. de Meijere, *Top. Curr. Chem.,* **1998**, *196*, 137–165; *cf.* T. Lendvai, T. Friedl, H. Butenschön, T. Clark, A. de Meijere, *Angew. Chem.,* **1986**, *98*, 734–735; *Angew. Chem. Int. Ed. Engl.,* **1986**, *25*, 719–720; for approaches to the tris-benzannelated derivative of **152** see D. Kuck, A. Schuster, B. Ohlhorst, V. Sinwell, A. de Meijere, *Angew. Chem.,* **1989**, *101*, 626–627; *Angew. Chem. Int. Ed. Engl.,* **1989**, *28*, 595–597; R. Haag, D. Kuck, X.-Y. Fu, J. M. Cook, A. de Meijere, *Synlett,* **1994**, 340–342.

107. R. Haag, B. Ohlhorst, M. Noltemeyer, R. Fleischer, D. Stalke, A. Schuster, D. Kuck, A. de Meijere, *J. Am. Chem. Soc.,* **1995**, *117*, 10474–10485.

108. R. Haag, F.-M. Schüngel, B. Ohlhorst, T. Lendvai, H. Butenschön, T. Clark, M. Noltemeyer, T. Haumann, R. Boese, A. de Meijere, *Chem. Eur. J.,* **1998**, *4*, 1192–1200.

109. R. Haag, D. Schröder, T. Zywietz, H. Jiao, H. Schwarz, P. v. R. Schleyer, A. de Meijere, *Angew. Chem.,* **1996**, *108*, 1413–1416; *Angew. Chem. Int. Ed. Engl.,* **1996**, *35*, 1317–1319.

110. R. Haag, R. Fleischer, D. Stalke, A. de Meijere, *Angew. Chem.,* **1995**, *107*, 1642–1644; *Angew. Chem. Int. Ed. Engl.,* **1995**, 34, 1492–1495.

111. H. Butenschön, A. de Meijere, *Tetrahedron,* **1986**, *42*, 1721–1729.

112. D. Bosse, A. de Meijere, *Chem. Ber.,* **1978**, *111*, 2223–2242; *cf.* D. Bosse, A. de Meijere, *Angew. Chem.,* **1974**, *86*, 706–707; *Angew. Chem. Int. Ed. Engl.,* **1974**, *13*, 663–664. For an alternate route to barretane (**163**) see D. Bosse, A. de Meijere, *Tetrahedron Lett.,*

1977, 1155–1158.

113. L. T. Scott, G. K. Agopian, *J. Am. Chem. Soc.*, **1974**, *96*, 4325–4326.

114. A possible photocyclization of the **173→170**-type has also been conceived by Prinzbach and co-workers; and they have proposed the trivial name *mitrane* to the hydrocarbon **165**: H. Prinzbach, D. Stusche, *Helv. Chim. Acta*, **1971**, *54*, 755–758.

115. D. Kaufmann, H.-H. Fick, O. Schallner, W. Spielmann, L.-U. Meyer, P. Gölitz, A. de Meijere, *Chem. Ber.*, **1983**, *116*, 587–609.

116. W. G. Dauben, C. H. Schallhorn, D. L. Whalen, *J. Am. Chem. Soc.*, **1971**, *93*, 1446–1452.

117. L. A. Paquette, J. C. Stowell, *J. Am. Chem. Soc.*, **1971**, *93*, 2459–2463. For an alternate synthesis of snoutene (**168**) see I. Erden, A. de Meijere, *Tetrahedron Lett.*, **1980**, *21*, 1837–1840.

118. P. K. Freeman, D. K. Kuper, V. N. Mallikarjuna Rao, *Tetrahedron Lett.*, **1965**, 3301–3304; *cf.* H. Prinzbach, M. Klaus, W. Mayer, *Angew. Chem.*, **1969**, *81*, 902–903; *Angew. Chem. Int. Ed. Engl.*, **1969**, *8*, 883–884.

119. D. Kaufmann, O. Schallner, L.-U. Meyer, H.-H. Fick, A. de Meijere, *Chem. Ber.*, **1983**, *116*, 1377–1385.

120. A. de Meijere, *Angew. Chem.*, **1979**, *91*, 867–884; *Angew. Chem. Int. Ed. Engl.*, **1979**, *18*, 809–826; *cf.* K. B. Wiberg in Houben-Weyl-Müller, *Methods of Organic Chemistry*, Vol. E17a A. de Meijere, (*Ed.*), Thieme Verlag, Stuttgart, **1997**, 1–27.

121. For the introduction of the terms 'homoaromaticity' and 'homoconjugation' see S. Winstein, *Quart. Rev.*, **1969**, *23*, 141–176; M. J. Goldstein, R. Hoffmann, *J. Am. Chem. Soc.*, **1971**, *93*, 6193–6204.

122. P. Bischof, R. Gleiter, E. Heilbronner, *Helv. Chim. Acta*, **1970**, *53*, 1425–1434.

123. H. Prinzbach, W. Eberbach, G. Philippossian, *Angew. Chem.*, **1968**, *80*, 910–911; *Angew. Chem. Int. Ed. Engl.*, **1968**, *7*, 887–888; *cf.* H. Prinzbach, D. Stusche, M. Breuninger, J. Markert, *Chem. Ber.*, **1976**, *109*, 2823–2848.

124. W. Spielmann, H.-H. Fick, L.-U. Meyer, A. de Meijere, *Tetrahedron Lett.*, **1976**, 4057–4060.

125. H. W. Whitlock, Jr, P. F. Schatz, *J. Am. Chem. Soc*, **1971**, *93*, 3837–3839.

126. W.-D. Braschwitz, Th. Otten, Chr. Rücker, H. Fritz, H. Prinzbach, *Angew. Chem.*, **1989**, *101*, 1383–1386; *Angew. Chem. Int. Ed. Engl.*, **1989**, *28*, 1348–1349; *cf.* G. Person, M. Keller, H. Prinzbach, *Liebigs Ann. Chem.*, **1996**, 507–527.

127. H. Prinzbach, D. Stusche, *Angew. Chem.*, **1970**, *82*, 836–838; *Angew. Chem. Int. Ed. Engl.*, **1970**, *9*, 799–800; *cf.* H. Prinzbach, D. Stusche, *Helv. Chim. Acta*, **1971**, *54*, 755–758.

128. R. Schwesinger, H. Prinzbach, *Angew. Chem.*, **1972**, *84*, 990–991; *Angew. Chem. Int. Ed. Engl.*, **1972**, *11*, 942–943; H. Prinzbach, R. Schwesinger, M. Breuninger, B. Gallenkamp, D. Hunkler, *Angew. Chem.*, **1975**, *87*, 349–350; *Angew. Chem. Int. Ed. Engl.*, **1975**, *14*, 347–348; H. Prinzbach, D. Stusche, M. Breuninger, J. Markert, *Chem. Ber.*, **1976**, *109*, 2823–2848; H. Prinzbach, D. Stusche, J. Markert, H.-J. Limbach, *Chem. Ber.*, **1976**, *109*, 3505–3626; R. Schwesinger, M. Breuninger, B. Gallenkamp, K.-H. Müller, D. Hunkler, H. Prinzbach, *Chem. Ber.*, **1980**, *113*, 3127–3160; C. Rücker, H. Prinzbach, *Tetrahedron Lett.*, **1983**, *24*, 4099–4102; S. Kakabu, C. Kaiser, R. Keller, P. G. Becher, K.-H. Müller, L. Knothe, G. Rihs, H. Prinzbach, *Chem. Ber.*, **1988**, *121*, 741–756; B. Zipperer, K.-H. Müller, B. Gallenkamp, R. Hildebrand, M. Fletschinger, D. Burger, M. Pillat, D. Hunkler, L. Knothe, H. Fritz, H. Prinzbach, *Chem. Ber.*, **1988**, *121*, 757–780 and references cited therein. For the synthesis of *cis*- and *trans*-tris-[2.2.2]-σ-homobenzenes see G. McMullen, M. Lutterbeck, H. Fritz, H. Prinzbach, C. Krüger, *Israel J. Chem.*, **1982**, *22*, 19–26.

129. E. Vogel, H.-J. Altenbach, C. D. Sommerfeld, *Angew. Chem.*, **1972**, *84*, 986–988; *Angew. Chem. Int. Ed. Engl.*, **1972**, *11*, 939–941; E. Vogel, H.-J. Altenbach, E. Schmidbauer, *Angew. Chem.*, **1973**, *85*, 862–864; *Angew. Chem. Int. Ed. Engl.*, **1973**, *12*, 838–840.

130. M. Engelhard, W. Lüttke, *Angew. Chem.*, **1972**, *84*, 346–347; *Angew. Chem. Int. Ed. Engl.*, **1972**, *11*, 310–311. Subsequently two alternate routes to **182** were published: R. T. Taylor, L. A. Paquette, *Angew. Chem.*, **1975**, *87*, 488–499; *Angew. Chem. Int. Ed. Engl.*, **1975**, *14*, 496; W. Spielmann, D. Kaufmann, A. de Meijere, *Angew. Chem.*, **1978**, *90*, 470–471; *Angew. Chem. Int. Ed. Engl.*, **1978**, *17*, 440–441.

131. W. R. Roth, B. Peltzer, *Liebigs Ann. Chem.*, **1965**, *685*, 56–74.

132. P. Binger, G. Schroth, J. McMeeking, *Angew. Chem.*, **1974**, *86*, 518; *Angew. Chem. Int. Ed. Engl.*, **1974**, *13*, 465; *cf.* P. Binger, J. McMeeking, *Angew. Chem.*, **1975**, *87*, 383–384; *Angew. Chem. Int. Ed. Engl.*, **1975**, *14*, 371.

133. D. L. Dalrymple, S. P. B. Taylor, *J. Am. Chem. Soc.*, **1971**, *93*, 7098.

134. H. Prinzbach, R. Schwesinger, *Angew. Chem.*, **1972**, *84*, 988–990; *Angew. Chem. Int. Ed. Engl.*, **1972**, *11*, 940–942.

135. G. Kaupp, K. Rösch, *Angew. Chem.*, **1976**, *88*, 185–186; *Angew. Chem. Int. Ed. Engl.*, **1976**, *15*, 163–164.

136. K. Menke, H. Hopf, *Angew. Chem.*, **1976**, *88*, 152–153; *Angew. Chem. Int. Ed. Engl.*, **1976**, *15*, 165–166. Besides **196**, a bis-cyclopropanation product is obtained, the two compounds being the first *cis*-bishomobenzene derivatives with unsubstituted or unbridged cyclopropane rings; *cf.* A. de Meijere, R. Näder, *Angew. Chem.*, **1976**, *88*, 153–154; *Angew. Chem. Int. Ed. Engl.*, **1976**, *15*, 166–167.

137. W. E. Billups, M. M. Haley, *J. Am. Chem. Soc.*, **1991**, *113*, 5084–5085.

138. H. E. Simmons, T. Fukunaga, *J. Am. Chem. Soc.*, **1967**, *89*, 5208–5215.

139. R. Hoffmann, A. Imamura, G. D. Zeiss, *J. Am. Chem. Soc.*, **1967**, *89*, 5215–5220; *cf.* R. Hoffmann, *Acc. Chem. Res.*, **1971**, *4*, 1–9; J. Kao, L. Radom, *J. Am. Chem. Soc.*, **1977**, *100*, 760–767; A. Tajiri, T. Nakajima, *Tetrahedron,* **1971**, *27*, 6089–6099.

140. M. J. S. Dewar, M. L. McKee, *Pure Appl. Chem.*, **1980**, *52*, 1431–1441.

141. A. Schweig, U. Weidner, J. G. Berger, W. Grahn, *Tetrahedron Lett.*, **1973**, 557–560; *cf.* P. Bischof, R. Gleiter, H. Dürr, B. Ruge, P. Herbst, *Chem. Ber.*, **1976**, *109*, 1412–1417.

142. C. Batich, E. Heilbronner, E. Rommel, M. F. Semmelhack, J. S. Foos, *J. Am. Chem. Soc.*, **1974**, *96*, 7662–7668.

143. Reviews: H. Dürr, R. Gleiter, *Angew. Chem.*, **1978**, *90*, 591–601; *Angew. Chem. Int. Ed. Engl.*, **1978**, *17*, 559–569; R. Gleiter, W. Schäfer, *Acc. Chem. Res.*, **1990**, *23*, 369–375.

144. W. E. Billups, W. Luo, M. Gutierrez, *J. Am. Chem. Soc.*, **1994**, *116*, 6463.

145. H. Dürr, B. Ruge, H. Schmidt, *Angew. Chem.*, **1973**, *85*, 616–617; *Angew. Chem. Int. Ed. Engl.*, **1973**, *12*, 577–578; *cf.* H. Dürr, B. Ruge, B. Weiß, *Liebigs Ann. Chem.*, **1974**, 1150–1161; T. Mitsuhashi, W. M. Jones, *J. Chem. Soc. Chem. Commun.*, **1974**, 103–104.

146. M. F. Semmelhack, J. S. Foos, S. Katz, *J. Am. Chem. Soc.*, **1972**, *94*, 8637–8638; *cf.* M. F. Semmelhack, J. S. Foos, S. Katz, *J. Am. Chem. Soc.*, **1973**, *95*, 7325–7336.

147. T. Haumann, J. Benet-Buchholz, R. Boese, *J. Mol. Struct.*, **1996**, *374*, 299–304.

148. M. F. Semmelhack, H. N. Weller, J. S. Foos, *J. Am. Chem. Soc.*, **1977**, *99*, 292–294.

149. D. Schönleber, *Angew. Chem.*, **1969**, *81*, 83; *Angew. Chem. Int. Ed. Engl.*, **1969**, *8*, 76; *cf.* D. Schönleber, *Chem. Ber.*, **1969**, *102*, 1789–1801; M. Jones, Jr, *Angew. Chem.,* **1969**, *81*, 83–84; *Angew. Chem. Int. Ed. Engl.*, **1969**, *8*, 76–77; H. Dürr, G. Scheppers, *Tetrahedron Lett.*, **1968**, 6059–6062.

150. H. Dürr, H. Kober, M. Kausch, *Tetrahedron Lett.*, **1975**, 1945–1948. At -10 °C the concentration of **209** amounts to approximately 20%. For the valence isomerization of several benzannelated derivatives of **209/210** see H. Dürr, H. Kober, *Tetrahedron Lett.*,

1975, 1941–1944; *cf.* W. Adam, H. Rebollo, H. Dürr, K.-H. Pauly, K. Peters, E.-M. Peters, H.-G. von Schnering, *Tetrahedron Lett.*, **1982**, *23*, 923–926.

151. R. Hoffmann, A. Imamura, W. J. Hehre, *J. Am. Chem. Soc.*, **1968**, 1499–1509.

152. R. Gleiter, *Angew. Chem.*, **1974**, *86*, 770–775; *Angew. Chem. Int. Ed. Engl.*, **1974**, *13*, 696–701; *cf.* M. N. Paddon-Row, *Acc. Chem. Res.*, **1982**, *15*, 245–251; M. N. Paddon-Row, K. D. Jordan in *Modern Models of Bonding and Delocalization*, J. F. Liebman, J. F. Greenberg, (*Eds.*), Verlag Chemie, Weinheim, **1989**.

153. P. Bischof, R. Gleiter, R. Haider, *J. Am. Chem. Soc.*, **1978**, *100*, 1036–1042; *cf.* R. Gleiter, A. Toyota, P. Bischof, G. Krennrich, J. Dressel, P. D. Pansegrau, L. A. Paquette, *J. Am. Chem. Soc.*, **1988**, *110*, 5490–5497; *cf.* L. A. Paquette, J. Dressel in *Strain and its Implications in Organic Chemistry*, A. de Meijere, S. Blechert, (*Eds.*), NATO ASI Series, Kluwer Academic Publishers, Dordrecht, **1989**, 77–107.

154. Review: O. Borzyk, Th. Herb, Chr. Sigwart, R. Gleiter, *Pure Appl. Chem.*, **1996**, *68*, 233–238.

155. M. Nakazaki, K. Naemura, H. Harada, H. Narutaki, *J. Org. Chem.*, **1982**, *47*, 3470–3474.

156. R. R. Sauers, K. W. Kelly, B. R. Sickles, *J. Org. Chem.*, **1972**, *37*, 537–543.

157. B. Kissler, R. Gleiter, *Tetrahedron Lett.*, **1985**, *26*, 185–188.

158. R. Gleiter, B. Kissler, C. Ganter, *Angew. Chem.*, **1987**, *99*, 1292–1294; *Angew. Chem. Int. Ed. Engl.*, **1987**, *26*, 1252–1254.

159. R. Gleiter, C. Sigwart, B. Kissler, *Angew. Chem.*, **1989**, *101*, 1561–1563; *Angew. Chem. Int. Ed. Engl.*, **1989**, *28*, 1525–1526; *cf.* R. Gleiter, C. Sigwart, *J. Org. Chem.*, **1994**, *59*, 1027–1038.

160. R. Gleiter, O. Borzyk, *Angew. Chem.*, **1995**, *107*, 1094–1095; *Angew. Chem. Int. Ed. Engl.*, **1995**, *34*, 1001–1003; R. Gleiter, G. Fritzsche, O. Borzyk, Th. Oeser, F. Rominger, H. Irngartinger, *J. Org. Chem.*, **1998**, *63*, 2878–2886; *cf.* R. Gleiter, B. Gaa, Chr. Sigwart, H. Lange, O. Borzyk, F. Rominger, H. Irngartinger, Th. Oeser, *Eur. J. Org. Chem.*, **1998**, 171–176; H. Lange, R. Gleiter, G. Fritzsche, *J. Am. Chem. Soc.*, **1998**, *120*, 6563–6568.

161. J. E. McMurry, G. J. Haley, J. R. Matz, J. C. Clardy, G. Van Duyne, R. Gleiter, W. Schäfer, D. H. White, *J. Am. Chem. Soc.*, **1986**, *108*, 2932–2938. See the synthesis and properties of the lower analog of **227**, tricyclo[4.2.2.2^{2,5}]dodeca-1,5-diene, in Section 7.4.

162. A. Krause, H. Musso, W. Boland, R. Ahlrichs, R. Gleiter, R. Boese, M. Bär, *Angew. Chem.*, **1989**, *101*, 1401–1402; *Angew. Chem. Int. Ed. Engl.*, **1989**, *28*, 1379–1380.

163. For a review of the interaction of C-H-bonds with π-system see M. Nishio, M. Hirota, Y. Umezawa, *The CH/π-Interaction–Evidence, Nature and Consequences*, Wiley-VCH, New York, N. Y., **1998**.

14 Extended Systems–I. From Benzene to Graphite Substructures

One of the main threads of this book has been the building-block approach to hydro-carbon chemistry, *i.e.* the use of smaller building units for the modular creation of larger hydrocarbon structures in synthesis, and the recognition of smaller structures in bigger hydrocarbons in analysis (retrosynthesis). And one of the most familiar build-ing blocks—and by far most important—is the benzene ring. We have used this ele-mentary unit already several times, as in the case of the 'construction' of the helicenes (see Section 12.1) and the circulenes (Section 12.2) by stepwise annelation beginning with a single benzene nucleus. I shall now return to this formally simple construction process on a broader scale, with the aim of proceeding as far as is presently known towards graphite, the ultimate molecule composed of 'benzene hexagons'. Obviously, the simplest mode of joining benzene rings is by way of *linear annelation* (**1**), with the centers of the constituting six-membered rings lying on a straight line. This class of hydrocarbons is called *linear acenes*. If, on the other hand, the centers lie on straight lines which form an angle of 120°, we speak of *angular annelation* (**2**). A *branched* polycyclic aromatic hydrocarbon (**3**, PAH) is finally encountered if annelated benzene rings extend from all three *ortho* positions of a central benzene core (**3**) (Scheme 1).

Two typical examples of angular annelation belonging to the already mentioned helicenes and circulenes are hexahelicene (**4**) and coronene (**5**). Formally the three-dimensional aromatic hydrocarbon **4** can be 'flattened' to **5** by removal of one ethyl-ene unit. I am returning to the circulenes here because earlier we discussed only three representative examples of this group of hydrocarbons—corannulene, coronene (**5**) and [7]-circulene, condensed aromatic compounds with central five-, six- and seven-membered rings, respectively. In principle, however, any other ring-size is possible—and hence much larger 'central cavities'—a most spectacular case being that of kekulene to be discussed below.

Of course, **5** can already be regarded as a (very small) section of graphite. Here, however, we want to progress to far bigger aggregates, symbolized by the general structure **7**, *ribbon* or *ladder structures*, **6**, being passed *en route*. Needless to say that there are many other ways of connecting benzene rings to each other. If we exclude functional groups as connecting elements (examples of this type of extended aromatic compound will be discussed in Section 15.2) the simplest way of joining benzene rings is by way of a single bond. The parent subsystem of this class is biphenyl (**8**), but again there is a large number of more complex systems, the *oligo-aryls* (or *aryl-enes*) **9** and **10**[1] being just two examples. The structural richness of the PAHs is en-hanced further by molecules not constructed exclusively from six-membered rings. If five-membered rings are also 'allowed' (*cf.* corannulene, Section 12.2), a large group

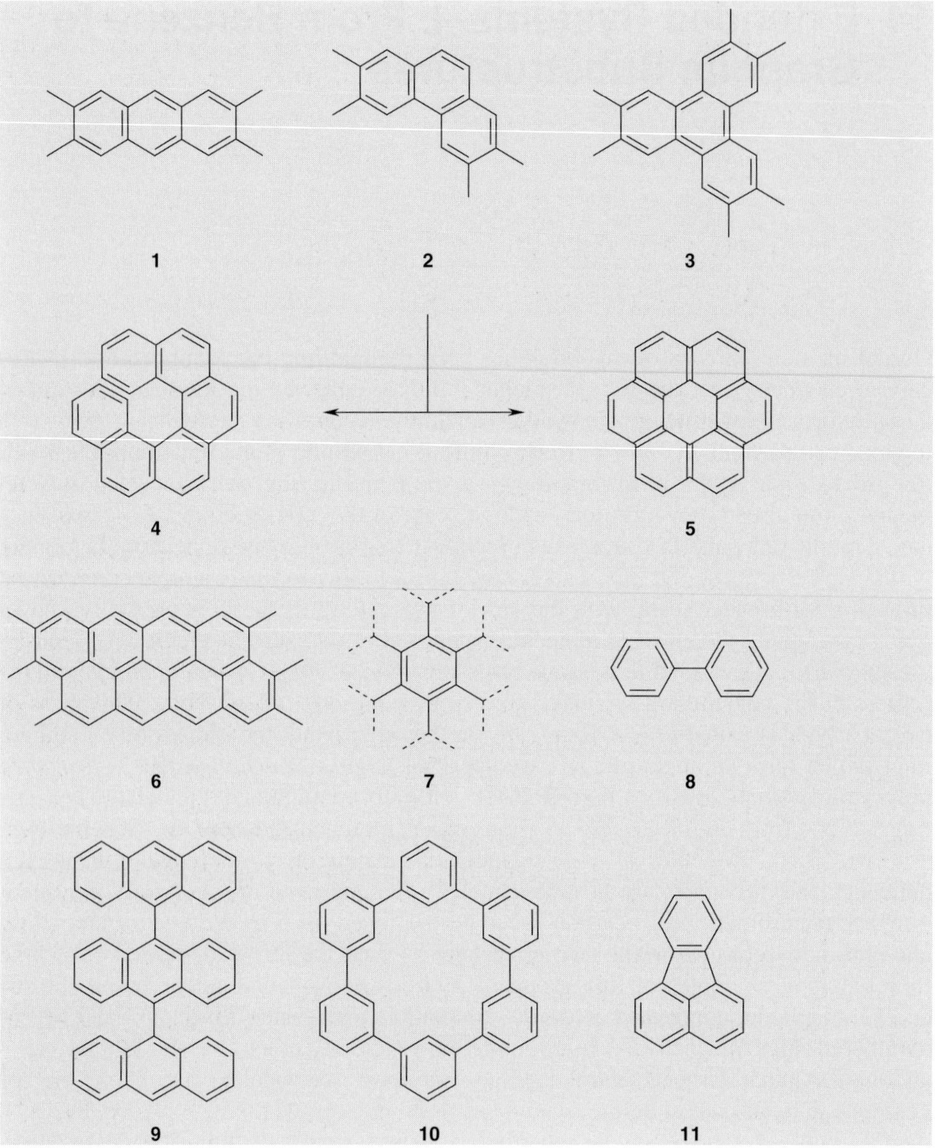

Scheme 1. Building with benzene rings.

of hydrocarbons becomes accessible, which can all be traced back to fluoranthene (**11**). Whereas hydrocarbons consisting solely of six-membered rings belong to the group of *alternant* hydrocarbons, those which also contain five-membered subsystems belong to the *non-alternant* category.

Condensed aromatic hydrocarbons are characterized by their extended π-systems, their well-defined geometric structures, and their relatively high thermal, photochemical and chemical inertness. If they do react they often undergo highly regioselective processes enabling the synthesis of specifically functionalized derivatives. Because of their overriding practical and academic importance there is a vast literature on PAHs. Fortunately, this has often been reviewed and summarized[2–9] allowing me to concentrate on the main lines of development here.

14.1 Linearly annelated polycyclic aromatic hydrocarbons

Largely for historical reasons the linear and the angular PAHs, **1** and **2** respectively, stood in the center of synthetic polyarene chemistry for a long time. Many hydrocarbons belonging to these two important categories were isolated during the 19th century from crude oil, coal, oil shale, and coal tar, and derivatives, such as the quinoid dyes, became of major industrial importance. The field received a further impetus when it was discovered in the mid-1930s that many condensed polyarenes are carcinogenic. To investigate the relationship between physical, chemical, and biological properties and the size and extension of these π-systems it was necessary to prepare so-called 'annelation-rows'. As far as the linear PAHs, the acenes, are concerned the homologous series could be extended to heptacene (see below), already a highly reactive hydrocarbon, difficult to obtain analytically pure. The parent systems with eight or more linearly annelated benzene rings are all unknown, only derivatives—largely hydroquinones and quinones—having been prepared.[2] Because the syntheses and the chemical behavior of the lower members of the series—naphthalene, anthracene, tetracene—have been extensively reviewed many times[2–9] the following discussion will focus on the preparation of penta- to heptacene. Many of the synthetic methods used to prepare these three hydrocarbon have also been applied for the preparation of other cata-condensed poylarenes[9] and are, therefore, exemplary. As with many other hydrocarbon syntheses encountered in this book it is of interest to compare different preparative schemes leading to the same target molecule.

Pentacene (**14**) was first obtained by Clar and John in 1929 using the so-called Elbs reaction for the preparation of the pentacyclic framework **13** from 4,6-dibenzoyl-*meta*-xylene (**12**) (Scheme 2).[10]

In this cyclodehydration process *ortho*-alkyl-diaryl ketones are heated to high temperatures (*ca* 400 °C) in the presence of copper or (more often) zinc. The dehydrogenation of the 6,13-dihydropentacene (**13**) formed can subsequently be achieved by sublimation over copper at 380 °C, heating under reflux in xylene with chloranil, *etc*. No yield was given in the original publication.

To employ the most important six-membered ring-forming process, the Diels–Alder reaction, in a pentacene synthesis is obvious, and, as was shown by Bailey and co-workers,[11] cycloaddition of 1,2-bismethylenecyclohexane (**17**) to the double dienophile *p*-benzoquinone (**18**) led to the 2:1 adduct **16** in good yield. The diketone was reduced to the pentacyclic diene **15** by a Mozingo reaction, and aromatization by catalytic dehydrogenation over Pd/C finally provided pentacene (**14**) in good yield.

In an newer Diels–Alder route to **14** (Scheme 3), the diene **17** was replaced by

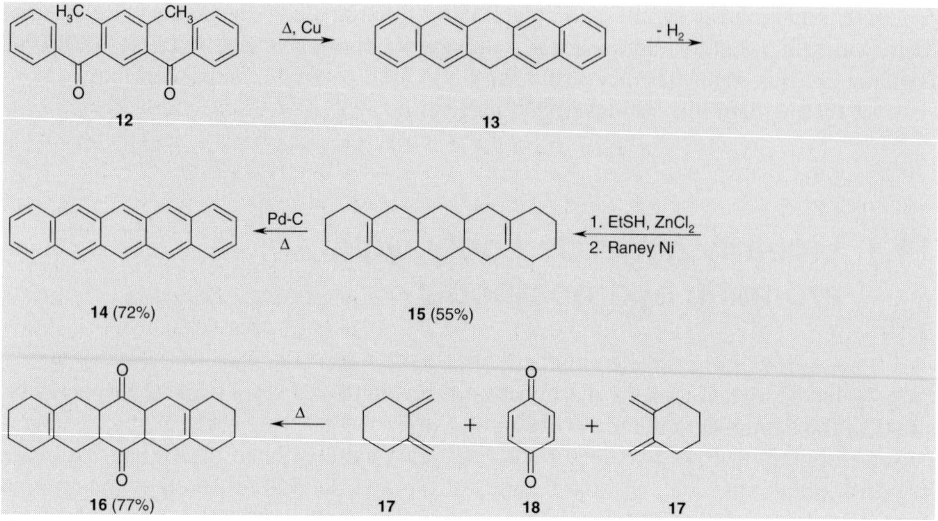

Scheme 2. Clar's and Bailey's classic routes ...

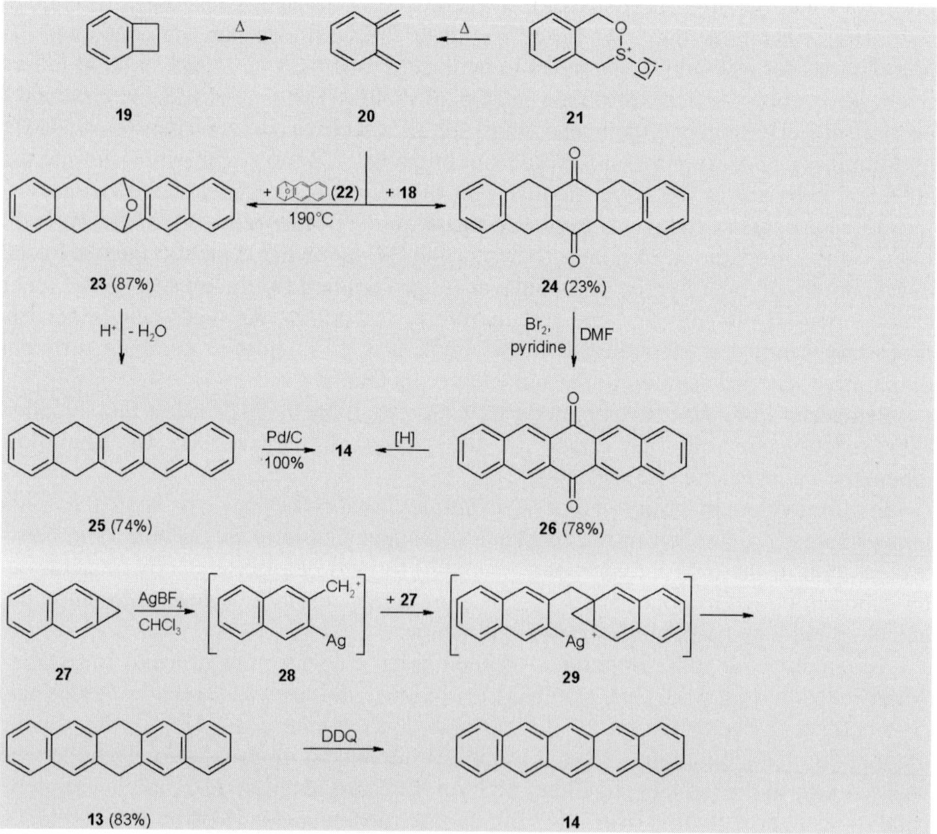

Scheme 3. ... and some modern routes to pentacene (**14**).

the more highly unsaturated (and reactive) *o*-xylylene (**20**, see Section 12.3). This was either generated by thermal ring-opening of cyclobutabenzene (**19**)[12] (see Section 16.3) or by fragmentation of 1,4-dihydro-2,3-benzoxathiin-3-oxide (**21**).[13]

Trapping of **20** with anthracene-1,4-endoxide (**22**) furnished adduct **23**, which was converted into **25** by acid-catalyzed dehydration.[12] Again, catalytic dehydrogenation concluded this route to **14**. Double [2+4]cycloaddition of **20** to *p*-benzoquinone (**18**) provided **24** which was oxidized to 6,13-pentacenequinone (**26**),[13] as yet another precursor for **14**. The most recent approach to pentacene exploits the remarkable behavior of cyclopropaaromatics (*cf.* Section 16.3) such as **27** towards silver ions. In the presence of silver tetrafluoroborate the hydrocarbon dimerized in excellent yield to the dihydropentacene **13**. The reaction is probably initiated by addition of Ag[+] to the strained three-membered ring and the thereby formed intermediate **28** subsequently adds to unreacted starting material **27** to provide the dimeric structure **29**, from which **13** could arise by demetalation and ring-closure. Oxidation of **13** with DDQ to **14** terminated this (quite general) synthesis for linearly annelated polyarenes.[14]

The quality and generality of Clar's and Bailey's concepts for the preparation of linear acenes is underlined by the successful syntheses of both hexacene (**34**) and heptacene (**35**) by these approaches. Whereas Clar's route to **34** begins with the Friedel–Crafts acylation of 1,5-hydroxynaphthalene with phthalic anhydride,[15] Bailey's Diels–Alder-route employs a sequence in which one equivalent of 1,2-dimethylenecyclohexane (**17**) was first added to *p*-benzoquinone (**18**). The resulting 1:1 adduct is still a dienophile which could be reacted with 2,3-dimethylenedecalin. The formed diketone has the σ-skeleton required for **34**, which was finally obtained from this precursor by the route presented in Scheme 2 for the synthesis of pentacene (**14**).[16]

In a more recent synthesis of **34**, base-induced condensation of 2,3-bis-formyl-naphthalene (**30**) and the dihydroxyanthracenedione **31** afforded the oxygenated hexacene derivative **32** which was subsequently hydrogenated over zinc to 6,15-dihydrohexacene (**33**) (Scheme 4).[17] Oxidation with cuprous oxide finally provided the desired **34**.

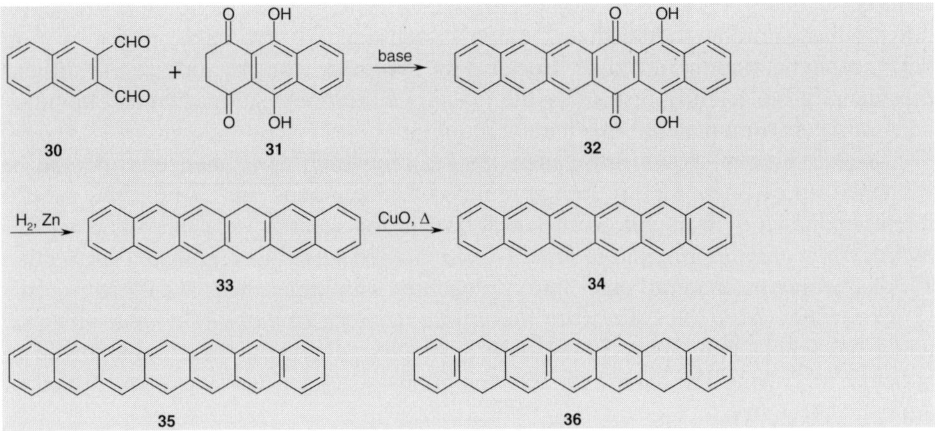

Scheme 4. Synthesis of hexacene (**34**) by a condensation route.

Clar, again, was the first to prepare the highly reactive condensed aromatic hydrocarbon heptacene (**35**).[18] The Bailey Diels–Alder route, now using the addition of two equivalents of 2,3-dimethylenedecalin to *p*-benzoquinone (**18**) in its first step is, however, more practical.[16] More recent approaches to heptacene derivatives have appeared in the chemical literature, and very often they proceed *via* hydro derivatives such as **36**, which are often not fully dehydrogenated in the last step, making the purification of the highly unstable **35** even more difficult.[12, 19] Even as late as 1997 information about its spectral, structural and chemical properties was lacking.[8] As already mentioned, heptacene marks the stability limit of the linearly annelated acenes; to extend this homologous series even further—and eventually to a polymeric acene[20]—will require some kind of protection of the reactive π-system, *e.g.* by bulky substituents. Alternatively ribbon structures might be made available by replacing at least some of the benzene rings of a molecule with the general structure **1** by other aromatic subsystems or by combining several of these simple linear ribbon structures with (formal) single bonds to bigger aggregates (see below).

14.2 Angularly annelated polycyclic aromatic hydrocarbons

As already seen in the section on helicenes (see Section 12.1) angular annelation can be extended considerably further than the linear connection of benzene rings. Furthermore, the number of methods of preparing PAHs which contain at least one phenanthrene unit—obviously the prerequisite for angular annelation—by far exceeds the number of ways leading to their linear isomers or analogs. Among these preparative methods the photochemical ring closure of stilbene derivatives, the only general method of synthesis for helicenes, is probably of greatest practical value, leading to a vast variety of angular annelated polyaromatics usually in short synthetic sequences starting from readily available substrate molecules.[21] According to the rules of orbital symmetry conservation the reaction begins with a (reversible) conrotatory 6π-electrocyclization of the *cis* isomer of the stilbene derivative to a *trans*-4a,4b-dihydrophenanthrene intermediate, which is subsequently oxidized, usually by air plus a catalytic amount of iodine. Because the *cis–trans* isomerization of the stilbene precursors is sufficiently fast under the reaction conditions, stereochemical homogeneity of the starting material is often not required.

As also already shown in Scheme 1 a helicene such as [6]-helicene (**4**) can be 'flattened' to coronene (**5**) by formal removal of one ethene unit. To discuss the different syntheses of **5** at this point is reasonable for 'generic' reasons—keeping the hydrocarbon *aufbau* principle in mind—especially because this circulene (see Section 12.1) constitutes a 'natural' entry into cycloarenes with larger internal cavities.

Coronene—from the Latin word for *corona* meaning ring or circle—was first prepared by Scholl and Meyer in 1932 in a lengthy, low-yield route summarized in Scheme 5,[22] using the harsh reaction conditions so typical for this early period of aromatic chemistry.

The synthesis began with a double Friedel–Crafts acylation of *m*-xylene (**38**) with

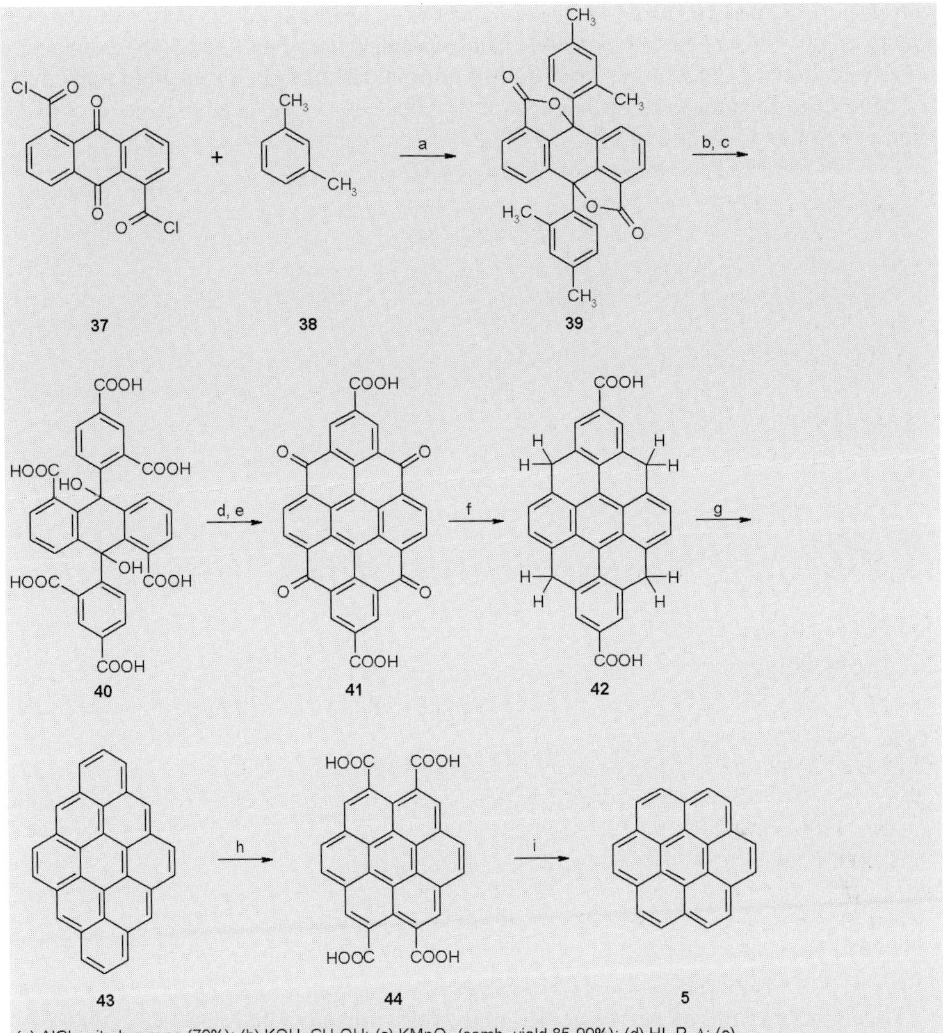

(a) AlCl$_3$, nitrobenzene (70%); (b) KOH, CH$_3$OH; (c) KMnO$_4$ (comb. yield 85-90%); (d) HI, P, Δ; (e) oleum, then P$_2$O$_5$ / H$_3$PO$_4$ (comb. yield 48%); (f) HI, P, Δ (no yield given); (g) soda lime, Cu, 500°C (10%); (h) HNO$_3$, Δ; (i) soda lime, 20 torr, 500°C (comb. yield 10%)

Scheme 5. The first synthesis of coronene (**5**) by Scholl and Meyer.

the diacid chloride **37** furnishing the bis lactone **39**, which already has all the carbon atoms of the final product plus the functionality required for the formation of the missing six-membered rings. Saponification and permanganate oxidation converted **39** into the dioxyhexacarboxylic acid **40**. When this was reduced with hydrogen io-dide and phosphorous in boiling acetic acid, a fully aromatic hexacarboxylic acid re-sulted which underwent ring-closure by loss of four equivalents of water to provide the dibenzocoronene-diquinone-diacid **41** under forcing conditions (oleum, then polyphosphoric acid). The keto functions were removed by another reduction step, and when the resulting diacid **42** was heated with soda lime in the presence of copper *anti*-di-*peri*-dibenzocoronene (**43**) was generated. Degradation to coronene (**5**) was finally achieved *via* the tetraacid **44**. The overall yield of this ten-step sequence is poor, with the two decarboxylations contributing most strongly to the yield reduction.

Much shorter and more efficient is a classical coronene synthesis developed by Clar and Zander (Scheme 6).[23]

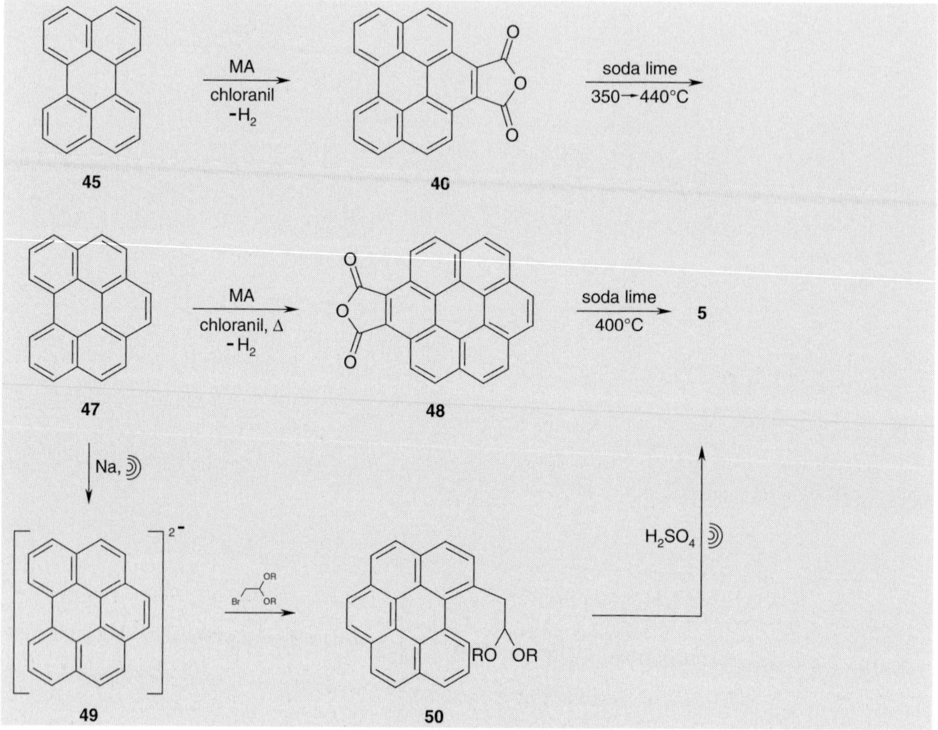

Scheme 6. The simple coronene (**5**) synthesis of Clar and Zander.

Heating a mixture of perylene (**45**) and excess maleic anhydride (MA) under re-flux in the presence of chloranil as oxidizing agent led to the Diels–Alder adduct **46**, which could be decarboxylated and decarbonylated to 1,12-benzopyrene (**47**). Re-peating this annelation sequence yielded—*via* **48**—coronene (**5**) in 25% overall yield. A still further simplification was realized[24] when **47** was first converted into its bis anion **49** by treatment of the hydrocarbon in tetrahydrofuran, with ultrasonic vibra-tion, and subsequent trapping of the latter with bromoacetaldehyde diethyl acetal, a reaction affording **50**. When this was treated with concentrated sulfuric acid, again with ultrasound activation, it cyclized to **5** in good yield. Because **47** could be pre-pared from **45** by the same approach, a method for the preparation of coronene is at hand which avoids forcing conditions altogether and can easily be performed on the multigram scale.

Several pathways to **5** have been reported which employ cyclophanes as sub-strates and reaction intermediates. At first sight the use of these archetypical three-dimensional aromatics (see Section 12.3) for a flat hydrocarbon *par excellence* might seem surprising. Actually this transformation is often quite easy because it can capitalize on strain reduction and gain in benzene resonance energy. Thus dehalogenation of 2,7-bis-bromomethylnaphthalene (**51**) with either phenyllithium[25] or sodium-dioxan[26] furnished [2.2](2,7)naphthalinophane (**54**) in moderate yields (15–20%, Scheme 7).

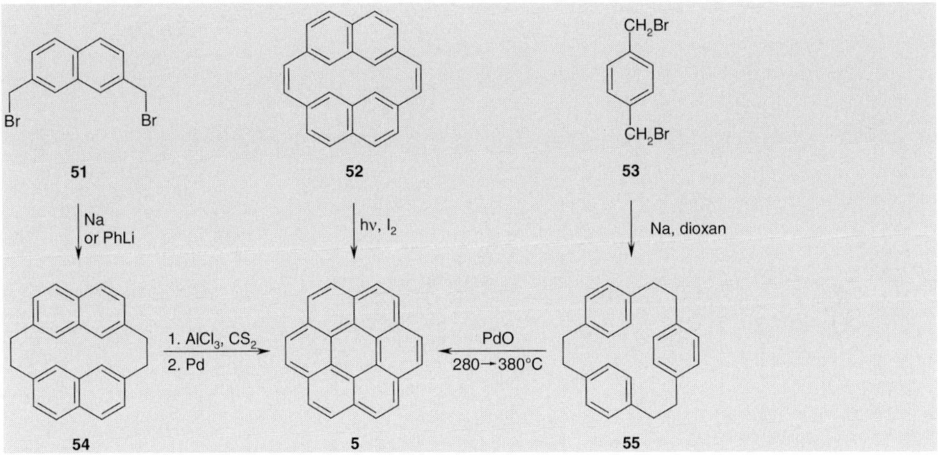

Scheme 7. Cyclophanes as precursors for coronene (**5**).

Whereas the direct dehydrogenation of **54** with classic aromatization reagents (S, Se, Pd/C, *etc*.) failed, **5** could be obtained by reacting **54** first with aluminum trichloride in carbondisulfide—the so-called Scholl reaction, a ring-forming process—followed by aromatization of the reaction mixture over palladium at 260 °C. Replacement of the ethano bridges of **54** by double bonds generates the cyclophandiene **52**, which formally is composed of two *cis*-stilbene units. Indeed, irradiation at 254 nm in the presence of iodine caused formation of coronene (**5**).[27] Finally, Wurtz–Fittig coupling of 1,4-bis(bromomethyl) benzene (**53**) with sodium in dioxan led—*inter alia*—to the trinuclear cyclophane tris-*p*-xylene (**55**), which on heating in the presence of palladium monoxide provided **5** again.[28]

Among the circulenes coronene is unique because it is the only planar compound in this series of condensed aromatic hydrocarbons (see the non-planar members corannulene ([5]-circulene) and [7]-circulene in Section 12.2). Circulenes beyond [7]-circulene have not been prepared; according to molecular models they are non-planar, as is also to be expected for [4]-circulene (see Section 12.2).

The situation changes, however, once linearly annelated subunits are incorporated into the circulene ring. A particular important and interesting representative of these 'extendedcirculenes' is kekulene (**64**), named after the discoverer of the cyclic structure of benzene, A. Kekulé and prepared by Staab and Diederich in 1983 (Scheme 8).[29]

The first important intermediate in the synthesis was 5,6,8,9-tetrahydrodibenzo[*a,j*]anthracene (**57**),[30] available from *m*-xylene (**38**) in four steps involving nitration of **38**, condensation of the resulting dinitro derivative with benzal-

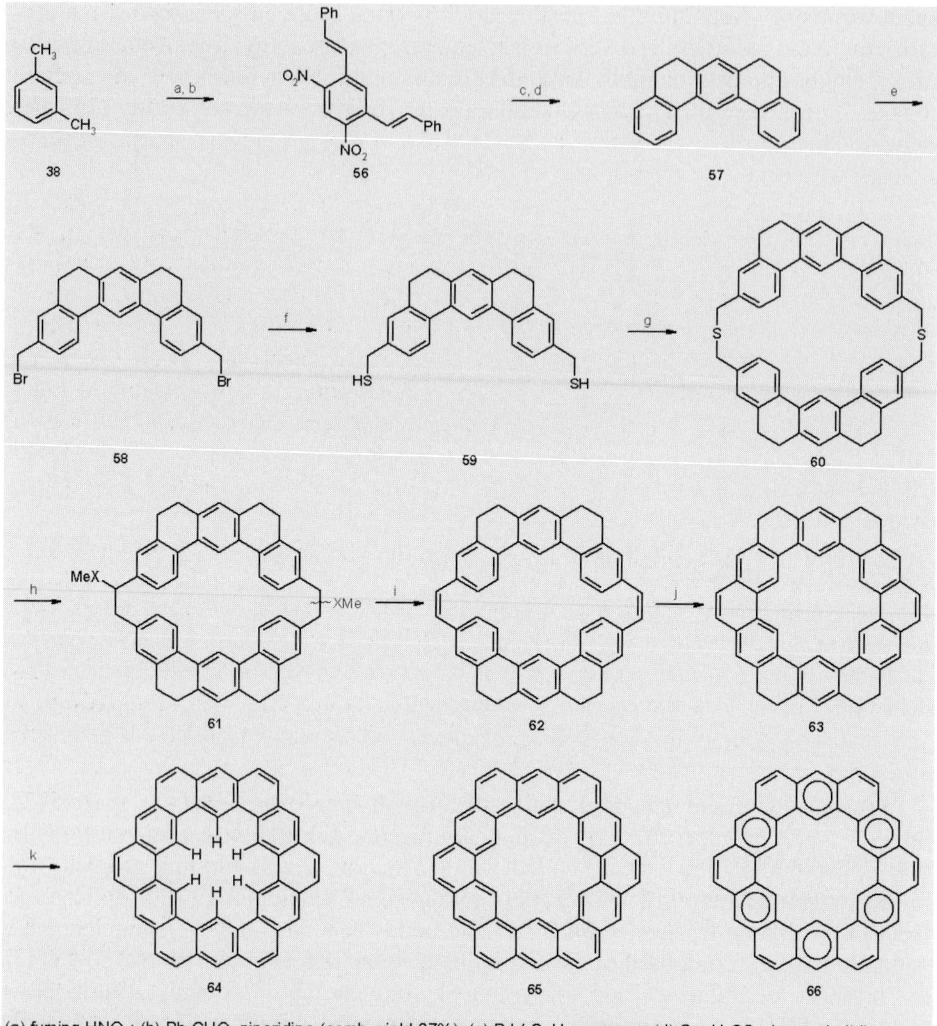

(a) fuming HNO₃; (b) Ph-CHO, piperidine (comb. yield 37%); (c) Pd / C, H₂, pessure; (d) Cu, H₂SO₄, isoamyl nitrite (comb. yield 7.4%); (e) (CH₂O)ₙ, H₃PO₄ / HOAc / HBr, then gas. HBr (51%); (f) thiourea, EtOH, Δ, then NaOH (76%); (g) + **58**, KOH, benzene / ethanol, high dilution (60%); (h) CF₃SO₃CH₃, then *tert*-BuOK (69%); (i) *m*-chloroperbenzoic acid, then 450°C (43%); (j) hv, I₂, benzene (70%); (k) DDQ in 1,2,4-trichlorobenzene, 100°C, 3 d (91%)

Scheme 8. The synthesis of kekulene (**64**) by Staab and Diederich.

dehyde in the presence of piperidine to the bis-styryl compound **56**, catalytic hydrogenation of the latter to the saturated diamine derivative, and then its twofold cyclization *via* the bis-diazonium compound. This last step especially is preparatively very costly giving the hydrocarbon **57** in 12% yield only. The subsequent steps owe much to paracyclophane chemistry (see Section 12.3). In fact, the next important synthetic intermediates are cyclophanes. Thus hydroxymethylation of **57** with paraformaldehyde then treatment with hydrogen bromide led to a cyclophane building block, the dibromide **58**, which was converted by reaction with thiourea and base to the other 'half', the dithiol **59**. Coupling of **58** and **59** by treatment with potassium hydroxide under high dilution conditions in benzene-ethanol provided the dithiaphane **60** in the surprisingly high yield of 60%. Although from here on several routes to **64** might be envisaged—and have, in fact, been tried—the one shown involving hydrogenated precursors proceeded most satisfactorily. Introduction of too much unsaturation at an too early stage caused severe losses in yield.

Thus in the second half of the synthesis **60** was first exhaustively methylated with methyl fluorosulfonate in dichloromethane and the bis-sulfonium salt formed was subjected to double Stevens rearrangement with potassium *tert*-butoxide in tetrahydrofuran. Oxidation of the produced thioether **61** (mixture of isomers) with *m*-chloroperbenzoic acid then led to the corresponding bis-sulfoxides which on heating to 450 °C yielded the diene **62** in acceptable yield; on irradiation this underwent the expected stilbene→phenanthrene ring closures (see above) furnishing hydrocarbon **63**, the first kekulene derivative with the complete σ-skeleton. The final dehydrogenation steps required extensive optimization. Success was eventually achieved with DDQ as the dehydrogenation reagent under the unusual and harsh reaction conditions shown in the scheme.

Kekulene (**64**) is of interest because on the one hand it can be regarded as a combination of two annulene perimeters connected by radial single bonds as indicated by the resonance structure **65**. These 'annulene rings' are both of the [4n+2]-type—the outer being a [30]annulene and the inner a [18]annulene (see Section 10.4).[31] On the other hand there could be localization of the π-electrons into benzene rings, as suggested by Clar in his *sextet concept*,[32] which in this compound could lead to **66** as an important or dominating resonance contributor (altogether 200 different resonance structures can be formulated for **64**!). Because kekulene—in contrast with the lower circulenes discussed above—has both external and internal hydrogen atom substituents, ^{1}H NMR spectroscopy should, in principle, enable a decision about the extent to which the diatropicity in the macrocyclic system can compete with ring-current induction within the benzenoid subunits. The molecular structure as determined by X-ray structural analysis and various spectroscopic methods enable clear distinction between these two alternatives. The bond lengths show a remarkable localization of aromatic sextets and double bonds, and the ^{1}H NMR spectrum also speaks clearly against a significant contribution from an annulenoid structure such as **65**. In 1,3,5-trichlorobenzene-d$_3$ at *ca* 200 °C the proton spectrum of the extremely insoluble **64** contains three signals in a 2:1:1 ratio at δ = 7.95, 8.37, and 10.45, and no signal shifted upfield, as would be expected for the internal protons should there be a pronounced diatropicity of the macrocyclic system. Absorption and emission spectra of kekulene are also in full agreement with the localized structure **66**.[33]

14.3 Condensed aromatic hydrocarbons with ribbon structures

As demonstrated by the linear acenes on the one hand and the angularly annelated polycyclic aromatic hydrocarbons on the other (see above), the latter mode of connecting benzene rings leads to far more stable compounds—a difference which already manifests itself in the anthracene-phenanthrene isomer pair and is caused by the higher number of resonance structures with localized benzene rings that can be drawn for the angular isomers (Clar's sextet rule).[32]

Thus if one wishes to synthesize polycyclic aromatic ribbons based exclusively on benzene rings, one way to surpass the stability/isolability limit set by heptacene (**35**) is by introducing phenanthrene units into these 'graphite ribbons'. In fact, the so-called [*n*]phenacenes, a series of PAHs beginning with phenanthrene itself and proceeding *via* chrysene to picene (**67**, [5]phenacene) and fulminene (**68**, [6]phenacene) consist of nothing but repeating phenanthrene units (Scheme 9).

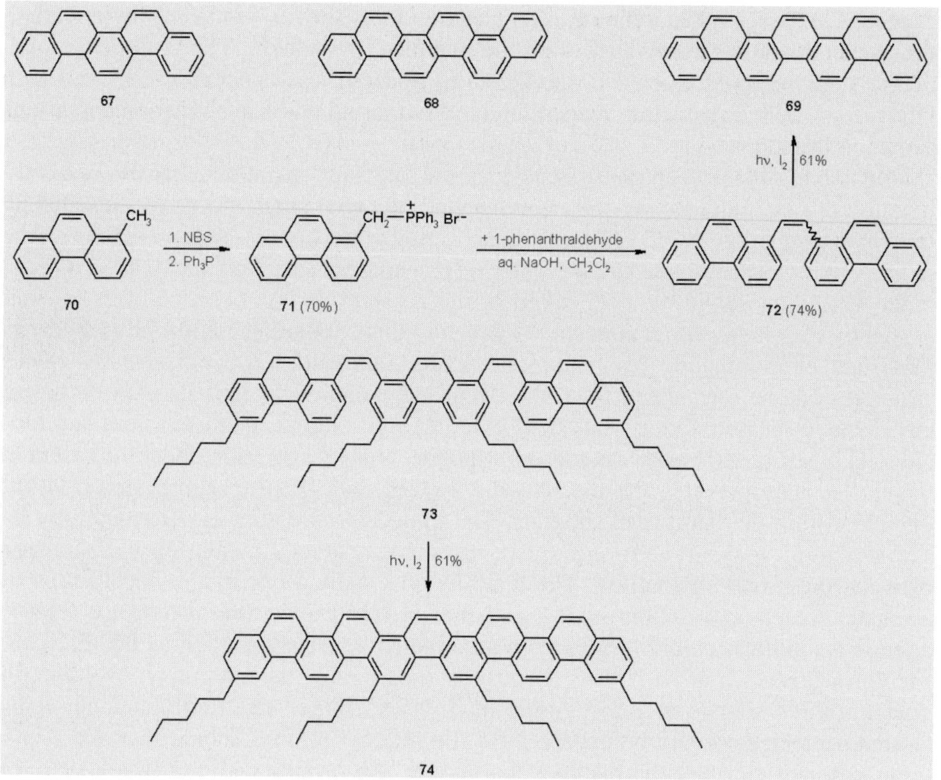

Scheme 9. Preparation of [*n*]phenacenes by stilbene→phenanthrene-type photocyclizations.

Various routes to individual members of this series have been reported,[34] the stilbene→phenanthrene-photocyclization being the most general method. In the parent hydrocarbon series it has been used to obtain the longest [*n*]phenacene known to date, [7]phenacene (**69**), recently synthesized by Mallory and co–workers.[35] The hydrocarbon was obtained in satisfactory overall yield by first preparing the phosphonium salt **71** from 1-methylphenanthrene (**70**), itself readily available by photocyclization of *o*-methylstilbene. Wittig reaction of the ylid prepared from **71** with 1-

phenanthraldehyde led to **72** (mixture of isomers) which cyclized to **69** on irradiation in the presence of iodine.[35] In principle analogous routes could provide even more extended [*n*]phenacene hydrocarbons, were it not for the extreme insolubility of these hydrocarbons. Thermal stability is no longer a problem—hydrocarbon **69** begins to decompose only at its melting point of 565 °C! To prepare longer ribbons of this type, which might have interesting and possibly useful properties as, *e.g.*, electrical conductors or as non-linear optical materials, solubilizing groups are hence mandatory. This is illustrated by the [11]phenacene derivative **74**, prepared by double photocyclization of the bis-(arylvinyl)phenanthrene derivative **73**.[36]

The width of these ribbon structures can be increased by adding 'another row of benzene rings'. This can be realized in different ways, for example with a homologous series of *peri*-condensed naphthalenes beginning with naphthalene and then progressing through perylene (**45**) and terrylene (**75**) to quaterrylene (**76**) (Scheme 10). Originally these so-called rylenes were prepared[2] by the typical brute-force methods of early aromatic chemistry, involving C–C coupling under harsh conditions (high temperature, the Lewis-acids melts of the Scholl reaction *etc.*).[37] These naphthalene oligomers are again highly insoluble, high-melting hydrocarbons. To investigate their chemical and physical properties in solution tetrakis-*tert*-butyl derivatives have been prepared by Müllen and co–workers applying the modern stepwise approach summarized in Scheme 10.[38]

(a) 3 mol% Pd(PPh₃)₄, K₂CO₃, toluene, 3 d, reflux (74%); (b) K, DME, 7 d, room temp., CdCl₂ (48%); (c) AlCl₃, CuCl₂, CS₂, 8 h, room temp. (48%)
Scheme 10. A modern route to rylenes by Müllen and co–workers.

To synthesize the *tert*-butyl derivative **81** of quaterrylene (**76**), 4,4'-dibromo-1,1'-binaphthyl (**77**) was coupled with two equivalents of the arylboronic acid **78** in the presence of Pd(0) to yield the tetranaphthyl intermediate **79**. Under the influence of potassium in 1,2-dimethoxyethane (DME) two of the three missing single bonds were then created, and the hydrocarbon **80** generated in acceptable yield. Scholl-type reaction conditions were employed to form the last connecting bond. By use of an analogous sequence the still larger rylene derivative **82**, the most extended hydrocarbon of this type presently known, could also be prepared. Modern metal-mediated coupling reactions have also been used to prepare several functional derivatives of **76**.[39]

The so-called *circumarenes* are another class of PAHs with a ribbon structure. The parent system is now coronene (**5**,'circumbenzene'), the next higher homolog has been termed ovalene (**83**, 'circumnaphthalene'), because of its egg-shaped appearance,[40] and the largest representative with a linear acene core[41] known to date is circumanthracene (**84**). Although the first synthesis of this hydrocarbon was claimed by Clar, who used a 'controlled graphitization process',[42] later work by the same group showed the original structure assignment to be erroneous.[43] Diederich and Broene, however, successfully prepared **84** in 1991 by the route described in Scheme 11.[44]

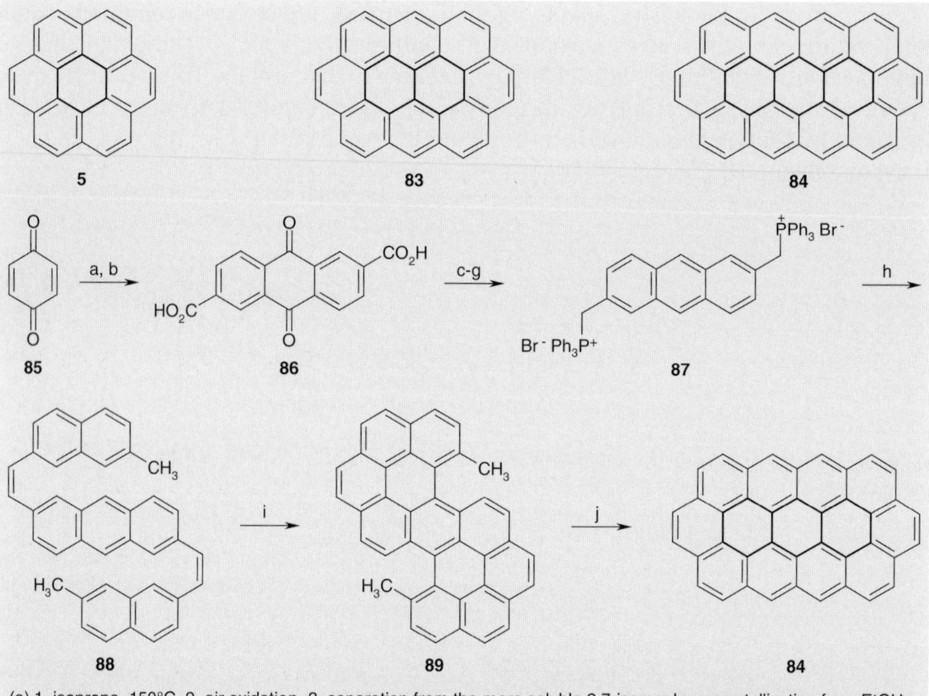

(a) 1. isoprene, 150°C, 2. air oxidation, 3. separation from the more soluble 2,7-isomer by recrystallization from EtOH (76%); (b) $Na_2Cr_2O_7$, H_2O, 250°C, 7 h (75%); (c) Zn, $CuSO_4$, NH_4OH (90%); (d) 1. $SOCl_2$, 2. EtOH (77%); (e) $LiAlH_4$, THF (98%); (f) PBr_3, CH_2Cl_2 (89%); (g) PPh_3, DMF (68%); (h) EtOH, EtONa, 1-methyl-2-naphthaldehyde (34%); (i) hv, toluene, Ar, I_2, 3 h (93%); (j) DDQ, 1,2,4-trichlorobenzene, Ar, 90°C, 15 h (ca. 50%)

Scheme 11. The synthesis of circumanthracene (**84**) by Diederich and Broene.

Starting from *p*-benzoquinone (**85**) the bis-phosphonium salt **87** was prepared *via* the diacid **86** employing standard reaction steps. When subsequently the bis-ylid generated from **87** was coupled with 7-methyl-2-naphthaldehyde the bis-styryl derivative **88** was formed; this underwent the expected photocyclization to **89** on irradiation with a 450-W mercury arc lamp. With the two methyl substituents properly oriented in their respective bay areas, dehydrogenation to the target hydrocarbon **84**, by 2,3-dichloro-5,6-dicyano-1,4-benzoquinone (DDQ), occurred under comparatively mild conditions in acceptable yield.

Although the chemical and physical properties of unsaturated polymers can often be predicted from the appropriate data of the corresponding oligomers[45]—their formal subunits—it is unsatisfactory that almost none of the hydrocarbons described above exceeds or even reaches 10 repeating units. The preparation of a fully unsaturated all-carbon ladder polymer was thus a considerable challenge. It was met by Schlüter and co-workers who used a Diels–Alder strategy to prepare truly polymeric unsaturated systems of a defined structure, employing either double dienes and double dienophiles as starting materials or substrates which are hybrids of dienes and dienophiles.[46] The latter approach is demonstrated by the preparation of the fully conjugated ladder polymer **92**, a 'molecular board', from the 'mixed' precursor **90** (Scheme 12).[47]

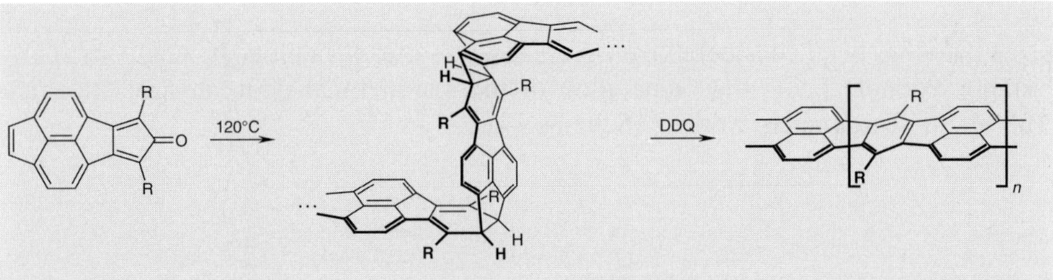

Scheme 12. The preparation of a molecular board, hydrocarbon **92**, by Schlüter and co-workers.

Besides the strain-activated double bond (the dienophile) and the cyclopentadienone unit (diene) the monomer **90** contains either a flexible alkyl loop (R–R = –$(CH_2)_{12}$–) or two ester substituents with long alkyl chains (R = R = –$CO_2C_{12}H_{25}$) which results in a dramatic increase in the solubility of the resulting polymer, growth of which, furthermore, is not impaired or prevented at an early stage by insolubility. Heating **90** in the presence of *o,o'*-di-*tert*-butyl-*p*-methylphenol as an antioxidant (which prevents the formation of byproducts) led to the still partially saturated polymer **91**, as a chloroform-soluble polyadduct—the initially formed [2+4]cycloadduct(s) having lost their carbonyl group as carbon monoxide by a retro reaction under the reaction conditions. The polymer **91** was dehydrogenated by exposure to DDQ in dichloromethane, the process taking place at room temperature already. Because each five-membered ring in **92** is joined to three benzene rings this board molecule should indeed be totally planar. Molecular weight determinations revealed that for the bridged polymer *n* is *ca* 15 whereas for the diester derivate 50 is exceeded.

An 'endless' structure need not only be realized by a ladder system but also by joining the ends of a linear acene-derived precursor, **93**, to a 'cycloacene' built exclusively from benzene rings, **94**, torands which have been termed beltenes,[48, 49] columnenes,[50] or collarenes[51] (Scheme 13).

Although fully unsaturated hydrocarbons of type **94** so far have not been synthe-

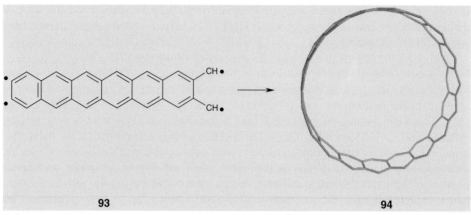

Scheme 13. From linear acenes (**93**) to molecular belts (**94**).

sized there has been considerable progress towards this difficult goal, a pa   rticularly striking example being the preparation of the symmetrical dodecahydro[12]beltene **101** by Stoddart and co–workers (Scheme 14).[51–53]

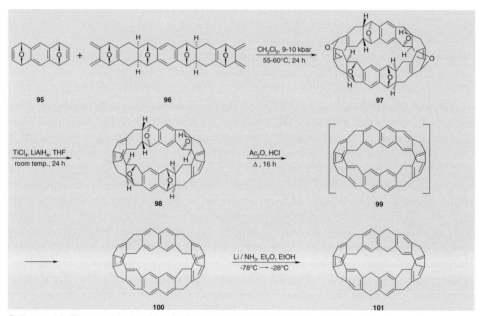

Scheme 14. The synthesis of the [12]collarene derivative **101** by Stoddart and co-workers.

In a most remarkable multiple cycloaddition sequence a belt-like intermediate, **97**, given the trivial name kohnkene, was obtained from the double dienophile **95** and the double diene **96** in *ca* 20% yield. Three factors account for the success of the trans-formation—both building blocks already have a concave structure, the Diels–Alder reaction prefers—as in countless other examples—the *endo* mode of addition, and, finally, the cycloaddition was performed under high-pressure (9-10 kbar), conditions long known to be rate-accelerating in the Diels–Alder process and which are particu-larly effective for reactions with a large negative volume of activation.[54] The torand **97** was characterized by X-ray structural analysis which also revealed that chloroform molecules are bound between layers of **97** in a clathrate fashion. When kohnkene (**97**) was treated with $TiCl_4$-$LiAlH_4$ in tetrahydrofuran at ambient temperature two oxygen atoms could be removed and the macropolycycle **98** was isolated in 43% yield. To remove the remaining four ether bridges, **98** was refluxed in a mixture of acetic anhydride and hydrochloric acid. Rather than isolating the expected 'symmetri-cal' hydrocarbon **99** with two anthracene and two benzene units the authors obtained the unsymmetrical compound **100** in 56% yield. Evidently this thermodynamically more stable isomer is generated by some kind of acid-catalyzed isomerization which occurs during dehydration. Both **100** and the symmetrical dodecahydro[12]beltene **101** formed from it by Birch reduction offer themselves as precursors for the fully conjugated cycloacene, but this transformation has not yet been accom-plished.[55]

Replacing the benzene rings in molecular belts such as **94** by larger aromatic sub-structures leads to tubular hydrocarbons with an obvious structural relationship to the carbon nanotubes which have become of prime importance in connection with the chemistry and physics of the fullerenes and other novel forms of carbon.[56] Because

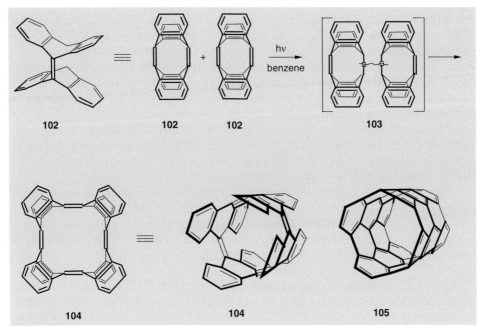

Scheme 15. The synthesis of the first tubular, fully conjugated hydrocarbon, **104**, by Herges and co–workers.

nanotubes are prepared by methods—*e.g.* the Krätschmer–Huffman process[57]—which differ fundamentally from the 'normal' synthetic procedures, the directed synthesis of a fully conjugated tubular hydrocarbon constitutes a major advance in preparative organic chemistry.

The first and very elegant solution of this problem was provided by Herges and co–workers, who subjected tetradehydrodianthracene (**102**)—a hydrocarbon which we already encountered in our discussion on pyramidalized double bonds (see Section 7.4)—to what amounts to a ring enlargement metathesis reaction (Scheme 15).[58]

Photolysis of a suspension of **102** in benzene in a quartz reactor furnished a dimer, the structure of which, **104**, was established by X-ray diffraction. To rationalize the formation of this tubular, fully conjugated hydrocarbon it has been suggested that the photodimerization of **102** is initiated by photoexcitation of the solvent benzene. Energy transfer could subsequently activate the starting material and trigger the addition to provide the cyclobutane intermediate **103**, which would subsequently loose part of its high strain by ring-opening (breaking of the four–membered ring at the marked positions) to **104**. The 'picotube' **104**, consisting of four anthracene units, has a diameter of 5.4 Å and a length of 8.2 Å. Dehydrocyclization should convert it into the 'buckytube' **105** with a length of three benzene rings and a circumference of eight.

14.4 Graphite substructures

The term 'graphite substructure' evades an exact definition. Of course, practically all of the polycyclic aromatics discussed so far in this chapter can be regarded as smaller sections of graphite. The hydrocarbons I want to present now, however, are considerably larger than any of the PAHs described so far (with the exception of molecular boards such as **92**). As graphite substrucures the so-called 'giant PAHs'[59] can be regarded, compounds with chemical and physical properties that are beginning to converge with those of graphite. Just as various cluster systems can be considered as an intermediate stage of matter between the atomic and the molecular level on the one hand and that of bulk materials on the other, the graphite subsystems assume an intermediate position between the traditional PAHs, as, *e.g.*, summarized in Clar's[2] or Harvey's monographs[8] and graphite itself.

The synthesis of giant PAHs constitutes a considerable conceptual and practical challenge. Although we would expect to use many of the concepts employed previously for the preparation of the smaller PAHs, *i.e.* the drastic routes of the classical period of polycyclic aromatic chemistry,[2–8] and the more recent Diels–Alder additions, photocyclizations, cyclodehydrogenations *etc.*, new approaches are also clearly needed.

To begin a synthesis with an already 'fairly extended' PAH precursor, *i.e.* to use the vast experience of our predecessors, is intuitively attractive, but might reach its limits quickly because of the low yield of many of the older syntheses or the poor solubility of such starting materials. Although this approach has been employed successfully (see below) most modern routes to very large PAHs favor a different strategy: A hydrocarbon is first assembled—often by use of modern metal-mediated coupling reactions—which does not yet have the highly symmetrical, planar structure of the intended target, because it is this feature that often causes the high insolubility. This precursor is then planarized towards the end of the synthesis or even in the very last step. As we shall see, for the practical synthetic work it is often very favorable for these precursors to have an internally 'twisted' structure.

An example for the first protocol is provided by the aluminum trichloride-catalyzed dimerization of coronene (**5**), an intermolecular Scholl reaction which has been studied by several authors (Scheme 16).[60–62]

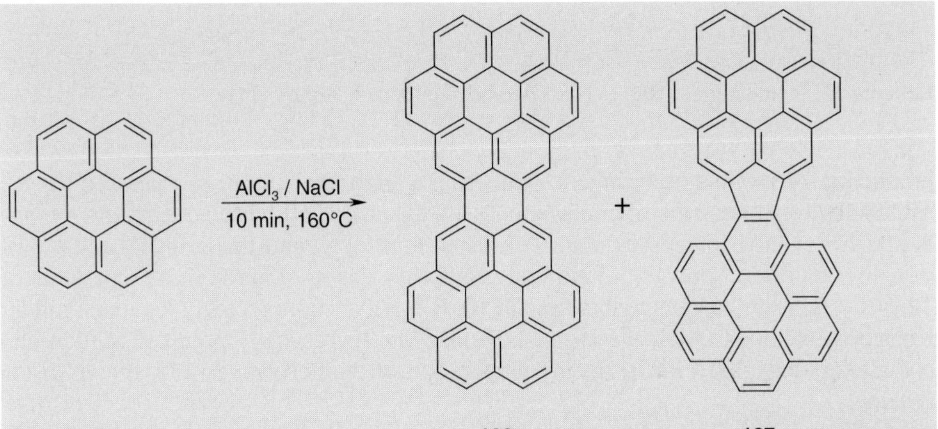

Scheme 16. Preparation of benzo[1,2,3-*b*,*c*:4,5,6-*b*',*c*']di-coronene (**106**) by Scholl dimerization of coronene (**5**).

This oxidative dimerization led to a mixture of the alternant hydrocarbon benzo[1,2,3,-*b*,*c*:4,5,6-*b*',*c*']dicoronene (**106**) and its non-alternant isomer cyclopenta[1,2-*a*:3,4,5-*b*',*c*']dicoronene (**107**). Although it has been claimed that the crude reaction mixture consists of approximately 80% **106** and 20% **107**,[62] the separation of these hydrocarbons to furnish analytically pure materials resulted in a dramatic reduction in yield.[63]

The second approach will be illustrated by several examples from the group of Müllen, who recently has made breathtaking advances *en route* to 'synthetic graphite'. As a first typical example we want to consider the preparation of various hexa-*peri*-hexabenzocoronenes **111** (R *inter alia* = *n*-alkyl, tert-butyl; Scheme 17).[64, 65]

To prepare the 'twisted' precursor **110**—hexaphenylbenzene (**110**, R = H) has a propeller structure[66]—the tolane derivative **108** was either trimerized using dicobaltoctacarbonyl as a catalyst[67] or **108** was cycloadded to the tetraphenylcyclopentadienone (or tetracyclone) derivative **109**. The latter approach, which involves a Diels–Alder addition between **108** and **109** then thermally-induced loss of CO from the primary adduct,[68] is synthetically more attractive than the former because the substituents in the two starting components can be varied independently (see below). In the cyclotrimerization route only uniform functionalization can be achieved–or the

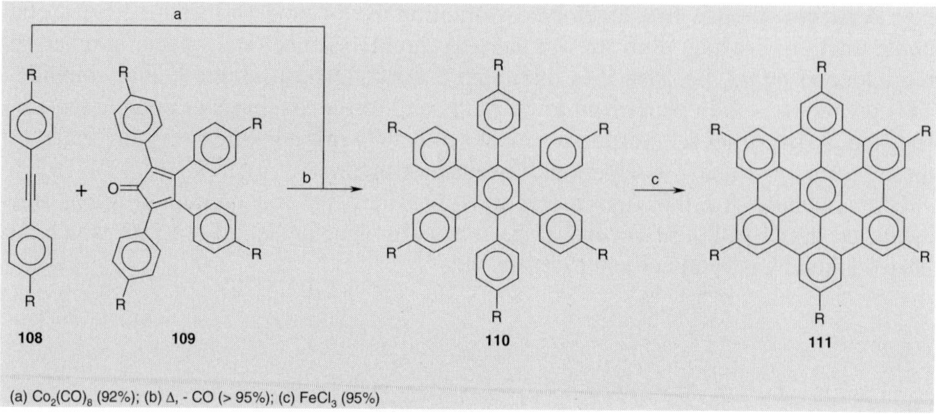

(a) Co$_2$(CO)$_8$ (92%); (b) Δ, - CO (> 95%); (c) FeCl$_3$ (95%)

Scheme 17. From tolanes (**108**) to hexa-*peri*-hexa-benzocoronenes (**111**).

production of mixtures of isomers. Numerous variations have been proposed for the cyclodehydrogenation of oligophenylenes to extended, all-benzenoid hydrocarbons, although oxidative processes, *e.g.* with vanadium(V)[69] and thallium(III) salts[70] or iron(III) chloride[71] prevail. Aluminum trichloride was used in the first preparation of **111** (R = H) from hexaphenylbenzene (**110**, R = H);[64] more recently the much milder Kovacic conditions (AlCl$_3$-FeCl$_3$)[72] have become increasingly popular. Müllen and co–workers have used FeCl$_3$ for the preparation of the derivatives **111** shown in the scheme.

Müllen and his group have demonstrated that tetracyclone-additions to tolanes of increasing complexity, followed by cyclodehydrogenations, do, indeed, currently constitute the *via regia* to very large PAHs. Treating octaphenylquinquephenyl (**113**), a hydrocarbon previously prepared by Ried and co-workers by double Diels–Alder addition of tetracyclone (**109**, R = H) to 1,4-bis(phenylethynyl)benzene (**112**), a double tolane with a 'shared' central benzene ring[73, 74] with copper(II) triflate-aluminum trichloride in carbon disulfide at room temperature resulted in quantitative formation of the C$_{78}$H$_{26}$ PAH **114**, as shown by laser desorption-time-of-flight-mass spectrometry (LD-TOFMS) (Scheme 18).[75]

Interestingly, the same hydrocarbon could be prepared from the *meta*-phenylene-bridged isomer **115**. This is surprising since intuitively one would not expect this cyclization to take place, **115** being a 'non-planarizable' (helical) precursor with two phenyl substituents directly on top of each other. The **115**→**114**-isomerization and cyclodehydrogenation has been rationalized by postulating a quantitative 1,2-phenyl shift taking place at the central *meta*-phenylene unit of **115**. It should come as no surprise, however, that **115** was obtained by tetracyclone addition to 1,3-bis(phenylethynyl)benzene.

The next step towards graphite involved the addition of tetracyclone (**109**, R = H) to the triyne **116** (Scheme 19).[76]

Addition of two equivalents of the diene in diphenylether at 190 °C occurred exclusively at the outer, sterically less shielded acetylene groups and yielded the tolane derivative **117**. Although highly hindered, with Co$_2$(CO)$_8$ catalyst this could be cyclotrimerized to **118** in good yield. Treatment of this soluble(!) hydrocarbon, which

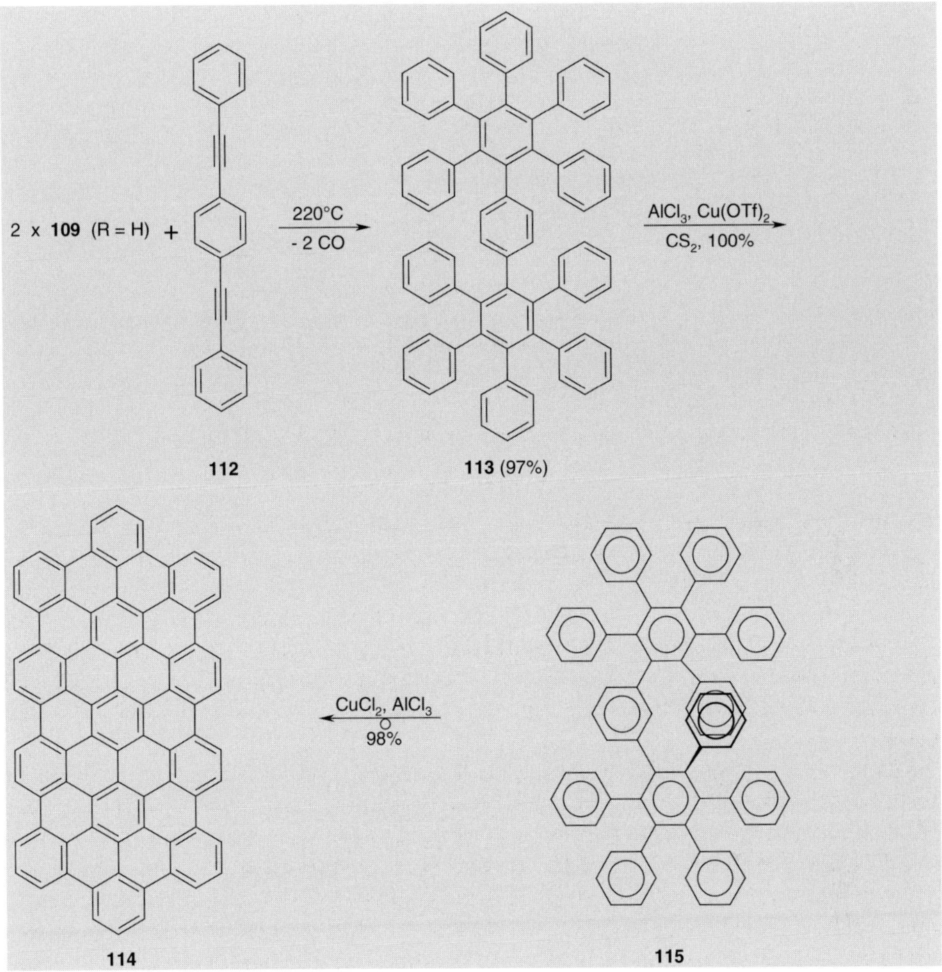

Scheme 18. Müllen's syntheses of superacenes from tolanes—the preparation of a $C_{78}H_{26}$ hydrocarbon, **114**, ...

contains 37 benzene rings, with AlCl$_3$-Cu(OTf)$_2$ in carbon disulfide resulted in a black solid completely insoluble in all common solvents. As indicated by LD-TOFMS analysis cyclodehydrogenation has occurred and signals could be recorded in the mass region expected for a C_{222} graphite substructure in which all bonds symbolyzed by dotted lines in **118** had indeed been generated.

Even with this, hitherto unimaginably large PAH, the above strategy has not been exhausted. One extension involved[77] the initial reaction of 3,3',5,5'-tetrakis(triisopropylsilylethynyl)biphenyl with 3,4-bis[4-(triisopropylsilylethynyl)phenyl]-2,5-diphenyl-cyclopentadiene at 180 to 200 °C; after deprotection with Bu$_4$NF in tetrahydrofuran at room temperature, the preceding sequence was repeated. When drawing the formulae for this synthesis the reader is adviced to use a sufficiently large piece of paper. Treatment with CuCl$_2$–AlCl$_3$ of the product isolated after the first cycle furnished a $C_{132}H_{34}$ hydrocarbon!

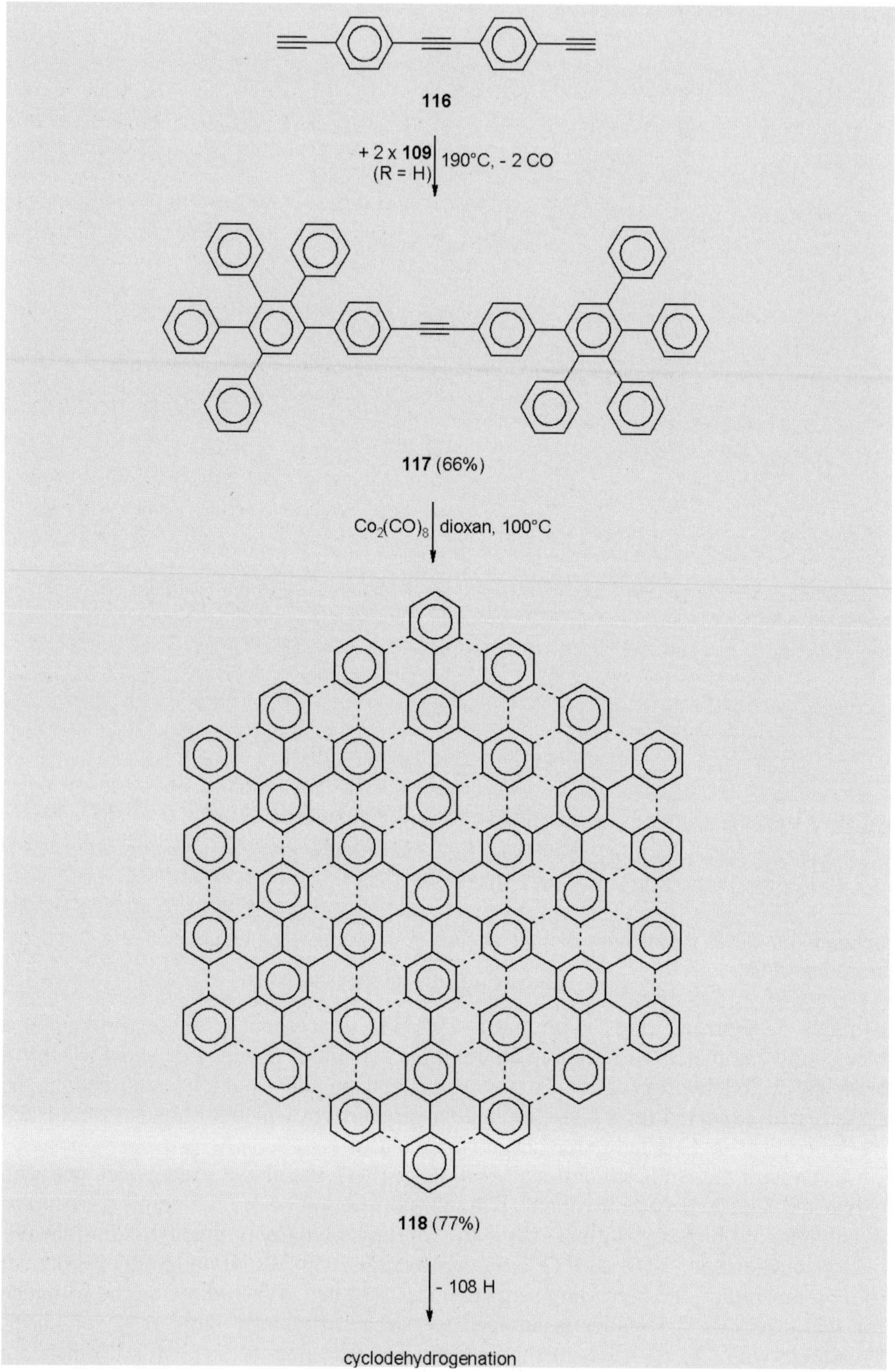

Scheme 19. ... and of a C_{222} hydrocarbon, **118**.

References

1. Penta-, hexa- (**10**), octa- and deca-*m*-phenylene have been synthesized by Staab and co-workers by metal-catalyzed cyclooligomerization of the bis Grignard reagent prepared from 3,3'-dibromo-biphenyl: H. A. Staab, F. Binnig, *Tetrahedron Lett.*, **1964**, 319–321; H. A. Staab, F. Binnig, *Chem. Ber.*, **1967**, *100*, 293–305; H. A. Staab, F. Binnig, *Chem. Ber.*, **1967**, *100*, 889–892; *cf.* H. A. Staab, H. Bräunling, *Tetrahedron Lett.*, **1965**, 45–49.

2. E. Clar, *Polycyclic Hydrocarbons*, Vol. I, II, Academic Press, New York, N. Y., **1964**.

3. A. Bjorseth, T. Ramdahl, *Handbook of Polycyclic Aromatic Hydrocarbons*, M. Dekker, New York, N. Y. Vol. I, **1983**, Vol II, **1985**.

4. H.-G. Franck, J. W. Stadelhofer, *Industrial Aromatic Chemistry*, Springer-Verlag, Berlin, **1988**.

5. J. R. Dias, *Handbook of Polycyclic Hydrocarbons*, Elsevier, Amsterdam, **1988**.

6. R. G. Harvey, *Polycyclic Aromatic Hydrocarbons, Chemistry and Carcinogenicity*, Cambridge University Press, Cambridge, **1991**.

7. M. Zander, *Polycyclische Aromaten*, Teubner Verlag, Stuttgart, **1995**.

8. R. G. Harvey, *Polycyclic Aromatic Hydrocarbons*, Wiley-VCH, New York, N. Y., **1997**.

9. For a recent summary of modern synthetic routes to extended aromatic compounds see S. Hagen, H. Hopf, *Top. Curr. Chem.*, **1998**, *196*, 45–89. When all quaternary carbon atoms of a polyarene are at the perimeter of the PAH under discussion, we speak of *cata*-condensation, in contrast to *peri*-condensation which is encountered in PAHs having both exterior and interior quaternary carbon atoms, as, *e.g.* in pyrene.

10. E. Clar, F. John, *Ber. Dtsch. Chem. Ges.*, **1929**, *62*, 3021–3029; *cf.* E. Clar, F. John, *Ber. Dtsch. Chem. Ges.*, **1930**, *63*, 2967–2977; E. Clar, F. John, *Ber. Dtsch. Chem. Ges.*, **1931**, *64*, 981–988.

11. W. J. Bailey, M. Madoff, *J. Am. Chem. Soc.*, **1953**, *75*, 5603–5604.

12. J. Luo, H. Hart, *J. Org. Chem.*, **1987**, *52*, 4833–4836.

13. N. Martin, R. Behnisch, M. Hanack, *J. Org. Chem.*, **1989**, *54*, 2563–2568.

14. W. E. Billups, D. J. McCord, B. R. Maughon, *Tetrahedron Lett.*, **1994**, *35*, 4493–4496.

15. E. Clar, *Ber. Dtsch. Chem. Ges.*, **1939**, *72*, 1817–1821; E. Clar, *Ber. Dtsch. Chem. Ges.*, **1952**, *75*, 1283–1287; E. Clar, *Ber. Dtsch. Chem. Ges.*, **1942**, *75*, 1330–1338; Ch. Marschalk, *Bull. Soc. Chim. France*, **1939**, *6*, 1112–1121. For a modification and improvement of Clar's original work see K. F. Lang, M. Zander, *Chem. Ber.*, **1963**, *96*, 707–711.

16. W. J. Bailey, C.-W. Liao, *J. Am. Chem. Soc.*, **1955**, *77*, 992–993; W. J. Bailey, C.-W. Liao, G. H. Coleman, *J. Am. Chem. Soc.*, **1955**, *77*, 990–991.

17. M. P. Satchell, B. E. Stacey, *J. Chem. Soc. (C)*, **1971**, 468–469; *cf.* A. Verine, Y. Lepage, *Bull. Soc. Chim. France*, **1973**, 1154–1159. For the preparation of pentacene-6,13-quinone by a condensation route see W. Ried, F. Anthöfer, *Angew. Chem.*, **1953**, *65*, 601.

18. E. Clar, *Ber. Dtsch. Chem. Ges.*, **1942**, *75*, 1330–1338. For a summary of the older routes to **35** see ref.[2]

19. J. Luo, H. Hart, *J. Org. Chem.*, **1988**, *53*, 1341–1343.

20. An elegant way towards this goal involves the cycloaddition between *p*-benzoquinone and 1,2,4,5-tetramethylenecyclohexane, *i.e.* between a double dienophile and a double diene. Although the Diels-Alder addition and the subsequent reduction of the ketone functionality of the primary product appear to be possible, the final aromatization step failed to proceed to completion before the material became intractable: W. J. Bailey, E. J. Fetter, E. Economy, *J. Org. Chem.*, **1962**, *27*, 3479–3482.

21. K. A. Muszkat, *Top. Curr. Chem.*, **1980**, *88*, 89–143; F. B. Mallory, C. W. Mallory, *Org. Reactions*, **1984**, *30*, 1–456; W. H. Laarhoven, *Org. Photochemistry*, **1987**, *9*, 129–224; D. H. Waldeck, *Chem. Rev.*, **1991**, *91*, 415–436; U. Mazzucato, F. Momiccholi, *Chem.*

Rev., **1991**, *91*, 1679–1719; H. Meier, *Angew. Chem.*, **1992**, *104*, 1425–1446; *Angew. Chem. Int. Ed. Engl.*, **1992**, *31*, 1399–1420.

22. R. Scholl, K. Meyer, *Ber. Dtsch. Chem. Ges.*, **1932**, *65*, 902–915; *cf.* R. Scholl, H. Dehnert, L. Manka, *Liebigs Ann. Chem.*, **1932**, *493*, 56–96; R. Scholl, H. K. Meyer, W. Winkler, *Liebigs Ann. Chem.*, **1932**, *494*, 201–224. For still another multi-step synthesis of coronene (**5**) see M. S. Newman, *J. Am. Chem. Soc.*, **1940**, *62*, 1683–1687.

23. E. Clar, M. Zander, *J. Chem Soc.*, **1957**, 4616–4619.

24. J. T. M. vanDijk, A. Hartwijk, A. C. Bleeker, J. Lugtenburg, J. Cornelisse, *J. Org. Chem.*, **1996**, *61*, 1136–1139.

25. W. Baker, J. F. W. McOmie, W. K. Warburton, *J. Chem. Soc.*, **1952**, 2991–2993.

26. W. Baker, F. Glockling, J. F. W. McOmie, *J. Chem. Soc.*, **1951**, 1118–1121.

27. J. R. Davy, J. A. Reiss, *J. Chem. Soc. Chem. Commun.*, **1973**, 806–807.

28. W. Baker, J. F. W. McOmie, J. M. Norman, *J. Chem. Soc.*, **1951**, 1114–1118.

29. F. Diederich, H. A. Staab, *Angew. Chem.*, **1978**, *90*, 383–385; *Angew. Chem. Int. Ed. Engl.*, **1978**, *17*, 372–374; *cf.* H. A. Staab, F. Diederich, *Chem. Ber.*, **1983**, *116*, 3487–3503; W. Jenny, R. Paioni, *Chimia*, **1969**, *23*, 41–42.

30. F. Vögtle, H. A. Staab, *Chem. Ber.*, **1968**, *101*, 2709–2716. For the preparation of related hydrocarbons see U. Meissner, B. Meissner, H. A. Staab, *Angew. Chem.*, **1973**, *85*, 957–958; *Angew. Chem. Int. Ed. Engl.*, **1973**, *12*, 916–917.

31. The two annulene perimeters must not be of the [4n+2]-type. Depending on the ring-size cycloarenes with [4n]-perimeters are also conceivable. - For the synthesis of a circulene formally composed of an outer [26]- and an inner [14]annulene ring see D. J. H. Funhoff, H. A. Staab, *Angew. Chem.*, **1986**, *98*, 757–759; *Angew. Chem. Int. Ed. Engl.*, **1986**, *25*, 742–744.

32. E. Clar, *The Aromatic Sextet*, John Wiley and Sons, London, **1972**.

33. C. Krieger, F. Diederich, D. Schweitzer, H. A. Staab, *Angew. Chem.*, **1979**, *91*, 733–735; *Angew. Chem. Int. Ed. Engl.*, **1979**, 16, 699–701; *cf.* H. A. Staab, F. Diederich, C. Krieger, D. Schweitzer, *Chem. Ber.*, **1983**, *116*, 3504–3512; D. Schweitzer, K. H. Hauser, H. Vogler, F. Diederich, H. A. Staab, *Mol. Physics*, **1982**, *46*, 1141–1153.

34. For a recent synthesis of [6]phenacene (**68**) involving alkylation of cyclohexanone-derived enamines or enamine salts by 1,5-bis(2′-bromoethyl)naphtalene followed by acidic cyclodehydration of the resulting diketone intermediate and final dehydrogenation see R. G. Harvey, J. Pataki, C. Cortez, P. Di Raddo, C. X. Yang, *J. Org. Chem.*, **1991**, *56*, 1210–1217.

35. F. B. Mallory, K. E. Butler, A. C. Evans, C. W. Mallory, *Tetrahedron Lett.*, **1996**, *37*, 7173–7176.

36. F. B. Mallory, K. E. Butler, A. C. Evans, E. J. Brondyke, C. W. Mallory, Y. Yang, A. Ellenstein, *J. Am. Chem. Soc.*, **1997**, *119*, 2119–2124.

37. For the synthesis of terrylene (**75**) see E. Clar, W. Kelly, R. M. Laird, *Monatsh. Chem.*, **1956**, *87*, 391; for a more specific and milder route to **75** see E. Buchta, H. Vates, H. Knopp, *Chem. Ber.*, **1958**, *91*, 228–241 and refs. cited therein. For the preparation and structural properties of quaterrylene (**76**) see E. Clar, J. C. Speakman, *J. Chem. Soc.*, **1958**, 2492–2494 and refs. to earlier work.

38. K.-H. Koch, K. Müllen, *Chem. Ber.*, **1991**, *124*, 2091–2100.

39. H. Quante, K. Müllen, *Angew. Chem.*, **1995**, *107*, 1487–1489; *Angew. Chem. Int. Ed. Engl.*, **1995**, *34*, 1323–1325.

40. E. Clar, *Nature*, **1948**, *161*, 238; E. Clar, *Chem. Ber.*, **1949**, *82*, 46–60; H. Inukuchi, *Bull. Soc. Chem. Jpn.*, **1951**, *24*, 222.

41. Other aromatic systems may also be totally 'surrounded' by condensed benzene rings. For the preparation of circum[34]para-terphenyl see M. Zander, W. Friedrichsen, *Chem.-*

Ztg., **1991**, *115*, 360–361.

42. E. Clar, W. Kelly, J. M. Robertson, M. G. Rossmann, *J. Chem. Soc.*, **1956**, 3878–3881.

43. E. Clar, J. M. Robertson, R. Schlögl, W. Schmidt, *J. Am. Chem. Soc.*, **1981**, *103*, 1320–1328.

44. R. D. Broene, F. Diederich, *Tetrahedron Lett.*, **1991**, *32*, 5227–5320.

45. *Electronic Materials: The Oligomer Approach* K. Müllen, G. Wegner, (*Eds.*), Wiley-VCH, Weinheim, **1998**.

46. Review: A.-D. Schlüter, M. Löffler, A. Godt, K. Blater in *Desk Reference of Functional Polymers - Syntheses and Applications* R. Arshady, (*Ed.*), American Chemical Society, Washington, D. C., **1997**, Chap. 1.5, 73–91.

47. A.-D. Schlüter, M. Löffler, V. Enkelmann, *Nature*, **1994**, *368*, 831–834; *cf.* B. Schlicke, A.-D. Schlüter, P. Hauser, J. Heinze, *Angew. Chem.*, **1997**, *109*, 2091–2093; *Angew. Chem. Int. Ed. Engl.*, **1997**, *36*, 1996–1998.

48. Review: A. Schröder, H.-B. Meckelburger, F. Vögtle, *Top. Curr. Chem.*, **1994**, *172*, 179–201.

49. For related, often computational studies see: S. Kivelson, O. L. Chapman, *Phys. Rev. B*, **1983**, *28*, 7236; M. A. Haase, R. W. Zoellner, *J. Org. Chem.*, **1992**, *57*, 1031–1033; F. Vögtle, *Top. Curr. Chem.*, **1983**, *115*, 157–159; R. W. Alder, R. B. Sessions, *J. Chem. Soc. Perkin Trans II*, **1985**, 1849–1854; *cf.* R. W. Alder, P. R. Allen, L. S. Edwards, G. I. Fray, K. E. Fuller, P. M. Gore, N. M. Hext, M. H. Perry, A. R. Thomas, K. S. Turner, *J. Chem. Soc. Perkin Trans. I*, **1994**, 3071–3077.

50. R. O. Angus, Jr, R. P. Johnson, *J. Org. Chem.*, **1988**, *53*, 314–317.

51. F. H. Kohnke, A. M. Z. Slawin, J. F. Stoddart, D. J. Williams, *Angew. Chem.*, **1987**, *99*, 941–943; *Angew. Chem. Int. Ed. Engl.*, **1987**, *26*, 892–894.

52. P. R. Ashton, N. S. Isaacs, F. H. Kohnke, A. M. Z. Slawin, C. M. Spencer, J. F. Stoddart, D. J. Williams, *Angew. Chem.*, **1988**, *100*, 981–983; *Angew. Chem. Int. Ed. Engl.*, **1988**, *27*, 966–968.

53. Reviews: J. P. Mathias, J. F. Stoddart, *Chem. Soc. Rev.*, **1992**, *21*, 215–225; J. F. Stoddart, *J. Inclusion Phenom. Mol. Recognit. Chem.*, **1989**, *7*, 227–2435.

54. In the meantime other (functionalized) concave molecules–both with open ('molecular tweezers') and closed structures–have been obtained by high-pressure Diels–Alder additions as primarily reported by Klärner: J. Benkhoff, R. Boese, F.-G. Klärner, A. E. Wigger, *Tetrahedron Lett.*, **1994**, *35*, 73–76; F.-G. Klärner, J. Benkhoff, R. Boese, U. Burkert, M. Kamieth, U. Naatz, *Angew. Chem.*, **1996**, *108*, 1195–1198; *Angew. Chem. Int. Ed. Engl.*, **1996**, *36*, 1130–1133; J. Benkhoff, R. Boese, F.-G. Klärner, *Liebigs Ann./Recueil*, **1997**, 501–516; M. Kamieth, F.-G. Klärner, F. Diederich, *Angew. Chem.*, **1998**, *110*, 3497–3500; *Angew. Chem. Int. Ed. Engl.*, **1998**, *37*, 3303–3306.

55. When **95** and **96** are replaced by comparable trisdienes and trisdienophiles, respectively, still more complex cage compounds result: P. R. Ashton, N. S. Isaacs, F. H. Kohnke, G. S. d´Alcontres, J. F. Stoddart, *Angew. Chem.*, **1989**, *101*, 1269–1271; *Angew. Chem. Int. Ed. Engl.*, **1989**, *28*, 1261–1263. For the preparation of smaller beltene precursors see: [4]beltene: W. Grimme, H. Geich, J. Lex, J. Heinze, *J. Chem. Soc. Perkin Trans. II*, **1997**, 1955–1958; [6]beltene: A. Godt, V. Enkelmann, A.-D. Schlüter, *Angew. Chem.*, **1989**, *101*, 1704–1706; *Angew. Chem. Int. Ed. Engl.*, **1989**, *28*, 1680–1682.

56. Review: S. Subramoney, *Adv. Mater.*, **1998**, *10*, 1157–1171.

57. W. Krätschmer, L. D. Lamb, K. Fostiropoulos, D. R. Huffman, *Nature*, **1990**, *347*, 354.

58. St. Kammermeier, P. G. Jones, R. Herges, *Angew. Chem.*, **1996**, *108*, 2834–2836; *Angew. Chem. Int. Ed. Engl.*, **1996**, *35*, 2669–2671; St. Kammermeier, P. G. Jones, R. Herges, *Angew. Chem.*, **1997**, *109*, 2317–2319; *Angew. Chem. Int. Ed. Engl.*, **1997**, *36*, 2200–2202; St. Kammermeier, P. G. Jones, R. Herges, *Angew. Chem.*, **1997**, *36*, 1757–1760;

Angew. Chem. Int. Ed. Engl., **1997**, *109*, 1825–1829; St. Kammermeier, R. Herges, *Angew. Chem.*, **1996**, *108*, 470–472; *Angew. Chem. Int. Ed. Engl.*, **1996**, *35*, 417–419.

59. M. Müller, Chr. Kübel, K. Müllen, *Chem. Eur. J.*, **1998**, *4*, 2099–2109; *cf.* J. R. Dias, *Acc. Chem. Res.*, **1985**, *18*, 241–248; St. E. Stein, *Acc. Chem. Res.*, **1991**, *24*, 350–356.

60. L. Boente, *Brennstoff-Chem.*, **1955**, *36*, 210.

61. M. Zander, W. Franke, *Chem. Ber.*, **1958**, *91*, 2794–2797.

62. H. J. Lempka, S. Obenland, W. Schmidt, *Chem. Phys.*, **1985**, *96*, 349–360.

63. M. Zander, W. Friedrichsen, *Z. Naturforsch.*, **1992**, *47b*, 1314–1318.

64. The parent system **111**, R = H was first prepared by Zn/ZnCl$_2$-induced cyclodehydrogenation of hexaphenylbenzene (**110**; R = H): A. Halleux, R. H. Martin, G. S. D. King, *Helv. Chim. Acta*, **1958**, *41*, 1177–1183. For later routes to **111** see: E. Clar, C. T. Ironside, M. Zander, *J. Chem. Soc.*, **1959**, 142–147; W. Hendel. Z. H. Khan, W. Schmidt, *Tetrahedron*, **1986**, *42*, 1127–1134; *cf.* M. Zander, *Angew. Chem.*, **1960**, *72*, 513–520. For the X-ray structural analysis of **106** and **111** see R. Goddard, M. W. Haenel, W. C. Herndon, C. Krüger, M. Zander, *J. Am. Chem. Soc.*, **1995**, *117*, 30–41.

65. P. Herwig, C. W. Kayser, K. Müllen, H. W. Spies, *Adv. Mater.*, **1996**, *8*, 510–513; *cf.* A. Stabel, P. Herwig, K. Müllen, J. P. Rabe, *Angew. Chem.*, **1995**, *107*, 1768–1770; *Angew. Chem. Int. Ed. Engl.*, **1995**, *34*, 1609–1611; *cf.* V. S. Iyer, K. Yoshimura, V. Enkelmann, R. Epsch, J. P. Rabe, K. Müllen, *Angew. Chem.*, **1998**, *110*, 2843–2846; *Angew. Chem. Int. Ed. Engl.*, **1998**, 37, 2696–2699.

66. For the determination of the structure of hexaphenylbenzene by X-ray structural analysis see: J. C. J. Bart, *Acta Crystallogr. Sect. B*, **1968**, *24*, 1277–1287; for the structure determination by gas phase electron diffraction see: A. Almenningen, O. Bastiansen, P. N. Skancke, *Acta Chem. Scand.*, **1958**, *12*, 1215–1220.

67. J. A. Hyatt, *Org. Prep. Proc. Int.*, **1991**, *23*, 460–463; *cf.* R. B. King, I. Haiduc, A. Efraty, *J. Organomet. Chem.*, **1973**, *47*, 145–151; R. Diercks, K. P. C. Vollhardt, *J. Am. Chem. Soc.*, **1986**, *108*, 3150–3152 and refs. cited therein.

68. W. Dilthey, G. Hurtig, *Ber. Dtsch. Chem. Ges.*, **1934**, *67*, 495–496; W. Dilthey, W. Schommer, W. Höschen, H. Dierichs, *Ber. Dtsch. Chem. Ges.*, **1935**, *68*, 1159–1162; L. F. Fieser, *Organic Experiments*, D. C. Heath & Co., 1st ed., Boston, **1964**, 307–308.

69. A. G. Brown, P. D. Edwards, *Tetrahedron Lett.*, **1990**, *31*, 6581–6584; *cf.* J. S. Bradshaw, L. Golic, M. Tisler, *Monatsh. Chemie*, **1988**, *119*, 327–332.

70. M. A. Schwartz, P. T. K. Pham, *J. Org. Chem.*, **1988**, *53*, 2318–2322; *cf.* E. C. Taylor, J. G. Andrade, G. J. H. Rall, A. McKillop, *J. Am. Chem. Soc.*, **1970**, *102*, 6513–6519.

71. N. Boden, R. J. Bushby, A. N. Cammidge, G. Headdock, *Synthesis*, **1995**, 31–32; *cf.* R. J. Bushby, C. Hardy, *J. Chem. Soc. Perkin Trans. I*, **1986**, 721–723.

72. Review: P. Kovacic, M. B. Jones, *Chem. Rev.*, **1987**, *87*, 357–379. Kovacic originally developed this method for the polymerization of benzene and its derivatives, *i.e.* for the synthesis of poly-*para*-phenylene.

73. W. Ried, K. H. Bönnighausen, *Chem. Ber.*, **1960**, *93*, 1769–1773; *cf.* W. Ried, D. Freitag, *Angew. Chem.*, **1968**, *80*, 932–942; *Angew. Chem. Int. Ed. Engl.*, **1968**, *7*, 835–845.

74. For a different route to **113** see: M. A. Ogliaruso, L. A. Shadoff, E. I. Becker, *J. Org. Chem.*, **1963**, *28*, 2725–2728; *cf.* M. A. Ogliaruso, E. I. Becker, *J. Org. Chem.*, **1965**, *30*, 3354–3360.

75. M. Müller, V. S. Iyer, C. Kübel, V. Enkelmann, K. Müllen, *Angew. Chem.*, **1997**, *109*, 1679–1682; *Angew. Chem. Int. Ed. Engl.*, **1997**, *36*, 1607–1610.

76. V. S. Iyer, M. Wehmeier, J. D. Brand, M. A. Keegstra, K. Müllen, *Angew. Chem.*, **1997**, *109*, 1676–1679; *Angew. Chem. Int. Ed. Engl.*, **1997**, *36*, 1604–1607.

77. F. Morgenroth, E. Reuther, K. Müllen, *Angew. Chem.*, **1997**, *109*, 647–649; *Angew. Chem. Int. Ed. Engl.*, **1997**, *36*, 631–634.

15 Extended Systems–II. Beyond the PAH Pattern

In Section 14.4 we used the benzene ring as a 'molecular tile'[1] to construct giant planar PAHs, which are indeed beginning to resemble graphite. The connection between these tiles was established by at least two shared carbon atoms.

As I pointed out, however, in the introductory paragraphs of Chapter 14 other connection modes are possible.[2] For example, a single bond may serve as the connector, and the resulting hydrocarbons—oligophenyls of increasing complexity—are, as we have seen, useful precursors for the preparation of very large PAHs. As in the preceding chapter I again want to present a selection of 'extended hydrocarbon systems' on the following pages, but I shall increase the variety of the 'allowed' building elements. Starting with additional single bonds as connectors of benzene rings, I shall move on to structures built from triple and double bonds and from triple bonds and benzene rings, respectively, and I shall conclude this overview with the synthesis of oligomeric hydrocarbons constructed from the cyclopropane ring as the repeating unit. Although this selection is somewhat arbitrary and certainly incomplete—many other combinations of our basic building units are conceivable and are beginning to be realized in the laboratory—it demonstrates very well the different synthetic approaches employed for the preparation of increasingly large and complex hydrocarbon systems.

15.1 From biphenylene to linear and angular [*n*]phenylenes

Having joined two benzene rings (**1**) by way of a single bond to biphenyl (**2**), there is only one way to introduce a second single bond between the two halves of **2**—removal of two *ortho* hydrogen atoms of biphenyl (**2**) and generation of biphenylene (**3**) (Scheme 1).

All other alternatives, *e.g.* closing a bridge between the *ortho* position of one phenyl ring of **2** and the *meta* position of the other results in extremely strained and deformed hydrocarbons, which would violate (*inter alia*) Bredt's rule (see Section 7.4) and therefore escape detection—not to speak of isolation—even under extreme conditions (liquid helium temperatures, matrix).

Continuing this process leads to the *linear* [*n*]phenylenes, with linear [3]- (**4**) and

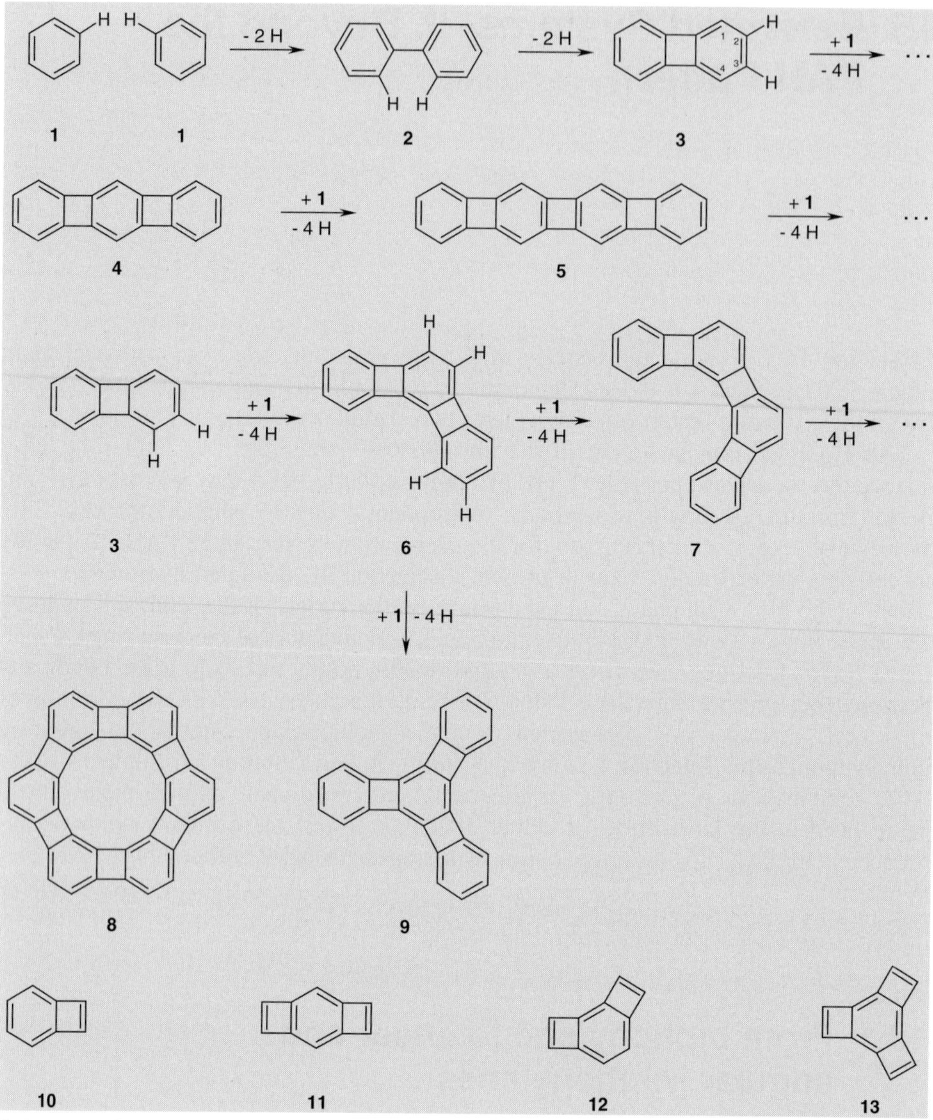

Scheme 1. Constructing [*n*]phenylenes by combining benzene rings with single bonds.

[4]phenylene (**5**) given as examples. Alternatively, connecting an additional benzene ring at the 3 and 4 positions of **3** leads to angular [3]phenylene (**6**), the first member of the *angular* [*n*]phenylenes. The angular connectivity is particularly attractive because in principle it enables—*via* **7** and further benzologs—the construction of cyclic phenylenes such as **8** with an internal cavity. And, of course, from **6** on *triangular* [*n*]phenylenes, such as **9**, could also become accessible. All these structures show an obvious relationship to the linear (Section 14.1) and angular acenes (Section 14.2) and the circulenes (Section 12.2) of which the above hydrocarbons are 'exploded' versions.

Another way of looking at the [*n*]phenylenes regards them as derivatives of the cyclobutarenes **10** (bicyclo[4.2.0]octa-1,3,5,7-tetraene, benzocyclobutene, commonly often referred to as benzocyclobutadiene), **11** (benzo[*a,d*]dicyclobutene), **12** (benzo[*a,c*]dicyclobutene), and **13** (benzo[*a,c,e*]tricyclobutene), a thoroughly studied and reviewed class of cyclobutannelated benzene derivatives[3-7] (see Section 16.3). According to molecular orbital calculations linear [*n*]phenylenes should be more aromatic than their angular isomers; because of their smaller HOMO-LUMO gap, however, the former should also be more reactive than the latter, a reactivity difference reminiscent of that encountered for linear and angular acenes (see Sections 14.1 and 14.2).[8]

The histrory of biphenylene (**3**), the parent molecule of the [*n*]phenylenes, dates back to the turn of the 20th century and follows—again—the typical hydrocarbon scenario. After many attempts, which were either unsuccessful immediately or later shown to have led to products with different structures, **3** was finally synthesized in 1941 by Lothrop, who obtained the hydrocarbon in poor yield (5%) by means of a modified Ullmann coupling on heating 2,2'-dibromobiphenyl (**14**, X = Br) with cuprous chloride at 350 °C (Scheme 2).[9]

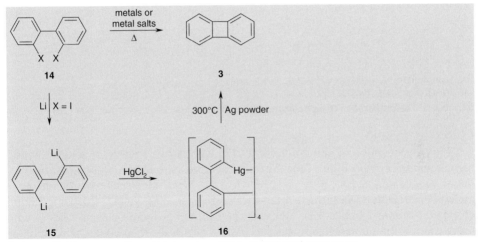

Scheme 2. The preparation of biphenylene (**3**) from biphenyl precursors.

Substituting the bromine substituents in **14** with iodine (X = I) led to a significant yield increase to 21%. A further improvement of the connection process was realized by Wittig and Herwig who first metalated **14** (X = I) to the dilithio biphenyl **15**, and then converted this to the organomercury tetramer **16**, which on heating (300 °C) in the presence of silver powder provided **3** in 54% yield.[10, 11] The Lothrop method has also been applied to prepare substituted biphenylenes, but acceptable yields are obtained only rarely,[3] and it is often preferable to use **3** as the starting material and, if derivatives are needed, functionalize it.

The other important route to **3** involves the generation of dehydrobenzene (**20**), available from a plethora of precursors (see Section 16.1), and its dimerization. Although the process is accompanied by trimer- (to triphenylene) and oligomer-formation, yields can occasionally be as high as 80% (Scheme 3).

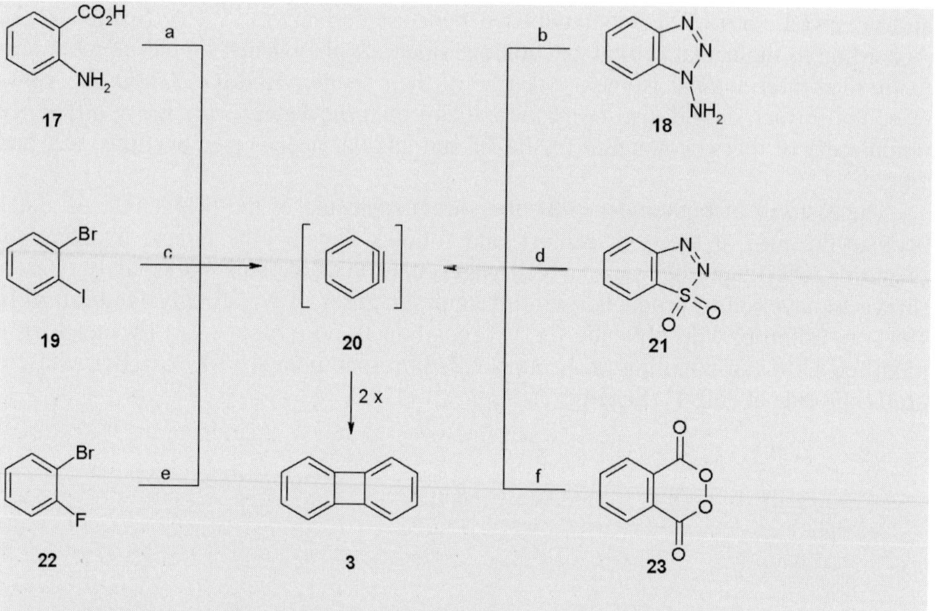

(a) isoamyl nitrite, H⁺, Δ (30%); (b) Pb(OAc)₄ (80%); (c) Mg (3.5%); (d) Δ (52%); (e) Li / Hg, ether (24%); (f) 600°C (27%)

Scheme 3. The preparation of biphenylene (**3**) by dimerization of dehydrobenzene (**20**).

Thus diazotization of anthranilic acid (**17**) with isoamyl nitrite in acidic solution then thermal decomposition of the diazonium carboxylate formed—the *Organic Syntheses* procedure—led to **3** in *ca* 30% yield.[12] Lead tetraacetate oxidation of 1-amino-benztriazole (**18**) furnished the hydrocarbon in 80% yield,[13] dehalogenation of either 1-bromo-2-iodo- (**19**)[14] or (better) 1-bromo-2-fluoro-benzene (**22**)[15] with magnesium or lithium amalgam was successful, as was the thermal decomposition of benzthiadiazol-1,1-dioxide (**21**)[16] and phthaloyl peroxide (**23**).[17] This selection is far from complete.[3] The dimerization route has also been of value for the preparation of simple alkyl derivatives of **3**, *e.g.* 1,5- and 1,8-dimethylbiphenylene or 1,4,5,8-tetramethylbiphenylene, and for functionalized biphenylenes.[18]

Although the two pathways to **3** summarized in Scheme 4 cannot compete in preparative simplicity with some of the routes just discussed, they are of mechanistic interest.

1,3,7,9-Cyclododecatetrayne (**24**), prepared by oxidative coupling of 1,5-hexadiyne (see Section 10.4) was isomerized to **3** in 25% yield at room temperature by treatment with potassium *tert*-butoxide. The mechanism of this rearrangement remains to be established; a byproduct formed under the reaction conditions used, the cyclododecatetraenediyne **25**, was shown not to be an intermediate.[19] The biphenyl diradical **27**, generated by flash vacuum-pyrolysis (FVP) of benzo[*c*]cinnoline (**26**) provided **3**

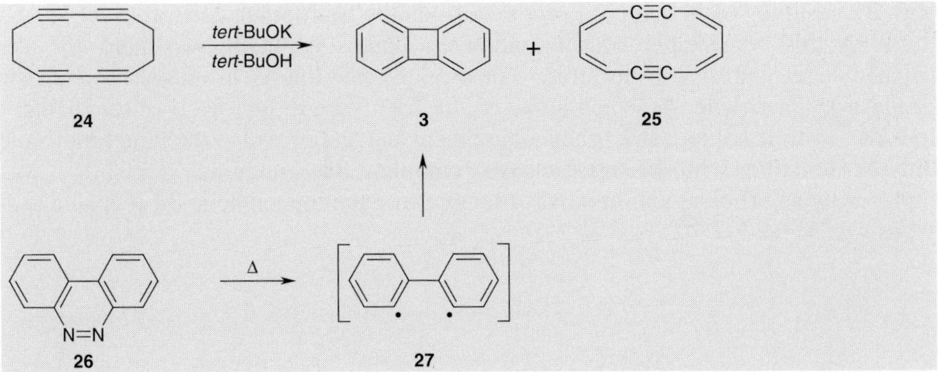

Scheme 4. Biphenylene (**3**) from non-aromatic (**24**) and heteroaromatic (**26**) precursors.

in 60% yield, by an intramolecular ring-closure.[20] Although this particular synthesis could be extended to the preparation of two higher homologs of **3** (see below) none of the (bi)phenylene syntheses yet discussed could be developed into a general approach towards the higher [*n*]phenylenes. The solution of this synthetic problem had to wait for a completely new phenylene synthesis discovered by Vollhardt and co–workers. The parent reaction of this ingenious approach consists in the catalysis by η^5-cyclopentadienylcobalt dicarbonyl (**29**, CpCo(CO)$_2$) of the cyclotrimerization of acetylenes **28** to aromatics **30**, a process widely used in the synthesis of numerous natural and unnatural products (Scheme 5).[21, 22]

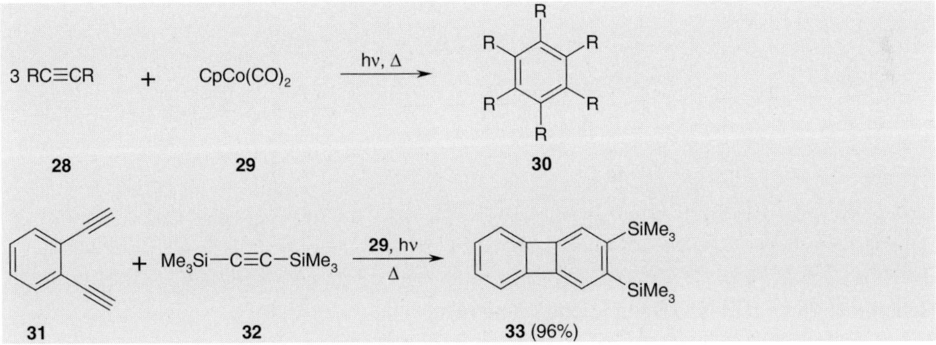

Scheme 5. The cobalt-catalyzed cyclotrimerization of acetylenes, **28**, to aromatics, **30**.

When the role of two molecules of **28** was taken over by a diacetylene, *viz.* 1,2-diethynylbenzene (**31**), the cycloaddition of monoalkynes such as bis-trimethylsilyl-acetylene (**32**) resulted in the near quantitative formation of 2,3-bis(trimethylsilyl-ethynyl)-biphenylenes (**33**).[23,24] In fact, because of the steric bulk of the TMS sub-stituents self-trimerization of **32** does not take place, and it can be used as a solvent for the reaction. Alkyl, aryl and certain functionalized acetylenes could also be co-cyclized with **31**, resulting—although in lower yields—in the appropriate derivatives of **3**.

With this remarkable transformation in hand, Vollhardt was able to take [*n*]phenylene chemistry in his stride. For example, the low-yield multistep synthesis of linear [3]phenylene (**4**) which involved the flash vacuum-pyrolysis of the dicinnoline **34**[25] (itself not an easily obtainable precursor) and provided the target molecule only in admixture with the mono-cleaved cinnoline **35** (combined yield 11%), was replaced by an efficient and iterative strategy using the biphenylene **33** prepared previously (Scheme 6).[24,26]

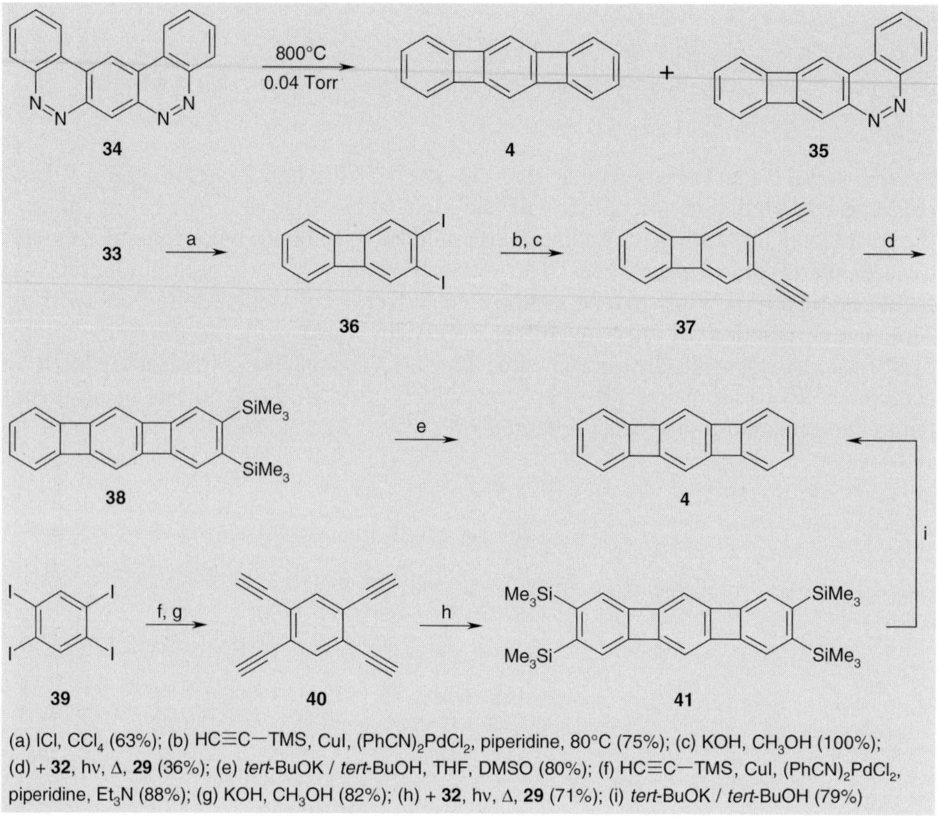

(a) ICl, CCl$_4$ (63%); (b) HC≡C—TMS, CuI, (PhCN)$_2$PdCl$_2$, piperidine, 80°C (75%); (c) KOH, CH$_3$OH (100%); (d) + **32**, hv, Δ, **29** (36%); (e) *tert*-BuOK / *tert*-BuOH, THF, DMSO (80%); (f) HC≡C—TMS, CuI, (PhCN)$_2$PdCl$_2$, piperidine, Et$_3$N (88%); (g) KOH, CH$_3$OH (82%); (h) + **32**, hv, Δ, **29** (71%); (i) *tert*-BuOK / *tert*-BuOH (79%)

Scheme 6. Three routes to linear [3]phenylene (**4**).

By iododesilylation with chloroiodide, 2,3-diiodobiphenylene (**36**) was first prepared from **33** and converted into the bisacetylene **37** by coupling with trimethylsilylethyne (TMSE), followed by desilylation. Repetition of the cobalt-mediated cocyclization step of **37** with **32** then led to **38** which in the concluding step was desilylated by treatment with base (the [3]phenylene system is acid-sensitive). An even shorter route to **4** was developed later in which the two four-membered-ring-forming processes were 'telescoped' into one.[26] Coupling of 1,2,4,5-tetraiodobenzene (**39**) with TMSE, then deprotection, furnished the (unstable) tetraethynylbenzene **40**, which

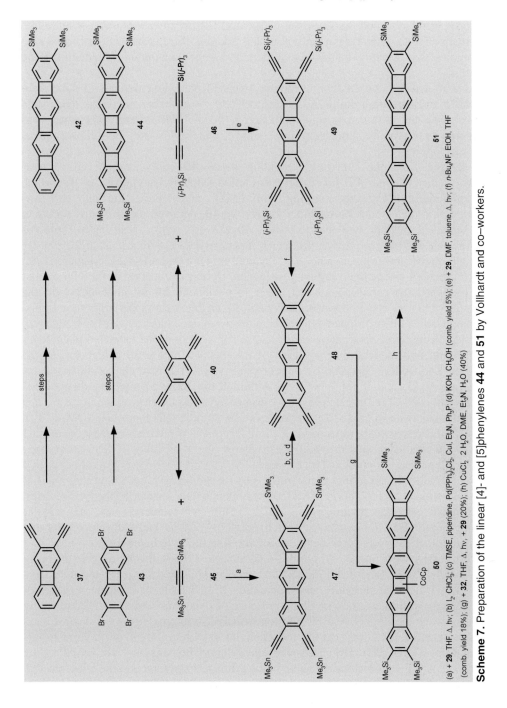

(a) + **29**, THF, Δ, hv; (b) I₂, CHCl₃; (c) TMSE, piperidine, Pd(PPh₃)₂Cl₂, CuI, Et₃N, Ph₃P; (d) KOH, CH₃OH (comb. yield 18%); (g) + **32**, THF, Δ, hv, + **29** (20%); (h) CuCl₂, 2 H₂O, DME, Et₃N, H₂O (40%)
(comb. yield 18%); (g) + **32**, THF, Δ, hv, + **29** (20%); (h) CuCl₂, 2 H₂O, DME, Et₃N, H₂O (40%)

Scheme 7. Preparation of the linear [4]- and [5]phenylenes **44** and **51** by Vollhardt and co-workers.

was converted to **4**, *via* **41**, by the now well-established cyclization methodology–the [3]phenylene being formed from **39** in an overall yield of 41%. Both [3]phenylene derivatives **4** and **41** were fully characterized by their spectroscopic data; the X-ray analysis of **41** revealed the interesting structural feature of considerable bond-length alternation in its outer rings (see below and Section 16.3), whereas the central ring resembles two strongly interacting allyl units rather than having alternating bond lengths.

I shall leave the preparation of the two linear [4]phenylene derivatives **42** and **44** from **37** and the tetrabromide **43**,[27] respectively, as an excercise to the reader and move on to the next benzocyclobutadienolog, linear [5]phenylene (**51**) which was prepared by the routes summarized in Scheme 7, which illustrates the full potential of this methodology.[28]

Starting from the 'universal' building block **40** co-cyclization with bis(tri-methylstannyl)ethyne (**45**) led to the tetrastannyl [3]phenylene derivative **47**, which by iodination, Pd-catalyzed coupling with TMSE, and deprotection with base was converted to the extended tetraethinyl derivative **48**. An even more effective route to the same intermediate was realized when **40** was co-cyclized with bis(triisopropyl-silyl)hexatriyne (**46**) to provide **49**, which was desilylated to **48** with fluoride ion. By this alternative the [3]phenylene was obtained in 18% overall yield, compared with 5% for the earlier described four-step sequence. Double cyclization of **48** with **32** and **29** furnished the cobalt-complexed linear [5]phenylene **50** in moderate yield; the amount obtained was, however, sufficient to enable decomplexation to tetrakis-(trimethylsilyl) linear [5]phenylene (**51**), a deep-red, air- and light-sensitive material. The increasingly reactive nature of the π-electron system of the extended phenylenes is not only indicated by the formation of the metal complex **50**, which breaks the through–conjugation, but also by the comparatively harsh reaction conditions to re-move the CpCo unit in the latter step. Although the above phenylenes alternate be-tween $4n\pi$- and $(4n+2)\pi$ electrons, they do not have the alternating antiaromatic and aromatic behavior so typical for the series of the simple annulenes (see Chapter 10). Rather, the compounds' spectroscopic and chemical properties suggest increasing paratropic character and growing reactivity of the π-electron system on passing through the series, as predicted by theory.[8]

A further prediction, that the angular isomers should be more stable than the lin-ear, was tested by preparation of various members of the former [n]phenylene series. The most extended hydrocarbon which could be prepared was the angular [5]phenylene (**61**). Its synthesis, which employs various steps developed for the preparation of its lower homologs[29] is presented in full in Scheme 8.[30]

Successive coupling of 1-bromo-2-iodobenzene (**19**) with two alkynes carrying two different silyl groups led initially to the unsymmetrical bis-protected derivative **52**, from which the sterically less demanding trimethylsilyl substituent was subse-quently removed by alcoholysis in the presence of potassium carbonate. The resulting **53** underwent regioselective cycloaddition with TMSE to provide the 1,2-disubstituted biphenylene **54**. This was readily iododesilylated with chloroiodide to furnish 1,2-diodobiphenylene (**55**). By successive ethynylation **55** was first converted to the monoiodide **56**, a crucial intermediate in the synthesis, and then into the bis-protected derivative **57**. Under basic conditions the latter was hydrolyzed at the less hindered silyl group to **58**, which on Pd-catalyzed coupling with **56** led to the disilylated triyne **59**. Deprotection with fluoride ion then yielded the free hydrocarbon **60**, which, how-ever, was not isolated but treated directly with CpCo(CO)$_2$ (**29**). The angular [5]phenylene (**61**) was isolated as an orange-yellow solid, and—as predicted—it was

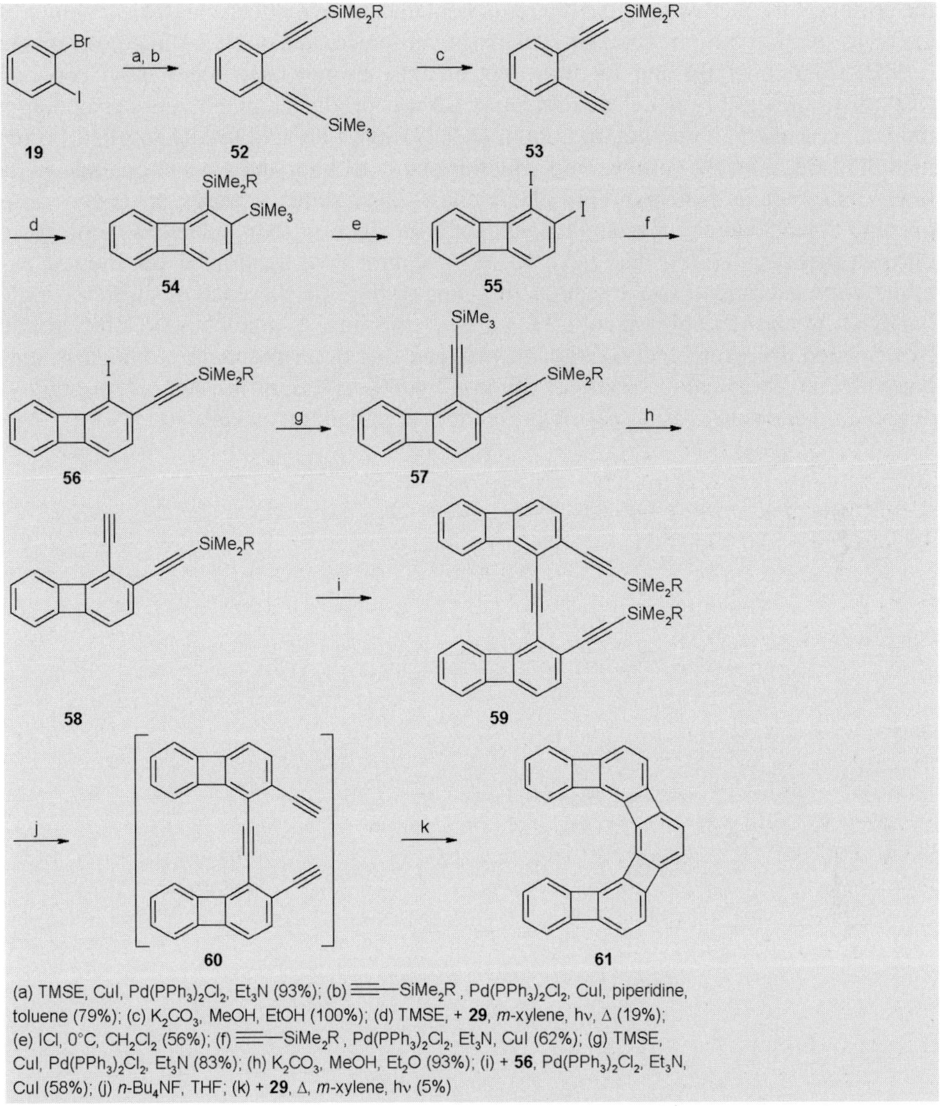

(a) TMSE, CuI, $Pd(PPh_3)_2Cl_2$, Et_3N (93%); (b) $\equiv$—$SiMe_2R$, $Pd(PPh_3)_2Cl_2$, CuI, piperidine, toluene (79%); (c) K_2CO_3, MeOH, EtOH (100%); (d) TMSE, + **29**, *m*-xylene, hν, Δ (19%); (e) ICl, 0°C, CH_2Cl_2 (56%); (f) $\equiv$—$SiMe_2R$, $Pd(PPh_3)_2Cl_2$, Et_3N, CuI (62%); (g) TMSE, CuI, $Pd(PPh_3)_2Cl_2$, Et_3N (83%); (h) K_2CO_3, MeOH, Et_2O (93%); (i) + **56**, $Pd(PPh_3)_2Cl_2$, Et_3N, CuI (58%); (j) *n*-Bu_4NF, THF; (k) + **29**, Δ, *m*-xylene, hν (5%)

Scheme 8. Vollhardt's synthesis of angular [5]phenylene (**61**; in all intermediates R = $C(CH_3)_2CH(CH_3)_2$).

less reactive than the linear [5]phenylene, one indication being, that no conjugation-interrupting CpCo-complex was formed.

That Vollhardt's route to the phenylenes had not yet reached its limits with the preparation of **61** was finally shown by the preparation of various triangular phenylenes, among them triangular [4]phenylene (**66**), for which, again, several syntheses were developed; the most efficient of these is reproduced in Scheme 9.[31]

Heck coupling of hexabromobenzene (**62**) with TMSE furnished the protected hexaalkyne **63**, a stable precursor molecule which on desilylation with fluoride ion in the presence of 18-crown-6 gave the remarkable hexaethynylbenzene (**64**),[33] which is stable in solution but polymerizes quickly in substance. When a 1,2-dimethoxyethane (DME) solution of **64** and **29** was injected into boiling bis-trimethylsilylacetylene (**32**) the triangular [4]phenylene derivative **65** was produced, in a process providing a total of six rings in a one-pot operation. Desilylation with acid finally resulted in formation of **66**, a bright yellow solid which is stable to light and air—in contrast to its linear and angular isomers. This phenylene is also stable towards acid, and ring-opening to hexadehydrotribenzo[12]annulene (see Section 10.4 and below) took place only on pyrolysis at 700 °C.[34] An X-ray structural investigation of **65** showed extreme bond alternation in the central ring, suggesting the presence of discrete single (*ca* 1.50 Å) and double bonds (1.33 Å). The ring thus resembles a cyclohexatriene (see Section 16.3 for bond-fixation in benzene rings), an observation which is supported by the chemical reactivity of **66** (*e.g.* hydrogenation of the central core by hydrogen or deuterium at atmospheric pressure over palladium on carbon).

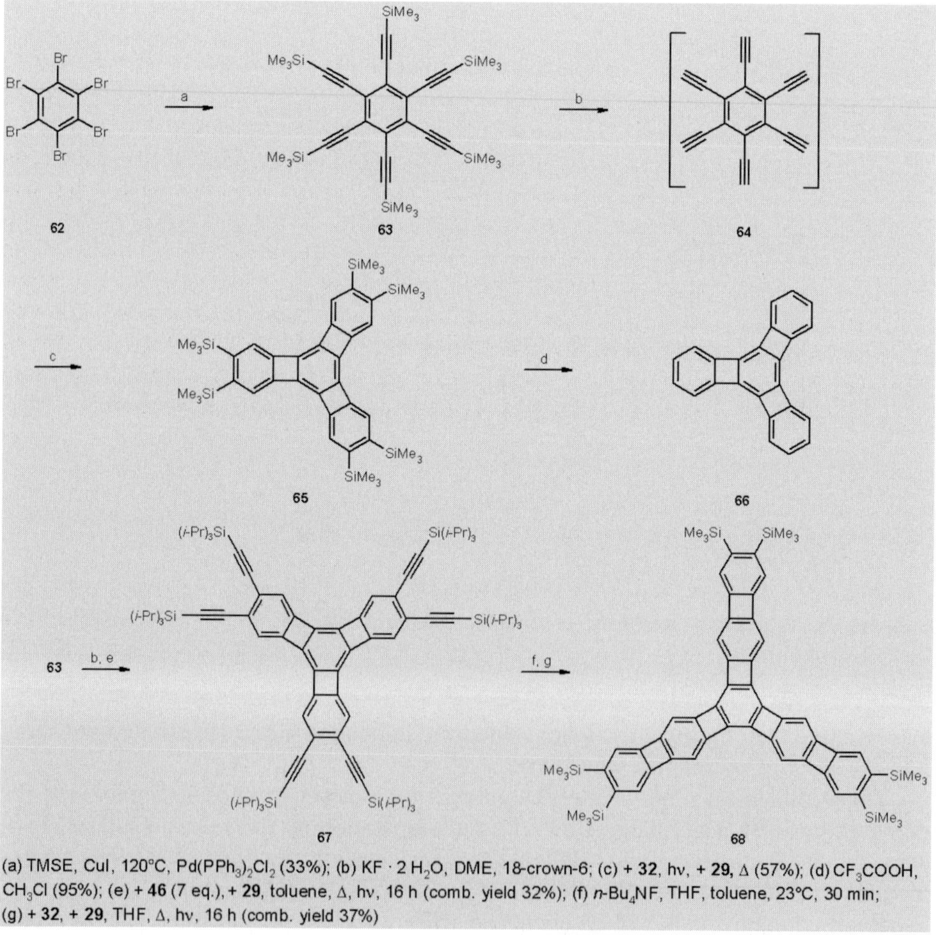

(a) TMSE, CuI, 120°C, Pd(PPh₃)₂Cl₂ (33%); (b) KF · 2 H₂O, DME, 18-crown-6; (c) + **32**, hv, + **29**, Δ (57%); (d) CF₃COOH, CH₃Cl (95%); (e) + **46** (7 eq.), + **29**, toluene, Δ, hv, 16 h (comb. yield 32%); (f) *n*-Bu₄NF, THF, toluene, 23°C, 30 min; (g) + **32**, + **29**, THF, Δ, hv, 16 h (comb. yield 37%)

Scheme 9. The synthesis of the triangular phenylenes **66** and **68**.

Still further elongation of triangular phenylene systems was achieved by applying the methodology developed for the preparation of the linear [3]phenylene derivative **49** (see above) to **64**, again generated *in situ* by deprotection of **63**. When **64** was intercepted by the triyne **46** under the usual co-cyclization conditions the functionalized triangular [4]phenylene **67** was obtained, ideally set up for yet another benzocyclobutadieno annelation step, resulting in the hexatrimethylsilyl derivative **68** of triangular [7]phenylene![35]

Comparison of the overall properties of linear and angular phenylenes suggests that the physical properties of the linear phenylenes are largely determined by their antiaromatic cyclobutadienoid circuits which they cannot avoid. In the angular series this effect is reduced by different amounts of bond alternation in the inner rings to enable maximization of the aromaticity of the outermost rings.[29]

15.2 Extended structures containing triple bonds

To use the carbon–carbon triple bond as an element for the construction of extended hydrocarbons is an obvious suggestion. It is readily available (see Chapter 8) and one of its most important reactions, the oxidative dimerization of terminal alkynes, already constitutes the first stage of elongation. In fact, we have already seen in Section 8.1 how polyacetylenes, the most elementary rod structures in organic chemistry, can be synthesized. However, we also noted the instability of the polyacetylenes, beginning even with as few as three consecutive triple bonds, and showed how this high reactivity can be tamed (*e.g.* by introduction of terminal *tert*-butyl or aryl substituents). Of course, if other than linear or angular structures are required, our building blocks need more than two terminal acetylene functions. Among the large number of potential acetylenic 'monomers' for the construction of extended hydrocarbon systems, those summarized in Scheme 10 have recently received growing attention.

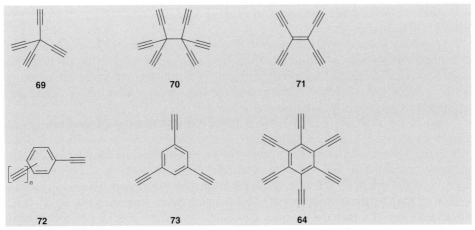

Scheme 10. A selection of acetylenic building units for the construction of extended hydrocarbon structures.

Beginning with tetraethynylmethane (**69**) and hexaethynylethane (**70**) a whole series of saturated hydrocarbons is conceivable, including cyclic variants such as the (unknown) hexaethynylcyclopropane and its higher homologs. The prototype for the unsaturated elements is tetraethynylethene (**71**, TEE). Obviously, the central double bond could be replaced by an allene group or even higher cumulogs.[36] Although all three parent systems **69**–**71** have been synthesized, none has yet been used directly for the construction of larger hydrocarbon structures so far, the main reason being their instability and the resulting difficulties in handling them in larger amounts. Some of the reported syntheses are, furthermore, too long to be of practical use. Still, since their preparation constitutes a notable achievement in hydrocarbon chemistry I shall discuss the routes leading to these very basic structures before progressing to extended hydrocarbon systems containing triple bonds.

It has long been known—*inter alia* from retinoid chemistry[37]—that the replacement of a double bond in a (larger) conjugated system by an aromatic ring often results in a very considerable increase in stability. If this observation were applied to polyacetylene chemistry we would expect that hydrocarbons derived from oligoacetylenes in which a number of triple bonds has been replaced by benzene rings (or other aromatic units) would be significantly more stable than their all-acetylenic model compounds. Taking triacetylene (see Section 8.1) as an example, substitution of its central triple bond by a benzene ring would lead to the isomeric diethynyl benzenes **72** ($n = 1$). For purposes of construction it is of importance that by placing the two triple bonds *ortho*, *meta*, and *para* to each other, different angles between them (60, 120, 180°) are attainable, enabling the building of circular, angular, or linear structures. Obviously, with a more complex aromatic core unit the number of orientations (angles) between the ethyne groups grows quickly. Increasing the number of triple bonds in these 'monomers' beyond two opens up numerous other modes of connection. For example, 1,3,5-triethynyl- (**73**) and even more so 1,2,3,4,5,6-hexaethynylbenzene (**64**) form a platform for the construction of large planar hydrocarbon sheets. Many of these simple ethynylbenzenes have long been known, the most recent addition being **64**, the synthesis of which has just been mentioned (see above).

Before proceeding to the preparation of large hydrocarbons consisting of triple and double bonds or triple bonds and aromatic systems, the syntheses of hydrocarbons **69**–**71** will be discussed.

After many unsuccessful attempts,[38–40] tetraethynylmethane (**69**) was finally prepared by Feldman and co–workers in 1993, by a multistep process which at first sight is in stark contrast to the structural simplicity of the target molecule, but which is, actually, necessitated by the difficulty of connecting the four most highly unsaturated substituents to one central carbon atom (Scheme 11).[41]

Starting from the diprotected 1,4-pentadiyne-3-one derivative **74**, olefination with carbon tetrabromide and triphenylphospine first yielded the dibromide **75**, which was transformed by metalation and quenching of the resulting vinyllithium intermediate with paraformaldehyde into the primary alcohol **76**. In the next step, decisive for the success of the synthesis, **76** was subjected to the acid-mediated Johnson orthoester variant of the Claisen rearrangement. This transformation furnished the highly functionalized ester **77** which not only has the central quaternary methane carbon atom of the target molecule but also two suitable functional groups for the preparation of the missing ethynyl functions. Lithium diisopropylamide (LDA)-induced dehydrobromination in the presence of trimethylsilylchloride next gave the triyne **78**, which by standard methodology was converted into the tosylhydrazone **79**, which, with its additional phenylsulfonyl group, is all set up for generation of the fourth triple bond by twofold elimination under basic conditions. Deprotection with poassium carbonate in methanol finally yielded tetraethynylmethane (**69**) as a colorless solid which rapidly decomposed at room temperature whether oxygen was excluded or not. As already stated, the route to **69** might seem overly complicated. However, the 'intuitive' route to this hydrocarbon *via* the tertiary alcohol **80**, its conversion into, *e.g.*, a halide, and ethynylation of the last, failed.[39] Although the preparation of **80** from **74** is an easy task, the introduction of the missing triple bond by substitution was impossible. In one of these experiments, however, a small amount of the hexaethynylethane derivative **81** was produced (Scheme 12).[38]

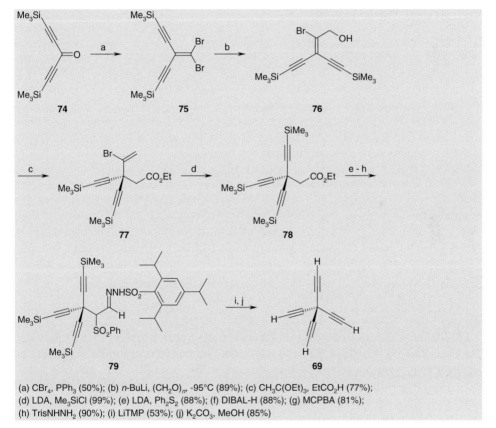

(a) CBr$_4$, PPh$_3$ (50%); (b) *n*-BuLi, (CH$_2$O)$_n$, -95°C (89%); (c) CH$_3$C(OEt)$_3$, EtCO$_2$H (77%);
(d) LDA, Me$_3$SiCl (99%); (e) LDA, Ph$_2$S$_2$ (88%); (f) DIBAL-H (88%); (g) MCPBA (81%);
(h) TrisNHNH$_2$ (90%); (i) LiTMP (53%); (j) K$_2$CO$_3$, MeOH (85%)

Scheme 11. The synthesis of tetraethynylmethane (**69**) by Feldman and co–workers ...

Whether **69** and **70** can be employed for the preparation of novel carbon nets—or 'only' oligomeric hydrocarbons incorporating these multifunctional building elements—remains to be seen.

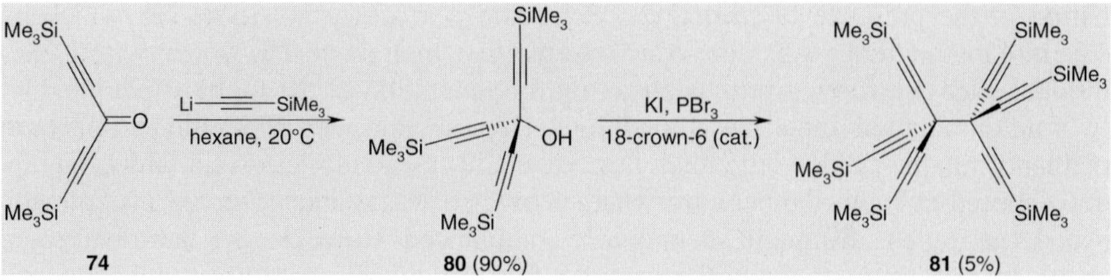

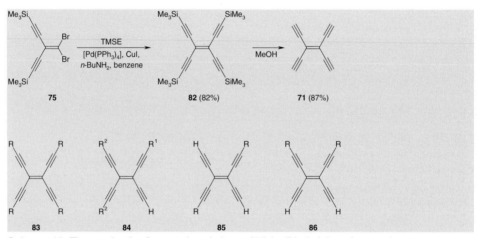

Scheme 12. ... and of **81**, a derivative of hexaethynylethane (**70**) by Alberts and Wynberg.

Far more advanced in this respect is the use of the olefinic building block tetra-ethynylethene (**71**, TEE) and several of its derivatives (see below), prepared by Diederich and co-workers in 1991, in a synthesis demonstrating again the crucial role of the ketone **74** and the dibromoolefin **75** derived from it in the preparation of pere-thynylated hydrocarbon systems (Scheme 13).[42]

Scheme 13. The synthesis of tetraethynylethene (**71**) by Diederich and co-workers.

Palladium-catalyzed coupling of **75** with trimethylsilylethyne (TMSE) provided the tetraprotected derivative **82** from which the parent hydrocarbon **71** was liberated by treatment with methanol. Although neither **82** nor **71** were the first tetra-ethynylethenes to be reported,[43] the above methodology (and variations thereof) is of general importance because it not only leads to the parent molecule but also enables the preparation of nearly any desired functionality and silyl protection,[44] as sym-bolyzed by the general structures **83** (fully substituted/protected), **84** (mono-deprotected), **85** (*trans*-bis-deprotected), and **86** (*cis*-bis-deprotected). To have access to these different substitution patterns is of importance in the construction of oli-gomeric structures from the TEEs. We have, for example, already discussed the preparation of the 'exploded' [*n*]radialenes from monomers of the type **84** in Section 11.5.

Whereas *cis*-bis-deprotected TEEs have been used as building blocks for cyclic structures, their *trans* isomers form the repeating unit for a new type of molecular wire with the poly(triacetylene) backbone.

Thus by oxidative Glaser–Hay macrocyclization Diederich and co-workers have obtained the per(silylethynyl)ated octadehydro[12]annulene (**88**) and dodecadehydro[18]annulene (**89**) from the precursor **87** (R = methyl and isopropyl in all cases) (Scheme 14),[45] several higher (unidentified) oligomers of **87** also being produced.

According to X-ray crystal structure analysis both annulene perimeters are completely flat and the electronic absorption spectra characterize the bright yellow derivatives **89** as Hückel-aromatic [18]annulenes, whereas their lower homologs **88** are antiaromatic.

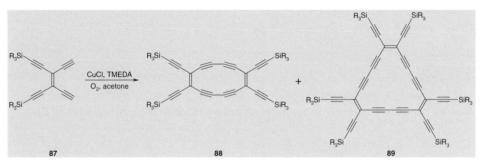

Scheme 14. From *cis*-bis-protected tetraethynylethenes, **87**, to dehydroannulenes, **88** and **89** ...

On the other hand end-capping polymerization of, *e.g.*, the bis-protected *trans* derivative **90** led to the oligomers **91** (*n* = 1-5), a series of highly stable conjugated molecular rods (Scheme 15).[46]

As expected the yields decrease with increasing rod length (*n* = 1 and 2, *ca* 30%; *n* = 5, 2%), but the new oligomers could be fully characterized and for several representatives X-ray crystal structure determinations have been conducted. The distance between the two extreme tips of the terminal phenyl groups increases from 19.4 Å (**91**, *n* = 1) to 49.2 Å (**91**, *n* = 5), *i.e.* one is dealing here with nanometer-sized molecular objects. In the meantime even more extended polyacetylenes such as **92** and **93** have been synthesized, the length of the polymers being determined by the solubility in the solvents in which they were produced.[47] Because, in principle, any desired substituent can be attached at the ends of these rods by changing the end-capping reagent, the attachment of these novel extended π-systems to substrates such as silicon wafers seems feasible.[48]

Turning to the aromatic acetylenes **72** and **73** it is at first sight not obvious that these structurally so simple hydrocarbons belong to the most versatile building blocks for extended π-structures presently known. Their preparative usefulness—availability, stability, structural rigidity—has already been discussed. As an additional bonus, the possibility of introducing either functional or solubilizing (usually alkyl) groups into the benzene core, must also be mentioned. In fact, the variety of oligomeric and macrocyclic compounds which have been prepared from and with phenylacetylenes is so

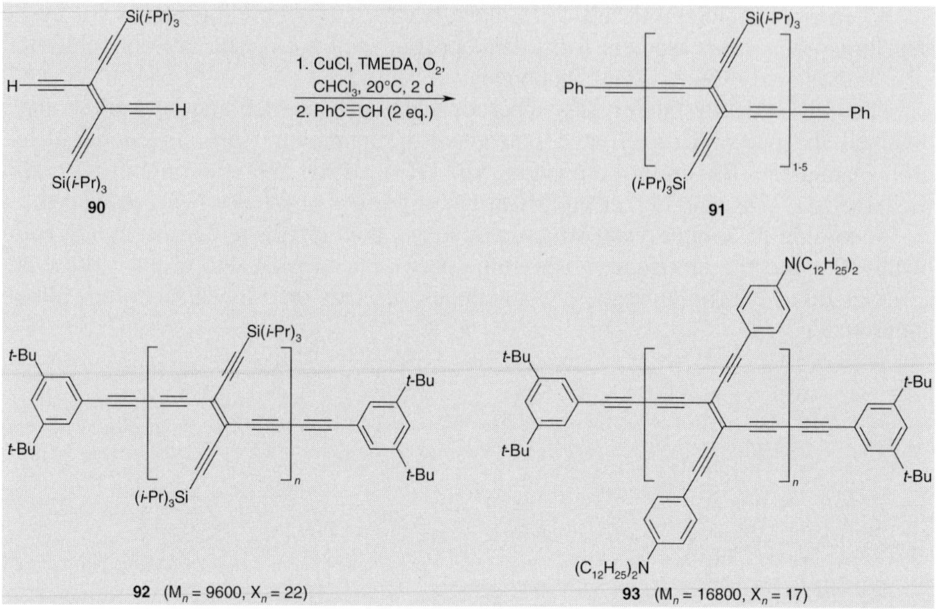

Scheme 15. ... and from *trans*-bis-protected tetraethynylethenes, **90**, to molecular rods, **91**.

extraordinary large, that the size restrictions of this monograph become particularly painful in this chapter. The examples which follow thus constitute only a small, highly subjective selection from a vast field.

Beginning with the diethynylbenzenes, obviously two basic motifs are possible— rod-like structures are accessible from 1,4-diethynylbenzene and angular and circular structures from its *ortho* and *meta* isomers. Although carbon rods such as **94** are known,[49] no systematic efforts have, apparently, been undertaken to prepare a complete series all the way up to, *e.g.*, the decamer of phenylacetylene, by a uniform concept. Rather, the synthesis of polymers such as **95** and **96** has been described by several groups (Scheme 16).[50–52]

These compounds, which can be regarded as extended versions of the widely studied poly(phenylenes),[53] are of importance in the synthesis of molecular wires to be used as components for novel electronic devices.[53, 54] Among the methods reported for the preparation of the π-conjugated polymers **99** Hagihara-type coupling of *para*-diethynylaromatics such as 1,4-diethynylbenzene (**97**) with various 1,4-dihaloaromatics (**98**) or the Sekiya–Ishikara coupling of bis Grignard components, **100**, with the dihalides **98** constitute the most general routes.

The use of 1,3-diethynylbenzene building units has been pioneered and developed to a very high level of complexity and sophistication particularly by Moore et al. in their efforts to create shape-persistent molecular architectures with nanoscale dimensions.[55]

To synthesize linear oligomers a highly efficient process was devised in which the high yield conversion of *N,N*-dialkylaryltriazenes to aryliodides by treatment with iodomethane at 100 °C plays a crucial rule, previous studies having shown that the

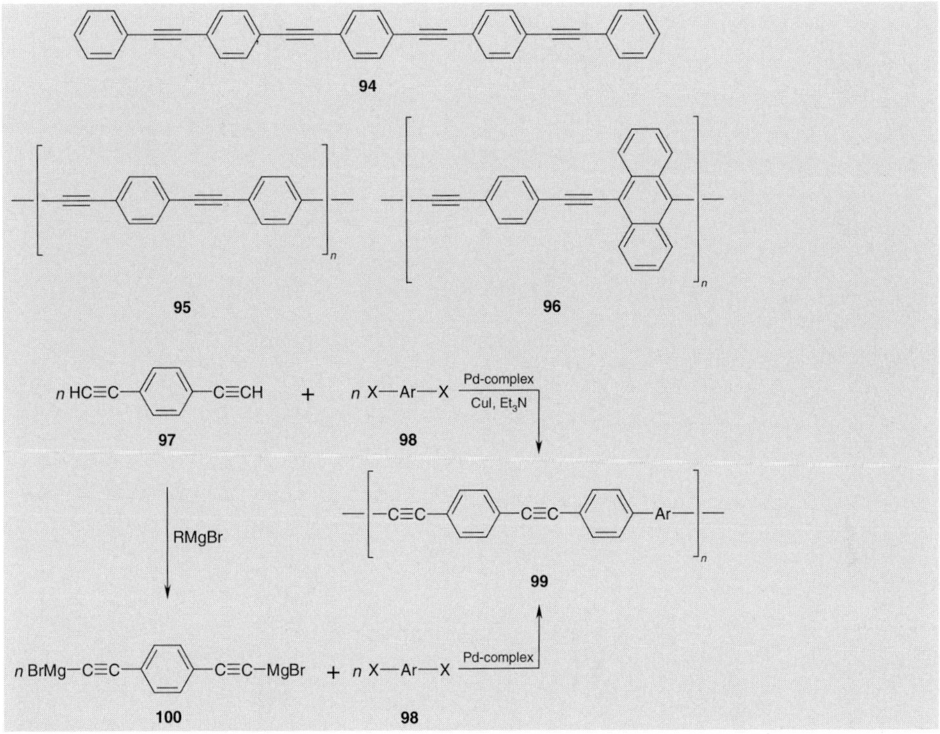

Scheme 16. Extended hydrocarbons structures from *para* diethynylaromatics.

reactivity of comparable arylbromides is not high enough to enable these iterative syntheses to proceed in sufficiently high yields. Furthermore, to simplify purification and increase the rate at which the different oligomers could be produced, solid-phase methods—a *novum* in hydrocarbon synthesis—were employed as shown in Scheme 17.[56, 57]

Reaction of Merrifield's resin (**101**) with neat propylamine furnished the propylbenzylamine **102** which on coupling with the diazonium salt **103** gave the triazene **104**. The *tert*-butyl groups will serve as solubilizing substituents in the future hydrocarbon oligomers, and the triazene group plays the role of a diazonium equivalent. It is stable to the conditions used in the subsequent cross-coupling steps and to those employed for deprotection of the terminal alkyne. Palladium-catalyzed coupling of **104** with trimethylsilylethyne (TMSE) then led to **105**, a polymer-bound bifunctional monomer used in the oligomer-building process next. Towards this end **105** was either desilylated by treatment with tetrabutylammoniumn fluoride (TBAF) to the terminal acetylene **106** or cleaved to the *meta*-iodophenylacetylene derivative **107** by reaction with methyl iodide at elevated temperature. Pd-mediated coupling of these two monomers yielded the dimer **108**, which, when subjected to a growth sequence by repetition of these steps, eventually provided oligomers such as **109** and **110**, the former having been set free from the resin by another substitution step with methyl iodide. By this methodology discrete oligomers containing up to 32 monomer units were obtained in high purity.

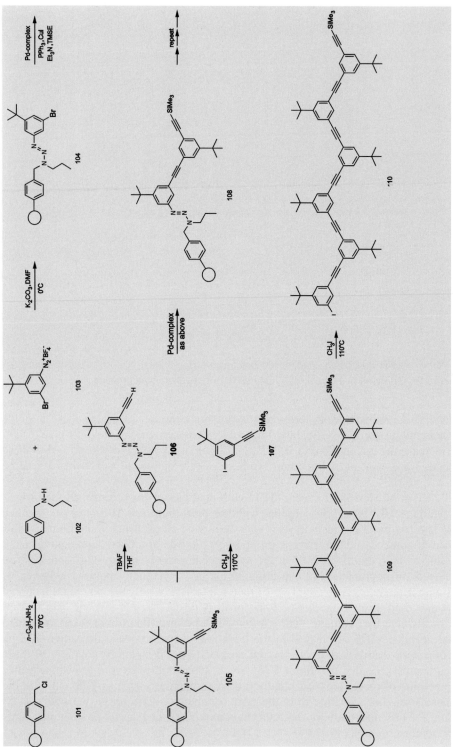

Scheme 17. The synthesis of *meta*-diethynylbenzene-derived hydrocarbon oligomers by Moore and co-workers...

Sequence-specific oligomers of the *meta*-substitition-type **110** are extremely valuable for the preparation of large cavity-containing hydrocarbons (Scheme 18).[58]

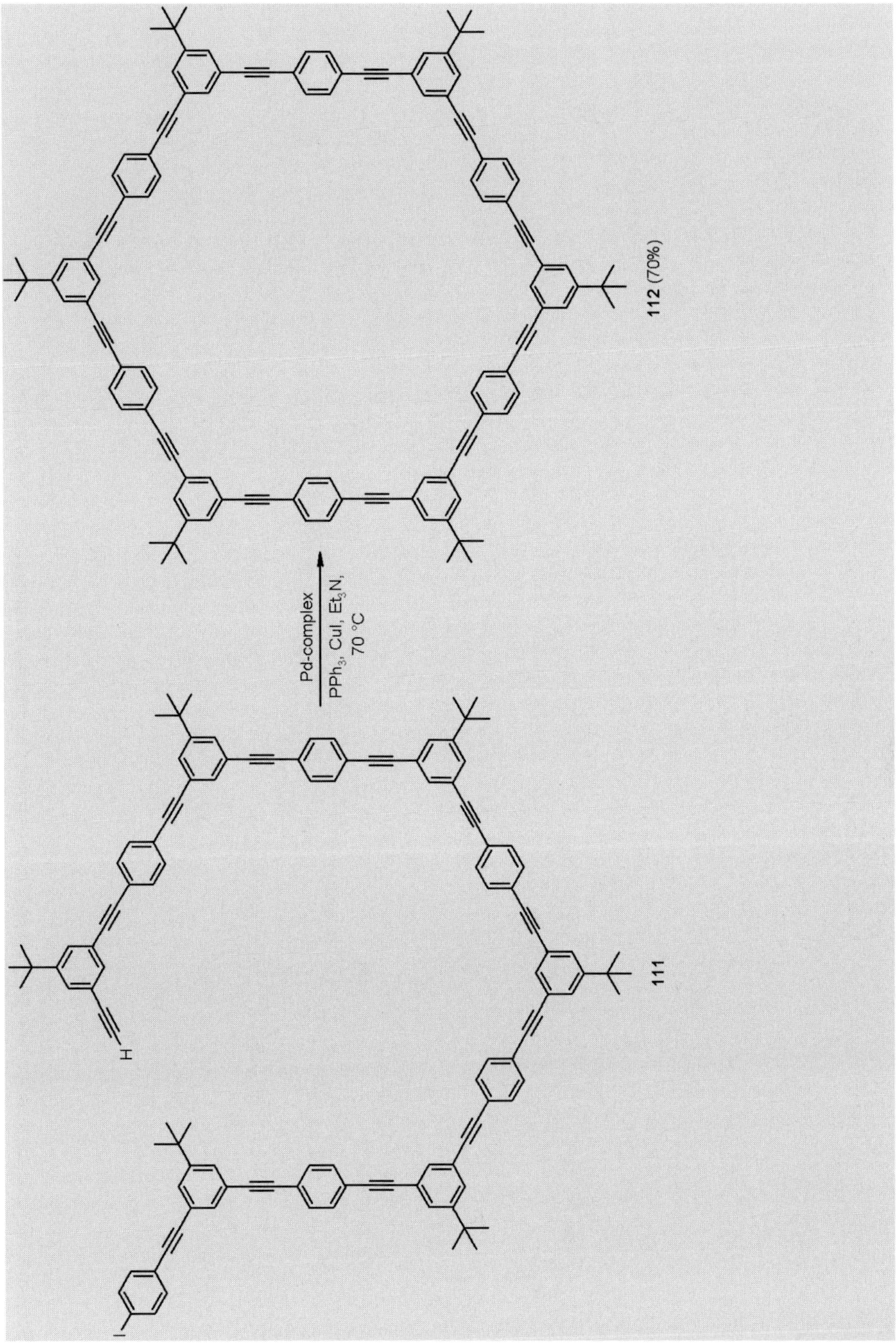

Scheme 18. .... and their use for the synthesis of cavity-containing hydrocarbons, **112**.

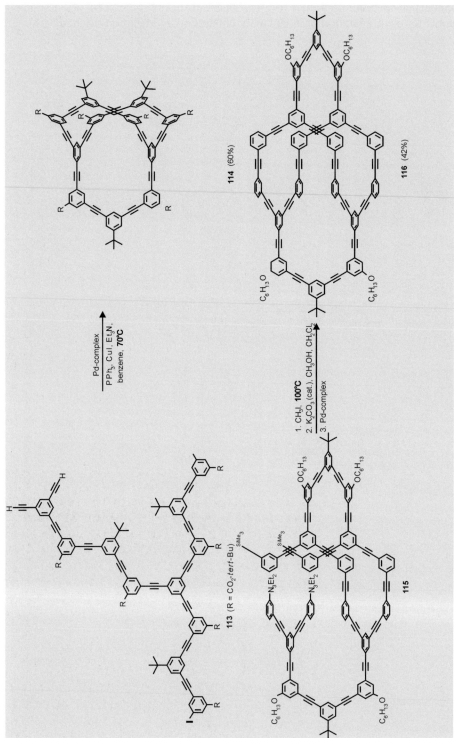

Scheme 19. The synthesis of the phenylacetylene-derived macropolycyclic systems **114** and **116**.

For example, macrocyclization of the 'extended' hexamer **111** furnished the cyclic hydrocarbon **112** in good yield. This amazing phenylacetylene cyclooligomer—formally **112** is composed of twelve phenylacetylene units—has an internal diameter of 22 Å as compared with *ca* 7 Å of kekulene (see Section 14.2). With appropriate functional groups these ring compounds could serve as modular units for the assembly of novel functional materials ('porous crystals').[54]

The introduction of a third triple bond into the building block **72**, for example as in 1,3,5-triethynylbenzene (**73**) causes a veritable explosion of construction possibilities. Macrobicycles and -tricycles such as **114**[59] and **116**,[60] respectively, were readily prepared in good yields by double cyclization of the branched precursor systems **113** and **115**, which were again assembled from smaller phenylacetylenes by Pd-catalyzed coupling reactions (Scheme 19).

Determination of the structure of these huge complex hydrocarbons has been called an 'arduous task';[54] it rests largely on ^{1}H NMR spectra—despite their high aromatic content these molecules are quite soluble in a variety of solvents including pentane (probably because of their spherical structure)—and, especially, the massive use of modern mass spectrometric techniques including MALDI-TOF and direct and silver chemical-ionization infrared laser desorption Fourier transform mass spectrometry (LD-FTMS and LDCI-FTMS).[61]

In **114** and **116** the 1,3,5-triethynylbenzene unit was used as a proxy for a bridgehead in polycyclic hydrocarbons. However, Moore has also exploited the 'branching properties' of **73** to synthesize dendrimer hydrocarbons of unprecedented complexity and size. By applying the above protection, deprotection, and coupling steps to suitable precursors, the monomers **117–119** and the core units **120–123**, respectively, dendrimers which combine high molecular weight with a very high degree of structural uniformity could be synthesized very efficiently. This is illustrated in Scheme 20,[62] in which the 1,4- or a 1,3,5-substituted benzene ring units and the connecting triple bonds are slowly beginning to 'vanish' in these most extended structures (Scheme 20).

In this repeating, non-linear *aufbau* process a set of increasingly branched trimethylsilylethynyl derivatives **124–128** was first prepared by a series of coupling steps beginning with the connection of 1-trimethylsilylethynyl-3,5-di-*tert*-butyl benzene (**124**) with **117** to **125** and ending with the monodendron **128**. In the next step the unprotected monodendrons were coupled to the core units **120–123** providing the tridendrons **129–133**. Whereas the structure of **132** was assigned unambiguously on the basis of ^{1}H NMR spectra (twodimensional H,H-COSY and *J*-resolved experiments), determination of the structure of **133** rests on elemental analysis, size-exclusion chromatography (SEC) data, and (unresolved) NMR spectra. With a formula of $C_{1398}H_{1278}$ and a molecular weight of 18.1 kDa the 127-mer **133**, which—according to space-filling molecular models—spans over 125 Å, is one of the biggest hydrocarbons ever prepared!

Compared with the acyclic oligo- or poly(phenyleneacetylenes) derived from building blocks with a *meta-* or *para*-substitution pattern at the benzene ring (see above), relatively little is known about the corresponding linear hydrocarbons derived from the appropriate *ortho*-substituted buildings blocks. Much more often the 1,2-diethynylbenzene unit has been used for the construction of cyclic hydrocarbons as we shall see below. Still, the parent systems **134** with *n* = 3-7 have recently been pre-

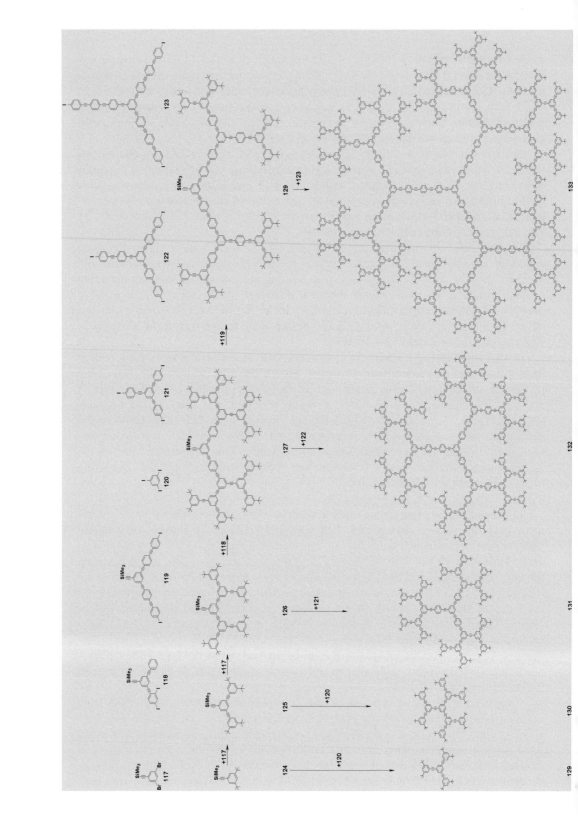

pared by Grubbs and Kratz, and, again, the success of their syntheses rests on metal-mediated coupling reactions using Pd[P(Ph)₃]₄ · CuI as the catalyst (Scheme 21).[63]

Starting from *ortho*-dibromobenzene (**135**) 1-phenyl-2-(2-ethynylphenyl)acetylene (**136,** 2-ethynyltolane) was first prepared by successive coupling with trimethyl-silylethyne (TMSE) and phenylacetylene, respectively, followed by deprotection with potassium fluoride. To introduce functionality for later coupling steps, **136** was reacted with *ortho*-diiodobenzene (**137**) to the monoiodide **138**, which was ethynylated in the usual way. Repetition of these last steps then led to **140**, which on coupling with **137** under suitable conditions, furnished **141**, the longest α-phenylethynyl-ω-phenylpoly[1,2-phenylene(2,1-ethynediyl)] known. X-ray analysis of two shorter

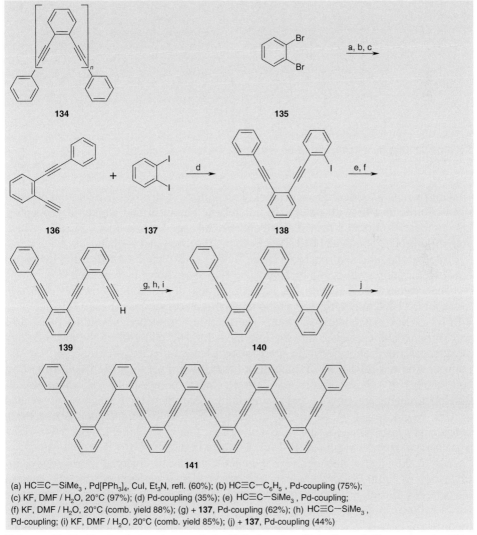

(a) HC≡C—SiMe₃ , Pd[PPh₃]₄, CuI, Et₃N, refl. (60%); (b) HC≡C—C₆H₅ , Pd-coupling (75%);
(c) KF, DMF / H₂O, 20°C (97%); (d) Pd-coupling (35%); (e) HC≡C—SiMe₃ , Pd-coupling;
(f) KF, DMF / H₂O, 20°C (comb. yield 88%); (g) + **137**, Pd-coupling (62%); (h) HC≡C—SiMe₃ ,
Pd-coupling; (i) KF, DMF / H₂O, 20°C (comb. yield 85%); (j) + **137**, Pd-coupling (44%)

Scheme 21. Synthesis of phenyl-capped oligo[1,2-phenylene(2,1-ethynediyl)] compounds, **134.**

members of the series (**134**, n = 2, 3) revealed a propensity for intramolecular π-stacking of the phenyl rings leading to a helical arrangement.

Combining triple bonds with aromatic cores not only opens the way to a sheer endless variety of hydrocarbons but can also give rise to novel forms of carbon. A case in point is 'graphyne' (**142**) proposed by Baughman and co–workers as a new planar form of carbon containing equal numbers of sp- and sp^2-hybridized carbon atoms (Scheme 22).[64]

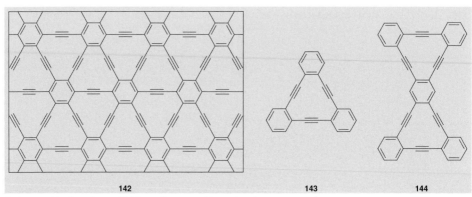

Scheme 22. Graphyne (**142**) and some of its subunits.

Our by now experienced view quickly recognizes many—known and unknown—substructures in **142** with potentially interesting chemical and structural properties, among them Vollhardt's hexaethynylbenzene (**64**, see Section 15.1), Staab's tribenzotrisdehydro[12]annulene (**143**) (see Section 10.4), and many others.

Another, larger dehydrobenzannulene (DBA) substructure of **142** is the hydrocarbon **144**, recently synthesized by Haley and Kehoe as shown in Scheme 23.[65]

The synthesis begins with the interconversion of the amino function of *ortho*-iodoaniline (**145**) into a triazene group and the replacement of the iodo substituent in **145** by a trimethylsilylethynyl group. The resulting bifunctional building block **146** was subsequently converted into the intermediates **147** and **148** by the methodology already described above for the preparation of *meta*-diethynylbenzene-derived oligomers. And the Pd-catalyzed coupling of these two units then furnished **149**. Two equivalents of these diacetylenes were next coupled to 1,5-dibromo-2,4-diiodobenzene as a core unit, and the resulting dihalo derivative was bis-ethynylated to **150**. Replacement of the triazene functions by iodide, deprotection of the triple bonds, and a last Pd-mediated C–C coupling step finally yielded the DBA **144**, which was characterized by means of IR, UV/Vis, and MS data.[65]

Just as the 'central hole' in, *e.g.*, the circulenes can be enlarged by successive introduction of additional *ortho*-annelated benzene rings (see Section 14.2) the internal diameter of the DBAs can be increased by replacement of the triple bonds in the above hydrocarbons by butadiyne, hexatriyne, *etc.*, rods. This has been accomplished by different groups,[66, 67] the examples shown in Scheme 24 again being from Haley and co-workers.[65, 68]

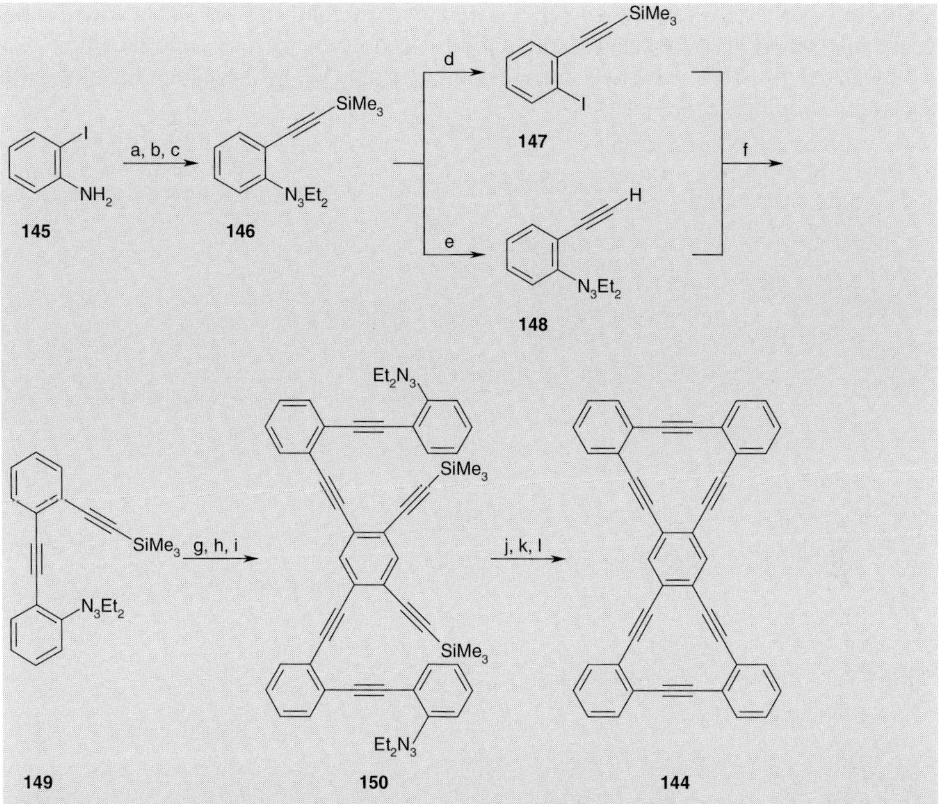

(a) HCl, NaNO$_2$; (b) K$_2$CO$_3$, Et$_2$NH; (c) TMSE, PdCl$_2$[PPh$_3$]$_2$, CuI, Et$_3$N (comb. yield 82%); (d) MeI, 120°C (96%); (e) K$_2$CO$_3$, MeOH (98%); (f) PdCl$_2$[PPh$_3$]$_2$, CuI, Et$_3$N (88%); (g) K$_2$CO$_3$, MeOH (>95%); (h) + 1,5-dibromo-2,4-diiodobenzene, PdCl$_2$[PPh$_3$]$_2$, CuI, Et$_3$N; (i) TMSE, PdCl$_2$[PPh$_3$]$_2$, CuI, Et$_3$N (comb. yield 63%); (j) MeI, 120°C; (k) K$_2$CO$_3$, MeOH; (l) Pd(d ba)$_2$, PPh$_3$, CuI, Et$_3$N (comb. yield 11%)

Scheme 23. The dehydrobenzannulene **144** as a substructure of graphyne (**142**).

Coupling of *ortho*-bromoiodobenzene (**19**), first with trimethylsilylbutadiyne and then with tris(isopropyl)silylethyne in the presence of the usual Pd-catalyst systems furnished the 1-butadiynyl-2-ethynylbenzene derivative **151**, which, after removal of its more reactive (trimethylsilyl) protective group, could be coupled with *ortho*-diiodobenzene (**137**) to a bis-coupled product in 71% yield. Desilylation with fluoride and oxidative dimerization at high dilution conditions then gave the 'stretched' DBA **152** as the sole product in moderate yield. Although the hydrocarbon is very poorly soluble in the common organic solvents (see above), all spectral data could be determined, and its structure was assigned unequivocally. When the intermediate **151** was desilylated in the presence of copper(II) acetate in pyridine-methanol, oxidative C–C-coupling took place *in situ*, and the the linear oligoacetylene **154** was produced in good yield. And again, deprotection with fluoride and ring closure by Cu(OAc)$_2$-CuCl in pyridine yielded the orange-colored 32-membered macrocycle **155** in acceptable yield. It is the largest tetrabenzo-DBA derivative yet characterized. That, finally, the acetylenic spacer units in these hydrocarbons need not be of the same length[67] was demonstated by **153**, which is also available from **19** by applying the powerful acetylene coupling techniques.[69]

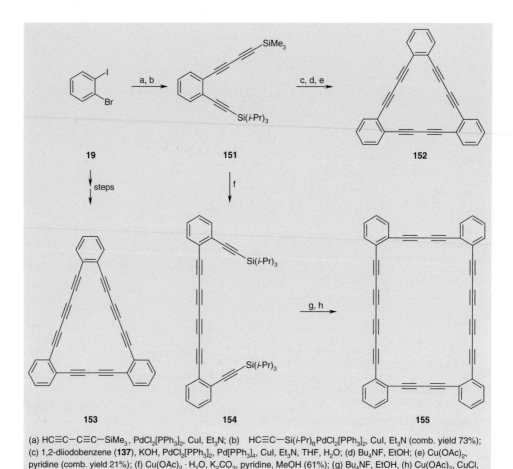

(a) HC≡C–C≡C–SiMe₃, PdCl₂[PPh₃]₂, CuI, Et₃N; (b)   HC≡C–Si(i-Pr)₆PdCl₂[PPh₃]₂, CuI, Et₃N (comb. yield 73%); (c) 1,2-diiodobenzene (**137**), KOH, PdCl₂[PPh₃]₂, Pd[PPh₃]₄, CuI, Et₃N, THF, H₂O; (d) Bu₄NF, EtOH; (e) Cu(OAc)₂, pyridine (comb. yield 21%); (f) Cu(OAc)₂ · H₂O, K₂CO₃, pyridine, MeOH (61%); (g) Bu₄NF, EtOH, (h) Cu(OAc)₂, CuCl, pyridine (51%)

Scheme 24. The preparation of di-, tri- and tetraacetylene dehydrobenzannulenes (DBAs) by Haley et al.

15.3 Building with cyclopropane rings

The usefulness of the benzene ring and of double and triple bonds as building blocks for extended structures rests largely on their ready availability, stability, and the numerous methods known to connect these construction units. In fact, nothing illustrates the importance of permanent improvement and invention of new preparative methodology better than the recent renaissance of acetylene chemistry, which—as we have seen above and in Chapter 8—is mainly because of the introduction of metal-mediated C–C coupling processes.

Although the cyclopropane ring has long stood in the shadow of these other building blocks, progress in this area of small ring chemistry has been so rapid during recent decades[70, 71] that today cyclopropanes carrying any type of substituent are now

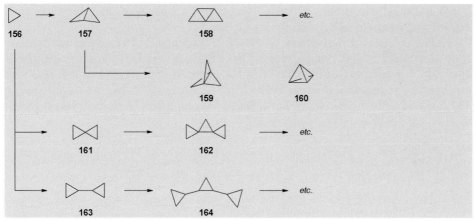

Scheme 25. Extended structures from cyclopropane (**156**).

available in multi-gram amounts by simple, reliable routes. And because of its steric, *i.e.* rigid, and electronic, double-bond-like properties, the importance of the three-membered ring as a building block is also growing rapidly. As shown in Scheme 25 there are several modes by which cyclopropane rings can be combined to produce more complex frameworks.

When two cyclopropane rings share one C–C bond bicyclo[1.1.0]butane (**157**) results. Annelation of a third ring could either lead to tricyclo[2.1.0.0^{1,3}]pentane (**158**) or to [1.1.1]propellane (**159**). Whereas the annelation process can, in principle, be extended beyond the linear structure **158**, its isomer **159** represents a dead-end. Another 'final' structure is reached with tetrahedrane (**160**) which is formally accessible from **157** by removal of two non-bridgehead hydrogen atoms. I have, in fact, already discussed the synthesis of **159** (see Section 6.1) and the different attempts to prepare **160** (see Section 5.1) at great length. With its inverted bridgehead carbon atom we would expect the zero-bridged spiropentane **158** to be highly strained, a property which would be even more pronounced in the higher homologs of this series. Extended structures of any size would, therefore, not be expected along this branch (and have not been prepared or generated, see below). This is, however, different if the two rings are connected by one carbon atom only, as shown by spiro[2.2]pentane (**161**), the simplest representative of this type of connection. This hydrocarbon also will be highly strained, but in its two constituting halves the strain should not significantly exceed that of cyclopropane itself, a perfectly stable hydrocarbon under normal laboratory conditions. We would, therefore, not expect serious synthetic problems on extension of this branch beyond dispiro[2.0.2.1]heptane (**162**). And indeed, the linear and angular oligo- and polytriangulanes to be discussed below can all formally be traced back to **161** and **162**. Finally, when the cyclopropane rings are joined by a single bond only, as in bicyclopropyl (**163**) as the parent hydrocarbon, they no longer have a common atom. A further extension—*via* **164**—seems feasible, and has, in fact, been realized. Interestingly, several natural products, isolated from the fermentation broth of *Streptoverticillium fervens* containing up to four consecutive cyclopropane rings have recently been identified and synthesized;[72, 73] these compounds have novel anti-fugal properties. Because of size and scope limitations bicyclopropyl-derived hydrocarbons will not be dealt with here.[74]

Although the first derivatives of bicyclo[1.1.0]butane (**157**) were discussed as early as the turn of the 19th century,[75] the first authentic bicyclobutane derivative, methyl bicyclobutane-1-carboxylate, was not obtained until more than half of a century later, by Wiberg and Ciula in 1959.[76, 77] The synthesis of the parent hydrocarbon **157** had to wait four more years, but was then accomplished by three different groups more or less simultaneously (Scheme 26).

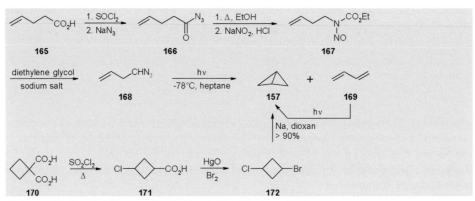

Scheme 26. The first three routes to bicyclo[1.1.0]butane (**157**).

The synthesis developed by Lemal and co–workers[78] started from 4-pentenoic acid (**165**) which was converted *via* its acid chloride into the azide **166**. When this was heated in ethanol the corresponding urethane was produced; this, on treatment with sodium nitrite and hydrochloric acid, gave the *N*-nitrosoallylcarbinyl urethane **167**. The immediate precursor for **157**, allyldiazomethane (**168**) was then generated by decomposition of **167** with the sodium salt of diethyleneglycol in diethyleneglycol at 0 °C. And when **168** was photolyzed in dilute heptane solution with a high pressure mercury lamp at –78 °C two hydrocarbons could be isolated by gaschromatography— **157** and 1,3-butadiene (**169**), formed in a 1:5 ratio. Preparatively more satisfactory was the synthesis of bicyclo[1.1.0]butane by Wiberg and Lampman.[79] When cyclobutane-1,1-dicarboxylic acid (**170**) was chlorinated by sulfuryl chloride, the 3-chloro-derivative was formed as the major product.[80] Decarboxylation subsequently furnished 3-chlorocyclobutane-1-carboxylic acid (**171**), which on treatment with mercuric oxide and bromine gave 1-bromo-3-chloro-cyclobutane (**172**). When this derivative was dehalogenated with sodium metal in dioxan, a mixture of hydrocarbons was obtained in nearly quantitative yield; **157** strongly dominated (93–96%), the remainder was a trace of cyclobutene (5–7%). Whether the *cis* or the *trans* isomer of **172** was employed had no effect on either yield or composition of the product mixture. Finally, Srinivasan has reported the synthesis of **157** by photoisomerization of 1,3-butadiene (**169**).[81] Although bicyclobutane is only produced as a side product—the main photoisomer of **169** is cyclobutene—this approach is extraordinarily simple. The photochemical synthesis was unsuccessful in the vapor phase, because any **157** produced was presumably in an excited state and could not be deactivated by collision before re-isomerizing to starting material. In solution the 'hot' bicyclobutane can collide with the inert solvent molecules and hence become deactivated. Today **157**, a gas with a boiling point of 8.3 °C, is a very well characterized hydrocarbon, both chemically and structurally. Its strain energy (66.5 kcal mol^{-1}) is considerably larger than the sum of the strain energies for two cyclopropane rings (2 x 28 kcal mol^{-1}).[82]

The synthesis of the next higher homolog of **157**, tricyclo[2.1.0.0^{1,3}]pentane (**158**), reported by Wiberg and co-workers in 1993 (Scheme 27),[83] was modeled on the Szeimies route to [1.1.1]propellane (**159**) (see Section 6.1).

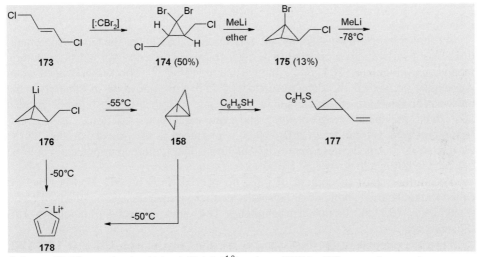

Scheme 27. The synthesis of tricyclo[2.1.0.0^{1,3}]pentane (**158**) by Wiberg and co–workers.

Addition of dibromocarbene (using phenyl(tribromomethyl)mercury as the carbene source) to *trans*-1,4-dichloro-2-butene (**173**) gave the tetrahalide **174** in *ca* 50% yield. When this adduct was treated with methyllithium in ether, 1-bromo-2-(chloromethyl)bicyclo[1.1.0]butane (**175**) was obtained, although, because of purification problems, only in a poor yield (13%). Metalation of **175** with methyllithium in THF at –78 °C provided the lithio derivative **176**; this was demonstrated by monitoring the reaction by NMR spectroscopy. Warming the solution to –50 °C caused the disappearance of the NMR signals of **176** and the formation of lithium cyclopentadienide (**178**) as shown by NMR analysis and trapping experiments. When the temperature was raised to only –55 °C, however, new signals appeared; these then disappeared as **178** was produced—a clear indication of the generation of a reaction intermediate. That this was probably the desired hydrocarbon **158** was not only supported by the ^{13}C NMR spectrum recorded at –55 °C but also by intercepting this highly reactive compound with phenylthiol to yield 2-vinyl-1-cycloropylphenylsulfide (**177**). The heat of formation of tricyclopentane **158** was estimated to be 132 kcal mol^{-1}, leading to a strain energy of 143 kcal mol^{-1} for this unique hydrocarbon; with *ca* 29 kcal mol^{-1} of strain per carbon atom this is one of the most highly strained organic compounds ever prepared.[82, 83]

Turning next to the spiro-condensed three-membered ring systems, the parent hydrocarbon, spiropentane (**161**) has been known since 1896 when it was prepared by Gustavson by debromination of tetrakis-(bromomethyl)methane (**179**) with zinc in ethanol (Scheme 28).[84]

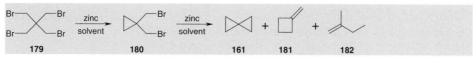

Scheme 28. The preparation of spiropentane (**161**) by Gustavson.

As shown by later investigators, the formation of **161** under the Gustavson conditions was invariably accompanied by the generation of several byproducts, including methylenecyclobutane (**181**) and 2-methyl-butene (**182**).[85] This only changed when Applequist performed the dehalogenation of **179** with zinc dust in ethanol in the presence of disodium ethylenediamine tetraacetate to trap the zinc bromide formed and thus prevented ring-opening and ring-expansion reactions of 1,2-bis-(bromomethyl)-cyclopropane (**180**), which is produced as the primary product of the bicyclization.[86] Under these optimized conditions nearly pure spiropentane could be obtained in 80% yield.

The further spiro-attachment of three-membered rings to **161** leads to the so-called triangulanes,[87] which can be subdivided into three classes: the linear or chain triangulanes **183**, the branched triangulanes **184**, and the cyclic triangulanes **185** (Scheme 29).[88]

Typical examples of the first category are the polyspiranes **162** and **186–188**; [3]rotane (**189**), the synthesis of which was already discussed in Section 11.6, is a branched triangulane, and the (unknown) 'davidane' (**190**)[89] is a [6]cyclic triangulane. Many of the higher triangulanes have interesting stereochemical properties. For example, proceeding from the achiral **162**, linear attachment of a further three-membered ring will lead to a racemic pair of hydrocarbons, one enantiomer being shown in **186**. For [5]unbranched triangulane three stereoisomers are posssible—the *meso* compound **187** and the D,L pair **188**.

The reductive ring-closure of a 1,3-dihalide, such as in the **180**→**161** cyclization, is a time-honored process to prepare cyclopropane rings. So is, of course, the addition of carbenes or carbenoids to C-C double bonds,[70, 71] and it is hence not surprising that this method has also been employed widely for the preparation of all types of triangulanes as shown predominantly be the groups of Binger,[90] Zefirov,[87] and de Meijere.[88]

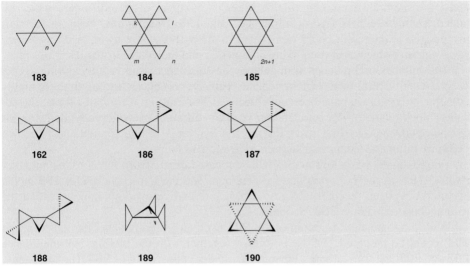

Scheme 29. Classification and stereoisomerism of the triangulanes.

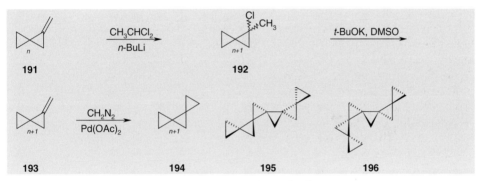

Scheme 30. Zefirov's general route to linear triangulanes (**194**).

To prepare the linear triangulanes the following general step-by-step extension developed by Zefirov has proven to be particularly valuable (Scheme 30).[91] Beginning with a methylenecyclopropane derivative **191**—for $n = 0$ this is methylenecyclopropane itself, of course—methylchlorocarbene (generated from 1,1-dichloroethane by treatment with *n*-butyllithium) was first added to produce the adduct **192**. When this was dehydrochlorinated with potassium *tert*-butoxide in DMSO a—now extended—methylenecyclopropane derivative **193** was (re)generated; this could either be resubmitted to the same reactive cycle or cyclopropanated to the final unbranched triangulane. By use of this route [3]-, [4]-, [5]-, and even [6]unbranched triangulanes have been prepared.[87, 88] Although elegant—and reminiscent of Fitjer's iterative route to the higher rotanes (see Section 11.6)—and proceeding in good to excellent yields, this approach has one drawback—its lack of stereoselectivity. For example, when linear [6]triangulane was prepared by Zefirov and co–workers[87] according to this scheme, mixtures of stereoisomers were obtained, among them the hydrocarbons **195** and **196**. Despite this, many of the thus synthesized triangulanes could be isolated in analytically pure form, enabling—*inter alia*—the investigation of their strain energies[92] and molecular structures.[93]

The number of branched triangulanes has also increased rapidly during recent years,[88] and several more or less general synthetic sequences leading to these highly complex polyspiranes habe been developed.[94] The most convergent of these, the so-called block-method,[88] is an elaboration of the addition of cyclopropylidenes to bicyclopropylidenes, as first used for the the synthesis of the triangulane [3]rotane (**189**) by Fitjer and Conia as discussed in Section 11.6.[95]

Employing diazocyclopropane (**198**), prepared *in situ* from the *N*-nitrosourea derivative **197**, as the cyclopropylidene precursor and the tetracyclic olefin **199** as the bicyclopropylidene receptor, Zefirov, de Meijere and co–workers have prepared the [6]branched triangulane **200** (Scheme 31).[94, 96]

With the 'extended' trapping olefin **201** even an [8]triangulane, the hydrocarbon **204**, could be prepared;[94, 96] it is formed together with the olefinic [6]triangulanes **202** and **203**, the former being the most likely direct precursor of **204**. Having been

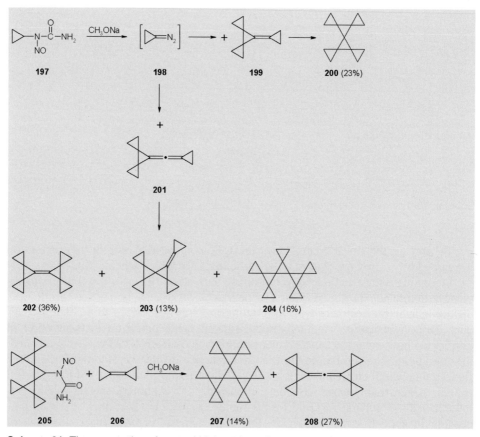

Scheme 31. The preparation of several higher triangulanes.

able to synthesize the highly symmetric [10]triangulane **207**, a spirocyclopropanated [3]rotane (D_{3h} symmetry) from **205** and **206**,[97] de Meijere and co-workers realized that a novel carbon network consisting only of spiro-linked three-membered rings might be constructed. Whether this further carbon allotrope can actually be prepared is a different matter. The allene **208**, produced as the main product in the above cyclopropanation, underwent the typical [2+2]cycloaddition of allenes (see Section 9.1) and yielded a cyclobutane derivative with a total of 12 cyclopropane rings. Clearly, construction of complex hydrocarbons from cyclopropane building blocks has come of age.

It was noted above that [1.1.1]propellane (**159**) constitutes a dead-end as far as further three-membered ring annelation is concerned. However, because of its very reactive central bond, which can be attacked both by ions and radicals,[98] the propellane can function as a building block *sui generis* for the construction of extended structures. For example, Michl and co-workers have been able to synthesize the so-called *staffanes* from **159** (Scheme 32).[99, 100]

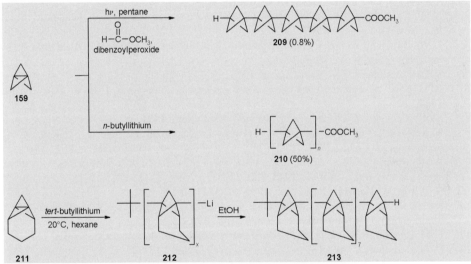

Scheme 32. The preparation of staffanes from [1.1.1]propellane (**159**).

These rod-like molecules were obtained either by irradiation of **159** in pentane in the presence of methyl formate and dibenzoylperoxide as free-radical initiator or under ionic conditions with *n*-butyllithium as the oligomerization-inducing reagent. In the former case the [5]staffane derivative **209** and its lower homologs were obtained, whereas anionic polymerization afforded the polymer **210** exclusively. [*n*]Staffanes of this type are 'straight molecular beams'[99] with a van der Waals radius of 2.37 Å and a length increment of 3.35 Å; *i.e.* derivative **209** is approximately 17 Å long.

That this oligomerization process is apparently typical for strained propellanes is borne out by the bridged [1.1.1]propellane **211**. When this was reacted with organolithium reagents such as *tert*-butyllithium at room temperature in hexane, a precipitate was formed after a short time to which structure **212** could be assigned.[101] Subsequent treatment with ethanol then led to a product mixture which contained 50% polymeric and 46% oligomeric material. From the latter a fraction could be isolated which according to ¹H NMR spectral analysis is the oligomer **213** consisting of nine monomer units.[101]

References

1.   For an excellent introduction to the art and science of tiling see B. Grünbaum, G. C. Shephard, *Tilings and Patterns*, W. H. Freeman and Comp., New York, N. Y., **1989**; *cf.* M. Senechal, G. Fleck (*Eds.*), *Shaping Space - A Polyhedral Approach*, Birkhäuser Boston Inc., Boston, **1988**.

2.   For the desgin of novel aromatics using graph theory see Y. Sritana-Anant, T. J. Seiders, J. S. Siegel, *Top. Curr. Chem.*, **1998**, *196*, 1–43.

3.   M. P. Cava, M. J. Mitchell, *Cyclobutadiene and Related Compounds*, Academic Press, Inc., New York, N. Y., **1967**.

4.   J. W. Barton in *Non-Benzenoid Aromatics* J. P. Snyder, (*Ed.*), Vol. I, Academic Press Inc., New York, N. Y., **1969**, 32–62.

5.   K. P. C. Vollhardt, *Top. Curr. Chem.*, **1975**, *59*, 113–136.

6.   D. I. Lloyd, *Non-Benzenoid Conjugated Carbocyclic Compounds*, Elsevier, Amsterdam, **1984**.

7.   M. K. Shepherd, *Cyclobutarenes*, Elsevier, Amsterdam, **1991**.

8.   N. Trinajstic, T. G. Schmalz, T. P. Zivkovic, S. Nikolic, G. E. Hite, D. J. Klein, W. A. Seitz, *Nouv. J. Chem.*, **1991**, *15*, 27–31.

9.   W. C. Lothrop, *J. Am. Chem. Soc.*, **1941**, *63*, 1187–1191. For a conformation of these results see W. Baker, *Nature*, **1942**, *150*, 210.

10. G. Wittig, W. Herwig, *Chem. Ber.*, **1954**, *87*, 1511–1512

11. For the thermal decomposition of other metal organic precursors see W. S. Rapson, R. G. Shuttleworth, J. N. van Niekerk, *J. Chem. Soc.*, **1943**, 326–327 and ref.[3], 266.

12. F. M. Logullo, A. H. Seitz, L. Friedman, *Org. Synth. Coll. Vol. V*, **1973**, 54–59.

13. C. D. Campbell, C. W. Rees, *J. Chem. Soc. (C)*, **1969**, 742–747; C. D. Campbell, C. W. Rees, *J. Chem. Soc. (C)*, **1969**, 752–756.

14. H. Heany, F. G. Mann, I. T. Millar, *J. Chem. Soc.*, **1957**, 3930–3938.

15. G. Wittig, L. Pohmer, *Chem. Ber.*, **1956**, *89*, 1334–1351; *cf.* G. Wittig, *Angew. Chem.*, **1957**, *69*, 245–251.

16. G. Wittig, R. W. Hoffmann, *Chem. Ber.*, **1962**, *95*, 2729–2734.

17. G. Wittig, H. F. Ebel, *Liebigs Ann. Chem.*, **1961**, *650*, 20–34; *cf.* G. Wittig, H. F. Ebel, *Angew. Chem.*, **1960**, *72*, 564.

18. J. F. Corbett, P. F. Holt, *J. Chem. Soc.*, **1961**, 4261–4263.

19. R. Wolovsky, F. Sondheimer, *J. Am. Chem. Soc.*, **1962**, *84*, 2844–2845. For a different route to **24** and its complete structural characterization see Chr. Werner, Ph. d. dissertation, Braunschweig, **1997**; *cf.* H. Hopf, Chr. Werner, P. Bubenitschek, P. G. Jones, *Angew. Chem.*, **1995**, *107*, 2592–2594; *Angew. Chem. Int. Ed. Engl.*, **1995**, *34*, 2367–2368.

20. J. A. H. MacBride, *J. Chem. Soc. Chem. Commun.*, **1972**, 1219–1220; *cf.* S. Kanok-

ranaporn, J. A. H. MacBride, *J. Chem. Res.*, **1980**, 203.

21. N. E. Schore, *Chem. Rev.*, **1988**, *88*, 1081–1119; *cf.* J. A. Casalnuovo, N. E. Schore, in *Modern Acetylene Chemistry* P. J. Stang, F. Diederich, (*Eds.*), VCH Verlagsgesellschaft, Weinheim, **1995**, 139–172.

22. K. P. C. Vollhardt, *Angew. Chem.*, **1984**, *96*, 525–541; *Angew. Chem. Int. Ed. Engl.*, **1984**, *23*, 539–555.

23. B. C. Berris, Y.-H. Lai, K. P. C. Vollhardt, *J. Chem. Soc. Chem. Commun.*, **1982**, 953–954.

24. B. C. Berris, G. H. Hovakeemian, Y.-H. Lai, H. Mestdagh, K. P. C. Vollhardt, *J. Am. Chem. Soc.*, **1985**, *107*, 5670–5687.

25. J. W. Barton, D. J. Rowe, *Tetrahedron Lett.*, **1983**, *24*, 299–302.

26. R. L. Hillard III, K. P. C. Vollhardt, *J. Am. Chem. Soc.*, **1977**, *99*, 4058–4069; *cf.* R. L. Funk, K. P. C. Vollhardt, *J. Am. Chem. Soc.*, **1980**, *102*, 5245–5253; R. L. Funk, K. P. C. Vollhardt, *J. Am. Chem. Soc.*, **1980**, *102*, 5253–5261.

27. M. Hirthammer, K. P. C. Vollhardt, *J. Am. Chem. Soc.*, **1986**, *108*, 2481–2482.

28. L. Blanco, H. E. Helson, M. Hirthammer, H. Mestdagh, S. Spyroudis, K. P. C. Vollhardt, *Angew. Chem.*, **1987**, *99*, 1276–1277; *Angew. Chem. Int. Ed. Engl.*, **1987**, *26*, 1246–1247.

29. K. P. C. Vollhardt, D. L. Mohler in *Advances in Strain in Organic Chemistry*, B. Halton, (*Ed.*), JAI Press, Inc. Greenwich, CT, Vol. V **1996**, 121–160.

30. R. H. Schmidt-Radde, M. K. Shepard, K. P. C. Vollhardt, *J. Am. Chem. Soc.*, **1992**, *114*, 9713–9715.

31. For a first, unsuccessful, attempt to prepare **66** by pyrolysis of a cinnoline precursor see J. W. Barton, M. K. Shepard, *Tetrahedron Lett.*, **1984**, *25*, 4967–4970.

32. R. Diercks, K. P. C. Vollhardt, *J. Am. Chem. Soc.*, **1986**, *108*, 3150–3152; *cf.* D. L. Mohler, Ph. d. thesis, University of California at Berkeley, **1992**. - For the synthesis of hexabutadiynylbenzene derivatives from **62** see: R. Boese, J. R. Green, J. Mittendorf, D. L. Mohler, K. P. C. Vollhardt, *Angew. Chem.*, **1992**, *104*, 1643–1645; *Angew. Chem. Int. Ed. Engl.*, **1992**, *31*, 1643–1645.

33. R. Diercks, J. D. Armstrong, R. Boese, K. P. C. Vollhardt, *Angew. Chem.*, **1986**, *98*, 270–271; *Angew. Chem. Int. Ed. Engl.*, **1986**, *25*, 268–269.

34. R. D. Bach, G. J. Wolber, H. B. Schlegel, *J. Am. Chem. Soc.*, **1985**, *107*, 2837–2841; *cf.* K. N. Houk, R. W. Gandour, R. W. Strozier, N. G. Rondan, L. A. Paquette, *J. Am. Chem. Soc.*, **1979**, *101*, 6797–6802.

35. R. Boese, A. J. Matzger, D. L. Mohler, K. P. C. Vollhardt, *Angew. Chem.*, **1995**, *107*, 1630–1633; *Angew. Chem. Int. Ed. Engl.*, **1995**, *34*, 1478–1481.

36. J.-D. van Loon, P. Seiler, F. Diederich, *Angew. Chem.*, **1993**, *105*, 1235–1238; *Angew. Chem. Int. Ed. Engl.*, **1993**, *32*, 1187–1189.

37. M. B. Sporn, A. B. Roberts, D. S. Goodman (*Eds.*), *The Retinoids*, Academic Press, Inc, Orlando, **1984**.

38. A. H. Alberts, H. Wynberg, *J. Chem. Soc. Chem. Commun.*, **1988**, 748–749.

39. M. Kreutzer, Diploma thesis, Technische Universität Braunschweig, **1989**.

40. U. Bunz, K. P. C. Vollhardt, J. S. Ho, *Angew. Chem.*, **1992**, *104*, 1645–1648; *Angew. Chem. Int. Ed. Engl.*, **1992**, *31*, 1678–1681.

41. K. S. Feldman, C. M. Kraebel, M. Parvez, *J. Am. Chem. Soc.*, **1993**, *115*, 3846–3847.

42. Y. Rubin, C. B. Knobler, F. Diederich, *Angew. Chem.*, **1991**, *103*, 708–710; *Angew. Chem. Int. Ed. Engl.*, **1991**, *30*, 698–700.

43. The first tetraethynylethene derivative **83** (R = Phenyl) was prepared in 1969 by Y. Hori, N. Noda, S. Kobayashi, H. Tamiguchi, *Tetrahedron Lett.*, **1969**, 3563–3566; the X-ray structure of this hydrocarbon being determined in 1991 by H. Hopf, M. Kreutzer, P. G.

Jones, *Chem. Ber.*, **1991**, *124*, 1471–1475. Ground-breaking work in this area was performed in the mid 1970s by H. Hauptmann, *Tetrahedron*, **1976**, *32*, 1293–1297, who also described the first preparation of **82** (by a different route than that above) and several peralkylated derivatives **83** (R = *tert*-Bu and Methyl): H. Hauptmann, *Angew. Chem.*, **1975**, *87*, 490–191; *Angew. Chem. Int. Ed. Engl.*, **1975**, *14*, 498–499; *cf.* H. Hauptmann, *Tetrahedron Lett.*, **1975**, 1931–1934.

44. R. R. Tykwinski, M. Schreiber, R. P. Carlón, F. Diederich, V. Gramlich, *Helv. Chim. Acta*, **1996**, *79*, 2249–2281.

45. J. Anthony, C. B. Knobler, F. Diederich, *Angew. Chem.*, **1993**, *105*, 437–440; *Angew. Chem. Int. Ed. Engl.*, **1993**, *32*, 406–409; *cf.* J. Anthony, A. M. Boldi, Y. Rubin, M. Hobi, V. Gramlich, C. B. Knobler, P. Seiler, F. Diederich, *Helv. Chim. Acta*, **1995**, *78*, 13–45; J. Anthony, A. M. Boldi, C. Boudon, J.-P. Gisselbrecht, M. Gross, P. Seiler, C. B. Knobler, F. Diederich, *Helv. Chim. Acta*, **1995**, *78*, 797–817.

46. A. M. Boldi, J. Anthony, V. Gramlich, C. B. Knobler, C. Boudon, J.-P. Gisselbrecht, M. Gross, F. Diederich, *Helv. Chim. Acta*, **1995**, *78*, 779–796.

47. M. Schreiber, F. Anthony, F. Diederich, M. E. Spahr, R. Nesper, M. Hubrich, F. Bommeli, L. Degiorgi, P. Wachter, Ph. Kaatz, Chr. Bosshard, P. Günter, M. Colussi, U. W. Suter, C. Boudon, J.-P. Gisselbrecht, M. Gross, *Adv. Mater.*, **1994**, *6*, 786–790.

48. F. Diederich, *Pure Appl .Chem.*, **1999**, in press.

49. S. Misumi, *Bull. Chem. Soc. Jpn.*, **1961**, *34*, 1827–1832: *cf.* S. Misumi, *Bull. Chem. Soc. Jpn.*, **1962**, *35*, 143–146.

50. K. Sanechika, T. Yamamoto, A. Yamamoto, *Bull. Chem. Soc. Jpn.*, **1984**, *57*, 752–755.

51. D. L. Trumbo, C. S. Marvel, *J. Polym. Sci., Part A: Polym. Sci.*, **1986**, *24*, 2311–2326.

52. M. E. Wright, *Macromolecules*, **1989**, *22*, 3256–3259.

53. *Electronic Materials*: *The Oligomer Approach*, K. Müllen, G. Wegner, (*Eds.*), Wiley-VCH, Weinheim, **1998**.

54. Review: K. Müllen, S. Valiyaveetil, V. Francke, V. S. Iyer in *Molecular Wires*, NATO-AST Series, Kluwer Int. Press, **1997**; *cf.* V. Francke, Ph. d. dissertation, University of Mainz, **1999**.

55. Reviews: J. K. Young, J. S. Moore in *Modern Acetylene Chemistry*, P. J. Stang, F. Diederich (*Eds.*), VCH Verlagsgesellschaft, Weinheim, **1995**, 415–442; J. F. Moore, *Acc. Chem. Res.*, **1997**, *30*, 402–413.

56. J. K. Young, J. C. Nelson, J. S. Moore, *J. Am. Chem. Soc.*, **1994**, *116*, 10841–10842; *cf.* J. Zhang, J. S. Moore, Z. Xu, R. A. Aguirre, *J. Am. Chem. Soc.*, **1992**, *114*, 2273–2274.

57. J. C. Nelson, J. K. Young, J. S. Moore, *J. Org. Chem.*, **1996**, *61*, 8160–8168.

58. J. S. Moore, J. Zhang, *Angew. Chem.*, **1992**, *104*, 873–874; *Angew. Chem. Int. Ed. Engl.*, **1992**, *31*, 922–923; *cf.* J. Zhang, D. J. Pesak, J. L. Ludwick, J. S. Moore, *J. Am. Chem. Soc.*, **1994**, *116*, 4227–4239.

59. Z. Wu, S. Lee, J. S. Moore, *J. Am. Chem. Soc.*, **1992**, *114*, 8730–8732; *cf.* T. C. Bedard, J. S. Moore, *J. Am. Chem. Soc.*, **1995**, *117*, 10662–10671.

60. Z. Wu, J. S. Moore, *Angew. Chem.*, **1996**, *108*, 320–322; *Angew. Chem. Int. Ed. Engl.*, **1996**, *35*, 297–299.

61. K. L. Walker, Z. Xu, J. S. Moore, *J. Am. Soc. Mass. Spectrometr.*, **1994**, *5*, 731.

62. Z. Xu, J. S. Moore, *Angew. Chem.*, **1993**, *105*, 1394–1396; *Angew. Chem. Int. Ed. Engl.*, **1993**, *32*, 1354–1356; *cf.* J. S. Moore, Z. Xu, *Macromolecules*, **1991**, *24*, 5893–5894.

63. R. H. Grubbs, D. Kratz, *Chem. Ber.*, **1993**, *126*, 149–157.

64. R. H. Baughman, H. Eckhadt, M. Kertesz, *J. Chem. Phys.*, **1987**, *87*, 6687–6699; *cf.* S. Housmans, H. P. Honnef (aka H. Musso, H. Hopf), *Nachr. Chem. Tech. Lab.*, **1984**, *32*, 379–381.

65. Review: M. Haley, *Synlett*, **1998**, 557–565.

66. Q. Zhou, P. J. Carroll, T. M. Swager, *J. Org. Chem.*, **1994**, *59*, 1294–1301.
67. K. P. Baldwin, A. J. Metzger, D. A. Scheimann, C. A. Tessier, K. P. C. Vollhardt, W. J. Youngs, *Synlett*, **1995**, 1215–1218.
68. M. M. Haley, M. L. Bell, J. J. English, C. A. Johnson, T. J. R. Weakley, *J. Am. Chem. Soc.*, **1997**, *119*, 2956–2957.
69. W. B. Wan, D. B. Kimball, M. M. Haley, *Tetrahedron Lett.*, **1998**, *39*, 6795–6798.
70. A. de Meijere (*Ed.*), *Top. Curr. Chem.*, **1986**, *133*;, **1987**, *135*;, **1988**, *144*;, **1990**, *155*;, **1996**, *178*.
71. Houben-Weyl, *Methods of Organic Chemistry, Carbocyclic Three- and Four-membered Ring Compounds* A. de Meijere, (*Ed.*), Thieme Verlag, Stuttgart, **1997**, Vol. E17a-c.
72. S. Stinson, *Chem. Eng. News*, **1995**, Apr. 17, 22.
73. A. G. M. Barrett, K. Kasdorf, G. J. Tustin, D. J. Williams, *J. Chem. Soc. Chem. Commun.*, **1995**, 1143–1144; *cf.* M. S. Kuo, R. J. Zielinski, J. I. Cialdella, C. K. Marschke, M. J. Dupuis, G. P. Li, D. A. Kloosterman, C. H. Spilman, V. P. Marshall, *J. Am. Chem. Soc.*, **1995**, *117*, 10629–10634.
74. For the conformational analysis of bicyclopropyl see: O. Bastiansen, A. de Meijere, *Acta Chem. Scand.*, **1966**, *20*, 516 521; K. Hagen, G. Hagen, M. Traetteberg, *Acta Chem. Scand.*, **1972**, *26*, 3649–3661; A. de Meijere, W. Lüttke, F. Heinrich, *Liebigs Ann. Chem.*, **1974**, 306–327; R. Stølevik, P. Bakken, *J. Mol. Struct.*, **1989**, *197*, 137–142 and references cited therein.
75. W. H. Perkin, Jr, J. L. Simonsen, *Proc. Chem. Soc.*, **1905**, *21*, 256; O. Döbner, G. Schmidt, *Ber. Dtsch. Chem. Ges.*, **1907**, *40*, 148–152; N. Zelinsky, J. Gutt, *Ber. Dtsch. Chem. Ges.*, **1907**, *40*, 4744–4749; M. Guthzeit, E. Hartmann, *J. Prakt. Chem.*, **1910**, *81*, 329–381; R. M. Beesley, J. F. Thorpe, *Proc. Chem. Soc.*, **1913**, *29*, 346.
76. K. B. Wiberg, R. P. Ciula, *J. Am. Chem. Soc.*, **1959**, *81*, 5261–5262.
77. For a summary of the work on small ring bicyclo[*n.m*.0]alkanes up to the late 1960s see K. B. Wiberg in *Advances in Alicyclic Chemistry*, Vol., 2. H. Hart, G. J. Karabatsos (*Eds.*), Academic Press, New York, N. Y., **1968**, 185–254.
78. D. M. Lemal, F. Menger, G. W. Clark, *J. Am. Chem. Soc.*, **1963**, *85*, 2529–2530.
79. K. B. Wiberg, G. M. Lampman, *Tetrahedron Lett.*, **1963**, 2173–2175.
80. W. A. Nevill, D. S. Frank, R. D. Trepka, *J. Org. Chem.*, **1962**, *27*, 422–428.
81. R. Srinivasan, *J. Am. Chem. Soc.*, **1963**, *85*, 4045–4046.
82. G. Haufe, G. Mann, *Chemistry of Alicyclic Compounds*, Elsevier, Amsterdam, **1989**, 84
83. K. B. Wiberg, N. McMurdie, J. V. McClusky, C. M. Hada, *J. Am. Chem. Soc.*, **1993**, *115*, 10653–10657.
84. G. Gustavson, *J. Prakt. Chem.*, **1896**, *54*, 97–107; *cf.* G. Gustavson, H. Bulatoff, *J. Prakt. Chem.*, **1897**, *56*, 93–95. Actually Gustavson had postulated the formation of vinylcyclopropane ('vinyltrimethylene') from **179**, a conclusion later corrected by H. Fecht, *Ber. Dtsch. Chem. Ges.*, **1907**, *40*, 3883–3891. The ultimate structure proof was accomplished by electron diffraction by F. Rogowski, *Ber. Dtsch. Chem. Ges.*, **1939**, *72*, 2021–2026.
85. A. I. Dyanchenko, E. L. Protasova, A. I. Ioffe, A. Ya. Steinschneider, O. M. Nefedov, *Tetrahedron Lett.*, **1979**, 2055–2058.
86. D. E. Applequsit, G. F. Fanta, B. E. Henrikson, *J. Org. Chem.*, **1958**, *23*, 1715–1716. For an electrochemical ring-closure of **179** see M. R. Rifi, *Org. Synth.*, **1972**, *52*, 22–33.
87. N. S. Zefirov, S. I. Kozhuskov, T. S. Kuznetsova, O. V. Kokoreva, K. A. Lukin, B. I. Ugrak, S. S. Tratch, *J. Am. Chem. Soc.*, **1990**, *112*, 7702–7707.
88. Reviews: A. de Meijere, S. I. Kozhushkov, *Adv. in Strain in Organic Chemistry* B. S. Halton (*Ed.*), Vol. 4, JAI Press, Inc. Greenwich, CT, **1995**, 225–282; K. A. Lukin, N. S. Zefirov in *The Chemistry of the Cyclopropyl Group* Z. Rappoport, (*Ed.*), J. Wiley & Sons, New York, N. Y., **1995**, Vol. 2, 861- 885.

89. A. Nickon, E. F. Silversmith, *Organic Chemistry - The Name Game*, Pergamon Press, New York, N. Y., **1987**, 34.
90. S. Arora, P. Binger, *Synthesis*, **1974**, 801–803.
91. N. S. Zefirov, K. A. Lukin, S. I. Kozhushkov, T. S. Kuznetsova, A. M. Domarev, L. M. Sosonkin, *Zh. Org. Khim.*, **1989**, *25*, 312–319; *J. Org. Chem. USSR*, **1989**, *25*, 278–284; *cf.* I. Erden, *Synth. Commun.*, **1986**, *16*, 117–121.
92. H.-D. Beckhaus, C. Rüchardt, S. I. Kozhushkov, V. N. Belov, S. .P. Verevkin, A. de Meijere, *J. Am. Chem. Soc.*, **1995**, *117*, 11854–11860.
93. R. Boese, Th. Haumann, E. D. Jemmis, B. Kiran, S. I. Kozhushkov, A. de Meijere, *Liebigs Ann. Chem.*, **1996**, 913–919.
94. N. S. Zefirov, S. I. Kozhushkov, B. I. Ugrak, K. A. Lukin, O. V. Kokoreva, D. S. Yufit, Y. T. Struchkov, S. Zoellner, R. Boese, A. de Meijere, *J. Org. Chem.*, **1992**, *57*, 701–708; *cf.* A. de Meijere, S. I. Kozhuskov, T. Spaeth, N. S. Zefirov, *J. Org. Chem.*, **1993**, *58*, 502–505.
95. L. Fitjer, J.-M. Conia, *Angew. Chem.*, **1973**, *85*, 349–350; *Angew. Chem. Int. Ed. Engl.*, **1973**, *12*, 334–335; *cf.* A. de Meijere, I. Erden, W. Weber, D. Kaufmann, *J. Org. Chem.*, **1988**, *53*, 152–161.
96. S. Zöllner, H. Buchholz, R. Boese, R. Gleiter, A. de Meijere, *Angew. Chem.*, **1991**, *103*, 1544–1546; *Angew. Chem. Int. Ed. Engl.*, **1991**, *30*, 15–18; *cf.* S. Zöllner, Ph. d. dissertation, Hamburg, **1991**.
97. S. I. Kozhushkov, T. Haumann, R. Boese, A. de Meijere, *Angew. Chem.*, **1993**, *105*, 426–429; *Angew. Chem. Int. Ed. Engl.*, **1993**, *32*, 401–403.
98. Review: K. B. Wiberg, *Chem. Rev.*, **1989**, *89*, 975–983.
99. P. Kaszynski, J. Michl, *J. Am. Chem. Soc.*, **1988**, *110*, 5225–5226; *cf.* C. Mazal, A. J. Paraskos, J. Miche, *J. Org. Chem.*, **1998**, *63*, 2116–2119. Rod-like molecules of this general type have previously been obtained by H. E. Zimmerman, T. D. Goldman, T. K. Hirzel, S. P. Schmidt, *J. Org. Chem.*, **1980**, *45*, 3933–3951. for building with the cubane-1,4-diylunit (construction of *p*–[*n*]cubyls) see P. E. Eaton, K. Pramod, . Emrich,R. Gilardi, *J. Am. Chem. Soc.*, **1999**, *121*, 4111–4123.
100. P. Kaszynski, A. C. Friedli, J. Michl, *J. Mol. Cryst. Liq. Cryst. Lett.*, **1988**, *6*, 27; J. Michl, P. Kaszynski, A. C. Friedli, G. S. Murthy, H.-C. Yang, R. E. Robinson, N. D. McMurdie, T. Kim in *Strain and Its Implications in Organic Chemistry* A. de Meijere, S. Blechert, (*Eds.*), Kluwer Academic Publishers, Dordrecht, **1989**, 463–482; P. Kaszynski, J. Michl in *Advances in Strain in Organic Chemistry* B. Halton (*Ed.*), The JAI Press, Inc., **1995**, Vol. 4, 283–331.
101. A. D. Schlüter, *Angew. Chem.*, **1988**, *100*, 283–285; *Angew. Chem. Int. Ed. Engl.*, **1988**, *27*, 296–298.

16 Classics in Hydrocarbon Synthesis— Three Examples from Physical Organic Chemistry

The main purpose of this book is to demonstrate to the reader the extraordinary structural richness of hydrocarbon chemistry, a diversity which many would probably not have expected when first looking at the elementary building blocks—the carbon-carbon single, double, and triple bonds (see Chapter 2)—and how these numerous different hydrocarbon systems have been synthesized in the laboratory. I have, furthermore, tried to establish generic relationships among and between different classes of hydrocarbons, and have attempted to show how many of our present day research projects are based on the ingenuity and the preparative efforts of our predecessors. Because of space restrictions I could not discuss the chemistry of radical, diradical, and carbene hydrocarbon intermediates to the extent which these important species doubtlessly deserve. In fact, whenever we encountered highly reactive ('unstable') hydrocarbons, for example the distorted alkenes in Section 7.4, the angle-strained cycloalkynes in Section 8.2, several of the cyclic allenes and cumulenes in Sections 9.3 and 9.4, respectively, or the cyclobutadienes in Chapter 10.1, to name but a few, these were discussed from the preparative viewpoint, the challenge to extend the limits of our preparative capabilities still further, not from the viewpoint of the electronic nature of these species. Realizing, however, that a text such as this cannot completely omit this important, although very heterogeneous group of hydrocarbons, I have decided to conclude the book with the presentation of three case-studies from physical-organic hydrocarbon chemistry. These will be concerned with the dehydrobenzenes, a selection of non-Kekulé hydrocarbons, and a discussion of various efforts to 'fix' the double bonds of a benzene ring, *i.e.* to synthesize a six-membered, six π-electron hydrocarbon which is not aromatic but polyolefinic, *i.e.* a cyclohexatriene. And again we will see how deeply these highly topical problems are rooted in the history of organic chemistry.

16.1 The dehydrobenzenes

Formally benzene (**1**) can be dehydrogenated by removal of a pair of *ortho*, *meta* and *para* hydrogen atoms. These possibilities lead to o-, m-, and p-dehydrobenzene (**2, 5,** and **8**, respectively, Scheme 1).

Although we would expect **2** to be a reactive hydrocarbon from what we have learned previously about cyclohexyne (see Section 8.2) we have no difficulty in

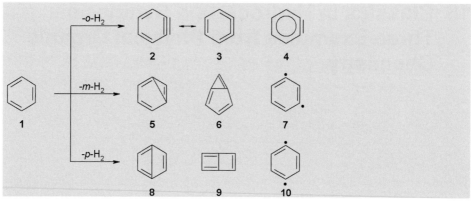

Scheme 1. The dehydrobenzenes.

drawing the new bond between the respective carbon atoms. The same is true for the cumulenic resonance structure of **2**, 1,2,3,5-cyclohexatetraene (**3**), because both cyclic cumulenes (Section 9.4) and even an allenic isomer of **1**, 1,2,4-cyclohexatriene (isobenzene) have been described (Section 9.3). Alternatively, the new π-bond of the dehydroaromatic compound might not effect the aromatic π-system, as indicated by the 'delocalized' formula **4**. In **5** and **8**, however, we would not expect the additional bond to connect the *meta* and the *para* carbon atoms because the bond lengths would be much too long. And even if we would draw these closed bicyclic structures more realistically, as in **6** and **9** (a hydrocarbon known as butalene), these molecules would be expected to be very highly strained. For *meta*- and *para*-dehydrobenzene the diradical structures **7** and **10** are thus much more likely.

Traditionally when talking about dehydrobenzene, the *ortho* isomer **2** was more or less automatically alluded to. In fact, this is the term introduced by Wittig in 1942 when he studied the decomposition of lithiated halobenzenes (see below) and rationalized their surprising chemical behavior by postulating **2** as a reactive intermediate.[1] Later (1953) work by Roberts and co-workers on the conversion of ^{14}C-labeled chlorobenzene to aniline strongly suggested the intermediacy of *o*-dehydrobenzene (**2**) in this[2] and related reactions. In more than half a century since then, **2** has not only become a cherished intermediate in preparative chemistry—see for example its use for the synthesis of triptycene and the iptycenes described in Section 13.2—but a thoroughly studied and comprehensively characterized hydrocarbon itself (see below). As a consequence there is a rich and frequently updated review literature on the generation, structural properties and synthetic applications of **2**.[3–7]

We have encountered several routes leading to **2** in the previous chapter already (see Section 15.3). Some of these are repeated in Scheme 2 which summarizes the three general approaches to *o*-dehydrobenzene which have emerged over the years.

The first mode of generation proceeds *via* 2-halophenyl anions, which can, for example, be prepared by treatment of *ortho*-bromofluorobenzene (**11**) with magnesium or a Grignard reagent or by subjecting chlorobenzene (**12**) to strong base such as sodamide or a lithium dialkylamide at or below room temperature. This route to **2** is essentially that first introduced and investigated by Wittig.[2] In the second approach

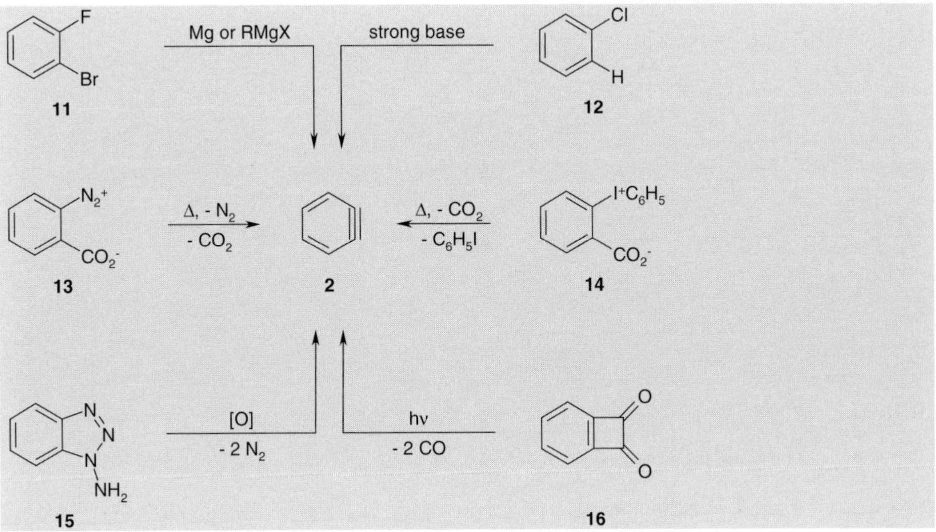

Scheme 2. Typical routes for the generation of o-dehydrobenzene (2).

internal salts such as benzenediazonium-2-carboxylate (13)[8] or diphenyliodonium-2-carboxylate (14)[9] are thermally decomposed with loss of nitrogen and carbon dioxide or of carbon dioxide and phenyliodide, respectively. The former reaction—usually performed between 40 and 80 °C—is the most common route to 2. Finally, o-dehydrobenzene has been generated by thermal, oxidative, and photochemical fragmentation of benzo-fused heterocyclic systems such as 1-aminobenzotriazole (15)[10] or benzocyclobutandione (16).[11] Whereas the decomposition of precursors such as 16 is ideally suited to the preparation of 2 in the gas phase or in a matrix, the ionic routes are employed preferentially in synthetic organic chemistry.

Today o-dehydrobenzene (2) is a very well characterized hydrocarbon. Its IR and UV spectra have been recorded,[11–16] as have its photoelectron,[17] mass,[18] and microwave spectra.[19] From the latter spectroscopic information it has been inferred that 2 is planar (C_{2v} symmetry), is a ground state singlet and has a gas-phase lifetime at low pressures considerably less than 1 s. For a long time NMR information on 2 was lacking; creative experimental techniques have, however, made this available recently. Grant and co–workers[20] prepared o-dehydrobenzene-1,2-$^{13}C_2$ from matrix-isolated phthalic anhydride-1,2-$^{13}C_2$ by UV irradiation at ca 20 K in argon and Warmuth has even been able to study 2 in solution ([D$_8$]THF at 77 K) by encapsulating this reactive hydrocarbon in a molecular container, making use of a hemicarcerand similar to that employed successfully to entrap cyclobutadiene (Section 10.1). Using benzocyclobutandione (16) as the dehydrobenzene precursor (see above), in the first step of the experiment this was inserted into the host molecule 17 by heating the two components at 145 °C. The hemicarceplex 18 was produced in 30–35% yield and could be separated by column chromatography from unreacted 17 (Scheme 3).[21]

Photolysis of 18 in degassed deuteriochloroform yielded the benzocyclopropenone complex 19 as the only product; the structure of this compound was secured not only by NMR spectroscopy but by X-ray crystal structure determination also. When,

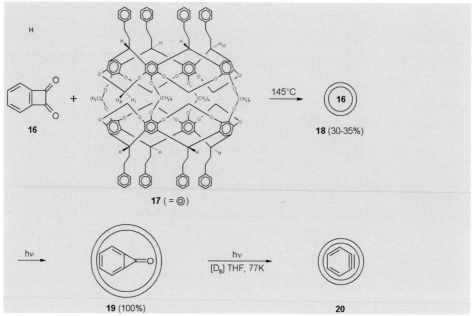

Scheme 3. The generation of *o*-dehydrobenzene (**2**) in a container molecule by Warmuth.

finally, **19** was irradiated in [D$_8$]THF at 77 K with a 300-W xenon lamp **20** was produced as demonstrated by signals at δ = 4.99 (H1, H1') and 4.31 (H2, H2') in the proton spectrum and by multiplets centered at δ = 181.33 (C1, C1'), 125.45 (C2, C2'), and 136.80 (C3, C3') in the carbon NMR spectrum. To estimate the chemical shift of 'free' *o*-dehydrobenzene the author assumed that benzene (**1**), which is comparable in form and size with **2**, experiences similar shielding from the surrounding host; he determined the chemical shift differences between free benzene and the complex between **1** and **17** which was prepared for this comparison. Applying the resulting Δδ values to **20** provided chemical shifts of δ = 7.69 and 7.01 and δ = 182.71, 126.86, and 138.13, respectively, for the uncomplexed **2**. The ^{13}C–^{13}C-coupling constants for **20**, determined by line-shape analysis of the ^{13}C multiplets and their comparison with the corresponding coupling constants of model compounds such as *cis*-1,3-hexadiene, 1,3-butadiene, and butatriene suggest a cumulenic structure for *o*-dehydrobenzene as indicated by resonance structure **3**. On the other hand both the solid-state NMR spectrum (see above) and extensive theoretical calculations concerning structure, aromaticity, and magnetic properties of *o*-dehydrobenzene favor the acetylenic alternative **2**.[22]

Once generated inside the cavity of **17**, *o*-dehydrobenzene (**2**) experiences an interesting fate—it undergoes an innermolecular Diels–Alder addition with the surrounding host in a new type of reaction at the borderline between intra- and intermolecular reactions.[23] Needless to say *o*-dehydrobenzene (**2**) has been the subject of numerous theoretical calculations.[24, 25]

Compared with **2** its *m* (**5–7**) and *p* isomers (**8–10**) for a long time have played only a minor role in mechanistic and preparative chemistry. Although these species could apparently be generated[18, 26, 27] and trapped,[28-33] these early attempts did not enable a detailed spectroscopic study of these highly reactive hydrocarbons. Usually decomposition of classical dehydrobenzene precursors such as the corresponding benzenediazonium carboxylates or diaryliodonium carboxylates was conducted out and detection was often performed by mass spectrometry or cycloaddition experiments. These precursors cannot be used in matrix-isolation experiments which normally require the sublimation of the substrate at elevated temperatures (and under high vacuum) on to a cold surface, where the intermediates under question are generated by irradiation. Whenever reactive intermediates are synthesized in the gas phase by thermal and/or photochemical decomposition of suitable precursor molecules their lifetimes must be long enough to enable trapping in the inert matrix at low temperatures. When thermochemical data of **7** and **10** had become available from negative-ion photoelectron spectroscopy[34] and kinetic experiments,[35] Sander and co–workers were able to design precursor systems enabling the matrix isolation of *m*- and *p*-dehydrobenzene (**7** and **10**, respectively).

For the parent *m*-dehydrobenzene **7** the [2.2]metaparacyclophandione **21** (see Section 12.3) and the bis(diacylperoxide) **22**[36] proved to be excellent precursors, both substrates being stable enough to survive the mentioned sublimation and—because of their high energy content—sufficiently reactive to enable ready fragmentation at extremely low temperatures (Scheme 4).[37, 38]

UV-photolysis of **21** in solid argon at 10 K led to its cleavage into two molecules of carbon monoxide, *p*-xylylene (**23**, see Sections 11.1 and 12.3), and the desired **7**.

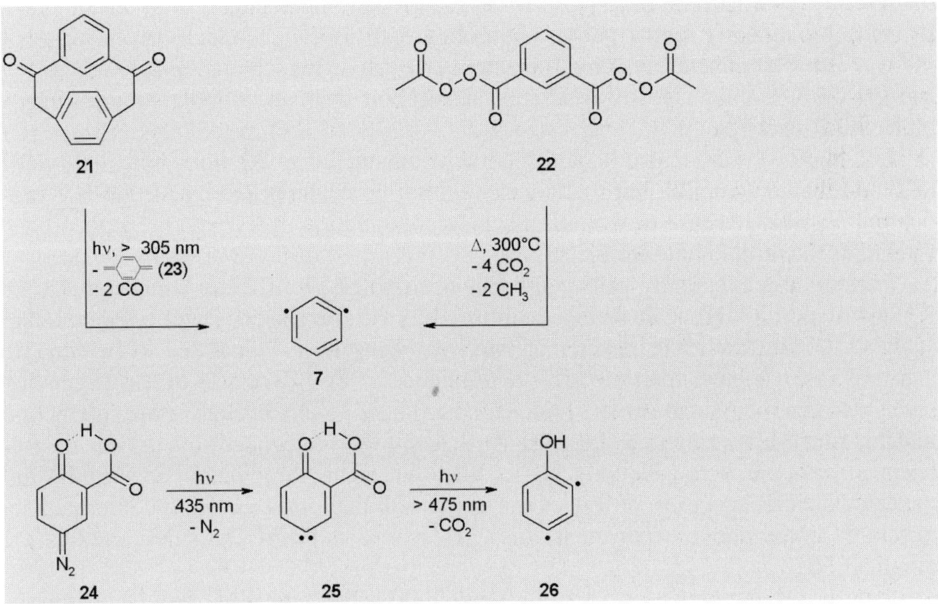

Scheme 4. The generation of *m*-dehydrobenzene (**7**) ...

Comparison of the experimentally determined IR spectrum and that calculated using CCSD(T) theory showed excellent agreement. Alternatively **7** could be prepared by flash-vacuum pyrolysis of **22** and subsequent trapping of the products—besides **7**, methyl radicals and carbon dioxide—in argon at 10 K.[37-39]

Careful analysis of the experimental and calculated IR spectra then led to a clear decision between the diradical, **7**, and the bicyclic structure, **6**, favoring the former unequivocally. The hydrocarbon does not, however, have a hexagonal geometry as would be intuitively expected. Compared with benzene (**1**) it is highly distorted: the distance between the radical centers is much shorter (2.1 Å) than the distance between two *meta* carbon atoms in **1** (2.41 Å), and the normal 120° C1–C2–C3 angle is reduced to 102° while the angle at the radical centers (C2–C1–C6 and C2–C3–C4) has increased to 136°.[38]

Besides the parent hydrocarbon **7**, Sander *et al.* have been able to prepare derivatives of *m*-dehydrobenzene. Thus the *m*-dehydrophenol **26**, which, in fact, was the first *m*-dehydrobenzene to be characterized by matrix spectroscopy, was obtained when the matrix-isolated *p*-benzoquinone diazide carboxylic acid **24** was irradiated with monochromatic blue light (435 nm). By nitrogen extrusion the carbene **25** was produced, which on irradiation with longer wavelength light (λ > 475 nm) yielded the phenol **26**.[38, 40]

Application of these decomposition routes to the isomers **27** and **30** of **21** and **22**, respectively, to produce *p*-dehydrobenzene (**10**) met with difficulties. Although the bis(diacylperoxide) **27** could be thermally decomposed at 300 °C, hydrocarbon **10** could not be detected by IR spectroscopy. Under these conditions the C_6H_4 hydrocarbon opens up to *cis*-1,5-hexadiyne-3-ene (**28**), in a process which is the reverse of the celebrated Bergman cyclization.[32] This unique ring-forming process is one of the main reasons why the chemistry of **10** has received so much attention in recent years. In many biologically highly active antibiotics such as calicheamicin cyclizations of the type **28**→**10** generate *p*-dehydrobenzene derivatives which are capable of causing irreversible DNA damage by hydrogen abstraction from the double-stranded DNA molecule[41] (Scheme 5).[42]

UV photolysis of matrix-isolated paracyclophanedione **30** furnished *p*-xylylene (**23**) and the diketene **29**, but the latter could not be decomposed by further UV irradiation, possibly because of a rapid in-cage recombination of CO and the ketocarbene formed as the initial photoproducts.[42]

Success was eventually achieved by matrix photolysis of **27** in solid argon at 10 K, the structure of **10** again being established by IR spectroscopy and computational studies.[43] The diradical character of *p*-dehydrobenzene is calculated to be *ca* 65% and thus much greater than the 20% determined for **7**.[44] Because of the larger distance between the radical centers (calculated to be *ca* 2.7 Å) through-space interaction in **10** is much less pronounced than in **7**. Because the (calculated) energy for the butalene structure **9** is considerably higher than that of the diradicaloid species **10** and the calculated IR spectrum of **9** does not agree with that observed for the intermediate generated in the above experiment, this C_6H_4 isomer is clearly best described as the diradical **10**.

Among the derivatives of **10**, 9,10-didehydroanthracene (**34**) has been investigated most thoroughly (Scheme 6).

It seems that this highly reactive hydrocarbon is accessible from precursors such as the bisketene **31** by photolysis (λ > 200 nm) in organic glasses (3-methylpentane, 77 K)[45] and by laser flash photolysis (LFP) of the anthracene adduct of **34**, the poly-

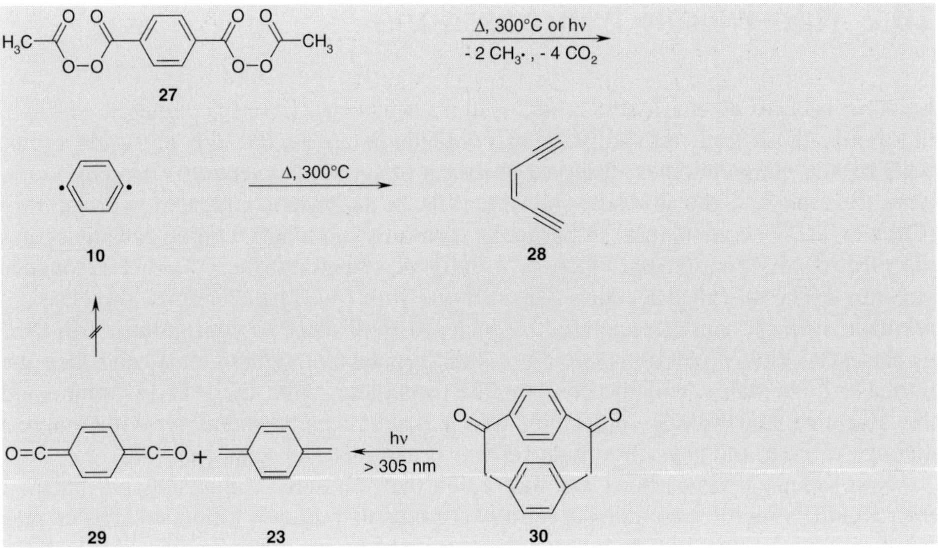

Scheme 5. ... and *p*-dehydrobenzene (10) by Sander and co-workers.

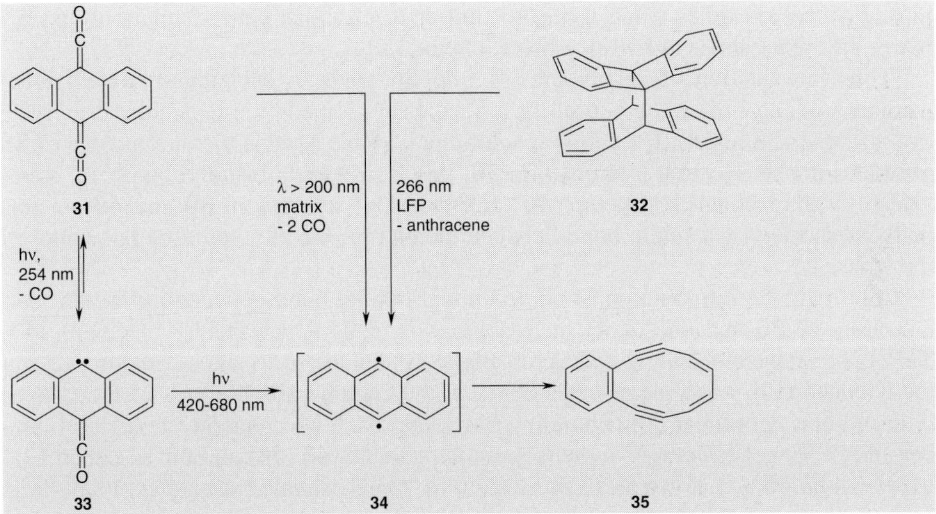

Scheme 6. The generation of 9,10-didehydroanthracene (34) from various precursors.

cyclic hydrocarbon 32,[46] but interpretation of the spectroscopic data is hampered by the extreme lability of 34, which easily opens up to 3,4-benzocyclodeca-3,7,9-trien-1,5-diyne (35), a bis dehydrobenzannulene (see Section 10.4). A recent reinvestigation of the diketene fragmentation experiment has *inter alia* shown that it proceeds in a stepwise manner *via* the novel ketocarbene 33.[47]

Whereas stabilization of 7 and 10 by metal complexation has not, apparently, been attempted, many metal complexes of *o*-dehydrobenzene (2) and its derivatives have been prepared or postulated as reaction intermediates.[48]

16.2 Non-Kekulé hydrocarbons

Looking back to all the hydrocarbon systems which we have encountered so far in this book, those with the property of conjugation are by far the most numerous, whether the conjugation is observed in linear polyolefins (Section 7.1), polyacetylenes (Section 8.1), the annulenes (Chapter 10), or the cross-conjugated hydrocarbons (Chapter 11). The annulenes, in particular, are fully, circularly conjugated molecules the electron distribution of which is usually described by their Kekulé resonance structures. The so-called Kekulé hydrocarbons with $(4n+2)$ π-electrons, the classical aromatic systems, are characterized by a closed-shell electron configuration. In their ground state their π-electrons occupy bonding molecular orbitals only, and they are normally quite stable, isolable compounds. Contrasting with these hydrocarbons are the so-called non-Kekulé molecules which have recently found growing interest among synthetic and physical organic chemists and material scientists.

Non-Kekulé hydrocarbons are molecules that are fully conjugated but each of their Kekulé structures contains at least two atoms that are not π-bonded.[49] The parent system of the non-Kekulé hydrocarbons is trimethylenemethane (**36**, TMM), from which more complex structures can be derived by either introduction of substituents or by application of the time-honored 'vinylog principle', which we have already employed numerous times when trying to find or to establish generic relationships between different classes of hydrocarbon (Scheme 7).

Thus introduction of substituents R into **36** leads to substituted trimethylene-methanes **37**, and when the end of these species are joined by a molecular bridge the cyclic TMMs **38** result. 'Inserting' an additional double bond at the olefinic end of **36** generates the six electron hydrocarbon **39**; this is, in fact, a bis-allylic system, as revealed by the resonance structure **40**. The two allyl subunits in **39** and **40** are formally connected by a single bond between the end of one allyl unit and the center of the other.

Alternatively, introduction of an additional double bond between the central carbon atom of **36** and one of its methylene ends leads to tetramethyleneethane (**41**, TME). Again we are dealing with a bis-allylic system, see resonance structure **42**, but the connection has now been made between the central carbon atoms of the two allylic subunits. Continuing this building block approach we can generate a large number and variety of structures; *m*-benzoquinodimethane (**43**, **44**) and the so-called Clar hydrocarbon **45** are shown as an illustration for more extended structures. In all these hydrocarbons we can, in principle, 'annihilate' the two radical centers by forming a bond between them. As shown for, *e.g.*, the closure of **36** to **46** (methylenecyclopropane, see Section 15.3) and **43** to **47**, however, we would have to pay a heavy price for it—not only would the delocalization be lost, but severe ring strain would be produced in the course of the C-C bond-forming process. In a sense **36** and **43** can be regarded as 'electronic isomers' of **46** and **47**, respectively, and we would expect the former to be generated readily from the latter once these have been made available by the synthetic chemist.

In the early 1950s Longuet-Higgins showed that in a non-Kekulé hydrocarbon containing no $4n$-membered rings, the number of atoms that cannot be assigned to π-bonds is equal to the number of π-non-bonding molecular orbitals (NBMOs) and to

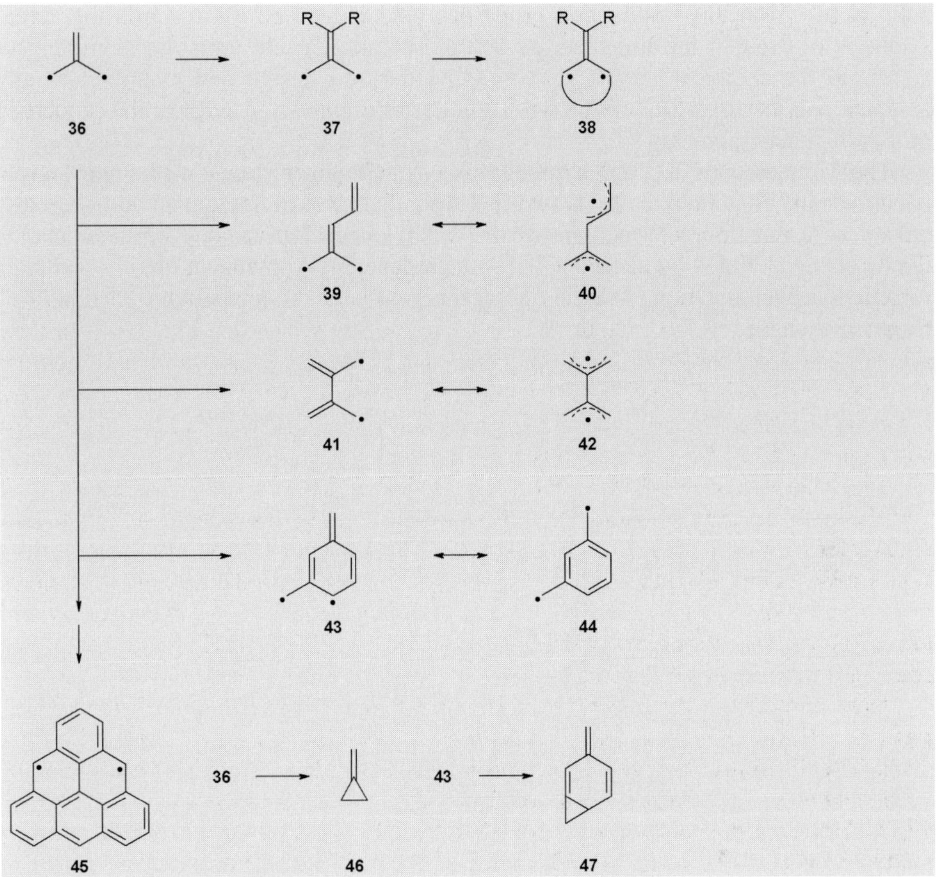

Scheme 7. The growing family tree of the non-Kekulé hydrocarbons.

the number of electrons that must be accommodated in these molecular orbitals. Thus hydrocarbons such as **36**, **41**, **43**, *etc.* have two π-NBMOs which contain a total of two electrons. In other words, each of the above molecules is a diradical.[50] Furthermore, Longuet-Higgins predicted on the basis of Hund´s rule that the ground state of a non-Kekulé hydrocarbon with n electrons in n π-NBMOs would have a spin quantum number of S = $n/2$. The diradicals shown in Scheme 7 should therefore have triplet ground states (S = 1), in which the two nonbonding electrons occupy different NBMOs and have parallel spins. Experimentally these predictions could be verified by detecting paramagnetic properties for these diradicals. On the other hand, should these species be singlets, and hence be diamagnetic, this would constitute a violation of Hund´s rule.

To test the limits of one of the very basic physical principles has been, and is, one of the reasons for the intense research effort on non-Kekulé hydrocarbons. Equally attractive is the use of these unusual species to develop new synthetic and analytical techniques and their potential use in material science. It has been argued that such units could serve as building blocks for the construction of polymers with ferromagnetic or electronically conductive properties, and indeed efforts towards the actual synthesis of organic ferromagnets by ligation of non-Kekulé units are beginning to appear in the chemical literature.[51, 52] Obviously, to design and manufacture any complex polymer structure of this type, requires thorough knowledge of the properties of its constituent elements.[53]

The simplest non-Kekulé hydrocarbon, trimethylenemethane (**36**, TMM) was unequivocally characterized by Dowd in 1966, the diradical having already been discussed as a reaction intermediate in the methylenecyclopropane rearrangement.[54] Dowd obtained TMM by photolysis of 4-methylenedihydropyrazole (**48**)[55] and later 3-methylenecyclobutanone (**49**) in an organic matrix at the temperature of liquid nitrogen (Scheme 8).[56]

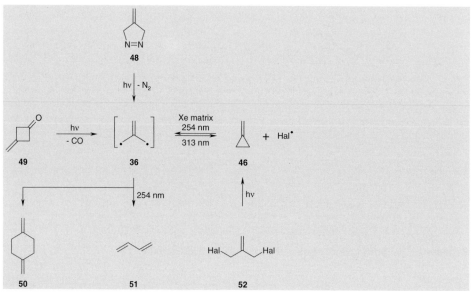

Scheme 8. The preparation of the parent non-Kekulé hydrocarbon, trimethylenemethane (**36**, TMM).

Alternatively, γ-irradiation of methylenecyclopropane (**46**) also led to **36**.[57] The triplet nature of the intermediate produced was established by ESR spectroscopy; at −196 °C the ESR spectrum of TMM is stable for extended periods of time, but at *ca* −150 °C the ESR lines vanish in a few minutes. According to theoretical calculations the triplet of **36** is 26.3 kcal mol^{-1} more stable than its singlet state.[58]

In all these experiments the concentration of **36** was too low for the hydrocarbon to be detected by spectroscopic methods other than ESR spectroscopy. This only changed when Maier and co-workers discovered a high-yield **46**→**36** ring-opening reaction—irradiation of the former in a halogen-doped xenon matrix at 10 K. Now an IR spectrum of **36** could be recorded for the first time.[59, 60] In fact, **46** is not even necessary as a precursor for **36**, the halides **52** (Hal = Br, I) work equally well. Irradiation experiments in the absence of the halogen atoms (either formed from **52** or deliberately added) showed **46** to be photostable. Furthermore, the matrix must consist of, or contain, xenon; experiments in a pure argon matrix were unsuccessful.

To explain the effect of the added halogen atoms $X^{\bullet}$ it has been postulated that on irradiation they form an exciplex with xenon; this immediately reacts further to a trisatomic charge transfer complex $[Xe_2^+X^-]$. The energy conserved in these species is subsequently transferred to **46** which undergoes ring-opening to **36**. The ring-opening of the cyclopropane to the diradical can be reversed by irradiation with 313-nm light; use of shorter wavelengths (254 nm), however, isomerizes **36** to 1,3-butadiene (**51**). Because of this photolability of **36** a yield of *ca* 25% of the non-Kekulé hydrocarbon could not be surpassed in the matrix. Additional chemical structure proof of the structure of **36** was found in its dimerization to 1,4-bismethylenecyclohexane (**50**), which occurred when the matrix was slowly warmed to room temperature. Comparison of the observed and the calculated IR spectra of **36** and several deuterated TMMs showed that TMM is planar with D_{3h} symmetry (out of the 24 fundamental vibrations only 14 are IR active resulting in 8 IR bands for **36**).

The success of the above approach to the synthesis of non-Kekulé hydrocarbons was underlined by additional examples from the same authors. Thus triplet 4-methylene-2-pentene-1,5-diyl (**39/40**, 1,2'-bisallyl diradical) has been generated by irradiation of 2-vinylmethylenecyclopropane (**53**) in a bromine-doped xenon matrix (Scheme 9).[61]

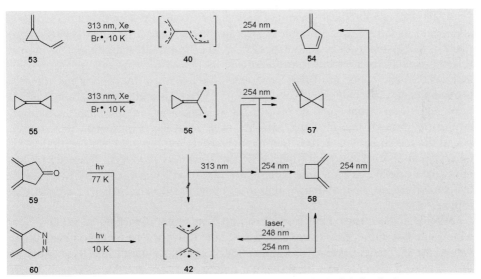

Scheme 9. The generation of the 1,2´- (**39/40**) and 2,2´- (**41/42**) bisallyl diradicals.

Again the starting hydrocarbon was photostable when irradiated at 313 nm in the absence of added bromine atoms. IR-spectroscopic structure assignment resulted in excellent agreement between the experimentally observed spectrum of triplet **39/40** and the calculated (UB3LYP/6-31G*) spectrum. On irradiation of **39/40** with 254-nm light intersystem crossing to the singlet diradical was followed by cylization to 3-

methylenecyclopentene (**54**). The thermal isomerization of **53** to **54** has been known for a long time.[62] Analogously, bicyclopropylidene (**55**, see Section 11.6) on exposure to light of wavelength 313 nm in a xenon matrix at 10 K was photostable, but readily opened to the diradical **56**, a cyclic derivative of **36** (see below), in a bromine-doped matrix under these conditions. As photoproducts 1,2-bismethylenecyclobutane (**58**, major product) and methylenespiropentane (**57**, trace) were characterized.[63] A further ring-opening of **56** to the 2,2'-bisallyl diradical (**42**, tetramethyleneethane, TME) could not be detected. Exposure of **56** to 254 nm light also caused the formation of **57** and **58**, but no ring-opening to **42**.

Unlike **55**, 1,2-bismethylenecyclobutane (**58**) was photostable at 313 nm in a bromine-doped xenon matrix. At a shorter wavelength (254 nm), however, **58** isomerized to **54**; neither bromine nor the xenon matrix was required to induce this transformation because **58** with its conjugated diene chromophor absorbs this wavelength directly. If, on the other hand, **58** was exposed to the light of a KF excimer laser (λ = 248 nm), although **54** was still the main photolysis product, new bands were recorded in the IR spectrum which indicated the formation of the desired TME diradical (**42**). That the generation of TME is reversible was shown by subsequent exposure of **42** to the light of a mercury low-pressure lamp at 254 nm and reformation of **58** under these conditions. The band position calculated for the most intense vibration of TME is in very good agreement with the experimentally determined absorption spectrum of the C_6H_8 hydrocarbon, thus giving supporting proof of the chemical structure.

TME (**42**) is a diradical which not only plays a central role in the dimerization of allene and its derivatives[64, 65] but is also an intermediate in the thermal rearrangement of 1,2-bismethylenecyclobutanes.[66, 67] The first deliberate attempt to prepare this non-Kekulé hydrocarbon and investigate its (ESR) spectroscopic properties was that of Dowd, who in 1970 generated this species by photolysis of 3,4-bismethylenecyclopentanone (**59**) at 77 K.[68] Although under these original conditions an ESR spectrum could be recorded it was impossible to establish the multiplicity of the ground state of **42**. When, however, in a later experiment the azo compound **60** was irradiated at 10 K in a methyltetrahydrofuran glass with a 1000-W xenon lamp at 265 nm a typical triplet ESR spectrum was produced. Because, furthermore, the Curie-Weiss plot of the intensity of the Δm = 2 line against $1/T$ over a broad temperature range gave a straight line it was concluded that TME is a ground state triplet.[69]

Why is the electronic nature of **42** so important as to merit these detailed investigations (and descriptions)? In contrast with TMM (**36**) for which experimental observation and theoretical predictions agree—**36** is a ground-state triplet (see above)—there has been disagreement for **41/42** for a long time. Tetramethyleneethane (TME) is a so-called even, alternant, disjoint hydrocarbon; it belongs to a group of molecules for which the non-bonding orbitals are 'geographically' isolated from each other.[53, 70] Therefore the electron repulsion that provides the basis for Hund's rule is minimized and the singlet could be the ground state.[69] *Ab initio* calculations on **41/42** all predict a singlet ground state for the *planar* diradical.[71] In fact, these calculations show that the singlet should be more stable than the triplet by *ca* 1.8 kcal mol^{-1}. Clearly, the ESR results of Dowd, indicating the paramagnetic nature of TME, are at odds with these predictions. To resolve this apparent contradiction it has been suggested that Dowd's TME prefers a non-planar geometry,[53] for which the triplet

Dowd's TME prefers a non-planar geometry,[53] for which the triplet configuration should be the more stable. 'Planarized' TMEs consequently became the target of an intense research effort. To prepare these species one can in principle use the trick so often employed in hydrocarbon chemistry if the fixation of a certain conformation is desired—the incorporation of the species or molecular structure under study into a (rigid) cyclic framework. This could be achieved for **41** by bridging its 'radical ends' by an ethano bridge and generation of the diyl **61** (Scheme 10).

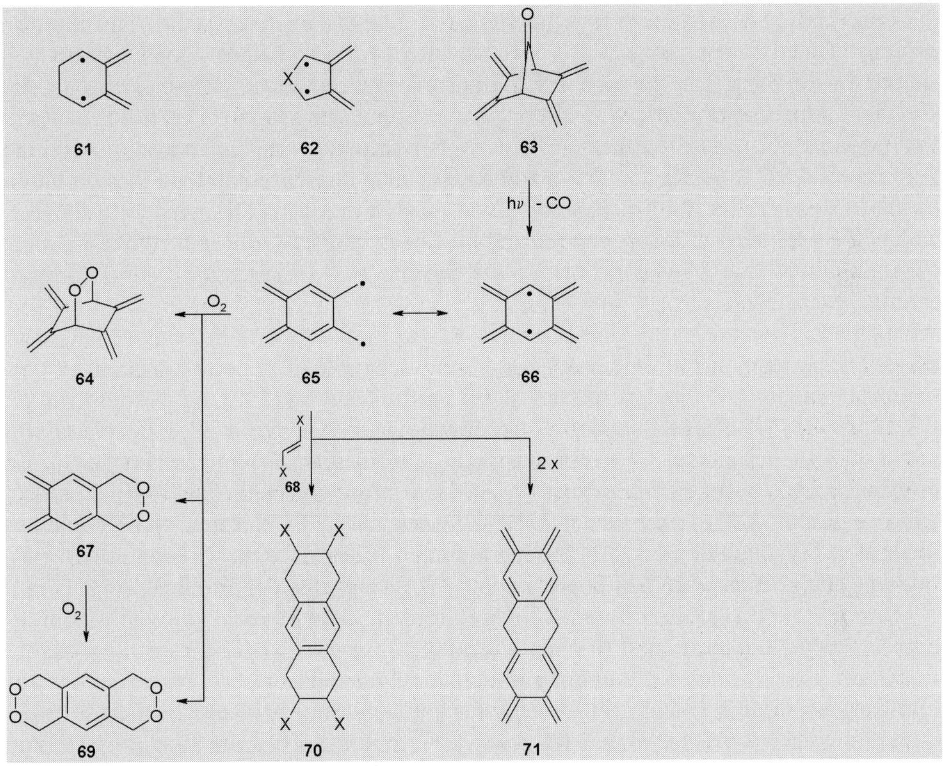

Scheme 10. Planarizing the TME diradical (**41/42**).

According to calculations, however, this cyclic diradical is non-planar and prefers a triplet ground state,[72] as is, in fact, observed experimentally.[73] Note that the synthesis of the non-Kekulé hydrocarbon **61** has already been described in Section 11.1. Exchange of the ethano bridge in **61** for a heteroatom generated the diradicals **62** (X = N, O, S), all of which could be prepared and be shown to be ground-state singlets.[74] Because the π lone pair of the heteroatom lifts the degeneracy between the NBMOs that exists in planar TME, the singlet ground state observed for **62** cannot be taken as evidence of a singlet ground state for **41/42**.

The next candidate to be studied as a diradical in which a violation of Hund's rule might be found experimentally was 1,2,4,5-tetramethylenebenzene (**66**, TMB). According to *ab initio* calculations singlet **66** should be more stable than the triplet by *ca*

5 kcal mol^{-1}.[75]

The non-Kekulé hydrocarbon **66** was first prepared by Roth, Maier, Sustmann and their co–workers from 2,3,5,6-tetramethylene-7-norbornanone (**63**), by photolytic extrusion of carbon monoxide in a matrix at low temperature.[76] From comprehensive spectroscopic data (UV/Vis, IR, ESR) the authors concluded that **66** was a triplet, that the theoretical predictions of a singlet ground state for TMB were incorrect, and, therefore, that the hypothesis that a disjoint non-Kekulé hydrocarbon could violate Hund's rule should be reexamined. A painstaking reinvestigation of this *experimentum crucis* by an interdisciplinary group headed by Berson[77] revealed however, that the observed ESR signals were generated by other ESR-active species and that the observed (intensely purple) color is not associated with these ESR signals. Rather it is caused by **66** which is, in fact, ESR-silent as required for a singlet. Proof of the chemical structure of TMB was obtained by trapping experiments in fluid medium. For example, dioxygen intercepted **66** and its resonance structure **65** to provide the peroxides **64**, **67**, and **69**, the mono adduct **67** being an intermediate in the formation of the diperoxide **69**. With olefins **68** (X = COOEt or CN) TMB reacted to the 2:1-adduct **70**, whereas in the absence of trapping reagent dimerization to the double *ortho*-quinodimethane **71** was observed (see Section 12.3 on preparative uses of these reactive intermediates).

Although the absence of an observable ESR spectrum supports the singlet nature of **66** positive experimental evidence was clearly desirable. This was provided by preparing a suitably ^{13}C labeled sample of **66** and investigating it by low-temperature, solid-state CP-MAS NMR spectroscopy. As shown by a single well-defined ^{13}C signal in the usual chemical shift region (δ = 113) **66** is indeed a singlet. For triplet **66** broadened and strongly shifted lines would have been expected. The TMB diradical, which is stable in rigid matrices at 77 K for weeks, is hence the first non-Kekulé hydrocarbon for which there is strong experimental evidence for the violation of Hund's rule—in accordance with theoretical predictions for this disjoint diradical.[78]

Besides these olefinic diradicals many aromatic non-Kekulé hydrocarbons have been prepared or postulated as reaction intermediates.[52, 79] The most thoroughly studied of these species is the *m*-quinodimethane diradical (**44**), a hydrocarbon also of historical importance in connection with the 'meta-quinone problem', which was discussed at great length in the late 19th century;[79] diradical **44** is the all-carbon version of this elusive quinone. Note also that **44** contains **39/40** as a subsystem.

The optical spectroscopy of **44** was first studied in 1968 by Migirdicyan who obtained the diradical either by photolysis of *m*-xylene (**72**) or α-chloro-*m*-xylene (**73**) in a methylcyclohexane glass at 77 K (Scheme 11).[80]

The first ESR observation of **44** was that of Wright and Platz who generated the bis carbene **75** by irradition of the bis diazo compound **74**[81] at cryogenic temperatures (22 K) and subsequently reduced this species with ethanol at 77 K.[82] That **44** was indeed produced could be detected by the fading of the quintet spectrum of **75**[81] and formation of a new triplet spectrum. Because its signal intensity was linearly dependent on 1/*T* between 30 and 77 K the triplet is the ground state or within a few calories of it, as required by theory. In yet another route to **44** involving a *m*-xylene derivative as a precursor, the dibromide **76** was photolyzed in the presence of diphenylamine.[82]

A completely different route to *m*-quinodimethane was taken by Berson and

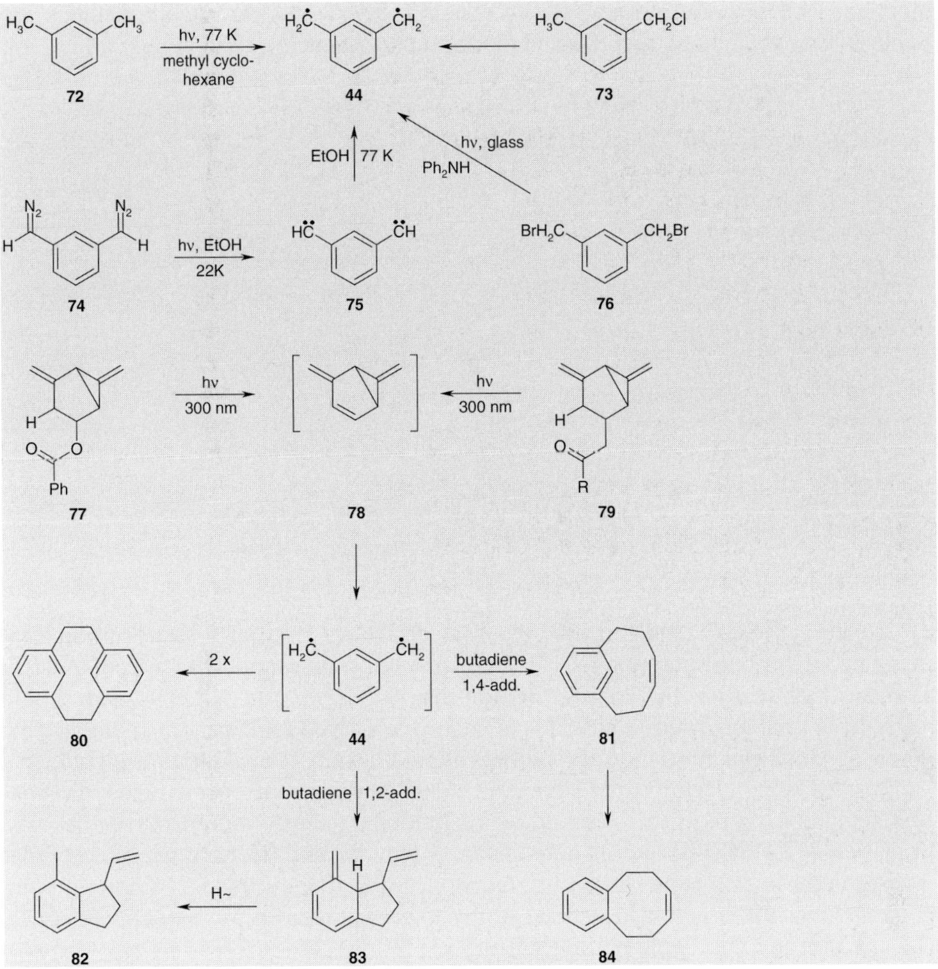

Scheme 11. The generation and trapping of *m*-quinodimethane (**44**).

Goodman.[83, 84] On photolysis of either the ester **77** or the ketones **79** (R = CH₃, phenyl) in 2-propanol glass at 77 K these underwent Norrish type-II photofragmentation to yield the covalent hydrocarbon **78** (an isomer of **47**, see above), and from it the parent **44**, as detected by its characteristic ESR signal.[83, 84] Upon warming of the matrix, dimerization of **44** to [2.2]metacyclophane (**80**) occurred (see Section 12.3). Photolysis of the methyl ketone **79** (R = CH₃) in the presence of a trapping agent such as 1,3-butadiene led to interesting trapping products. By 1,4-addition the [6]metacyclophene **81** was produced (see Section 12.3) and from it the *ortho* isomer **84** by a secondary photoprocess.[85] Alternatively, 1,2-addition of **44** to the diene first yielded a prearomatic adduct, **83**, from which the indane derivative **82** was formed by a hydrogen-shift process. Besides **82** a regioisomeric hydrocarbon was isolated; the yields of the four 1:1-trapping products amounted to 40–60%.

Many other aromatic diradicals have been described in the chemical literature, including the naphthalene-derived species **85–87**[86] and the stable Schlenk (or Schlenk-Brauns) hydrocarbon **88** and its derivatives (Scheme 12).[87]

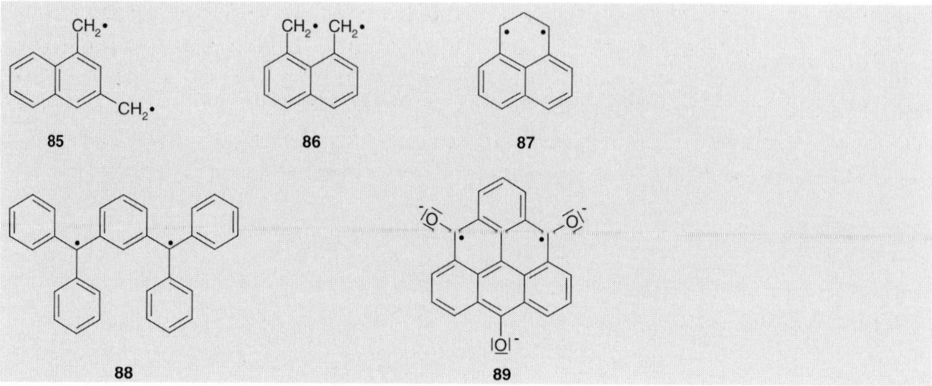

Scheme 12. A selection of aromatic diradicals.

Recently trioxytriangulene, **89**, the first derivative of Clar's hydrocarbon (**45**, triangulene) has been described.[88] Triangulene, the simplest non-Kekulé polynuclear aromatic hydrocarbon, has a pair of non-disjoint degenerate NBMOs and should hence have a triplet ground state. This is, indeed, observed for **89**, a remarkably stable diradical which can be stored for months without intensity loss of the ESR signal.

Besides being electronically important model and reference compounds, the non-Kekulé hydrocarbons are also beginning to attract the interest of preparative chemists. In particular TMM (**36**) and its derivatives of type **37** and **38** have become popular intermediates for the construction of complex carbon frameworks.[89] A case in point is the alcohol **92**, a key intermediate in the total synthesis of the anti-cancer agents D,L-coriolin and D,L-hypnophilin (Scheme 13).[90]

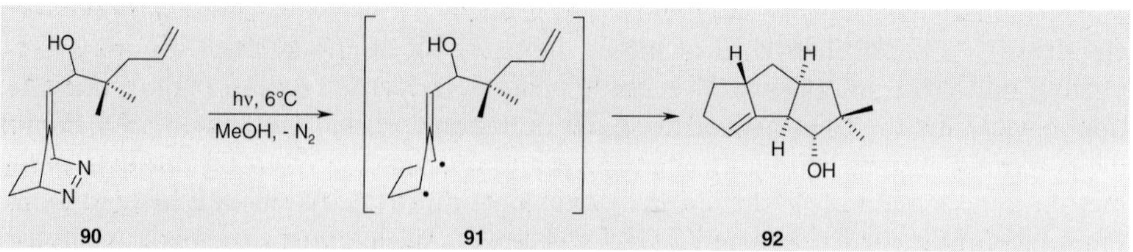

Scheme 13. Use of the TMM-intermediate **91** in natural product synthesis.

To prepare **92**, the diazene **90** was irradiated through a Pyrex filter at 6 °C in methanol. The TMM intermediate **91** was generated from **90** by extrusion of nitrogen and subsequently intramolecularly trapped by the side chain double bond; with 84% the yield of the tricyclic alcohol was excellent.

16.3 Bond fixation in benzene rings

In the discussion of the cyclobutadiene problem (Section 10.1) it was pointed out that square-planar cyclobutadiene (D_{4h} symmetry) with delocalized π-electrons can be viewed as the transition state of the interconversion between the rectangular cyclobutadiene valence tautomers (D_{2h} symmetry) in which the π-electrons are localized into distinct double bonds. As shown experimentally this prototypical antiaromatic hydrocarbon prefers the localized structures. In benzene (**1**), the aromatic hydrocarbon *par excellence*, exactly the opposite situation is encountered (Scheme 14).

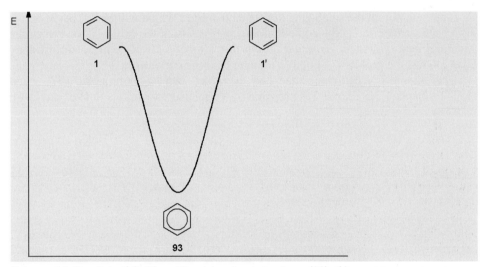

Scheme 14. The Kekulé (**1,1'**) and the delocalized structures (**93**) of benzene.

Here, the delocalized structure (**93**) represents the energy minimum, and the localized Kekulé structures **1** and **1'** with alternating single and double bonds are energetically unfavorable. As is generally known Kekulé favored a rapid oscillation between the 'limiting structures' **1** and **1'** to account for certain chemical properties of benzene. With the development of modern bonding theory in the 1920s and 1930s Kekulé's oscillation theory was shown to be erroneous, **1** and **1'** had to be abandoned and were replaced by **93** as representing the π-electron distribution in benzene. Despite this, the question of whether the benzene nucleus can be induced to give up its delocalized structure in favor of a 1,3,5-cyclohexatrienoid structure has attracted experimental and theoretical chemists until the present day.

One way to coerce a benzene nucleus into a cyclohexatriene structure consists in bridging the aromatic core. As we have seen in the chapter on cyclophanes (Section 12.3) the introduction of a bridging unit such as a short polymethylene chain in the *meta* or in *para* positions causes a severe deformation of the benzene ring; this deformation increases rapidly with decreasing bridge length. Although deformation angles in the 20° range are easily obtained, no bond alternation in the benzene nucleus could, however, be detected by X-ray crystallography. Unfortunately, the most extremely distorted [*n*]cyclophanes with only a tetra- and pentamethylene bridge are too reactive to be isolated. That bridging of the *ortho* positions of a benzene ring might lead to 'bond fixation' was put forward as early as 1930 by Mills and Nixon,[91] who interpreted the different selectivities observed in the electrophilic aromatic chemistry of indanes and tetralins as an indication of bond fixation. Whereas the tetralin derivatives were assumed to prefer the tautomer with a double bond at the ring junction, in the indane series the tautomer should be preferred in which the double bonds of the benzene ring would be semicyclic to the five-membered ring. Although this interpretation could not be maintained, the postulate of the so-called Mills–Nixon effect, *i.e.* that annelation of small rings to a benzene nucleus should induce bond-length alternation, survived and remained a challenge for synthetic chemists for a long time.[92]

Although not all the small-ring annelated benzenes which will be discussed below have been synthesized in the context of the controversy of the Mills–Nixon effect, the structural data ultimately collected during this synthetic effort contributed to finally laying to rest the problem of bond fixation in benzene derivatives.[92]

The simplest small-ring annelated benzene, benzocyclopropene (**97**, bicyclo[4.1.0]hepta-1,3,5-triene) was prepared by Vogel and co-workers in 1965 as shown in Scheme 15.[93]

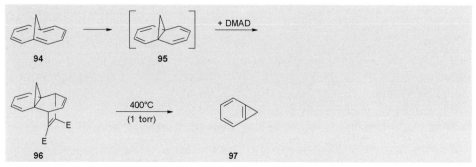

Scheme 15. Vogel's synthesis of benzocyclopropene (**97**, E = COOCH₃).

Diels–Alder addition of dimethyl acetylenedicarboxylate (DMAD) to 1,6-methano[10]annulene (**94**, see Section 10. 5) at 120 °C furnished the 1:1 adduct **96**, evidently by trapping of the norcaradiene valence isomer **95** of the substrate hydrocarbon. When **96** was subjected to flash-vacuum pyrolysis at 400 °C and 1 torr it suffered Alder–Rickert cleavage and yielded dimethyl phthalate and the C₇H₆ hydrocarbon **97** in 45% yield. The benzocyclopropene is an unpleasant smelling liquid which is stable at room temperature for weeks but polymerizes on distillation.

Although several derivatives of **97** such as the 1,1-difluoro compound,[94] cyclopropa[*a*]naphthalene,[95] and the very reactive cyclopropa[*l*]phenanthrene[96] have been prepared by the retro-Diels–Alder route, more convenient syntheses to cycloproparenes are known (Scheme 16). Among these the base-induced dehydrohalogenation of different dihalo bicyclo[4.2.0]hept-2-enes, **98**–**100**, are most often used, because not only are these substrates usually readily available (see below), and the elimination can be performed relatively simply, but also because these aromatization reactions have the widest range of application.

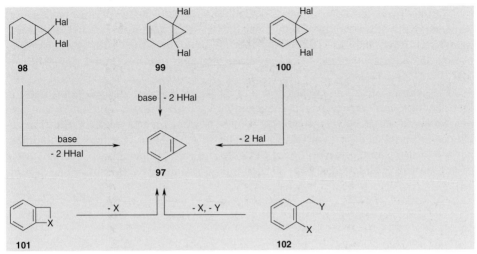

Scheme 16. Different routes to cycloproparenes.

Alternative approaches starting from substrates lacking the three-membered ring, as observed in the photochemical or thermal fragmentation of the benzannelated precursors **101**, or the closure of one of the lateral cyclopropane single bonds in the *ortho*-substituted benzylic derivatives **102**, lack generality.

The dihalides **98** are easily obtainable by dichlorocarbene addition to 1,4-cyclohexadiene (**103**) or synthetic equivalents. Thus the dichloride **104** has been prepared from **103** by dichlorocyclopropanation under phase-transfer conditions (Scheme 17).[97] Treatment of **104** with potassium *tert*-butoxide in dimethyl sulfoxide subse-

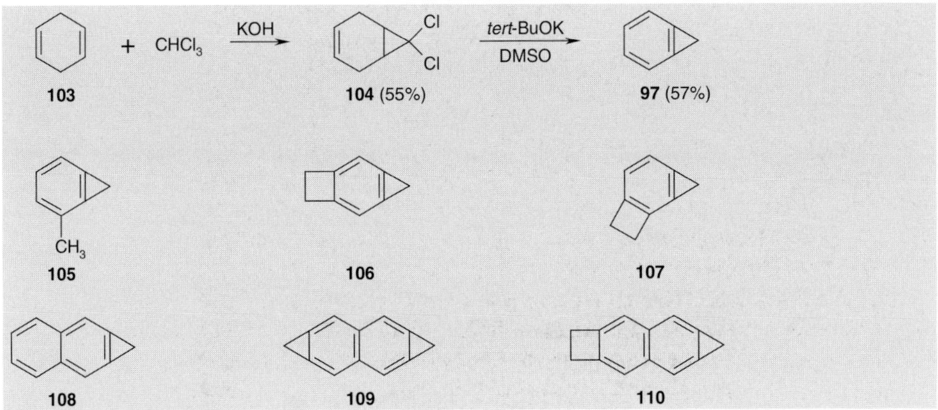

Scheme 17. The Billups route to cycloproparenes.

quently caused twofold dehydrochlorination and migration of the double bonds into the six-membered ring to give benzocyclopropene (**97**).[98] This sequence, first described by Billups and co-workers in 1971, could be scaled up to the 0.5 molar level and is presently one of the most widely used routes to cycloproparenes.

Starting from the appropriate 7,7-dihalobicyclo[4.1.0]heptenes this method has provided many other cycloproparenes whether these contain just one annelated three-membered ring such as **105**[99] or **108**[100] or two as in **109** (1,4-dihydrodicyclopropa[*b,g*]naphthalene[101] or one three- and one four-membered ring, **106**,[102] **107**,[102, 103] and **110**.[102, 104]

The isomeric dihalides **99** are as useful for the preparation of cycloproparenes; in fact, the structural variety which is accessible by base-induced elimination and aromatization of these bicyclo[4.1.0]hept-3-enes is even greater than for **98**. The dihalides are produced by Diels–Alder addition of 1,2-dihalo- or tetrahalocyclopropenes—1-bromo-2-chlorocyclopropene (**112**) being used frequently—to a large variety of acyclic and cyclic dienes **111**, as shown predominantly by Billups *et al.* (Scheme 18).

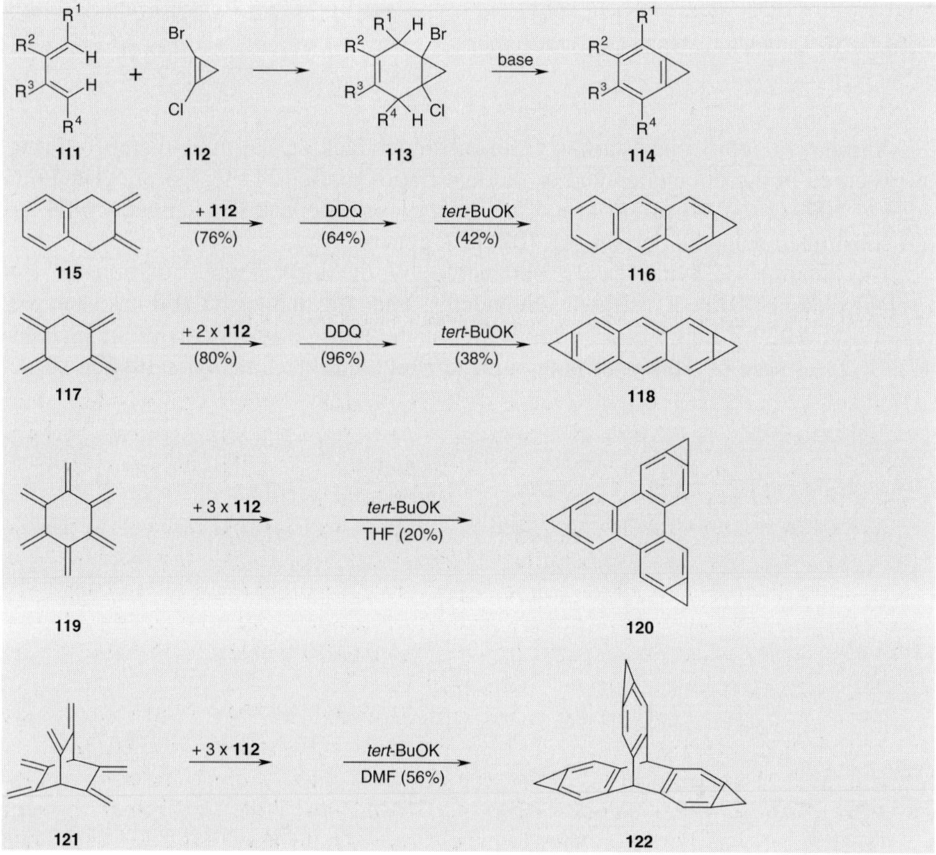

Scheme 18. Cycloproparenes *via* the Diels–Alder route.

Because the halogen substituents in the Diels–Alder adduct **113** are now positioned β to hydrogen atoms, the elimination to the cycloproparenes **114** can proceed directly and no halocyclopropenes are required as reaction intermediates.

The examples in Scheme 18 leading from **115** to **116**,[105] **117** to **118**,[106] from hexaradialene (**119**, see Section 11.5) to the triangular cycloproparene **120**,[107] and from [2.2.2]hericene (**121**, see Section 11.5) to the triptycene derivative (see Section 13.2) **122**[107] are typical of the versatility of this approach.

The aromatization step in these syntheses was abbreviated when the diene component **111** was replaced by an *ortho*-quinodimethane (*ortho*-xylylene), a cycloaddition partner which had proven its worth already in cyclophane chemistry (see Section 12.3). The highly reactive parent system **125** may be generated by various routes— one possibility, the base-catalyzed isomerization of 1,7-octadiyne-4-ene (**123**, see Section 8.2), is shown in Scheme 19.[108]

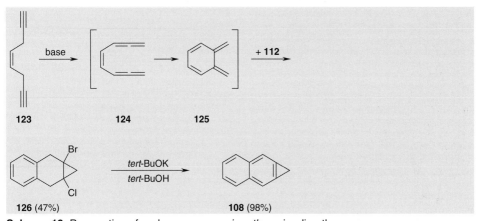

Scheme 19. Preparation of cycloproparenes *via ortho*-quinodimethanes.

Treatment of **123** with potassium *tert*-butoxide in *tert*-butanol initially furnished the conjugated bisallene **124** (see Section 9.1), which immediately cyclized to **125**. Trapping with **112** to **126** followed by β-elimination of the latter with base then led to 1-*H*-cyclopropa[*b*]naphthalene (**108**) in *ca* 50% total yield.

Halides of the general structure **100** have been used far less often in cycloproparene synthesis than their isomers **98** and **99**; one reason is that these norcaradiene precursors (which usually exist in their 1,3,5-cycloheptatriene form) are normally obtained from benzocyclopropenes by halogen addition in the first place.[109] Still, **108**, for example, could be prepared from 4,4,6,6-tetrabromobenzocycloheptene-2,7-dione by ring contraction according to this protocol.[110]

Likewise, fragmentation of the benzannelated substrates **101** constitutes a means of preparing benzocyclopropenes, but it is unattractive preparatively because its scope is too limited. Mechanistically the high-temperature[111] and photochemical decomposition[112] of benzocyclobutenone (**127**) and 3*H*-indene-1,2-dione (**129**) have been investigated thoroughly; both processes yield only small amounts of **97** and other C_7H_6 hydrocarbons such as fulvenallene (**128**) (Scheme 20).[113]

The photofragmentation of the 3*H*-indazole **130** to the ester **131** is noteworthy because Anet and Anet in 1964 prepared the first benzocyclopropene derivative ever reported, the ester **131**, by this approach.[114] Again, the major product of this nitrogen extrusion is the unstrained isomer **132**.

Finally, the closure of lateral cyclopropane single bonds in monocyclic precursors of type **102** has not yet been exploited to any significant extent and apparently seems to be restricted to the preparation of the very elementary cycloproparenes only. Thus, benzocyclopropene (**97**) has been obtained by lithiation of 1-bromo-2-methoxymethylbenzene in THF with *n*-butyllithium at –20 °C,[115] and 3,4-dihydro-1*H*-cyclobuta[3,4]benzocyclopropene (**106**) has been prepared similarly from 4-

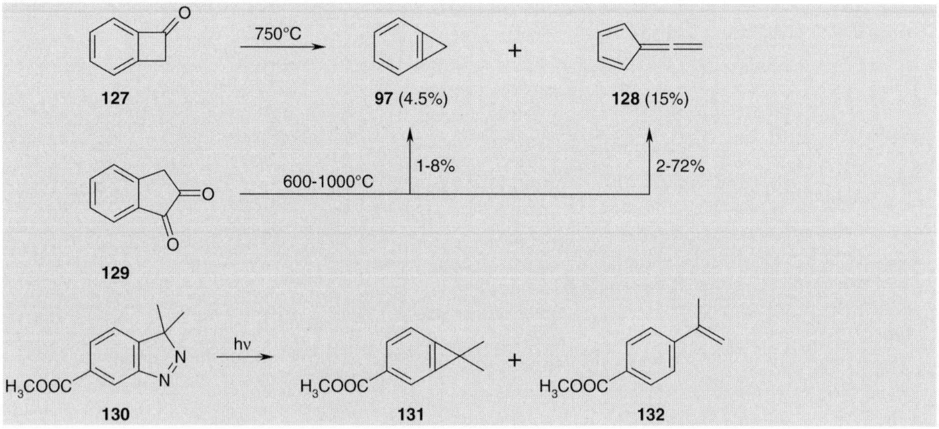

Scheme 20. The preparation of benzocyclopropenes by fragmentation reactions.

bromo-5-methoxymethyl-1,2-dihydrocyclobutabenzene.[116]

Today, cycloproparenes present themselves as a readily available and carefully studied class of highly strained hydrocarbon[117]—a selection of their structural properties will be discussed below after description of the synthesis of their next higher homologs, the cyclobutarenes.

Traditionally benzocyclobutenes, as these hydrocarbons are also called, are prepared by more or less the same methods as have been used for their lower homologs, *i.e.* either by closure of a lateral C–C single bond and formation of the four-membered ring or by use of [2+4]cycloaddition reactions with carefully chosen diene components (see below). In contrast with the cycloproparenes, however, the former approach is much more popular than the latter. The reason for this preference is simple—the starting materials for the first strategy, suitably functionalized *o*-xylenes, are readily available and many four-membered ring-forming reactions are known. In this approach the substrates normally have the number of carbon atoms eventually needed in the product, whereas in the cycloaddition route the σ-skeleton is first assembled and the target molecule subsequently produced by aromatization steps. For the synthesis of functionalized benzocyclobutenes a third approach, in which the six- and the four-membered ring are created simultaneously in a catalyzed co-oligomerization step has recently become increasingly important.

The parent hydrocarbon **137** (bicyclo[4.2.0]octa-1,3,5-triene) was first synthe-sized by Cava and Napier in 1956[118] who treated α,α,α'α'-tetrabromo-*o*-xylene (**133**) with excess sodium iodide in ethanol under reflux, following a procedure which Finkelstein had reported nearly half a century earlier.[119] Like he they obtained the dibromide **135**, the first benzocyclobutene derivative ever reported, presumably formed by cyclization of the initially generated *o*-xylylene intermediate **134** (Scheme 21).

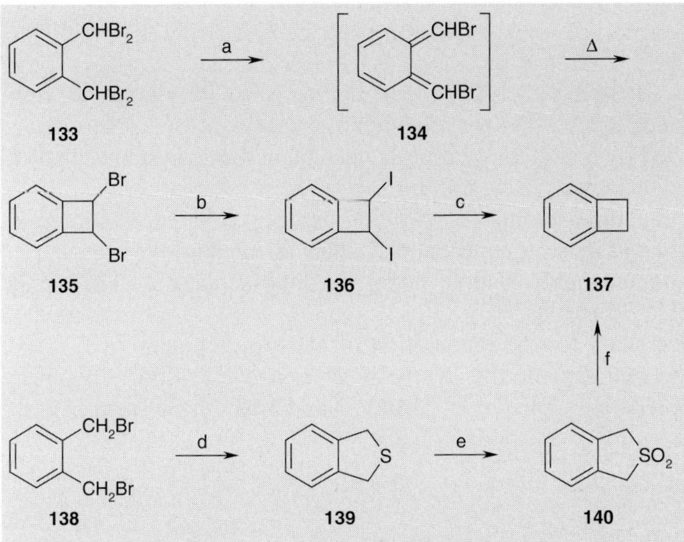

(a) NaI, EtOH, reflux, 2 d (60%); (b) NaI, EtOH, reflux, 8 d (78%); (c) H₂, Pd / C, NaOEt (55%); (d) Na₂S, EtOH, reflux (44%); (e) CH₃CO₃H, CH₃CO₂H (89%); (f) 770°C, 2 torr (67%)

Scheme 21. The preparation of benzocyclobutene (**137**) by Cava and Napier.

Extended exposure of **135** to sodium iodide in ethanol under reflux resulted in the production of the diiodide **136**, which on hydrogenolysis over palladium on charcoal furnished the hydrocarbon **137**.[120] Soon afterwards it was discovered that α,α'-dibromo-*o*-xylene (**138**) could be 'debrominated' to **137** by a sequence involving its conversion to 1,3-dihydroisothianaphthene (**139**), oxidation of the latter, and gas-phase pyrolysis of the obtained sulfone **140** at 770 °C.[121] *o*-Xylene derivatives have been used as substrates in benzocyclobutene synthesis ever since;[122] a small collec-tion of successful eliminations and cyclizations is assembled in Scheme 22.

Thus 1,4-elimination of hydrogenchloride from 2-methylbenzylchloride (**141**) under flash vacuum pyrolysis conditions enables a particularly easy access to **137**.[123] The method is successful with many other precursor aromatics as long as they contain **141** as a substructure. It has found an impressive application in Boekel-heide's celebrated superphane synthesis (see Section 12.3). Likewise, thermal frag-mentation of the tellurophene **142**,[124] 3-isochromanone (**143**),[125] and 1,2-bis(phenylselenomethyl) benzene (**144**)[126] yielded benzocyclobutene in acceptable yields.

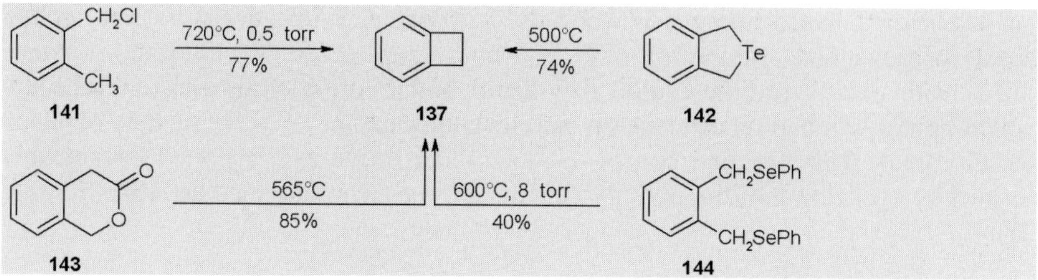

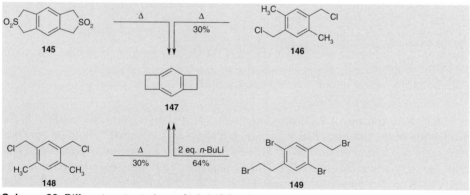

Scheme 22. Further routes to benzocyclobutene (**137**) from *o*-xylene-derived precursors.

The transformations of **142** and **143** can, of course, also be viewed as ring contraction processes, and, indeed, benzocyclobutene and several of its derivatives have also been synthesized by contracting larger benzannelated ring systems. Typical examples are 2-diazo-indan-1-one[127] and tetralin.[128]

A large number of reactions yielding or involving benzocyclobutenes as intermediates have been investigated more for mechanistic than for preparative reasons,[129, 130] a particularly well-documented example being the intermediacy of **137** in the thermal isomerization of arylcarbenes.[131]

The methods just described for the preparation of benzocyclobutene (**137**) have also been employed successfully for the synthesis of benzo[1,2:4,5]dicyclobutene (**147**) starting from the precursors **145**,[132, 133] **146**,[134] and **148**[134] (Scheme 23).

Scheme 23. Different routes to benzo[1,2:4,5]dicyclobutene (**147**).

Because of the thermal lability of benzocyclobutenes in general (see below), and of **147** in particular,[135] pyrolytic 1,4-eliminations—besides requiring special apparatus—reach a natural limit when applied to the synthesis of preparative amounts of multiply cyclobutannelated aromatics—the strain of these hydrocarbons increases rapidly with the number of condensed small rings. Less forcing reaction conditions are thus needed for these molecules and were found for **147** in the so-called Parham cyclization of the tetrabromide **149**. When this precursor was first metalated with two equivalents of *n*-butyl lithium at -100 °C and the temperature then raised to room temperature **147** was formed in 64% yield, enabling the preparation of gram quantities of the hydrocarbon.[136]

The isomeric benzo[1,2:3,4]dicyclobutene (**155**) has also been prepared by a conventional pyrolytic route and by a process avoiding high temperatures (Scheme 24).

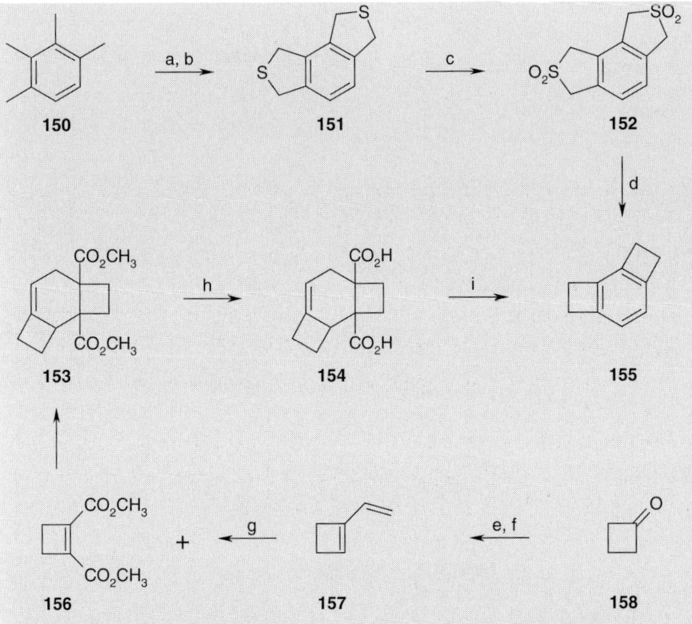

(a) Br$_2$, CCl$_4$ (65%); (b) Na$_2$S, EtOH (20%); (c) CH$_3$CO$_3$H, CH$_3$CO$_2$H (60%); (d) 280°C (5%); (e) H$_2$C=CHMgBr (66%); (f) Δ, I$_2$ (72%); (g) + **156**, Δ; (h) KOH, aq. MeOH; (i) Pb(OAc)$_4$, DMSO (50%)

Scheme 24. The synthesis of benzo[1,2:3,4]dicyclobutene (**155**).

Thus bromination of prehnitene (**150**) with elemental bromine and cyclization of the resulting tetrabromide with sodium sulfide provided 1,3,6,8-tetrahydrodithiopheno[1,2-c:3,4-c']benzene (**151**). After oxidation of this to the bis sulfone **152**, its gas phase pyrolysis yielded **155**.[137] Earlier the same cyclobutarene had been prepared by Thummel in what seems to be the first application of the the Diels–Alder reaction in this area of hydrocarbon chemistry.[138] The required diene for this latter approach, 1-vinylcyclobutene (**157**) was first produced from cyclobutanone (**158**) by reaction with vinyl magnesium bromide and dehydration of the resulting alcohol. When **157** was treated with dimethyl cyclobutene-1,2-dicarboxylate (**156**) as the dienophile, 1,8-bis(methoxycarbonyl)tricyclo[6.2.0.0^{4,5}]dec-5-ene (**153**) was formed as a mixture of isomers. Hydrolysis with potassium hydroxide in methanol under reflux subsequently provided the corresponding diacid **154**, which was decarboxylated and aromatized to the target hydrocarbon **155** by treatment with lead tetraacetate.

Diels–Alder reactions were also the key steps in the synthesis of tricyclobutabenzene (**164**). In an earlier attempt to prepare this hydrocarbon Vollhardt and co-workers had studied the rearrangement of 1,5,9-cyclodecatriyne at 650 °C, but had isolated only its ring-opened isomer, [6]radialene (see Section 11.5).[139] As the diene component in the first successful synthesis of **164**[140] Thummel and co-workers used α,α'-dicyclobutenyl (**161**) prepared from 1-bromocyclobutene (**159**) via copper chloride promoted coupling of its Grignard derivative **160** according to a procedure by Lüttke and Heinrich (Scheme 25).[141]

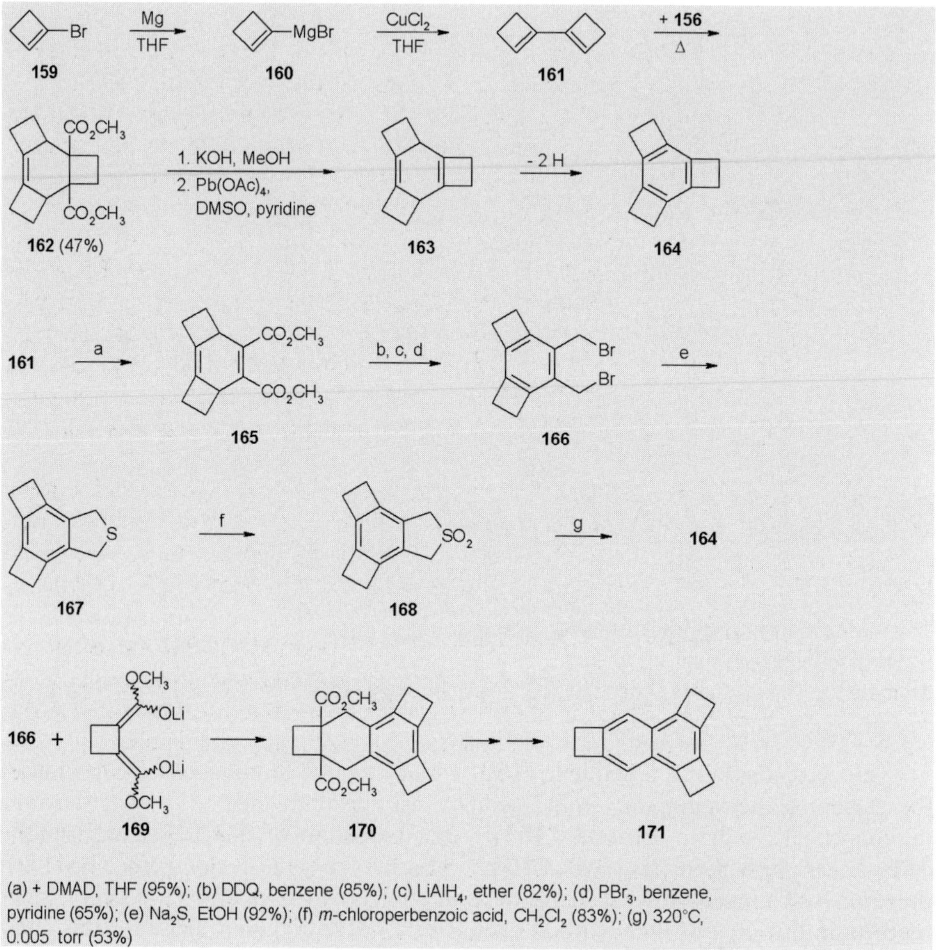

(a) + DMAD, THF (95%); (b) DDQ, benzene (85%); (c) LiAlH₄, ether (82%); (d) PBr₃, benzene, pyridine (65%); (e) Na₂S, EtOH (92%); (f) m-chloroperbenzoic acid, CH₂Cl₂ (83%); (g) 320 °C, 0.005 torr (53%)

Scheme 25. The synthesis of tricyclobutabenzene (**164**) according to Thummel and Garratt.

Cycloaddition with **156** at 110 °C led to the tetracyclic diester **162** (mixture of isomers), which was saponified with potassium hydroxide in methanol in quantitative yield to the corresponding diacid. Disappointingly, however, the yield of the last step—decarboxylation as described above for the synthesis of **155**—was very poor. The hydrocarbon **164** was only formed in 3.6% yield, and it was contaminated by traces of the dihydro derivative **163**, evidently its immediate precursor which aromatizes under the reaction conditions. These disadvantages were circumvented by Garratt and co-workers who replaced **156** by dimethyl acetylenedicarboxylate (DMAD) in the [2+4]cycloaddition step. The expected adduct, **165**, was isolated in good yield, and because its aromatization and the conversion of the two ester substituents into bromomethyl groups could also be accomplished readily the dibromide **166** was available in sufficient amounts. Preparation of the sulfide **167** and its oxidation to the sulfone **168** posed no problems, and the final thermal fragmentation and ring-contraction step led to **164** in 53% yield, the overall yield from **166** to **164** amounting to a comfortable 36%.[142]

The importance of **166** as a synthetic intermediate for the preparation of cyclobutannelated aromatic hydrocarbons is underlined by the synthesis of tricyclobuta[1,2:3,4:6,7]naphthalene (**171**), obtained by first coupling **166** with **169**, dilithiated dimethyl cyclobutane-1,2-dicarboxylate, and then subjecting the diester formed, **170**, to the above decarboxylation and aromatization conditions.[142, 143]

Triple-bond-containing precursors have been used in benzocyclobutene synthesis on several occasions. The obvious possibility of adding alkynes to dehydrobenzene (which, after all, dimerizes to biphenylene, a benzannelated derivative of **137**, as described in Section 15.1) has met with limited success. It is restricted to the parent *o*-dehydrobenzene (see above) and requires functionalized alkenes as addition partners.[144, 145] Despite this, sometimes, as with vinylacetate, the [2+2] adduct is formed in acceptable yield (*ca* 50%).[146] Better results were obtained when dehydrobenzene was added to dienes such as **157**, enabling the preparation of naphthocyclobutenes.[147]

Acylic acetylenic precursors are involved in Vollhardt's co-oligomerization route in which the complete skeleton of benzocyclobutenes is generated by (η^5-cyclopentadienyl)cobalt dicarbonyl-catalyzed addition of substituted acetylenes to 1,5-hexadiyne, a procedure which we already encountered in Section 15.1 in connection with the synthesis of the [*n*]phenylenes.[148]

Benzocyclobutene (**137**) and its derivatives are a thoroughly studied class of hydrocarbons.[144, 149] Their most important reaction is ring-opening to *o*-punodimethanes (*o*-xylylenes), a process which has been studied extensively from the mechanistic viewpoint[150] and been exploited numerous times in synthetic organic chemistry, because these highly reactive intermediates are excellent partners in cycloaddition reactions in which they serve as the diene component.[151]

Having acquainted ourselves with the most important preparative routes to cyclopropa- and cyclobutabenzenes we now return to the question of bond fixation in these hydrocarbons. Fortunately, highly reliable single crystal X-ray structural analyses are available for most of the hydrocarbons mentioned above.[152]

Beginning with cyclopropabenzene (**97**) Neidlein, Boese, Gieren and co-workers have shown that at 120 K this hydrocarbon is non-planar: the six- and the three-membered ring are tilted by 2.4° towards each other. Although the C1–C6 bond is somewhat shorter than the other C–C bonds in the benzene ring, **97** shows no pronounced bond localization (Scheme 26).[153]

Excluding the C–C bond shared by the two rings, the average bond length of the other five C–C bonds of the six-membered ring is 1.378 Å, and is thus only 0.04 Å longer than an average C=C bond length. The bond angles of cyclopropabenzene, on the other hand, deviate substantially from those encountered in the subsystems of the molecule, *viz.* benzene and cyclopropene. For example C4–C5–C6 = 113.2°, C1–C6–

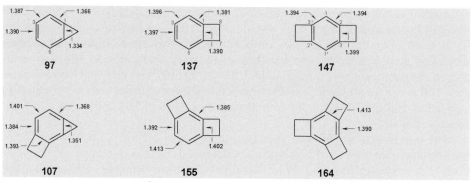

Scheme 26: Bong lengths (in Å) in cyclopropa- and cyclobutabenzene derivatives.

C5 = 124.5°, C5–C6–C7 = 171.7°, and C1–C7–C6 = 52.9°. For **137** (measured at 100 K) and **147** (100 K) the situation is similar—there is no bond alternation in the aromatic rings and bond angle deformation is comparable with that of **97**.[154] Because no double bond fixation could be discovered for several other small-ring fused aromatic hydrocarbons, including **107**, **155** and **164** (for which it amounts to < 0.025 Å, if present at all), the impression arose that a Mills–Nixon effect cannot be induced by annelation.[155] The sheer endless variability of organic structures proved this conclusion to be premature, and, indeed, when the monocyclic annelation used in all the above examples is replaced by bicyclic bridging, bond alternation can be induced in benzene rings.

When the 'annelation element' is comparatively large—and hence unstrained—as the anthracene unit in heptiptycene (**172**, see also Section 13.2)[156] bond alternation in the central ring of the hydrocarbon becomes barely discernible—the bonds semicyclic to the annelating molecular bridges are about 2.2 pm shorter than the six-membered ring bonds which are part of the bridging element.[157] For trisbicyclo[2.2.2]octabenzene (**173**), prepared by Komatsu and co-workers by reacting 2,3-dibromobicyclo[2.2.2]oct-2-ene with *n*-butyllithium in THF at –78 °C, and presumably produced *via* the intermediately generated bicyclo[2.2.2]oct-2-yne (see Section 8.2), the situation is similar—bond alternation amounting to *ca* 1.5 pm.[158, 159] These effects would probably hardly provoke deeper comment did they not mark the beginning of a trend.

In the next lower homolog of **173**, trisbicyclo[2.2.1]heptabenzene (**174**) averaging of the C–C bond lengths semi- and endocyclic to annelation leads to a bond alternation of the order of 4 pm. As in the preceding cases the central benzene ring is essentially planar.[160] That, indeed, bond fixation is a manifestation of semicyclic bond angle strain produced by bicyclic annelation, was finally demonstrated by the most strained member of this series, trisbicyclo[2.1.1]hexabenzene (**178**) (Scheme 27).

This hydrocarbon was prepared by Siegel and co-workers in 1995[161] by first treating bicyclo[2.1.1]hexan-2-one (**175**) with phosphorus pentachloride in phosphorus trichloride to furnish the dichloride **176**. When this was dehydrochlorinated with potassium *tert*-butoxide in THF under reflux, the vinylchloride **177** was produced as the only product. To induce the second dehydrochlorination step, **177** was treated

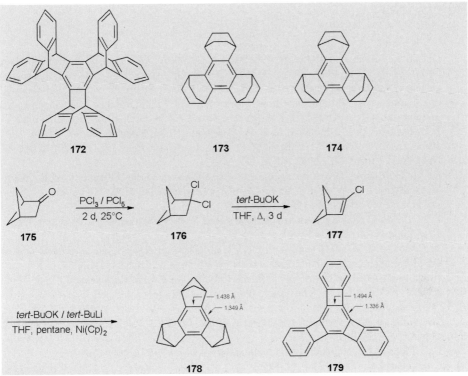

Scheme 27. Siegel's trisbicyclo[2.1.1]hexabenzene (**178**), a benzene derivative with a cyclo-hexatriene structure.

with a mixture of *tert*-butyllithium and potassium *tert*-butoxide in a pentane-THF so-lution at –78 °C, followed by addition of nickelocene and warming to reflux. After work-up, **178** was isolated in < 1% yield. According to single crystal X-ray structural analysis **178** is the first mononuclear benzenoid hydrocarbon with a true cyclo-hexatriene structure.[162] The two non-equivalent C–C bonds of the benzene ring differ by 8.9 pm, yet the ring is planar and its internal angle is 120°. The bond lengths corre-spond to Pauling bond orders of 1.86 and 1.39 for the shorter and the longer bond, respectively. We have already encountered an even larger bond-order difference (0.75) in hydrocarbon **179** (starphenylene, see Section 15.1). In this polynuclear con-densed system, however, π-effects play a dominating role[163, 164] as in triphenylene (see Chapter 14) in which some bond alternation is also present.[165]

   With 1,3,5-cyclohexatriene derivatives having become a reality the next interest-ing problem which must be solved by hydrocarbon chemists is the actual demonstra-tion of the phenomenon of 'oscillation' between cyclohexatrienoid valence tautomers (see above)—as had been suggested by Kekulé in his classical studies on the structure of benzene in 1872.[166] Raising this possibility is asking the question whether hydro-carbon chemistry has come full circle with the above developments. If the circle is an open one and recognizing that its radius has increased tremendously during these 125 years the answer is yes.

References

1. G. Wittig, *Naturwiss.*, **1942**, *30*, 696–703; *cf.* G. Wittig, G. Pieper, G. Fuhrmann, *Ber. Dtsch. Chem. Ges.*, **1940**, *73*, 1193–1197. Dehydroaromatics have apparently been suggested as reactive intermediates as early as 1902 when Stoermer und Kahlert reported that treatment of 3-bromobenzofuran with strong base in the presence of ethanol yielded 3-ethoxyfuran: R. Stoermer, B. Kahlert, *Ber. Dtsch. Chem. Ges.*, **1902**, *35*, 1633–1640.
2. J. D. Roberts, H. E. Simmons, Jr, L. A. Carlsmith, C. W. Vaughan, *J. Am. Chem. Soc.*, **1953**, *75*, 3290–3291; *cf.* J. F. Bunnett, R. E. Zahler, *Chem. Rev.*, **1951**, *49*, 273–412.
3. H. Heaney, *Chem. Rev.*, **1962**, *62*, 81–97.
4. R. W. Hoffmann, *Dehydrobenzene and Cycloalkynes*, Verlag Chemie, GmbH, Weinheim, **1967**; *cf.* R. W. Hoffmann in *Chemistry of Acetylenes*, H. G. Viehe, (*Ed.*), Marcel Dekker, New York, N. Y., **1969**, 1063–1148.
5. E. K. Fields in *Organic Reactive Intermediates*, S. P. McManus, (*Ed.*), Academic Press, New York, N. Y., **1973**, 449–508.
6. T. L. Gilchrist in *The Chemistry of Triple-bonded Functional Groups, Supplement C, Part I* S. Patai, Z. Rappoport, (*Eds.*), J. Wiley and Sons, Chichester, **1983**, 383–419.
7. H. Hart in *The Chemistry of Triple-bonded Functional Groups, Supplement C2*, S. Patai, (*Ed.*), J. Wiley and Sons, Chichester, **1994**, 1017–1131.
8. F. M. Logullo, A. H. Seitz, L. Friedman, *Org. Synth. Collect. Vol. V*, **1973**, 54–59; *cf.* L. Friedman, F. M. Logullo, *J. Org. Chem.*, **1969**, *34*, 3089–3092; P. Crews, J. Beard, *J. Org. Chem.*, **1973**, *38*, 522–528;
9. L. F. Fieser, M. J. Haddadin, *Org. Synth. Collect. Vol. V*, **1973**, 1037–1047.
10. C. D. Campbell, C. W. Rees, *J. Chem. Soc. (C)*, **1969**, 742–747; *cf.* C. D. Campbell, C. W. Rees, *J. Chem. Soc. (C)*, **1969**, 748–751.
11. O. L. Chapman, K. Mattes, C. L. McIntosh, J. Pacansky, G. V. Calder, G. Orr, *J. Am. Chem. Soc.*, **1973**, *95*, 6134–6135.
12. O. L. Chapman, C.-C. Chang, J. Kolc, N. R. Rosenquist, H. Tomioka, *J. Am. Chem. Soc.*, **1975**, *97*, 6586–6588.
13. C. Wentrup, R. Blanch, H. Briehl, G. Gross, *J. Am. Chem. Soc.*, **1988**, *110*, 1874–1880.
14. A. Schweig, N. Münzel, H. Meyer, A. Heidenreich, *Struct. Chem.*, **1990**, *1*, 89–100.
15. G. G. Simon, N. Münzel, A. Schweig, *Chem. Phys. Lett.*, **1990**, *170*, 187–192.
16. J. G. Radziszewski, B. A. Hess, Jr, R. Zahradnik, *J. Am. Chem. Soc.*, **1992**, *114*, 52–57.
17. N. H. Werstiuk, C. D. Roy, J. Ma, *Can. J. Chem.*, **1995**, *73*, 146–149 and refs. cited therein.
18. R. S. Berry, G. N. Spokes, M. Stiles, *J. Am. Chem. Soc.*, **1962**, *84*, 3570–3577; *cf.* R. S. Berry, J. Clardy, M. E. Schafer, *J. Am. Chem. Soc.*, **1964**, *86*, 2738–2739; M. E. Schafer, R. S. Berry, *J. Am. Chem. Soc.*, **1965**, *87*, 4497–4501.
19. R. D. Brown, P. D. Godfrey, M. Rodler, *J. Am. Chem. Soc.*, **1986**, *108*, 1296–1291.
20. A. M. Orendt, J. C. Facelli, J. G. Radziszewski, W. J. Horton, D. M. Grant, J. Michl, *J. Am. Chem. Soc.*, **1996**, *118*, 846–852.
21. R. Warmuth, *Angew. Chem.*, **1997**, *109*, 1406–1409; *Angew. Chem. Int. Ed. Engl.*, **1997**, *36*, 1347–1350.
22. H. Jiao, P. v. R. Schleyer, B. R. Beno, K. N. Houk, R. Warmuth, *Angew. Chem.*, **1997**, *109*, 2929–2933; *Angew. Chem. Int. Ed. Engl.*, **1997**, *36*, 2761–2764.
23. R. Warmuth, *J. Chem. Soc. Chem. Commun.*, **1998**, 59–60.
24. A. C. Scheiner, H. F. Schaefer III, *Chem. Phys. Lett.*, **1991**, *177*, 471–476 and references cited therein.
25. R. Lindh, M. Schütz, *Chem. Phys. Lett.*, **1996**, *258*, 409–415.

26. I. P. Fischer, F. P. Lossing, *J. Am. Chem. Soc.*, **1963**, *85*, 1018–1019.
27. R. S. Berry, J. Clardy, M. E. Schafer, *Tetrahedron Lett.*, **1965**, *15*, 1003–1010; R. S. Berry, J. Clardy, M. E. Schafer, *Tetrahedron Lett.*, **1965**, *15*, 1011–1017; *cf.* C. F. Logan, P. Chen, *J. Am. Chem. Soc.*, **1996**, *118*, 2113–2114.
28. H. E. Bertorello, R. A. Rossi, R. Hoyos de Rossi, *J. Org. Chem.*, **1970**, *35*, 3332–3338.
29. F. Gaviña, S. V. Luis, V. S. Safont, P. Ferrer, A. M. Costero, *Tetrahedron Lett.*, **1986**, *27*, 4779–4782; S. V. Luis, F. Cavina, P. Ferrer, V. S. Safont, M. C. Torres, M. I. Burguete, *Tetrahedron*, **1989**, *45*, 6281–6296.
30. W. E. Billups, J. D. Buynak, D. Butler, *J. Org. Chem.*, **1978**, *44*, 4218–4219.
31. W. N. Washburn, R. Zahler, *J. Am. Chem. Soc.*, **1976**, *98*, 7828–7830 and references cited therein.
32. R. R. Jones, R. G. Bergman, *J. Am. Chem. Soc.*, **1972**, *94*, 660–661. Review: R. G. Bergman, *Acc. Chem. Res.*, **1973**, *6*, 25–31.
33. R. Breslow, J. Napierski, T. C. Clarke, *J. Am. Chem. Soc.*, **1975**, *97*, 6275–6276.
34. P. G. Wenthold, R. R. Squires, W. C. Lineberger, *J. Am. Chem. Soc.*, **1998**, *120*, 5279–5290; *cf.* P. G. Wenthold, J. Hu, R. R. Squires, *J. Am. Chem. Soc.*, **1996**, *118*, 11865–11871 and references cited therein; J. J. Nash, R. R. Squires, *J. Am. Chem. Soc.*, **1996**, *118*, 11872–11883.
35. W. R. Roth, H. Hopf, C. Horn, *Chem. Ber.*, **1994**, *127*, 1765–1779; *cf.* W. R. Roth, O. Adamczak, R. Breuckmann, H.-W. Lennartz, R. Boese, *Chem. Ber.*, **1991**, *124*, 2499–2521.
36. Matrix-isolated radical pairs of this type separated by carbon dioxide molecules have been reported previously: J. Pacansky, W. Koch, M. D. Miller, *J. Am. Chem. Soc.*, **1991**, *113*, 317–328 and previous papers in this series; J. G. Radziszewski, M. R. Nimlos, P. R. Winter, G. G. Ellison, *J. Am. Chem. Soc.*, **1996**, *118*, 7400–7401.
37. Review: W. Sander, *Acc. Chem. Res.*, **1999**, *32*, 669–676 R. Marquardt, W. Sander, E. Kraka, *Angew. Chem.*, **1996**, *108*, 825–827; *Angew. Chem. Int. Ed. Engl.*, **1996**, *35*, 746–748.
38. W. Sander, G. Bucher, H. Wandel, E. Kraka, D. Cremer, W. S. Sheldrick, *J. Am. Chem. Soc.*, **1997**, *119*, 10660–10672; *cf.* E. Kraka, D. Cremer, G. Bucher, H. Wandel, W. Sander, *Chem. Phys. Lett.*, **1997**, *268*, 313–320.
39. W. Sander, R. Marquardt, G. Bucher, H. Wandel, *Pure Appl. Chem,*.**1996**, *68*, 353–356.
40. G. Bucher, W. Sander, E. Kraka, D. Cremer, *Angew. Chem.*, **1992**, *104*, 1225–1228; *Angew. Chem. Int. Ed. Engl.*, **1992**, *31*, 1230–1233.
41. K. C. Nicolaou, A. L. Smith in *Modern Acetylene Chemistry* P. J. Stang, F. Diederich (*Eds.*), VCH Verlagsgesellschaft, Weinheim, **1995**, 203–283.
42. R. Marquardt, W. Sander, T. Laue, H. Hopf, *Liebigs Ann. Chem.*, **1995**, 1643–1648.
43. R. Marquard, A. Balster, W. Sander, E. Kraka, D. Cremer, J. G. Radziszewski, *Angew. Chem.*, **1998**, *110*, 1001–1005; *Angew. Chem. Int. Ed. Engl.*, **1998**, *37*, 955–958.
44. E. Kraka, D. Cremer, *Chem. Phys. Lett.*, **1993**, *216*, 333–340.
45. O. L. Chapman, C. C. Chang, J. Kolc, *J. Am. Chem. Soc.*, **1976**, *98*, 5703–5705.
46. M. J. Schottelius, P. Chen, *J. Am. Chem. Soc.*, **1996**, *118*, 4896–4903.
47. C. Kötting, W. Sander, S. Kammermeier, R. Herges, *Eur. J. Org. Chem.*, **1998**, 799–803; *cf.* H.-H. Wenk, W. Sander, *Eur. J. Org. Chem.*, **1999**, 57–60.
48. Reviews: M. A. Bennett, H. P. Schwemlein, *Angew. Chem.*, **1989**, *101*, 1349–1373; *Angew. Chem. Int. Ed. Engl.*, **1989**, *28*, 1296–1320; S. L. Buchwald, R. B. Nielsen, *Chem. Rev.*, **1988**, *88*, 1047–1058; S. L. Buchwald, R. A. Fisher, *Chem. Scr.*, **1989**, *29*, 417; *cf.* D. Peña, S. Escudero, D. Pérez, E. Guitián, L. Castedo, *Angew. Chem.*, **1998**, *110*, 2804–2806; *Angew. Chem. Int. Ed. Engl.*, **1998**, *37*, 2659–2661.
49. The term was introduced by M. J. S. Dewar in *The Molecular Orbital Theory of Organic*

Chemistry, McGraw Hill, New York, N. Y., **1969**, *p.* 232, the concept, however, is much older, see ref.[50].

50.   H. C. Longuet-Higgins, *J. Chem. Phys.*, **1950**, *18*, 265–274.
51.   C. Kollmar, O. Kahn, *Acc. Chem. Res.*, **1993**, *26*, 259–265.
52.   A. Rajca, *Chem. Rev.*, **1994**, *94*, 871–893.
53.   For reviews on non-Kekulé hydrocarbons see: W. Th. Borden (*Ed.*), *Diradicals*, J. Wiley and Sons, New York, N. Y., **1982**; J. A. Berson, *Acc. Chem. Res.*, **1978**, *11*, 446–453; W. Th. Borden, E. R. Davidson, *Acc. Chem. Res.*, **1981**, *14*, 69–76; W. Adam, S. Grabowski, R. M. Wilson, *Acc. Chem. Res.*, **1990**, *23*, 165–172; D. A. Dougherty, *Acc. Chem. Res.*, **1991**, *24*, 88–94; H. Iwamura, N. Koga, *Acc. Chem. Res.*, **1993**, *26*, 346–351; W. Th. Borden, H. Iwamura, J. A. Berson, *Acc. Chem. Res.*, **1994**, *27*, 109–116; J. A. Berson, *Acc. Chem. Res.*, **1997**, *30*, 238–244; D. A. Hrovat, W. Th. Borden. *J. Mol. Struct. (Theochem.)*, **1997**, *398–399*, 211–220; C. J. Cramer, *J. Chem. Soc. Perkin Trans. 2*, **1998**, 1007–1013.
54.   M. G. Ettlinger, *J. Am. Chem. Soc.*, **1952**, *74*, 5805–5806; E. F. Ullmann, *J. Am. Chem. Soc.*, **1960**, *82*, 505–506; J. P. Chesick, *J. Am. Chem. Soc.*, **1963**, *85*, 2720–2723.
55.   P. Dowd, *J. Am. Chem. Soc.,* **1966**, *88*, 2587–2589.
56.   P. Dowd, K. Sachdev, *J. Am. Chem. Soc.*, **1967**, *89*, 715–716.
57.   K. Takeda, H. Yoshida, K. Hayashi, S. Okamura, *Bull. Inst. Chem. Res. Kyoto Univ.*, **1967**, *45*, 55–62; *Chem. Abstr.*, **1967**, *67*, 77761 q.
58.   O. Claesson, A. Lund, T. Gillbro, T. Ichikawa, O. Edlund, H. Yoshida, *J. Chem. Phys.*, **1980**, *72*, 1463–1470. According to more recent calculations the singlet/triplet energy gap amounts to 15.3 kcal mol^{-1}: R. Janoschek, *Chem. Unserer Zeit*, **1991**, *25*, 59–66; *cf.* also ref.[59], footnote 16.
59.   G. Maier, H. P. Reisenauer, K. Lanz, R. Tross, D. Jürgen, B. A. Hess, Jr, L. J. Schaad, *Angew. Chem.*, **1993**, *105*, 119–121; *Angew. Chem. Int. Ed. Engl.*, **1993**, *32*, 74–76.
60.   G. Maier, D. Jürgen, R. Tross H. P. Reisenauer, B. A. Hess, Jr, L. S. Schaad, *Chem. Phys.,* **1994**, *189*, 383–399; *cf.* G. Maier, H. P. Reisemaner, T. Preiss, H. Pacl, D. Jürgen, R. Tross, St. Senger, *Pure Appl. Chem.*, **1997**, *69*, 113–118.
61.   G. Maier, St. Senger, *J. Am. Chem. Soc.*, **1999**, *121*, in press.
62.   T. C. Shields, W. E. Billups, A. R. Lepley, *J. Am. Chem. Soc.*, **1968**, *90*, 4749–4751; *cf.* W. E. Billups, K. H. Leavell, E. S. Lewis, S. Vanderpool, *J. Am. Chem. Soc.*, **1973**, *95*, 8096–8102.
63.   G. Maier, St. Senger, *Eur. J. Org. Chem.*, **1999**, 1291–1294.
64.   J. D. Roberts, C. M. Sharts, *Org. Reactions*, **1962**, *12*, 1–56.
65.   H. Hopf in *The Allenes* S. R. Landor, (*Ed.*), Academic Press, London, **1982**, Vol. 2, 525–562.
66.   J. J. Gajewski, *Hydrocarbon Thermal Isomerizations*, Academic Press, New York, N. Y., **1981**.
67.   J. J. Gajewski, C. N. Shih, *J. Am. Chem. Soc.*, **1967**, *89*, 4532–4533.
68.   P. Dowd, *J. Am. Chem. Soc.*, **1970**, *92*, 1066–1068.
69.   P. Dowd, W. Chang, Y. H. Paik, *J. Am. Chem. Soc.*, **1986**, *108*, 7416–7417.
70.   A. A. Ovchinnikov, *Theor. Chim. Acta*, **1978**, *47*, 297–304.
71.   W. T. Borden, E. R. Davidson, *J. Am. Chem. Soc.*, **1977**, *99*, 4587–4594; P. Du, W. T. Borden, *J. Am. Chem. Soc.*, **1987**, *109*, 930–931; P. Nachtigall, K. D. Jordan, *J. Am. Chem. Soc.*, **1992**, *114*, 4743–4747; P. Nachtigall, K. D. Jordan, *J. Am. Chem. Soc.*, **1993**, *115*, 270–271.
72.   J. J. Nash, P. Dowd, K. D. Jordan, *J. Am. Chem. Soc.*, **1992**, *114*, 10071–10072; *cf.* W. R. Roth, U. Kowalczik, G. Maier, H. P. Reisenauer, R. Sustmann, P. Müller, *Angew. Chem.*, **1987**, *99*, 1330–1331; *Angew. Chem. Int. Ed. Engl.*, **1987**, *26*, 1285–1286.

73. P. Dowd, W. Chang, Y. H. Paik, *J. Am. Chem. Soc.*, **1987**, *109*, 5284–5285.

74. M. M. Greenberg, S. C. Blackstock, J. A. Berson, R. A. Merrill, J. C. Duchamp, K. W. Zilm, *J. Am. Chem. Soc.*, **1991**, *113*, 2318–2319.

75. P. Du, D. A. Hrovat, W. T. Borden, P. M. Lahti, A. R. Rossi, J. A. Berson, *J. Am. Chem. Soc.*, **1986**, *108*, 5072–5074.

76. W. R. Roth, R. Langer, M. Bartmann, B. Stevermann, G. Maier, H. P. Reisenauer, R. Sustmann, W. Müller, *Angew. Chem.*, **1987**, *99*, 271–272; *Angew. Chem. Int. Ed. Engl.*, **1987**, *26*, 256–257; *cf.* W. R. Roth, R. Langer, T. Ebbrecht, A. Beitat, H.-W. Lennartz, *Chem. Ber.*, **1991**, *124*, 2751–2760.

77. J. H. Reynolds, J. A. Berson, K. K. Kumashiro, J. C. Duchamp, K. W. Zilm, J. C. Scaiano, A. B. Berinstain, A. Rubello, P. Vogel, *J. Am. Chem. Soc.*, **1993**, *115*, 8073–8090 and previous publications.

78. Violation of Hund´s rule have also been found experimentally in dicarbenes, dinitrenes, and dinitroxides, see ref.[53] and K. Itoh, *Pure Appl. Chem.*, **1978**, *50*, 1251–1259; *cf.* Y. Teki, T. Takui, M. Kitano, K. Itoh, *Chem. Phys. Lett.*, **1987**, *142*, 181–186.

79. Review: J. A. Berson in *The Chemistry of Quinonoid Compounds* S. Patai, Z. Rappoport (*Eds.*), J. Wiley and Sons, Chichester, **1988**, Vol. 2, part 1, 455–536.

80. E. Migirdicyan, *C. R. Hebd. Séances Acad. Sci.*, **1968**, *266*, 756–759; *cf.* W. Th. Borden, E. R. Davidson, *J. Am. Chem. Soc.*, **1977**, *99*, 4587–4594; V. Lejeune, A. Despres, E. Migirdicyan, J. Baudet, G. Berthier, *J. Am. Chem. Soc.*, **1986**, *108*, 1853–1860.

81. The bis carbene **75** was first generated from **74** by A. M. Trozzolo, R. W. Murray, G. Smolinsky, W. A. Yager, E. Wasserman, *J. Am. Chem. Soc.*, **1963**, *85*, 2526–2527.

82. B. B. Wright, M. S. Platz, *J. Am. Chem. Soc.*, **1983**, *105*, 628–630; *cf.* M. S. Platz, *J. Am. Chem. Soc.*, **1979**, *101*, 3398–3399.

83. J. L. Goodman, J. A. Berson, *J. Am. Chem. Soc.*, **1984**, *106*, 1867–1868.

84. J. L. Goodman, J. A. Berson, *J. Am. Chem. Soc.*, **1985**, *107*, 5409–5424.

85. J. L. Goodman, J. A. Berson, *J. Am. Chem. Soc.*, **1985**, *107*, 5424–5428.

86. S. L. Buchwalter, G. L. Closs, *J. Am. Chem. Soc.*, **1975**, *97*, 3857–3858; J.-F. Muller, D. Muller, H. J. Dewey, J. Michl, *J. Am. Chem. Soc.*, **1978**, *100*, 1629–1630; S. L. Buchwalter, G. L. Closs, *J. Am. Chem. Soc.*, **1979**, *101*, 4688–4694; R. Jain, G. J. Snyder, *J. Am. Chem. Soc.*, **1984**, *106*, 7294–7295; J. J. Fisher, J. Michl, *J. Am. Chem. Soc.*, **1987**, *109*, 583–584; K. Haider, M. S. Platz, A. Despres, V. Lejeune, E. Migirdicyan, T. Bally, E. Haselbach, *J. Am. Chem. Soc.*, **1988**, *110*, 2318–2320; R. Jain, M. B. Sponsler, F. D. Coms, D. A. Dougherty, *J. Am. Chem. Soc.*, **1988**, *110*, 1356–1366; M. C. Biewer, C. R. Biehn, M. S. Platz, A. Despres, E. Migirdicyan, *J. Am. Chem. Soc.*, **1991**, *113*, 616–620.

87. W. Schlenk, M. Brauns, *Ber. Dtsch. Chem. Ges.*, **1915**, *48*, 661–669; W. Schlenk, M. Brauns, *Ber. Dtsch. Chem. Ges.*, **1915**, *48*, 716–728; *cf.* V. D. Sholle, E. G. Rozantsev, *Russ. Chem. Rev.*, **1973**, *42*, 1011–1019; M. Ballester, *Acc. Chem. Res.*, **1985**, *18*, 380–387; N. A. Porter, D. J. Hogenkamp, F. F. Khouri, *J. Am. Chem. Soc.*, **1990**, *112*, 2402–2407.

88. G. Allinson, R. J. Bushby, J.-L. Pailland, D. Oduwole, K. Sales, *J. Am. Chem. Soc.*, **1993**, *115*, 2062–2064; *cf.* for the sybthesis of 2,5,8-tri-*tert*-butyl-phenalenyl see K. Goto, T. Kubo, K. Yamamoto, K. Makasuji, K. Sato, D. Shiomi, T. Takui, M. Kubota, T. Kabayashi, K. Yakusi, J. Ouyang, *J. Am. Chem. Soc.*, **1999**, *121*, 1619–1620.

89. Reviews: R. D. Little, *Chem. Rev.*, **1986**, *86*, 875–884; A. K. Allan, G. L. Caroll, R. D. Little, *Eur. J. Org. Chem.*, **1998**, 1–12.

90. L. Van Hijfte, R. D. Little, *J. Org. Chem.*, **1985**, *50*, 3940–3942; *cf.* L. Van Hijfte, R. D. Little, J. L. Petersen, K. D. Moeller, *J. Org. Chem.*, **1987**, *52*, 4647–4661.

91. W. H. Mills, I. G. Nixon, *J. Chem. Soc.*, **1930**, 2510–2524.

92. For recent reviews on the Mills-Nixon effect see: J. S. Siegel, *Angew. Chem.*, **1994**, *106*,

1808–1810; *Angew. Chem. Int. Ed. Engl.*, **1994**, *33*, 1721–1723; N. L. Frank, J. S. Siegel in *Advances in Theoretically Interesting Molecules* R. P. Thummel, (*Ed.*), The JAI Press, Inc., **1995**, Vol. 3, 209–260; K. K. Baldridge, J. S. Siegel, *Angew. Chem.*, **1997**, *109*, 765–768; *Angew. Chem. Int. Ed. Engl.*, **1997**, *36*, 745–748; *cf.* A. Stanger, *J. Am. Chem. Soc.*, **1998**, *120*, 12034–12040.

93. E. Vogel, W. Grimme, S. Korte, *Tetrahedron Lett.*, **1965**, 3625–3631. Apparently benzocyclopropene derivatives were first discussed by W. H. Perkin, Jr, *J. Chem. Soc.*, **1888**, 1–20.

94. E. Vogel. S. Korte, W. Grimme, H. Günther, *Angew. Chem.*, **1968**, *80*, 279–280; *Angew. Chem. Int. Ed. Engl.*, **1968**, *7*, 289–290.

95. S. Tanimoto, R. Schäfer, J. Ippen, E. Vogel, *Angew. Chem.*, **1976**, *88*, 643–644; *Angew. Chem. Int. Ed. Engl.*, **1976**, *15*, 613–614.

96. B. Halton, B. R. Dent, S. Böhm, D. L. Officer, H. Schmickler, F. Schophoff, E. Vogel, *J. Am. Chem. Soc.*, **1985**, *107*, 7175–7176.

97. W. E. Billups, A. J. Blakeney, W. Y. Chow, *J. Chem. Soc. Chem. Commun.*, **1971**, 1461–1462; *cf.* W. E. Billups, A. J. Blakeney, W. Y. Chow, *Org. Synth.*, **1976**, *55*, 12–15; R. Okazaki, M. O-oka, N. Tokitoh, N. Inamoto, *J. Org. Chem.*, **1985**, *50*, 180–185.

98. For the details of the mechanism of this process which involves the intermediate formation of chlorocyclopropene derivatives see J. Prestien, H. Günther, *Angew. Chem.*, **1974**, *86*, 278–279; *Angew. Chem. Int. Ed. Engl.*, **1974**, *13*, 276–277.

99. L. K. Bee, P. J. Garratt, M. M. Mansuri, *J. Am. Chem. Soc.*, **1980**, *102*, 7076–7079.

100. W. E. Billups, W. Y. Chow, *J. Am. Chem. Soc.*, **1973**, *95*, 4099–4100; *cf.* F. A. Cotton, J. M. Troup, W. E. Billups, L. P. Liu, C. V. Smith, *J. Organomet. Chem.*, **1975**, *102*, 345–351.

101. J. Ippen, E. Vogel, *Angew. Chem.*, **1974**, *86*, 780–781; *Angew. Chem. Int. Ed. Engl.*, **1974**, *13*, 736; *cf.* B. Halton, R. Boese, D. Bläser, Q. Lu, *Austr. J. Chem.*, **1991**, *44*, 265–276.

102. D. Davalian, P. J. Garratt, W. Koller, M. M. Mansuri, *J. Org. Chem.*, **1980**, *45*, 4183–4193; *cf.* D. Davalian, P. J. Garratt, *J. Am. Chem. Soc.*, **1975**, *97*, 6883–6884.

103. D. Davalian, P. J. Garratt, M. M. Mansuri, *J. Am. Chem. Soc.*, **1978**, *100*, 980–981.

104. D. Davalian, P. J. Garratt, *Tetrahedron Lett.*, **1976**, 2815–2818.

105. W. E. Billups, E. W. Casserly, B. E. Arney, *J. Am. Chem. Soc.*, **1984**, *106*, 440–441.

106. W. E. Billups, M. M. Haley, R. C. Claussen, W. A. Rodin, *J. Am. Chem. Soc.*, **1991**, *113*, 4331–4332.

107. W. E. Billups, D. J. McCord, B. R. Maughon, *J. Am. Chem. Soc.*, **1994**, *116*, 8831–8832.

108. M. Bartmann, Ph. d. thesis, University of Bochum, **1980**; *cf.* P. Müller, D. Rodriguez, *Helv. Chim. Acta*, **1983**, *66*, 2540–2542.

109. P. Müller, J.-P. Schaller, *Helv. Chim. Acta*, **1989**, *72*, 1608–1617; *cf.* P. Müller, J.-P. Schaller, *Chimia*, **1986**, *40*, 430–431.

110. J. A. Barltrop, A. J. Johnson, D. D. Meakins, *J. Chem. Soc.*, **1951**, 181–185; *cf.* G. L. Buchanan, J. K. Sutherland, *J. Chem. Soc.*, **1956**, 2620–2628.

111. P. O. Schissel, M. E. Kent, D. J. McAdoo, E. Hedaya, *J. Am. Chem. Soc.*, **1970**, *92*, 2147–2149; E. Hedaya, *Acc. Chem. Res.*, **1969**, *2*, 367–373.

112. D. R. Arnold, E. Hedaya, V. Y. Merritt, L. A. Karnischky, M. E. Kent, *Tetrahedron Lett.*, **1972**, 3917–3920.

113. E. Hedaya, M. E. Kent, *J. Am. Chem. Soc.*, **1970**, *92*, 2149–2151; *cf.* M. Tokuda, L. L. Miller, A. Szabo, H. Suhr, *J. Org. Chem.*, **1979**, *44*, 4504–4508.

114. R. Anet, F. A. L. Anet, *J. Am. Chem. Soc.*, **1964**, *86*, 525–526.

115. P. Radlick, M. T. Crawford, *J. Chem. Soc. Chem. Commun.*, **1974**, 127; *cf.* A. T. Nichols, P. J. Stang, *Synlett*, **1992**, 971–972 for a modern, higher-yielding variant of this process.

116. C. J. Saward, K. P. C. Vollhardt, *Tetrahedron Lett.*, **1975**, 4539–4542.

117. Reviews on cyclopropanes: W. E. Billups, *Acc. Chem. Res.*, **1978**, *11*, 245–251; Y. Apeloig, M. Karni, D. Arad in *Strain and its Implications in Organic Chemistry*, A. de Meijere, S. Blechert (*Eds.*), Kluwer Academic Publishers, Dordrecht, **1989**, 457–462; B. Halton, *Chem. Rev.*, **1989**, *89*, 1161–1185; P. Müller in *Advances in Theoretically Interesting Molecules* R. P. Thummel, (*Ed.*), The JAI Press, Inc., Vol. 3, **1995**, 37–107; P. Müller in Houben-Weyl, *Methods of Organic Chemistry*, Thieme Verlag, Stuttgart, **1997**, Vol. E17d, 2866–2914.

118. M. P. Cava, D. R. Napier, *J. Am. Chem. Soc.*, **1956**, *78*, 500; *cf.* M. P. Cava, D. R. Napier, *J. Am. Chem. Soc*, **1957**, *79*, 1701–1705.

119. H. Finkelstein, *Ber. Dtsch. Chem. Ges.*, **1910**, *43*, 1528–1532; *cf.* H. Finkelstein, Ph. d. dissertation, Univ. Straßburg, **1910**.

120. M. P. Cava, D. R. Napier, *J. Am. Chem. Soc.*, **1958**, *80*, 2255–2257.

121. M. P. Cava, A. A. Deana, *J. Am. Chem. Soc.*, **1959**, *81*, 4266–4268; *cf.* J. A. Oliver, P. A. Ongley, *Chem. Ind.* (*London*), **1965**, 1024–1025; R. A. Aitken, P. K. G. Hodgson, A. O. Oyewale, J. J. Morrison, *J. Chem. Soc. Chem. Commun.*, **1997**, 1163–1164.

122. An early synthesis of **137** not starting from a o-xylene precursor has been described by A. P. ter Borg, A. F. Bickel, *Proc. Chem. Soc.*, **1958**, 283, who obtained the benzocyclobutene skeleton by thermal isomerization of the dichlorocarbene adduct of cycloheptatriene.

123. P. Schiess, M. Heitzmann, S. Rutschmann, R. Stäheli, *Tetrahedron Lett.*, **1978**, 4569–4572; P. Schiess, S. Rutschmann, V. V. Toan, *Tetrahedron Lett.*, **1982**, *23*, 3665–3668; K. A. Walker, L. J. Markoski, J. S. Moore, *Synthesis*, **1992**, 1265–1268.

124. E. Cuthbertson, D. D. MacNicol, *Tetrahedron Lett.*, **1975**, 1893–1894.

125. R. J. Spangler, B. G. Beckmann, *Tetrahedron Lett.*, **1976**, 2517–2518 and references cited therein.

126. H. Higuchi, Y. Sakata, S. Misumi, T. Otsubo, F. Ogura, H. Yamaguchi, *Chem. Letters*, **1981**, 627–630.

127. L. Horner, W. Kirmse, K. Muth, *Chem. Ber.*, **1958**, *91*, 430–437.

128. M. R. Berman, P. B. Comita, C. B. Moore, R. G. Bergman, *J. Am. Chem. Soc.*, **1980**, *102*, 5692–5694; W. S. Trahanovsky, K. E. Swenson, *J. Org. Chem.*, **1981**, *46*, 2984–2985; W. Tsang, J. P. Cui, *J. Am. Chem. Soc.*, **1990**, *112*, 1665–1671.

129. W. S. Trahanovsky, P. W. Mullen, *J. Am. Chem. Soc.*, **1972**, *94*, 5911–5913; *cf.* J. M. Riemann, W. S. Trahanovsky, *Tetrahedron Lett.*, **1977**, 1863–1866; J. M. Riemann, W. S. Trahanovsky, *Tetrahedron Lett.*, **1977**, 1867–1870.

130. H. Meier, N. Hanold, H. Kolshorn, *Angew. Chem.*, **1982**, *94*, 67–68; *Angew. Chem. Int. Ed. Engl.*, **1982**, *31*, 66; *cf.* N. Hanold, H. Meier, *Chem. Ber.*, **1985**, *118*, 198–209.

131. O. L. Chapman, R. J. McMahon, P. R. West, *J. Am. Chem. Soc.*, **1984**, *106*, 7973–7974; P. P. Gaspar, J.-P. Hsu, S. Chari, M. Jones, Jr, *Tetrahedron*, **1985**, *41*, 1479–1507; M. Rahman, P. B. Shevlin, *Tetrahedron Lett.*, **1985**, *26*, 2959–2960; O. L. Chapman, U.-P. E. Tsou, J. W. Johnson, *J. Am. Chem. Soc.*, **1987**, *109*, 553–559; O. L. Chapman, J. W. Johnson, R. J. McMahon, P. R. West, *J. Am. Chem Soc.*, **1988**, *110*, 501–509; W. Kirmse, W. Konrad, D. Schnitzler, *J. Org. Chem.*, **1994**, *59*, 3821–3829.

132. M. P. Cava, A. A. Deana, K. Muth, *J. Am. Chem. Soc.*, **1960**, *82*, 2524–2525; *cf.* H. Rapoport, G. Smolinski, *J. Am. Chem. Soc.*, **1960**, *82*, 1171–1180.

133. The first derivative of **147** was prepared by a photo Wolff ring contraction: L. Horner, K. Muth, H.-G. Schmelzer, *Chem. Ber.*, **1959**, *92*, 2953–2957.

134. R. Gray, L. G. Harruff, J. Krymowski, J. Peterson, V. Boekelheide, *J. Am. Chem. Soc.*, **1978**, *100*, 2892–2893.

135. Gas-phase pyrolysis of **147** in a nitrogen stream at 425 °C gives [2₄](1,2,4,5)cyclophane (see Chapter 12.3), 14,20-dimethyl[2₅](1,2,4,5)(1,3,4)(1,3,4)-cyclophan-17-ene and

[2₆](1,2,4,5)cyclophane, the so-called deltaphane, a hydrocarbon of interest in view of its cavity which is capable of binding metal ions: H. C. Kang, A. W. Hanson, B. Eaton, V. Boekelheide, *J. Am. Chem. Soc.*, **1985**, *107*, 1979–1985.

136. C. K. Bradsher, D. A. Hunt, *J. Org. Chem.*, **1981**, *46*, 4606–4610.
137. E. Giovannini, H. Vuilleumier, *Helv. Chim. Acta*, **1977**, *60*, 1452–1455.
138. R. P. Thummel, *J. Am. Chem. Soc.*, **1976**, *98*, 628–629; *cf.* R. P. Thummel, W. Nutakul, *J. Org. Chem.*, **1977**, *42*, 300–305.
139. A. J. Barkovich, E. S. Strauss, K. P. C. Vollhardt, *J. Am. Chem. Soc.*, **1977**, *99*, 8321–8322; *cf.* A. J. Barkovich, K. P. C. Vollhardt, *J. Am. Chem. Soc.*, **1976**, *98*, 2667–2668.
140. W. Nutakul, R. P. Thummel, A. D. Taggart, *J. Am. Chem. Soc.*, **1979**, *101*, 770–771. For the determination of the molecular structure of tris(tricarbonylironcyclobutadieno)benzene see A. Stanges, N. Ashkenazi, R. Boese, *J. Org. Chem.*, **1998**, *63*, 247–253.
141. F. Heinrich, W. Lüttke, *Liebigs Ann. Chem.*, **1978**, 1880–1886.
142. C. W. Doecke, P. J. Garratt, H. Shahriari-Zavareh, R. Zahler, *J. Org. Chem.*, **1984**, *49*, 1412–1417. For the synthesis of a hexabromotricyclobutabenzene derivative see A. Stanger, N. Ashkenazi, R. Boese, D. Bläser, P. Stellberg, *Chem. Eur. J.*, **1997**, *3*, 208–211.
143. For the synthesis of various polycyclic biphenylenes structurally related to the hydrocarbons discussed in this chapter see: J. W. Barton, M. K. Shephard, R. J. Willis, *J. Chem. Soc. Perkin Trans. 1*, **1986**, 967–971.
144. Review: I. L. Klundt, *Chem. Rev.*, **1970**, *70*, 471–487.
145. T. A. Christopher, R. H. Levin, *Tetrahedron Lett.*, **1976**, 4111–4114; M. A. O'Leary, M. B. Stringer, D. Wege, *Austr. J. Chem.*, **1978**, *31*, 2003–2012; O. Abou-Teim, M. C. Goodland, J. F. W. McOmie, *J. Chem. Soc. Perkin Trans I*, **1983**, 2659–2662;
146. J. H. Markgraf, S. J. Basta, P. M. Wedge, *J. Org. Chem.*, **1972**, *37*, 2361–2363; for an intramolecular cycloaddition involving a dehydrobenzene intermediate see J. J. Bunnett, J. A. Skorzc, *J. Org. Chem.*, **1962**, *27*, 3836–3843.
147. R. P. Thummel, W. E. Cravey, W. Nutakul, *J. Org. Chem.*, **1978**, *43*, 2473–2477.
148. K. P. C. Vollhardt, *Acc. Chem. Res.*, **1977**, *10*, 1–8.
149. R. P. Thummel, *Acc. Chem. Res.*, **1980**, *13*, 70–76.
150. W. R. Roth, B. P. Scholz, *Chem. Ber.*, **1981**, *114*, 3741–3750; W. R. Roth, Th. Ebbrecht, A. Beitat, *Chem. Ber.*, **1988**, *121*, 1357–1358; *cf.* W. R. Roth, R. Langer, Th. Ebbrecht, A. Beitat, H.-W. Lennartz, *Chem. Ber.*, **1991**, *124*, 2751–2760. For the photochemistry of benzocyclobutene see N. J. Turro, Z. Zhang, W. S. Trahanovsky, C.-H. Chou, *Tetrahedron Lett.*, **1988**, *29*, 2543–2546.
151. Typical recent examples are described by J. Luo, H. Hart, *J. Org. Chem.*, **1987**, *52*, 4833–4836; M. Toda, K. Okada, M. Oda, *Tetrahedron Lett.*, **1988**, *29*, 2329–2332; S. V. D'Andrea, J. P. Freeman, J. Szmuszkovicz, *J. Org. Chem.*, **1990**, *55*, 4356–4358; A. Gügel, A. Kraus, J. Spickermann, P. Belik, K. Müllen, *Angew. Chem.*, **1994**, *106*, 601–603; *Angew. Chem. Int. Ed. Engl.*, **1994**, *33*, 559–560.
152. For the photoelectron spectra of the above small-ring fused aromatic hydrocarbons see C. Santiago, R. W. Gandour, K. N. Houk, W. Nutakul, W. E. Cravey, R. P. Thummel. *J. Am. Chem. Soc.*, **1978**, *100*, 3730–3737; E. Heilbronner, B. Kovac, W. Nutakul, A. D. Taggart, R. P. Thummel, *J. Org. Chem.*, **1981**, *46*, 5279–5284; for their ^{1}H and ^{13}C NMR spectra see R. P. Thummel, W. Nutakul, *J. Org. Chem.*, **1978**, *43*, 3170–3173.
153. R. Neidlein, D. Christen, V. Pognée, R. Boese, D. Bläser, A. Gieren, C. Ruiz-Pérez, T. Hübner, *Angew. Chem.*, **1988**, *100*, 292–293; *Angew. Chem. Int. Ed. Engl.*, **1988**, *27*, 294–295.
154. R. Boese, D. Bläser, *Angew. Chem.*, **1988**, *100*, 293–295; *Angew. Chem. Int. Ed. Engl.*, **1988**, *27*, 304–305. For an earlier determination of the structure of **147** by X-ray structural analysis see J. Lawrence, S. G. G. MacDonald, *Acta Crystallogr. Sect. B*, **1969**, *B35*,

978–981.

155. R. Boese, D. Bläser, W. E. Billups, M. M. Haley, A. H. Maulitz, D. L. Mohler, K. P. C. Vollhardt, *Angew. Chem.*, **1994**, *106*, 321–325; *Angew. Chem. Int. Ed. Engl.*, **1994**, *33*, 313–317. For a review on the structural parameters of small-ring fused aromatics see R. Boese in *Advances in Strain in Organic Chemistry* B. Halton, (*Ed.*), The JAI Press, Greenwich, CT, **1992**, Vol. 2, 191–254.

156. H. Hart, S. Shamouilian, Y. Takehira, *J. Org. Chem.*, **1981**, *46*, 4427–4432.

157. The crystal structure of **172** was determined as a 1:1-complex with chlorobenzene: P. Venugopalan, H.-B. Bürgi, N. L. Frank, K. K. Baldridge, J. S. Siegel, *Tetrahedron Lett.*, **1995**, *36*, 2419–2422.

158. K. Komatsu, Y. Jinbu, G. R. Gillette, R. West, *Chem. Lett.*, **1988**, 2029–2032; *cf.* K. Komatsu, Sh. Aonuma, Y. Jinbu, R. Tsuji, Ch. Hirosawa, K. Takeuchi, *J. Org. Chem.*, **1991**, *56*, 195–203. Hydrocarbon **173** is also interesting from the chemical viewpoint since it can be converted to the corresponding tropylium ion, an all-hydrocarbon carbocation of extraordinary stability: K. Komatsu, H. Akamatsu, Y. Jinbu, K. Okamoto, *J. Am. Chem. Soc.*, **1988**, *110*, 633–634; *cf.* K. Komatsu, *Pure Appl. Chem.*, **1993**, *65*, 73–80; K. Komatsu, *Eur. J. Org. Chem.*, **1999**, 1495–1502.

159. The first bicyclically annelated hydrocarbon apparently was trisbicyclo[2.2.1]heptabenzene prepared by trimerization of bicyclo[2.2.1]hept-2-yne (norbornyne, see Chapter 8.2) by P. G. Gassman, I. Gennick, *J. Am. Chem. Soc.*, **1980**, *102*, 6863–6864.

160. N. L. Frank, K. K. Baldridge, P. Gantzel, J. S. Siegel, *Tetrahedron Lett.*, **1995**, *36*, 4389–4392.

161. N. L. Frank, K. K. Baldridge, J. S. Siegel, *J. Am. Chem. Soc.*, **1995**, *117*, 2102–2103.

162. H.-B. Bürgi, K. K. Baldridge, K. Hardcastle, N. L. Frank, P. Gantzel, J. S. Siegel, J. Ziller, *Angew. Chem.*, **1995**, *107*, 1575–1577; *Angew. Chem. Int. Ed. Engl.*, **1995**, *34*, 1454–1456.

163. R. Dierks, K. P. C. Vollhardt, *J. Am. Chem. Soc.*, **1986**, *108*, 3150–3152; *cf.* D. L. Mohler, K. P. C. Vollhardt, St. Wolff, *Angew. Chem.*, **1990**, *102*, 1200–1202; *Angew. Chem. Int. Ed. Engl.*, **1990**, *29*, 1151–1153.

164. K. K. Baldridge, J. S. Siegel, *J. Am. Chem. Soc.*, **1992**, *114*, 9583–9587.

165. G. Filippini, *J. Mol. Struct.*, **1985**, *130*, 117–124.

166. A. Kekulé, *Liebigs Ann. Chem.*, **1872**, *162*, 77–124.

Author Index

Alberts, A. H. 460
Alder, K. 87, 121
Allinger, N. L. 341
Anet, F. A. L. 506
Anet, R. 506
Applequist, D. E. 476
Arens, J. F. 178
Askani, R. 394

Bailey, W. J. 253, 423, 425, 426
Ball, W. J. 184
Bally, Th. 291
Barth, W. E. 331
Bartlett, P. D. 385
Bates, R. B. 347
Baughman, R. H. 470
Berson, J. A. 349, 498, 499
Bertelli, D. J. 286
Bickelhaupt, F. 341, 343, 344, 345, 349
Billups, W. E. 207, 261, 263, 407, 503, 504
Binger, P. 477
Blomquist, A. T. 158, 184, 253
Boekelheide, V. 235, 236, 352, 354, 355, 357, 507
Boese, R. 511
Bohlmann, F. 106, 151, 154
Borden, W. Th. 134
Böttger, O. 24
Brandsma, L. 178
Bredt, J. 128
Broene, R. D. 434
Brown, C. J. 351
Brown, M. 357

Cahn, R. S. 131
Capraro, H.-G. 33
Cava, M. P. 507
Chapman, O. 199
Chow, H. S. 359
Christl, M. 185, 206

Ciula, R. P. 474
Clar, E. 386, 423, 425, 426, 428, 431, 434, 438
Cole Jr, T. W. 60, 61
Conia, J. M. 301, 302, 304, 478
Cook, J. M. 91, 92, 400
Cope, A. C. 124, 210
Cram, D. J. 339, 352, 354, 356
Criegee, R. 198

Dauben, H. J. 286
De Lucchi, O. 382
de Meijere, A. 304, 305, 307, 401, 407, 477, 478
de Wolf, W. H. 341
Demyanov (Demjanoff), N. Y. 112, 173
Deslongchamps, P. 33, 399, 401
Dewar, M. J. S. 396
Diederich, F. 298, 429, 430, 434, 461
Doering, W. von E. 172, 266, 267, 271, 390, 391, 392
Dopper, J. H. 330
Dorko, E. A. 291
Dowd, P. 494, 496
Dürr, H. 408

Eaton, P. E. 46, 47, 60, 61, 65, 214
Eglinton, G. 226
El-Tamany, S. 358
Engelhard, M. 406
Erden, I. 301
Eugster, C. H. 111

Faraday, M. 197
Farthing, A. C. 351
Favorskii, A. E. 182
Fawcett, F. S. 353
Feldman, K. S. 458, 459
Finkelstein, H. 507
Fitjer, L. 301, 302, 306, 308, 477, 478

Ganter, C. 33
Garratt, P. J. 268, 511
Gieren, A. 511
Ginsburg, D. 82
Glaser, C. 152
Gleiter, R. 63, 161, 367, 410, 411
Goodman, J. L. 499
Grant, D. M. 487
Greene, F. D. 136
Griffin, G. W. 291, 293
Grubbs, R. H. 468
Grunewald, G. L. 383
Gustavson, G. 173, 476

Hafner, K. 265, 276, 279, 280, 281, 282
Hagen, St. 335
Haley, M. M. 207, 407, 470, 472
Hart, H. 388
Harvey, R. G. 437
Haselbach, E. 291
Hedaya, E. 271
Heinrich, F. 510
Hellmann, H. 205
Herges, R. 437, 438
Herwig, W. 449
Hine, J. 380
Hirano, S. 348
Hisatome, M. 367
Hoffmann, R. 396, 408, 410
Hollins, R. A. 356
Hopf, H. 110, 185, 293, 347, 354, 356, 358
Hopff, H. 297
Hückel, E. 202
Huisgen, R. 214, 339
Hunsmann, W. 152

Ingold, C. K. 131
Iyoda, M. 292, 296,

Jenneskens, L. W. 341
Jenny, W. 361
John, F. 423
Johnson, R. P. 139
Jones Jr, M. 342, 343
Jones, E. R. H. 155, 175

Karrer, P. 103, 111
Katz, T. J. 44, 45, 206, 268, 324
Keese, R. 90
Kekulé, A. 198, 203, 513
Kloster-Jensen, E. 178
Knox, G. R. 339

Köbrich, G. 292
Komatsu, K. 292, 512
Krantz, A. 199
Kratz, D. 468
Krause, N. 97
Krebs, A. W. 140, 201, 217
Kuck, D. 92, 93
Kuhn, R. 107, 171, 181
Kumada, M. 347

Ladenburg, A. 41
Lampman, G. M. 474
Landa, S. 24
Landor, S. R. 184
Lawton, R. G. 331
Lednicer, D. 326, 329
LeGoff, E. 279
Lehn, J. M. 95
Lemal, D. M. 474
Longone, D. T. 359
Longuet-Higgins, H. C. 198, 492
Lothrop, W. C. 449
Ludwig, R. 386
Lüttke, W. 140, 406, 510
Lüttringhaus, A. 337

Machacek, S. 24
Maier, G. 55, 261, 494, 498
Mallory, F. B. 432
Marshall, J. A. 129, 131
Martin, R. H. 329
Masamune, S. 219, 228, 237
McMurry, J. E. 98
Meerwein, H. 24
Mehta, G. 48
Meinwald, J. 394
Meyer, K. 426
Michl, J. 479
Migirdicyan, E. 498
Mills, W. H. 502
Misumi, S. 339, 341, 360, 363
Mitchell, R. H. 236, 362
Moore, J. S. 462, 464
Moore, W. R. 173
Mozart, W. A. 1
Müllen, K. 109, 238, 433, 439, 440, 441
Musso, H. 35, 36

Nakagawa, M. 225
Nakazaki, M. 131, 187
Napier, D. R. 507
Natta, G. 119

Neidlein, R. 511
Neuenschwander, M. 263, 268, 271, 272, 281
Newman, M. S. 326, 329
Nishida, S. 340, 346
Nixon, I. G. 502
Nozaki, H. 348

Oda, M. 292
Orgel, L. 198
Oth, J. F. M. 393
Otsubo, T. 341
Overberger, C. G. 210

Pappas, S. P. 204
Paquette, L. A. 65, 66, 68, 74, 286, 394
Park, C. H. 95
Pascal Jr, R. A. 367
Paufler, R. M. 380
Pellegrin, M. 361
Perkin Jr, W. H. 16, 198
Peterson, L. I. 293
Pettit, R. 61, 62, 198
Pfau, A. S. 281
Plato, 54
Plattner, P. A. 281
Platz, M. S. 498
Plieninger, H. 257
Prelog, V. 24, 131
Prinzbach, H. 69, 71, 74, 216, 272, 273, 275,
    406
Proksch, E. 305

Quast, H. 396

Rabideau, P. W. 336
Raphael, R. A. 226
Reppe, W. 119, 210
Ried, W. 440
Ripoll, J. L. 179, 294, 304
Roberts, J. D. 158, 486
Roberts, W. P. 65
Roth, W. R. 498
Rücker, C. 216

Sander, W. 489, 490, 491
Saunders, M. 97
Schäfer, W. 205
Schleyer, P. v. R. 25, 135
Schlüter, A.-D. 435
Scholl, R. 426
Scholler, K. L. 181
Schönleber, D. 410

Schrock, R. 108
Schröder, G. 198, 221, 390
Schubert , N. M. 178
Scott, L. T. 164, 237, 332, 336
Scwarc, M. 351
Seiwerth, R. 24
Sekine, Y. 357
Semmelhack, M. 408
Serratosa, F. 391
Siegel, J. S. 512
Simmons, H. E. 95, 408
Skattebøl, L. 173
Sondheimer, F. 105, 222
Srinivasan, R. 474
Staab, H. A. 226, 352, 429, 430, 470
Staley, S. W. 44
Stang, P. J. 264
Staudinger, H. 7, 256
Steinberg, H. 352
Stetter, H. 25
Stoddart, J. F. 436
Sustmann, R. 498
Szeimies, G. 83, 128, 135, 188, 475

Tashiro, M. 367
Thiele, J. 198, 260, 264
Thorpe, J. F. 58
Thummel, R. P. 509, 510
Tobe, Y. 344
Tochtermann, W. 344
Trabert, L. 293
Traetteberg, M. 125
Trost, B. M. 66
Truesdale, E. A. 356
Tsuji, T. 340, 346, 364, 365

van Tamelen, E. E. 204
van´t Hoff, J. H. 171, 181
Viehe, H. G. 43, 205
Vogel, E. 121, 219, 230, 234, 238, 286, 287,
    406, 502
Vogel, P. 299
Vögtle, F. 352
Vollhardt, K. P. C. 451, 452, 453, 455, 470,
    509, 511
von Auwers, K. 257
von Baeyer, A. 16, 152

Walton, D. R. M. 152, 153
Warmuth, R. 487
West, R. 292
Whiting, M. C. 105

Whitlock Jr, H. W. 31, 32
Wiberg, K. B. 83, 133, 137, 474, 475
Wick, A. K. 297
Wilke, G. 119
Willstätter, R. 198, 210
Wilzbach, K. E. 206
Winberg, H. E. 353
Wirz, J. 178
Wiseman, J. R. 128, 129
Wittig, G. 158, 182, 386, 449, 486
Woodward, R. B. 65, 397

Wright, B. B. 498
Wynberg, H. 330, 460

Yamamoto, K. 334

Zander, M. 428
Zecher, D. 292
Zefirov, N. S. 477, 478
Ziegler, K. 119, 124, 282, 283
Zimmerman, H. E. 214, 267, 380, 383, 394
Zimmermann, G. 185, 335

Subject Index

A page number followed by a second number in parentheses refers to information given in the reference section of the respective chapter , *e.g.*, 21(30) refers to reference 30 given on page 21.

σ-π interaction 412
π-π interaction intraannular 361
α,α,α',α'-tetrabromo-*o*-xylene 507
α,α'-dicyclobutenyl 510
β-carotene 8, 103, 111
(CH)$_{10}$ hydrocarbons isomerizations 392
(CH)$_{10}$ map 220, 392
(CH)$_8$ hydrocarbons 394
α-chloro-*m*-xylene photolysis 498
σ-complex 346
σ-homobenzenes 405
β-ionone 112
π-isomer 405
α-methylbenzylamine 125
α-phenylethynyl-ω-phenylpoly[1,2-phenylene (2,1-ethynediyl)] 469
α-pyrone 380
π-spacers 290
π-stacking 469
π-systems [4*n*]- 337
π-systems non-planar 321
[1.1.1]propellane 83, 85, 99(15), 473, 479
[1.1.1]propellane bridged 479
[1.1.1]propellane strain energy 77(21)
[1.1.1]propellane structural properties 83
[10][10]-betweenanene optically active 132
[10][8]-betweenanene 131
[14]annulene ethano-bridged 237
[2.1.1]propellane 82, 84
[2.2.1]propellane 82, 85
[2.2.2]hericene 17, 298, 299, 319(208), 505
[2.2.2]propellane 85, 86
[2.2.2]propellane heat of hydrogenation 85
[2.2]metacyclophane 235
[2+2]cycloaddition 479
[2+2]cycloaddition intramolecular 404
[4+2]cycloaddition 506
[3.1.1]propellane 82, 85
[3.2.1]propellane 86
[3]cumulenes alkyl substituted 179

[4.1.1]propellane 85, 136
[4.3.3]propellane 401
[4.4.1]propellane 86
[4.4.4]propellane 87
[4.4.4]propellane X-ray structure 88
[4*n*+2]annulenes methano-bridged 231
[5.5.5.5]fenestrane 69, 89, 90
[5.5.5.5]fenestratetraene 92
[5]pericyclyne 164
[8][8]-betweenanene 131
[*n*]metacyclophanes 337, 338
[*n*]paracyclophanes 337, 338
[*n*]paracyclophanes AM1 calculations 339
[*n*]phenacenes 432
[*n*]phenylenes angular 447, 448
[*n*]phenylenes linear 447, 448
[*n*]phenylenes triangular 448
1,16-dimethyldodecahedrane 64
1,1-dibromo-2-methyl-1-propene 292, 296
1,1-diethynyl-cyclopropane 164
1,1-dimethyl-propargyl alcohol 164
1,2,3,4,5-hexapentaene permethyl 180
1,2,3,4,5-tetradecapentaene 190
1,2,3,4-pentatetraene 179, 180
1,2,3,4-tetrachloro-5,5-dimethoxycyclopentadiene 47
1,2,3,4-tetramethylenecyclohexane 259
1,2,3,5-cyclohexatetraene 486
1,2,3-butatriene 178, 193(64)
1,2,3-cycloalkatrienes 188
1,2,3-cyclodecatriene 188, 189
1,2,3-cycloheptatriene 188, 189
1,2,3-cyclohexatriene 188, 189
1,2,3-cyclononatriene 189
1,2,3-cyclooctatriene 189
1,2,3-trifluoro-4,5,6-tris-*tert*-butyl-benzene 43
1,2,4,5-cyclooctatetraene 186
1,2,4,5-hexatetraene 176, 351, 353, 357
1,2,4,5-tetrabromobenzene 389
1,2,4,5-tetramethylenebenzene (TMB) 497, 498

1,2,4,6,7,9-cyclodecahexaene 187
1,2,4,6,7-cyclooctapentaene 186
1,2,4,6-cycloheptatetraene 186, 271
1,2,4-cyclohexatriene 184, 209, 486
1,2,4-pentatriene 174
1,2-bis(phenylselenomethyl) benzene 507
1,2-bis(phenylsulfonyl)ethene 382
1,2-bismethylenecyclobutane 133, 496
1,2-bismethylenecycloheptane 349
1,2-bismethylenecyclooctane 349
1,2-bis-trifluoromethyl benzene 109
1,2-cyclodecadiene 182
1,2-cycloheptadiene 184
1,2-cycloheptadiene *tert*-butyl 184
1,2-cyclohexadiene 183
1,2-cyclononadiene 184
1,2-cyclononadiene structural properties 184
1,2-cyclooctadiene 184
1,2-cyclopentadiene 182
1,2-dibromocyclohexene 158
1,2-dibromocyclopentene 182
1,2-diethynylbenzene 451, 467
1,2-diethynylbenzene oligomers 469
1,2-dihalocyclopropenes 504
1,2-dihydropentalene 251, 252, 280
1,2-hexadien-5-yne 176, 223
1,3,5,7,9-cyclodecapentaene 218
1,3,5-cycloheptatriene 233
1,3,5-cyclohexatriene 16, 501
1,3,5-cyclooctatriene 210
1,3,5-hexatriene 104
1,3,5-trienes 176
1,3,5-triethynylbenzene 467
1,3,6,8-tetrahydro-dithiopheno[1,2-*c*:3,4-*c'*]
  benzene 509
1,3,7,9,13,15-hexakisdehydro[18]annulene 225
1,3,7,9-cyclododecatetrayne 180, 225
1,3-bis(bromomethyl)benzene 361
1,3-bis(mercaptomethyl)benzene 363
1,3-butadiene 495
1,3-butadiene cyclooligomerization 120
1,3-butadiene photoisomerization 474, 475
1,3-cyclobutadiene 47, 48, 198, 263
1,3-cyclobutadiene spectroscopic properties
  200
1,3-cyclobutadiene 1,2,3-tri-*tert*-butyl-4-
  (trimethylgermyl)- 77(26)
1,3-cyclobutadiene bis-homo- 116
1,3-cyclobutadiene chloro- 199
1,3-cyclobutadiene dimerization 62, 63
1,3-cyclobutadiene dynamic NMR
  spectroscopy 202

1,3-cyclobutadiene equivalent 47, 70
1,3-cyclobutadiene fluoro- 199
1,3-cyclobutadiene iron tricarbonyl complex 61
1,3-cyclobutadiene structural studies 202
1,3-cyclobutadiene tetrakis-*tert*-butyl- 200
1,3-cyclobutadienes kinetically stabilized 202
1,3-cyclobutadienes push-pull 199
1,3-cyclohexadiene 188
1,3-cyclopentadiene dimer 264
1,3-dihydroisothianaphthene 507
1,3-dipolar cycloadditions 116
1,3-enynes 172
1,3-hexadien-5-yne 208, 209, 223
1,3-*trans*-5-cycloheptatriene 128
1,4,5,8-tetrahydronaphthalene 231, 286
1,4-bismethylenecyclohexane 495
1,4-cycloheptadiene 396
1,4-cyclohexadiene 503
1,4-cyclohexadiyne 163
1,4-dibromo-2-butyne 178
1,4-dichloro-2-butene 109
1,4-dihydrobenzoic acid chloride 257
1,4-diidobicyclo[2.1.1]hexane 83
1,4-pentadiyne-3-one 155, 458
1,5,7,11-dodecatetrayne 223
1,5,9-cyclodecatriyne 509
1,5,9-cyclododecatriene 119
1,5,9-trisdehydro[12]annulene 225
1,5-cyclooctadiene 120
1,5-cyclooctadiyne 160, 178, 294
1,5-dibromo-bicyclo[3.1.1]heptane 85
1,5-dihydropentalene 281
1,5-hexadiyne 11, 209, 223, 266, 293, 294
1,5-hexadiyne starting material for annulenes
  224
1,5-hexadiyne-3-ene 490, 491
1,5-methano[10]annulene 237
1,6-cyclodecadione 161
1,6-cyclodecadiyne 161, 367
1,6-dibromo-2,4-hexadiyne 152
1,6-didehydro[10]annulene 247(194)
1,6-dihydroheptalene 286
1,6-dimethyldodecahedrane 69
1,6-imino[10]annulene 231
1,6-methano[10]annulene 230, 286, 502
1,6-methano[10]annulene structural properties
  231
1,6-methano[12]annulene 238
1,6-oxido[10]annulene 231
1,7-octadiyne-4-ene 505
1,8-diazabicyclo[5.4.0]undec-7-ene (DBU) 253
15,16-dihydropyrene 235

1-acetoxy-1-bromomethane 268
1-adamantene 130
1-aminobenzotriazole 487
1-bromo-2-chlorocyclopropene 504
1-bromo-3-chloromethyl-bicyclo[1.1.0]butane
    83
1-bromo-4-chloro-bicyclo[2.2.0]hexane 133
1-bromocyclobutene 510
1-bromocyclohexene 183
1-chloro-1,3,5-cycloheptatriene 186
1-chloro-2,3-butadiene 256, 257
1-chlorocyclohexene 158
1-chlorotricyclo[3.1.0.0^{2,6}]hexane 136
1-chlorovinyl substituents 332
1-cyclopropyl-cyclopropene 303
1-diazo-2,4-cyclopentadiene 271
1-H-cyclopropa[b]naphthalene 505
1-nonen-6,8-diyne 195(139)
1-norbornene 130
1-penten-4-yne 175
1-vinyl-cyclobutene 253, 509
2-(2,4,5,7-tetranitro-9-fluorenylideneaminoxy)
    propionic acid (TAPA) 324
2,3,5,6-tetramethylene-7-norbornanone 498
2,3-dehydronaphthalene 389
2,3-dibromobicyclo[2.2.2]oct-2-ene 512
2,3-dichloro-5,6-dicyano-1,4-benzoquinone
    (DDQ) 231, 293, 435
2,3-dimethylene-1,4-cyclohexandiyl 256
2,4,6,8,10,12-tetradecahexaene 105
2,4,6-heptantrione 332, 333
2,4,6-trimethylenetricyclo[3.3.0.0^{3,7}]octane 411
2,4,6-tris(chloromethyl)mesitylene 298
2,4-pentadienal 105
2,5-dibromobenzoquinone 61
2,5-dihydro-thiophen-1,1-dioxide 253
2,5-diphenyl-3,4-isobenzofuran (DIBF) 126
2,6-diazocyclohexanone 157
2,6-dimethylenetricyclo[3.3.0.0^{3,7}]octane
    410, 411
2-bromo-cyclohexen-3-one 188
2-bromo-cyclopentadiene 60
2-butyne-1,4-diol 105
2-butyne 408, 409
2-butyne trimerization 205
2-ethinyl-1,3-butadiene 251, 252, 253
2-formyl-4,4-dimethylcyclohexadienone 259
2-halophenyl anions 486
2-oxabicyclo[2.2.0]hex-5-en-3-one 199
2-vinyl-1-cylopropylphenylsulfide 475
2-vinylmethylenecyclopropane 495
3,3′-bicyclopropenyl 42, 204, 207

3,3′-bicyclopropenyl derivatives 207
3,4,5-trimethylene-1,6-heptadiene 253
3,4-benzocyclodeca-3,7,9-trien-1,5-diyne 491
3,4-bismethylenecyclobutene 11, 209, 293
3,4-bismethylenecyclopentanone 495, 496
3,4-dimethylene-1,5-hexadiene 253
3,8-dihydroheptalene 287
3-chlorocyclobutane-1-carboxylic acid 474
3H-indene-1,2-dione 505
3-isochromanone 507
3-methylene-1,4-cyclohexadiene 257
3-methylene-1,4-pentadiene 251, 252, 253, 290
3-methylene-1,4-pentadiyne 251, 252
3-methylene-1-penten-4-yne 251, 252
3-methylenecyclobutanone 494
3-methylenecyclobutene 175
3-methylenecyclohexene 258
3-methylenecyclopentene 496
3-methylenespiro[5.6]dodeca-1,4-diene 342
3-silacyclopropyne 167(48)
3-sulfolene as a 1,3-butadiene equivalent
    253, 254
4,4′-bipyridyl 283
4,4-dimethoxy-cyclohexa-2,5-diene-1-one 47
4,9-methano[11]annulenone 274
4,9-twistadiene 34
4-methylene-2-pentene-1,5-diyl 495
4-methylenedihydropyrazole 494
4-pentenoic acid 474
4-thia-3,3,5,5-tetramethylcyclopentyne 167(55)
4-vinyl-1-cyclohexene 120
5,10-dibromobenzocyclooctatetraene 218
5,11,17-tris-dehydro-tribenzo[a,e,i]-
    [12]annulene- 226
5-chloro-3-penten-1-yne 175
5-cyclopentadienyl-2,4-pentadienal 282, 283
6,6′-biazulenyl 283
6-dimethylamino-pentafulvene 265
7,8-bis(trifluoromethyl)tricyclo[4.4.4.0^{2,5}]deca-
    3,7,9-triene 108
7,8-dehydropentalane 132
7-acetoxynorbornadiene 271
7-methylene-dispiro[2.0.2.1]heptane 303
7-phenyl-heptatrienal 107
9,10′-dibromodianthracene 136
9,10-[1′,2′]benzeno-9,10-dihydroanthracene
    385
9,10-didehydroanthracene 490, 491
9,10-dihydrofulvalene 220, 271
9,9′,10,10′-tetrahydrodianthracene 136
9,9′,10,10′-tetrahydrodianthracene X-ray
    structure 136

9,9′-didehydrodianthracene 136

acenaphthene 331
acenaphthenequinone 332, 333
acenaphthyne 157
acenes 9, 231
acenes linear 421, 423
acepentalene 397, 401
acepentalene metal complexes 403
acepentalenediide dilithium 402
acepentalenediide dipotassium 401, 402
acetylallene 411
acetylene [2+2]dimerization 201
acetylene coupling 471, 472
acetylene cycloaddition 511
acetylene equivalent 323, 332, 333, 382
acetylene tetramerization 210
acetylenes catalytic trimerization 451
acetyltetrafluoroborate 266
acyclic allenes 171
acylation Friedel-Crafts 326, 338, 356, 425, 427
acyloin condensation 131, 303, 332, 338, 341
adamantane 23, 209
adamantane higher analogs 29
adamantane thermodynamic stability 27
adamantane-land 28, 209
addition 1,4- 42
addition Diels-Alder 65, 71, 96, 136, 299, 323,
    332, 333, 344, 345, 346, 408, 424, 425, 440,
    502, 509
Alder-Rickert cleavage 113, 179, 502
aldol reaction intramolecular 348, 400, 401
aldol reaction retrograde 69
alkenes 103
alkenes fully substituted 138
alkylation Meerwein 355
alkynes 151
all-*cis*-[10]annulene 392
all-*cis*-1,4,7-cyclononatriene 405
allene chlorocarbene addition 261
allene dimerization 183, 186, 256, 479, 496
allene tetrakis-*tert*-butyl- 173, 174
allenes 1,1-diaryl- 172
allenes 8, 171
allenes acyclic 171
allenes aryl 173
allenes chiral 192(57)
allenes cyclic 182
allenes cycloadditions 174, 175
allenes optically active 177, 178
allenes tetrasubstituted 174
allenes thermal racemization 194(95)

allenes vinyl 174, 175
allenyl magnesium bromide 176
allyl chloride 113
allyldiazomethane 474
alternant hydrocarbons 422
alternation bond 221
amphotericin B 103
angle deformation 227
angle-strained cycloalkynes 156
angular annelation 321, 421, 426
annelation 255
annelation angular 321, 421, 426
annelation bicyclic 512
annelation linear 421
annelation Robinson 342
annulene [10]- 219, 278
annulene [12]- 89, 221
annulene [12]- octadehydro- 461
annulene [12]- short-circuited 285
annulene [14]- 12, 223
annulene [16]- 222, 223
annulene [18]- 222, 223
annulene [18]- dodecadehydro- 461
annulene [20]- 223
annulene [22]- 223
annulene [24]- 223
annulene [30]- 223
annulenes 151, 197
annulenes higher members 218
annulenes short-circuited 251
annulenes zero-bridged 277
ansa compounds 337
anthracene 386, 402, 423
anthracene as leaving group 180
antiaromatic 278
antibiotics endiyne 161
anti-Bredt hydrocarbons 122, 122
anti-Bredt system double 130
anti-van't Hoff geometry 16, 82
anti-van't Hoff geometry metalorganic systems
    82
arachidonic acid 162
Arndt-Eistert homologation 87
aromatic character 323
aromatic compound non-planar 321
aromatic compounds three-dimensional 350
aromatic systems bridged 229
aromaticity 197
aromaticity NMR criterion 223
aromaticity theory 202
arylallenes 173
arylenes 421, 422

aryliodides 462, 463
asteranes 35
atropisomerism 310(26)
aufbau principle 6, 11, 269
automerization 11, 395
azasnoutanes 395
azasnoutenes 395
azulene 237, 251, 252, 277, 281, 282, 402
azulene 6-methyl- 284
azulene history 314(114)
azulenophane [2.2]- 366

Baeyer strain 156
Baeyer-Villiger oxidation 67
barbaralane 415(74)
barbaralone 391
barrelene 10, 209, 210, 380, 381
barrelene homoaromaticity 385
barretane 221, 402, 403
Barton-Kellog fragmentation 138, 138
basketane diaza- 394
basketene 45, 69, 220, 404
bay area 435
Beckmann rearrangement 381
beltene 435, 445(55)
belts molecular 436
bending out-of-plane antisymmetric 124
bending out-of-plane symmetric 124
benzamide 251, 252
benzannelation 274, 278
benzene 203
benzene bond fixation 501
benzene diethynyl- 458
benzene dimer 221
benzene oxide 382
benzene photoaddition of acetylenes 212
benzene photochemistry 243(101)
benzene photoisomerization 206
benzene tetraphenyl- 208
benzenediazonium-2-carboxylate 487
benzo[1,2:3,4]dicyclobutene 509
benzo[1,2:4,5]dicyclobutene 508
benzo[c]cinnoline 450
benzo[c]phenanthrene 321, 323
benzo[g,h]perylene 324
benzobarrelene 384
benzocyclobutadiene 449
benzocyclobutandione 487
benzocyclobutene 358, 449, 507
benzocyclobutenone 505
benzocyclooctene 344
benzocyclopropene 502, 504

benzocyclopropene substituted 306
benzocyclopropenone 487, 488
Benzolfest 203
benzopyrene 428
benzotriazole 1-amino- 487
benzvalene 43, 204, 206, 242(87)
benzvalene as starting material in hydrocarbon
    chemistry 206
benzvalene tri-fluoro-tris-*tert*-butyl 205
Bergman cyclization 228, 229, 490, 491
betweenanenes 122, 123
biallenyl 176, 351, 353, 357
bicyclo[5.5.5]heptadecane 95
bicyclic annelation 512
bicyclic bridging 512
bicyclo[1.1.0]-1(3)-butene 118, 133, 135
bicyclo[1.1.0]butane 473, 474
bicyclo[1.1.0]butane strain energy 475
bicyclo[1.1.0]butanes isomerization to
    conjugated dienes 287
bicyclo[2.1.1]hexan-2-one 512
bicyclo[2.2.0]hex-1(4)-ene 134
bicyclo[2.2.0]hex-1(4)-ene dimerization 137
bicyclo[2.2.0]hex-1(4)-ene X-ray structure 137
bicyclo[2.2.0]hexa-2,5-diene 203, 204
bicyclo[2.2.1]hept-2-en-5-yne 167(55)
bicyclo[2.2.1]hept-2-yne 521(159)
bicyclo[2.2.2]oct-2-ene 382
bicyclo[2.2.2]oct-2-yne 512
bicyclo[2.2.2]octa-2,5,7-triene 380, 381
bicyclo[3.2.2]nonane 95
bicyclo[3.3.0]octan-3,7-dione 91
bicyclo[3.3.0]octane 400
bicyclo[3.3.1]-1(2)-nonene 129
bicyclo[3.3.1]-1(2)-nonene structural properties
    129, 130
bicyclo[4.1.0]hepta-1,3,5-triene 502
bicyclo[4.1.0]heptene 7,7-dihalo- 504
bicyclo[4.2.0]hept-2-ene dihalo- 502
bicyclo[4.2.0]oct-3-ene 121
bicyclo[4.2.0]octa-1,3,5,7-tetraene 449
bicyclo[4.2.0]octa-1,3,5-triene 507
bicyclo[4.2.0]octa-2,4,7-triene 209, 210
bicyclo[4.3.0]nona-3,7-diene 406
bicyclo[4.4.0]-1(6)-decene 132
bicyclo[4.4.3]tridecane 95
bicyclo[5.3.0]decane 282
bicyclo[6.1.0]nonatriene 268
bicyclo[6.2.0]deca-2,4,6,9-tetraene 220
bicyclo[6.5.1]tetradecane 97
bicyclo[6.6.6]eicosane 95
bicyclo[7.5.1]pentadecane 95

bicyclo[*n.m.*0]alkanes 483
bicyclononylidene-2,4,6,8,2′,4′,6′,8′-octaene 272
bicyclopropyl 473, 483(74)
bicyclopropylidene 117, 301, 302, 303, 496
Binor-S 29
Binor-S tetrahydro 30
biphenyl 421, 422, 447
biphenyl dilithio 449
biphenylene 2,3-diiodo- 452
biphenylene 447, 450
biphenylenes alkyl 450
Birch reduction 69, 231, 286, 437
bis(bromomethyl)benzene 429
bis(phenylethynyl)benzene 440
bisallene 209, 505
bisallenes conjugated 176
bisallenes cyclic stereochemistry 186, 187
bisallyl diradicals 495
bis-biphenyleneallene 173, 174
biscarbenes 158, 498, 499
bis-Dewar-benzene 365
bissecododecahedradiene 73
bis-trimethylsilylethynylacetylene 451
bivalene 64
block-method for triangulane synthesis 478
board molecular 435
bond alternation 202, 221, 456
bond fixation 501, 511, 512
bond orders Pauling 513
bowtiediene 407
branched polycyclic aromatic hydrocarbons (PAH) 421
branched triangulanes 476
Bredt's rule 27, 32, 47, 128, 447
Bredt's rule modern interpretation 128
bridged [1.1.1]propellane 479
bridged annulenes 227
bridged aromatic systems 229
bridgehead functionalization 47
bridgehead olefins structural data 147(143)
bridgehead-distorted hydrocarbons 81
bridging σ- 44
bromocyan 283
bromomethyl magnesium bromide 268
buckytube 438
bullvalene 12, 220, 389, 390, 402
bullvalene concept 392, 393
bullvalene dynamic behavior 393
bullvalene structural properties 392, 415(59)
bullvalone 391
butadiene photodimerization 121

butadiene co-oligomerization with 2-butyne 146(98)
butadiene cyclo-co-oligomerization 122
butadiene dimerization 212
butadiene dimerization to 2-methylenevinylcyclopentane 145(92)
butadiyne 152, 470
butalene 486
butane 30
butatriene 8, 160
butatriene dimerization 293, 294
butatriene perchloro- 294
butatriene permethyl- 292, 295
butatriene tetraphenyl- 296

$C_{10}H_{10}$ isomers total number 245(150)
C_6H_6 isomers total number 241(73)
C_8H_8 land 213, 214
cage hydrocarbons 6
calculation molecular mechanics 27
calicene 269
calicene sterically protected 275
calicheamicin 490
carbanions 3
carbenacyclopropane 173
carbene 1,4-adddition 349
carbene addition 291
carbene cyclopropenyl 59, 201
carbene cyclopropylidene 286
carbene methoxycarbonyl 266
carbene spiro 342
carbene vinylidene 156, 333
carbenes 42, 58, 59, 230, 409, 477, 508
carbenes addition to acetylenes 264
carbenes from 1,1-dichloroethane 302
carbocations 3
carbon allotrope sp 155
carbon insertion 191(17)
carbon planar 16
carbon rods 151
carbon-carbon bond double 6
carbon-carbon bond single 6
carbon-carbon bond triple 6
cascade process 308, 309
cata-condensation 443(9)
cata-condensed polyarenes 423
catalyst chiral 525
catalyst tuning 120
catalyst Wilkinson 190
catenane 17, 18
cavities molecular 465, 519(135)
centropolyquinanes 92

chair conformation 309
chamazulene 281, 282
chamomille 281
charge transfer (CT) complexes 270, 361, 379
cheletropic ring-opening 253
chemical shift reversal 238
chemistry designer 23
chemistry non-natural product 23
chiral solvents 329
chirality in iptycenes 389
chirality plane 124, 337
chlorobenzene 486
chlorocarbons 6
chlorocycloheptatriene isomers 271
chlorodurene 357
chloroiodide 452
chloromethylene(triphenyl)phosphorane 233
chrysene 432
circularly polarized light 329
circulene [4]- 429, 331
circulene [7.7]- 336
circulene [7]- 334, 335
circulene [7]- molecular structure 334, 335
circulene extended 429
circulenes 44(31), 321, 330, 470
circum[34]para-terphenyl 444(41)
circumanthracene 434
circumarenes 434
circumbenzene 434
cis,trans-1,5-cyclododecadiene 121
cis,trans-1,5-cyclooctadiene 128
cis-1,2-cyclobutanediacetic acid 121
cis-1,2-divinylcyclobutane 120, 121
cis-1,3-hexadien-5-yne 185
cis-1,5-dimethylbicyclo[3.3.0]octan-2,6-dione 396
cis-15,16-dimethyldihydropyrene 236
cis-3,4-dichloro-1-cyclobutene 198, 212
cis-9,10-dihydronaphthalene 219, 391
cis-benzene-trioxide 216
cis-bis-σ-homobenzene 407
cis-tris-σ-homobenzene 405
Claisen rearrangement 458
Clar's hydrocarbon 492, 493, 500, 517(88)
Claus benzene 204
cleavage Alder-Rickert 113, 179, 502
Clemmensen reduction 339
collarene 435
columnene 435
concave molecules 445(54)
condensation acyloin 131, 303, 332, 338, 341
condensation Dieckmann 87

condensation Knoevenagel 261, 332, 333
condensation Thiele 277
condensed aromatic hydrocarbons properties 422, 423
conductivity electric 270
conformation boat 31
conformation chair 31, 309
conformation half-chair 31
conformation twist-boat 31
conformational analysis 27, 30
congressane 38(22)
conjugated molecular rods 461
conjugated polyenes 103
container molecule 200, 201
contrathermodynamic reactions 67
co-oligomerization 506
Cope elimination 124, 293, 410, 411
Cope rearrangement 121, 209, 389, 412
corannulene 9, 297, 321, 330, 331, 332
corannulene derivatives 332
corannulene reduction to anions 334
core units 467, 470
Corey-Winter fragmentation 125, 134
coriolin 500
coronanes 308
coronene 321, 330, 421, 422, 426, 427, 428, 434
coronene dimerization 439
corset effect 55
coumalic acid 299, 380
coupling Eglinton 223, 225, 298, 299, 339
coupling Glaser 105, 152, 153
coupling Glaser-Hay 461
coupling Hagihara 462
coupling Hay 153
coupling Heck 455, 456
coupling Kumada 347
coupling McMurry 97, 111, 132, 138, 260, 344, 345
coupling metal-mediated 434, 463
coupling of acetylenes 471, 472
coupling Pd-catalyzed 467, 468
coupling Sekiya-Ishikara 462
coupling Stephens-Castro 226
coupling Stille 109
coupling Ullmann 449
coupling Wurtz 356, 361
coupling Wurtz-Fittig 429
cross-breeding pyrolysis 360
cross-conjugated hydrocarbons 251
cross-conjugation 6
crown ethers 337

cryptands 95
CT complexes 270, 361, 379
cubane 42, 54, 60, 209, 210
cubane metalation 63
cubane octamethyl- 63
cubane perfluorooctamethyl- 63
cumulenes 171
cumulenes acyclic 178
cumulenes aryl 181
cumulenes cyclic 188
cuneane 209, 210, 215
cuprates 296
Curie-Weiss plot 496
Curtius degradation 399
cyanoacetylene 201
cyanocarbons 6
cyclic allenes 182
cyclic cumulenes 188
cyclic triangulanes 476
cyclic trimethylenemethanes 492, 493
cyclization Bergman 11, 228, 229, 490, 491
cyclization Dieckmann 90
cyclization Friedel-Crafts 92, 332, 409
cyclization Haworth 327
cyclization Parham 508
cyclization Thorpe-Ziegler 90
cyclo[3]cumulenes 189, 190
cycloacene 435
cycloaddition [2+2]- 96
cycloaddition [4+2]- 506
cycloaddition [6+4]- 284
cycloaddition 1,3-dipolar 116
cycloaddition allenes 174, 175
cycloaddition carbene 291
cycloadditions 91
cycloalkadienes 160
cycloalkenes large-ring 118
cycloalkynes 10
cycloalkynes angle-strained 156
cycloalkynes large-ring 160
cycloalkynes medium-ring 160
cycloallenes 160
cycloallenes doubly-bridged 187
cycloallenes optically active 187
cyclobutabenzene 425
cyclobutane-1,1-dicarboxylic acid 474
cyclobutanes by photodimerization of alkenes 525
cyclobutanone 157, 509
cyclobutarenes 449, 506
cyclobutene dicarboxylic acid anhydride 199
cyclobutene ring-opening 304

cyclobutyne 156
cyclocarbon 9
cyclo-co-oligomerization 119
cyclocumulenes 136, 486
cyclodehydrogenation 440, 442
cyclododeca-1,4-dione 348
cyclododecahexaene 221
cyclododecatetraenediyne 450, 451
cyclododecayne 121
cycloheptatriene 392
cycloheptatriene-1,6-dicarboxaldehyde 233
cycloheptatrienone 261, 266
cycloheptatrienyl anion 261
cycloheptatrienylidene 186, 271
cycloheptatrienylidene ketene 274
cycloheptyne 3,3,7,7-tetramethyl- 159
cyclohexadienone 342
cyclohexane 30
cyclohexatriene 323, 337, 456, 513
cyclohexyne 158, 159, 485
cyclohexyne trapping 160
cyclononatetraenide lithium 268
cyclooctatetraene 48, 119, 186, 209, 210, 221, 268, 281, 382, 390, 394
cyclooctatetraene bond-switching 217
cyclooctatetraene derivatives 211
cyclooctatetraene dimers 221, 390
cyclooctatetraene isomerization reactions 213
cyclooctatetraene planar derivatives 217
cyclooctatetraene planar, transient 245(140)
cyclooctatetraene Reppe synthesis 211
cyclooctatetraene ring inversion 217
cyclooctatetraenophane [2.2]- 366
cyclooctatrienyne 217
cycloocten-3-one 97
cyclooctyne 284, 285, 344, 345
cycloolefins 112
cycloolefins *trans*- 9
cyclooligomerization 119
cyclopentadiene 116, 402, 411
cyclopentadienide lithium 44
cyclopentadienide sodium 65
cyclopentadienyl anion 261
cyclopentadienylcobalt dicarbonyl 451, 511
cyclopentadienylidene 408, 409, 410
cyclopentyne 156, 157, 182
cyclopentyne stabilization 157
cyclopentyne tetramer 63
cyclophandiene 429
cyclophane [2₃]- 355, 356
cyclophanes 321, 337, 428, 429
cyclophanes [2.2]-isomers 377(186)

cyclophanes [2₄]- 357
cyclophanes binuclear 337, 338
cyclophanes chirality 124
cyclophanes multi-bridged 69, 322, 337, 338, 355
cyclophanes multi-layered 322, 337, 338, 359
cyclophanes multi-stepped 338, 363
cyclophanes *pseudo*-geminally substituted 356
cyclophanes short-bridged 205
cyclophanes structural richness 366
cyclopropa[*a*]naphthalene 502
cyclopropa[*l*]phenanthrene 502
cyclopropaaromatics 425
cyclopropabenzene 118
cyclopropacarbene 173
cyclopropanation Simmons-Smith 302, 303, 305
cyclopropanes as building blocks 472
cycloproparene triangular 505
cycloproparenes 503
cyclopropene 9, 112, 113
cyclopropene 1,2-dicyclopropyl- 117
cyclopropene 3-cyclopropyl- 115
cyclopropene 3-methyl- 113
cyclopropene acidity 117
cyclopropene derivatives as natural products 145(74)
cyclopropene oligomerization 115
cyclopropene pyrolysis 117, 118
cyclopropene reactions 113, 115, 116
cyclopropene spectral properties 144(66)
cyclopropene strain energy 112
cyclopropenium cation 114, 117
cyclopropenone 261
cyclopropenyl anion 114, 261
cyclopropenyl carbenes 201
cyclopropenyl radical 117
cyclopropyl bromide 113
cyclopropyl-1-nitroso urea 301
cyclopropylidene 172
cyclopropylidene carbene 286
cyclopropylidenecyclopropane 117
cyclopropylidenedispiroheptane 306, 307
cyclopropylmethylketone 302
cyclopropyne 10, 156, 163
cyclotetradeca-1,6-diyne 121
cyclotrimerization of acetylenes 451
cylopentadienide lithium 206, 475

davidane 477
DBAs 470, 471
DBU 253

DDQ 231, 293, 435
decacyclene 158, 336
decanonaene 181, 182
decapentaene 105, 109
decapentayne 152
decarboxylation 36, 72
dechlorination 294
dechlorosilylation 207
deformation angle 15
deformation bond-length 15, 16
deformations non-planar 124
degradation Curtius 399
degradation Hunsdiecker 25, 36, 83, 84
dehalogenation by sodium metal 474
dehydro[10]annulene 228
dehydro[8]annulene 217
dehydro[8]annulene dibenzo 218
dehydroannulenes 224
dehydroannulenes large-ring 225
dehydrobenzannulenes (DBAs) 470, 471
dehydrobenzene 136, 353, 449, 485, 486, 511
dehydrobenzene non-benzenoid 223
dehydrobromination 458, 459
dehydrochlorination 298, 409
dehydrogenation 71
dehydrohalogenation 502
deltaphane 519(135)
demetalation 425
dendralene [3]- 253, 254, 290
dendralene [3]- conformation 255
dendralene [3]- spectroscopic properties 256
dendralene [4]- 177, 253
dendralene [5]- 253
dendralenes 7, 253
dendralenes chemical reactivity 255
dendralenes conformation 257
dendralenes planarized 259
dendralenes sterically fixed 258, 259
dendrimeric hydrocarbons 467, 468
desilylation 455
desulfurization Raney nickel 364
Dewar benzene 42, 43, 199, 203, 204
Dewar benzene 1,4-pentamethylene- 345, 346
Dewar benzene 1,4-tetramethylene- 345
Dewar benzene bis- 365
Dewar benzene bridged 343, 364, 365
Dewar benzene hexamethyl- 205
Dewar benzene tris-*tert*-butyl- 204, 205
di(2-methyl-9-triptycyl) 414(28)
di-π-methane rearrangement 214, 215, 383, 384
diacetylene 152
diacetylenes α,ω- 161

diademane 220, 403
diallenes α,ω- 176, 177
diamantane 29
diamonds explosive 152
diamonds industrial 39(28)
diaryne equivalent 389
diatropic 217
diazabasketane 394
diazamacrobicyclics 94
diazocyclopentadiene 408, 409, 410
diazocyclopropane 478
diazo-group transfer 72, 85
diazomethane 302
dibenzo[c,g]phenanthrene 321, 323
dibenzocoronene 428
dibenzocyclooctadiene 361
dibenzylideneacetone 256
DIBF 126, 130, 183, 188
dibromocarbene 475
dichlorocarbene 503
dicinnoline 452
dicoronene derivatives 439
dicyclopentadiene 89
dicyclopropylketone 303
Dieckmann condensation 87, 90
Diels-Alder addition 65, 71, 87, 96, 136, 175,
    183, 299, 323, 332, 333, 344, 345, 346, 408,
    424, 425, 440, 502, 509
Diels-Alder addition diene-transmissive 254
Diels-Alder addition domino 65, 71, 381
Diels-Alder addition innermolecular 488
Diels-Alder addition intramolecular 79(69), 129
Diels-Alder addition retro- 179, 264, 294, 351
Diels-Alder hetero 116
diene double 435
dienes oligomerization 118
dienophile double 423, 424, 435, 436
diethynylbenzene 458, 462
dihalides vicinal 173
dihalo-1,3-butadienes 295, 296
dihydroacepentalene derivatives 401, 402
dihydrofulvalene 71
dihydrohexacene 425
dihydropentacene 423
dihydropentalenes 213
dihydropyrenes X-ray structural analysis 236
diketene 491
dimerization acetylenes 201
dimerization allene 183, 256, 496
dimerization oxidative 270
dimerization reductive 270
dimerization solid-state 288

dimerization topologically-controlled 288
dimethoxycarbonium tetrafluoroborate 236
dimethyl acetylenedicarboxylate (DMAD) 65,
    199, 255, 285, 358, 502, 511
dimethyl cyclobutene-1,2-dicarboxylate 509
dimethyl pagodane 73
dimethyl trans-1,2-dihydrophthalate 406
dimethylsulfonium methylide 268
diphenyliodonium-2-carboxylate 487
diphenylisobenzofuran (DIBF) 126, 130, 183,
    188
diphenylketene 256
diradicals bisallyl 495
diradicals triplet 494
disjoint hydrocarbon 496, 497
dispiro[2.0.2.1]heptane 473
distance intramolecular 16
distorted olefins 122
distortions out-of-plane 123
di-tert-butyl-3-oxoglutarate 400
di-tert-butyl-acetylene 55
di-tert-butylicosadecaene 110
dithia[3.3]metacyclophanes 236, 248(216), 363
dithia[3.3]metacyclophanes chromium
    complexes 362, 363
dithiacyclophane 351
divinylcyclopropane 392, 396
divinylether 251, 252
divinylmethane 384
diynes cyclic 121
DMAD 65, 199, 255, 285, 358, 502, 511
DMS method 172, 184, 189, 302
DNA double stranded 18
DNA duplex circular 18
DNA single-stranded 18
dodecahedrane 54, 63, 271, 397
dodecahedrane 1,16-dimethyl- X-ray structural
    analysis 80(90)
dodecahedrane derivatives 75
dodecahedrane heat of formation 74
dodecahedrane hexaene 75
dodecahedrane retrosynthesis 63
dodecahedrane spectroscopic properties 73, 74
dodecahedrane X-ray structure 74
dodecahedrene 75
dodecahydro[12]beltene 437
dodecahydrotriphenylene 159
dodecamethyl[6]radialene 40(49)
Doering-Moore-Skattebøl reaction (DMS) 172,
    184, 189, 302
domino Diels-Alder addition 65, 381
double bonds semicyclic 512

double diene 435
double dienophile 435, 436
double-bond semicyclic 290
durene 359
dynamic NMR spectroscopy 97

Eglinton coupling 223, 225, 298, 299
Eglinton coupling intramolecular 339
Elbs reaction 423, 424
electrocyclic isomerization 283, 285, 289
electrocyclic reactions 21(30)
electrolysis 406
electronic devices 462
electronic interactions over long distances 379
electronically conductive materials 494
elimination Cope 124, 293, 410, 411
elimination Hofmann 112, 121, 124, 129, 236,
    253, 256, 258, 264, 267, 280, 293, 339, 353,
    355, 359, 381
endocyclic 6
endohedral 397
endohedral fullerenes 367
enediyne antibiotics 490
episulfides 138
equivalent diaryne 389
Eschweiler-Clarke methylation 399
ESR spectroscopy 494
ester pyrolysis 256, 391
ethane hexaethynyl 458, 460
ethene tetraethynyl 458, 460, 461, 481(43)
ethyl diazoacetate 391
ethylene 116
exohedral 397
exploded radialenes 290, 297, 298, 299, 460
explosive diamonds 152
extended structures 421, 447, 457
extrusion nitrogen 307

Favorskii ring contraction 47, 48, 60, 92
felicene 17
fenestranes 88
fenestranes ab initio calculations 88
fenestranes X-ray structure 92, 94
fenestrindane 92, 93
ferrocenophane 367
ferromagnetism 494
fidecene 273, 274
flash vacuum pyrolysis (FVP) 263, 280, 336,
    342, 349, 356, 450
fluoranthene 422
fluoranthene 7,10-diacetyl- 332, 333

fluxional systems 12, 384, 391, 395, 396,
    415(74)
formylation Rieche 358
fragmentation Barton-Kellog 138, 139
fragmentation Corey-Winter 125, 134
fragmentation Grob 36
Friedel-Crafts acylation 326, 338, 356, 425, 427
Friedel-Crafts cyclization 92, 332, 409
fullerenes 9, 152, 331
fullerenes endohedral 367
fullerenes partial structures 336
fulminene 432
fulvadienes 275
fulvalenes 118, 269
fulvalenes symmetrical 269
fulvalenes unsymmetrical 269
fulvenallene 288, 505
fulvene 6,6-dimethyl- 265
fulvene 6,6-diphenyl- 265
fulvene 6-ethynyl- 265
fulvene 6-methyl- 266
fulvene 6-vinyl- 265
fulvenes 6, 208, 260
fulvenes open 7, 256
fulvenes retrosynthesis 260
fulvenes Thiele synthesis 265
FVP 263, 280, 336, 342, 343, 349, 356, 450

gas-phase pyrolysis 178
gedankenexperiment 282
giant PAHs 438
Glaser coupling 105, 152, 153
Glaser-Hay coupling 461
glutacondialdehyde 283
glyoxal 276, 400
glyoxal sulfate 276
grail molecule 142
graphite 9, 16, 421
graphite ribbons 432
graphite substructures 421, 438
graphite synthetic 439
graphitization 434
graphyne 470
Grob fragmentation 36
guaiazulene 281, 282

Hafner-Ziegler synthesis 282, 283
Hagihara coupling 462
Haworth cyclization 327
Hay coupling 153
Heck coupling 455, 456
helicene [14]- 328, 329

helicene [5]- 323, 325
helicene [6]- 326, 327
helicene [7]- 328
helicene [9]- 328
helicenes 321, 323
helicenes by Diels-Alder additions 325
helicenes charge-transfer complexes 330
helicenes chirality 321
helicenes configurational stability 330
helicenes optically active 329
helicenes thermal racemization 330
helicenes X-ray structure determination 330
helvetane 50(3)
hemicarceplex 487, 488
hemicarcerand 200, 487, 488
heptacene 425, 426
heptafulvalene 269, 271
heptafulvalene chemical behavior 272
heptafulvene 260, 266, 267
heptahendecafulvalene 274
heptalene 277, 286, 288
heptalenedicarboxylate 287
heptapentafulvalene 269, 272
heptiptycene 512, 513
hericene 17, 298, 299, 319(208), 505
heterocirculenes 330
heterotrishomoaromatics 406
hexacene 425
hexacyclo[4.4.0.0^{2,10}.0^{3,5}.0^{4,8}.0^{7,9}]decane 403
hexaethynylbenzene 456, 458
hexaethynylethane 458, 460
hexahelicene 326, 327, 421, 422
hexahelicene resolution 329
hexahelicene reactivity 371(49)
hexahydroanthracene 231
hexamethyl[3]radialene 292
hexamethyldisilazane 90
hexapentaenes 181, 182
hexa-*peri*-hexabenzocoronene 439, 440
hexaphenylisobenzofuran 368
hexaprismane 47
hexaquinacene 64
hexaradialene 505
hexatriyne 152, 470
hexphenylbenzene 446(64), 446(66)
high-temperature pyrolysis 340
high-temperature synthesis 333, 334
Hofmann degradation 293
Hofmann elimination 112, 121, 124, 129, 236,
    253, 256, 258, 264, 267, 280, 293, 339, 355,
    359, 381
Hofmann elimination bis-vinylogous 353

homoacetylene 113
homoaromaticity 116, 395, 397, 417(102)
homobenzvalene 207
homoconjugation 163, 380
homocyclopropyne 133
homodiademane 405
homologation Arndt-Eistert 87
homopentaprismane 47
homophthalaldehyde 233
homotropylidene 12, 392
Horner reaction 90
Hückel ketone 282
Hückel rule 114, 202
Hückel rule and planarity 232
Hückel systems 461
Hund's rule 493, 497, 498
Hunsdiecker degradation 36, 83, 84
hydroboration asymmetric 178
hydrocarbon acyclic 19(5)
hydrocarbon cage 6, 24
hydrocarbon charged 2
hydrocarbon disjoint 496, 497
hydrocarbons C$_6$H$_6$ 184, 185, 203
hydrocarbons *out,out-, out,in-, in,in-* 81, 82, 94
hydrocarbons (CH)$_{10}$ 392
hydrocarbons (CH)$_6$ 203, 204
hydrocarbons alternant 422
hydrocarbons anti-Bredt 122, 123öö
hydrocarbons bicyclic 95
hydrocarbons bridgehead-distorted 81
hydrocarbons C$_6$H$_6$ land 208
hydrocarbons C$_7$H$_6$ 186
hydrocarbons C$_8$H$_8$ photochemical
isomerizations 214, 215
hydrocarbons C$_8$H$_8$ metal-catalyzed
isomerizations 214, 215
hydrocarbons cavity-containing 465
hydrocarbons cross-conjugated 177, 251
hydrocarbons dendrimeric 467, 468
hydrocarbons extended 447, 457
hydrocarbons fluxional systems 395, 396
hydrocarbons macropolycyclic 465
hydrocarbons non-alternant 422
hydrocarbons non-Kekulé 256, 349, 492
hydrocarbons non-planar 368(1)
hydrocarbons platonic 53
hydrocarbons saturated 6
hydrocarbons shape-persistent 462
hydrocarbons skipped 162
hydrocarbons spiro 294
hydrocarbons tubular 437
hydrogenation catalytic 67

hydrogenation Lindlar 106, 112, 253
hydrogen-hydrogen repulsion 227
hydroxylamine *O*-mesitylene sulfonate 136
hyperstable olefin 73, 79(86), 97
hypnophilin 500
hypostrophene 45, 220

iceane 37
in-[3^{4,10}][7]metacyclophane 98
indeno[2.1-*a*]indene 278
indigo 253
innermolecular Diels-Alder addition 488
insertion carbon 191(17)
interaction non-bonding 95
interaction bowsprit-flagpole 35
interaction through-space 408
intramolecular [2+2]cycloaddition 404
intramolecular aldol reaction 400, 401
inverted tetrahedron 81
iododecarboxylation 72
iododesilylation 452
iodolactonization 66
iptycenes 385, 388
iptycenes chiral 389
iptycenes dendritic 388
iptycenes properties 389
isobenzene 184, 209, 486
isobenzvalene 135
isobullvalene 220
isobutylidenes addition to acetylenes 264
isomerism homeomorphic 95, 97, 98
isomerism *out, out-, out, in-, in, in-* 81, 82
isomerization intramolecular 13
isomerization electrocyclic 283, 285, 289
isomerization metal-induced 273
isomerization photochemical 273
isomerization propargylallene 175
isomerization valence 279
isonaphthalene 185
isotetralin 231
isotriquinacene 401
israelane 50(3)

Jones oxidation 66, 91

Kekulé benzene 337
Kekulé resonance structures 492
Kekulé structure 204, 396, 501
kekulene 331, 421, 429, 430
kekulene structural properties 431
ketene 56, 274
Knoevenagel condensation 261, 332, 333

kohnkene 436
Kolbe electrolysis 406
Kovacic conditions 440
Krätschmer-Huffman process 438
Kumada coupling 347

ladder structure 421, 431, 443(20)
Ladenburg benzene 41
laser desorption 467
laser flash photolysis (LFP) 490, 491
LDA 411, 458
lead tetraacetate 382
LFP 490, 491
Lindlar hydrogenation 106, 112, 153
linear acenes 421, 423
linear annelation 421, 423
linear polyenes 103
linear triangulanes 476, 477
linearmycin A1 103
liquid crystals cholesteric 329
lithium diisopropylamide (LDA) 411, 458
Lochmann-Schlosser base 347, 401
loop knotted 17
lycopene 103

macrocyclic diacetylenes 161
macrocyclization 461
macropolycyclic hydrocarbons 466
maleic anhydride 121, 255, 323
malononitrile 261, 266
Mannich reaction 266
manxane 17
mass spectrometry time-of-flight 440, 467
material science 290, 297, 492, 493, 525
matrix 447, 487
matrix halogen-doped 494, 495
matrix isolation 157, 186, 199, 200, 515
m-benzoquinodimethane 492, 493
m-chloroperbenzoic acid (MCPBA) 341
McMurry coupling 97, 111, 132, 138, 260, 344, 345
MCPBA 341
m-dehydrobenzene 489
m-dehydrophenol 490
Meerwein alkylation 355
Meerwein ester 24
Meerwein reagent 264, 363
Meldrum acid 159
Merrifield resin 463
metacyclophane [2.2]- , conformational behavior 362
metacyclophane [2.2]- 361, 362, 499

metacyclophane [4]- 349
metacyclophane [5]- 349
metacyclophane [6]- 348, 349
metacyclophane [6]-, 3-ene- 349
metacyclophane [7]- 348
metacyclophanes 229
metacyclophene [6]- 499
meta-diethynylbenzene as building unit for
extended structures 464
metalation 63
metal-catalyzed rearrangement 344, 345, 395,
404
metallocenes 65
metallocenes chirality 124
metal-mediated coupling 434, 463
metaparacyclophandione [2.2]- 489
methane tetraethynyl 458, 459
methane tetrakis-(bromomethyl) 476
methano[10]annulene 12
methano[10]annulenophane 366
methanoannulene as a subunit 273
methano-bridged [4n+2]annulenes 231
methano-bridged annulenes aufbau sequences
234
methyl benzoate triplet sensitizer 126
methyl bicyclobutane-1-carboxylate 474
methyl propiolate 199, 357
methyl vinyl ketone 381
methylation Eschweiler-Clarke 399
methylchlorocarbene 477
methylene(triphenyl)phosphorane 233
methylenecyclobutane 253, 254, 476
methylenecyclopropane 113, 447, 492, 493
methylenecyclopropane rearrangement 494
methylenespiropentane 496
Michaelis-Arbuzov reaction 233
Mills-Nixon effect 502, 512
mitrane 221, 417(114)
Möbius system 384
molecular beams 479
molecular belts 436
molecular board 435
molecular cavities 519(135)
molecular container 200, 201
molecular electronics 104
molecular ribbon 9
molecular rods conjugated 461
molecular sieves 115
molecular tile 447
molecular tweezers 445(54)
molecular wires 9, 104, 107, 155, 461, 482(54)
Mozingo reduction 423, 424

m-quinodimethane 492, 493, 498, 499
mucodialdehyde 277
Müller-Röscheisen conditions 361
multi-bridged cyclophanes 355, 356
multi-layered cyclophanes 359, 360
multi-layered cyclophanes charge-transfer
complexes 361
m-xylene photolysis 498
m-xylylene 492, 493

N,N-dialkylaryltriazines 462, 463
naked nickel 120, 121
nanochemistry 104, 461, 462, 479
nanotubes 438
naphthalene 187, 230, 404, 423
naphthalene bis-formyl 425
naphthalenes *peri*-condensed 433
naphthalinophane [2.2]- 429
naphthobarrelene 384
natural products containing cyclopropane rings
473, 474
natural products containing triple bonds 151,
152
Nenitzescu's hydrocarbon 220, 382, 402
neutralization-reionization 401, 402
nickel naked 120, 121
nickel Raney 140
nickelocene 65, 271
nitrogen extrusion 307
NMR spectroscopy dynamic 97
nonadecaiptycene 389
nonafulvene 268, 272
nonaiptycene 389
non-alternant hydrocarbons 422
non-Kekulé hydrocarbons 349, 492
non-Kekulé hydrocarbons aromatic 498
norbornadiene 332, 333
norbornadiene 7-methoxycarbonyl- 382
norbornyne 167(55), 521(159)
norcaradiene 266, 502, 505
norcarene 116
Norrish type-II photofragmentation 499
N-phenyl-triazolindione 44
nylon 119

o-chinodimethane 511
octabisvalene 209, 210, 216
octachloro[4]radialene 294
octaheptaene peralkylated 181
octaheptaenes 182
octalin 132, 282
octaphenyldibenzo[*a,c*]naphthacene 367, 368

octatetrayne 152, 153
octavalene 207, 209, 210
o-dehydrobenzene 217, 487
o-dehydrobenzene metal complexes 491
o-dehydrobenzene NMR spectrum 488
o-dehydrobenzene structural properties 487,
    488
o-fluorobromobenzene 386, 486
olefins diastereomers 122
olefins distorted 122
olefins hyperstable 73, 79(86), 97
olefins pyramidalized 122
oligoaryls 421, 422, 443(1)
oligoenes conjugated 6
oligomerization metal-catalyzed 118
oligomers sequence-specific 465
oligoquinanes 400
oligotriangulanes 473
open fulvenes 256
o-quinodimethane 177, 358, 425, 498, 505
organic ferromagnets 494
organocuprates 296
ortho-bromofluorobenzene 386, 486
ortho-dehydrobenzene 217, 487
ortho-diidobenzene 471
orthogonalization 124
ortho-magnesation 63
oscillation theory 501
oxaquadricyclane 344
oxepin 382
oxepin bridged 345
oxidation Baeyer-Villiger 67
oxidation cerium ammonium nitrate 198, 199
oxidation Jones 66, 91
oxidation manganese dioxide 112
oxidation Sarett 303
oxidation Sharpless 132
oxidative dimerization 270
oxocarbons 6
o-xylylene 177, 358, 425, 505, 507, 511
ozonolysis 121

p-[*n*]cubyls 484(99)
paddlanes 94
pagodane 69, 70
pagodane di-cation 73
pagodane dimethyl 73
pagodane force-field calculations 69
pagodane heat of formation 69
pagodane radical cation 73
pagodane reactivity 73
pagodane structure 73

PAH 421
PAHs giant 438
paracyclene 9
paracyclophane [2.2]- strain energy 351
paracyclophane [1.1]- 363, 365
paracyclophane [1.1]- strain 365
paracyclophane [1.1]- structural properties 365
paracyclophane [1]- 230
paracyclophane [2.2]- 349
paracyclophane [2.2]-, intraannular distance
    354
paracyclophane [2.2]- chirality 350
paracyclophane [2.2]- retrosynthesis 351
paracyclophane [2.2]-, tetramethyl 358
paracyclophane [4]- 345, 346
paracyclophane [5]- 344, 346
paracyclophane [6]-, 8-methoxycarbonyl- 344,
    345
paracyclophane [6]-, irradiation 347
paracyclophane [7]-, 3-carboxy- 341
paracyclophane [7]- 342
paracyclophane [8]- 339, 340
paracyclophane [8]-, 4-carboxy- 341
paracyclophane intramolecular charge-transfer
    complexes 350
paracyclophanes 229
paracyclophyne [2.2]- 377(203)
parallel phanes 354
paratropic 217
Parham cyclization 508
Paterno-Büchi reaction intramolecular 411
patterns reactivity 11
patterns structure 11
Pauling bond orders 513
p-benzoquinone 255, 386, 423, 424
p-benzoquinone diazide carboxylic acid 489,
    490
Pd-catalyzed coupling 467, 468
p-dehydrobenzene 490, 491
pentacene 423, 424
pentacyclo[3.3.0.0^{2,4}.0^{3,7}.0^{6,8}]octane 209, 210
pentacyclo[4.2.0.0^{2,5}.0^{3,8}.0^{4,7}]octane 60
pentacyclo[5.1.0.0^{2,4}.0^{3,5}.0^{6,8}]octane 209, 210
pentadecaene 108
pentafulvadiene 276
pentafulvalene 1,2:5,6-dibenzo 271
pentafulvalene 251, 252, 269, 270
pentafulvene 251, 252, 260, 264
pentafulvene 6-chloro- 287, 288
pentahelicene 323, 325
pentahelicene configuration 323
pentahelicene racemization 324

pentahendecafulvalene 273, 274
pentalene 1,3,5-tri-*tert*-butyl- 279, 284
pentalene 1,3-di-*tert*-butyl- 280
pentalene 251, 252, 277
pentalene perphenyl 279
pentalenes metal complexes 315(127)
pentalenes stabilized 278
pentaprismane 221
pentaprismane carboxylic acid 47
pentatetraene dimerization 295
pentatriafulvalene 269
pentiptycene 387, 389
perfluoro-2-butyne 96
peri-condensation 443(9)
pericyclynes 163
pericyclynes exploded 164
pericyclynes extended 164
pericyclynes permethylated 163, 164
peristylane 64
permethylcyclooctatetraene 415(61)
permethylsemibullvalene 415(61)
perylene 428, 433
phase-transfer catalysis 503
phenacene [11]- 433
phenacene [6]- 444(34)
phenacene [7]- 432
phenanthrene 321, 432
phenanthrene 4,5-dimethyl- 323
phenanthro[3,4-*c*]phenanthrene 326, 327
phenyl(tribromomethyl) mercury 475
phenylacetylene building block for
macropolycyclic hydrocarbons 466, 468
phenylene [4]- 453
phenylene [4]- triangular 455, 456
phenylene [5]- angular 454, 455
phenylene [5]- linear 453, 454
phenylene [7]- triangular 457
phosgene 287
photocleavage 281, 400
photocleavage Norrish type-II 349
photocyclization 327
photocyclization asymmetric 329
photocyclization homo-Norrish 69
photocycloaddition [2+2]- 42, 44
photodehydrocyclization of stilbenes 21(30), 327
photofragmentation Norrish type-II 499
photoisomerization 383, 384
photolysis laser flash 490, 491
photolysis low-temperature 280
phthalic acid 204
physical organic chemistry 485

picene 432
picotubes 137, 438
pincer effect 79(71)
p-isotoluene 257, 258
pivaldehyde 108
planar carbon 81
planarization 81
p-nitrobenzenezenesulfonylazide 306
poly(phenylenes) 462
polyacetylene 8, 104, 151
polyacetylenes α,ω-di-aryl 155
polyacetylenes α,ω-di-*tert*-butyl 154, 155
polyacetylenes bis rhenium complexes 155
polyacetylenes of natural origin 151
polyacetylenes with even number of triple
bonds 153
polyacetylenes with odd number of triple bonds
154
polyarenes *cata*-condensed 423
polyenepolyynes 152
polyenes α,ω-dimethyl 106, 107
polyenes α,ω-diphenyl 107
polyenes α,ω-di-*tert*-butyl 108
polyenes α,ω-tetra-*tert*-butyl 110
polyenes conjugated 6
polyenes linear conjugated 103
polyethylene 119
polymerization ring-opening 108
poly-*para*-phenylene 446
polypropylene isotactic 119
polyquinane 69, 91
polyspiranes 300, 476
polyspiranes photoelectron spectra 319(224)
polyspiranes strain energies 319(211)
polytriangulanes 473
pomegranate 210
porous crystals 467
p-quinodimethane 290, 339, 340, 351
prebullvalene 220
prehnitene 509
Prins reaction 253, 254
prismane [3]- 41
prismane [4]- 60
prismane [5]- 44
prismane [6]- 47
prismane bridged 349
prismanes 41
propadiene 8, 173
propadienylidene 156
propargylallene 176, 223
propargylbromide 354

propellanes 69, 82, 133, 231, 344, 345, 348, 349
propellanes heat of hydrogenation 85
propellanes, higher homologs 87
propellaprismane 63
propene-1,3-diyl 117
propenylidene 117
prostaglandins 162
Pschorr reaction 323
pseudo-pelleterine 210
pterodactyladiene 17, 221
p-xylene 351
p-xylylene 290, 339, 340, 351, 489
p-xylylene bishomo 339, 340
p-xylylene mechanism of dimerization 375(162)
p-xylylenophanes 359
pyramidalization 75, 133
pyramidalized olefins 122
pyridinium chloride 283
pyrolysis 178, 286, 333, 334, 340, 351
pyrolysis cross-breeding 360
pyrolysis ester 253, 256, 391
pyrolysis flash vacuum (FVP) 263, 280, 342, 343, 349, 356, 450
pyrolysis sulfone 342, 347, 353, 361, 507
pyrylium salts 284

quaterrylene 433, 444(37)

radialene [3]- hexaethynyl- 293
radialene [3]- hexamethyl- 292
radialene [4]- 6, 189, 294
radialene [4]- octachloro- 294
radialene [4]- permethyl- 295
radialene [4]- perphenyl- 296
radialene [5]- decamethyl- 296
radialene [6]- 510
radialene [6]- hexaethyl- 298
radialene [6]- hexamethyl- 297
radialenes substrates for molecular magnets 317(164)
radialenes [3]- substituted 291
radialenes bridged 295
radialenes exploded 290, 460
Ramberg-Bäcklund rearrangement 158
Raney nickel 140
Raney nickel desulfurization 364
rearomatization 358
rearrangement divinylcyclopropane 11
rearrangement [3,3]sigmatropic- 392
rearrangement 1,5-hydrogen shift 21(30)

rearrangement Beckmann 381
rearrangement Claisen 13, 458
rearrangement Claisen ester 13
rearrangement Cope 11, 121, 209, 389, 412
rearrangement cyclobutene 21(30)
rearrangement degenerate 11
rearrangement di-π-methane 21(30), 214, 215, 383, 384
rearrangement divinylcyclobutane 11
rearrangement metal-catalyzed 344, 345, 395, 404
rearrangement methylenecyclopropane 494
rearrangement oxy-Cope 13
rearrangement Ramberg-Bäcklund 158
rearrangement Saito-Myers 11
rearrangement semibenzene 258, 341
rearrangement sigmatropic 410
rearrangement Stevens 236, 353, 355, 361, 363, 431
rearrangement vinylcyclopropane 21(30)
rearrangement vinylidene to acetylene 156
rearrangement Wagner-Meerwein 26, 27
rearrangement Wolff 72, 200, 341
reduction Birch 69, 231, 286, 437
reduction Clemmensen 339
reduction diisobutylaluminum hydride 69
reduction Mozingo 423, 424
reduction Wolff-Kishner 32, 87, 327
reductive dimerization 270
relationships generic 13
relay system 412
Reppe chemistry 181
repulsion intraannular 227
resolution kinetic 178
retro-Diels-Alder reaction 179, 351, 264
retrosynthesis 18
ribbon structures 421, 431, 443(20)
Rieche formylation 358
Rigisolve 56, 58
ring expansion annulene-to-annulene 221
ring-contraction 72, 341
ring-contraction Favorskii 60, 92
ring-opening cheletropic 253
ring-opening polymerization 108
Robinson annelation 342
rod-like structures 484(99)
rods molecular 461
rotane [3]- 301, 477
rotane [4]- 303
rotane [4]- octachloro- 304, 305
rotane [5]- 304, 305
rotane [6.3]- 308

rotane [6.4]- 308
rotane [6]- 305, 306
rotanes 300
rotanes structural properties 306
rotanes exploded 164, 309, 309
rotanes mixed 320(228)
rotanes universal synthesis 306, 307
rotaradialenes 304
rylenes 433

Sarett oxidation 303
Schlenk hydrocarbon 500
Schlenk-Brauns hydrocarbon 500
Scholl reaction 429, 433, 439
seco-dodecahedrane 69
seco-hexaprismane 48
Sekiya-Ishikara coupling 462
selenadiazoles 161
self-trapping 134, 183, 186
semibenzene rearrangement 258, 341
semibuckminsterfullerenes 335, 336
semibullvalene 209, 210, 215, 384, 389, 394, 415(69)
semibullvalene 2,6-dicyano-1,5-dimethyl-4,8-diphenyl- 396, 397
semibullvalene dialkyl derivatives 416(79)
semibullvalenes functionalized 396
semicyclic 6, 290, 512
sesquifulvalene 269, 272, 273
sextet concept 431, 432
shape-persistent hydrocarbons 462
Sharpless oxidation 132
Siegrist reaction 327
sigmatropic rearrangement [3,3]- 392
sigmatropic rearrangements 410
Simmons-Smith cyclopropanation 302, 303, 305
s-indacene 288, 289
skipped double bonds 384, 403, 405
snoutanes aza- 395
snoutene 220, 404, 418(117)
snoutenes aza 395
solid state NMR spectroscopy 498
solid-phase reactions 463
solvolysis 87
sparteine butyllithium complex 187
spiro compounds 263, 340
spiro[2.2]hepta-1,4,6-triene 408, 409
spiro[2.2]pentane 473
spiro[4.4]nona-1,3,7-triene 409
spiro[4.4]nona-1,3-diene 409
spiro[4.4]nona-2,6-diene-1,5-dione 409

spiro[4.4]nonatetraene 408, 409
spiro[4.4]nonatetraene spectroscopic properties 409
spiro[4.4]nonatetraene thermal isomerization 409
spirocarbene 342
spiro compounds 263, 340
spiroconjugation 408
spirodienone 339, 340
spiropentadiene 407
spiropentane 476, 483(84)
spiropolyenes 407
stability kinetic 55
stability thermodynamic 55
stabilomer 27, 69, 203
staffanes 83, 479
starphenylene 513
staurane 92, 101(64)
stelladiene 410
stellapolyenes 407, 410, 411
stellatetraene 412
stellatriene 411
Stephens-Castro coupling 226
stereoselectivity 14
Stevens rearrangement 236, 353, 355, 361, 363, 431
Stille coupling 109
strain Baeyer 156
strain release 205
strained olefins metal complexes 133
structures non-planar 9
substituents bulky 274
substitution S_N2 32
sulfone pyrolysis 342, 347, 353, 361, 507
sumanene 9
superacenes 441, 442
superphane 137, 322, 338, 354, 357, 358, 378(214), 507
superphane generalized concept 367
superphane heteroaromatic 367
superphane hexahomo 367
superphane X-ray structure 359
superphanes metal-capped 367
superthiophenophane 367
supertriptycene 389
Symmetrel 27
syn-1,6:8,13-bismethano[14]annulene 232, 234
syn-pyramidalization 134
synthetic graphite 439
syn-tricyclo[8.2.0.0^{2,9}]dodeca-3,5,7,11-tetraene 221
syn-tricylo[4.2.0.0^{2,5}]octa-3,7-diene 63

tandem reaction 91, 344
tandem-annelating reagents 254
TAPA 324
tautomerism valence 202
tautomerization 283
TCNE 361
tellurophene 507
template effect 119
terrylene 433
tert-butyl groups stabilizing effect on
    polyacetylenes 155
tert-butyl groups as protecting groups 274
tert-butyl maleic anhydride 55
tert-butylacetylene 109
tert-butylcyclopentadienide lithium 279
tert-butyl-fluoroacetylene 205
tetraasterane 34
tetracene 423
tetrachlorocyclopropane 293
tetrachlorocyclopropene 293
tetrachlorothiophene-1,1-dioxide 70, 92
tetracyanoethylene (TCNE) 361
tetracyclo[2.2.0.0^{2,6}.0^{3,5}]hexane 42, 203, 204
tetracyclo[6.1.0.0^{2,4}.0^{5,7}]nonane 406
tetracyclone 159, 439, 440
tetradehydro[18]annulene 226
tetraethynylbenzene 452, 453
tetraethynylethene (TEE) 298, 299, 458, 460,
    461, 481(43)
tetraethynylmethane 458, 459
tetrahalocyclopropenes 504
tetrahedrane 54, 199, 473
tetrahedrane 1,2,3-tri-*tert*-butyl-4-
    (trimethylgermyl)- 77(26)
tetrahedrane tetrakis-*tert*-butyl- 200
tetrahedrane tetralithio- 58
tetrahedrane tetramethyl- 58
tetrahedrane trimethylsilyl substituted 58
tetrahydrodibenzopentafulvalene 272
tetraiodobenzene 452
tetraisopropylethene 138
tetrakis-(bromomethyl)methane 476
tetrakis-*tert*-butylethene 138
tetrakis-*tert*-butylethene ab initio and DFT
    calculations 141, 142
tetramantane 30
tetramethyleneethane planarized 497
tetramethyleneethane structure 496, 497
tetramethyleneethane (TME) 492, 496, 497
tetraphenylallene 173, 174
tetraphenylcyclopentadienone 217, 439
tetra-*tert*-butyl-tetrahedrane 55

tetra-*tert*-butyl-tetrahedrane strain 57
tetra-*tert*-butyl-tetrahedrane X-ray structure 57
tetrathiafulvalene 270
thiacycloheptyne 140
Thiele acid 399
Thiele condensation 277
thiodiazolines 138
thiophene-1,1-dioxide 284
Thorpe-Ziegler cyclization 90
three-phase test 199
through-space interaction 408
tied-back approach 139
tied-back substituents 386
tiling 447, 480(1)
TMB 497, 498
TME 492, 494, 496, 497
TMM 492, 494
tolanes as building blocks for superacenes 441,
    442
torand 435
trans-[2.2]metacyclophane-1,9-diene 236
trans-1,3,5,7-octatetraene 104
trans-15,16-dimethyldihydropyrene 236
trans-1-phenyl-cyclohexene
trans-9,10-dihydronaphthalene 219
trans-cycloalkenes 122, 123
trans-cyclodecene optically active 125
trans-cyclododecene 96
trans-cycloheptene 125, 126
trans-cycloheptene structural properties 127
trans-cyclohexene 127
trans-cyclononene optically active 125
trans-cyclooctene 124, 125
trans-cyclooctene racemization barrier 125
trans-cyclooctene resolution 125
trans-cycloolefins self-trapping 130
trans-cyclopentene 128
trans-cyclopentenone 128
trans-tris-σ-homobenzene 406
trans-tropylidene 128
triacenaphthotriphenylene 336
triacetylene 152
triafulvalene 269
triafulvene 118, 251, 252, 260, 262
triafulvene spectroscopic properties 263
triafulvene trapping by cyclopentadiene 263
triamantane 30
triangulane [10]- 478
triangulane [5]- unbranched 477
triangulane [6]- branched 478
triangulane [6]- linear 477
triangulane [8]- 478

triangulanes 476
triangulanes branched 476
triangulanes cyclic 476
triangulanes linear 476, 477
triangular [4]phenylene X-ray structure 456
triangular [7]phenylene 457
triangulene 500
triazines N,N-dialkylaryl 462, 463
tribenzobicyclo[2.2.2]octatriene 385
tribromomethyl phenylmercury 302
tricyclo[2.1.0.0^{1,3}]pentane 473, 475
tricyclo[2.1.0.0^{2,5}]pentan-3-one 199
tricyclo[3.1.0.0^{2,6}]hex-1(6)-ene 136
tricyclo[3.1.0.0^{2,6}]hex-3-ene 204
tricyclo[3.3.0.0^{2,6}]octa-3,7-diene 394
tricyclo[3.3.0.0^{2,6}]octa-3,8-diene 209, 210
tricyclo[3.3.0.0^{2,8}]octa-3,6-diene 384
tricyclo[3.3.1.0^{2,8}]nona-3,6-dien-9-one 391
tricyclo[3.3.2.0^{4,6}]deca-2,7,9-triene 390
tricyclo[4.1.0.0^{2,7}]hept-4-en-3-one 217
tricyclo[4.2.0.0^{2,5}]octa-3,7-diene 209, 210
tricyclo[4.2.2.2^{2,5}]dodeca-1,5-diene 137
tricyclo[4.4.0.0^{3,8}]decane 31
tricyclo[4.4.2.2^{2,5}]dodeca-1,5-diene 420 (161)
tricyclo[5.2.1.0^{4,10}]deca-2,5,8-triene 399
tricyclobuta[1,2:3,4:6,7]naphthalene 511
tricyclobutabenzene 509
tricyclopenta[$def;jkl;pqr$]triphenylene 9
tricycloproparene 297
trimethylene methane structural properties 495
trimethylenemethane (TMM) 492, 494
trimethylenemethanes cyclic 492, 493
trimethylsilylbutadiyne 470
trimethylsilylethyne 463
trimethylenemethanes in natural product
    synthesis 500
triangular [4]phenylene 456
trioxyangulene 500
triphenylene 450
triphenylmethane dyes 253
triphenyltinhydride 349
triple bond dienophiles 351
triplet diradicals 494
triprismane 42, 203, 204
triptycene 384, 385, 505
triptycene bridgehead reactivity 386
triquinacene 64, 65, 220, 397, 398, 399, 400
triquinacene bridged 401
triquinacene homoaromaticity 401
triquinacene photochemistry 403
tris-(2-adamantylidene)cyclopropane 292
tris(triphenylphosphine)platinum(0) 159

trisbicyclo[2.1.1]hexabenzene 512
trisbicyclo[2.2.1]heptabenzene 512, 513,
    521(159)
trisbicylo[2.2.2]octabenzene 512, 513
trishomobenzene 116
trispirane 37
trispiro[2.0.2.0.2.1]decan-10-one 304
tris-p-xylene 429
tris-$tert$-butylcyclopropenium ion 208
tris-$tert$-butylethene 141
tri-$trans$-di-cis-[10]annulene 237
tritriptycene 389
trityl anion 386
trityl cation 386
trityl tetrafluoroborate 275, 286
tropone 261, 266, 272
tropone valence isomer 273
tropylidene 12
tropylium bromide 271
tropylium cation 266
tropylium tetrafluoroborate 268
TTF 270
tubular hydrocarbons 437
tweezers molecular 445(54)
twistane 30
twistane configuration 34
twistanone 32

Ullmann coupling 449
urea 251, 252
urotropine 24

vacuum-gas-solid reaction (VGSR) 207, 261,
    407
valence isomerization 279
valence tautomerism 202
vestamide 119
vetivazulene 281, 282
VGSR technique 207, 261, 407
vibrationally excited intermediates 213
vinylallenes substituted 175
vinylallenes reactivity 175, 176
vinylcyclopropane 384
vinylidene carbene 156, 188, 333
vinylidene-acetylene rearrangement 156
vinylidenes 156, 188, 333
vinylog principle 492
vinylphosphorane 172
vitamin A 176
vitamin D 176

Weiss reaction 91, 400

Wilkinson catalyst 190
windowpane 100(51)
wire molecular 461
Wittig reaction 105, 172, 222, 226, 294, 303, 327, 341, 391, 412
Wittig reaction intramolecular 129
Wittig-Horner double condensation reaction 260
Wittig-Horner reaction 190, 233
Wolff rearrangement 72, 85, 200, 341
Wolff-Kishner reduction 87, 327
wormwood 281
Wurtz coupling 356, 361

Wurtz-Fittig coupling 429
wurtzitane 39(46)

yarrow 281

Y-conjugation 251, 252
ylid diphenylcyclopropylsulfonium 66

Zeise salt 125
zero-bridged annulenes 277
Ziegler catalyst 119
Zincke aldehyde 283
zirconocenechlorohydride 308